Natural Disasters

Fourth Edition

Patrick L. Abbott
San Diego State University

Boston Burr Ridge, IL Dubuque, IA Madison, WI New York San Francisco St. Louis
Bangkok Bogotá Caracas Kuala Lumpur Lisbon London Madrid Mexico City
Milan Montreal New Delhi Santiago Seoul Singapore Sydney Taipei Toronto

NATURAL DISASTERS, FOURTH EDITION

Published by McGraw-Hill, a business unit of The McGraw-Hill Companies, Inc., 1221 Avenue of the Americas, New York, NY 10020.

 This book is printed on recycled, acid-free paper containing 10% postconsumer waste.

1 2 3 4 5 6 7 8 9 0 QPD/QPD 0 9 8 7 6 5 4 3

ISBN 0–07–252809–5

Publisher: *Margaret J. Kemp*
Sponsoring editor: *Thomas C. Lyon*
Developmental editor: *Lisa A. Leibold*
Executive marketing manager: *Lisa L. Gottschalk*
Lead project manager: *Jill R. Peter*
Production supervisor: *Sherry L. Kane*
Lead media project manager: *Judi David*
Media technology producer: *Renee Russian*
Coordinator of freelance design: *Rick D. Noel*
Cover/interior designer: *Elise Lansdon*
Cover image: © *The Image Bank, image number 10125754, Tornado*
Senior photo research coordinator: *John C. Leland*
Photo research: *Mary Reeg*
Compositor: *ElectraGraphics, Inc.*
Typeface: *10/12 Times Roman*
Printer: *Quebecor World Dubuque, IA*

Interior design images: © *PhotoDisc Volume 44 Nature, Wildlife, and the Environment: #44221 (ocean waves), #44225 (hurricane), #44272 (drought), #44285 (grass fire), #44287 (forest fire aftermath), #44328 (volcano), and #44335 (lightning)*

The credits section for this book begins on page 449 and is considered an extension of the copyright page.

Library of Congress Cataloging-in-Publication Data

Abbott, Patrick L.
Natural disasters / Patrick L. Abbott.—4th ed.
p. cm.
Includes bibliographical references.
ISBN 0–07–252809–5 (alk. paper)
1. Natural disasters. I. Title.
GB5014.A24 2004
904'.5—dc21 2003001976
CIP

www.mhhe.com

Clement L. And Constance V. Abbott

for their lifetime of interest and support of my activities

Contents

Chapter 3

Basic Principles of Earthquake Geology, Seismology, and Tsunami 59

Chapter 4

Some Earthquakes in Western North America 96

Chapter 5

More United States and Canadian Earthquakes 128

Chapter 6

Volcanic Eruptions: Plate Tectonics and Magmas 151

Chapter 7

Volcanic Eruptions Continue 180

Chapter 8

Mass Movements 208

Chapter 9

Climate Change 242

Chapter 10

Severe Weather 273

Chapter 11

Hurricanes and the Coastline 302

Chapter 12

Floods 334

Chapter 13

Fire 365

Chapter 14

The Great Dyings 388

Chapter 15

Impacts with Space Objects 407

Chapter 16

Population Growth 427

Preface

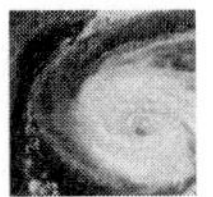

Why the Book Was Written

In the early 1970s, Bill Ganus and I developed an environmental geology course at San Diego State University. The growing awareness of the environment and the availability of good textbooks made it natural to offer a general education course looking at geological hazards, resource utilization and disposal, and intelligent planning in concert with the environment. The course had moderately successful enrollments, chugging along at 25 to 35 students per semester for over a decade.

In 1987, Tom Rockwell and I were discussing the environmental geology course and speculating on why it never attracted large enrollments. We agreed that the natural disasters portions of the course were the most popular. So, I formally changed the name of the course to "Natural Disasters" but did not change the course description or textbook, or advertise the change in any way. Yet almost instantly, students reading through the fine print of semester course offerings saw the "Natural Disasters" listing and enrollments skyrocketed. Now we offer multiple sections filling more than 4,500 classroom seats per academic year and still do not satisfy demand.

San Diego State University students do not have to take Natural Disasters. They can select from over 30 courses among 10 departments with offerings such as Biology of Sex, Evolution, Origin of Life, The Oceans, Dinosaurs, and Confronting AIDS. But more students opt for Natural Disasters than any other course. If your department could benefit from higher enrollments of non-major students, I strongly recommend offering a Natural Disasters course. Earthquakes, hurricanes, tornadoes, and other high-energy processes of our active Earth affect children's lives. As students, they want to understand why these natural disasters happen. The students' high level of interest can be channeled by the instructor into some significant understanding about how the Earth works.

About the Book

This book focuses on natural disasters: how the normal processes of the Earth concentrate their energies and deal heavy blows to humans and their structures. It largely ignores the numerous case histories describing human actions and resultant environmental responses; these topics are left to the excellent textbooks on environmental geology. Nor does this book address resource extraction, utilization, and disposal; these subjects are covered by fine textbooks on earth resources, minerals, energy, soils, and water. This book is concerned with how the natural world operates and, in so doing, kills and maims humans and destroys their works.

Throughout the book, certain themes are maintained:

- Energy sources underlying disasters
- Plate tectonics and climate change
- Earth processes operating in rock, water, and atmosphere
- Significance of geologic time
- Complexities of multiple variables operating simultaneously
- Detailed and readable case histories

The text aims to explain important principles about the Earth and then develop further understanding through numerous case histories. I hope that students will actually enjoy reading most of this book.

The primary organization of the book is based on an energy theme. Chapter 1 leads off with data describing death and destruction, then examines the energy sources underlying disasters: 1) Earth's internal energy from its formative impacts and continuing decay of radioactive elements; 2) gravity; 3) external energy from the Sun, and; 4) impacts with asteroids and comets.

Disasters fueled by Earth's internal energy are addressed in Chapters 2 through 7 and are organized on a plate-tectonics theme. Chapter 2 provides the basic description of plate tectonics and its relationship to earthquakes. Chapter 3 covers the basic principles of earthquake geology, seismology and tsunami, and assumes no prior knowledge. Chapter 4 uses plate tectonics and historic and prehistoric records to explain earthquakes along western North America. Chapter 5 examines the history and potential for earthquakes throughout the rest of North America. The intent is to cover every geographic area and major historic earthquake. Chapters 6 and 7 discuss volcanoes; their characteristic magmas are organized around the 3 Vs—viscosity, volatiles, and volume. Eruptive behaviors are related to plate-tectonic setting. As throughout, case histories are employed to enliven the text.

Disasters powered primarily by gravity are covered in Chapter 8 on mass movements. Many types are discussed and illustrated, from falls to flows and slides to subsidence.

Disasters fueled by the external energy of the Sun are examined in Chapters 9 through 13. Chapter 9 looks at climate change and provides some basis for succeeding chapters. Climate principles governing energy transfer over time scales of millions, thousands, hundreds, and several years are discussed. The time focus shrinks through the chapter, leading to Chapter 10 on severe weather phenomena, such as thunderstorms, lightning, and tornadoes. Chapter 11 examines hurricanes and the coastline. The emphasis on water continues in Chapter 12 on floods and how human activities increase flood damage. Chapter 13 on fire examines the liberation of ancient sunlight captured by photosynthesis and stored in organic material.

Before moving to the fourth energy source (impacts), Chapter 14 examines the great dyings encased in the fossil record. The intent is to document the greatest of all natural disasters and to use multiple variables in analyzing their causes. Specific mass extinctions are examined using causative factors, such as continental unification and separation, climate change, flood-basalt volcanism, sea-level rise and fall, impacts, biologic processes, and the role of humans in the latest mass dying. Chapter 15 examines impact mechanisms in greater detail and includes plans to protect Earth from future impacts.

Chapter 16 looks at population growth, the unprecedented exponential increase in the human population.

There is a lot of material in this book, probably too much to cover in one semester. But the broad range of natural disasters topics allows each instructor to select those chapters that cover their interests and local hazards. The goal is to involve the students for a lifetime in understanding the Earth, Atmosphere, Oceans and Skies—Observe; Think; Explain; Discuss.

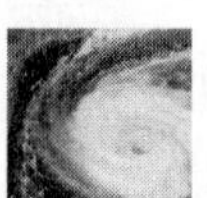

New to This Edition

For the fourth edition, *all* chapters have been revised and updated, and 24 new pieces of line art, 16 new photos, and 19 new tables have been added. Changes include major reorganizations and expansions. Chapter 1 has more physical and economic data on disasters and hazards, and more discussion of energy flow. Chapters 6 and 7 on volcanism are unified into a continuous sequence based on plate tectonics and magma characteristics with eruptions explained using

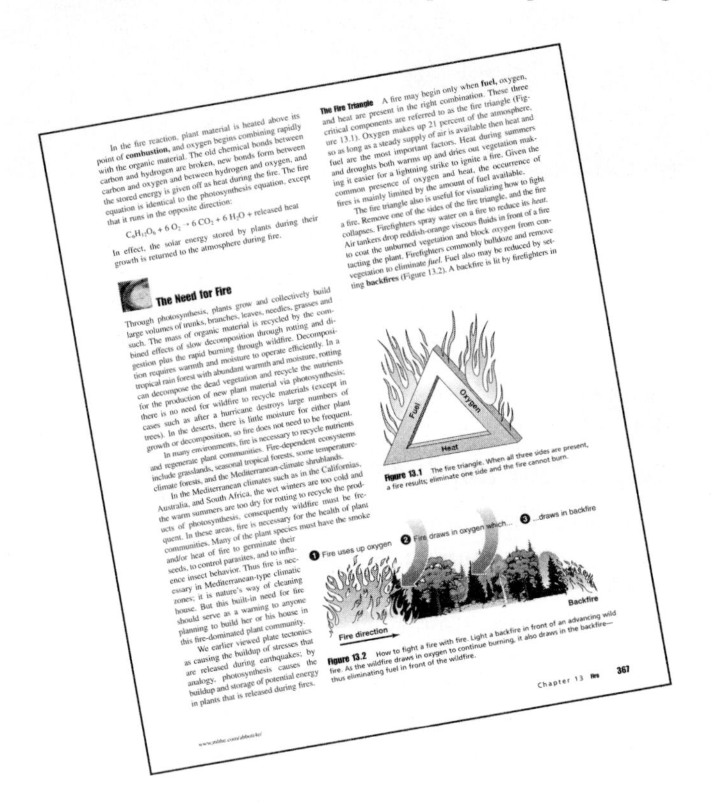

The Fire Triangle

The Need for Fire

Figure 13.1 The fire triangle. When all three sides are present, a fire results; eliminate one side and the fire cannot burn.

Figure 13.2 How to fight a fire with fire. Light a backfire in front of an advancing wild fire. As the wildfire draws in oxygen to continue burning, it also draws in the backfire—thus eliminating fuel in front of the wildfire.

Chapter 13 Fire 367

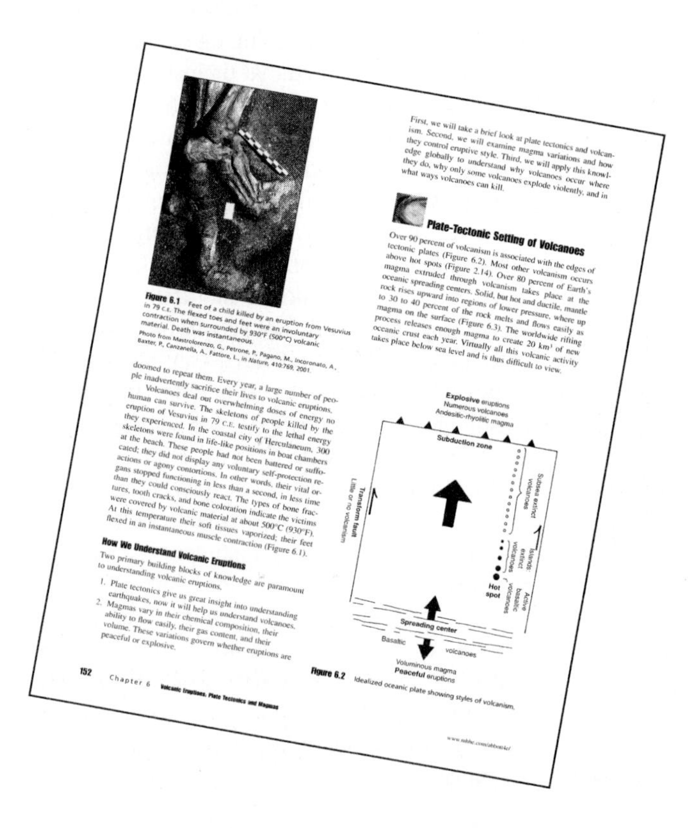

First, we will take a brief look at plate tectonics and volcanism. Second, we will examine magma variations and how they control eruptive style. Third, we will apply this knowledge globally to understand why volcanoes occur where they do, why only some volcanoes explode violently, and in what ways volcanoes can kill.

Plate-Tectonic Setting of Volcanoes

Over 90 percent of volcanism is associated with the edges of tectonic plates (Figure 6.2). Most other volcanism occurs above hot spots (Figure 2.14). Over 80 percent of Earth's magma extruded through volcanism takes place at the oceanic spreading centers. Solid, but hot and ductile, mantle rock rises upward into regions of lower pressure, where up to 30 to 40 percent of the rock melts and flows easily as magma on the surface (Figure 6.3). The worldwide rifting process releases enough magma to create 20 km³ of new oceanic crust each year. Virtually all this volcanic activity takes place below sea level and is thus difficult to view.

How We Understand Volcanic Eruptions

Figure 6.1 Feet of a child killed by an eruption from Vesuvius in 79 C.E.

Figure 6.2 Idealized oceanic plate showing styles of volcanism.

152 Chapter 6 Volcanic Eruptions, Plate Tectonics and Magmas

the 3 Vs—viscosity, volatiles, and volume. Chapter 13 on fire has added 7 new figures and 5 new tables on wildland fire data and how fires work. Chapter 9 on climate change has new sections on the last thousand years, and on global warming. Chapter 10 on severe weather has expanded discussion of energy flow and how thunderstorms work.

Supplements

For the Student

Online Learning Center at http://www.mhhe.com/abbott4e

This site gives you the opportunity to further explore topics presented in the book using the Internet. The site contains interactive quizzing with immediate feedback, interactive key term flashcards, web links, a career center, and more.

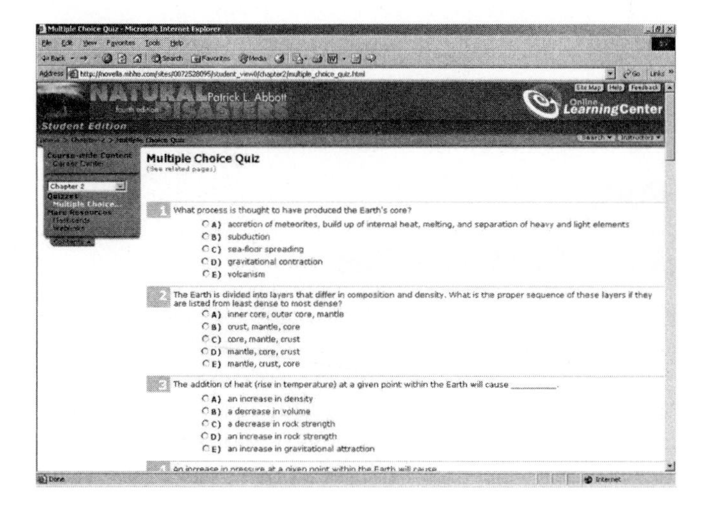

For the Instructor

Online Learning Center at http://www.mhhe.com/abbott4e

Take advantage of the instructor's manual, PowerPoint lecture outlines, and access to PageOut—McGraw-Hill's course management tool.

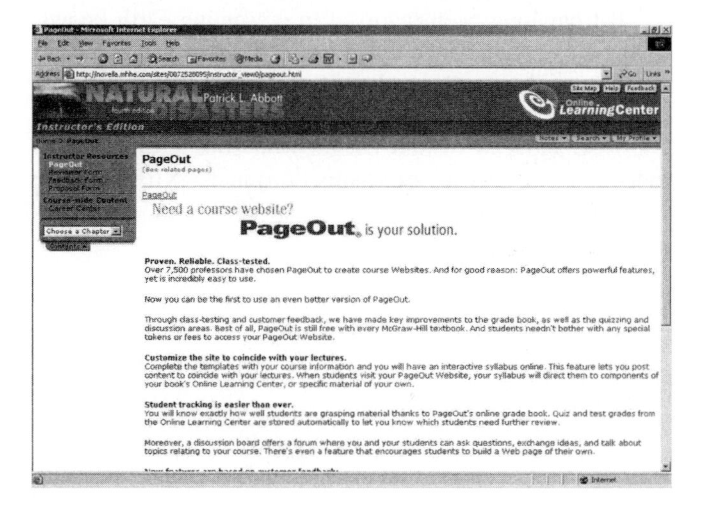

Digital Content Manager CD-ROM

This CD-ROM contains **all of the line art, photographs, and tables** from the text to make customizing your multimedia presentation easy. You can organize figures in any order you want; add labels, lines, and your own artwork; integrate materials from other sources; edit and annotate lecture notes, and then have the option of placing your media lecture into a presentation program, such as PowerPoint.

Instructor's Testing and Resource CD-ROM

This cross-platform CD-ROM provides a wealth of resources for the instructor. Supplements featured on this CD-ROM include a computerized test bank utilizing Brownstone Diploma ® testing software to quickly create customized exams. This user-friendly program allows instructors to search for questions by topic, format, or difficulty level; edit existing questions or add new ones; and scramble questions and answer keys for multiple versions of the same test.

Other assets on this CD are grouped within easy-to-use folders. The Instructor's Manual and Test Item File are available in both Word and PDF formats. Word files of the test bank are included for those instructors who prefer to work outside of the test-generator software.

Acknowledgments

I am deeply appreciative of the help given by others to make this book a reality. The photograph collection in the book is immeasurably improved by the aerial photographs generously given by John S. Shelton, the greatest geologist photographer of them all. The collection of John Shelton photographs in this book is second in number only to his classic book *Geology Illustrated.* Many of the figures were drafted or drawn by Rene Wagemakers of San Diego State University. Rene's talent and ready willingness to help are invaluable.

I am indebted to other geologists who provided photographs: Alan Mayo of GeoPhoto Publishing Company on the Winter Park sinkhole and Tucson flooding; Gerald G. Kuhn of San Diego from his space shuttle image collection; Michael W. Hart of San Diego on mass movements; Al Boost of Caltrans on the San Fernando earthquake; Peter Weigand of California State University Northridge, Greg Davis of University of Southern California, and Kerry Sieh of Caltech on the Northridge earthquake; Anne Jennings of the University of Colorado on climate; José Aguirre on the Berkeley fire; and the photo libraries of the USGS, NOAA, and NASA.

For the first edition several chapters benefited from helpful reviews by San Diego State University colleagues: Michael J. Walawender on volcanism, J. David Archibald and Richard H. Miller on great dyings, and David L. Kimbrough on impacts.

The quality of the book was significantly improved by the insights provided by comments from the following reviewers of the third and earlier editions:

Judson Ahern, *University of Oklahoma*
Wang-Ping Chen, *University of Illinois at Urbana-Champaign*
Patrick Colgan, *Northeastern University*
Michael Conway, *Arizona Western University*
John Dunbar, *Baylor University*
Michael Forrest, *Rio Hondo Junior College*
Kevin P. Furlong, *Pennsylvania State University*
David Gonzales, *Fort Lewis College*
Paul K. Grogger, *University of Colorado–Colorado Springs*
John Hidore, *University of North Carolina–Greensboro*
George Hupper, *University of Wisconsin–LaCrosse*
Ernest L. Kern, *Southeast Missouri State University*
Alan Lester, *University of Colorado*
Jon Nourse, *California State Polytechnic University–Pomona*
Peter Sadler, *University of California–Riverside*
Bingming Shen-Tu, *Indiana University*
Don Steeples, *University of Kansas*
Donald J. Stierman, *University of Toledo*
Philip Suckling, *University of Northern Iowa*

The expanded coverage of volcanoes was much improved by the advice of Victor E. Camp of San Diego State University.

The fourth edition benefited greatly from detailed reviews by:

Sandra Allen, *Lindenwood University*
Cathy Busby, *University of California–Santa Barbara*
Stanley Dart, *University of Nebraska–Kearney*
John Dooley, *North Hennepin Community College*
John Dunbar, *Baylor University*
Sue Morgan, *Utah State University–Logan*
Leslie Sonder, *Dartmouth College*

I am grateful for the help of others at San Diego State University: Jacobe Washburn for his original line drawings, Tony Carrasco for invaluable aid on the computer with photos and line drawings, and Marie Grace for forming many of the tables.

I sincerely appreciate the talents and accomplishments of the McGraw-Hill professionals in Dubuque who took my manuscript and produced it into this book. For the shortcomings that remain in the book, I alone am responsible. I welcome all comments, pro and con, as well as suggested revisions.

Pat Abbott
pabbott@geology.sdsu.edu

About the Author

Patrick Abbott is a native San Diegan lucky enough to live and work in his hometown. Pat earned his M.A. and Ph.D. degrees in geology at The University of Texas at Austin. He benefited greatly from the depth and breadth of the faculty in the Department of Geological Sciences at Austin; this was extended by their requirement to take five additional graduate courses outside the department. Developing interests in many topics helped lead to writing this textbook.

Pat's research has concentrated on the Mesozoic and Cenozoic sedimentary rocks of the southwestern United States and northwestern Mexico. Studies have focused on reading the history stored within the rocks—depositional environments, provenance, paleoclimate, palinspastic reconstructions, and high-energy processes.

Pat has long been involved in presenting earth knowledge to the public, primarily through local TV news. At present, he has embarked on producing videos for TV broadcast in a series called Written in Stone. The first video, The Rise and Fall of San Diego, won 2002 awards in the Videographers (Award of Distinction) and AXIEM (Silver Axiem) competitions.

Chapter 1

Natural Disasters and Their Energy Sources

In 2001, more than 35,000 people lost their lives to **natural disasters.** The 16 deadliest events of the year were **earthquakes,** floods, typhoons (hurricanes), and landslides (Table 1.1). All the disasters were the result of extreme events of natural phenomena operating at the high end of the **energy** scale for a short time in a restricted area. The killer events were spread around the world.

On Friday morning, 26 January 2001, Hidendre Barot was at home with his wife in their apartment on the top floor of a ten-story building in Ahmedabad, in the state of Gujarat in west-central India. At 8:46 A.M. the apartment building began shaking and Barot and his wife fled up onto the roof, held hands, and waited for the violent motions to stop. But the building failed, and Barot fell ten

Table 1.1 The 16 Deadliest Natural Disasters in 2001

Fatalities	Date	Event	Country
20,103	26 Jan	Earthquake (Gujurat)	India
886	10 Nov	Flood	Algeria
844	13 Jan	Earthquake (landslide)	El Salvador
396	15 Aug	Flood (Mekong River)	Vietnam
360	25 Jul	Typhoon Toraji	China
350	10 Aug	Flood	Iran
320	7 Nov	Typhoon Lingling	Philippines
302	27 Aug	Flood	Nigeria
277	31 Jul	Storm (flood, landslide)	Indonesia
274	13 Feb	Earthquake	El Salvador
196	4 Jul	Typhoon Utor	Philippines
178	18 Jun	Flood	China
177	10 Aug	Typhoon Usagi	China
169	24 Jun	Typhoon Chebi	China
146	7 Sep	Flood	India
145	23 Jun	Earthquake (tsunami)	Peru
25,123 Total			

Source: Data from Swiss Reinsurance Company (2002).

stories with the collapsing structure, ending up surrounded by debris but with no serious injuries. For days afterward, he helped search the building wreckage looking for his wife, but she was one of the 20,103 people killed by this major earthquake, the deadliest natural disaster of 2001 (Figure 1.1; and see Chapter 2). This earthquake was a **great natural disaster,** an event that so overwhelms a region that outside assistance is needed in rescuing and caring for people, in helping clean up the destruction, and in beginning the process of reconstruction. We need to learn more about earthquakes and building behavior to better protect ourselves; these events happen all around the world.

On the other side of our planet, 13 days earlier, a similar-sized earthquake struck El Salvador in Central America. This time most of the deaths occurred when the shaking earth caused a steep hillside to fail and send a landslide through the middle-class neighborhood of Santa Tecla, a suburb of the capital San Salvador (Figure 1.2). We need to learn more about recognizing steep slopes that are liable to fail; landslides occur around the world.

It is not just shaking or sliding earth that kills people, but also water. In 2001, 12 of the 16 deadliest events were storms and floods (Table 1.1). In November, Algeria was suffering through a severe drought. In Algiers, the capital city, water supplies were so limited that harsh restrictions were imposed on water usage and some religious leaders were calling on people to pray for rain. And rain it did. During 9–10 November, heavy rains fell during a 36-hour period and raging torrents of water, mud, and debris flowed down from the surrounding hills into this Mediterranean city, killing 886 people (Figure 1.3). Rainfall and water runoff are familiar to everyone, yet we need to learn more about floods; every year they kill thousands of people around the world.

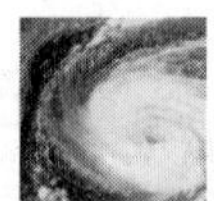

Human Fatalities in Natural Disasters

The numbers of natural disaster fatalities between 1980 and 2001 vary markedly from year to year (Figure 1.4). The saw-tooth shape of the curve is created by the largest or great natural disasters that kill so many people in one event. As horrifying as the year 2001 life losses are, that number of natural disaster fatalities is about average for the past 22 years.

Figure 1.1 A crowd gathers outside a collapsed building in Ahmedabad, India on the day of the earthquake, Friday, 26 January 2001.

Figure 1.2 This earthquake-triggered landslide destroyed 300 homes killing 680 people in Santa Tecla, El Salvador on Saturday, 13 January 2001.

Figure 1.3 Masked rescuers carry a dead body from an Algiers building collapsed by debris-carrying flood water five days earlier, 15 November 2001.

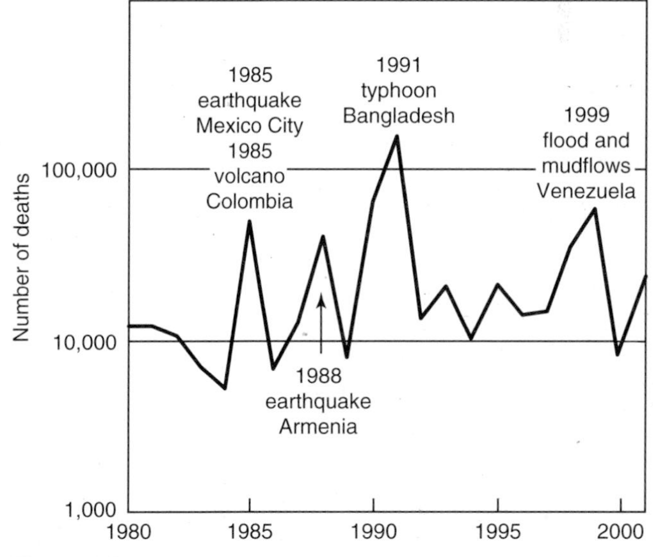

Figure 1.4 Deaths due to natural disasters, 1980–2001.

Data from Swiss Reinsurance Company.

Table 1.2 **Fatalities from Natural Disasters, 1947–1980**

	Earthquake	Tsunami	Volcanic Eruption	Flood	Hurricane	Tornado	Other Severe Weather	Landslide/ Avalanche	Life Loss Totals by Geographic Area
Number of Killing Events	**180**	**7**	**18**	**333**	**210**	**119**	**147**	**45**	
North America	77	60	96	1,633	1,997	4,568	5,003	323	13,757
Caribbean and Central America	30,613	—	151	2,575	16,541	26	510	260	50,676
South America	38,837	—	440	4,396	—	—	340	5,262	49,275
Europe	7,750	—	2,000	11,199	250	39	6,816	640	28,694
Asia	354,521	4,459	2,805	170,664	478,574	4,308	34,403	4,356	1,054,090
Africa	18,232	—	—	3,891	864	548	5	—	23,540
Oceania	18	—	4,000	77	290	—	117	—	4,502
Life loss totals by Earth process	450,048	5,519	9,492	194,435	498,516	9,489	47,194	10,841	Life loss total = 1,225,534

Source: Data from Shah (1983).

Going back to data from 1947 to 1980 gives more insight into the what, where, and who of natural disaster types versus geographic area and fatalities (Table 1.2). The numbers presented are quite conservative and actually understate the number of deaths, yet the patterns in the data are quite instructive. Notice that during this 34-year-long period the biggest killers worldwide were hurricanes and earthquakes, and that the water-related phenomena of severe weather and floods killed more people than **volcanoes** and landslides.

The 40 deadliest disasters in the 32-year-long period from 1970 to 2001 are shown in Table 1.3. Notice that 37 of the 40 disasters were due to natural causes. The most frequent mega-killers were earthquakes (21 of 40) and hurricanes (9 of 40). However, the 9 storms killed more people (599,344) than the 21 earthquakes (571,146). Notice that 21 of the 40 worst disasters occurred in a belt running from China and Bangladesh through India and Iran to Turkey. Only one mega-killer disaster happened in western Europe and none in the United States and Canada.

What is the correlation between human population density and the number of natural-disaster deaths? The data of Table 1.2 paint a clear picture: densely populated Asia dominates the list with 86 percent of the fatalities. The Asian experience offers a sobering view of what may befall the global population of humans if we continue our rapid growth. Where humans are concentrated, disasters kill many more people during each high-energy event.

Disasters kill many people but what are the effects upon survivors? One effect is an increase in suicides. A study of suicide rates among almost 20 million people in 377 U.S. counties following a natural disaster showed a 13.8 percent increase in the four years after a flood, a 31 percent increase in the two years after a hurricane, and a 62.9 percent increase in the first year after an earthquake. The suicide rate in the whole United States increased by less than 1.3 percent during these time intervals.

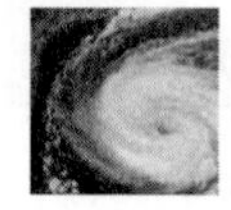

Economic Losses from Natural Disasters

The deaths and injuries caused by natural disasters grab our attention and squeeze our emotions, but, in addition, there are the economic losses. The destruction and disabling of buildings, bridges and roads, of power-generation plants, of transmission systems for electricity, natural gas, and water, plus all the other built works of our societies add up to a huge dollar cost. But the economic losses are greater than just damaged structures; there are the industries and businesses knocked out of operation, causing losses in productivity and resulting in lost wages for employees left without places to work.

In 2001, there were about 700 loss events globally with known economic losses of $36 billion (U.S.). And yet, the year 2001 was only an average year (Figure 1.5). The worldwide economic losses for 2001 have been exceeded by the economic losses from individual events in the recent past such as the Northridge, California earthquake in 1994 ($44 billion). The trend in economic losses becomes more evident when the dollar figures are averaged over longer time intervals such as decades (Table 1.4). The increasing numbers of great natural disasters and economic losses are evident. Do these increases mean the Earth is experiencing more earthquakes and hurricanes? Or are these increasing economic losses related to the global population of humans doubling

Table 1.3 **The 40 Deadliest Disasters, 1970–2001**

Fatalities	Date/Start	Event	Country
400,000	14 Nov 1970	Hurricane	Bangladesh
250,000	28 Jul 1976	Earthquake (Tangshan)	China
165,000	26 Apr 1986	Nuclear power plant accident	Ukraine
140,000	30 Apr 1991	Hurricane Gorky	Bangladesh
60,000	31 May 1970	Earthquake and landslide (Nevados Huascaran)	Peru
50,000	15 Dec 1999	Flooding and mudslides	Venezuela
50,000	21 Jun 1990	Earthquake (Gilan)	Iran
25,000	7 Dec 1988	Earthquake	Armenia
25,000	16 Sep 1978	Earthquake (Tabas)	Iran
23,000	13 Nov 1985	Volcanic eruption and mudflows (Nevado del Ruiz)	Colombia
22,000	4 Feb 1976	Earthquake	Guatemala
20,103	26 Jan 2001	Earthquake (Gujarat)	India
19,118	17 Aug 1999	Earthquake (Izmit)	Turkey
15,000	19 Sep 1985	Earthquake (Mexico City)	Mexico
15,000	11 Aug 1979	Dam failure (Morvi)	India
15,000	1 Sep 1978	Flood (Monsoon rains in north)	India
15,000	29 Oct 1999	Hurricane (Orissa)	India
11,000	22 Oct 1998	Hurricane Mitch	Honduras
10,800	31 Oct 1971	Flood	India
10,000	25 May 1985	Hurricane	Bangladesh
10,000	20 Nov 1977	Hurricane (Andhra Pradesh)	India
9,500	30 Sep 1993	Earthquake (Marashtra state)	India
8,000	16 Aug 1976	Earthquake (Mindanao)	Philippines
6,425	17 Jan 1995	Earthquake (Kobe)	Japan
6,304	5 Nov 1991	Typhoons Thelma and Uring	Philippines
5,300	28 Dec 1974	Earthquake	Pakistan
5,000	10 Apr 1972	Earthquake (Fars)	Iran
5,000	23 Dec 1972	Earthquake (Managua)	Nicaragua
5,000	30 Jun 1976	Earthquake (West Irian)	Indonesia
4,800	23 Nov 1980	Earthquake (Campagna)	Italy
4,500	10 Oct 1980	Earthquake (El Asnam)	Algeria
4,375	21 Dec 1987	Boat collision	Philippines
4,000	15 Feb 1972	Storm; snow	Iran
4,000	24 Nov 1976	Earthquake (Van)	Turkey
4,000	30 May 1998	Earthquake (Takhar)	Afghanistan
3,840	1 Nov 1997	Hurricane Linda	Vietnam
3,800	8 Sep 1992	Floods (Punjab)	Pakistan
3,656	1 Jul 1998	Flood (Yangtze River)	China
3,400	20 Sep 1999	Earthquake (Nantou)	Taiwan
3,200	16 Apr 1978	Hurricane	Reunion
1,443,121 Total			

Source: Data from Swiss Reinsurance Company (2002).

from three billion in 1960 to six billion in 1999 with increasing percentages of the population living in cities?

Insured Portion of Economic Losses

The 40 greatest natural disasters between 1970 and 2001 from the insurance company perspective of dollar losses are listed in Table 1.5. Notice that 36 of the 40 most expensive disasters were due to natural processes. The list of most expensive events is dominated by storms (26 of 40), whereas earthquakes contributed only four events. Compare the events on the 40 deadliest disasters list for 1970 to 2001 (Table 1.3) with Table 1.5. The list based on dollar losses

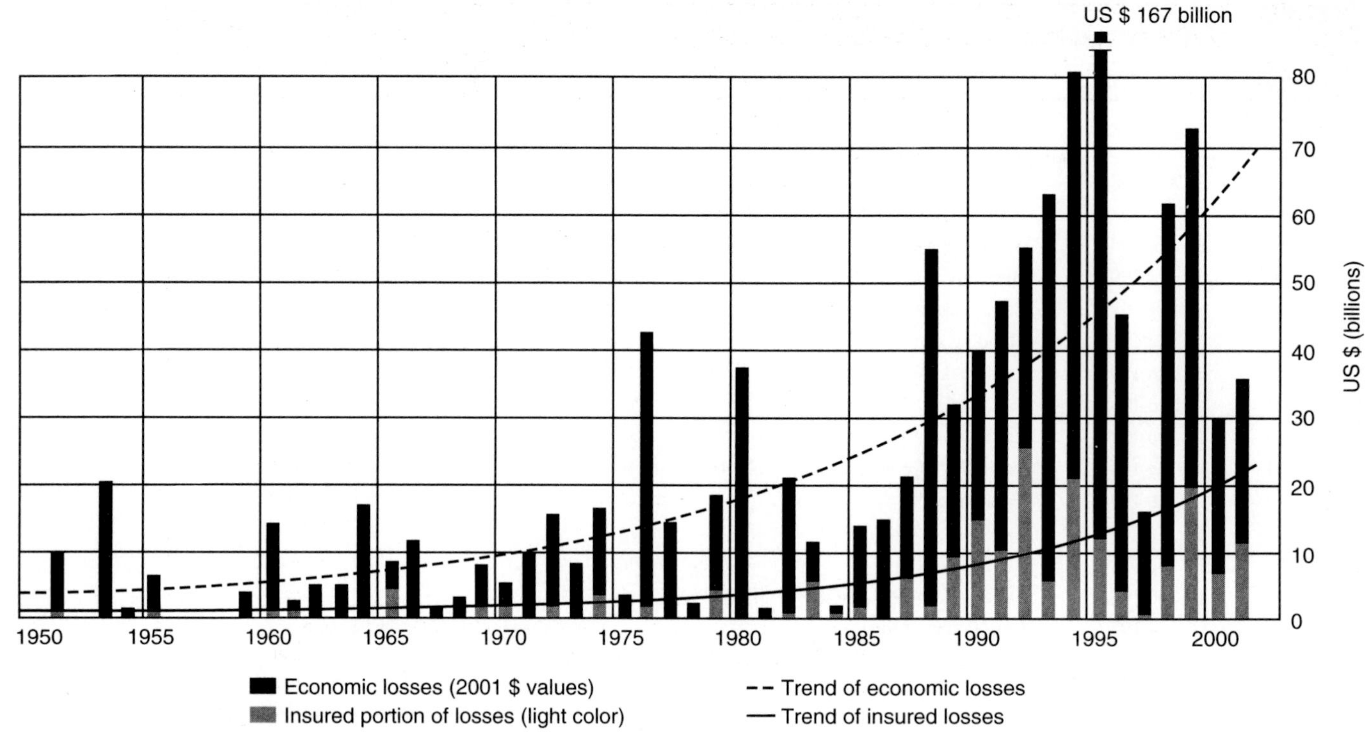

Figure 1.5 Economic and insured losses from natural disasters, 1950–2001.
Data from Munich Reinsurance Company.

Table 1.4 Dollar Losses from Great Natural Disasters, 1950–1999 (Billions of 2001 U.S. $)

Years	1950–59	1960–69	1970–79	1980–89	1990–99
Number Great Events	20	27	47	63	89
Economic Losses	42.2	75.7	136.1	211.3	652.3
Insured Losses	—	7.2	12.4	26.4	123.2

Source: Data from Munich Reinsurance Company (2002).

(Table 1.5) is quite different from the one based on fatalities (Table 1.3).

The locations of the worst dollar-loss disasters for the insurance industry (Table 1.5) are different from the worst locations for fatalities (Table 1.2). The highest insurance dollar losses occurred in the United States (24 of 40), Europe (8), and Japan (4). Wealthy countries are better insured and their people live in safer buildings.

The extent of economic and insured losses may take years to become known. For example, the insured losses from the January 1994 Northridge earthquake were listed at $2.8 billion in February 1994, but grew to $10.4 billion in January 1995, and continued increasing to total $15.3 billion in April 1998.

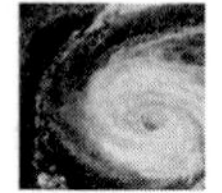

Natural Hazards

Many sites on Earth have not had a natural disaster in recent time, but nonetheless they show clear signs of danger; they have **natural hazards.** For example, people migrate and build next to rivers that will have a big flood, on the shoreline of the sea awaiting a powerful storm, and on the slopes of volcanoes that will erupt. Decades, or even centuries, may pass with no great disasters, but the hazard remains.

Sites with natural hazards need to be studied and understood. Their risks must be evaluated. Then we can design actions that try to prevent natural hazards from causing natural disasters. In this process of **mitigation,** we make plans and take actions to eliminate or reduce the threat of future death and destruction when natural hazards suddenly increase their activity and become great threats. The mitigating actions taken to protect us may be engineering, physical, social, and political.

Another need for mitigation occurs after great disasters occur, because people around the world tend to reoccupy the same site after a disastrous event is done. Earthquakes knock cities down, then the survivors may use the same bricks and stones to rebuild on the same site. Floods inundate towns,

Table 1.5 **The 40 Most Costly Insurance Disasters, 1970–2001**

Losses in Millions of 2001 U.S. $	Fatalities	Date/Start	Event	Country
20,185	38	24 Aug 1992	Hurricane Andrew	USA
19,000	3,000	11 Sep 2001	Terrorist attack	USA
16,720	60	17 Jan 1994	Earthquake (Northridge)	USA
7,338	51	27 Sep 1991	Typhoon Mireille	Japan
6,221	95	25 Jan 1990	Winter Storm Daria	Europe
6,164	80	25 Dec 1999	Winter Storm Lothar	Europe
6,008	61	15 Sep 1989	Hurricane Hugo	USA
4,933	63	17 Oct 1989	Earthquake (Loma Prieta)	USA
4,674	13	15 Oct 1987	Storm	Europe
4,323	64	26 Feb 1990	Winter Storm Vivian	Europe
4,293	26	22 Sep 1999	Typhoon Bart	Japan
3,833	600	20 Sep 1998	Hurricane Georges	USA, Caribbean
3,150	33	5 Jun 2001	Tropical Storm Allison	USA
2,994	167	6 Jul 1988	Explosion on Piper Alpha offshore oil rig	UK
2,872	6,425	17 Jan 1995	Earthquake (Kobe)	Japan
2,551	45	27 Dec 1999	Winter Storm Martin	France
2,508	70	10 Sep 1999	Hurricane Floyd	USA & Bahamas
2,440	59	4 Oct 1995	Hurricane Opal	USA
2,144	246	10 Mar 1993	Storm (East Coast)	USA
2,080	19,118	17 Aug 1999	Earthquake	Turkey
2,019	41	11 Sep 1992	Hurricane Iniki	USA
1,900	—	6 Apr 2001	Storms (tornado/hail)	USA
1,892	23	23 Oct 1989	Explosion at Phillips Petroleum	USA
1,834	—	12 Sep 1979	Hurricane Frederic	USA
1,806	39	5 Sep 1996	Hurricane Fran	USA
1,795	2,000	18 Sep 1974	Hurricane Fifi	Honduras
1,743	116	3 Sep 1995	Hurricane Luis	Caribbean
1,665	350	12 Sep 1988	Hurricane Gilbert	Jamaica
1,594	20	3 Dec 1999	Winter Storm Anatol	Europe
1,578	54	3 May 1999	Tornadoes	USA
1,564	500	17 Dec 1983	Storms (snow/frost)	USA
1,560	26	20 Oct 1991	Fire—into urban area, drought	USA
1,546	350	2 Apr 1974	Tornadoes in 14 states	USA
1,475	—	25 Apr 1973	Flood (Mississippi River)	USA
1,461	—	15 May 1998	Tornadoes	USA
1,413	31	4 Aug 1970	Hurricane Celia	USA
1,386	12	19 Sep 1998	Typhoon Vicki	Japan
1,357	30	21 Sep 2001	Explosion (fertilizer factory)	France
1,337	46	5 Jan 1998	Ice storm	Canada/USA
1,319	21	5 May 1995	Storm (wind, hail, floods)	USA
$156.675 billion	33,973 total			

Source: Data after Swiss Reinsurance Company (2002).

but people return to refurbish and again inhabit the same buildings. Volcanic eruptions pour huge volumes of magma and rock debris onto the land, burying cities and killing thousands of people, yet survivors and new arrivals build new towns and cities on top of their buried ancestors. Why do people return to a devastated site and rebuild? What are their thoughts and plans for the future? For a case history of a natural hazard, let's visit Popocatépetl in Mexico.

Popocatépetl Volcano, Mexico

Popocatépetl is a 17,883-ft (5,452-m) high volcano sitting between the huge populations of Mexico City (largest city in the world) and Puebla (fourth largest city in Mexico) (Figure 1.6). The volcano has had numerous small eruptions over thousands of years, thus its Nahuatl name of Popocatépetl, or Popo as it is affectionately called, which means smoking mountain. But sometimes Popo blasts forth with huge eruptions that destroy cities and alter the course of civilizations. Around the year 822, Popo had large eruptions that buried significant cities. Even its smaller eruptions have affected the course of human affairs. In 1519, Popo was in an eruptive sequence as Hernan Cortez and about 500

Figure 1.6 Popocatépetl in eruption on 19 December 2000. The cathedral was built by the Spanish on top of the great pyramid at Cholula, an important religious site in a large city that was mostly buried by an eruption around 822 C.E.

Spanish conquistadors marched westward toward Tenochtitlan, the Aztec capital city. The superstitious Aztec priest-king Montezuma interpreted the eruptions as omens and they affected his thinking on how to deal with the invasion.

Popocatépetl has helped change the path of history, but what is the situation now? Have people returned to rebuild and raise families? Today, about 100,000 people live at the base of the volcano; they have been attracted by the rich volcanic soil, lots of sunshine, and fairly reliable rains. Millions more people live in the danger zone extending 25 miles (40 km) away. The Nahuatl people consider El Popo to be divine—a living, breathing being. In their ancient religion, God, rain, and volcano are intertwined. Most do not fear the volcano. Commonly expressed thoughts are that God decides events, and that with faith things will work out. Thus, good opportunities for farming, coupled with faith and fatalism, bring people back.

Volcanic activity resumed on 21 December 1994 with eruptions of ash and gases. The sequence of intermittent eruptions continues today. How do we evaluate this hazard? Is this just one of the common multi-year sequences of small eruptions that gave the volcano its name? Or are these little eruptions the forewarnings of a giant killing eruption that will soon blast forth? The answer to these questions is: we don't know. How would you handle the situation? Would you order the evacuation of 100,000 people to protect them, and in so doing, have them abandon their homes, sell their livestock, and leave their independent way of life for an unknown length of time that could be several years? Or would you explain the consequences of an unlikely but possible large eruption and let them decide whether to stay or go? If they decide to stay, and then die during a huge volcanic blast, would this be your fault?

It is relatively easy to identify natural hazards, but as the Popocatépetl case history shows, it is not easy deciding how to answer the questions presented by this volcanic hazard. We are faced with the same types of questions about hazards again and again, for earthquakes, landslides, tornadoes, hurricanes, floods, fire, and **meteorite** impacts.

Magnitude, Frequency, and Return Period

Earth is not a quiet and stable body. Our planet is dynamic and has major flows of energy. Everyday the Earth experiences earthquakes, volcanic eruptions, landslides, storms, floods, fires, meteorite impacts, and extinctions. These energy-fueled events are common, but their **magnitudes** vary markedly over space and time.

Natural hazards and disasters are not spaced evenly about the Earth. For example, some areas experience gigantic earthquakes and some areas are hit by powerful hurricanes; some areas are hit by both, while other areas receive neither.

During a time period of several years, or even several decades, a given area may experience no natural disasters.

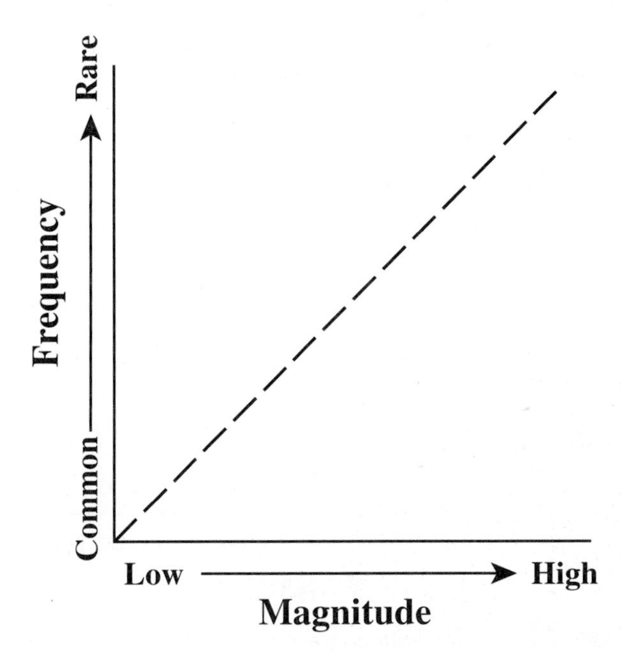

Figure 1.7 Relationship between disaster magnitude and frequency. The greater the magnitude, the rarer the event.

But given a long enough time, some very powerful, high-energy events will occur in every area. It is the concentrated pulses of energy that concern us here, for they are the cause of natural disasters; but how frequent are the big ones? In general, there is an inverse correlation between the **frequency** and the magnitude of a process. The frequent occurrences are low in magnitude, involving little energy in each event. As the magnitude of an event increases, its frequency of occurrence decreases (Figure 1.7). For all hazards, small-scale activity is common, but big events are rarer.

Another way of understanding how frequently the truly large events occur is to match a given magnitude event with its **return period**, or recurrence interval, which is the number of years between same-sized events (Figure 1.8). In general, the larger and more energetic the event, the longer the return period.

A U.S. Geological Survey mathematical analysis of natural-disaster fatalities in the United States assessed the likeliness of killer events. Table 1.6 shows the probabilities of 10- and 1,000-fatality events for earthquakes, hurricanes, floods, and tornadoes for 1-, 10- and 20-year intervals, and estimates the return times for these killer events. On a yearly basis, the majority of low-fatality events are due to floods and tornadoes, and their return times are brief, less than one year. High-fatality events are dominantly hurricanes and earthquakes, and their return times for mega-killer events are much shorter than for floods and tornadoes.

Knowing the magnitude, frequency, and return period for a given event in a given area provides us useful information, but it does not answer all our questions. There are still the cost/benefit ratios of economics to consider. For example,

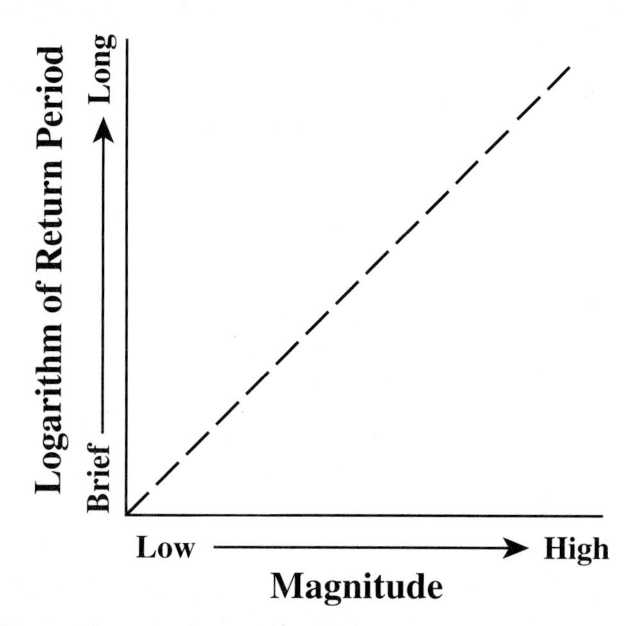

Figure 1.8 Relationship between disaster magnitude and return period. The greater the magnitude, the longer the return period.

Table 1.6 Probability Estimates for 10- and 1,000-Death Natural Disasters in the United States

	Likeliness of a 10-Fatality Event			
	During 1 Year	During 10 Years	During 20 Years	Return Time (in Years)
Earthquakes	11%	67%	89%	9
Hurricanes	39	99	>99	2
Floods	86	>99	>99	0.5
Tornadoes	96	>99	>99	0.3
	Likeliness of a 1,000-Fatality Event			
	During 1 Year	**During 10 Years**	**During 20 Years**	**Return Time (in Years)**
Earthquakes	1%	14%	26%	67
Hurricanes	6	46	71	16
Floods	0.4	4	8	250
Tornadoes	0.6	6	11	167

Data from U.S. Geological Fact Sheet (unnumbered).

given an area with a natural hazard that puts forth a dangerous pulse of energy with a return period of about 600 years, how much money should you spend constructing a building that will be used about 50 years before being torn down and replaced? Will your building will be affected by a once-in-600-year disastrous event during its 50 years? Do you spend the added money necessary to guarantee that your building will withstand the rare destructive event? Or do economic considerations suggest that your building be constructed to the same standards as similar buildings in nearby non-hazardous areas?

Energy Sources of Disasters

Disasters occur where and when the Earth's natural processes concentrate energy and then release it, killing life and causing destruction. Our interest is especially peaked when this energy deals heavy blows to humans. As the growth of the world's population accelerates, more and more people find themselves living in close proximity to Earth's most hazardous places. The news media increasingly present us with vivid images and stories of the great losses of human life and destruction of property caused by natural disasters. As Booth Tarkington remarked: "The history of catastrophe is the history of juxtaposition."

To understand the natural hazards that kill and maim unwary humans, one must know about the energy sources that fuel Earth processes. Four primary energy sources make the Earth an active body: 1) the impact of extraterrestrial bodies, 2) **gravity,** 3) the Earth's internal heat, and 4) the Sun.

An energy source for disasters arrives when visitors from outer space—**asteroids** and **comets**—impact the Earth. Impacts were abundant and important early in Earth's history. In recent times, collisions with large bodies have become infrequent, although when they hit, their effects on life can be global.

Gravity is an attractional force between bodies. At equal distances, the greater the mass of a body, the greater its gravitational force. The relatively great mass of the Earth has powerful effects on smaller masses such as ice, causing **glaciers** to flow and hillsides to fail in landslides.

Internal energy, stored inside the Earth, flows unceasingly toward the surface. Over short time spans, internal energy is released as eruptions from volcanoes and by earthquakes; over longer intervals of geologic time, it has caused the formation of **continents,** oceans, and **atmosphere.** On a planetary scale, this outward flow of internal energy causes continents to drift and collide, thus constructing mountain ranges and elevated plateaus.

About a quarter of the Sun's energy that reaches Earth evaporates and lifts water into the atmosphere. At the same time, the constant pull of gravity helps bring atmospheric moisture down as snow and rain. Gravity powers the agents of **erosion**—glaciers, streams, underground waters, winds, ocean waves and currents—which wear away the continents and dump their broken pieces and dissolved remains into the seas. Thus, solar radiation is the most important external energy source because it evaporates and elevates water, but gravity is the immediate force that drives the agents of erosion.

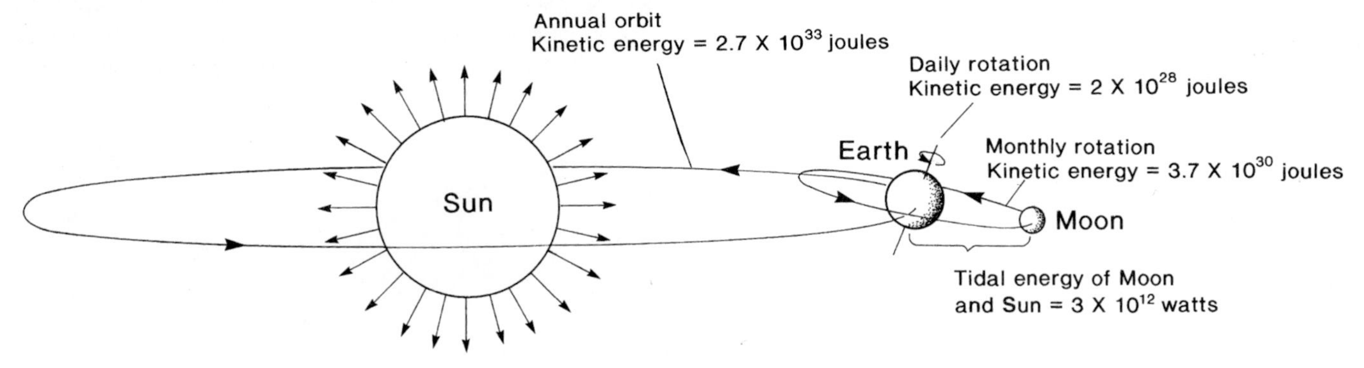

Figure 1.9 The rotations and orbits of the Earth-Moon-Sun system result in tremendous amounts of energy.

Impacts with Asteroids and Comets

The Earth moves through space at a high rate of speed, as do asteroids and comets. When their paths intersect, there are explosive impacts. The Earth travels over 950 million km (590 million mi) around the Sun each year—an orbital speed in excess of 108,000 km/hr (67,000 mph)(Figure 1.9). The kinetic energy of this orbital motion is about 2.7×10^{33} joules. When this tremendous amount of energy is involved in a head-on collision with a large asteroid moving 65,000 km/hr (40,000 mph) or comet traveling 150,000 km/hr (93,000 mph), the effects on life are catastrophic and worldwide.

Additional sources of energy lie in the rotational motions of the Earth—the daily rotation of the Earth about an axis that pierces its center, and the monthly rotation of the Earth-Moon system about its common center of gravity lying about 4,680 km (2,900 mi) away from the center of the Earth toward the Moon (Figure 1.9).

Origin of the Sun and Planets

Impacts of material are not rare and insignificant events in the history of our Solar System; they probably were responsible for its formation. The most widely accepted hypothesis of the origin of the Solar System was stated by the German philosopher Immanuel Kant in 1755. He thought the Solar System formed by growth of the Sun and planets through collisions of matter within a rotating cloud of gas and dust.

The early stage of growth began about 4.6 billion years ago within a rotating spherical cloud of gas, ice, dust, and other solid debris (Figure 1.10a and b). Gravity acting upon matter within the cloud attracted particles, bringing them closer together. Small particles stuck together and grew in size resulting in greater gravitational attraction to nearby particles and thus more collisions. As matter drew inward and the size of the cloud decreased, the speed of rotation increased and the mass began flattening into a disk (Figure 1.10c). The greatest accumulation of matter occurred in the center of the disk, building toward today's Sun (Figure 1.10 d and e). The two main constituents of the Sun are the lightweight elements hydrogen (H) and helium (He). As the central mass grew larger, its internal temperature increased to about 1,000,000 degrees **centigrade** (C) or 1,800,000 degrees **Fahrenheit** (F) and the process of **nuclear fusion** began. In nuclear fusion, the smaller hydrogen atoms combine (fuse) to form helium with some mass converted to energy. We Earthlings feel this energy as **solar radiation** (sunshine).

The remaining rings of matter in the revolving Solar System formed into large bodies as particles continued colliding and fusing together to create the planets (Figure 1.10f). Late-stage impacts between ever-larger objects would have been powerful enough to melt large volumes of rock with some volatile elements escaping into space. The inner planets (Mercury, Venus, Earth, Mars) formed so close to the Sun that solar radiation drove away most of their volatile gases and easily vaporized liquids, leaving behind rocky planets. The next four planets outward (Jupiter, Saturn, Uranus, Neptune) are giant icy bodies of hydrogen, helium, and other frozen materials from the beginning of the Solar System.

Impact Origin of the Moon

Large impacts can generate enough heat to vaporize and melt rock; they can produce amazing results. For example, the dominant hypothesis on the origin of Earth's Moon involves an early impact of the young Earth with a Mars-size body. The resultant impact generated a massive vapor cloud, part of which condensed to form the Moon. This theory suggests the Moon is made mostly from the Earth's rocky **mantle.** The theory accounts for the lesser abundance of iron on the Moon (iron on the Earth is mostly in the central **core**) and the Moon's near absence of lightweight materials (such as gases and water), which would have been lost to space.

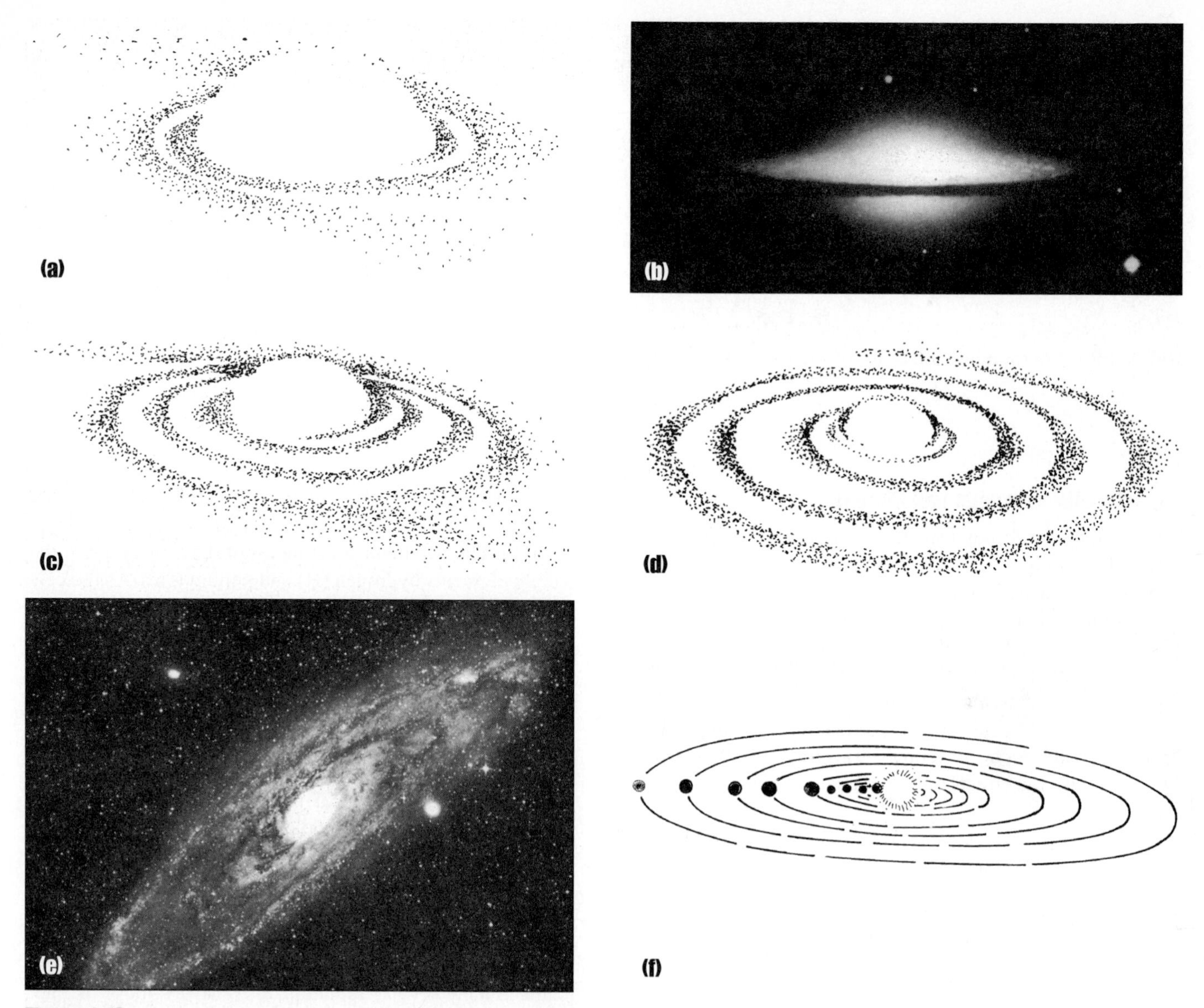

Figure 1.10 Hypothesis of the origin of the Solar System. **(a and b)** Initially, a huge, rotating spherical cloud of ice, gas, and other debris forms. **(c)** Spinning mass contracts into a flattened disk with most mass in the center. **(d and e)** Planets grow as masses collide and stick together. **(f)** Ignited Sun is surrounded by planets. Earth is the third planet from the Sun.

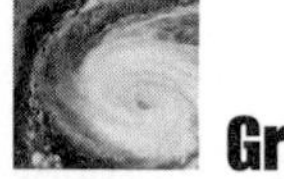

Gravity

The existence of gravity was first discussed scientifically by Isaac Newton (1642–1727), one of the true geniuses in history. It is our good fortune that Newton isolated himself in the countryside during his twenties to avoid the bubonic plague that was ravaging the city-dwelling population. To be successful in any generation, one must avoid the plague of one's time. Newton's accomplishments were many, including inventing calculus and determining the laws of motion and the universal law of gravitation. The importance of fundamental laws was underscored by Ralph Waldo Emerson in 1841, when he wrote: "Nature is an endless combination and repetition of a very few laws. She hums the old well-known air through innumerable variations."

Gravity is an attraction between objects. It is a force that humans are unable to modify; it cannot be increased, decreased, reversed, or reflected. The law of gravity states that two bodies attract each other with a force directly proportional to the product of their masses and inversely proportional to the square of the distance between them:

$$\text{gravity (g)} = \frac{\text{G} \times \text{mass 1} \times \text{mass 2}}{\text{distance} \times \text{distance}}$$

where G = a universal constant

Sidebar

Energy, Force, Work, Power, and Heat

The effectiveness of agents and events is measured using the related terms of energy, **force, work, power,** and **heat.** Energy is the capacity to do work; it may be potential or kinetic. **Potential energy** is poised and ready to go to work. For example, a house-sized boulder resting precariously high on a steep slope has the potential to roll and bounce downhill and do a lot of damage (Figure 1.11a). The potential energy (PE) of the huge boulder is equal to its mass (m) times the force of gravity (g) times its height (h) above a certain level, which in this case is the elevation of the valley floor below it:

$$PE = mgh$$

If the boulder starts to roll, its potential energy now becomes kinetic—the energy of motion (Figure 1.11b). **Kinetic energy** (KE) is determined by half the product of mass (m) times the velocity (v) squared:

$$KE = 1/2\ mv^2$$

The kinetic energy of the bouncing boulder adds to its work if its downslope collisions cause other boulders to move and the resultant moving mass brings soil, trees, and other debris downhill with it. The work (w) done on the sliding mass is determined as force (F) times distance (d), where force equals mass (m) times acceleration (a):

$$\text{Force (F)} = ma$$
$$\text{Work (w)} = Fd = mad$$

Force may be measured using a unit called a dyne. Each dyne equals a mass of one gram (gm) accelerated one centimeter (cm) per second squared:

$$\text{dyne} = 1\ \text{gm cm/sec}^2$$

Energy, as well as work, may be expressed in dyne-centimeters (= dyne × cm). Work is defined as force acting over a distance and is dimensionally equivalent to energy. A force of one dyne acting over a distance of one centimeter is called an erg:

$$\text{erg} = \text{dyne} \times \text{cm} = \text{gm cm}^2/\text{sec}^2$$

A dyne is a small unit of force, and an erg is a small unit of work. Thus, for large-scale phenomena, a larger unit of measurement called a joule is used. A joule is 10 million ergs:

$$\text{joule} = 10^7\ \text{erg}$$

The landslide triggered by the bouncing mega-boulder may move rapidly (faster than a human can run) or slowly. Whether fast or slow, if the same amount of material ends up at the bottom of the slope, then the amount of work is the same. However, the power is different. Power is defined as the rate at which work occurs and is measured in watts:

$$\text{power} = \text{work/time} = \text{joule/second} = \text{watt}$$

After the boulder caused the landslide to move, what happened to slow it down and stop it? Friction—friction with the underlying ground and friction among the boulders, sand grains, trees, and other debris inside the moving mass. As you know from sliding into second base or across a dance floor, friction generates heat. Thus, kinetic energy is related to heat. The definition of heat is the capacity to raise the temperature of a mass; it is expressed in calories. The relationship of work to heat is:

4.185 joules = 1 **calorie**

Calories are units of energy. One calorie is defined as the amount of heat needed to raise the temperature of one gram of water by one degree centigrade under a pressure of one atmosphere. One gram of water is equivalent to about 10 drops. The calorie commonly discussed with food is actually a kilocalorie, or 1,000 of the calories defined here.

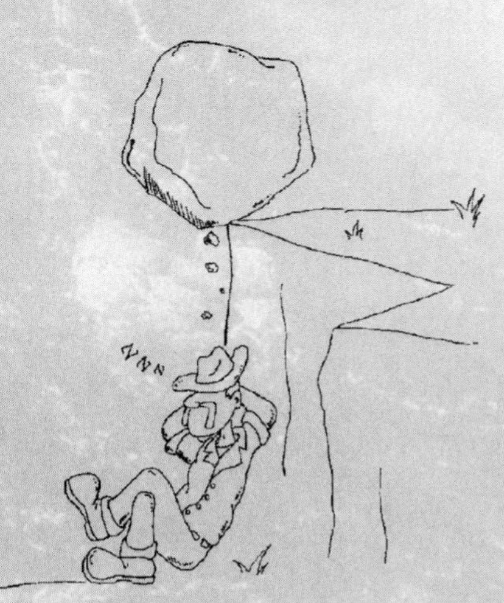

Figure 1.11 (a) When the boulder is poised and ready to move, it has potential energy. **(b)** When the boulder is rolling, its energy is kinetic.
Drawings by Jacobe Washburn.

In a planetary body, the attracting mass can be considered as concentrated at its center. In the discussion of the Earth's internal energy, it was seen that gravity caused the melted iron-rich materials to collapse toward the Earth's center to form the core, thus liberating internal heat and helping create the Earth's density layering.

The gravitational system of the Earth, Moon, and Sun, and their interactions generates tidal energy. Using Newton's equation to assess the gravitational effects of the Sun and Moon on the Earth requires knowledge of masses and distances. The Moon has a diameter of about 3,500 km (2,160 mi), whereas the Earth has an average diameter of about 12,800 km (7,926 mi). For comparison, if the Earth were reduced to the size of a basketball, then the Moon would be slightly smaller than a tennis ball.

The volume of the Moon is only about one-forty-ninth that of the Earth, and the Moon's lower average density of 3.34 means its mass is only about one-eightieth that of the Earth. By comparison, the Sun's diameter is about 1,395,000 km (864,000 mi), and even though its density is only about one-fourth that of the Earth, its mass is still about 332,000 times greater. Since gravitational attraction is directly proportional to mass, the gravitational pull of the massive Sun would absolutely dwarf that of the Moon, if it were at the same distance from the Earth. But this gravitational attraction is reduced by dividing by distance times distance, and the Sun is 150 million km (93 million mi) away. Thus, the Sun's massive gravitational pull is so diminished by distance that our Moon, only about 386,000 km (239,000 mi) away, exerts more force on the Earth. Calculations show that

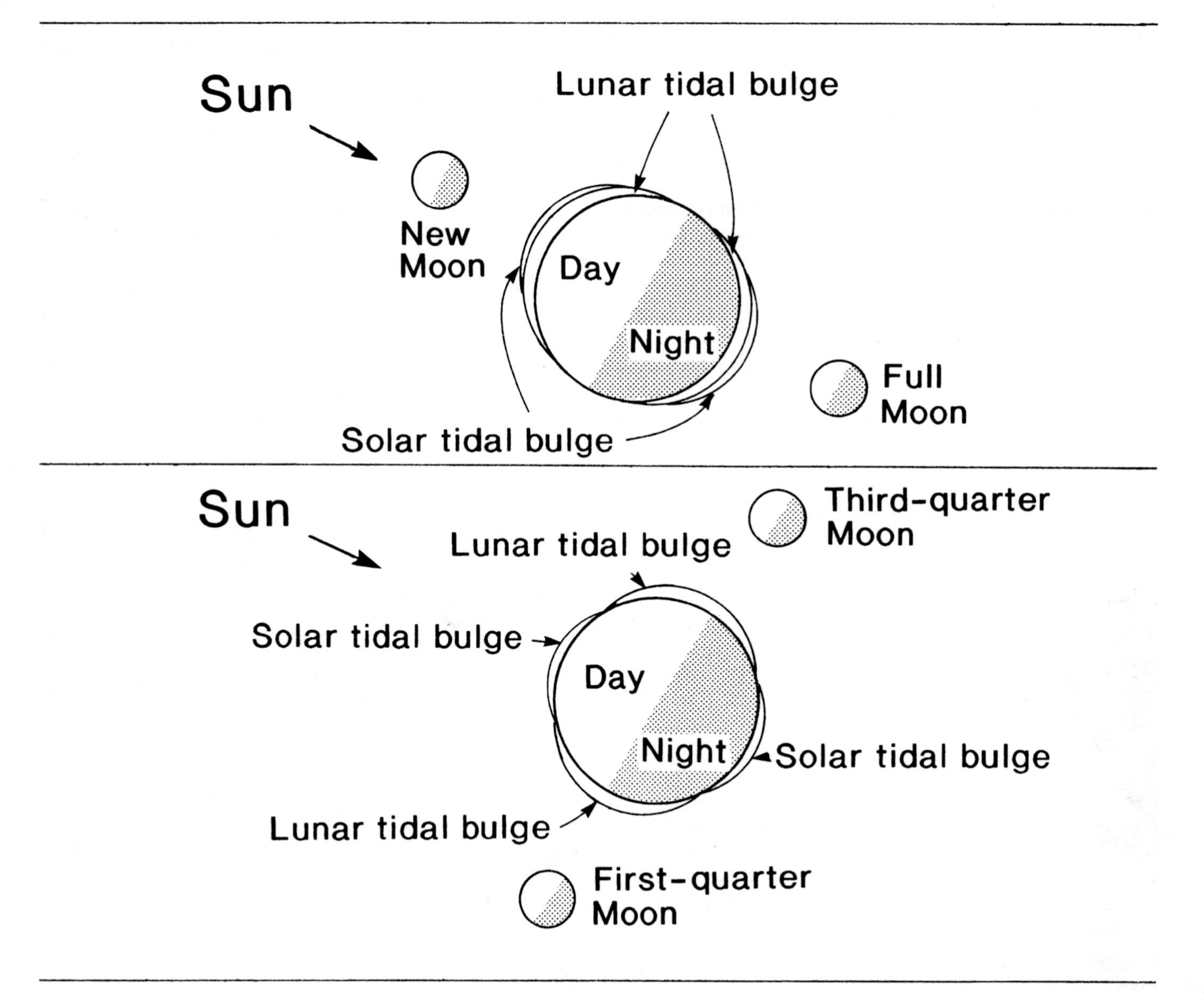

Figure 1.12 Earth tides are caused by the gravitational attractions of the Sun and the Moon. The greatest daily range of tides occurs at the new and full Moons; the lowest daily range occurs at first- and third-quarter Moons.

the gravitational effect of the Sun on the Earth is only 46 percent as strong as the pull from the Moon; that is, the Moon's pull on Earth is more than double that of the Sun.

The Earth has rather unique tidal effects because: 1) 71 percent of its surface is covered by oceans; 2) it has a long period of rotation compared to many other planets; and 3) its relatively large Moon is nearby. The gravitationally attracted bulges we call tides affect the land, water, and air but are most visible in the daily rises and falls of the ocean surface. The Sun appears overhead once every 24 hours, while the Moon takes about 24 hours and 52 minutes to return to an overhead position. Thus, the Moon appears to move in the sky relative to the Sun. So, too, will the tidal bulges attracted by the Moon move in relation to the tidal bulges caused by the Sun. The two sets of tidal bulges will coincide twice a month, at the new and full moons, when the Sun and Moon align with the Earth (Figure 1.12). These highest tides of the month are called spring tides. In the first and third quarters of the Moon, the Sun and Moon are at right angles to Earth, thus producing the lowest tidal ranges, called neap tides.

The tidal bulges moving across the face of the Earth, and within mobile intervals in the Earth's interior, cause a frictional braking of the Earth's rotation. Following Newton's laws of motion, as the rotations of the Earth and Moon slow down, they move farther apart, days become longer, and the years have fewer days. At present, the Earth and Moon are separating an additional 3.8 cm (1.5 in) per year. Substantiation of the lengthening days is evident in the fossil record. For example, careful counting of growth ridges in the skeletons of corals (broadly similar to tree rings) shows daily additions that vary in size according to the season of the year. A study of 370-million-year-old corals has shown that each day on Earth during their life was about 22 hours long, and a year had 400 days.

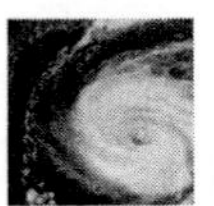

Internal Sources of Energy

To understand the origin and character of Earth's internal energy, one must know the early history of our planet. Studying early history is difficult because the Earth is a dynamic planet; it recycles its rocks and thus removes much of the record of its early history. The older the rocks, the more time and opportunities there have been for their destruction. Nonetheless, the remaining early Earth rocks, along with our growing knowledge of the processes in the Earth's interior and in the Solar System, allow us to build an increasingly sophisticated approximation of early Earth history.

Earth appears to have begun as an aggregating mass of particles and gases from a rotating cloud some 4.57 billion years ago. During a 50- to 100-million-year period, bits and pieces of metal-rich particles (similar to iron-rich meteorites), rocks (similar to stony meteorites), and ices (of water, carbon dioxide, and other compounds) accumulated to form the Earth. As the ball of coalescing particles enlarged, the gravitational force may have pulled more of the metallic pieces toward the center, while some of the lighter-weight materials may have concentrated near the exterior. Nevertheless, the Earth in its infancy probably grew from random collisions of debris that formed a more or less homogeneous mixture of materials.

But the Earth did not remain homogeneous. The very processes of planet formation (Figure 1.13) created tremendous quantities of heat, which fundamentally changed the young planet. The heat that transformed the Earth came primarily from: 1) impact energy, 2) gravitational energy, and 3) decay of **radioactive** elements.

As the internal temperature of the Earth rose beyond 1,000° centigrade (C) or 1,800° Fahrenheit (F), it passed the melting points of iron at various depths below the surface. Iron forms about one-third of the Earth's mass, and although it is much denser than ordinary rock, it melts at a much lower temperature. The buildup of heat caused immense masses of iron-rich meteorites to melt. The high-density liquid iron was pulled by gravity toward the Earth's center. As these gigantic volumes of liquid iron moved inward to form the Earth's core, they released a tremendous amount of gravitational energy that converted to heat and probably raised the Earth's internal temperature by another 2,000° C. The release of this massive amount of heat would have produced widespread melting likely to have caused low-density materials to rise and form: 1) a primitive **crust** of low-density rocks at the surface of the Earth; 2) large oceans; and 3) a heavier atmosphere. The formation of the iron-rich core was a unique event in the history of the Earth. The planet was changed from a somewhat homogeneous ball into a density-stratified mass with the heavier materials in the center and progressively lighter materials outward to the atmosphere.

Impact Energy and Gravitational Energy

The impact energy of masses colliding with the growing Earth produced heat. Tremendous numbers of large and small asteroids, meteorites, and comets hit the early Earth, with their energy of motion converted to heat on impact.

Gravitational energy was released as the Earth pulled into an increasingly dense mass during its first 50 to 100 million years. The ever-deeper burial of material within the growing mass of the Earth caused an increasingly greater gravitational pull that further compacted the interior. This gravitational energy was converted to heat.

The immense amount of heat generated during the formation of the Earth did not readily escape because heat conducts very slowly through rock. Some of this early heat still is flowing to the surface today.

Radioactive Elements

Energy is released from radioactive elements as they decay. Radioactive atoms are unstable and must kick out subatomic

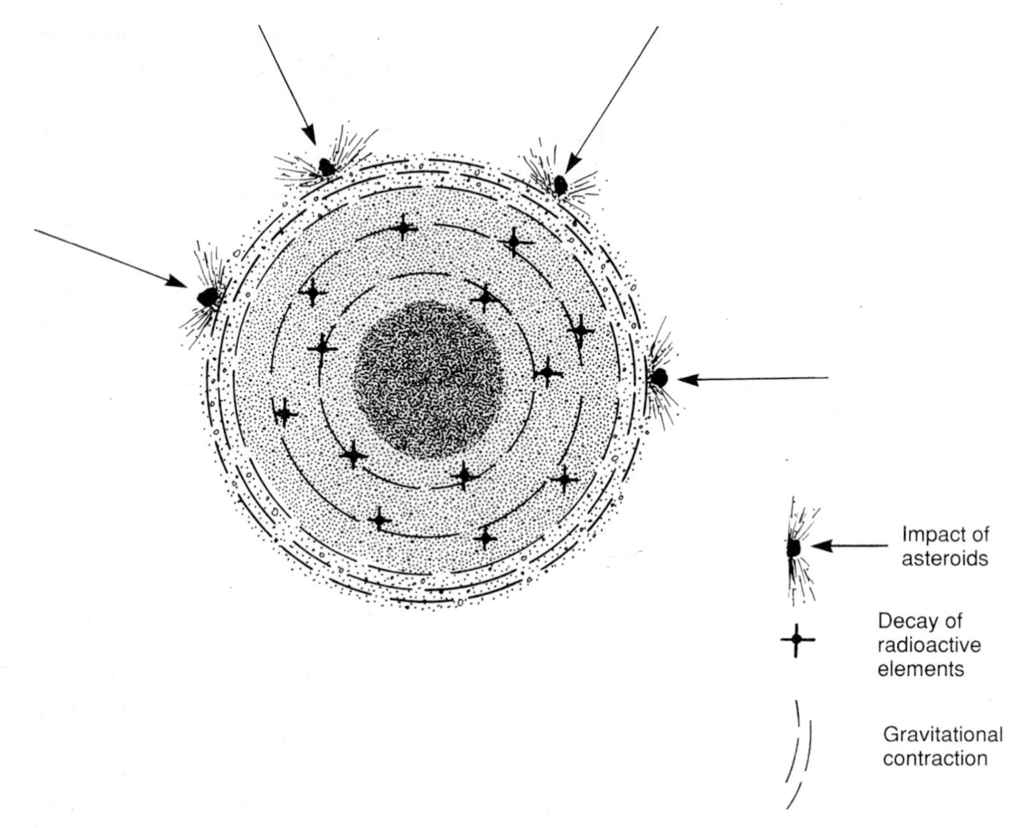

Figure 1.13 Heat-generating processes during the formative years of the Earth include: 1) impact of asteroids, 2) decay of radioactive elements, and 3) gravitational contraction.

particles to attain stability. As radioactive atoms decay, heat is released.

In the beginning of the Earth, there were abundant, short-lived radioactive **elements,** such as aluminum-26, that are now effectively extinct, as well as long-lived radioactive elements, many of which have now expended much of their energy (Table 1.7). The young Earth had a much larger complement of radioactive elements and a much greater heat production from them than it does now (Figure 1.14). With a declining output of radioactive heat inside the Earth, the flow of energy from the Earth's interior is on a slow decline curve heading toward zero.

The radioactive-decay process is measured by the **half-life,** which is the length of time needed for half the present number of atoms of a radioactive element (parent) to disintegrate to a decay (daughter) product. As the curve in Figure 1.15 shows, during the first half-life, one-half of the radioactive atoms decay. During the second half-life, one-half of the remaining radioactive atoms decay (equivalent to 25 percent of original parent atoms). The third half-life witnesses the third halving of radioactive atoms present (12.5 percent of the original parent atom population), and so forth. Half-lives plotted against time produce a negative exponential curve (Figure 1.15); this is the opposite direction of a positive exponential curve, such as interest being paid on money in a savings account.

Table 1.7 Some Radioactive Elements in the Earth

Parent	Decay Product	Half-Life (billion years)
Aluminum-26	Magnesium-26	0.00072 (720,000 years)
Uranium-235	Lead-207	0.71
Potassium-40	Argon-40	1.3
Uranium-238	Lead-206	4.5
Thorium-232	Lead-208	14
Rubidium-87	Strontium-87	47
Samarium-147	Neodymium-147	106

The sum of the internal energy from impacts, gravity, and radioactive elements, plus additional energy produced by tidal friction, is very large. The greater abundance of radioactive elements at the Earth's beginning plus the early gravitational compaction and more frequent meteorite impacts combined to elevate the Earth's internal temperature during its early history. It is noteworthy that this heat buildup inside the Earth reached a maximum early in the Earth's history and has declined significantly since then. Nonetheless, the flow of internal heat toward the Earth's surface today is still great enough to provide the energy for continents to drift, volcanoes to erupt, and earthquakes to shake.

Age of the Earth The Earth is inferred to be about 4.57 billion years old; this is 4,570 million years—time for many changes to occur. The 4.57 billion-year age has been measured using radioactive elements and their decay products collected from Moon rocks and meteorites. The oldest Earth rocks found to date are 4.055 billion years old in northwest Canada and 3.9 billion years old in Greenland. These rocks are of crustal composition, implying that they were recycled and formed from even older rocks. The oldest ages obtained on Earth materials are 4.4 billion years, measured on zircon sand grains collected from a 3.1 billion-year-old sandstone in western Australia.

How can we infer that the Earth is 4.57 billion years old if the oldest known Earth rocks are slightly more than 4 billion years old and the oldest known minerals are 4.4 billion years old? Earth is such an energetic planet that surface rocks are continually being formed and destroyed. Because of these active Earth processes, truly old materials are rarely preserved; there have been too many events, over too many years, that destroy rocks. We look instead to the oldest rocks on the Moon, which is no longer geologically active, and to the meteorites that arrive from the refrigerator of space, noting that they have consistent ages of about 4.57 billion years. Then, from the hypothesis of a common origin for the

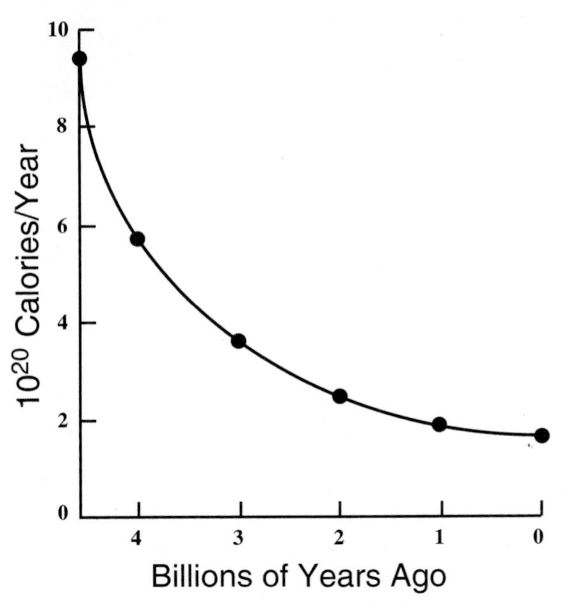

Figure 1.14 Rate of heat production from decay of radioactive atoms has declined throughout the history of the Earth.

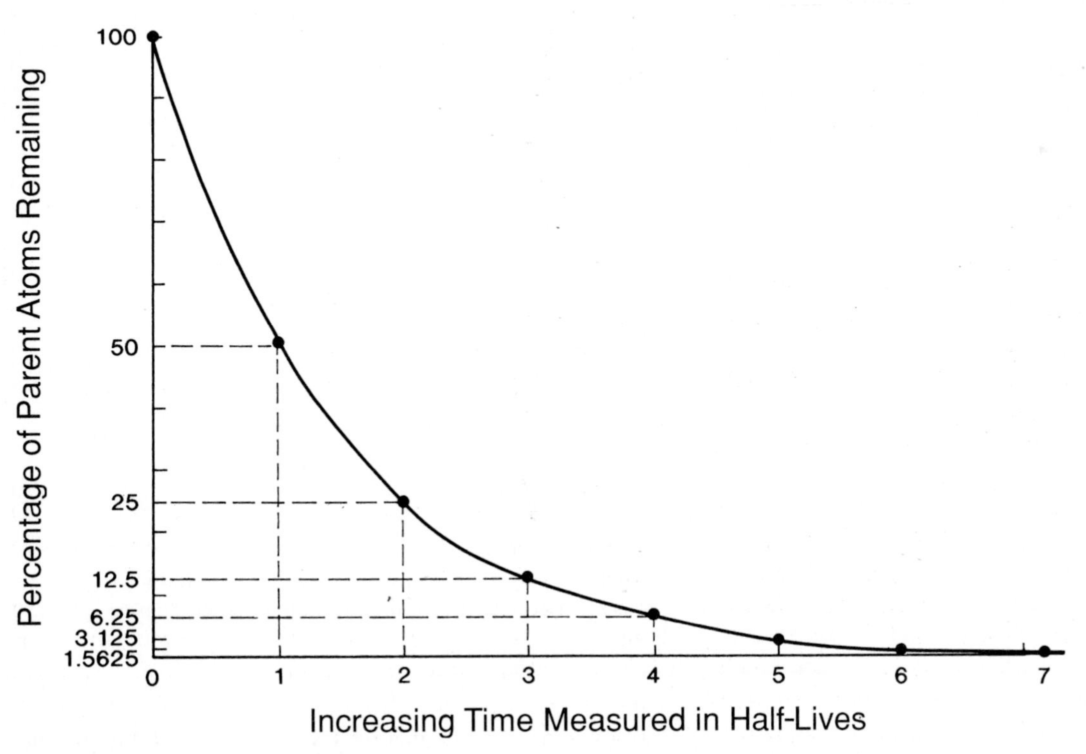

Figure 1.15 Negative exponential curve showing decay of radioactive parent atoms to stable daughter atoms over time. Each half-life witnesses the disintegration of half the remaining parent atoms.

Sidebar

Radioactive Elements

Energy is released from radioactive elements in the process of **nuclear fission** when unstable, radioactive parent atoms shed excess subatomic particles, reducing their weight and becoming smaller daughter atoms (Figure 1.16). The nuclei of radioactive atoms are unstable and contain too many subatomic particles, both positively charged protons and neutral neutrons. The overly heavy radioactive atoms slim down to a stable weight by emitting: 1) alpha particles, consisting of two protons and two neutrons (effectively, the nucleus of a helium atom); 2) beta particles, which are electrons freed upon a neutron's splitting; and 3) gamma radiation, which is similar to X rays but with shorter wavelength. As the rapidly expelled particles are slowed and absorbed by surrounding matter, their energy of motion is transformed into heat.

Dating the Events of Earth History

The same decaying radioactive elements producing heat inside the Earth also may be read as clocks that date events in Earth history. For example, uranium-238 decays to lead-206 through numerous steps involving different isotopes and new elements (Figure 1.17). By emitting alpha and beta particles, 32 of the 238 subatomic particles in the U-238 nucleus are lost, leaving the 206 particles of the Pb-206 nucleus. Laboratory measurements of the rate of the decay process have given us the U-238-to-Pb-206 half-life of 4.5 billion years. These facts may be applied to quantifying Earth history by reading the radiometric clocks preserved in some minerals. For example, some **igneous rocks** (crystallized from **magma**) can be crushed, and the very hard mineral zircon (from which zirconium, the diamond substitute in jewelry, is synthesized) separated from it. Zircon crystals contain uranium-238 that was locked into their atomic structure when they crystallized from magma, but they originally contained virtually no lead-206. Thus, the lead-206 present in the crystal must have come from decay of uranium-238.

The collected zircon crystals are crushed into a powder and dissolved with acid under ultraclean conditions. The sample is placed in a mass spectrometer to measure the amounts of parent uranium-238 and daughter lead-206 present. Then with three known values—1) the amount of U-238, 2) the amount of Pb-206, and 3) the half-life of 4.5 billion years for the decay process—it is easy to calculate how long the U-238 has been decaying into Pb-206 within the zircon crystal. In other words, the calculation tells us how long ago the zircon crystal formed and consequently the time of formation of the igneous rock.

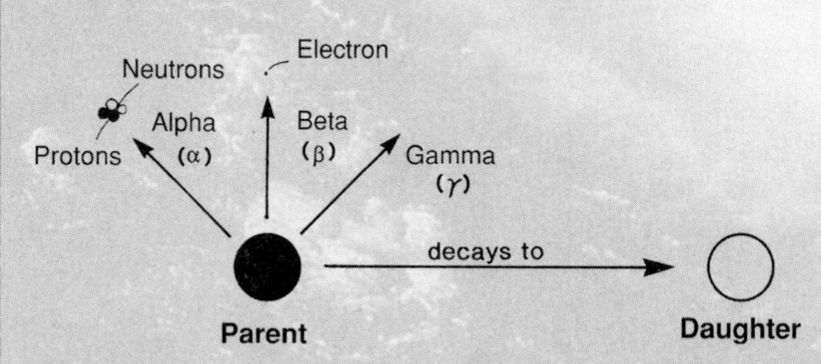

Figure 1.16 A radioactive parent atom decays to a smaller daughter atom by emitting alpha particles (such as the nucleus of a helium atom, i.e., two protons and two neutrons), beta particles (electrons), and gamma radiation (such as X rays).

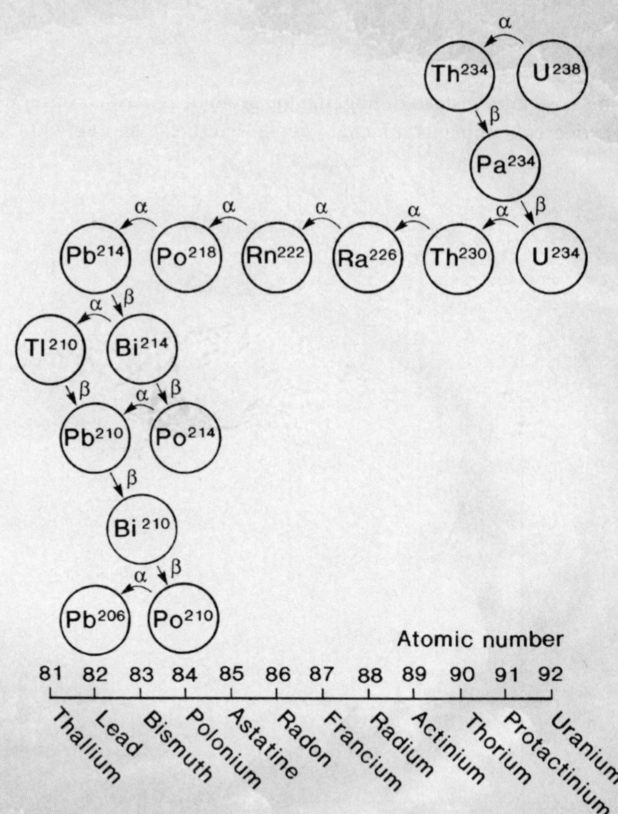

Figure 1.17 Radioactive uranium-238 (U^{238}) decays to stable lead-206 (Pb^{206}) by steps through many intermediate radioactive atoms. The atomic number is the number of protons (positively charged particles) in the nucleus.

Earth, Moon, meteorites, and the rest of our Solar System, the 4.57 billion-year age measured on Moon rocks and meteorites is used as the age for all. Thus, we can conclude from direct evidence that the Earth is 4+ to 4.4 billion years old and infer another 0.17 to 0.52 billion years, bringing Earth's age into line with that of the Moon and meteorites.

Sidebar

Radioactivity Disasters

Radioactivity disasters—the term brings to mind the meltdown of the uranium-rich core of a nuclear power plant, such as happened at Chornobyl in Ukraine, part of the former Soviet Union, on 26 April 1986. This human-caused disaster occurred when the night-shift workers made a series of mistakes that unleashed a power surge so great that the resultant explosions knocked off the 1,000-ton lid atop the nuclear reactor core, blew out the building's side and roof, triggered a partial meltdown of the reactor core's radioactive fuel, and expelled several tons of uranium dioxide fuel and fission products, such as cesium-137 and iodine-131, in a 5-km-(3-mi)-high plume. As much as 185 million **curies** of dangerous radioactive atoms were released. (The worst U.S. incident released 17 curies from the Three Mile Island nuclear power plant in Pennsylvania during 1979.) After the 1:24 A.M. explosion, people near Chornobyl were at least fortunate that they were indoors and thus somewhat sheltered, there was no rain in the area, and the contaminant plume rose high instead of hugging the ground. The cloud of radioactive contaminants affected people, livestock, and agriculture from Scandinavia to Greece. At the Chornobyl power plant, 31 workers were killed. But most deaths came later from cancer and other diseases. At the end of 1999, there were 165,000 deaths attributed to this nuclear accident by Swiss insurance companies. And many more will die in upcoming years.

An earthquake may have helped trigger this disaster. It is widely reported in Europe that the Chornobyl power-plant workers were having difficulties in the early morning hours of 26 April and then a magnitude 3 earthquake occurred 12 km (7 mi) away. The panicked supervisor thought the shaking meant the power plant was losing control and he quickly made emergency maneuvers, but they jammed the internal works of the reactor leading to the fateful explosion 22 seconds after the earthquake. Can the Chornobyl meltdown be considered an earthquake disaster?

But Chornobyl was a human-caused disaster. What can happen under natural conditions? Today, on Earth and the Moon, uranium is present mostly as the heavier U-238 **isotope,** which has a combined total of 238 protons and neutrons in each uranium atom nucleus. The lighter-weight uranium isotope, U-235, makes up only 0.7202 percent of all uranium atoms. In nuclear power plants, the uranium ore fed to nuclear reactors is enriched to 2 to 4 percent U-235 to promote more potent reactions. Remember from Table 1.7 that U-235 has a half-life of 0.71 billion years, whereas the half-life of U-238 is 4.5 billion years. Because U-235 decays more rapidly, it would have been relatively more abundant in the geologic past. In fact, at some past time, the U-235 natural percentage relative to U-238 would have been like the U-235 percentage added to U-238 and fed as ore to nuclear reactors today.

Have natural nuclear reactors operated in the geologic past? Yes. A well-documented example has been exposed in the Oklo uranium mine near Franceville in southeastern Gabon, a coastal country in equatorial West Africa. At Oklo 2.1 billion years ago, sands and muds accumulated along with organic carbon from the remains of fossil bacteria. These carbon-bearing **sediments** were enriched in uranium; U-235 was then 3.16 percent of total uranium. The sand and mud sediments were buried to shallow depths, and at least 800 m^3 of uranium ore sustained nuclear fission reactions that generated temperatures of about 400°C regionally with much higher local temperatures. At Oklo, 17 sites started up as natural nuclear reactors about 1.85 billion years ago; they ran for at least 500,000 years (and maybe as long as 2 million years). Nine of the natural reactors that have been carefully studied are estimated to have produced at least 17,800 megawatt years of energy.

External Sources of Energy

Energy generated by nuclear fission flows to the Earth's surface constantly via **conduction** from the Earth's interior and by **convection** through magma in volcanoes and water in hot springs. These internal energy flows accomplish impressive geologic work yet their total amount of energy is miniscule compared to the energy radiated from the nuclear fusion occurring in the Sun (Figure 1.18). Only a minute percentage of the radiant energy of the Sun reaches the Earth, yet it is more than 5,300 times greater than the heat flow from the Earth's interior (Table 1.8).

Energy is also supplied externally via gravitational attractions between the Earth, Moon, and Sun that add tidal energy to the Earth. In addition, incoming meteorites, asteroids, and comets still impact upon our planet.

The Sun

The Sun emits or radiates energy across a broad spectrum of wavelengths ranging from radio waves that are tens of kilometers apart to gamma rays spaced closer together than one billionth of a centimeter. Most of the solar radiation is concentrated in the part of the wavelength spectrum visible to humans (light) or nearly visible (infrared and ultraviolet). Visible light is about 43 percent of the solar radiation received on

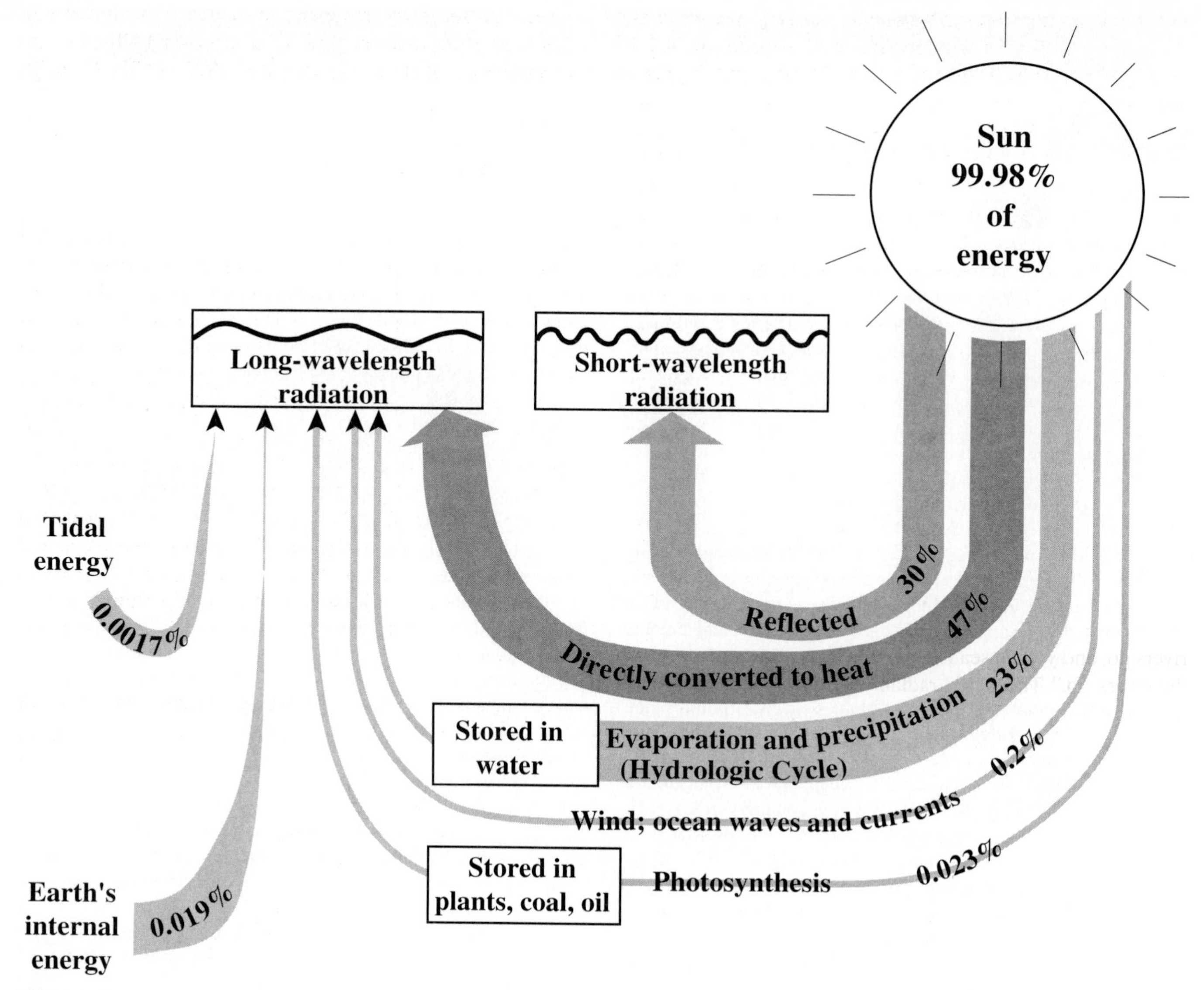

Figure 1.18 The flow of energy on Earth from the Sun, tides (gravity), and Earth's interior. After M. King Hubbert (1971).

Table 1.8 Power Flow to and from Earth

($\times 10^{12}$ Joules per Second)		
Solar radiation	173,410	
Direct reflection		52,000
Direct conversion to heat		81,000
Evaporation		40,000
Water transport in oceans and atmosphere		370
Photosynthesis		40
Heat flow from interior	32.3	
General heat flow by conduction		32
Volcanoes and hot springs		0.3
Tidal energy	3	

Source: Data from Hubbert (1971).

Earth; it ranges in wavelengths from 0.0004 cm (violet) to 0.0007 cm (red). The almost-visible solar energy is received 49 percent in near-infrared wavelengths we can detect as heat and 7 percent in the ultraviolet (UV) wavelengths that give us sunburn.

All objects radiate energy. The hotter the object, the more energy it radiates and increasingly more is at shorter wavelengths. The Sun radiates hundreds of thousands times more energy than does the Earth and mostly at shorter wavelengths. Solar radiation commonly is referred to as short wavelength, and radiation from the Earth is called long wavelength.

Not all of the Sun's energy reaching the Earth's surface is involved in accomplishing work (creating activity in and among Earth's systems). About 30 percent is directly reflected back to space as short-wavelength radiation (Figure 1.18).

Reflectivity is known as **albedo** and is usually measured as the percentage of solar radiation reflected. Albedo can be 80 to 85 percent from fresh snow, 20 to 25 percent off grass, and 10 percent off wet earth. Another 47 percent of incoming solar radiation is absorbed as heat by the air, sea, and land. Water has an amazing ability to absorb and hold solar energy. The global ocean is such an enormous body of water that the solar heat it stores influences weather around the world. On an average day, the amount of solar energy absorbed by the ocean and then reradiated back as long-wavelength radiation would be enough heat to raise the temperature of the whole atmosphere almost 2°C (3°F). However, the atmosphere does not have as great a capacity to store heat; its total heat storage is equivalent to that held in the upper 3 m (10 ft) of the ocean. The remaining 23 percent of solar radiation is expended to evaporate water and begin the hydrologic cycle.

The Hydrologic Cycle Evaporated water rises convectively, due to its lower density, up into the atmosphere, performing the critical initial work of the **hydrologic cycle.** The hydrologic cycle was in part recognized in the third century B.C.E. in Ecclesiastes 1:7, where it is stated: "Into the sea all the rivers go, and yet the sea is never filled, and still to their goal the rivers go." The Sun's radiant energy evaporates water, primarily from the oceans, which then drops on the land and flows back to the sea both above and below ground (Figure 1.19). The same water, for the most part, has run through this same cycle, time and time again, for over 4 billion years. Percy Bysshe Shelley described it in 1820 in "The Cloud":

> I am the daughter of Earth and Water,
> And the nursling of the sky;
> I pass through the pores of the oceans and shores;
> I change, but I cannot die.

The hydrologic cycle is still the subject of study. Figure 1.20 depicts the volumes of water in temporary storage in ice, lakes, underground, and elsewhere, as well as the water moving about. Volumes of water in thousands of cubic kilometers are hard to visualize (4.1 km^3 equals 1 mi^3), but the relative abundances are easy to compare. The hydrologic cycle is a continuously operating, distilling-and-pumping system. The heat from the Sun evaporates water, while plants transpire (evaporate from living cells) water into the atmosphere. The atmospheric moisture condenses and precipitates as snow and rain. Some falls on the land and then is pulled back to the sea by gravity as glaciers, rivers, and via underground water flow, i.e., as the agents of erosion. The system is over 4 billion years old and will continue to operate as long as the Sun shines and water lies on the surface of the Earth.

Heat Capacity, Latent Heat, and Energy Transfer The Sun not only powers the hydrologic cycle, its energy is stored in

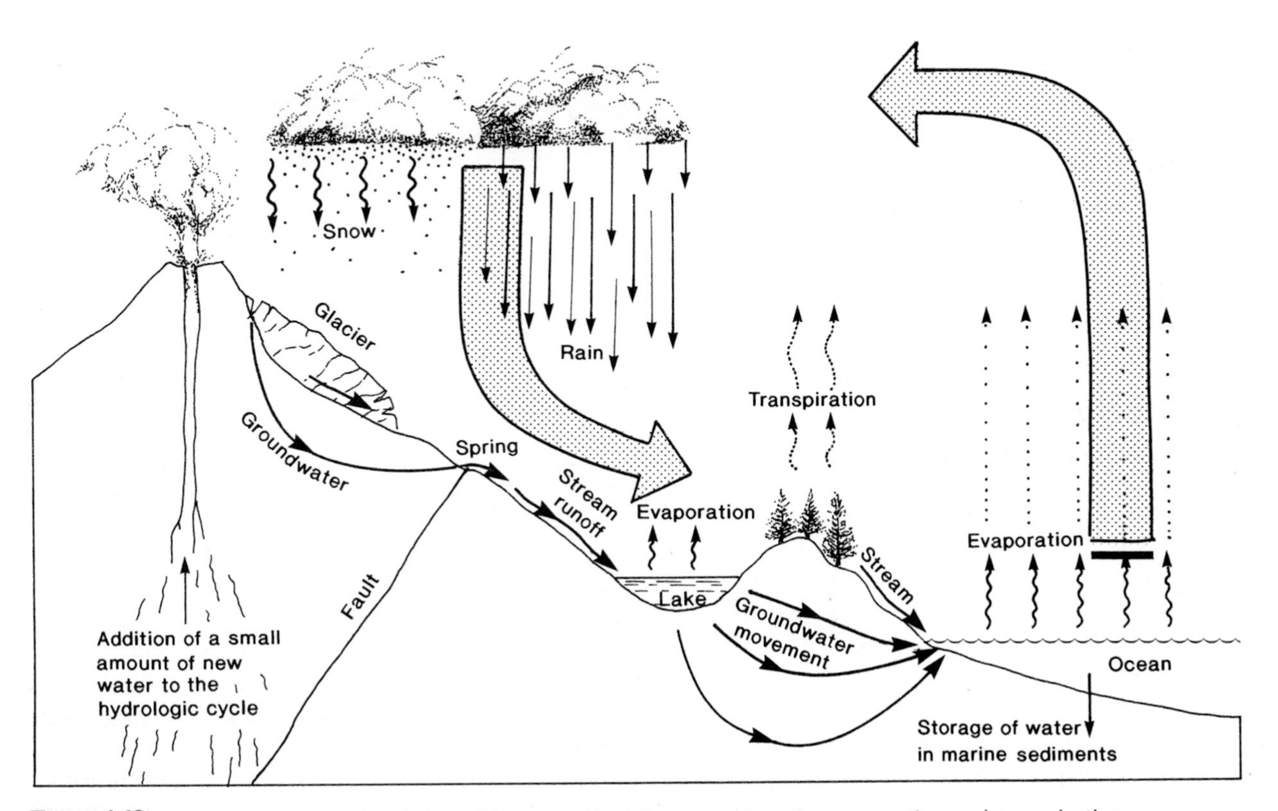

Figure 1.19 The hydrologic cycle. The Sun lifts water into the atmosphere by evaporation and transpiration. Atmospheric water condenses and falls under the pull of gravity. The water then flows as glaciers, streams, and groundwater, returning to the seas.

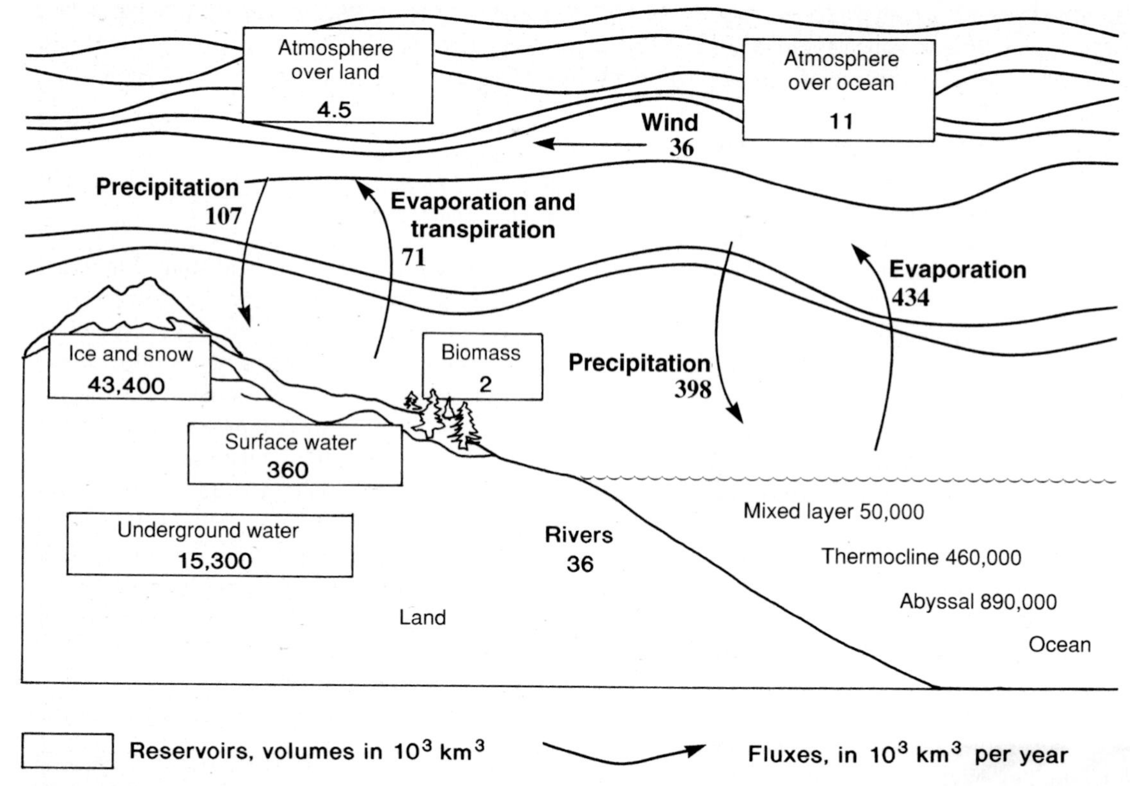

Figure 1.20 Water moving through the hydrologic cycle at global scale.

water and water vapor. Where is solar energy delivered to the Earth? Different areas of the Earth's surface receive different amounts of solar radiation; the amounts received are measured as **insolation.** The equator and low latitudes receive excess solar radiation, but moving toward the North and South Poles, energy loss begins to exceed gains around 32° to 35° latitude and the losses progressively increase to the poles (Figure 10.3). The imbalances in heat between the tropical and polar latitudes help cause ocean currents and winds that transfer heat from the tropics toward the poles. The ocean and the atmosphere act like gigantic heat engines that transfer energy around the world.

Where is the water on Earth held? The oceans hold the greatest share: 97.1 percent of Earth's water covering 71 percent of Earth's surface. Of the remaining 2.9 percent of all water, 68 percent is locked up in glaciers, 21 percent moves slowly underground, 10 percent is in the atmosphere, and the remaining less than 1 percent is shared by rivers, lakes, and inland seas and as moisture in the soil. Table 1.9 shows the relative percentages of the Earth's water storage sites and locations when they are proportionately reduced to the contents of a 55-gallon barrel. The water so vital to land plants and animals is relatively minuscule in volume compared to that of the oceans.

Satellite photos of Earth show the abundance of water in the oceans and moving as moisture in the atmosphere (Figure 1.21). Water is uniquely qualified to absorb and release solar energy (see Sidebar on Water). Water has an exceptionally high heat capacity allowing storage of great amounts of heat (calories) in the ocean. Water gains and loses more heat than any other common substance on Earth. The unceasing motion of ocean water transfers its stored solar energy throughout the world. Winds blowing across the water surface cause circulation of surface water (Figure 9.2). Deep-ocean circulation is mostly driven by density differences caused by colder and/or saltier water masses sink-

Table 1.9 Proportional Distribution of Water on Earth Reduced to a 55-Gallon Barrel

World ocean	53 gallons,	1 quart,	1 pint,	3	ounces
Glaciers	1 gallon,	—	—	12	—
Groundwater	—	1 quart,	—	11.4	—
Atmosphere	—	—	1 pint,	4.5	—
Freshwater lakes	—	—	—	0.5	—
Saline lakes, inland seas	—	—	—	0.34	—
Soil moisture	—	—	—	0.25	—
Rivers	—	—	—	0.01	—

Figure 1.21 Satellite view showing atmosphere and ocean; two fluid masses in continual motion transporting energy about the Earth. Photo courtesy of NASA/JPL.

ing and flowing at depth in a global circuit (Figure 9.3). The ocean currents also are affected by Earth's rotation that adds energy to water movements, and by continents that block and divert the flow of warm ocean water sending it up to colder latitudes. The solar energy stored in the oceans acts as a thermal regulator strongly influencing global climate.

The evaporation of water to vapor (gas) requires about 600 calories per gram of water. This heat energy is absorbed and stored in the water vapor as latent (hidden) heat, or specifically, as the latent heat of vaporization. When water vapor condenses, or changes back to a liquid such as rain or fog, it releases its stored latent heat at the same 600 calories per gram. Although only 0.29 percent of water near the Earth's surface is in the atmosphere, this water vapor is important because of the solar energy it holds, transports, and releases. On a broad scale, the latent heat of vaporization (evaporation) is an important factor in global climate, and on a local scale it is the energy, the power, the juice behind severe weather such as hurricanes and tornadoes.

The atmosphere may be viewed as a heat engine that uses solar radiation to produce the mechanical energy of winds. The warm air masses of the equatorial region are less dense, rising buoyantly and flowing toward the poles where they cool and sink (Figure 10.5). Cold, dense polar air masses flow away from the poles toward the equator. The rotation of the Earth beneath its low-density fluid shell of atmosphere adds complexities to this simplistic model (Figure 10.6). The circulation of the atmosphere distributes heat around the Earth.

Sidebar

Water—The Most Peculiar Substance on Earth?

It is an understandable human trait to consider things that are common and abundant as being ordinary, whereas those items that are uncommon and rare are regarded as extraordinary. The most common substance about the surface of the Earth is water. It is so much a part of our daily lives that it is all too easy to regard water as being ordinary. But such a bias will lead us astray. Water is a truly extraordinary chemical compound. Were it not such an odd substance, everything on Earth, from weather to life, would be radically different. Let's examine some of the characteristics of this most peculiar substance.

1. Water is the only substance on Earth that is present in vast quantities in solid, liquid, and gaseous states.
2. Water has the highest **heat capacity** of all solids and liquids except liquid ammonia. Because water stores so much heat, the circulation of water in the ocean transfers immense quantities of heat.
3. Water has the highest heat conduction of all liquids at normal Earth surface temperatures.
4. Water has the highest **latent heat of vaporization** of all substances. It takes about 600 calories to evaporate a gram of water. This latent heat is carried by water vapor into the atmosphere and is released when water vapor condenses to liquid rain. Much heat is transported about the atmosphere as air masses circulate.
5. Water has the second highest **latent heat of fusion,** exceeded only by ammonia. When ice melts, it absorbs about 80 calories per gram. When water freezes, it releases about 80 calories per gram.
6. Water is a bipolar molecule. The negative oxygen and positive hydrogen atoms bond together yielding a molecule with a negative and a positive side (Figure 1.22). This + and – polarity allows water to readily bond with charged **ions.**
7. Water has the highest **dielectric constant** of all liquids. This property tends to keep ions apart and prevent their bonding, thus maintaining a solution. This is why water has been called the universal solvent.
8. Water has the highest **surface tension** of all liquids.
9. Water expands about 9 percent when it freezes. This is anomalous behavior. Usually, as a substance gets colder, it shrinks in volume and becomes denser. The maximum density for water occurs at about 4°C (39°F). Imagine what lakes and oceans would be like if ice were heavier than liquid water and ice sank to the bottom.

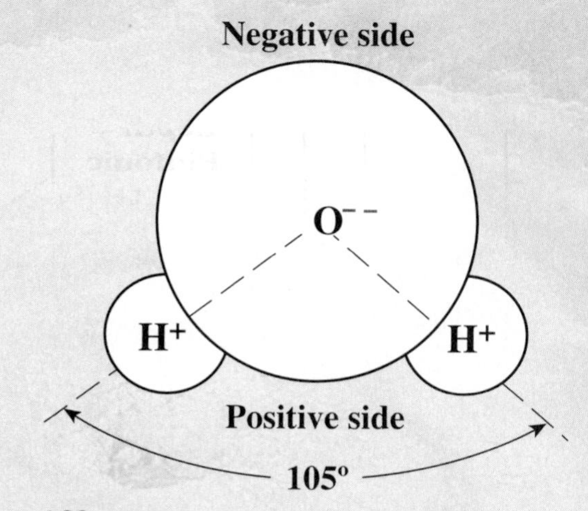

Figure 1.22 Water is a bipolar molecule exhibiting a negative and a positive side. This bipolarity greatly increases the activity of water.

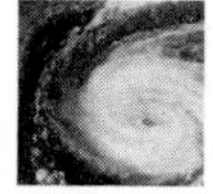

Forces of Construction Versus Destruction

Another way to visualize the amounts of energy flow on the Earth is by understanding the rock cycle and the construction and destruction of land (continents). Energy flowing up from Earth's interior melts rock that rises as magma and then cools and crystallizes to form the igneous rocks; they are plutonic rocks if they solidify at depth or volcanic rocks if they cool and harden at the surface. These newly formed rocks help create new land (Figure 1.23). Igneous-rock formation is part of the internal energy-fed **Forces of Construction** that create and elevate land masses.

At the same time, the much greater flow of energy from the Sun is driving the hydrologic cycle that weathers the igneous rocks exposed at or near the surface and breaks them down into sediments. Physical weathering disintegrates rocks into gravel and sand, while chemical weathering decomposes rock into clay. The sediments are eroded, transported mostly by water, then deposited in topographically low areas, ultimately the ocean. These processes are part of the **Forces of Destruction** that work to erode the lands and dump the debris into the oceans. These land-building and land-destroying forces result from Earth's energy flows that create, transform, and destroy rocks as part of the rock cycle (Figure 1.23).

Think about the incredible amount of work done by the prodigious flows of energy operating over the great age of the Earth. There is a long-term conflict raging between the internal-energy powered forces of construction that create and elevate landmasses and the external-energy powered forces of destruction that erode the continents and dump the

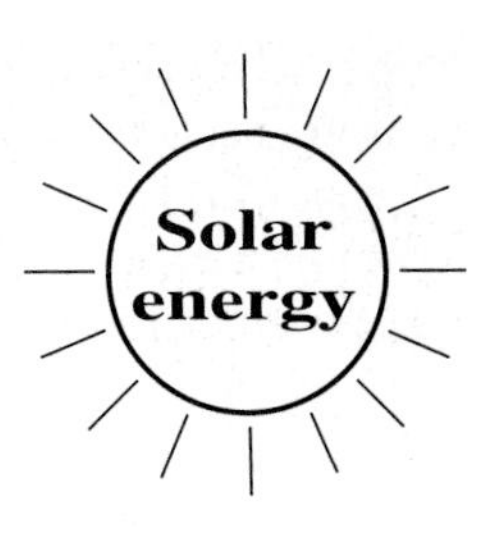

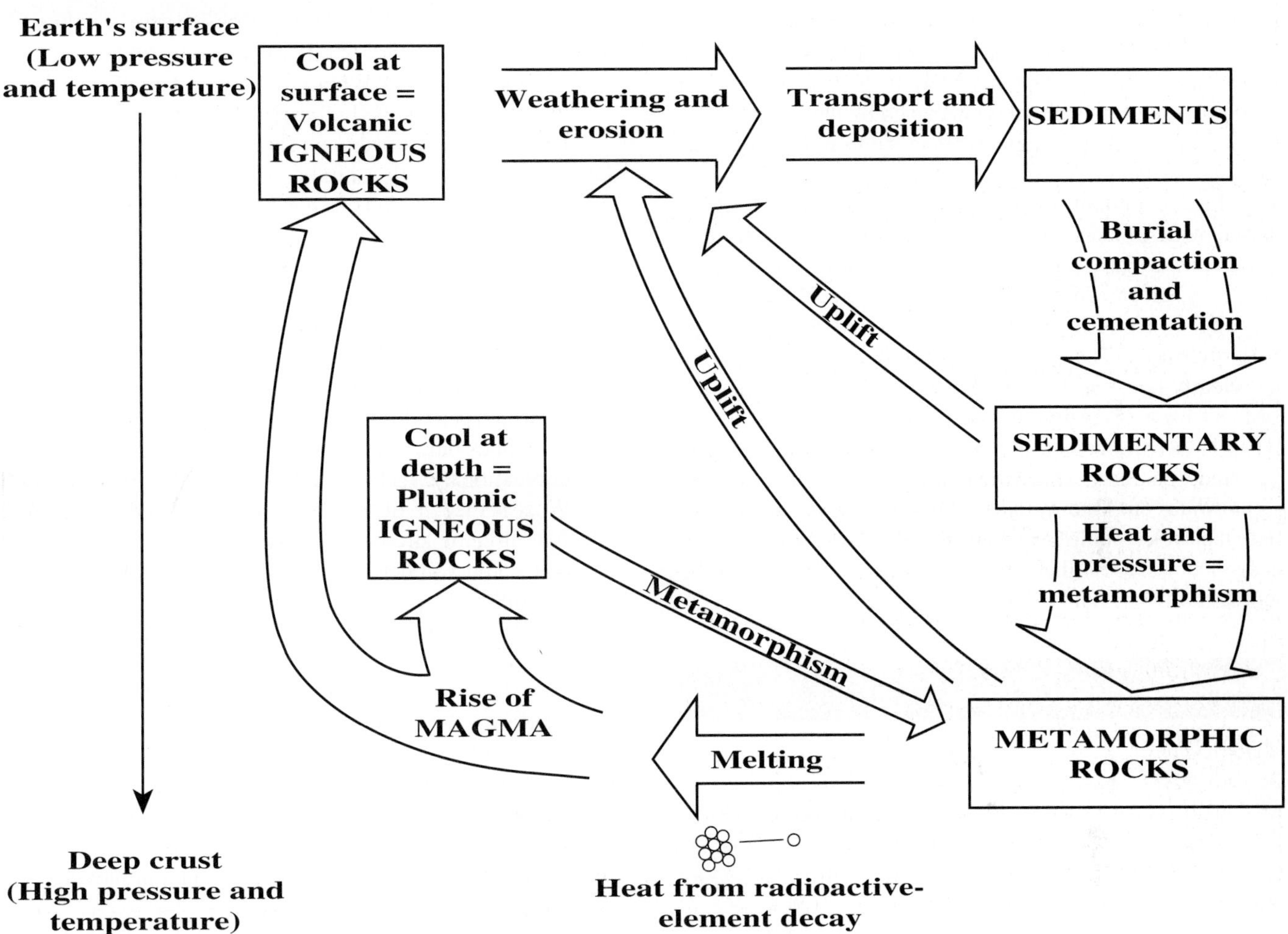

Figure 1.23 The rock cycle. Magma cools and solidifies to form igneous rocks. Rocks exposed at the Earth's surface break down and decompose into sediments (e.g., gravel, sand, clay) which are transported, deposited, and hardened into sedimentary rock. With increasing burial depth, temperature and pressure increase causing changes or metamorphosis of rocks into metamorphic rocks.

continental debris into the ocean basins. Visualize this: If the interior of the Earth cooled and the flow of internal energy stopped, then mountain building and uplift also would stop; then the ongoing solar-powered agents of erosion would reduce the continents to sea level in just 45 million years. There would be no more continents, only an ocean-covered planet.

Think about the time scales involved in eliminating the continents. At first reading, 45 million years of erosion may seem like an awfully long time, but remember that the Earth is 4.57 billion years old. The great age of the Earth indicates that erosion is powerful enough to have leveled the continents about 100 times. This shows the power of the internal forces of construction to keep elevating old continents and adding new landmasses. And woe to humans and other life forms that get too close to these forces of construction and destruction, for this is where disasters occur.

Summary

The flow of energy through the Earth's land, ocean, and atmosphere has significant impacts on humans. The biggest killers of the twentieth century have been hurricanes and earthquakes with deaths especially high in Asia and the Mideast. The greater the magnitude of an event, the lesser its frequency of occurrence.

Disasters occur where the Earth unleashes its concentrated energy. The main energy sources are: 1) impacts with large asteroids and comets, 2) gravitational contraction within the Earth, which helps heat the interior—and gravitational attractions between the Earth, Moon, and Sun that generate tidal energy, 3) the Earth's interior releasing heat generated by the ongoing decay of radioactive elements in the process of nuclear fission, and 4) the Sun, where nuclear fusion produces radiant energy received by the Earth's exterior at a rate 5,300 times greater than the heat flow from the interior.

Massive amounts of internal heat within the early Earth caused widespread melting. The radioactive elements that help heat the Earth's interior by their decay do so at measurable rates known as half-lives. Elements that radioactively decay act as clocks that can be used to date the events of Earth history. The Earth is 4.57 billion years old.

Nearly one-fourth of the Sun's energy that reaches the Earth is used to evaporate water to begin the hydrologic cycle. Under the pull of gravity, snow and rain fall back to the land, then run downslope as glaciers, streams, and groundwater until the water is returned to the ocean to complete the cycle. While in motion, the ice, water, and wind act as agents of erosion that wear down the land and dump the debris into the ocean basins.

Terms to Remember

albedo 21
asteroid 10
atmosphere 10
B.C.E. 19
calorie 13
centigrade 11
comet 10
conduction 19
continent 10
convection 19
core 11
crust 15
curie 19
dielectric constant 24
earthquake 1
element 16
energy 1
erosion 10
Fahrenheit 11
force 13
forces of construction 24
forces of destruction 24
frequency 9
glacier 10
gravity 10
great natural disaster 2
half-life 16
heat 13
heat capacity 24
hydrologic cycle 19
igneous rocks 18
insolation 22
ion 24
isotope 19
kinetic energy 13
latent heat of fusion 24
latent heat of vaporization 24
magma 18
magnitude 9
mantle 11
meteorite 9
mitigation 6
natural disaster 1
natural hazard 6
nuclear fission 18
nuclear fusion 11
potential energy 13
power 13
radioactive 15
return period 9
sediment 19
solar radiation 11
surface tension 24
volcano 4
work 13

Questions for Review

1. What natural disasters killed the most people in the twentieth century? Where in the world are deaths from natural disasters the highest? Where in the world are insurance losses from natural disasters the highest?
2. What is the relationship between the magnitude of a given disaster and its frequency of occurrence?
3. What is the difference between a natural disaster and a natural hazard? How do economic losses differ from insured losses?
4. What energy sources caused the interior of the early Earth to heat up?
5. How does nuclear fusion differ from nuclear fission?
6. At what speed does the Earth travel around the Sun?
7. Why does our relatively small Moon have a greater gravitational effect on the Earth than the gigantic Sun?
8. What is the age of the Earth? How is this determined?
9. After freeing zircon crystals from an igneous rock, how could you determine when the rock formed (solidified)?
10. Where are the oldest known Earth rocks found? How old are they?
11. How does the amount of energy flowing from the interior of the Earth compare with the energy received from the Sun?
12. Explain how the hydrologic cycle operates. What are the roles of the Sun and gravity?
13. What properties make water so peculiar?
14. What is latent heat? How is it important in moving energy through the atmosphere?
15. Describe the effects on the Earth's surface from the "internal forces of construction" versus the "external forces of destruction."

Questions for Further Thought

1. Would we call a large earthquake or major volcanic eruption a natural disaster if no humans were killed or buildings destroyed?
2. Are new natural nuclear reactors likely to spring into action on Earth?
3. Your lifetime will be what percentage of geologic time?
4. If the heat flow from Earth's interior ceased, what would happen to the landmasses? After the internal heat flow had stopped for 100 million years,

how would the Earth appear to a future visitor from space?

5. Could global building designs be made disaster proof thus reducing the large number of fatalities?

Suggested Readings and References

Alexander, D. (1993). *Natural Disasters.* New York: Chapman & Hall.

Calder, N. (1972). *The Restless Earth.* New York: Viking Press.

Cowan, G. A. (1976, July). A natural fission reactor. *Scientific American,* 36–47.

Dalrymple, G. B. (1991). *The Age of the Earth.* Palo Alto, Calif.: Stanford University Press.

Hubbert, M. K. (1971, September). The energy resources of the Earth. *Scientific American,* 225, 61–70.

Levin, H. L. (1999). *The Earth through Time* (6th ed.). New York: Saunders College.

Menard, H. W. (1974). *Geology, Resources, and Society.* New York: W. H. Freeman.

National Research Council. (1986). *Global Change in the Geosphere-Biosphere.* Washington, D.C.: National Academy Press.

Press, F., and Siever, R. (1986). *Earth* (4th ed.). New York: W. H. Freeman.

Press, F., and Siever, R. (2001). *Understanding Earth* (3rd ed.). New York: W. H. Freeman.

Schmidt, V. A. (1986). *Planet Earth and the New Geoscience.* Dubuque, Iowa: Kendall/Hunt. (See also the seven-part television series "Planet Earth.")

Shah, B. V. (1983). Is the environment becoming more hazardous? *Disasters,* 7, 202–209.

Skinner, B. J., and Porter, S. C. (2000). *The Dynamic Earth* (4th ed.). New York: John Wiley and Sons.

Van Andel, T. H. (1994). *New Views on an Old Planet* (2nd ed.). New York: Cambridge University Press.

Zanetti, A., and Enz, R. (2002). Natural catastrophes and man-made disasters in 2001. *Sigma No. 1,* Swiss Reinsurance Company.

Videos

The Earth Revealed—Geologic Time. (1992). Annenberg/CPB Project (30 min.).

Geologic Time. (1986). Encyclopedia Britannica/American Geological Institute (24 min.).

The Earth Has a History. (1988). Geological Society of America (20 min.).

The Miracle Planet—The Heat Within. (1987). Nova/KCST (60 min.).

Chapter 2

Plate Tectonics and Earthquakes

Such superficial parts of the globe seemed to me unlikely to happen if the Earth were solid to the centre. I therefore imagined that the internal parts might be a fluid more dense, and of greater specific gravity than any of the solids we are acquainted with; which therefore might swim in or upon that fluid. Thus the surface of the globe would be a shell, capable of being broken and disordered by the violent movements of the fluid on which it rested.

—Benjamin Franklin, 1780

Figure 2.1 Earthquakes in Turkey on 17 August and 12 November 1999 killed almost 20,000 people. On this block in Duzce, seven multi-story buildings collapsed. Photo by Roger Bilham, courtesy of NOAA.

Earthquakes destroy our buildings and kill us (Figure 2.1). We begin protecting ourselves by learning how the Earth operates. The mobility and activity of our planet cannot be fully appreciated until we understand how the Earth differentiated into a series of floating layers and that the outermost rocky layers move in the process of plate tectonics.

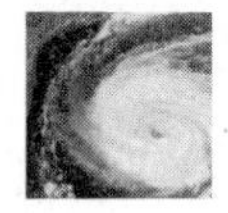

Earth History

Our present understanding of Earth's origin has an accreting mass of dust and larger particles orbiting the Sun about 4.6 billion years ago. Internal heat rose for the first several hundred million years during and after accretion of these particles due to meteorite impacts, gravitational contraction, and the decay of radioactive elements. During this time, heat continued to build inside the Earth until the critical temperature was reached at which iron melted. Gravity then pulled the liquid iron inward to form the metallic core. The process of core formation released gravitational energy and much more heat, which in turn caused widespread melting. Low-density materials (magmas, waters, and gases) freed by the melting rose and accumulated on Earth's exterior as continents, oceans, and atmosphere. We know that large oceans and small continents existed by 3.9 billion years ago, life was present as photosynthetic bacteria 3.5 billion years ago, large continents were present 2.5 billion years ago, and the outer layers of the Earth were active in the process of plate tectonics by at least 1.5 billion years ago.

The history of the Earth has been metaphorically contrasted with the life history of a 46-year-old woman by Nigel Calder in his book *The Restless Earth.* In this metaphor, each of "Mother" Earth's years equals 100 million years of geologic time. The first seven of her years are wholly lost to the biographer. Like the human memory, the early rock record on Earth is distorted; it emphasizes the more recent events in both number and clarity. Most of what we know of "Mother" Earth happened in the past six years of her life. Her continents had little life until she was 42. Flowering plants did not appear until her forty-fifth year. Her pet dinosaurs died out eight months ago. In the middle of last week, some ancestors of present apes evolved into human ancestors. Yesterday, modern humans (*Homo sapiens*) evolved and began hunting other animals, and in the last hour discovered agriculture and settled down. Fifteen minutes ago, Moses led his people to safety, and five minutes later, Jesus was preaching along the same fault line. In the last minute, the Industrial Revolution began, and the number of humans has increased enormously.

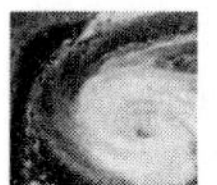

The Layered Earth

The Earth today is differentiated into layers of varying densities. Much of the densest materials have been pulled toward the center, and some of the least dense substances have escaped to the surface (Figure 2.2). At the center of the Earth is a dense, iron-rich core measuring about 7,000 km (4,350 mi) in diameter. The inner core is a 2,450-km (1,520-mi)-diameter solid mass with temperatures up to 4,300ºC (7,770ºF). The outer core is mostly liquid, and the **viscous** movements of convection currents within it are responsible for generating

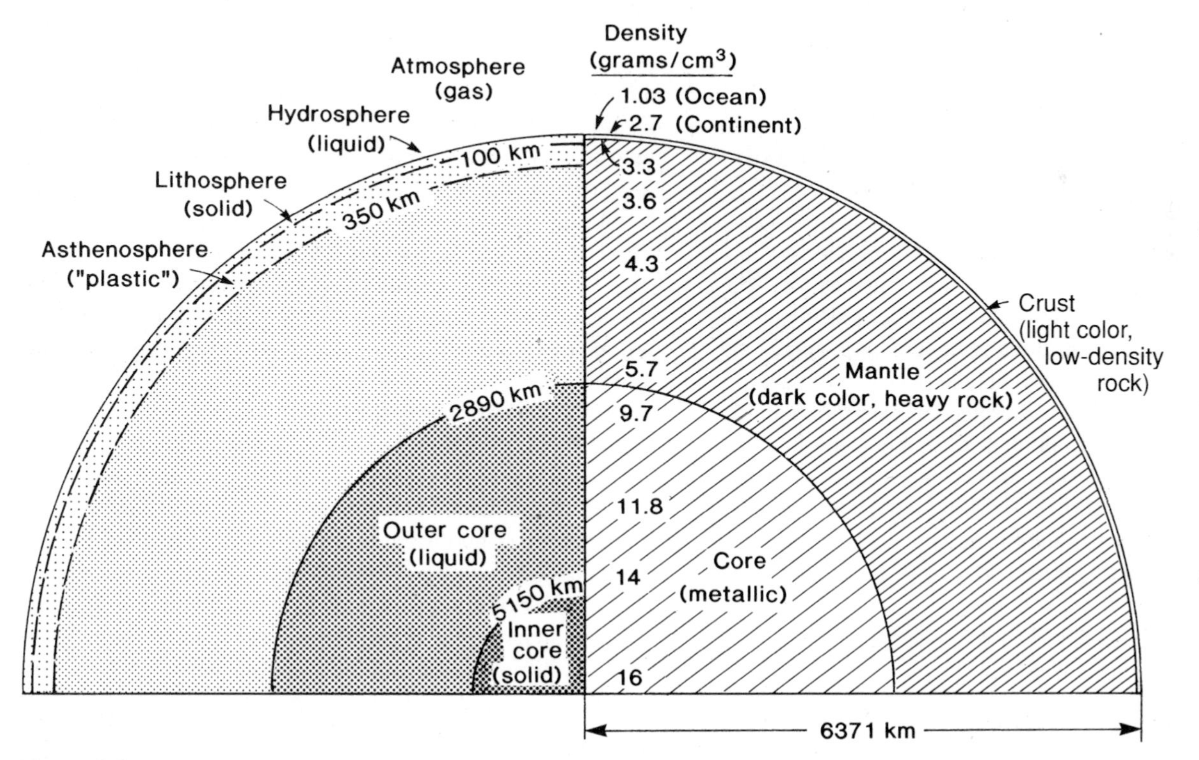

Figure 2.2 Density stratification within the Earth, i.e., lower density materials float atop higher density materials. Pressure and temperature both increase from surface to center of Earth. Layers illustrated on the left show the differences in physical properties and strengths. Layers on the right emphasize different mineral and chemical compositions.

the Earth's magnetic field. The entire iron-rich core is roughly analogous in composition to a melted mass of metallic meteorites. Recent work has suggested that the solid inner core is spinning in the same direction as the surrounding Earth, only faster. The inner core may be moving about 3 degrees farther per year, suggesting it will do an extra rotation about once every 120 years. The cause of the faster rotation of the inner core is hotly debated but it is likely due to a lesser slowing by tidal friction with the Moon. It takes time for the tide-caused deceleration (slowing) to pass through the spinning Earth, especially the liquid outer core, so the inner core keeps racing ahead of the rest of our planet.

Wrapped around the core is a nearly 2,900-km (1,800-mi)-thick, rocky **mantle** with compositions similar to **chondritic,** stony meteorites. The mantle comprises 83 percent of Earth's volume and 67 percent of its mass. Floating atop the hot, buoyant rock of the mantle is a mosaic crust of less dense rocks, and above these are the oceans and the atmosphere.

The uppermost 700-km (435-mi)-thickness of mantle is depleted in light elements and thus differs from the lower 2,200-km (1,365-mi)-thick zone. The differences can be approximated by melting a chondrite meteorite in the laboratory. Melting produces a separation in which an upper froth rich in low-density elements rises above a residue of denser minerals/elements. The differentiation is similar to continental crust that by melting and separation has risen above the uppermost mantle. All the years of heat flow toward the Earth's surface have "sweated out" many low-density elements to form a continental crust. Today, the continents make up only 0.1 percent of Earth's volume.

From a broad perspective, it may be seen that the Earth is not a homogeneous, solid ball but rather it is a series of floating layers where less-dense materials successively rest upon layers of more-dense materials. The core with densities up to 16 gm/cm^3 supports the mantle with densities ranging from 5.7 to 3.3 gm/cm^3. Atop the denser mantle float the continents with densities around 2.7 gm/cm^3, which in turn support the salty oceans with a density about 1.03 gm/cm^3, and then the least dense layer of them all—the atmosphere. The concept of floating layers holds true on smaller scales as well. For example, the oceans are made of layered masses of water of differing densities. Very cold, dense Antarctic waters flow along the ocean bottoms and are overlain by cold Arctic water, which is overlain by extra salty waters, which in turn are overlain by warmer, less dense seawater. The Earth is comprised basically, from core to atmosphere, of density-stratified layers.

Isostasy

The Earth's layering can be described either as 1) separations based on differing densities due to varying chemical and mineral compositions or 2) layers with different strengths (Figure 2.2). Both temperature and pressure increase continuously from the Earth's surface to the core, yet their effects on materials are different. Increasing temperature causes rock to expand in volume and become less dense and more capable of flowing under pressure and in response to gravity. Increasing pressure causes rock to decrease in volume and become more dense and more rigid. Visualize tar at the Earth's surface. On a cold day, it is solid and brittle, but on a hot day, it can flow as a viscous fluid. Similar sorts of changes in physical behavior mark different layers of the Earth. In fact, from a perspective of geological disasters, the crust-mantle boundary is not as important as the boundary between the rigid **lithosphere** (from the Greek word *lithos,* meaning "rock") and the fluidlike **asthenosphere** (from the Greek word *asthenes,* meaning "weak") (Figure 2.3). The differences in strength and mechanical behavior between solid and fluid states are partly responsible for earthquakes and volcanoes.

The concepts of gas, liquid, and solid are familiar. A gas is a fluid capable of indefinite expansion. A liquid is a fluid, a substance that flows readily and has a definite volume but no definite shape. A solid is firm; it offers resistance to pressure and does not easily change shape. What is not stated but is implicit in these definitions is the effect of time. All of these definitions describe behavior at an instant in time, but what is the behavior when viewed over a longer time scale? Specifically, some solids yield to long-term pressure such that at any given moment, they are solid, yet internally they are deforming and flowing, that is, behaving as a fluid. A familiar example is the ice in a glacier. When a glacier is hit with a rock hammer, solid chunks of brittle ice are broken off. Yet inside the glacier, atoms are changing positions within the ice and moving to downhill positions of lower stress. At no instant in time does the glacier fit our everyday concept of a liquid, yet over time, the glacier is flowing downhill as an ultrahigh-viscosity fluid.

When materials are subjected to sufficient **stress** or force, they will deform or undergo **strain** in different ways. Stress may produce **elastic** or recoverable deformation such as when you pull on a spring. The spring deforms while you pull or stress it, but when you let go, the spring recovers and returns to its original shape.

If greater stress is applied for a longer time, or at higher temperatures, then **ductile** or plastic deformation may occur and the change is permanent. An example occurs deep within glaciers where the ice deforms and moves with ductile flow.

If stress is applied rapidly to a material, then it may abruptly fracture or break into pieces in **brittle** deformation. Take a chunk of ice from your refrigerator and drop it or hit it; it will shatter with brittle failure. Notice that the ice in a glacier exhibits both brittle and ductile behavior. Near the surface there is little pressure on the rigid ice and it abruptly fractures when stressed. Deep within the glacier where the weight of overlying ice creates a lot of pressure, the ice deforms and moves by ductile flow. The style of ice behavior depends on the amount of pressure confining it.

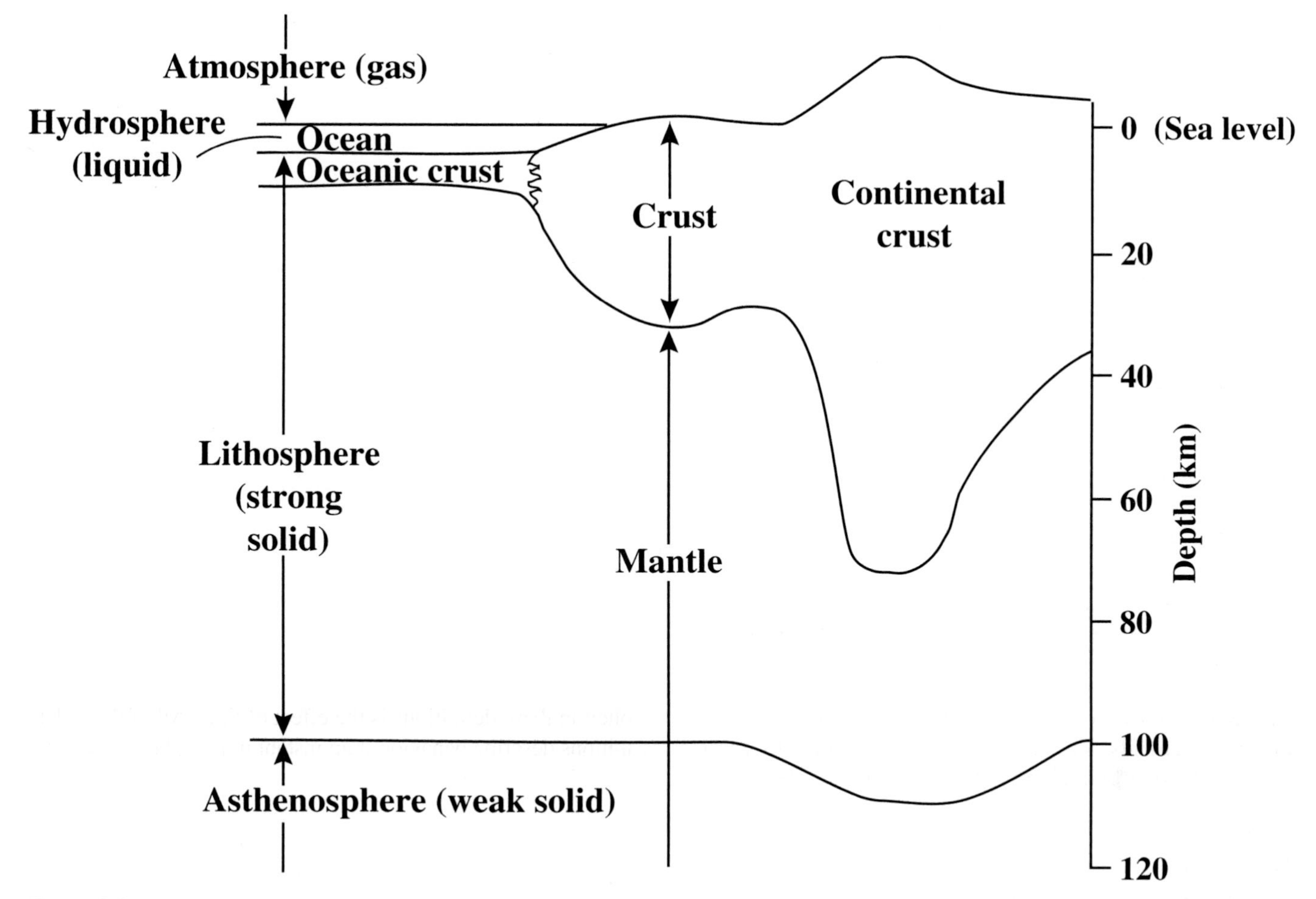

Figure 2.3 Upper layers of the Earth may be recognized 1) compositionally, as lower density crust separated from the underlying higher density mantle, or 2) on the basis of strength, as rigid lithosphere riding atop "plastic" asthenosphere.

The type of mechanical behavior illustrated by ice deep within a glacier typifies that of the rock within the Earth's asthenosphere. This layer of rock is "plastic" in the sense used by William James in his 1890 *Principles of Psychology.* He defined *plastic* as "possession of a structure weak enough to yield to an influence, but strong enough not to yield all at once." The top of the asthenosphere comes to the surface at the ocean's volcanic mountain chains but lies more than 100 km (about 60 mi) below the surface in other areas. It has gradational upper and lower boundaries and is about 250-km (155-mi) thick. What are the effects of having this "plastic" or ductile zone so near the Earth's exterior? There is a lot of flowage of rock within the asthenosphere that helps cause the Earth's surface to rise and fall. For example, the Earth is commonly described as a sphere, but it is not. The Earth may be more properly described as an oblate ellipsoid that is flattened at the poles (nearly 30 km or 19 mi) and bulged at the equator (nearly 15 km or 9 mi). The Earth is neither solid enough nor even strong enough to spin and maintain a spherical shape. Rather, the Earth deforms its shape in response to the spin force.

The concept of **isostasy** was developed in the nineteenth century. It applies a **buoyancy** principle to the low-density continents and mountain ranges that literally float on the denser mantle below. Just as an iceberg juts up out of the ocean while most of its floating mass is beneath sea level, so does a floating continent jut upward at the same time it has a thick "root" beneath it (Figure 2.3). An example of this buoyancy effect or isostatic equilibrium was defined by carefully surveying the landscape before and after the construction of Hoover Dam across the Colorado River east of Las Vegas, Nevada (Figure 2.4). On 1 February 1935, the impoundment of Lake Mead began. By 1941, about 24 million **acre feet** of water were detained, placing a weight of 40,000 million tons over an area of 232 square miles. Although this is an impressive reservoir on a human scale, what effect can you imagine it having on the whole Earth? In fact, during the 15 years from 1935 to 1950, the central region beneath the reservoir subsided up to 175 millimeters (7 inches) (Figure 2.4). The relatively simple act of impounding water triggered an isostatic adjustment that caused the area to depress.

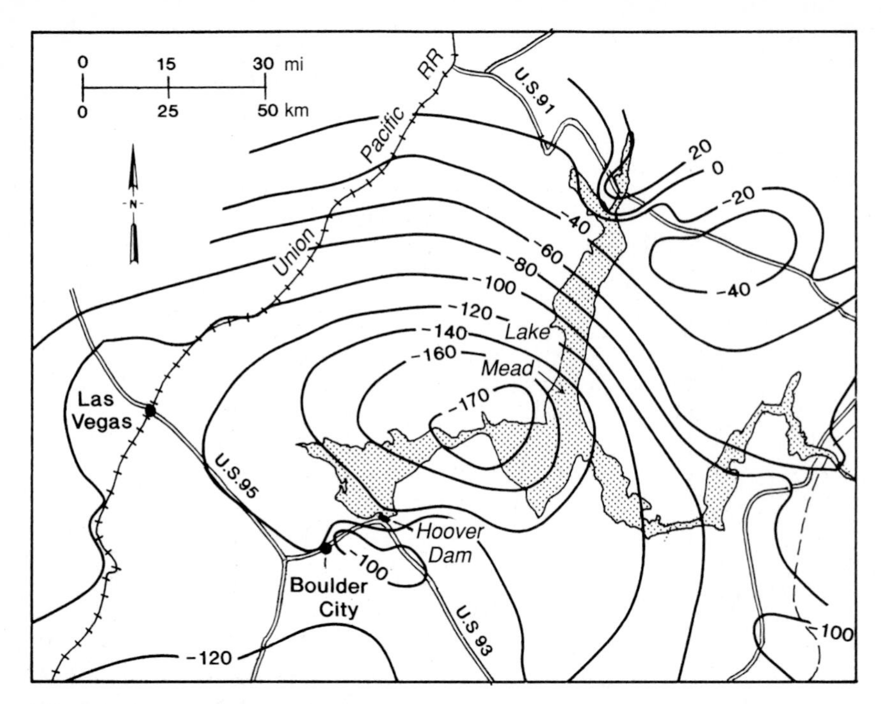

Figure 2.4 Isostatic downwarping caused by the weight of Lake Mead, from 1935 to 1950. Contour lines show depression (or uplifting) of surface in millimeters.

Source: Smith, W. O., et al., "Comprehensive Survey of Sedimentation in Lake Mead, 1948–49," in *U.S. Geological Survey Professional Paper 295,* 1960.

Just how solid and firm is the surface of the Earth we live on? Larger-scale examples are provided by the great ice sheets of the recent geologic past. The continental glacier that buried the Finland-Sweden region was up to 3-km (2-mi) thick less than 20,000 years ago. The land was depressed beneath this great weight. By 10,000 years ago, the ice sheet had retreated and melted, and the water returned to the ocean. The long-depressed landmass, now freed from its heavy load, is rebounding upward via isostatic adjustment. In the last 10,000 years, northeastern coastal Sweden and western Finland have risen about 200 m (650 ft). This upward movement was vividly shown during excavation for a building foundation in Stockholm, Sweden. Workers uncovered a Viking ship that had sunk in the harbor and been buried with mud. The ship had been lifted above sea level, encased in its mud shroud, as the harbor area rose during the ongoing isostatic rebound. Gravity measurements of this region show a negative anomaly indicating another 200 m (650 ft) of isostatic uplift is yet to come. The uplift will add to the land of Sweden and Finland and reduce the size of the Gulf of Bothnia between them.

Some of the early uplifting of land after ice-sheet removal occurred in rapid movements that ruptured the ground surface generating powerful earthquakes. In northern Sweden, there are ground ruptures up to 160-km (100-mi) long with parallel cliffs up to 15-m (50-ft) high. The rocks in the region are ancient and rigid suggesting that ruptures may go 40-km (25-mi) deep and that they generated truly large earthquakes.

Vertical movements of the rigid lithosphere floating on the flexible asthenosphere are well documented. If we add a load on the surface of the Earth we can measure the downward movement. When nature buries a continent beneath ice, the surface sinks hundreds of meters as rock in the asthenosphere flows away under the load. If we remove a load on the surface, such as a hill, we can measure the resulting uplift. The surface of the Earth clearly is in a delicate vertical balance. Do major adjustments and movements also occur horizontally? Yes, there are horizontal movements between lithosphere and asthenosphere, which brings us into the realm of plate tectonics.

Plate Tectonics

Adding the horizontal components of movements on Earth allows us to understand the **tectonic cycle.** Ignoring complexities for the moment, the tectonic cycle can be simplified as follows (Figure 2.5). First, melted asthenosphere flows upward as magma and cools to form new ocean floor/lithosphere. Second, the new lithosphere slowly moves laterally away from the zones of oceanic crust formation on top of the underlying asthenosphere (**seafloor spreading**). Third, when the leading edge of a moving slab of oceanic

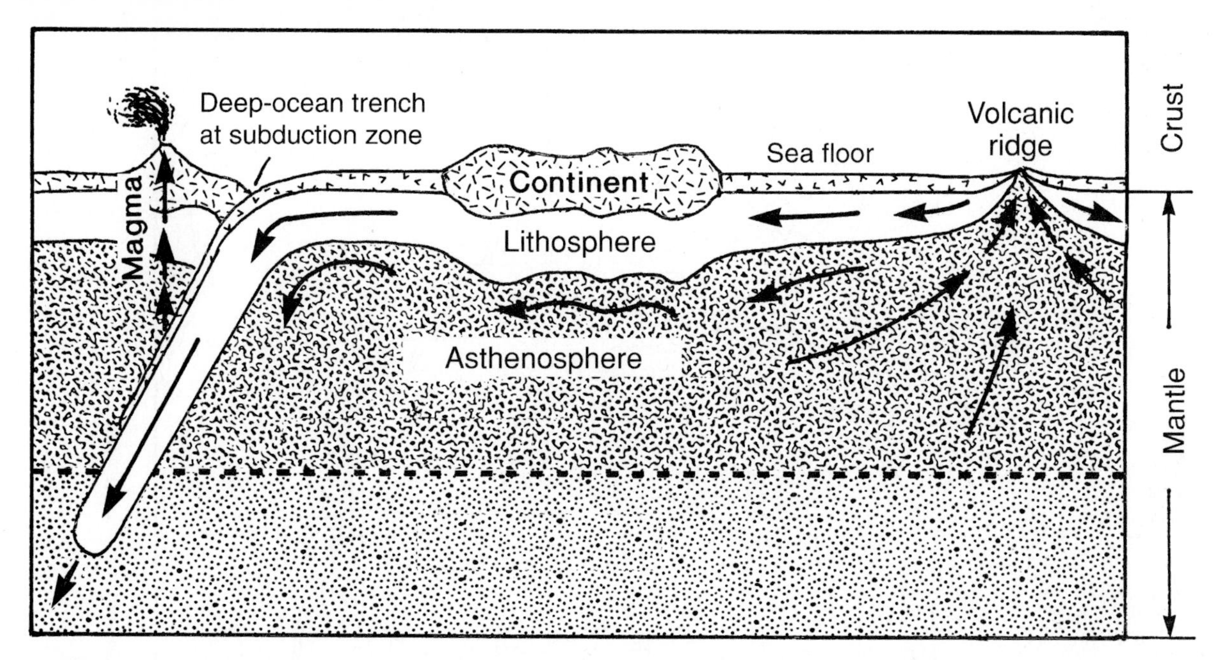

Figure 2.5 Schematic cross section of the tectonic cycle. Magma rises from the asthenosphere to the surface at the oceanic volcanic ridges where it solidifies and adds to the plate edges. As the igneous rock cools, the plate subsides and gravity pulls the plates from their topographic highs. The plate continues to cool, grows thicker at its base, becomes denser, collides with a less-dense plate, and turns down into the mantle, where it is ultimately reassimilated.

lithosphere collides with another slab, the older, colder, denser slab turns downward and is pulled by gravity back into the asthenosphere (**subduction**), while the less-dense, more buoyant slab overrides it. Last, the slab pulled into the mantle is reabsorbed. The time needed to complete this cycle is long, commonly in excess of 250 million years.

The grand recycling of the upper few hundred kilometers of the Earth is called the tectonic cycle. The Greek word *tekton* comes from architecture and means "to build"; it has been adapted by geologists as the term **tectonics,** which describes the building of **topography** and the deformation and movement within the Earth's outer layers. If we adopt the perspective of a geologist-astronaut in space and look down upon the tectonic cycle, we see that the lithosphere of the Earth is broken into pieces called **plates** (Figure 2.6). The study of the movements and interactions of the plates is known as **plate tectonics.** The gigantic pieces of lithosphere (plates) pull apart during seafloor spreading at **divergence zones,** slide past at **transform faults,** or collide at **convergence zones.** These plate-edge interactions are directly responsible for most of the earthquakes, volcanic eruptions, and mountains on Earth.

Another way that plate tectonics can be visualized is by using a hard-boiled egg as a metaphor for the Earth. Consider the hard-boiled egg with its brittle shell as the lithosphere, the slippery inner lining of the shell as the asthenosphere, the egg white (albumen) as the rest of the mantle, and the yolk as the core. Before eating a hard-boiled egg, we break its brittle shell into pieces that slip around as we try to pluck them off. This hand-held model of brittle pieces being moved atop a softer layer below is a small-scale analogue to the interactions between Earth's lithosphere and asthenosphere.

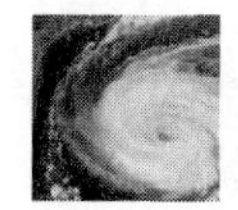

Development of the Plate Tectonics Concept

Human thought about the Earth has long been limited by the smallness of our bodies and the restricted range of our travels in relation to the Earth's gigantic size; it also has been limited by the shortness of our life spans compared to the age of the Earth. The Earth is so large and so old that the combined efforts of many geologists and philosophers during the last few hundred years have been required to amass enough observations to begin understanding how and why the Earth changes as it has and does. The first glimpse of our modern understanding began after the European explorers of the late 1400s and 1500s made maps of the shapes and locations of the known continents and oceans. These early world maps raised intriguing possibilities. For example, in 1620, Francis Bacon of England noted the parallelism of the Atlantic coastlines of South America and Africa and suggested that these continents had once been joined. During the late 1800s, the Austrian geologist Eduard Suess presented abundant evidence in support of **Gondwanaland,** an ancient southern supercontinent

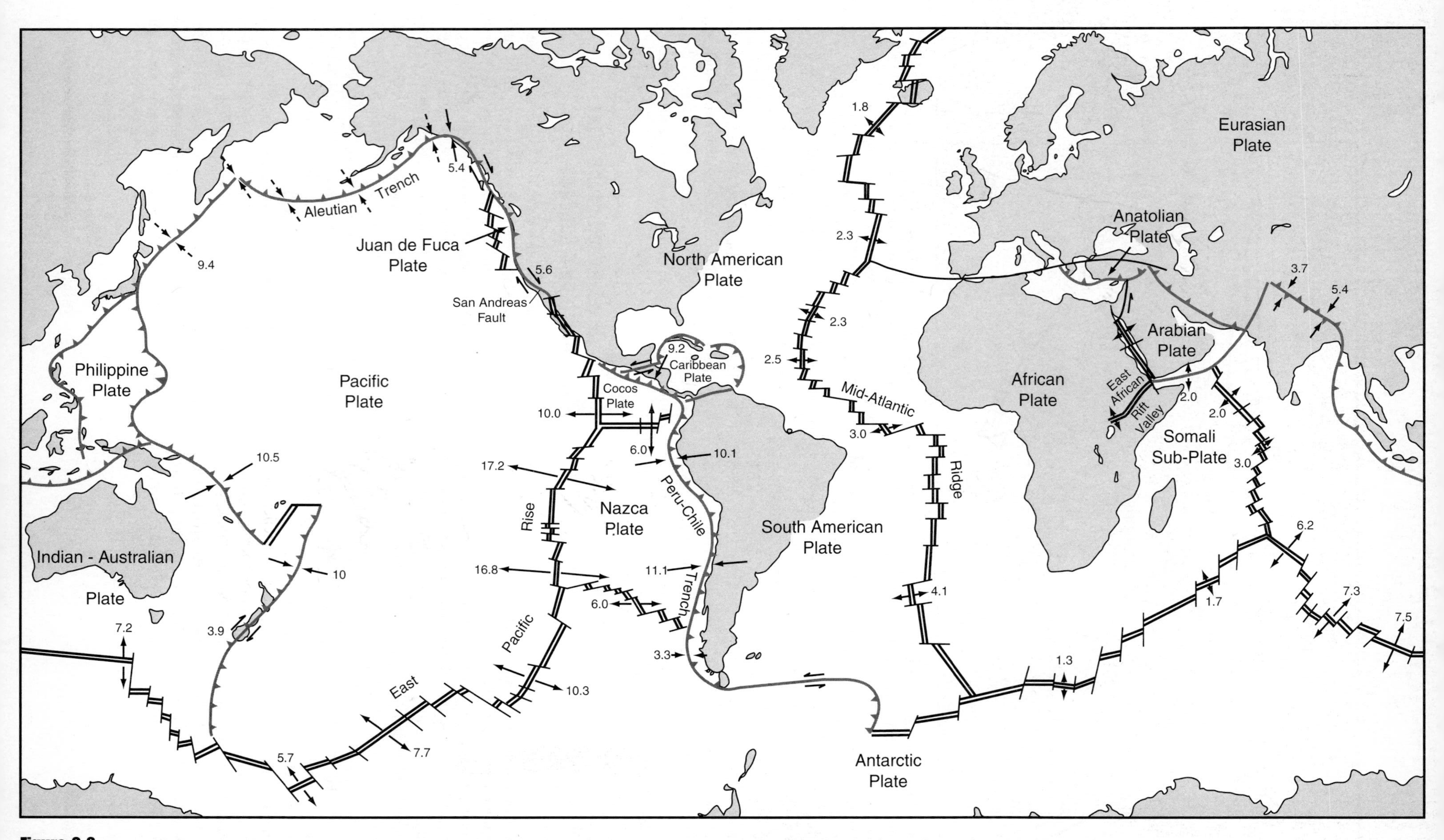

Figure 2.6 Map of the major tectonic plates with arrows showing directions of movement. Amounts of movement are shown in centimeters per year.

Sidebar

Earth's Magnetic Field

Anyone who has ever held a compass and watched the free-turning needle point toward the north has experienced the **magnetic field** that surrounds the Earth. The Chinese invented and were the first to use magnetic compasses. They in turn taught fourteenth-century European travelers who brought this knowledge back to Europe, where it was developed into the navigational tool that helped late fifteenth-century explorers make their voyages of discovery.

The Earth's magnetic field operates as if a gigantic bar magnet is located in the core of the Earth at an inclination of 11°. The **magnetic pole** and geographic north pole do not coincide but the magnetic pole axis has apparently always been near the rotational pole axis. Notice in Figure 2.7 that the inclination of the magnetic lines of force with respect to the Earth's surface varies with **latitude.** At the magnetic equator, the magnetic lines of force are parallel to the Earth's surface (inclination of 0°). Toward the poles, either northward or southward, the angle of inclination continuously increases until it is perpendicular to the surface at both the North and South Poles (inclinations of 90°). Notice also that the lines of force are inclined downward and into the Earth's surface near the North Pole and upward and out of the Earth's surface near the South Pole.

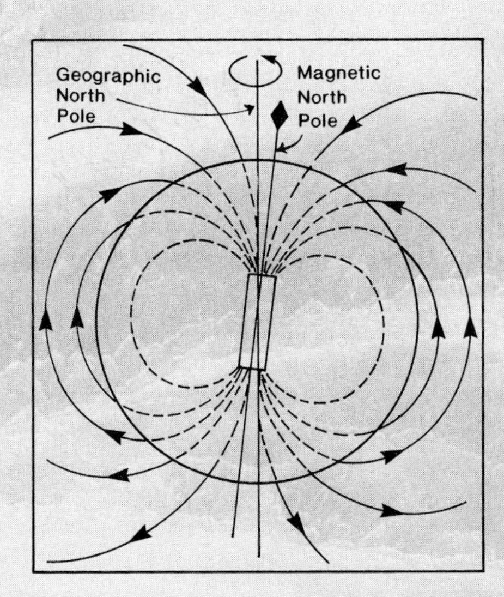

Figure 2.7 Schematic diagram of Earth's magnetic field. The bar magnet pictured does not exist, but it would create the same magnetic field achieved by the electrical currents in the Earth's liquid, iron-rich outer core. Notice that 1) the magnetic pole and the rotational pole do not coincide, 2) the magnetic lines of force are parallel to the Earth's surface at the magnetic equator and perpendicular at the magnetic poles, and 3) the lines of force go into the Earth at the North Pole and out at the South Pole.

In reality, the interior of the Earth is much too hot for a bar magnet to exist. **Magnetism** in rocks is destroyed by temperatures above 550°C, and temperatures in the Earth's core are estimated to reach 5,800°C (10,470°F). The origin of Earth's magnetic field involves movements of the iron-rich fluid in the outer core, which generate electric currents that in turn create the magnetic field. Fluid iron is an excellent conductor of electricity. The molten iron flowing around the solid inner core is a self-perpetuating dynamo deriving its energy both from the rotation of the Earth and from the convection of heat released by the crystallization of minerals at the boundary of the inner and outer cores.

A closer look at the Earth's magnetic field yields several problems awaiting resolution. The simplified magnetic field portrayed in Figure 2.7 does not show the complexities that occur over years and centuries as the magnetic field's strength waxes and wanes. In addition, the magnetic pole moves about the geographic North Pole region. Every several thousand to tens of millions of years, a highly dramatic change occurs in the magnetic field: the magnetic polarity reverses. In a reversal, the orientation of the magnetic field flip-flops from a north (normal) polarity to a south (reverse) polarity or vice versa. After the next reversal, your hand-held compass will point toward the South Pole region. The change in orientation of the magnetic field leaves its imprint in rocks, where geologists (paleomagnetists) can read it. The paleomagnetic history contained in the rocks has provided the most important evidence of seafloor spreading; it also has allowed charting of the paths of continents as they have moved through different latitudes. In addition, the record of magnetic reversals provides the data for a magnetic time scale, a third geologic time scale. (The first time scale is based on the irreversible sequence of occurrence of fossils in sedimentary rocks, and the second time scale is founded on the decay of radioactive elements.)

composed of a united South America, Africa, Antarctica, Australia, India, and New Zealand, which later split apart. The most famous and outspoken of the early proponents of **continental drift** was the German meteorologist Alfred Wegener. In his 1915 book, *The Origin of Continents and Oceans,* he collected all available evidence, such as similar rocks, fossils, and geologic structures, on opposite sides of the Atlantic Ocean. Wegener suggested that all the continents had once been united in a supercontinent called **Pangaea** (*pan* meaning "all" and *gaea* meaning "earth").

Much is made of the fact that during his lifetime Wegener's hypothesis of continental drift gathered more ridicule than acceptance. But why were his ideas not widely accepted? Wegener presented an intriguing hypothesis well supported with observations and logic, but his mechanism of continental drift was deemed impossible. When Wegener presented his evidence for continental drift, geologists and geophysicists were faced with trying to visualize how a continent could break loose from the underlying rocks and plow a path over them. It seemed physically impossible then. The breakthrough in understanding came when the ocean floors were studied and the data were best explained by the formation of new sea floor that spread apart and then was consumed by subduction. When it became known that the lithosphere decouples from the asthenosphere and moves laterally, then the relatively small, low-density continents, set within the oceanic crust, were seen to be carried along as incidental passengers (Figure 2.5).

In the mid-1960s, evidence abounded, mechanisms seemed plausible, the plate tectonic theory was developed and widely accepted, and Wegener was restored to an elevated status. Scientific understanding grew with the addition of new data, old hypotheses were modified, and new theories were created. Science is never static; it is a growing, evolving body of knowledge that creates ever-better understanding of how the Earth works.

It is rare in science to find widespread agreement on a large-scale hypothesis such as plate tectonics. But when data from the Earth's magnetic field locked inside seafloor rocks were widely understood, skeptics around the world became convinced that seafloor spreading occurs and that the concept of plate tectonics is valid. These paleomagnetic data are so powerful that we need to understand their story so that plate tectonics can be seen as real.

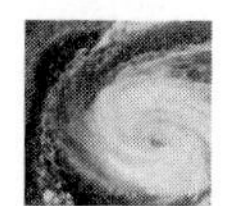

Magnetization of Volcanic Rocks

Lava erupts from a volcano, flows outward as a sheetlike mass, slows down, and stops. Then as the lava cools, minerals begin to grow as crystals. Some of the earliest formed crystals incorporate iron into their structures. After the lava cools below the **Curie point,** about 550°C, atoms in iron-

Figure 2.8 A stratified pile of former lava flows of the Columbia River Basalt exposed in the east wall of Grand Coulee, Washington. The oldest flow is on the bottom and is overlain by progressively younger flows.
Photo by John S. Shelton.

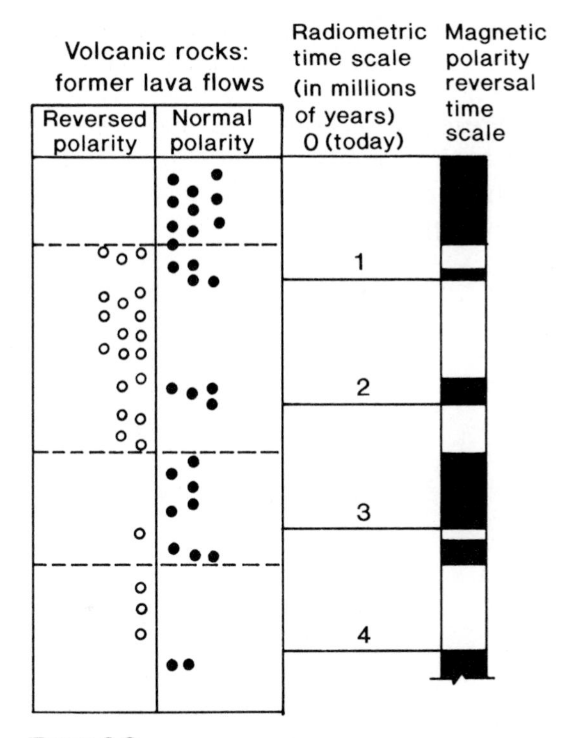

Figure 2.9 A portion of the magnetic polarity time scale. Magnetic polarity measurements in volcanic rocks combined with radiometric ages determined from the same rocks allow formation of a time scale based on magnetic polarity reversals. Notice the unique and nonrepetitive pattern of the polarity reversals.

From P. J. Wyllie, *The Way the Earth Works.* Copyright © 1976 John Wiley & Sons, Inc., New York. Reprinted with permission of John Wiley & Sons, Inc.

bearing minerals become magnetized in the direction of the Earth's magnetic field at that time and place. The lined-up atoms in the iron-rich crystals behave like compass needles pointing toward the magnetic pole of their time (measured as declination or "compass bearing"); they also become inclined at the same angle as the lines of force of the magnetic field (measured as inclination or dip). Ancient magnetic fields have been measured in rocks as old as 3.5 billion years.

Lava flows pile up as sequences of stratified (layered) rock, and the magnetic polarity of each rock layer can be measured (Figure 2.8). Many of the volcanic rocks also contain minerals with radioactive elements that allow determination of the age of the volcanic rock, i.e., how long ago the lava flow solidified. When this information is plotted together in a vertical column, a time scale of magnetic polarities emerges (Figure 2.9). It is interesting to note that the timing of polarity reversals appears random. There is no discernible pattern to the lengths of time the magnetic field was oriented either to the north or to the south. The processes that reverse the polarity of the magnetic field are likely related to changes in the flow of the iron-rich liquid in the outer core. The reversal-causing mechanism does not occur at any mathematically definable time interval.

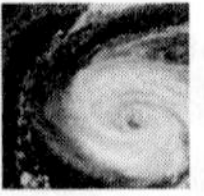

Magnetization Patterns on the Sea Floors

Since the late 1940s, some oceanographic research vessels crisscrossing the Atlantic Ocean have towed magnetometers to measure the magnetization of the sea floor. As the number of voyages grew and more data were obtained, a striking pattern began to emerge (Figure 2.10). The floor of the Atlantic Ocean is striped by parallel bands of magnetized rock that show alternating polarities. The pattern is symmetrical and parallel with the mid-ocean volcanic **ridge** (spreading center). That is, each striped piece of sea floor has its twin on the other side of the oceanic mountain range.

A remarkable relationship exists between the time of reversals of magnetic polarity, as dated radiometrically from a sequence of solidified lava flows (Figure 2.9), and the widths of alternately polarized sea floor (Figure 2.10)—they are comparable. How stunning it is that the widths of magnetized seafloor strips have the same ratios as the lengths of time between successive reversals of the Earth's magnetic field. This means that distance in kilometers is proportional to time in millions of years. Now, if the Earth's magnetic field is reversing polarity in a known time scale and if that time scale reappears in distances, then the relationship must take the form of a velocity. That is, magma is injected into the ocean ridges where it is imprinted by the Earth's magnetic field as it cools to form new rock. Then the sea floor/ocean crust/lithosphere is physically pulled away from the oceanic ridges as if they were parts of two large conveyor belts going in opposite directions (Figure 2.11).

The evidence provided by the paleomagnetic time scale and the magnetically striped sea floors is compelling. These phenomena are convincing evidence that seafloor spreading occurs and that plate tectonics is valid. Let us briefly consider other evidence for plate tectonics.

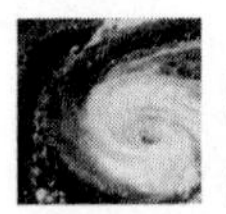

Some Other Evidence of Plate Tectonics

Earthquake Epicenters Outline Plates

The map of earthquake **epicenters** (Figure 2.12) can be viewed as a connect-the-dots puzzle. Each epicenter represents a place where one major section of rock has moved past another section. Take your pen or pencil, connect the dots (epicenters), and you will outline and define the edges of the tectonic plates, the separately moving pieces of lithosphere. Remember that these plates are about 100-km (60+-mi) thick and thousands of kilometers across.

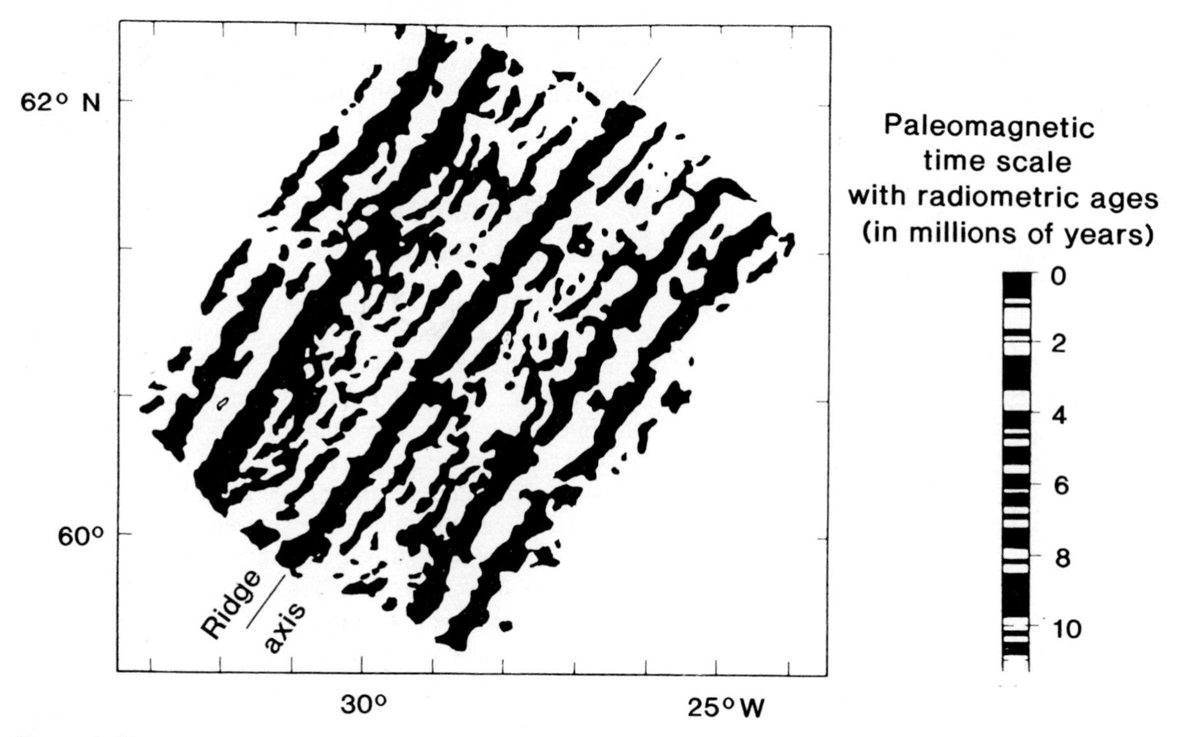

Figure 2.10 Map of the magnetically striped Atlantic Ocean floor southwest of Iceland. Black areas are magnetized pointing to a north pole and white areas to a south pole. Notice the near mirror images of the patterns on each side of the volcanic ridge (spreading center).

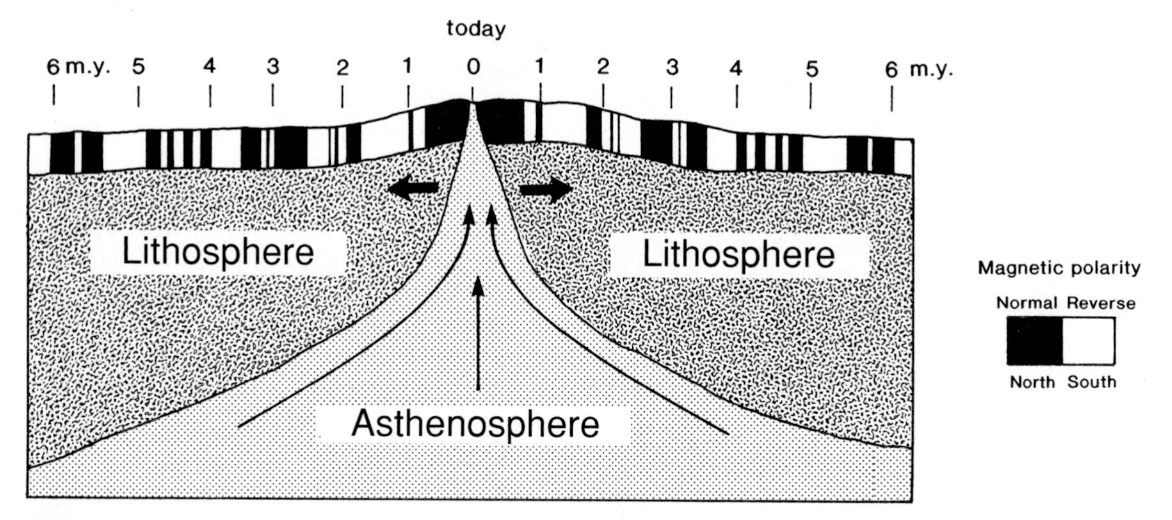

Figure 2.11 Cross section of magnetically striped sea floor. Numbers above the sea floor are radiometrically determined ages in millions of years. The near mirror-image magnetic pattern is like a tape recorder that documents "conveyor belt" movements away from volcanic ridges.

Oceanic Mountain Ranges and Deep Trenches

The greatest mountain ranges on Earth lie on the ocean bottoms and extend more than 65,000 km (40,000 mi). These long and continuous volcanic mountains are seen to form at **spreading centers** where plates pull apart and magma rises to fill the gaps.

The ocean bottom has an average depth of 3.7 km (2.3 mi), yet depths greater than 11 km (nearly 7 mi) exist in elongate, narrow **trenches.** The long and deep trenches were known since the Challenger oceanographic expedition in the 1870s but they were not understood until the 1960s when it

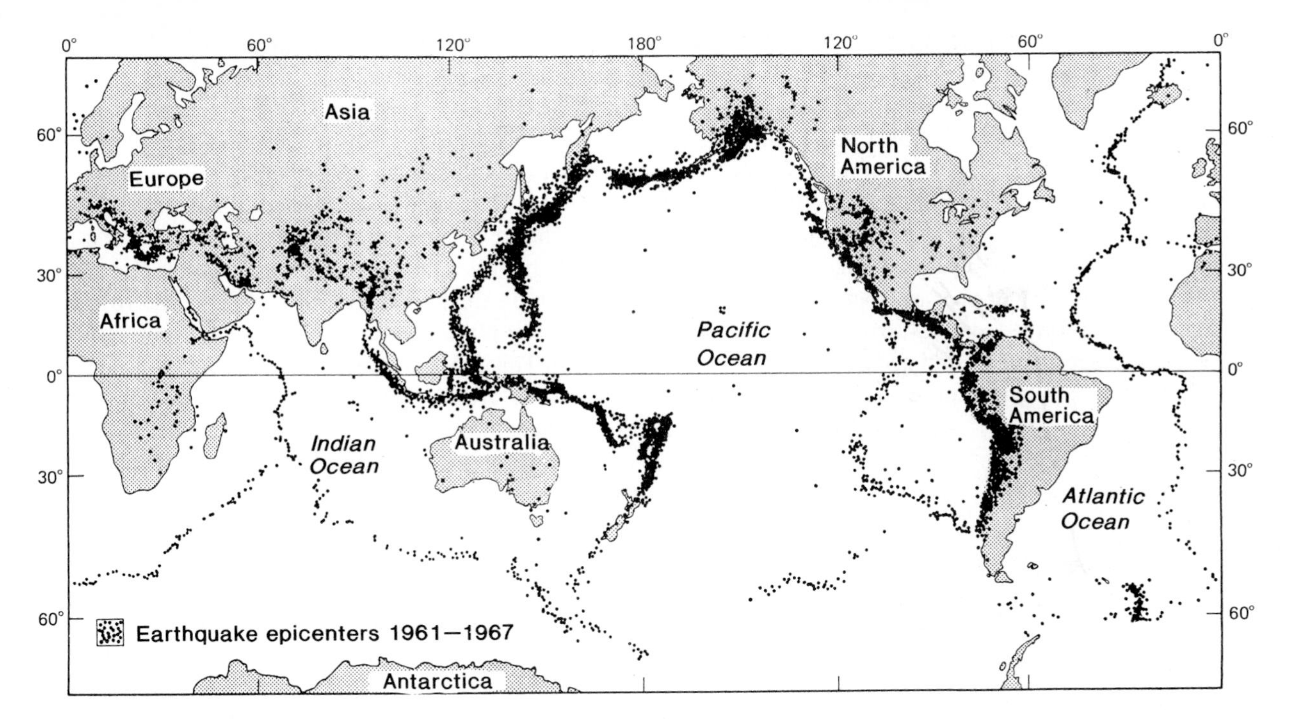

Figure 2.12 Map of the earthquake epicenters recorded by the U.S. Coast and Geodetic Survey, 1961–1967. Notice that the epicenters are concentrated in linear belts.

Source: M. Barazangi and J. Dorman, in *Bulletin of Seismological Society of America,* 59:369, 1969.

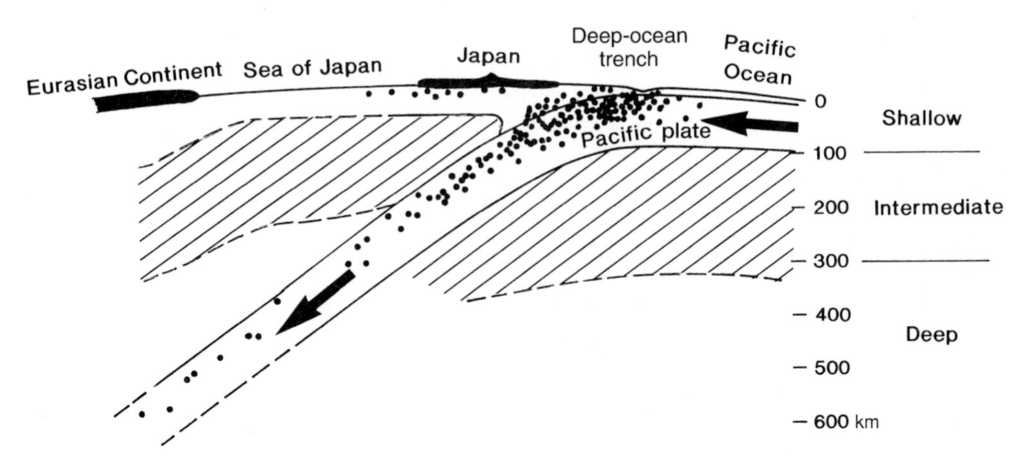

Figure 2.13 Cross section showing earthquake (fault movement) locations at depth; notice the inclined plane defined by the earthquake sites. The earthquake locations define the subducting plate beneath Japan. At shallow depths, earthquakes are generated in brittle rocks in both subducting and overriding plates. At greater depths, only the interior of the subducting Pacific plate is cold enough to maintain the rigidity necessary to produce earthquakes. Striped areas are hot rocks defined by relatively lower-velocity seismic waves.

was recognized that they are the tops of the subducting plates turning downward to reenter the mantle.

Deep Earthquakes

Earthquakes at depth commonly occur along inclined planes (Figure 2.13) adjacent to deep-ocean trenches. These deep earthquakes define the subducting plates being pulled forcefully back into the mantle.

Ages from the Ocean Basins

One of the most stunning facts discovered during the recent exploration of the oceans is the youthfulness of the ocean basins. The oldest rocks on the ocean floors are about 200 million years in age; this is less than 5 percent of the age of the Earth. Remember that some continental rocks are over 4,000 million years old. Meteorites are over 4,500 million years old. Some Moon rocks are over 4,500 million years old, and none are younger than 3,100 million years. But the ocean basins (not the water in them) and their contained volcanic mountains, sediments, and fossils are all much, much younger. Why? Because the ocean basins are young features that are continuously being formed and destroyed.

Along the oceanic ridges, volcanism is active, and new seafloor/oceanic crust is forming (Figure 2.11). Moving away from the ridges, the seafloor volcanic rocks and islands become progressively older. The oldest seafloor rocks are found at the edges of the ocean basins.

At certain locations, deep-seated **"hot spots"** generate magma that rises due to its lower density; these **plumes** of buoyant hot rock pass upward through the mantle, begin to melt near the top of the overlying asthenosphere, and pass up through the lithosphere as magma. Hot spots have active volcanoes above them on the Earth's surface. The volcanoes rest on moving plates that carry them away from their hot-spot source. This process forms lines of extinct volcanoes on the ocean floor, from youngest to oldest, pointing in the direction of plate movement (Figure 2.14). In other words, the ages of the former volcanoes increase with their distance from the hot spot.

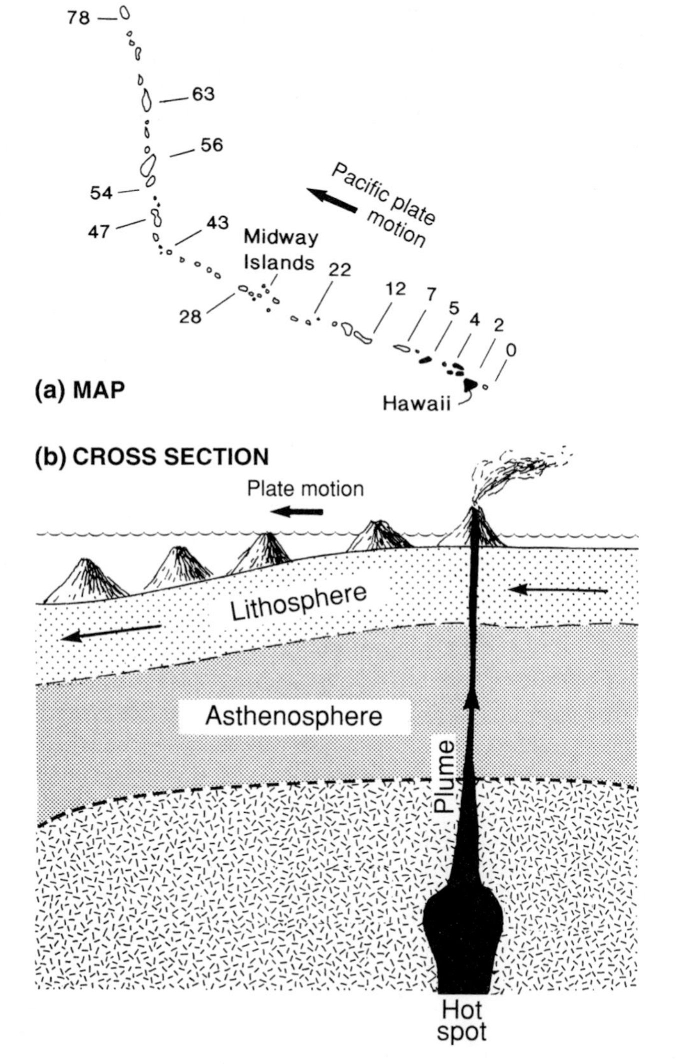

Figure 2.14 A hot spot and its path. **(a)** Map shows the Hawaiian Islands–Emperor Seamount chain of hot spot-fed volcanoes with plots of their radiometric ages. The map pattern of volcano ages testifies to movement of the Pacific plate through time. **(b)** Cross section shows a hot spot at a depth where hot mantle rock rises up through the asthenosphere and passes through the lithosphere as a plume of magma supplying a volcano. Because the lithospheric plate keeps moving, new volcanoes are formed.

The blanket of sediment on the sea floor ranges from very thin to nonexistent at the volcanic ridges and thickens toward the ocean margins (Figure 2.15). The older the sea floor, the more time it has had to accumulate a thick cover of sand, silt, clay, and **fossils.**

Systematic Increases in Seafloor Depth

Above the oceanic ridges, the ocean water depths are relatively shallow. However, moving progressively away from the ridges, the ocean water depths increase systematically with seafloor age (Figure 2.15). This is due to the cooling and contraction of the oceanic crust with a resultant increase in density. Also, there is some isostatic downwarping due to the weight of sediments deposited on the sea floor. The progressive deepening of the sea floor with increasing age also testifies to the existence of seafloor spreading.

The Fit of the Continents

If the continents have really drifted apart, then one should be able to take a map, cut out the continents to make puzzle pieces, and then reassemble them in their former configuration. In fact, this can be done if we know where to cut the map. On two-dimensional world maps, the landmasses occupy about 29 percent of the Earth's surface and the oceans the other nearly 71 percent. If we cut the puzzle pieces at the land-sea shoreline and then attempt to reassemble them, the fit will not be good. The problem here is that the significant boundary is not between land and water but instead lies at the real edge of the continent—the change from low-density continental rocks to higher-density oceanic rocks. This change occurs at about 1,800-m (6,000-ft)-water depth. If we remove the oceans, we find that the continental masses cover 40 percent of the Earth's surface and the ocean basins the other 60 percent. If we cut the puzzle pieces at the 1,800-meter water-depth line, then the continental puzzle pieces fit together quite well. There are some overlaps and gaps, but these are reasonably explained by changes during the last 220 million years, since the last major split of the continents. Examples of the changes include deformation dur-

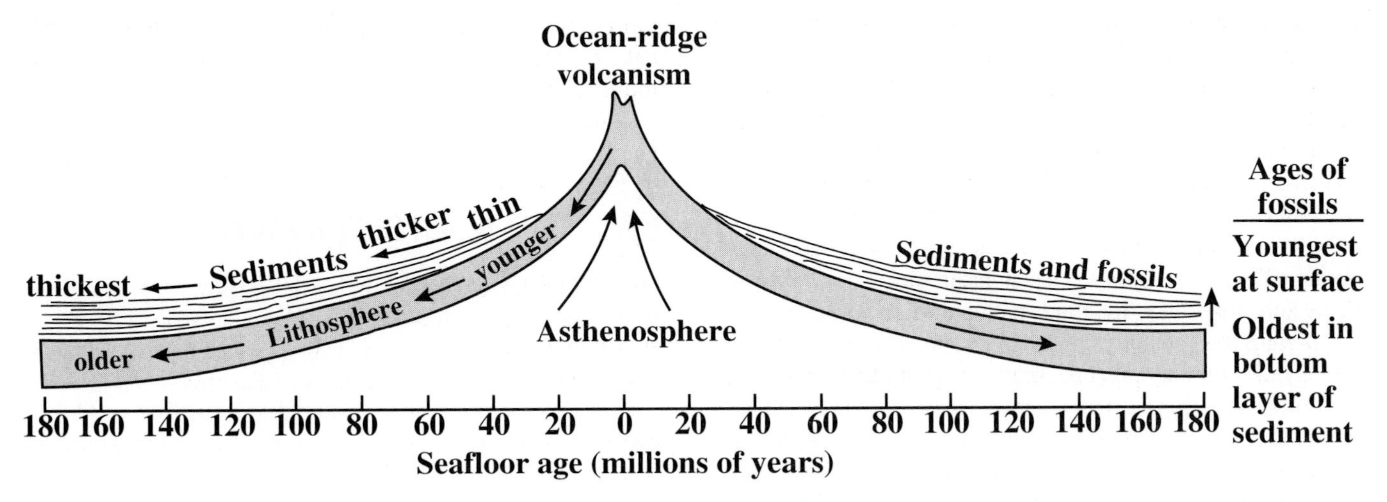

Figure 2.15 Schematic cross section through oceanic lithosphere perpendicular to a volcanic ridge. Moving away from the ridge: 1) radiometric ages of oceanic lithosphere increase, 2) thicknesses of accumulated sediments increase, and 3) ages of fossils in the sediments increase. The systematic increases in water depth are due to cooling, shrinking, and increase in density of the aging seafloor rocks.

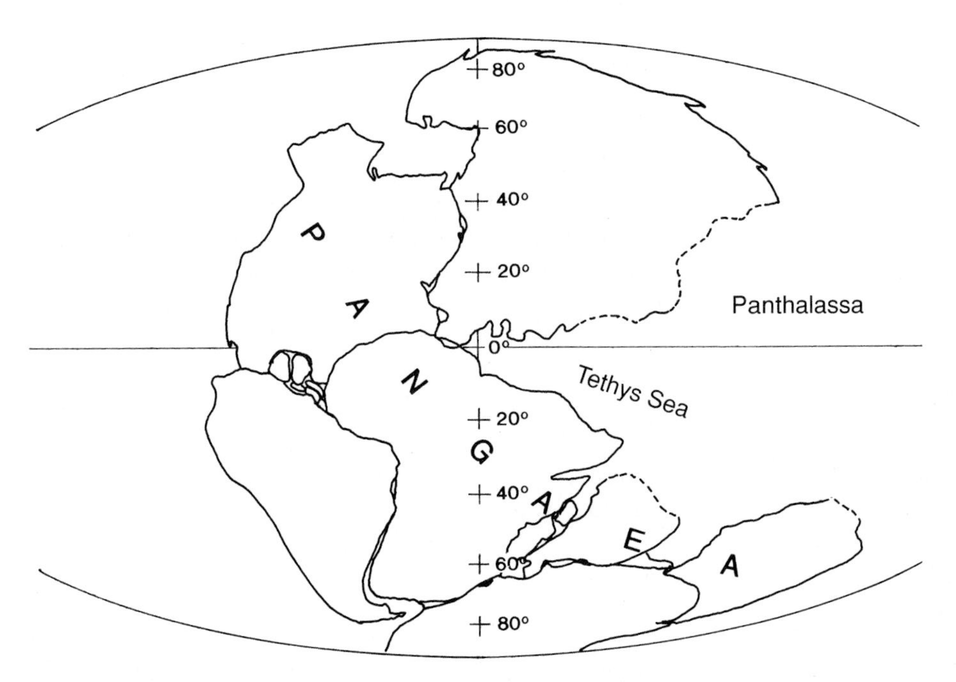

Figure 2.16 Pangaea, the supercontinent, 220 million years before present. The modern continents are drawn to be recognizable in this restoration. The superocean of the time (Panthalassa) exists today in shrunken form as the Pacific Ocean.

After R. S. Dietz and J. C. Holden, "Reconstruction of Pangaea: Breakup and Dispersion of Continents, Permian to Present" in *Journal of Geophysical Research,* 75:4, 939–56, 1970. Copyright © 1970 American Geophysical Union.

ing the process of rifting; growth of river deltas, volcanoes, and coral **reef** masses; erosion of the continents; and land movements.

Changing Positions of the Continents Undoing the seafloor spreading of the last 220 million years restores the continents of today into the supercontinent Pangaea that covered 40 percent of the Earth (Figure 2.16). Although the present continents had yet to form, this figure shows their relative positions within Pangaea before its breakup. The remaining 60 percent of the Earth's surface was a massive ocean called **Panthalassa** (meaning "all oceans").

Figure 2.17a shows the breakup of Pangaea at 180 million years before present. An equatorial spreading center separated the northern supercontinent **Laurasia** from the southern supercontinent **Gondwanaland.** Much of the sediment deposited in the Tethys Sea at that time has since been uplifted to form mountain ranges from the Himalayas to the Alps. Another spreading center began opening the Indian Ocean and separating Africa–South America from Antarctica–Australia.

At 135 million years ago, seafloor spreading had begun opening the North Atlantic Ocean, India was moving toward Asia, and the South Atlantic Ocean was a narrow sea similar to the Red Sea today (Figure 2.17b).

By 65 million years ago, seafloor spreading had opened the South Atlantic Ocean and connected it with the North Atlantic, and Africa came into contact with Europe cutting off the western end of the Tethys Sea to begin the Mediterranean Sea (Figure 2.17c). Although the modern world had become recognizable, note that North America and Eurasia were still connected and that Australia had not yet left Antarctica.

Nearly half of the present ocean floor was created during the last 65 million years (Figure 2.17d). India has rammed into Asia, continued opening of the North Atlantic has split Eurasia from North America, and Australia has moved a long way from Antarctica.

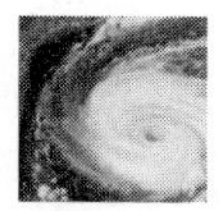

The Grand Unifying Theory

Figure 2.18 shows a part of the world and how the Earth's outer layers are operating today after 220 million years in the latest round of plate-tectonic action. Magma from the Earth's interior rises. The buildup of magma and heat can cause expansion and topographic elevation of the overlying lithosphere/continent, which then may fracture because of the uplift and begin to be pulled apart laterally by gravity. The pulling apart causes a reduction in pressure on the superheated asthenosphere rock, which liquefies and rises upward to fill the fractures and create new oceanic lithosphere via seafloor spreading. The continuing elevation of the volcanic mountain chain (**ridge**) forms a setting for gravity to keep pulling material downward and outward

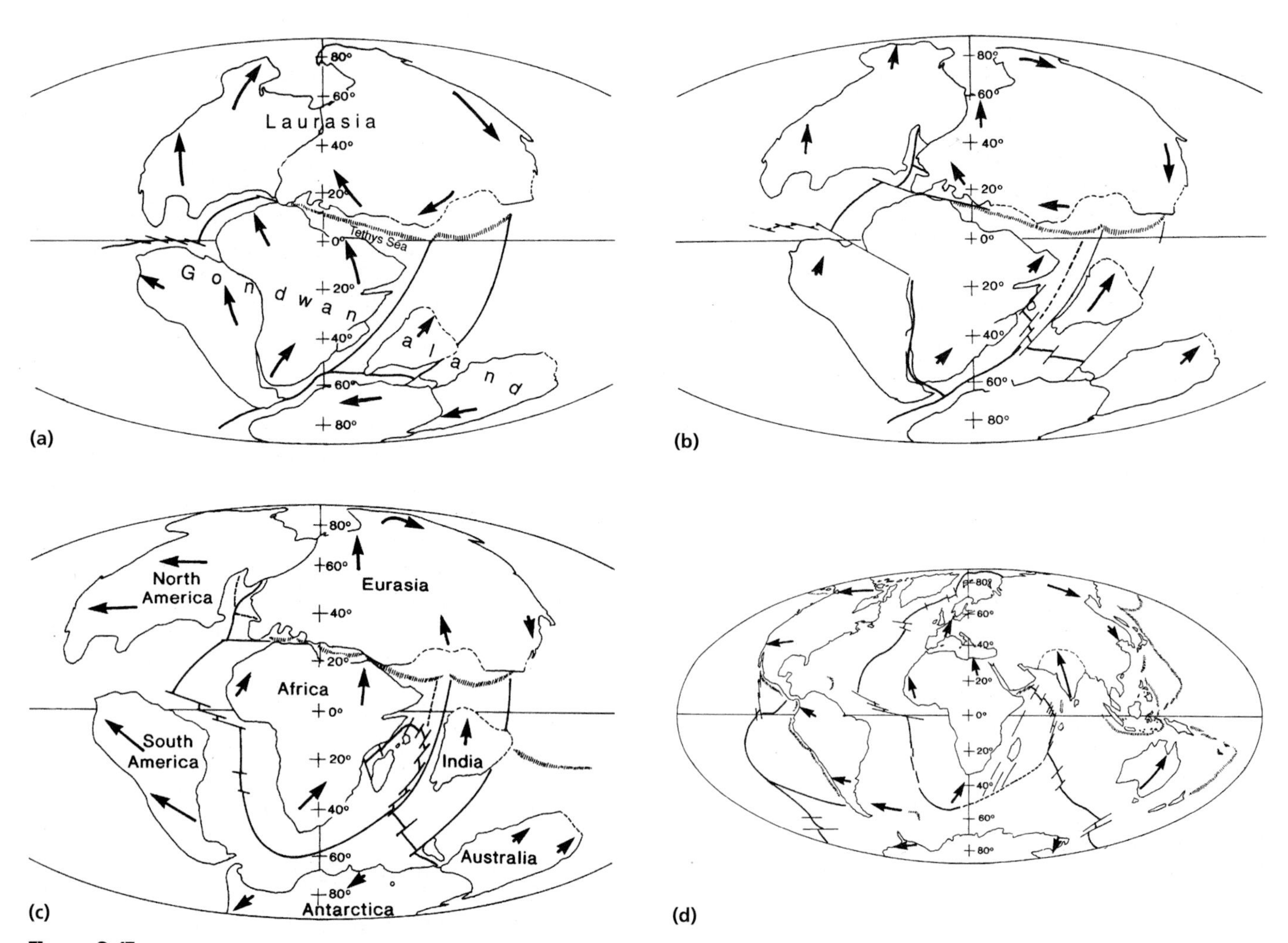

Figure 2.17 Changing positions of the continents. **(a)** 180 million years ago. **(b)** 135 million years ago. **(c)** 65 million years ago. **(d)** Today.

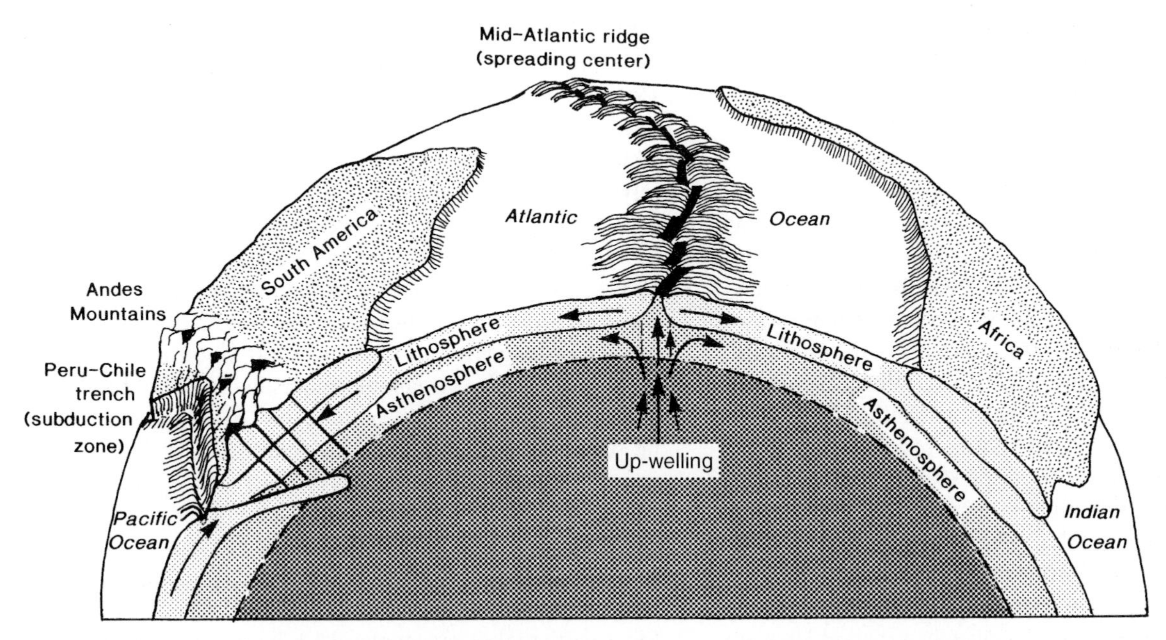

Figure 2.18 Schematic diagram showing plate-tectonic processes for part of the modern world. Atlantic Ocean is opening due to active spreading at its mid-ocean volcanic mountain chain. Westward-moving South America is running over a subducting oceanic plate beneath the Pacific Ocean; the downward bend of the subducting plate creates a deep-ocean trench.

From P. J. Wyllie, *The Way the Earth Works.* Copyright © 1976 John Wiley & Sons, Inc., New York. Reprinted with permission of John Wiley & Sons, Inc.

(spreading). The lateral spreading may be aided by **convection** cells of mantle heat, which rise and move laterally beneath the lithosphere before descending. As the lithosphere spreads, cools, and becomes denser, it is pulled ever more strongly by gravity. When oceanic lithosphere collides with another plate, the heavier, denser (older, colder) plate goes beneath the lighter (younger, warmer) plate in the process of subduction. As the leading edge of the negatively buoyant subducting plate turns downward, gravity exerts an even stronger pull on it, which helps tear the trailing edge of the plate away from the spreading center. The combination of gravity pulling on elevated spreading-center mountains and on denser, down-going plates at subduction zones (slab pull) plus convection in the mantle (mantle drag) keep the lithospheric plates moving. Thus, an ongoing tectonic cycle operates where each moving part stimulates and maintains motions of the others in a large-scale, long-term, recycling operation. Subducted plates are reassimilated into the mantle as physical slabs that remain solid enough to be recognized by their effects on travel velocities of seismic waves. Plate movements are now so well understood due to the magnetic record of seafloor rocks that the plates are not only outlined, but their rates of movement are defined as well (Figure 2.6).

Plate tectonics is a great scientific concept. It provides us with new perspectives about the Earth that are quite different from those encountered in our life or historical experiences. Because the Earth is so much older and so much larger than a human being, we must set aside our personal time and size scales. Our lives are measured in decades, and our personal measuring rods are our 5- to 6-feet-tall bodies; with these as reference guides, we can be only mystified by the Earth. However, if we change our time perspective to millions and billions of years and our size scales to continents and plates, then, and only then, can we begin to understand the Earth. An active plate may move 1 cm (0.4 in) in a year; this is only 75 cm (30 in) in a human lifetime. The rates of plate movement are comparable to those of human fingernail growth.

From the time perspective of a human life span, how can one explain the building of mountains or the formation of ocean basins? One is left waving her or his arms and speaking wildly about catastrophic upheavals or massive sinkings and other fantasized interpretations that make no sense. But when we consider the Earth over its own time span of 4,570 million years, then there is plenty of time for small events to add up to big results. The plate moving 1 cm/yr travels 10 km (6+ mi) in just one million years. The 1 cm/yr process is fast enough to uplift a mountain in a small amount of geologic time. This style of thought is key to understanding how the Earth behaves; we must think of repeated small changes occurring for great lengths of time to create large features like mountains. This style of thought has been called **Uniformitarianism.**

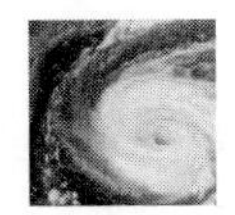

Uniformitarianism

Thousands of years ago, human thought had already made great advances in topics such as philosophy, government, religion, drama, and engineering. But our understanding of the

Earth was insignificant until Earth's great age was realized. This recognition came late in human history; it started with James Hutton in the 1780s. Hutton carefully observed his Scottish landscape and thought deeply about it. For example, he saw rock walls built by the Romans that had stood for 15 centuries with only slight change. If 1,500 years was not long enough to break down a wall, then Hutton wondered how much time had been required to break down some of the hard rock masses of Scotland into the abundant pebbles and sand grains he saw? And how much more time had been necessary to lift the pebbly and sandy sedimentary rocks to form hills? All the active processes Hutton observed worked slowly so his answer to the questions was that great lengths of time were required. In 1788, Hutton described the history of the Earth with: "The result, therefore, of our present inquiry is that we find no vestige of a beginning, no prospect of an end." And this was Hutton's great gift to human thought: time is long, and everyday changes on Earth add up to major results.

Hutton's thought pattern revolutionized our understanding of the Earth. Uniformitarianism implies that natural laws are uniform through time and space. Physical and biological laws produce certain effects today, as they have in the past, and will in the future. If we can understand how the Earth works today, we can use this knowledge to read the rock and fossil record to understand Earth history. The present is the key to the past.

The term Uniformitarianism has come under attack by some who assume it says that Earth processes have always acted at a uniform and slow rate, but we all know that rates can vary. For example, seafloor spreading has operated at slower rates in the past, and it has run at faster rates, but the laws governing how seafloor spreading operates do not change just because the rates vary. Some suggest using the term **Actualism** instead of Uniformitarianism, but the concept is basically the same. Actualism tells us to understand the processes actually operating on and in the Earth today, and to use these known and testable processes to interpret the past; do *not* invent unknown and untestable processes to explain problems.

How do we go about understanding the Earth? We study the present to understand the past, and then make probabilistic forecasts about the future.

Plate Tectonics and Earthquakes

Most earthquakes are explainable based on plate tectonics theory. The lithosphere is broken into rigid plates that move away from, past, and into other rigid plates. These global-scale processes are seen on the ground as individual **faults** where the Earth ruptures and the two sides move past each other in earthquake-generating events. Figure 2.19 shows an idealized tectonic plate and assesses the varying earthquake hazards that are concentrated at plate edges:

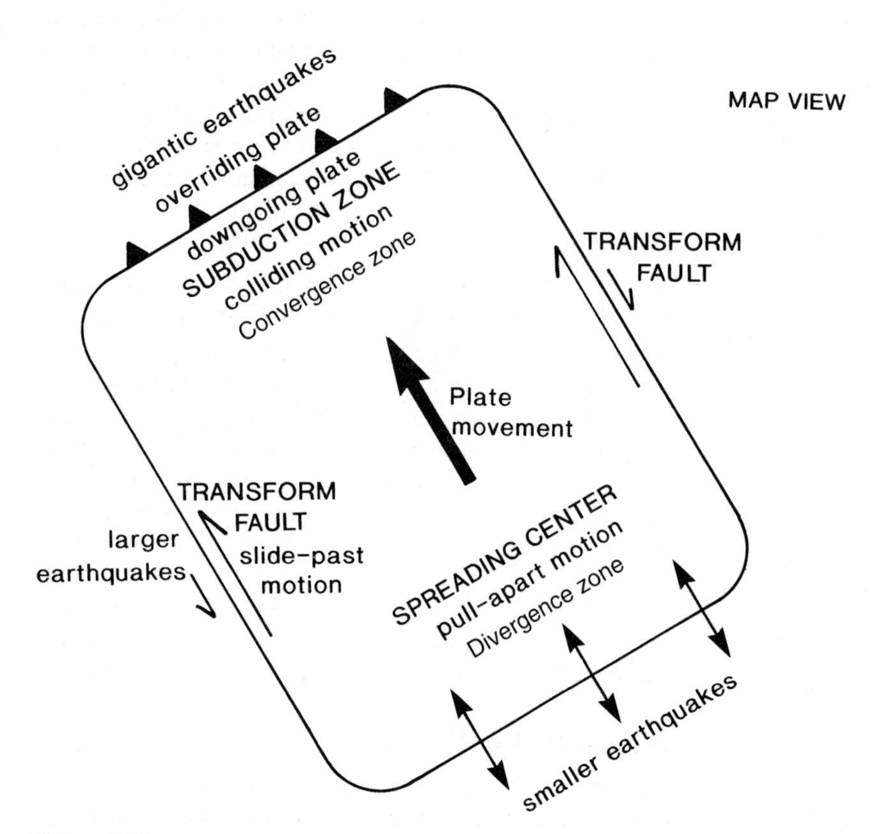

Figure 2.19 An idealized plate and the earthquake potential along its edges.

1. The divergent or pull-apart motion at spreading centers causes rocks to fail in **tension.** Rocks rupture relatively easily when subjected to tension. Thus, this process yields mainly smaller earthquakes that do not pose an especially great threat to humans.
2. The slide-past motion occurs as the rigid plates fracture and move around the curved Earth. The plates slide past each other in the dominantly horizontal movements of transform faults. This process creates large earthquakes as the irregular plate boundaries retard movement because of irregularities along the faults. It takes a lot of stored energy to overcome the rough surfaces, nonslippery rocks, and bends in faults. When these impediments are finally overcome, a large amount of seismic energy is released.
3. The convergent motions that occur at subduction zones and in continent-continent collisions store immense amounts of energy that are released in Earth's largest tectonic earthquakes. The very processes of pulling a 70- to 100-km (45- to 60-mi)-thick oceanic plate back into the mantle via a subduction zone or of pushing continents together—such as India slamming into Asia to uplift the Himalayas—involve incredible amounts of energy. This results in Earth's greatest earthquakes.

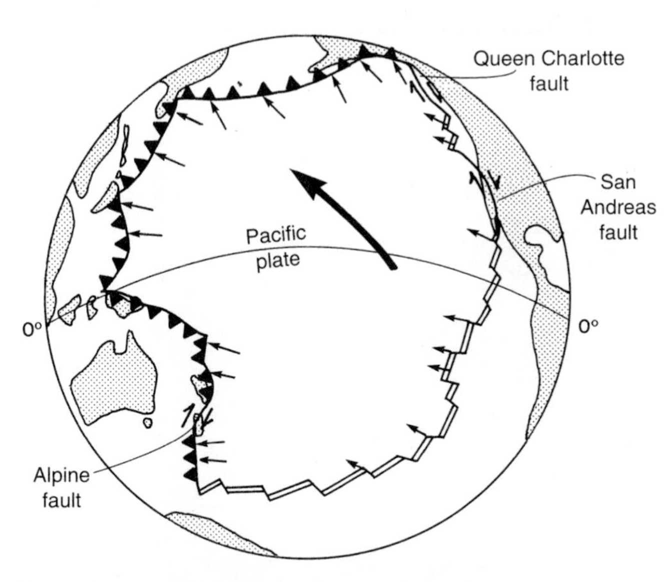

Figure 2.20 The Pacific plate is the largest in the world; it underlies part of the Pacific Ocean. Its eastern and southern edges are mostly spreading centers characterized by small- to intermediate-size earthquakes. Three long transform faults exist along its sides in Canada (Queen Charlotte), California (San Andreas), and New Zealand (Alpine); all are marked by large earthquakes. Subduction zones lie along the northern and western edges, from Alaska to Russia to Japan to the Philippines to Indonesia to New Zealand; all are characterized by gigantic earthquakes.

From P. J. Wyllie, *The Way the Earth Works.* Copyright © 1976 John Wiley & Sons, Inc., New York. Reprinted with permission of John Wiley & Sons, Inc.

Moving from an idealized plate, let us examine an actual plate—the Pacific plate. Figure 2.20 shows the same type of plate-edge processes and expected earthquakes. The Pacific plate is created at the spreading centers along its eastern and southern edges. The action there produces smaller earthquakes that also happen to be located away from major human populations.

The slide-past motions of long transform faults occur: 1) in the northeastern Pacific as the Queen Charlotte fault, located near a sparsely populated region of Canada; 2) along the San Andreas fault in California with its famous earthquakes; and 3) at the southwestern edge of the Pacific Ocean where the Alpine fault cuts across the South Island of New Zealand (Figures 3.5 and 3.6).

The Pacific plate subducts along its northern and western edges and creates enormous earthquakes, such as the 1964 Alaska event, the 1923 Tokyo seism, and the 1931 Napier quake on the North Island of New Zealand.

Our main emphasis here is to understand plate-edge effects as a means of forecasting where earthquakes are likely to occur and what their relative sizes may be.

Spreading Centers and Earthquakes

Inspection of earthquake epicenter locations around the world (see Figure 2.12) reveals that earthquakes are not as common in the vicinity of spreading centers or divergence zones as they are at transform faults and at subduction/collision zones. The expanded volumes of warm rock in the oceanic ridge systems have a higher heat content and a resultant decrease in rigidity. These heat-weakened rocks do not build up and store the huge stresses necessary to create great earthquakes.

Iceland

The style of spreading-center earthquakes can be appreciated by looking at the earthquake history of Iceland, a nation that exists solely on a hot-spot-fed, volcanic island portion of the mid-Atlantic ridge spreading center (Figures 2.21 and 2.22). The Icelandic geologist R. Stefansson reported on catastrophic earthquakes in Iceland and stated that in the portions of the country underlain by north-south-oriented spreading centers, stresses build up to cause earthquakes too small to destroy buildings or kill people. These moderate-sized earthquakes tend to occur in swarms, as is typical of volcanic areas where magma is on the move. Iceland does have large earthquakes, but they are associated with east-west-oriented transform faults between the spreading-center segments.

Red Sea and Gulf of Aden

Iceland has been built on a mature spreading center that has been opening the North Atlantic Ocean basin for about 180 million years. What would a young spreading center and

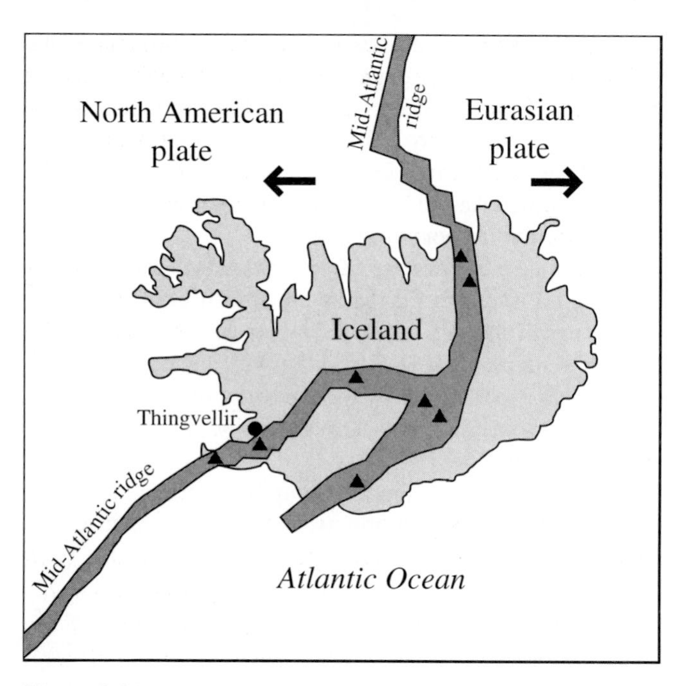

Figure 2.21 Iceland sits on top of a hot spot and is being pulled apart by the spreading center in the Atlantic Ocean.

new ocean basin look like? Long and narrow. In today's world, long and narrow ocean basins exist in northeast Africa as the Red Sea and the Gulf of Aden (Figure 2.23). The northeastern portion of Africa sits above an extra hot area in the upper mantle. The heat contained within this mantle hot zone is partially trapped by the blanketing effect of the overlying African plate and its embedded continent (Figure 2.24a). The hot rock expands in volume and some liquefies to magma. This volume expansion causes doming of the overlying rocks, with resultant uplift of the surface to form topography (Figure 2.24b). The doming uplift sets the stage for gravity to pull the raised landmasses downward and apart, thus creating pull-apart faults with centrally located, downdropped **rift** valleys, also described as pull-apart basins (Figure 2.24c). As the fracturing/faulting progresses, magma rises up through the cracks to build volcanoes. As rifting and volcanism continue, seafloor spreading processes take over, the downdropped linear rift valley becomes filled by the ocean, and a new sea is born (Figure 2.24d).

Figure 2.23 reveals another interesting geometric feature. Three linear pull-apart basins meet at the south end of the Red Sea at a **triple junction,** a point where three plate

Figure 2.22 Looking south along the fissure at Thingvellir, Iceland. This is the rift valley being pulled apart in an east-west direction by the continuing spreading of the Atlantic Ocean basin.
Photo by John S. Shelton.

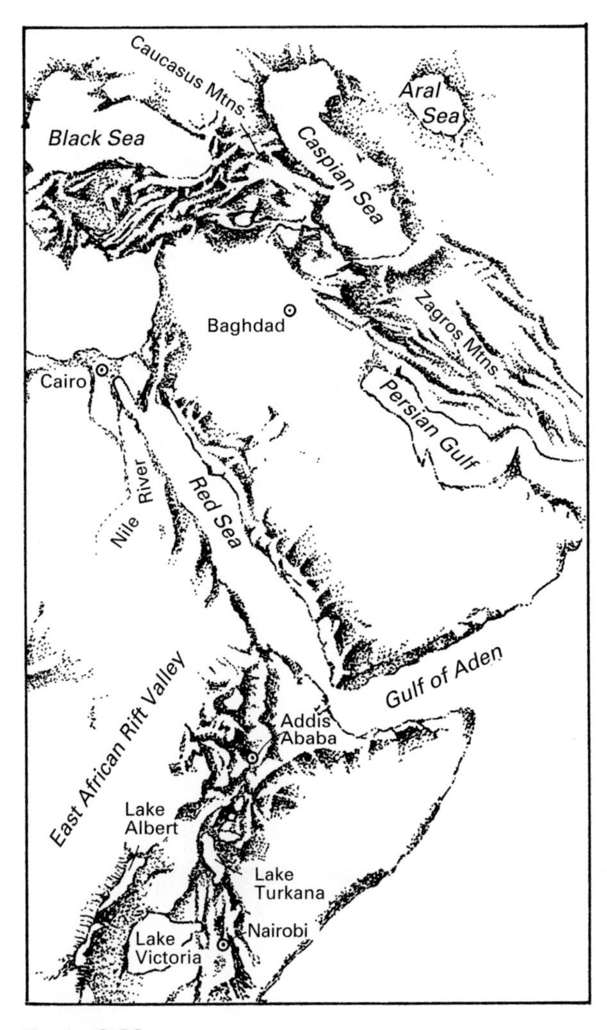

Figure 2.23 Topography in northeastern Africa and Arabia.

edges touch. A concentration of heat in the upper mantle began the process of creating this triple junction. The Earth's surface bulged upward into a dome, causing the elevated rocks to fracture into a radial pattern (Figure 2.25). Gravity pulled the dome apart, magma welled up to fill three major fracture zones, and the spreading process was initiated.

The triple junction in northeast Africa is geologically young, having begun about 25 million years ago. To date, spreading in the Red Sea and Gulf of Aden has been enough to split off northeast Africa and create an Arabian plate and to allow seawater to flood between them. But the East African Rift Valley has not yet been pulled far enough apart for the sea to fill it (Figure 2.26). The East African Rift Valley is a truly impressive physiographic feature. It is 5,600-km (3,500-mi) long and has steep escarpments and dramatic valleys. Beginning at the Afar triangle at its northern end and moving southwest are the domed and stretched highlands of Ethiopia, beyond which the rift valley divides into two major branches. The western rift is markedly curved and has many deep lakes, including the world's second deepest lake, Lake Tanganyika. The eastern rift is straighter and holds shallow, alkaline lakes and volcanic peaks, such as Mount Kilimanjaro, Africa's highest mountain. The rift valley holds the oldest humanoid fossils found to date and is the probable homeland of the first human beings. Will the spreading continue far enough to split a Somali plate from Africa? It is simply too early to tell.

How severe are the earthquakes in the geologically youthful Red Sea and Gulf of Aden? Significant, but spreading-center earthquakes are not as large as the earthquakes on the other types of plate edges.

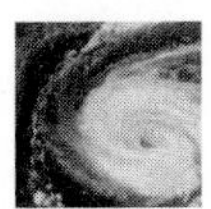

Convergent Zones and Earthquakes

The greatest earthquakes in the world occur where plates collide. Three basic classes of collisions are: 1) oceanic plate versus oceanic plate; 2) oceanic plate versus continent-bearing plate; and 3) continental plate versus continental plate. These collisions result in either subduction or continental upheaval. If oceanic plates are involved, subduction will occur. The younger, warmer, less-dense plate edge will override the older, colder, denser plate, which will then bend downward and be pulled back into the mantle. If a continent is involved, it cannot subduct because its huge volume of low-density, high-buoyancy rocks simply cannot sink to great depth and cannot be pulled into the denser mantle rocks below. The fate of oceanic plates is destruction via subduction and reassimilation within the mantle, whereas continents float about on the asthenosphere in perpetuity. Continents are ripped asunder and then reassembled into new configurations via collisions, but they are not destroyed by subduction.

Subduction Zones

Subduction zones are the sites of great earthquakes. Imagine pulling a 100-km (62-mi)-thick, rigid plate into the weaker, deformable rocks of the mantle that resist the plate's intrusion. This process creates tremendous stores of energy, which are released periodically as great earthquakes. Subduction occurs on a massive scale. At the present rates of subduction, oceanic plates with an area equivalent to the entire surface area of the Earth will be pulled into the mantle in only 180 million years.

A descending slab of oceanic lithosphere is defined by an inclined plane of deep earthquakes or fault-rupture locations (Figure 2.13). Earthquakes at subduction zones result from different types of fault movements in shallow versus deeper realms. At shallow depths (less than 100 km or 62 mi), the two rigid lithospheric plates are pushing against each other. Earthquakes result from compressive movements where the overriding plate moves upward and the subducting plate moves downward. Pull-apart fault movements also

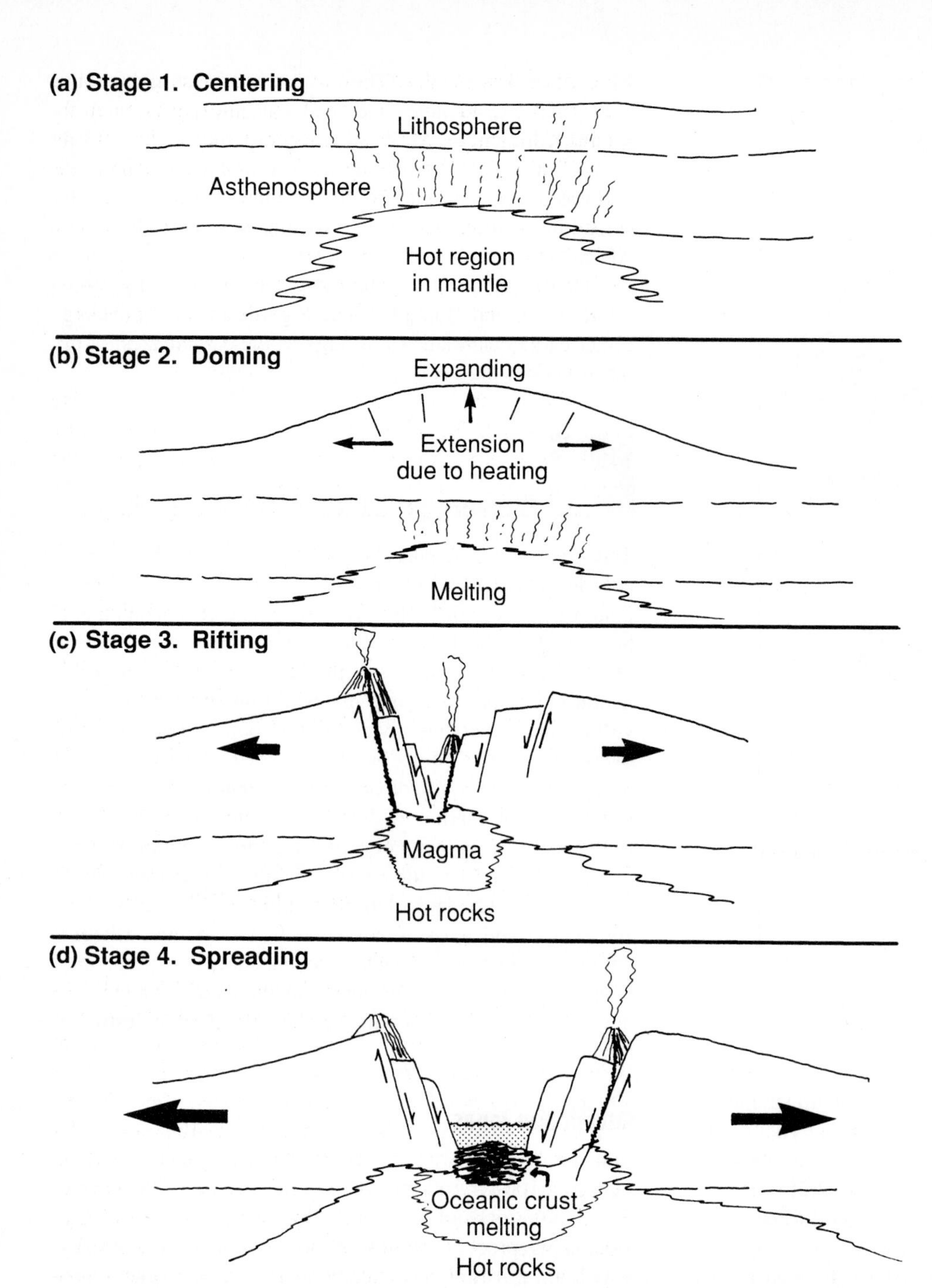

Figure 2.24 Stages in the formation of an ocean basin. **(a) Stage 1, Centering:** Moving lithosphere centers over an especially hot region of the mantle. **(b) Stage 2, Doming:** Mantle heat causes melting and the overlying lithosphere/continent extends. The increase in heat causes surface doming through uplifting, stretching, and fracturing. **(c) Stage 3, Rifting:** Volume expansion causes gravity to pull the uplifted area apart; fractures fail and form faults. Fractures/faults provide escape for magma; volcanism is common. Then, the dome's central area sags downward, forming a valley such as the present East African Rift Valley. **(d) Stage 4, Spreading:** Pulling apart has advanced, forming a new sea floor. Most magmatic activity is seafloor spreading, as in the Red Sea and Gulf of Aden.

occur near the surface within the subducting plate as it is bent downward and snaps in tensional failure and as the overriding plate is lifted up from below. Notice in Figure 2.13 that the shallow earthquakes occur: 1) in the upper portion of the down-going plate, 2) at the bend in the subducting plate, and 3) in the overriding plate.

Compare the locations of the shallow earthquake sites to those of intermediate and deep earthquakes (Figure 2.13).

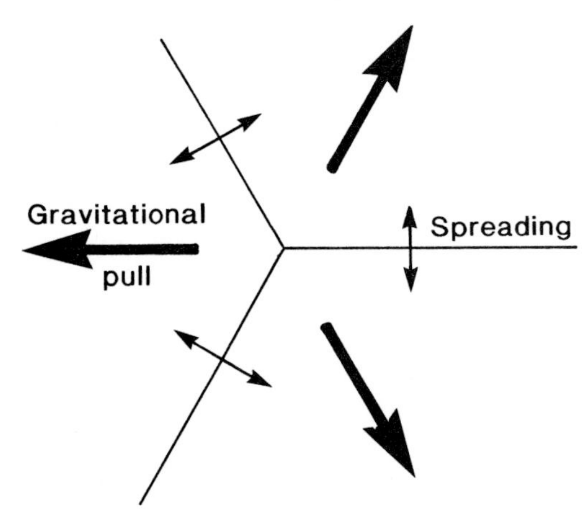

Figure 2.25 Schematic map of a triple junction formed by three young spreading centers. Heat concentrates in mantle and rises in a magma plume, doming the overlying lithosphere and causing fracturing into a radial set with three rifts. Gravity then pulls the dome apart, initiating spreading in each rift.

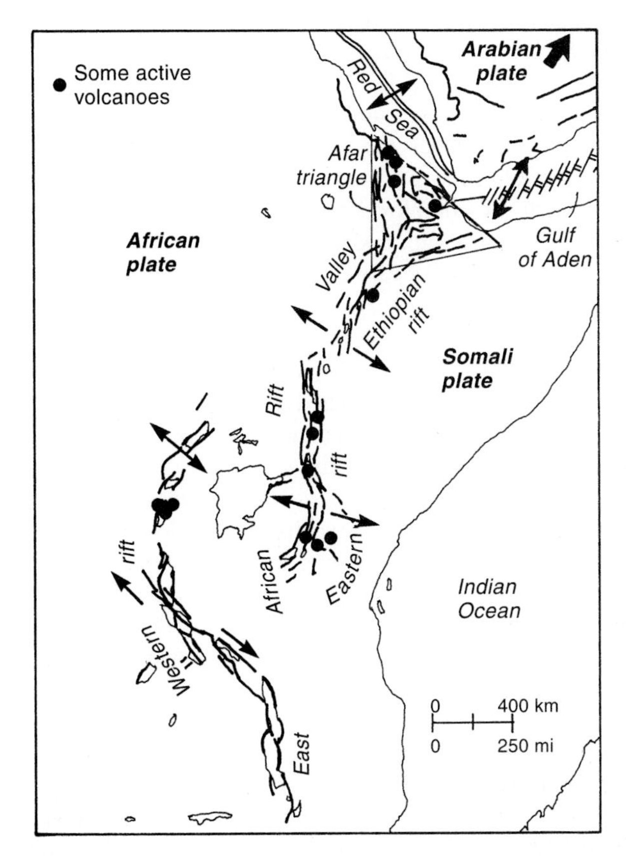

Figure 2.26 Northeastern Africa is being torn apart by three spreading centers which meet at the triple junction in the Afar triangle.

At depths below 100 km, earthquakes occur almost exclusively in the interior of the colder oceanic lithosphere, the heart of the subducting slab. The high temperatures of rock in the upper mantle cause it to yield more readily to stresses and thus not build up the stored energy necessary for gigantic earthquakes. At depth, the upper and lower surfaces of the subducting slabs are too warm to generate large earthquakes. Thus, the earthquakes occur in the cooler interior area of rigid rock, where stress is stored as gravity pulls against the mantle resistance to slab penetration. In the areas of most rapid subduction, the down-going slab may remain rigid enough to spawn large earthquakes to depths in excess of 700 km (435 mi). A great earthquake with a deep hypocenter has much of its seismic energy dissipated while traveling to the surface. Thus, the biggest disasters are from the great earthquakes with shallow hypocenters.

Most of the subduction-zone earthquakes of today occur around the rim of the Pacific Ocean. This is shown by the presence of most of the deep-ocean trenches and by the dense concentrations of earthquake epicenters (Figure 2.12).

A popular way of forecasting the locations of future earthquakes uses the **seismic-gap method.** If segments of one fault have moved recently, then it seems reasonable to expect that the unmoved portions will move next, and thus fill the gaps. For example, some of the abundant earthquake sites in Figure 2.12 clearly define the subducting Pacific plate. Figure 2.27 gives a closer look at an area of Pacific plate subduction and the location, frequency, and severely felt areas of some recent earthquakes. Looking at Figure 2.27 where would you forecast the next large earthquakes to occur? It is easy to see the gaps in earthquake locations and although seismic-gap analysis is logical, it yields only expectations, not guarantees. One segment of a fault can move two or more times before an adjoining segment moves once.

Tokyo, Japan, 1923 Early on Saturday morning, 1 September 1923, the cities of Tokyo and Yokohama were drenched by the last squalls of a waning storm. Later that morning, the skies cleared and the sun beamed down as the residents prepared their midday meal. Moments later, this tranquil scene was shattered by a deadly series of earthquakes. The principal shock was powerful; the Earth averages less than one earthquake a year that releases this much energy. It occurred beneath Sagami Bay southwest of the big cities. The floor of Sagami Bay dropped markedly and sent 11-m (36-ft)-high seismic sea waves (**tsunami**) crashing against the shore. The waves washed away hundreds of homes. Yet fishermen spending their day out on the open ocean were unaware of the monster waves. At day's end, as they sailed toward home through Sagami Bay, they were sickened to find the floating wreckage of their houses and the bodies of their families. Devastation on land was great. Houses were destroyed, bridges fell, tunnels collapsed, and landslides destroyed both forested slopes and terraced hillsides created for agriculture. The wreckage of Tokyo and Yokohama buildings begun by the shaking earth was completed by the ensuing fires. The shaking caused the collapse of flammable house materials onto cooking fires, and the flames, once liberated, quickly raced out of control throughout both cities. Little could be done to stem their spread because the earthquake had broken the water mains.

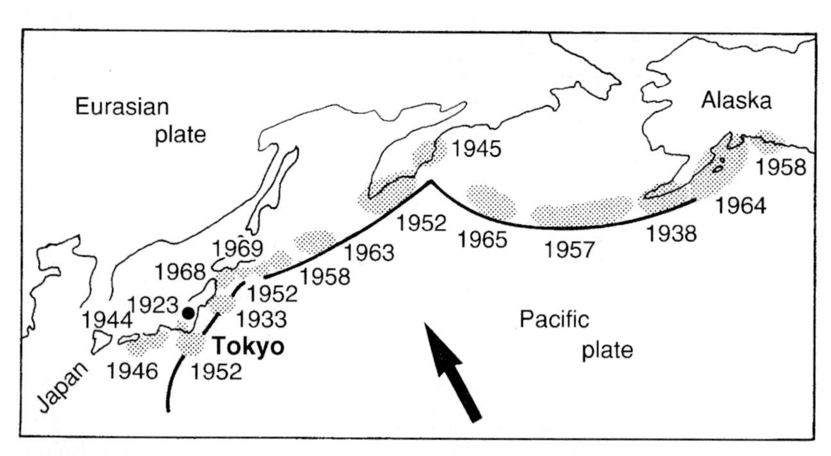

Figure 2.27 Dotted patterns show severely shaken areas, with dates, from some recent earthquakes caused by Pacific plate subduction. Using the seismic-gap method, where are the next earthquakes most likely to occur?

Shifting winds advanced the fires through Tokyo for two and a half days, destroying 71 percent of the city's houses. Infernos in Yokohama gutted the city, a 100 percent loss.

Possibly the most tragic event in this disaster occurred when 40,000 people, clutching their personal belongings, attempted to escape the flames by crowding into a 250-acre garden owned by a wealthy banker on the edge of the Sumida River. People packed themselves into this open space so densely that they were barely able to move. At about 4 P.M., several hours after the earthquake, the roaring fires approached on all three landward sides of the crowd. Suddenly the fire-heated winds spawned a tornado that carried flames onto the huddled masses and their combustible belongings. After the flames had died, 38,000 people lay dead, either burned or asphyxiated. The usual instinct to seek open ground during a disaster was shockingly wrong this time.

The combined forces of earthquakes, seismic sea waves, and fires killed 99,331 people and left another 43,476 missing and presumed dead. In Tokyo, irreplaceable records were lost, and 2,000 years of art treasures were destroyed. Yet, despite this immense catastrophe, the morale of the Japanese people remained high. They learned from the disaster. They have rebuilt their cities with wider streets, more open space, and less use of combustible construction materials.

When will the next large earthquake strike? Figure 2.27 helps frame a very general estimate; the gaps in historic seismicity probably will be filled next. The historical record of earthquakes in the region is also instructive. The region 80 km (50 mi) southwest of Tokyo has been rocked by five very strong earthquakes in the last 400 years. The seisms have occurred roughly every 73 years, the most recent in 1923.

Continent-Continent Collisions

The grandest continental pushing match in the modern world is the ongoing ramming of Asia by India. When Gondwanaland began its breakup, India moved northward toward Asia. The 5,000 km (more than 3,000 mi) of sea floor (oceanic plate) that lay in front of India's northward path had all subducted beneath Asia by about 40 million years ago. Then, with no sea floor left to separate them, India punched into the exposed underbelly of Asia (Figure 2.28). Since the initial contact, the assault has remained continuous, and India has moved another 2,000 km (1,250 mi) farther north causing complex accommodations within the two plates as they shove into, under, and through each other accompanied by folding, overriding, and stacking of the two continents into the huge mass of the Himalayas and the Tibetan Plateau. The precollision crusts of India and Asia were each about 35-km (22-mi) thick. Now, after the collision, the combined crust has been thickened to 70 km (44 mi) to create the highest-standing continental area on Earth. The Tibetan Plateau dwarfs all other high landmasses. In an area the size of France, the average elevation exceeds 5,000 m (16,400 ft). But what does all of this have to do with earthquakes? Each year, India continues to move about 5 cm (2 in) into Asia along a 2,000-km (1,250-mi) front. This ongoing collision jars a gigantic area with great earthquakes. The affected area includes India, Pakistan, Afghanistan, the Tibetan Plateau, much of eastern Russia, Mongolia, and most of China.

A relatively simple experiment shows how earthquake-generating faults may be caused by continental collision (Figure 2.29). The experiment uses a horizontal jack to push into a pile of plasticine, deforming it under the force. The experimental deformation is similar to the tectonic map of the India-Asia region (Figure 2.30). The northward wedging of India seems to be forcing Indochina to escape to the southeast and is driving a large block of China to the east.

Gujarat, India, 2001 The ongoing push of India into Asia caused another big earthquake on 26 January 2001 near Bhuj in the State of Gujarat (Figure 2.30). The event killed 20,103 people, destroyed 348,000 houses, damaged an addi-

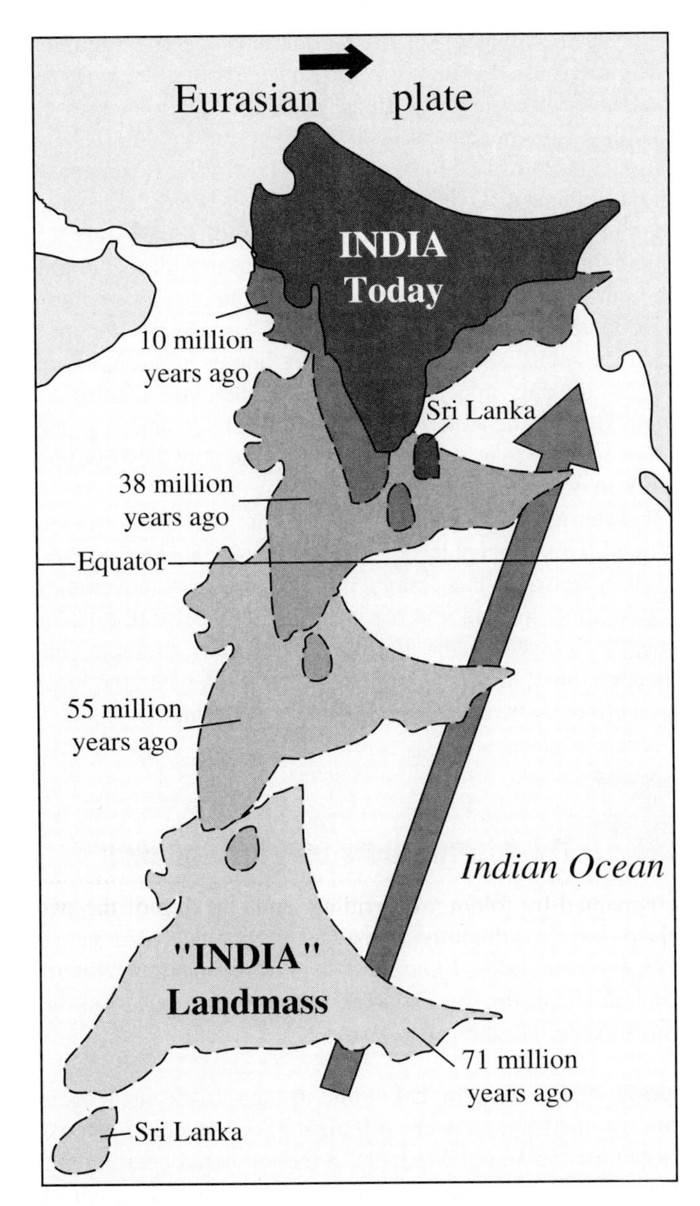

Figure 2.28 Map showing the movement of India during the last 71 million years. India continues to shove into Eurasia creating great earthquakes all the way through China.

Figure 2.29 Simulated collision of India into Asia. Wedge is slowly jacked into layered plasticine confined on its left side but free to move to the right. From top to bottom of figure, notice the major faults that form and the masses that are compelled to move to the right. Compare this pattern to tectonic map of India and Asia in Figure 2.30.
(After P. Tapponier, et al. (1982). *Geology,* 10, 611–16.)

tional 844,000 houses, and killed more than 20,000 cattle. Economic losses are estimated at $5 billion.

This earthquake essentially was a repeat of the 1819 Rann of Kutch earthquake; the two quakes had the same cause, same size, same area, and same results. The 1819 earthquake, and the nineteenth-century geological studies that followed, clearly spelled out the risks for the future. A building code was implemented requiring that buildings be constructed to withstand earthquakes, yet the 2001 earthquake killed the same percentage of the population as the 1819 event. There were ten times as many deaths in the 2001 earthquake, but the population had grown. During the first week after the earthquake, local police registered 37 cases of criminal conspiracy and homicide against builders, architects, and engineers of some collapsed buildings.

Shaanxi Province, China, 1556 The deadliest earthquake in history occurred in 1556 when about 830,000 Chinese were killed in and near Xi'an on the banks of the mighty Huang River (once known as the Yellow River). The region has numerous hills composed of deposits of windblown silt and fine sand (**loess**) that have very little **cohesion** or ability to stick together. Because of the ease of digging in these loose sediments, a tremendous number of the homes in the region were caves dug by the inhabitants. Most of the residents were in their cave homes at five o'clock on the wintry morning of

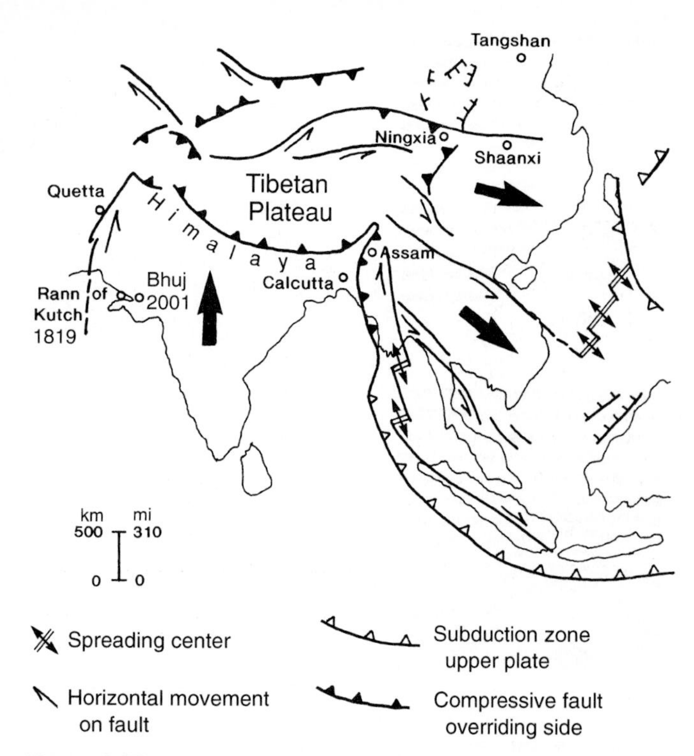

Figure 2.30 Tectonic map showing India pushing into Asia. The ongoing collision causes devastating earthquakes including two in the state of Gujarat—in 1819 at Rann of Kutch and in 2001 near Bhuj.

23 January, when the seismic waves rolled in from the great earthquake. The severe shaking caused much of the soft silt and sand sediments of the region to vibrate apart and literally behave like fluids. Most of the cave-home dwellers were entombed when the once-solid walls of their homes liquefied and collapsed.

Tangshan, China, 1976 The deadliest earthquake in recent times occurred directly beneath the city of Tangshan when a fault ruptured at a depth of 11 km (7 mi) in a local response to the regional stress created by the ongoing collision of India with Asia. The earthquake was much larger than local officials expected. Building codes were lenient—fatally so. This poor decision was instrumental in the deaths of 240,000 people.

In 1976, Tangshan was an industrial and mining city with 2 million residents. It contained the largest coal mines in China, so heavy industry found a home there also. Its coal, steel, electricity, and diesel- and locomotive-engine industries combined to create about 1 percent of China's gross national product. For Tangshan residents, the night of 27 July was unusually warm with rain and wind. But the most unusual happenings that night were the fireballs and lightning that rolled through the sky in all the colors of the rainbow. At 3:42 A.M. on 28 July, the ground began the rumbling that reduced the city to almost total rubble. Most residents were at home in densely packed houses made of mud-bricks held together by poor-quality mortar and covered with mud-and-lime roofs that had grown heavier through the years as new layers were added. Home was not a good place to be that day as 93 percent of residential buildings collapsed. Industrial buildings fared somewhat better; still, 78 percent of them collapsed. Overall, older buildings performed better than newer ones. Some of the "luckier" individuals were the night-shift coal miners hard at work thousands of feet below the surface during the earthquake. Although 13 of these 15,000 miners died, as a whole, they fared far better than their day-shift comrades. Collapsing homes killed 6,500 of 85,000 off-duty miners. Through it all, the human spirit remained unquenchable. Tangshan was rebuilt and is again home to more than a million residents, but now they live and work in better-designed buildings.

Earthquake weather is a term one hears tossed around. This concept has no validity because there is no connection between the energy released by fault movements miles below ground and the weather, which is due to solar energy received at the Earth's surface. For example, the Shaanxi and Tangshan earthquakes occurred when the local temperatures were at opposite ends of the scale.

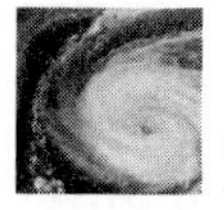

Transform Faults and Earthquakes

The transform faults forming the sides of some tectonic plates have dominantly horizontal movements that cause major earthquakes. Examples include the Alpine fault of New Zealand, the San Andreas fault in California, and the North Anatolian fault in Turkey.

Turkey, 1999 A warm and humid evening made sleep difficult, so many people were still up at 3:01 A.M. on 17 August 1999 near the Sea of Marmara in the industrial heartland of Turkey. They were startled by a ball of flame rising out of the sea, a loud explosion, sinking land along the shoreline, and a big wave of water. Another big rupture moved along the North Anatolian fault as a magnitude 7.4 earthquake. This time the fault ruptured the ground surface for 120 km (75 mi) with the south side of the fault moving westward up to 5 m (16.5 ft) (Figure 2.31). Several weeks later, after evening prayers for Muslims, a segment of the North Anatolian fault to the east ruptured in a 7.1 magnitude earthquake. The two devastating events combined to kill over 19,000 people and cause an estimated $20 billion in damages.

Why were so many people killed? Bad buildings collapsed (Figures 2.32 and 2.33). Industrial growth in the region attracted hordes of new residents who, in turn, caused a boom in housing construction. Unfortunately, many residential buildings were built on top of soft, shaky ground and some building contractors cut costs by increasing the percentage of sand in their concrete causing it to crumble during the ground shaking.

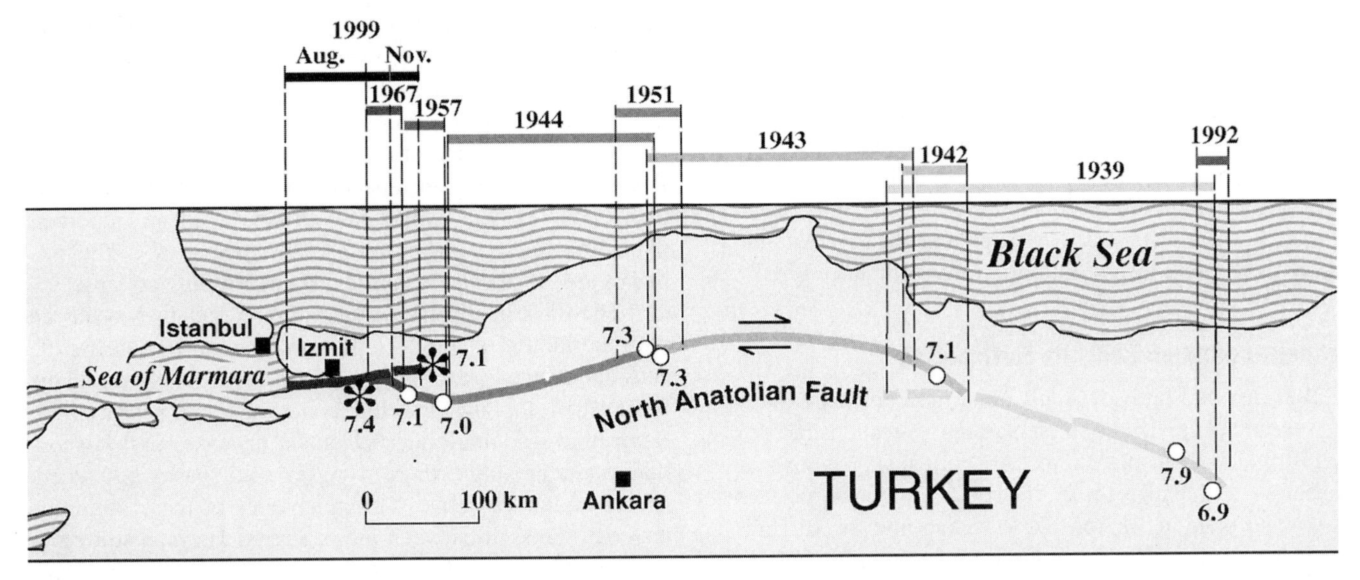

Figure 2.31 The North Anatolian fault accommodates the movement of Turkey westward into the Mediterranean basin (see Figure 2.34). Note the time sequence of the fault ruptures. What does the near future hold for Istanbul?

Figure 2.32 Ground ruptured by movement of the North Anatolian fault from center right to center of photo in Kaynasli, Turkey, on 12 November 1999.
Photo by Roger Bilham, courtesy of NOAA.

Figure 2.33 A six-story building pancaked in Duzce, Turkey, on 12 November 1999 when its supporting columns failed.
Photo by Roger Bilham, courtesy of NOAA.

The North Anatolian fault is not on the Arabian plate, but it is caused by it (Figure 2.34). As the Arabian plate pushes farther into Eurasia, Turkey is forced to move westward and slowly rotate counterclockwise in **escape tectonics.** Bounded by the North Anatolian fault in the north and the east Anatolian fault in the southeast, Turkey is squeezed westward like a watermelon seed from between your fingers.

The North Anatolian fault is a 1,400 km-(870 mi)-long fault zone made of numerous subparallel faults that split and combine, bend and straighten. A remarkable series of earthquakes began in 1939 near the eastern end of the fault with the magnitude 7.9 Erzincan earthquake that killed 30,000 people. Since 1939, 11 earthquakes with magnitudes greater than 6.7 have occurred as the fault ruptures westward in a semi-regular pattern that is unique in the world (Figure 2.31). At intervals ranging from 3 months to 32 years, over 1,000 km (620 mi) of the fault has moved in big jumps.

What is likely to happen next? There is every reason to expect the fault rupture to keep moving to the west, ever closer to Istanbul. The Sea of Marmara fills a basin partly created by movements along subparallel strands of the North Anatolian fault. The next big earthquake will likely occur near Istanbul, a city of 13 million people and growing rapidly. In the last 15 centuries, Istanbul has been heavily damaged by 12 earthquakes. Calculations indicate the next big earthquake affecting Istanbul has a 62 (+/– 15) percent probability of occurring within the next 30 years.

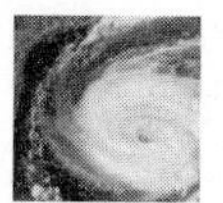

The Arabian Plate

The emergence of the geologically young spreading centers in the Red Sea and Gulf of Aden has cut off the northeast tip of the African continent (Figures 2.26 and 2.34) and created the Arabian plate. Analysis of the movement of the Arabian plate gives us good insight into different earthquake types.

Continent-Continent Collision Earthquakes

The Red Sea and Gulf of Aden areas may not have many large earthquakes, but their spreading is responsible for shoving the Arabian plate into Eurasia, causing numerous devastating earthquakes there. The rigid continental rocks of the Arabian plate are driven like a wedge into the stiff underbelly of Eurasia. The force of this collision uplifts mountain ranges (e.g., Caucasus and Zagros in Figure 2.23) and moves many faults that create the killer earthquakes typical of this part of the world.

Armenia, December 1988 Winter is cold in the Caucasus Mountains. On Wednesday morning, 7 December 1988, temperatures were freezing. The 700,000 people living within a 50-km (30-mi)-diameter area near Spitak, Armenia, were up and about their work or in school. At this time, the region provided shelter to about 100,000 refugees from the battles with Azerbaijan; many of whom were poorly housed. All lives changed dramatically at 11:41 A.M., when the ground began shaking from about 30 seconds of fault movements that generated a magnitude 6.9 earthquake. At 11:45 A.M., a magnitude 5.9 aftershock added to the devastation and despair. Within just five minutes, at least 25,000 people were dead, and over 31,000 were seriously injured, including 4,000 who lost limbs. The damage to buildings was so relentless that over 500,000 people were left homeless, and the partial or total destruction of 83 schools and 88 kindergartens left innumerable children in need of medical care. But these same earthquake waves also ruined 84 hospitals and killed or injured 80 percent of the doctors and nurses just when they were needed most.

What caused this killer quake? Arabia is pushing into Eurasia at a rate of about 4.5 cm (1.8 in) per year, creating compression and shortening. This continent-continent collision is causing uplift of part of the broad mountain belt that stretches across southern Europe to Asia, from the Alps to the Himalayas. In this earthquake the surface of the Earth rose 2 m (6.5 ft) in a dominantly vertical, compressive movement as part of Armenia rode upward and onto the Arabian plate.

Why were so many people killed in the Armenian earthquake? The magnitude 6.9 main shock and magnitude 5.9 aftershock killed more than 25,000 of the 700,000 residents within a 50-km (30-mi) radius of the epicenter. Compare this to the World Series (Loma Prieta) earthquake south of San Francisco that occurred ten months later. It was a magnitude 7.1 event that killed fewer than 25 of the more than 1.5 million people who lived within a 30-mile radius of its epicenter. Why did a California earthquake that was more than twice as powerful and affected more than twice the population than the Armenian seism kill fewer than 0.1 percent as many people? The explanation brings us to an important punchline—*earthquakes don't kill, buildings do.* The primary cause of deaths in earthquakes is building collapse.

The most horrifying of the building collapses occurred with the failure of 9- or 12-story-tall structures. During the earthquake, the horizontal beams and vertical columns came apart, then the unsupported concrete floors simply fell. Some nine-story buildings became a "pancake" of floors less than one story high; there were no void spaces left where people could stay alive. The inhabitants of these buildings were compressed between the pancaked floors, resulting in the incredibly high death tolls.

Transform Fault Earthquakes

The Arabian plate is moving away from Africa and is pushing into Eurasia but what is happening along the sides of the Arabian plate? The slide-past movements of transform faults. On the eastern side, the plate-boundary fault occurs beneath the Indian Ocean and has scant effect on humans. But look where the slide-past fault movements occur along the western side of the Arabian plate (Figures 2.34, 2.35, 2.36).

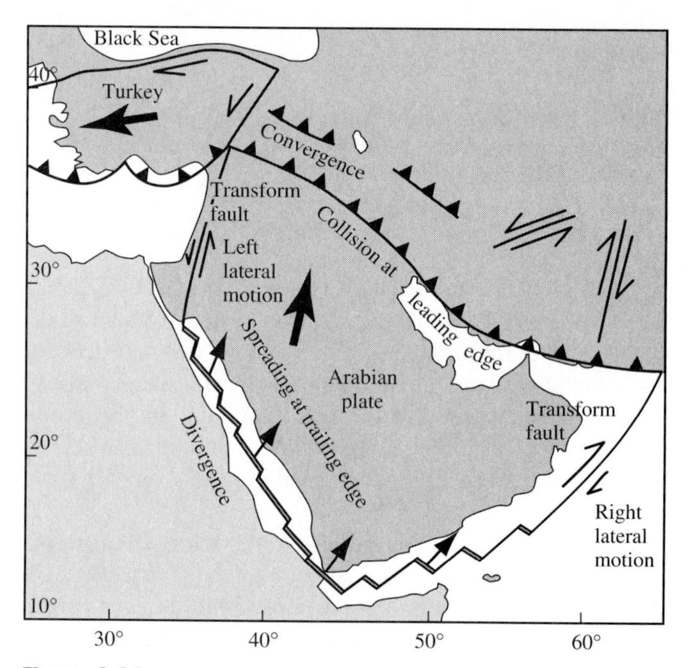

Figure 2.34 The Arabian plate pulls away from Africa, pushes into Eurasia, slices through the Holy Land with a transform fault, and squeezes Turkey westward.

Sidebar

Historical Perspective

It is interesting to ponder the effects of earthquakes in the Holy Land on the thinking of the religious leaders in this region, which is the birthplace of Judaism and Christianity and an important area to Islam. Imagine the great early leaders living in stiff, mud-block and stone buildings along one of the world's major strike-slip faults. They understood little about the workings of the Earth, yet they had to explain and interpret events that destroyed entire cities and killed many thousands of people. It is not surprising that many of them interpreted the disastrous events of their times as directly due to "the hand of God."

Let us use today's understanding of plate tectonics and fault movements to think about past events. For example, how might one interpret the account of Joshua leading the Israelites into the promised land, specifically the famous event when the walls of the oasis city Jericho came tumbling down? Is it possible that during the long siege of Jericho, an earthquake knocked down the walls of the city, killing and disabling many of the residents and allowing Joshua's army to enter and take over? Residents of cities suffer injuries and deaths from the collapse of buildings during an earthquake, not troops camped in the surrounding fields. Recent historical and archaeological investigations in the Holy Land have shown that many of the destroyed buildings and cities of the past did not meet their ends by time or humans alone; many fell to earthquakes.

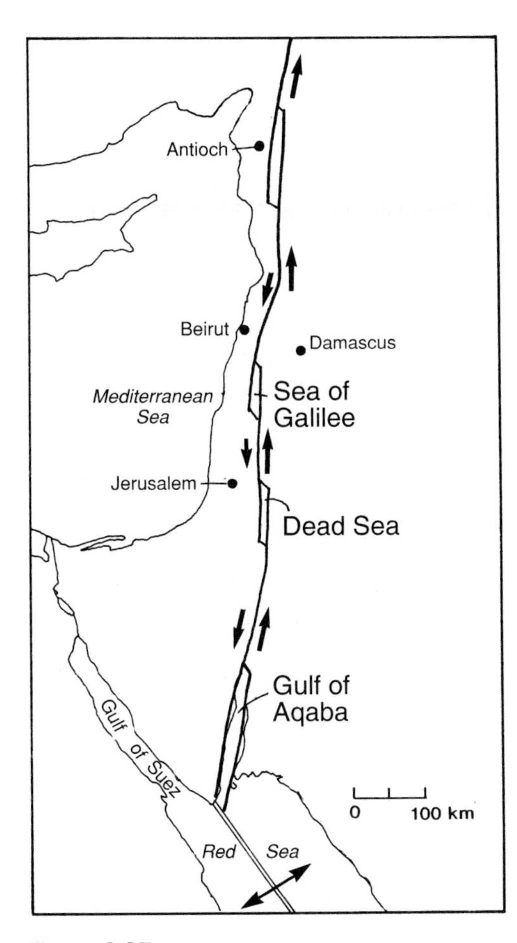

Figure 2.35 Map of the Dead Sea fault zone. Notice the subparallel faults have pull-apart basins in the steps between faults. The Dead Sea basin is deep; it has a 7-km (greater than 4-mi)-thick infill of sediments below its water. On 21 November 1995, a magnitude 7.2 earthquake in the Gulf of Aqaba killed people as far away as Cairo, Egypt.

Figure 2.36 Space shuttle view of the Dead Sea fault, 14 June 1991. The Dead Sea fault runs from lower right to upper left. Photo courtesy of NASA.

Dead Sea Fault Zone The Dead Sea fault zone is an Eastern Hemisphere analogue of the San Andreas fault in California. It not only runs right through the Holy Land but has also created much of the area's well-known topography. Notice in Figure 2.35 that there are four prominent overlaps or steps in the Dead Sea fault zone. Fault movements on both sides of these steps have created pull-apart basins that are filled by historically famous water bodies, such as the Dead Sea and

Table 2.1 Some Earthquakes in the Holy Land

Year	Magnitude	Year	Magnitude
1927	6.5	1068	6.6
1834	6.6	1033	7.0
1759	6.5	749	6.7
1546	6.7	658	6.2
1293	6.4	363	7.0
1202	7.2	31 B.C.E.	6.3
		759 B.C.E.	7.3

Source: "When the Walls Came Tumbling Down." (1991). [Video] Amos Nur, Stanford University.

the Sea of Galilee. The Dead Sea fault zone has been operating for as long as the Red Sea has been opening. During that time, there has been 105 km (65 mi) of offset, and 40 km (25 mi) of this movement has happened in the last 4.5 million years. This computes to an average **slip** (movement) rate of more than 5 mm/yr over the longer time frame or 9 mm/yr over the more recent time span. However, the rough, frictionally resistant faults do not easily glide along at several millimeters each year. The rocks along the fault tend to store stress until they can't hold any more, then they rupture in an earthquake-producing fault movement. How often do these earthquakes occur? Table 2.1 is a partial list put together by Amos Nur of Stanford University.

Summary

Gravity has pulled the Earth into layers of differing density, ranging from a heavy metallic core outward through layers of decreasing density through the mantle to the continents, then the ocean, and finally, the atmosphere. These layers exist in a state of flotational equilibrium known as isostasy. Up-and-down movements of the land due to isostatic adjustments are readily measurable.

The outer layers of the Earth are involved in a grand recycling known as the tectonic cycle. Hot buoyant rock and magma rise up from the mantle, through the lithosphere, to build world-encircling mountain ranges of volcanoes (oceanic ridges). The injection of magma elevates ridges that are pulled apart by gravity (slab pull) in gigantic slabs (plates) to form ocean basins in the process known as seafloor spreading. When these moving lithospheric plates collide, if one plate is composed of denser rock, it will turn back down into the mantle in the process known as subduction, to become melted and reabsorbed. The entire lithosphere is fractured into plates that pull apart (diverge), slide past, and collide (converge) with each other. The plate collisions cause mountains to rise, sea floors to bend down forming trenches that are elongate and deep, volcanoes to erupt, and earthquakes to be generated; this cyclic process is the topic of plate tectonics. Continents are composed of lower-density rock that rides on top of the denser rock of the moving plates.

The evidence for plate tectonics is overwhelming. Ancient magnetic fields locked into iron-bearing minerals in rocks point toward former south or north magnetic poles in patterns indicating that seafloor spreading and continental drift occur. The ages of rocks, sediments, and fossils, as well as the depth of water, all increase away from the oceanic ridges, indicating that oceanic crust/lithosphere is continuously forming and spreading apart. The oldest rocks and fossils in the ocean basins are less than 5 percent of the Earth's age, indicating that oceanic material is destroyed by recycling into the mantle.

Most earthquakes are caused by fault movements associated with tectonic plates. Plates have three types of moving edges: 1) divergent at spreading centers, 2) slide-past at transform faults, and 3) convergent at collision zones. The tensional (pull-apart) movements at spreading centers do not produce very large earthquakes. The dominantly horizontal (slide-past) movements at transform faults produce large earthquakes. The compressional movements at subduction zones and continent-continent collisions generate the largest tectonic earthquakes and they affect the widest areas.

Deaths from earthquakes are mostly due to building failures. For example, for thousands of years, humans have built stone and mud-block houses along the Dead Sea fault zone (a major transform fault), and for thousands of years, these rigid houses have collapsed during earthquakes causing many deaths. These geologic disasters have affected the teachings of Judaism, Christianity, and Islam.

Continent-continent collisions produce great earthquakes throughout Asia and Asia Minor. The 1988 Armenia earthquake killed more than 25,000 people, mainly in the collapse of multistory concrete buildings where upright supports and walls were not securely tied to floors. The 1976 Tangshan, China, earthquake killed 242,419 people as mud-block houses with heavy roofs collapsed along with multistory buildings. The deadliest earthquake in history occurred in 1556 in Shaanxi Province, China, when the loess (loose, silty sediment) into which cave homes had been dug collapsed and flowed, killing 830,000 people.

Subduction zones produce the largest number of great earthquakes. In 1923, a subduction movement of the Pacific plate destroyed nearly all of Tokyo and Yokohama; much of the devastation was caused by fires unleashed during building collapses.

The idea that earthquakes occur during certain weather conditions is flawed. Earthquakes are powered by the outflow of Earth's internal energy toward the surface; this is not affected by whether it is hot or cold, dry or humid, day or night, or any other surface condition.

Terms to Remember

acre feet 31
Actualism 44
asthenosphere 30
brittle 30
buoyancy 31
chondrite 30
cohesion 51
continental drift 35
convection 43
convergence zone 33
Curie point 36
divergence zone 33
ductile 30
earthquake weather 52
elastic 30
epicenter 37
escape tectonics 53
fault 44
fossil 40
Gondwanaland 33
hot spot 40
isostasy 31
latitude 35
Laurasia 42
lava 36
lithosphere 30
loess 51
magnetic field 35
magnetic pole 35
magnetism 35
mantle 30
Pangaea 35
Panthalassa 41
plate 33
plate tectonics 33
plume 40
reef 41
ridge 37
rift 46
seafloor spreading 32
seismic-gap method 49
slip 56
spreading center 38
strain 30
stress 30
subduction 33
tectonic cycle 32
tectonics 33
tension 45
topography 33
transform fault 33
trench 38
triple junction 46
tsunami 49
Uniformitarianism 43
viscous 29

Questions for Review

1. Describe how the Earth became segregated into layers of differing density.
2. How did Earth's continents, ocean, and atmosphere form?
3. Draw and label a cross section that explains the tectonic cycle.
4. What are the ages of the oldest 1) rocks on the Moon, 2) meteorites, 3) rocks on the continents, and 4) rocks making up the ocean floor?
5. Explain some evidence indicating that sea floors spread.
6. Describe a deep-ocean trench. How does one form?
7. Why do deep earthquakes tend to occur within inclined bands?
8. Explain several lines of evidence indicating that the continents move about the Earth.
9. Draw a cross section that shows a hot spot and plume. How do they help determine the directions of plate motions?
10. How is the Earth's magnetic field formed?
11. Describe some examples of isostasy.
12. Draw a map of an idealized tectonic plate and explain the earthquake hazards along each type of plate edge.
13. What are the differences between brittle, ductile, and elastic behavior?
14. Explain why earthquakes at subduction zones are many times more powerful than spreading-center earthquakes.
15. Sketch a sequence of cross sections that show how a continent is separated to accommodate an ocean basin.
16. Draw a map view of a triple junction made of spreading centers.
17. Sketch a southwest-northeast cross section across the Arabian plate and explain the origin of the Armenian and Turkey earthquakes.
18. Draw a cross section and explain why the Himalayas are the world's largest mountain range.
19. Explain the seismic-gap method of forecasting earthquakes.
20. Evaluate the concept of earthquake weather.
21. Explain the concept of Uniformitarianism.

Questions for Further Thought

1. The Earth is commonly called "terra firma." Does this make good geologic sense?
2. Are the type of earthquakes that ripped through Sweden after the glaciers melted likely to occur again in your lifetime?
3. How can the magnetic record inside a volcanic rock be used to determine the latitude at which the lava cooled?
4. How can the rate of motion of a plate be calculated?
5. Why does the polarity of the Earth's magnetic field switch from north to south and back again?
6. How much does the ground sink under a load, or rise after removal of a load, during isostatic adjustments?
7. How might you use food to create a plate-tectonics model in your kitchen?
8. Is East Africa likely to pull away from the rest of Africa to form a Somali plate?
9. How might people with no geologic knowledge, living in stone houses next to a major fault, explain a disastrous earthquake?
10. Why are fires in cities so commonly associated with major earthquakes?

Suggested Readings and References

Ambraseys, N. N., and Adams, R. D. (1989, 7 March). Long-term seismicity of north Armenia. *Eos,* 151–54.

Bolt, B. A., ed. (1980). *Earthquakes and Volcanoes: Readings from* Scientific American. New York: W. H. Freeman.

Boore, D. M. (1977, December). The motion of the ground in earthquakes. *Scientific American,* 69–78.
Cox, A., ed. (1973). *Plate Tectonics and Geomagnetic Reversals.* New York: W. H. Freeman.
Cox, A., and Hart, R. B. (1986). *Plate Tectonics: How It Works.* Palo Alto and Oxford, England: Blackwell Scientific Publications.
Dietz, R. S., and Holden, J. C. (1970). The breakup of Pangaea. *Scientific American,* 223, 30–41.
Glen, W. (1975). *Continental Drift and Plate Tectonics.* Indianapolis: Merrill.
Kearey, P., and Vine, F. J. (1990). *Global Tectonics.* Oxford, England: Blackwell Scientific Publications.
Molnar, P., and Tapponnier, P. (1977, April). The collision between India and Eurasia. *Scientific American.*
Press, F., and Siever, R. (1974). *Planet Earth: Readings from* Scientific American. New York: W. H. Freeman.
Richter, C. F. (1958). *Elementary Seismology.* New York: W. H. Freeman.
Stefansson, R. (1979). Catastrophic earthquakes in Iceland. *Tectonophysics,* 53, 273–78.
Tapponnier, P., Peltzer, G., LeDain, A. Y., and Armijo, R. (1982). Propagating extrusion tectonics in Asia. *Geology,* 10, 611–16.
Tarling, D., and Tarling, M. (1971). *Continental Drift: A Study of the Earth's Moving Surface.* Garden City, N.Y.: Doubleday.
Wyllie, L. A., Jr., and Filson, Jr., eds. (1989). Armenia earthquake reconnaissance report. *Earthquake Spectra,* special supplement.
Wyllie, P. J. (1976). *The Way the Earth Works.* New York: John Wiley and Sons.
Yong, C., Tsei, K., Feibi, C., Zhenhuan, G., Qujia, Z., and Zhangli, C., eds. (1988). *The Great Tangshan Earthquake of 1976.* New York: Pergamon Press.

Videos

Continental Drift and Plate Tectonics. (1988). University of California—Santa Barbara (20 min.).
The Earth Revealed—Birth of a Theory. (1992). Annenberg/CPB Project (30 min.).
The Earth Revealed—Plate Dynamics. (1992). Annenberg/CPB Project (30 min.).
Planet Earth—The Living Machine. (1986). Annenberg/CPB Project (60 min.).
When the Walls Came Tumbling Down—Earthquakes in the Holy Land. (1991). Amos Nur, Stanford University (56 min.).
Tibet: Where Continents Collide. (1989). Earth Vision (47 min.).

CD-ROM

Jones, A., Siebert, L., Kimberly, P., and Luhr, J. F. (2000). Earthquakes and Eruptions: Temporal and spatial display of earthquake hypocenters, seismic-wave paths, and volcanic eruptions. Smithsonian Institution, Global Volcanism Program, Digital Information Series, GVP-2.

A must have for classroom instruction and personal computer study.

Chapter 3

Basic Principles of Earthquake Geology, Seismology, and Tsunami

Diseased nature
oftentimes breaks forth
In strange eruptions:
oft the teeming earth
Is with a kind of colic
pinch'd and vex'd
By the imprisoning
of unruly wind
Within her womb;
which, for enlargement
striving,
Shakes the old
beldam earth, and
topples down
Steeples, and moss-
grown towers.
—William Shakespeare, 1598,
King Henry IV

The earth beneath our feet moves, releasing energy that shifts the ground and sometimes topples cities. Some earthquakes are so immense that their energy is equivalent to thousands of atomic bombs exploded simultaneously. The power of earthquakes to destroy human works, to kill vast numbers of people, and to alter the very shape of our land has left an indelible mark on many civilizations. Earthquake unpredictability instills an uneasy respect and fear in humankind that, through the millennia, have helped shape thought about life and our place in it.

Ancient accounts of earthquakes tend to be quite incomplete. Instead of providing rigorous descriptions of Earth behavior, they emphasize interpretations. For over 2,000 years, based on Aristotle's ideas, many explanations of earthquakes were based on winds rushing beneath the Earth's surface. Even Leonardo da Vinci wrote in his *Notebooks* about 1500 C.E. that:

> When mountains fall headlong over hollow places they shut in the air within their caverns, and this air, in order to escape, breaks through the Earth, and so produces earthquakes.

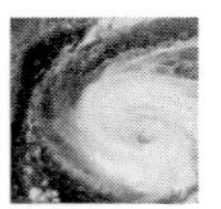

The Lisbon Earthquake of 1755

Even closer to our time, the scientific method was not applied to earthquake descriptions. Portugal in the eighteenth century, and especially its capital city of Lisbon, was rich with the wealth its explorers brought from the New World. Yet Portugal's decline probably began with a set of earthquakes. On the morning of 1 November 1755—All Saints Day—Lisbon rocked under the force of closely spaced earthquakes originating offshore in the Atlantic Ocean to the southwest. On this day of religious observance, the churches were full of worshippers. About 9:40 A.M., a thunderous underground sound began, followed by increasingly violent ground shaking. The severe ground movement lasted two to three minutes and caused widespread damage to houses, churches, and public buildings in this city of over 250,000 people. Most of Lisbon's churches were built of masonry; they collapsed into the narrow streets, killing thousands of trapped and fleeing people. Tapestries fell onto candles and lamps—all lit on this holy day—and started fires that burned unchecked for six days. Before an hour had passed, crippled Lisbon was rocked by a second earthquake, more violent but shorter-lived than the first. In a panic, many of the frightened survivors of the first earthquake rushed to the shore for safety, only to be swept away by quake-caused sea waves up to 10 m high (more than 30 ft). These walls of water spilled onto the land, carrying boats and cargo over one-half km inland. As the seawater withdrew, it dragged people and debris from the earthquake-shattered structures back to the ocean. About noon, Lisbon suffered yet another shaking, the effects of a major earthquake that occurred 550 km (340 mi) away, near Fez, Morocco. The third quake caused as many casualties and as much destruction in North Africa as had the earlier events in Lisbon. Fez and numerous smaller cities were destroyed, and Algiers and Tangier were severely wracked. Back in Lisbon, the first two earthquakes had already killed almost 70,000 people and had destroyed or seriously damaged about 90 percent of the houses and churches. So the third quake could do little more harm except to further frighten the survivors (Figure 3.1).

The effects of the Lisbon earthquake series extended far beyond the demolished city. According to the 1914 estimates of American seismologist Harry Fielding Reid, a little more than 3 percent of the Earth's surface was sensibly shaken. Structures located 600 km (375 mi) from the source of the earthquake suffered damage. Giant sea waves generated by the earthquake not only hit the Portugal coast but were also recorded at lesser heights in North Africa, the British Isles, and the Netherlands. Even in the West Indies, across the Atlantic Ocean, the sea retreated and advanced for over two and one-half hours with waves up to 4 m (13 ft) high.

Figure 3.1 The Lisbon earthquake.

At the time the earthquakes occurred, Lisbon was staggeringly rich in bullion, jewels, and merchandise, and it had great commercial and cultural importance. The city was also home to the dreaded Inquisition. The destruction of this famous city by earthquakes and their attendant sea waves and fires was an unparalleled shock to Western civilization. Not only were the losses of lives and buildings staggering, but the fires also incinerated irreplaceable libraries, maps and charts of the Portuguese voyages of discovery, and paintings of such masters as Titian, Correggio, and Rubens.

It seemed incomprehensible to many that such devastation should occur on a high holy day, when the churches were filled with the devout. Yet to others, such as the church reformer John Wesley, the earthquakes were just punishment for sin. He roared:

> It comes! The Roof trembles! The Beams crack! The Ground rocks to and fro! Hoarse Thunder resounds from the Bowels of the Earth! And all these are but the Beginning of Sorrows. Now what Help? What Wisdom can prevent? What strength resist the Blow? What Money can purchase, I will not say Deliverance, but an Hour's Reprieve? Poor honourable Fool, where are now thy Titles? Wealthy Fool, where is now thy golden God? If any Thing can help, it must be Prayer. But what wilt thou pray to? Not to the God of Heaven: you suppose Him to have nothing to do with Earthquakes?

To calm the terrified people of Lisbon, Sebastiao de Carvalho, the emergency governor supervising the cleanup and reconstruction, had to seek the bishop's aid in ending the insistent warnings of the priests that more evidences of God's wrath were forthcoming.

The German philosopher Immanuel Kant regarded the earthquakes as a natural phenomenon and wisely suggested that people note where earthquakes occur and then not build cities there. The French intellectual Jean Jacques Rousseau also viewed earthquakes as natural occurrences and pointed out how the artificialities of civilization (e.g., stone buildings) caused the deaths. That is, people living the simple outdoors life would not have been affected by the earthquakes. The French writer Voltaire wrote an impassioned poem on the earthquake; an excerpt follows:

> **The Lisbon Earthquake**
> Say what advantage can result to all,
> From wretched Lisbon's lamentable fall?
> Are you then sure, the power which could create
> The universe and fix the laws of fate,
> Could not have found for man a proper place,
> But earthquakes must destroy the human race?
> Will you thus limit the eternal mind?
> Should not our God to mercy be inclined?

Cannot then God direct all nature's course?
Can power almighty be without resource?
Humbly the great Creator I entreat,
This gulf with sulfur and with fire replete,
Might on the deserts spend its raging flame,
God my respect, my love weak mortals claim;
When man groans under such a load of woe,
He is not proud, he only feels the blow.
Would words like these to peace of mind restore
The natives sad of that disastrous shore?

The Lisbon disaster was an especially notable earthquake series because of its severity, the amount of damage and number of lives lost, the large area in which it was felt, the far-ranging destructive sea waves it spawned, the long distances to which surface waters were agitated, and the place and time in which it occurred. Before this event, Europe had an air of prosperity and tranquillity in an ordered world. The Lisbon earthquakes did more than devastate a city; they changed the prevailing philosophies of the era. All was not well in the world after all.

Despite the profound effects that earthquakes have had on civilizations for so many centuries, scientific observations did not begin until the early nineteenth century, when good descriptions were made of earthquake effects on the land. Today, less than two centuries later, our knowledge of earthquakes has increased enormously. We have a fairly comprehensive understanding of what earthquakes are, why and where they happen, and how big and how often they occur at a given site. Our scientific data and theories allow us to understand phenomena that even the greatest minds of the past could not have glimpsed. Such are the rewards from the pyramidal building of knowledge we call science.

What Is an Earthquake?

The word *earthquake* is effectively a self-defining term—the Earth quakes, the Earth shakes, and we feel the vibrations. Earthquakes may be created by volcanic activity, meteorite impacts, undersea landslides, explosions of nuclear bombs, and more; but most commonly, they are caused by sudden earth movements along faults. A **fault** is a **fracture,** a crack in the Earth across which the two sides move relative to each other (Figure 3.2). Pressures build up in near-surface rocks until the **stress** is so great that the rocks fracture and shift along a fault. The shock waves sent off as the rock ruptures and moves are what we experience as an earthquake. A pictorial analogy may be made with the familiar sight of a rock thrown into a still body of water. The rock hitting the water creates a brief, violent disturbance that sends shock waves radiating through the water away from the impact site. Likewise, a fault rupture releases concentrated energy where the fault moves, plus it sends several types of shock waves radiating long distances away from the disturbance.

Figure 3.2 Offset of tilled farmland by 1979 movement of the Imperial fault, southernmost California. View is to the east; the west side of the fault (closest to you) has moved northward (to your left).

Photo courtesy of Pat Abbott.

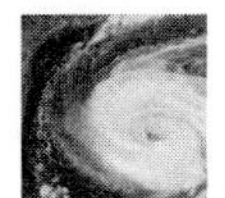

Faults and Geologic Mapping

The nineteenth-century recognition that fault movements cause earthquakes was a fundamental advance that triggered a whole new wave of understanding. With this relationship in mind, geologists go into the field to map faults, which in turn identifies earthquake-hazard belts. Because a fault moves formerly continuous rock layers apart, the careful mapping of different rock masses can define sharp lines that separate offset segments of single rock masses. These sharp lines are the surface expression of faults.

The principles that help us understand faults begin with some of the earliest recognized relationships about rocks that are still useful today. Several were formalized by the Danish physician Niels Steensen, working in Italy and known by his Latinized name of Steno. In 1669, he set forth several laws that are fundamental in interpreting geologic history. His **Law of Original Horizontality** explains that sediments (sands, gravels, and muds) are originally deposited or settled out of water in horizontal layers. This is important because some older sedimentary rock layers are found at angles ranging from horizontal to vertical. But since we know they started out as horizontal layers, their postdepositional history of deformation can be unraveled by mentally returning their orientations back to horizontal (Figures 3.3 and 3.4).

In the **Law of Superposition,** Steno stated that in an undeformed sequence of sedimentary rock layers, each successive layer is deposited on top of a previously formed, and hence older, layer. Thus, each sedimentary rock layer is younger than the bed beneath it, but older than the bed above it (Figures 3.3 and 3.4).

Steno's **Law of Original Continuity** states that sediment layers are continuous, ending only by butting up against a

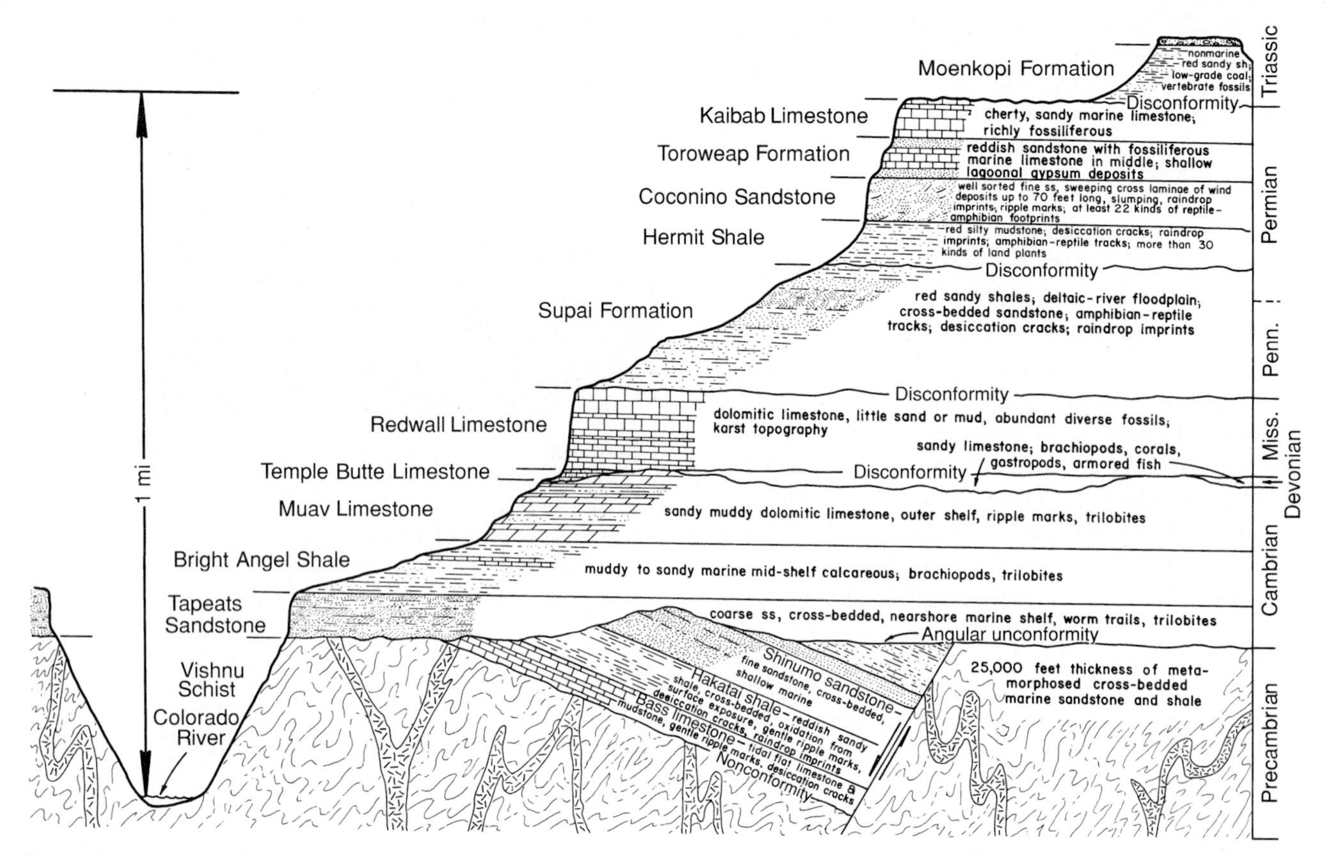

Figure 3.3 Representational cross section of the Grand Canyon, Arizona. Different geologic histories are shown at different levels. 1) The lowermost rocks are the oldest; they were deposited as sediments in marine water, then buried to great depths and altered by high temperature and pressure (metamorphism), then exposed at the surface again by uplift and erosion. 2) Younger shallow-marine sediments were deposited as horizontal layers on top of the old, deformed rocks (e.g., Shinumo Sandstone); these layers were tilted, lifted above sea level, and eroded. 3) The seas returned, and horizontal sediment layers were again deposited, building a thick sedimentary sequence interrupted by intervals of erosion and deposition of nonmarine sedimentary layers. 4) The region was uplifted, raising it above sea level. The Colorado River has cut down to form the Grand Canyon.

From John Shelton, 1966.

topographic high, such as a hill or a cliff, by pinching out due to lack of sediment, or by gradational change from one sediment type to another. This relationship allows us to appreciate the incongruity of a sedimentary rock layer that abruptly terminates. Something must have happened to terminate it. For example, a stream may have eroded through it, or a fault may have truncated it. Geologists spend a lot of time locating and identifying offsets of formerly continuous rock layers. In this way, we can determine the lengths of faults and estimate the magnitude of earthquakes they produce.

On a much broader scale, we can find large offsets on long-acting, major faults. Figure 3.5 shows a pronounced line cutting across the land in a northeast-southwest trend; this is the Alpine fault on the South Island of New Zealand. The west (left) side has been moved 480 km (300 mi) toward the north. In Otago province in the southern part of the South Island, gold was discovered in 1861 in stream gravels (Figure 3.6). This set off a gold rush that brought in prospectors and miners from all over the world. The gold fever that had attracted so many fortune seekers to California in 1849 now moved to New Zealand. Prospectors panned the streams and worked their way upstream into bedrock hills to find the source of the gold. Yet much of the wealth lay 480 km to the northeast in Nelson province, where the same gold-bearing rock masses had been offset along the Alpine fault by more than 23 million years of fault movements. As this example shows, fault studies also can have tremendous implications for locating mineral wealth.

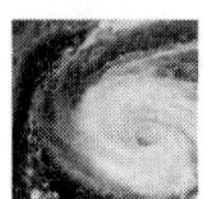

Types of Faults

As tectonic plates split apart and collide, as mountains are elevated and basins are warped downward, the brittle rocks of the lithosphere respond by fracturing (also called **jointing** or cracking). When regional forces create a large enough stress differential in rocks on either side of a fracture, then movement occurs and the fracture becomes a fault. Accumulated movements of rocks along faults range from millimeters to hundreds

Figure 3.4 North wall of the upper Grand Canyon. At the canyon bottom, the once horizontal sedimentary rock layers are now dipping to the east. Their uptilted ends have been eroded and buried by the Tapeats Sandstone and younger rock layers.
Photo by John S. Shelton.

of kilometers. These movements can cause originally horizontal sedimentary rock layers to be tilted and folded into a wide variety of orientations. To describe the location in three-dimensional (3-D) space of a deformed rock layer, a fault surface, or any other planar feature, geologists make measurements known as **dip** and **strike.** Dip is seen in the two-dimensional (2-D) vertical view (**cross section**) as the angle of inclination from the horizontal of the tilted rock layer as well as the direction of the dip (Figure 3.7). Strike is viewed in the 2-D horizontal view (**map**) as the compass bearing of the rock layer where it pierces a horizontal plane.

The classification of faults uses some terminology of early miners. Many ore veins were formed in ancient fault zones. Thus, many mines consist of adits (passages) dug along old, inactive faults. Ores are common along faults because when one block of rocks moves past another in a fault zone, there are tremendous frictional forces that tend to shatter and pulverize the rocks in the fault zone. The broken rock in the fault zone creates an avenue of **permeability** through which water can flow. If the underground water carries a concentration of dissolved metals, they may precipitate as valuable elements or minerals within the fault zone. Early miners working in excavated fault zones called the floor beneath their feet the **footwall** and the rocks above their heads the **hangingwall** (Figure 3.8). This terminology is used to define the two major types of faults dominated by vertical movements—the **dip-slip faults.**

Dip-Slip Faults

Faults with the major amounts of their offset in the dip or vertical direction are caused by either a pulling or a pushing force. Faults where the dominant force is extensional are recognized by the separation of the pulled-apart rock layers in a zone of omission (Figure 3.9). A **normal fault** occurs when the hangingwall moves down relative to the footwall. The selection of the word *normal* as a name for this type of fault was unfortunate because it carries a connotation of normalcy, as if this were the standard or regular mode of fault movement; such is not the case. Extensional- or normal-style faults are typical of the faults at seafloor spreading centers and in regions of continents where the plates are pulled apart.

If the dominant force that creates a fault movement is compressional, then the rock layers are pushed together, or repeated, when viewed in a cross section (Figure 3.10). With compressional forces, the hangingwall moves upward relative to the footwall; this type of fault is referred to as a

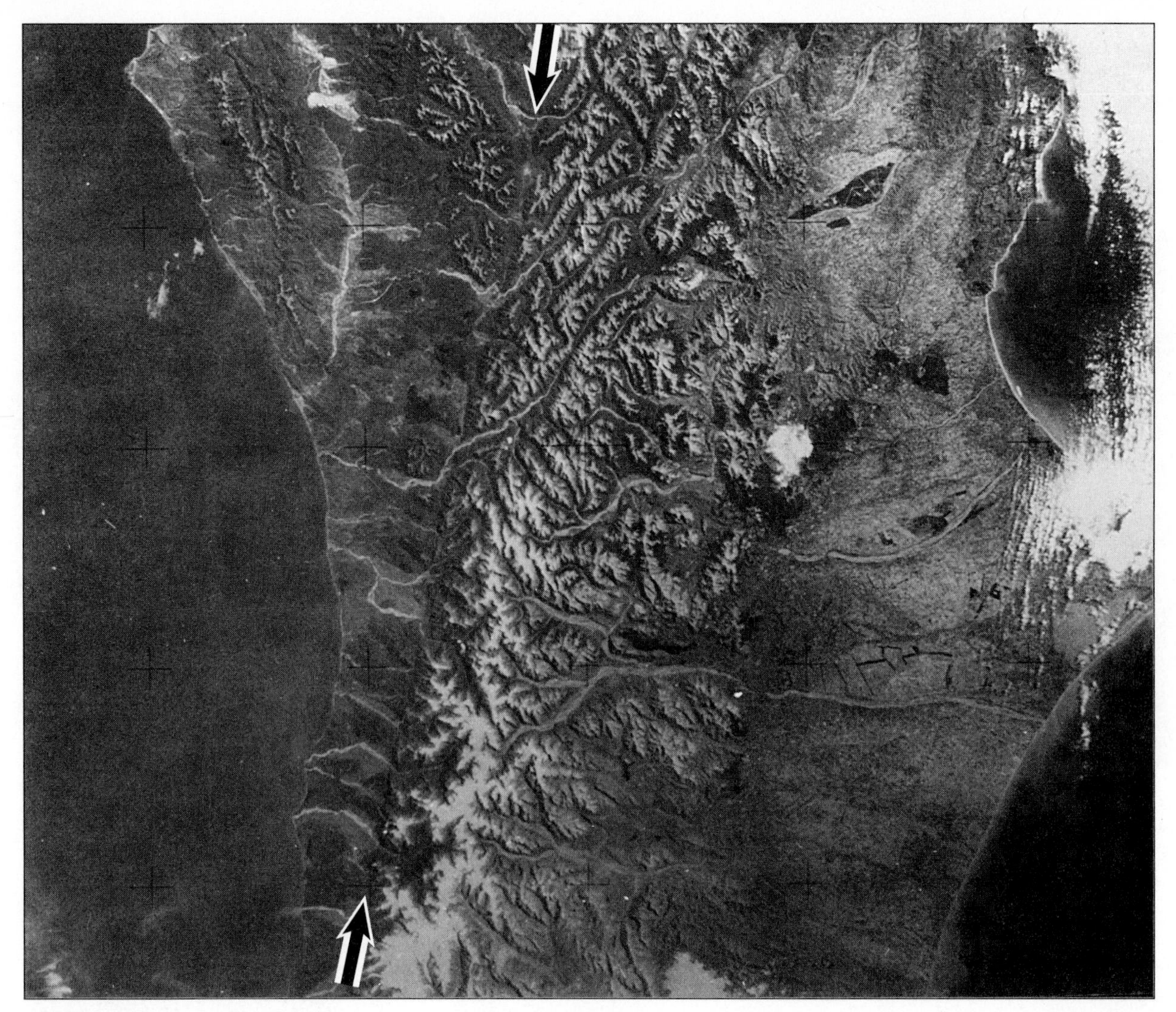

Figure 3.5 Aerial photo of part of South Island, New Zealand (see Figure 3.6 for location). The Alpine fault cuts a prominent slash from near the lower left (southwest) corner of the photo to the top center (northeast). Arrowheads line up with fault.
Photo courtesy of Pat Abbott.

reverse fault. The compressional motions of reverse faults are commonly found at areas of plate convergence where subduction or continental collision occurs.

The extensional versus compressional origins of movement can have enormous economic implications. Look again at Figures 3.9 and 3.10. Visualize the emphasized rock layer in each figure as being an oil reservoir. Now imagine yourself to be the landowner above either the zone of omission or the zone of repetition. In one case, it could mean poverty; in the other, great wealth.

Strike-Slip Faults

When most of the movement along a fault is horizontal (parallel to the strike direction), the fault is referred to as a **strike-slip fault.** These fault offsets are seen in map view as if we were in a balloon or airplane looking down on the Earth's surface. Strike-slip faults are further classified on the basis of the relative movement directions of the fault blocks (Figure 3.11). If you straddle a fault and the block on your right-hand side has moved relatively toward you, then it is called a **right-lateral,** or dextral, **fault.** Notice that this convention for naming the fault works no matter which way you are straddling the fault; try it facing both directions with Figure 3.11. Similarly, if features on the left-hand side of the fault have moved closer to you, then it is a **left-lateral,** or sinistral, **fault.**

We have looked at a large strike-slip fault in New Zealand, the Alpine fault, but the most famous strike-slip fault in the world is the San Andreas in California. This right-lateral fault is more than 1,300 km (800 mi) long. On 18 April 1906, a 430 km (265 mi) long segment of the San Andreas fault ruptured and moved horizontally as much as 6.5 m (20 ft) in 60 seconds. The great burst of energy gen-

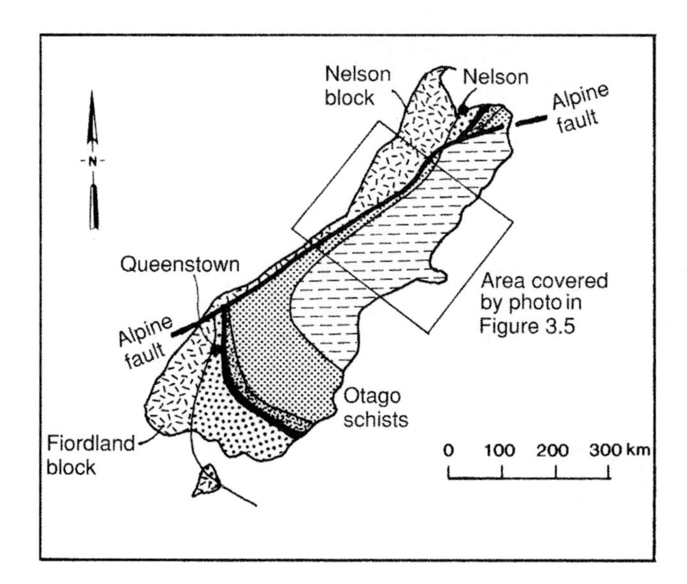

Figure 3.6 Generalized geologic map of South Island, New Zealand. Each map pattern records a different type of rock. Locate the Alpine fault, then match up the rock patterns across the fault. The gold-bearing rocks near Queenstown have been offset 480 km to near Nelson.

erated by the fault movement was actually the release of elastic energy that had built up and been stored in the rocks for many decades. The San Andreas fault is discussed in Chapter 4.

Faults are not simple planar surfaces that glide readily when subjected to stress. Instead, faults are complex zones of breakage where rough and interlocking rocks are held together over an irregular surface that extends many miles below the ground. Stress must build up over many years before enough potential energy is stored to allow a rupture on a fault. The initial break occurs at a weak point on the fault and then propagates rapidly along the fault surface. Much of the energy stored in the rocks is released as radiating seismic waves that humans call earthquake. The point where the fault first ruptures is known as the **hypocenter** or focus. The point on the Earth's surface directly above the hypocenter is called the epicenter (Figure 3.12).

A fault rupture is not a simple, one-time movement that produces "the earthquake." In fact, we never have just one earthquake. The stresses that build up in the rocks in an area are released by a series of movements along the fault, or several faults, that continue for weeks to months to years. Each fault movement generates an earthquake; the biggest one is called "the earthquake," the smaller ones before it are known as foreshocks, and the smaller ones after it are called aftershocks. Realistically, there are no differences between these earthquakes other than size; they are all part of the same series of stress release on the fault. When a large earthquake

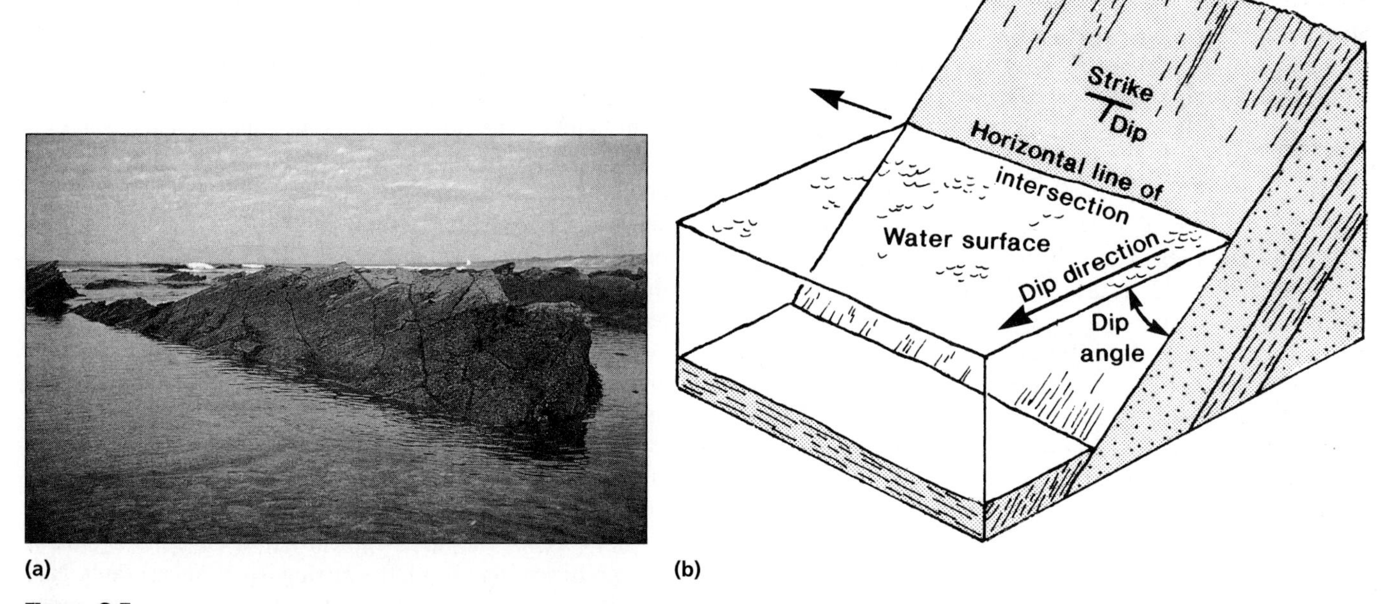

Figure 3.7 **(a)** A 75 million-year-old sandstone layer at La Jolla Bay, California, exposed at a moderately high tide. The sea surface forms a horizontal plane against the inclined sandstone bed. **(b)** The strike of a rock layer is the compass bearing of the "shoreline." The dip angle is the number of degrees below horizontal that the rock layer is inclined.

Photo by Pat Abbott.

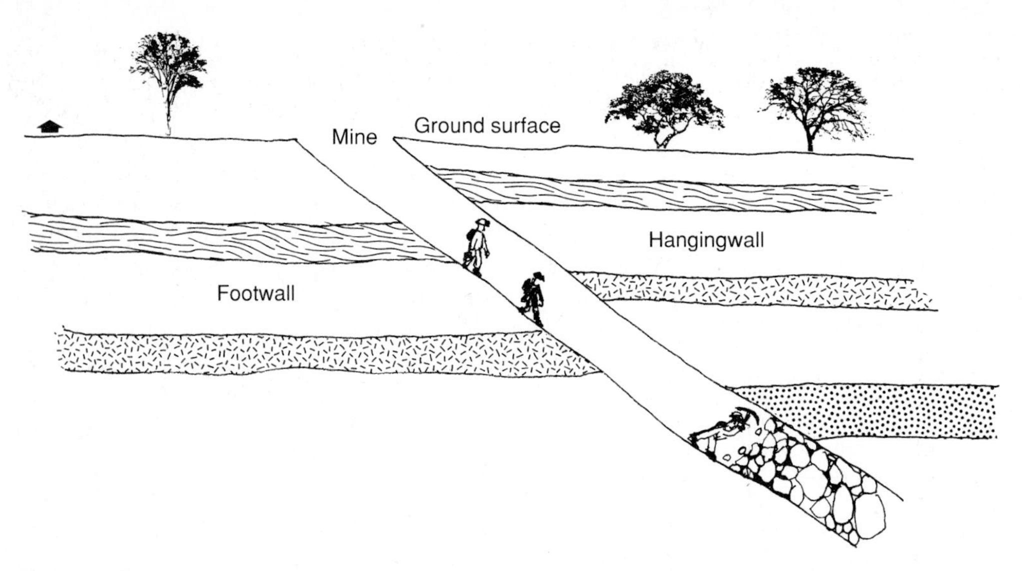

Figure 3.8 Schematic cross section of miners excavating ore that precipitated in broken rock within an old fault zone. Notice that the rock layers in the footwall and hangingwall are no longer continuous; this gives evidence of the movements that occurred along the fault in the past.

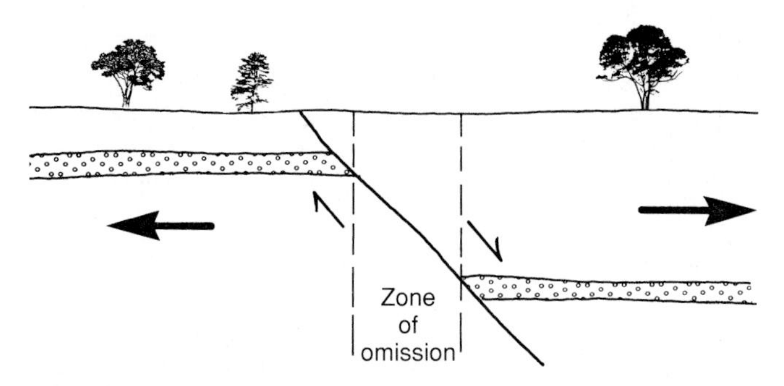

Figure 3.9 Schematic cross section of a normal fault, i.e., the hangingwall has moved downward (in a relative sense). Extensional forces are documented by the zone of omission, where the originally continuous rock layers are missing. The small arrows indicate movement; the larger arrows show force.

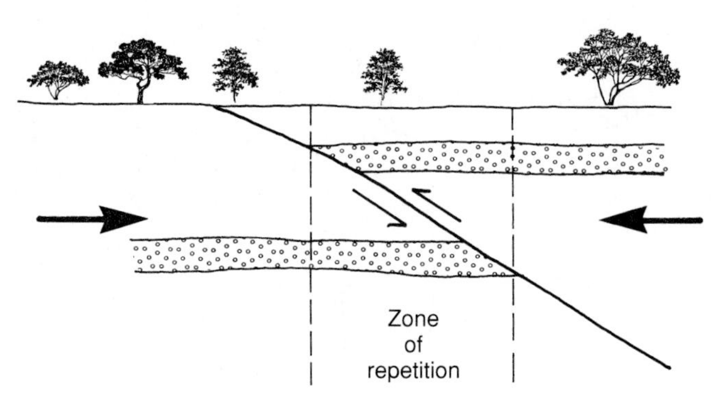

Figure 3.10 Schematic cross section of a reverse fault, i.e., the hangingwall has moved upward (in a relative sense). Compressional forces are documented by the zone of repetition, where the originally continuous rock layers have been split, shoved together, and stacked above each other.

occurs, it may be announced that there is a 6 percent probability that another, even larger earthquake may occur in the next three days. This statistical statement comes because it is not yet known how to distinguish a foreshock from the main earthquake. We simply wait until the series of earthquakes concludes, then label all quakes before the biggest one as foreshocks and all after as aftershocks.

Steps in Strike-Slip Faults Strike-slip faults do not simply split the surface of the Earth along perfectly straight lines. The rupturing fault tears apart the rocks along its path in numerous subparallel breaks that stop and start, bend left and bend right. For analogy, visualize a sheet cake. Put your right hand on the upper right corner and your left hand on the lower left corner. Now pull toward you with your right hand and push away with your left. Do you visualize the cake ripping along one straight line? Or along several breaks that stop and start, bend left and bend right? And so it is with the Earth when it ruptures during an earthquake-generating fault movement. Normal and reverse faults also have bends; we just don't see them as easily on the surface.

The bends along a fault have profound implications for the creation of topography and the starting and stopping of fault rupture events. Figure 3.13a is a sketch of a right-lateral fault with a bend (step) in it—a left-stepping bend. Stand to either side of the fault and look at the region of the bend. Note that the fault segment left of the bend is closest to you; hence, this is a left-stepping, right-lateral fault. Notice what occurs at the bend in the fault when the two sides slide past each other—compression, pushing together, collision, constraint. The photo in Figure 3.13b shows a left step in the right-lateral Superstition Hills fault west of Brawley,

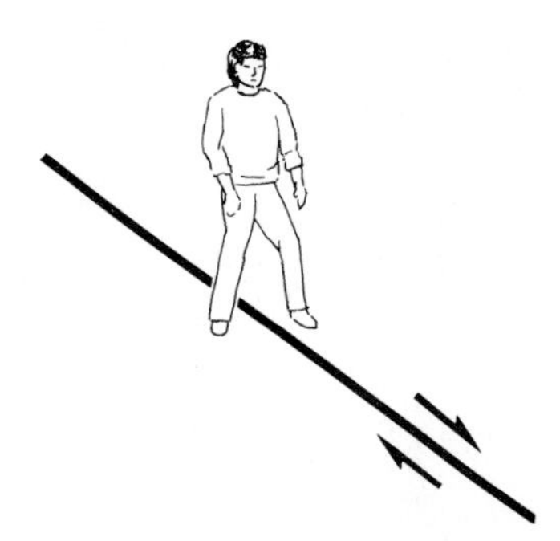

Right-lateral fault
Straddle the fault; right-hand side moves toward you.

Figure 3.11 Map of a right-lateral, strike-slip fault. As the man straddles the fault, the right-hand side of the fault has moved relatively closer to him. If he turns around, will his right-hand side still have moved closer to him?

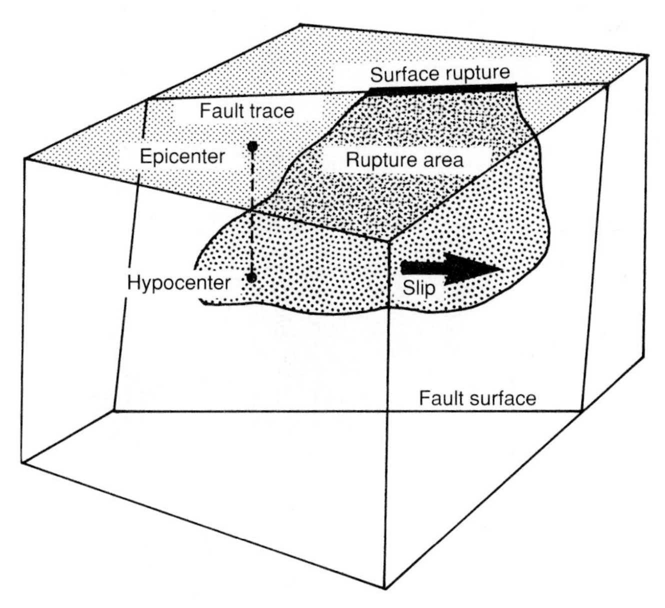

Figure 3.12 Block diagram of a fault surface. Hypocenter is place where rupture began; epicenter is point on Earth's surface above hypocenter. Notice that because the fault surface is inclined (it dips), the epicenter does not plot on the trace of the fault at the surface.

Source: J. Ziony, ed., "Earthquakes in the Los Angeles Region." *U.S. Geological Survey.*

California, that was created on 16 November 1987. Notice how the compression at the bend produced a little hill. What size could this hill attain if movements at this left step were to occur for millions of years? We will see examples in Chapter 4.

Similarly, Figure 3.14a depicts a right step along a right-lateral fault. Visualize what happens at the bend in the fault. In this case, the two sides pull apart from each other, extend, diverge, release. The photo in Figure 3.14b is from the same earthquake, along a different length of the same fault, as in Figure 3.13b. At this right step, the two sides pulled apart and created a wide crack, a little basin, a downdropped area. Steps have effects on fault rupture initiation, propagation, and cessation (see Landers earthquake, Chapter 4).

Transform Faults

Transform faults are a special type of horizontal-movement fault recognized first by the Canadian geologist J. Tuzo Wilson in 1965. Figure 3.15 depicts the basic components of plate tectonics. Seafloor crust forms at oceanic volcanic ridges and is pulled apart by gravity. When plates collide, the denser plate subducts. But what happens along the sides of the plates? They slide past each other at transform faults. Visualize this process in three dimensions. The spreading plates are rigid slabs of oceanic rock, tens of kilometers thick, that are being wrapped around a near-spherical Earth. How does a rigid plate move about a curved surface? The plates must fracture, and these fractures are transform faults. In fact, transform faults must link spreading centers or connect spreading centers with subduction zones.

In Figure 3.15, notice that in the region between the two spreading centers, the relative motions of the two plates are in opposite directions in typical strike-slip fault fashion. However, passing both to the right and left of the spreading centers, notice that the two slabs are moving in the same direction and there they are called fracture zones. Can the same fault be classified as both a strike-slip and a transform fault? Yes.

Development of Seismology

The study of earthquakes is known as **seismology** (after **seism,** meaning "earthquake"). The earliest earthquake-indicating device known was invented in China in 132 C.E. by Chang Heng. The modern era of seismologic instrumentation began about 1880. Instrumentation continues to evolve through many different styles, but a basic need is to

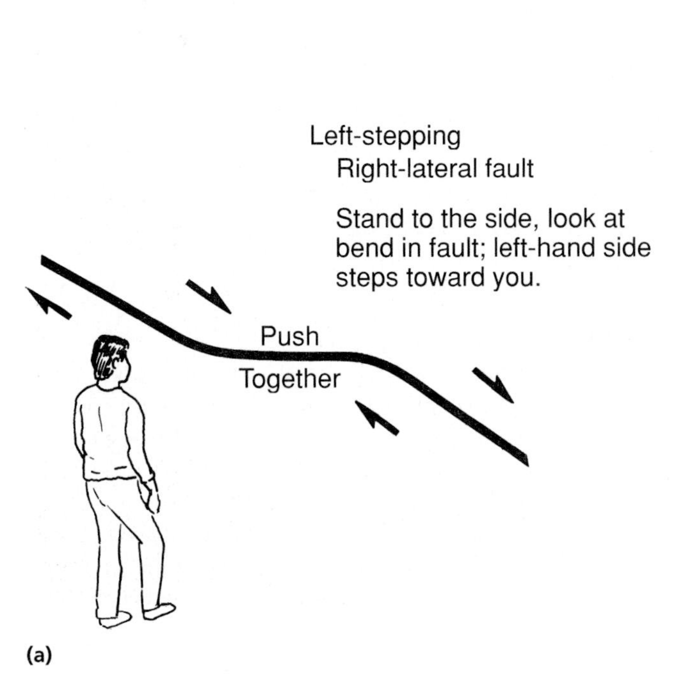

(a)

(b)

Figure 3.13 **(a)** Left step in right-lateral fault. Notice that the land is pushed together at the fault bend whenever the fault moves. Movements will create a hill, which could grow to a mountain if the fault remains active for a long enough time. **(b)** Land offset along the Superstition Hills right-lateral fault during its 16 November 1987 earthquake. See the left step and the uplift at the bend. Photo by Pat Abbott.

record the 3-D movement of earthquake waves. This is achieved by having instruments detect Earth motions (**seismometers**) and record them (**seismographs**) as north-south horizontal movements, east-west horizontal movements, and vertical movements. To accurately record the passage of seismic waves, a seismometer needs to have a part that remains as stationary as possible while the whole Earth beneath it vibrates. One way to accomplish this is by building a frame that suspends a heavy mass (Figure 3.16). The support frame rests on the Earth and moves as the Earth does, but the mass suspended by a wire must have its **inertia** overcome before it moves. The differences between motions of the frame and the hanging mass are recorded on paper by pen and ink or, increasingly, as digital data. Visualize the process this way: hold an ink pen steady in your hand, then vibrate the entire Earth beneath your pen to make an inked line.

Other important pieces of information to record include the arrival times and the durations of the various seismic waves. This is accomplished by having time embedded in the seismographic record either as inked tick marks on the paper graph or within the digital data. Time is standardized in the United States by the national clock in Boulder, Colorado.

First-order analysis of the seismic records allows seismologists to identify the different kinds of seismic waves generated by the fault movement, to estimate the amount of energy released (magnitude), and to locate the epicenter/hypocenter (where the rock hit the water, so to speak).

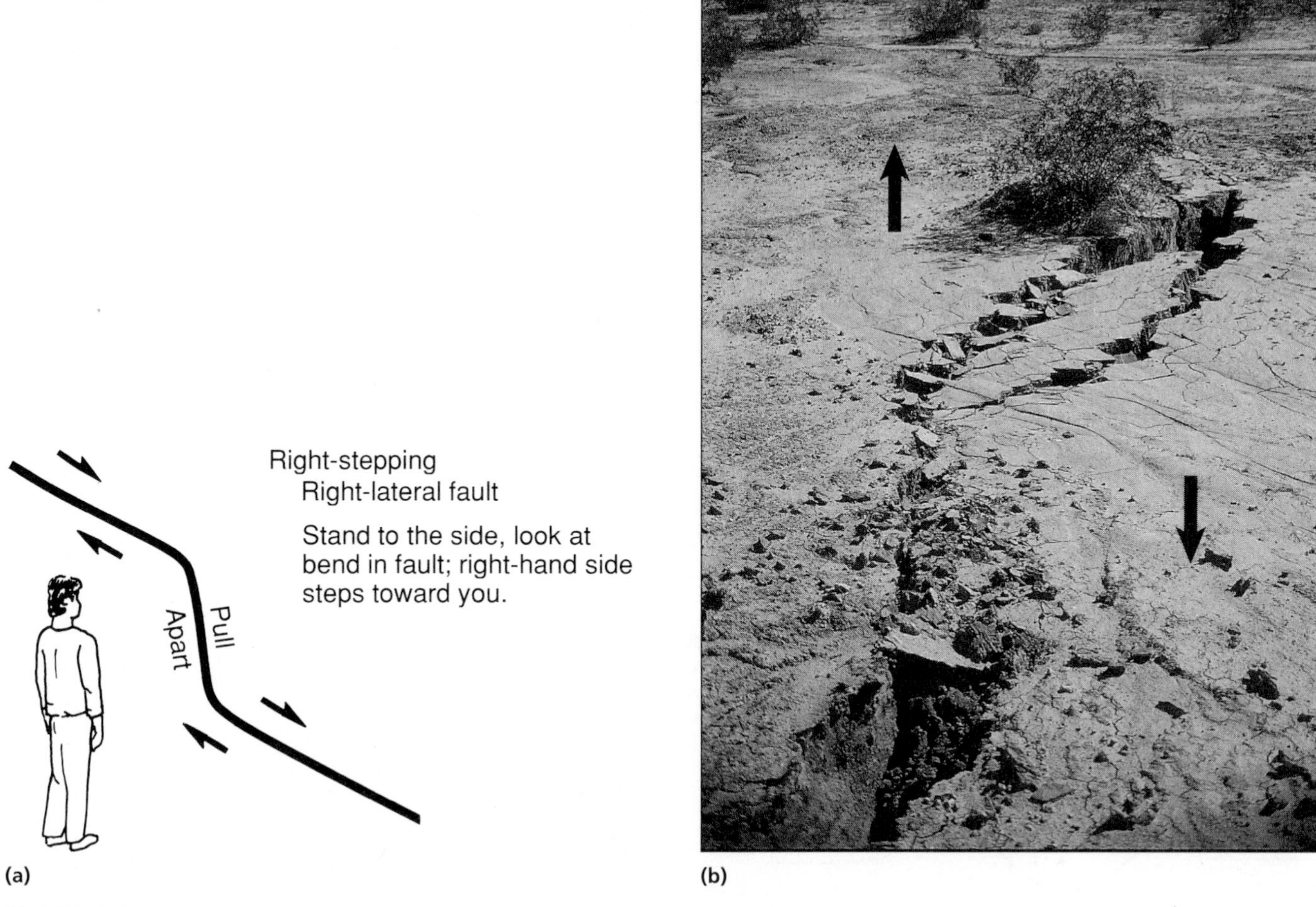

Figure 3.14 **(a)** Right step in right-lateral fault. Notice that the land is pulled apart at the fault bend whenever the fault moves. Movements will create a hole, which could become a basin if the fault stays active for a geologically long time. **(b)** Land offset along the Superstition Hills right-lateral fault during its 1987 rupture. See the right step and the pull apart at the bend.

Photo by Pat Abbott.

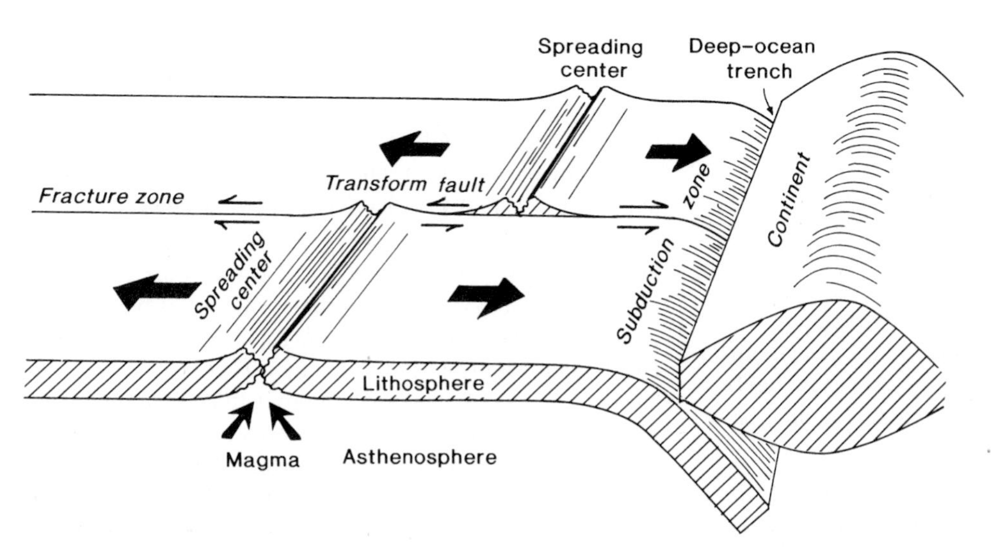

Figure 3.15 Plate-tectonics model of transform faults. Notice that the transform fault connects the two separated spreading centers; sea floor moves in opposite directions here. Beyond the spreading centers, the two plates move in the same direction and are separated by a fracture zone; there is no transform fault here.

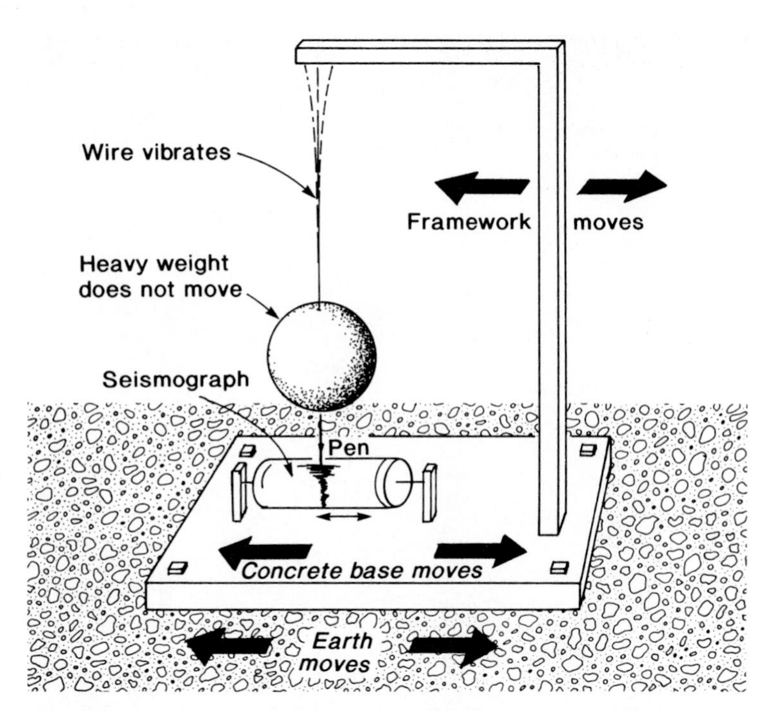

Figure 3.16 A basic seismograph. The Earth moves, the seismograph framework moves, the hanging wire vibrates, but the suspended heavy mass and pen beneath it remain relatively steady. Ideally, the pen holds still while the Earth moves beneath the pen to produce an inked line.

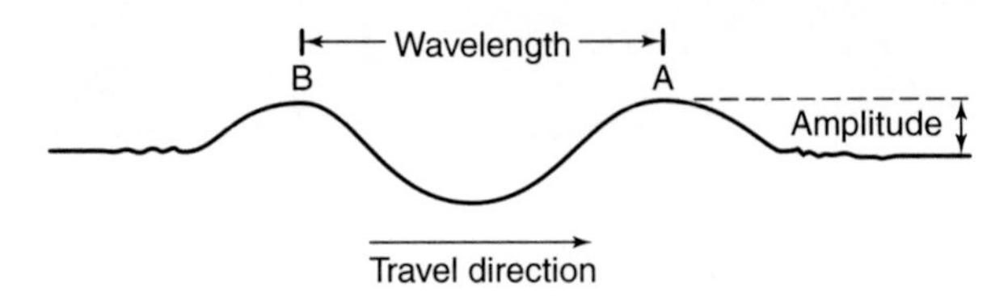

Figure 3.17 Wave motion. Amplitude is the height of the wave above the starting point. Wavelength is the distance between wave crests B & A. Period is the amount of time in seconds for wave crest B to travel to site A.

Waves

Throw a rock into a pond, blow air into a trombone, or experience a fault movement, and the water, the air, or the earth will transmit waves of energy that travel away from the initial disturbance. All these waves have similarities (Figure 3.17); they have displacement called **amplitude,** distance between successive waves measured as **wavelength,** a **period** of time between waves measured in seconds, and **frequency** measured as number of waves passing a given point during one second. Frequencies are measured in **hertz** where one hertz equals one cycle per second. Note that period and frequency are inversely related:

$$\text{period} = \frac{1}{\text{frequency (in hertz)}}$$

For example, if five waves passed a given point in one second, then the frequency is 5 hertz (Hz) and the period of time between each wave is 0.2 second.

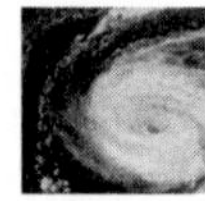

Seismic Waves

When the Earth shakes, it releases energy in **seismic waves** that pass through the whole body of the planet (**body waves**) and others that move near the surface only (**surface waves**).

Body Waves

Body waves are the fastest and are referred to as either primary or secondary waves. Body waves are most abundant at high frequencies of 0.5 to 20 hertz, that is, one-half to 20 cycles per second, and are called short-period waves. The high-frequency, short-period waves are most energetic for short distances close to the hypocenter/epicenter.

Primary Waves The **primary** or **P wave** is the fastest and thus is the first to reach a recording station. P waves move in a push-pull fashion of alternating pulses of compression (push) and extension (pull); this motion is probably best visualized using a Slinky toy (Figure 3.18a). P waves radiate outward from their source in an ever-expanding sphere, like a rapidly inflating balloon. They travel through any material, be it solid, liquid, or gas. Their speed depends on the density and compressibility of the materials through which they pass. The greater the density and resistance to compression, the greater the speed of the seismic waves passing through packed atomic lattices. Representative velocities for P

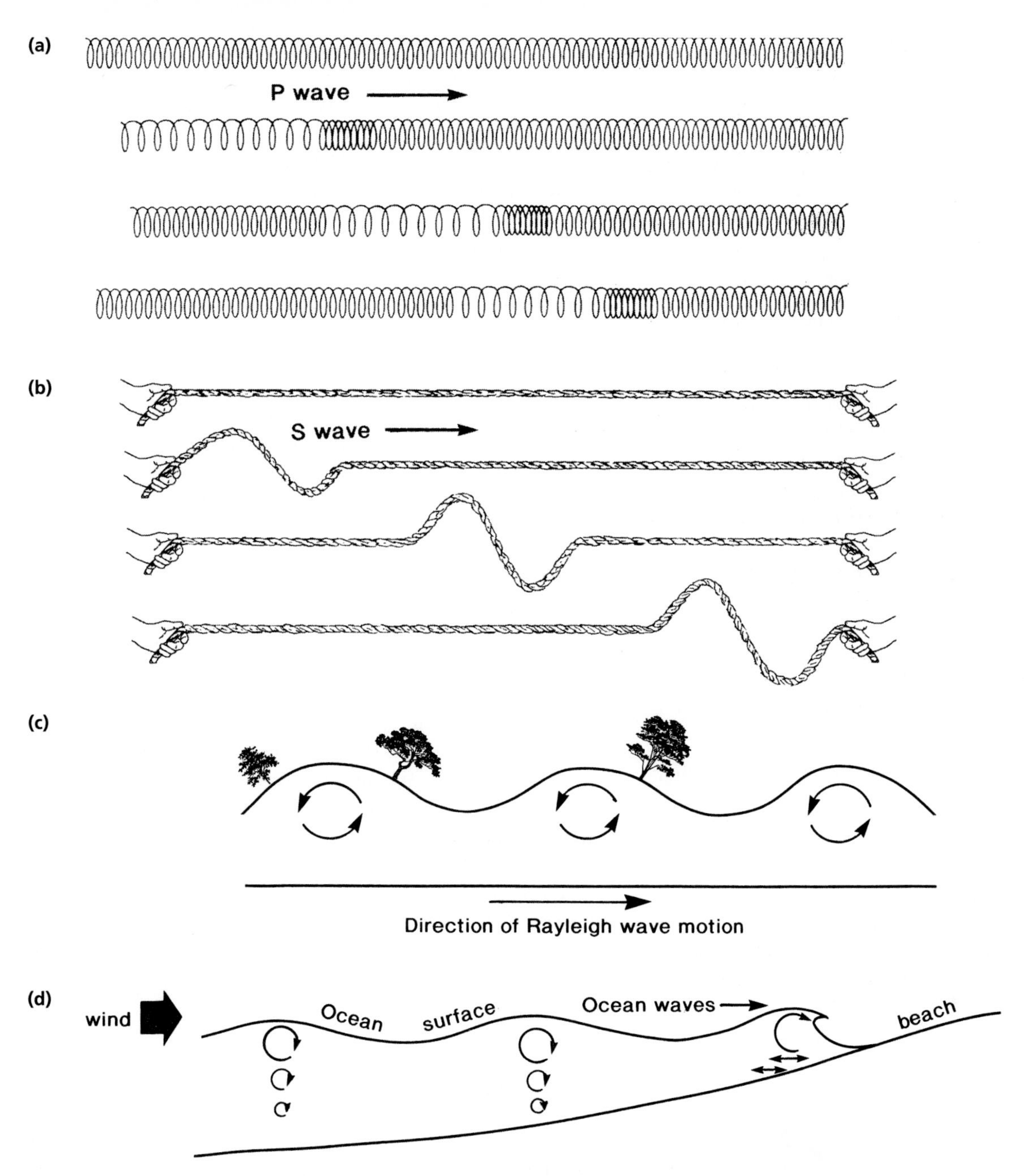

Figure 3.18 Types of seismic waves. **(a)** P waves move in the push-pull motion of a Slinky toy. **(b)** S waves move in up-and-down motions perpendicular to the direction of advance, like a shaken jump rope. **(c)** Rayleigh waves advance in a backward-rotating motion, as opposed to **(d)** wind-blown ocean waves, which cause water to move in forward-rotating circles.

waves in dense rocks (e.g., **granite**) are about 4.8 km/sec (about 10,700 mph). P waves in water slow to 1.4 km/sec (about 3,100 mph). Because P waves are similar to sound waves, they can travel through air. P waves may emerge from the ground, and if you are near the epicenter, you may be able to hear those P waves pulsing at around 15 cycles per second as low, thunderous noises. The arrival of P waves at your home or office is similar to a sonic boom with the rattling of windows.

Secondary Waves The **secondary** or **S wave** is the second wave to reach a recording station. S waves are transverse waves that propagate by shearing or shaking particles in their path at right angles to the direction of advance. This motion is probably most easily visualized by considering how a jump rope moves when you shake one end up and down (Figure 3.18b). S waves travel only through solids. On reaching liquid or gas, the S wave energy is reflected back into rock or is converted to another form. The velocity of an S wave depends on the density and resistance to shearing of materials. Liquids and gases do not have shear strength and thus cannot transmit S waves. Representative velocities for S waves in dense rocks (e.g., granite) are about 3 km/sec (about 6,700 mph). Passing into liquid (e.g., water), the shear waves stop. With their up-and-down and side-to-side motions, S waves shake the ground surface and can do severe damage to buildings.

Seismic Waves and the Earth's Interior

Large earthquakes generate body waves energetic enough to be recorded on seismographs all around the world. These P and S waves do not follow simple paths as they pass through the Earth; they speed up, slow down, and change direction, and S waves even disappear. Analysis of the travel paths of the seismic waves gives us our models of the Earth's interior (Figure 3.19). The Earth is not homogeneous. Following the paths of P and S waves from the Earth's surface inward, there is an initial increase in velocity but then a marked slowing occurs at about 100 km depth; this is the top of the asthenosphere. Passing farther down through the mantle, the velocities vary but generally increase until about 2,900 km depth; there, the P waves slow markedly and the S waves disappear. This is the mantle-core boundary zone. The stopping of S waves indicates that the outer core is mostly liquid. Moving into the core, P wave velocities gradually increase until a jump is reached at about 5,150 km depth, suggesting that the inner core is solid.

Surface Waves

Seismic waves that travel near the Earth's surface are of two main types—Love and Rayleigh waves. Surface waves are created by body waves disturbing the surface. You can visualize these types of seismic waves by analogy. Throw a rock into water and watch the circular ripple trains flow outward from the point of impact (epicenter). Both Love and Rayleigh waves are referred to as L waves (long waves) because they take longer periods of time to complete one cycle of motion and are the slowest moving. The frequencies of surface waves are low, less than one cycle per second. The low-frequency, long-period waves carry significant amounts of energy for much greater distances away from the epicenter.

Love Waves Love waves were recognized on seismograms and first explained by the British mathematician A. E. H. Love. Their motion is similar to S waves, except it is from side-to-side in a horizontal plane roughly parallel to the Earth's surface. As with S waves, their shearing motion is at right angles to the direction of advance; that is, visualize the jump rope in Figure 3.18b lying on the ground. Love waves generally travel faster than Rayleigh waves. Like S waves, they do not move through water or air.

Rayleigh Waves Rayleigh waves were predicted to exist by Lord Rayleigh 20 years before they were actually recognized on seismograms. They advance in a backward-rotating, elliptical motion (Figure 3.18c) similar to the orbiting paths of water molecules in wind-blown waves of water, except that waves in water are forward-rotating (Figure 3.18d). The shaking produced by Rayleigh waves causes both vertical and horizontal movement. The shallower the hypocenter, the more P and S wave energy will hit the surface, thus putting more energy into Rayleigh waves. The rolling waves pass through both ground and water. The often-heard report that an earthquake feels like one is rocking in a boat at sea well describes the passage of Rayleigh waves. These waves have long periods and once started, they go a long way.

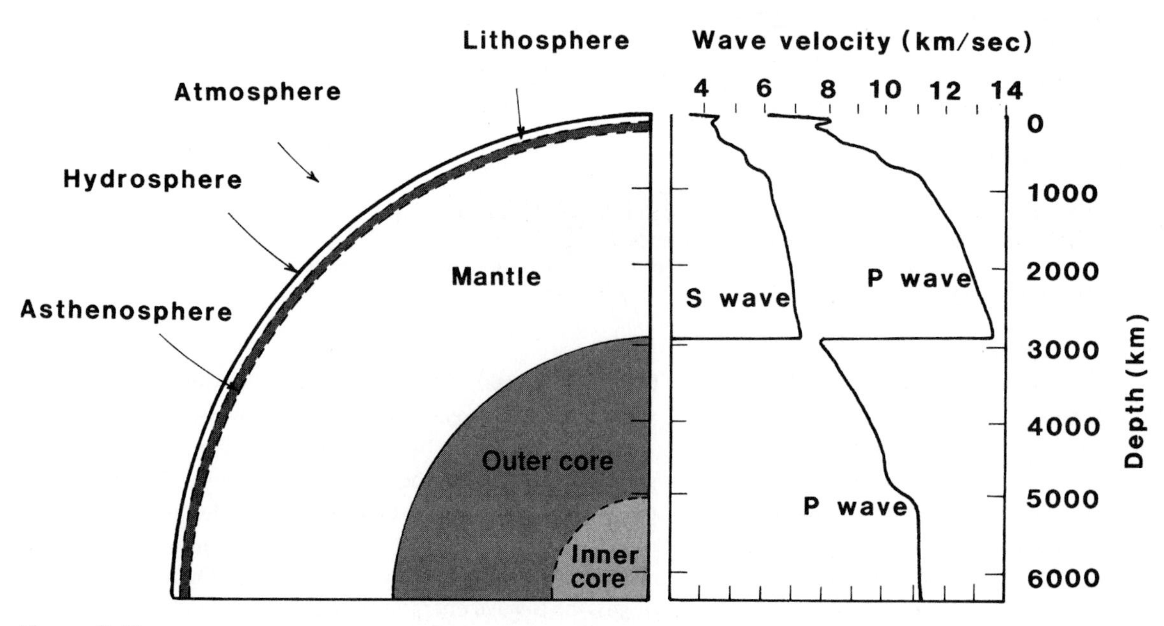

Figure 3.19 Varying velocities of P and S waves help define the internal structure of the Earth.

Sound Waves and Seismic Waves

Waves are fundamental to both music and seismology. Musicians use instruments to produce the sound waves we hear as music. For example, trombone players control the amount of sound with their breath and the frequencies of the sound waves by extending and retracting the slide on their trombones. Seismologists record the body and surface waves generated during an earthquake and analyze the wave patterns to determine where the earthquake occurred and how much energy was released during the event.

Music is a common part of our lives and we are familiar with hearing sound waves. Seismic waves are usually presented to us in an unfamiliar visual form. Waveforms for a trombone and a moderate-size earthquake are shown in Figure 3.20. Both trombone and earthquake have more higher frequency waves if a shorter path is traveled, that is, the trombone is retracted and the earthquake occurred nearby. As the travel paths become longer for both trombone (extended) and earthquake (distant), the amount of high-frequency waves decreases. Musically, as the path through the trombone lengthens, the vibrations per second decrease, the frequencies are lower, the tone is lower. Seismically, a nearby earthquake has more frequent, sharp jolting waves, and at greater distances the waves are less frequent and more gently rolling. The ranges of some common frequencies are listed in Table 3.1.

Table 3.1 Some Common Frequencies (in hertz)

Some Common Frequencies (in hertz)
Audio
30,000 Hz—heard by dogs
15–20 Hz to 15,000–20,000 Hz—range of human hearing
15–20 Hz—P waves in air heard by humans near epicenter
Earthquake
0.5–20 Hz—body waves
0.005–0.1 Hz—surface waves

Locating the Source of an Earthquake

Using the lengths of time the various seismic waves take to reach a seismograph, the locations of the epicenter and hypocenter can be determined. P waves travel about 1.7 times faster than S waves. Thus, the farther away from the earthquake origin, the greater is the difference in arrival times between P and S waves (Figure 3.21). When a seismograph records an earthquake, the difference in arrival times of P and S waves (S–P) is determined. Inspection of the **seismogram** in Figure 3.22 shows that S waves arrived 3 minutes and 45 seconds after P waves. Figure 3.21 indicates that an S–P arrival time difference of 3 minutes and 45 seconds corresponds to an earthquake about 2,250 km (1,400 mi) away. But in what direction?

Epicenters can be located using seismograms from three recording stations. As an example, S–P wave arrival time differences yield distances to the epicenter of 164 km (102 mi) from Memphis State University in Tennessee, 236 km (146 mi) from St. Louis University in Missouri, and 664 km (412 mi) from Ohio State University in Columbus. Now if the distance from each station is plotted as the radius of a

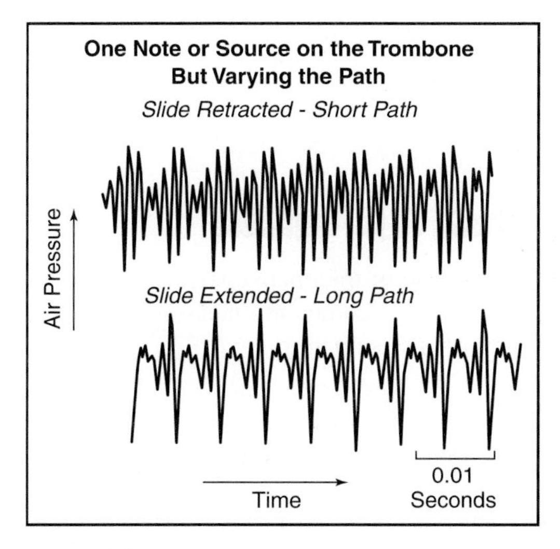

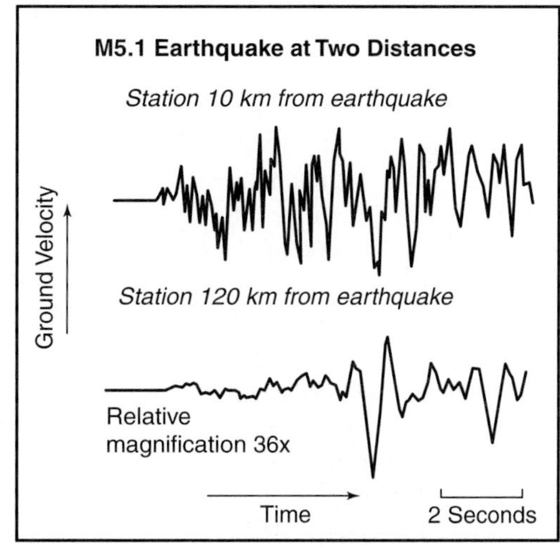

Figure 3.20 Wave patterns for trombone and earthquake for short- and long-distance travel paths.

Source: A. Michael, S. Ross, and D. Schaff, "The Music of Earthquakes: Waveforms of Sound and Seismology" in *American Scientist* (2002).

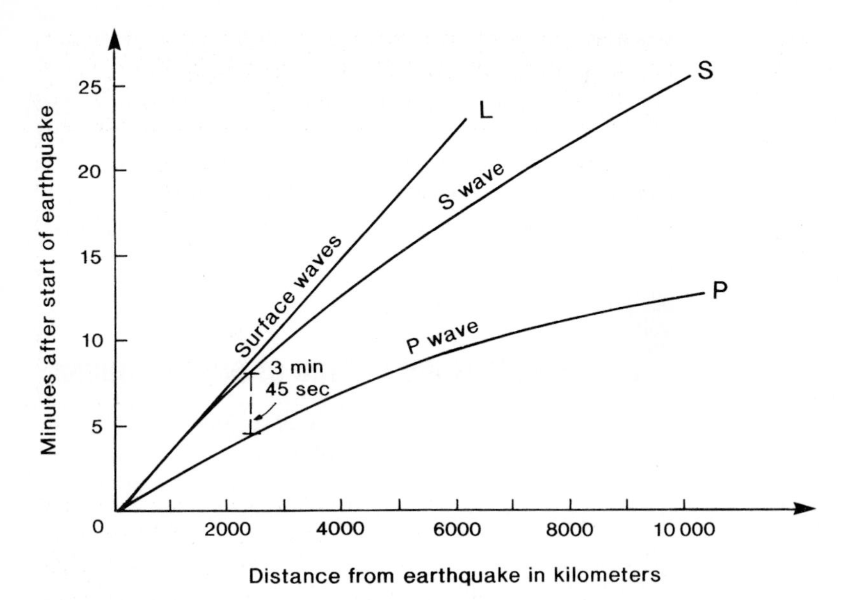

Figure 3.21 Plot of travel time versus distance for seismic waves. Note that the arrival time difference for P and S waves of 3 minutes and 45 seconds in Figure 3.22 corresponds to a distance of about 2,250 km.

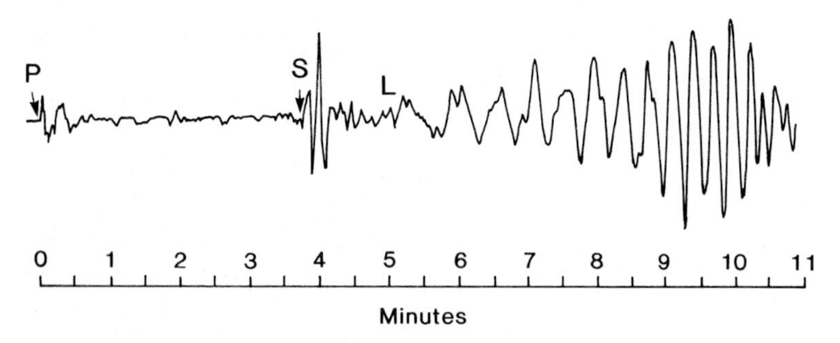

Figure 3.22 Seismogram recorded in Russia in 1909 of an Asia Minor earthquake. Notice that the difference in arrival times of P and S waves (S–P) is 3 minutes and 45 seconds.

circle, then the three circles will intersect at one unique point—an epicenter at New Madrid, Missouri (Figure 3.23). Computers usually make the calculations to determine epicenter locations; however, a better mental picture of the process is gained via the hand-drawn circles.

The difference in arrival times of P and S waves (S–P) actually measures the distance from the recording station to the hypocenter (or focus) of the earthquake, the site of initial fault movement (Figure 3.12). If the hypocenter is at the Earth's surface, then the hypocenter and epicenter are the same. However, if the hypocenter is deep below the surface, it will affect the arrival time of surface (L) waves because L waves do not begin until P waves strike the Earth's surface. The depth to a hypocenter is best determined where an array of seismometers is nearby, thus allowing careful analysis of P wave arrival times.

Magnitude of Earthquakes

Magnitude is an estimate of the relative size or energy release of an earthquake. It is commonly measured from the seismic wave traces on a seismogram.

Richter Scale

Several systems are available to assess the magnitudes of earthquakes. The best-known scheme is the Richter scale. In 1935, Charles Richter of the California Institute of Technology devised a quantitative scheme to describe the magnitude of Californian earthquakes, that is, of events with shallow hypocenters that are located near (less than 300 mi from) the seismometers. Richter based his scale on the idea that the bigger the earth-

Figure 3.23 Location of an earthquake epicenter. S–P arrival time difference calculations gave a radius of 164 km from Memphis, 236 km from St. Louis, and 664 km from Columbus. The circles plotted with these values intersect uniquely at New Madrid, Missouri—the epicenter.

quake, the greater the shaking of the Earth, and thus the greater the amplitude (swing) of the lines made on the seismogram. To standardize this relationship, he defined magnitude as:

> the logarithm to the base ten of the maximum seismic wave amplitude (in thousandths of a millimeter) recorded on a standard seismograph at a distance of 100 kilometers from the earthquake center.

Since not all seismographs would be sitting 100 km from the epicenter, corrections are made for distance. Richter assigned simple, whole numbers to describe magnitudes; for every tenfold increase in the amplitude of the recorded seismic wave, the Richter magnitude increases one number, e.g., from 4 to 5. The energy released by earthquakes increases even more rapidly than the tenfold increase in amplitude of the seismic wave trace. For example, if the amplitude of the seismic waves increased 10,000 times ($10 \times 10 \times 10 \times 10$), the Richter magnitude would move up from a 4 to an 8. However, the energy release from 4 to 8 increases by 2,800,000 times (Table 3.2).

What does this increase mean in everyday terms? If you feel a magnitude 4 earthquake while sitting at your dinner table, then a magnitude 8 comes along while you are still at the table, would you really be shaken 2,800,000 times as hard? No. The greater energy of the magnitude 8 earthquake would be spread out over a much larger area, and the magnitude 8 event would dissipate its energy over a time interval about 20 times longer (e.g., 60 seconds as opposed to 3 seconds). The actual shaking in earthquakes above magnitude 6 does not increase very much more (maybe three times more); it certainly does not increase as much as the values in Table 3.2 might lead one to think. *In effect, the bigger earthquake means that more people in a larger area and for a longer time will experience the intense shaking.*

Table 3.2 Energy of Richter Scale Earthquakes

Richter Magnitude		Energy Increase	Energy Compared to Magnitude 4
4			1
5	=	48 Mag 4 EQs	48
6	=	43 Mag 5 EQs	2,050
7	=	39 Mag 6 EQs	80,500
8	=	35 Mag 7 EQs	2,800,000

Computing a Richter magnitude for an earthquake is quickly done, and this is one of the reasons for its great popularity with the deadline-conscious print and electronic media. Upon learning of an earthquake, usually by phone calls from reporters, one can rapidly measure: 1) the amplitude of the seismic waves and 2) the difference in arrival times of P and S waves. Figure 3.24 has reduced Richter's equation to a nomograph that allows easy determination of magnitude. Take a couple of minutes to figure out the magnitude of the earthquake whose seismogram is printed above the nomograph.

Each year, the Earth is shaken by millions of quakes that are recorded on seismometers. Most are too small to be felt by humans (Table 3.3). Notice the distinctive "pyramidal" distribution of earthquakes by size—the smaller the earthquake magnitude, the greater their numbers. Yet the fewer than 20 major and great earthquakes (magnitudes of 7 and higher) each year account for more than 90 percent of the energy released by earthquakes. This fact underscores the logarithmic nature of the Richter scale; each step up the scale has major significance.

Other Measures of Earthquake Size

An earthquake is a complex event, and more than one number is needed to assess its magnitude. Although the Richter scale is useful for assessing moderate-sized earthquakes that occur nearby, the 0.1-to-2-second-period waves it uses do not work well for distant or truly large earthquakes. The short-period waves do not become more intense as an earthquake

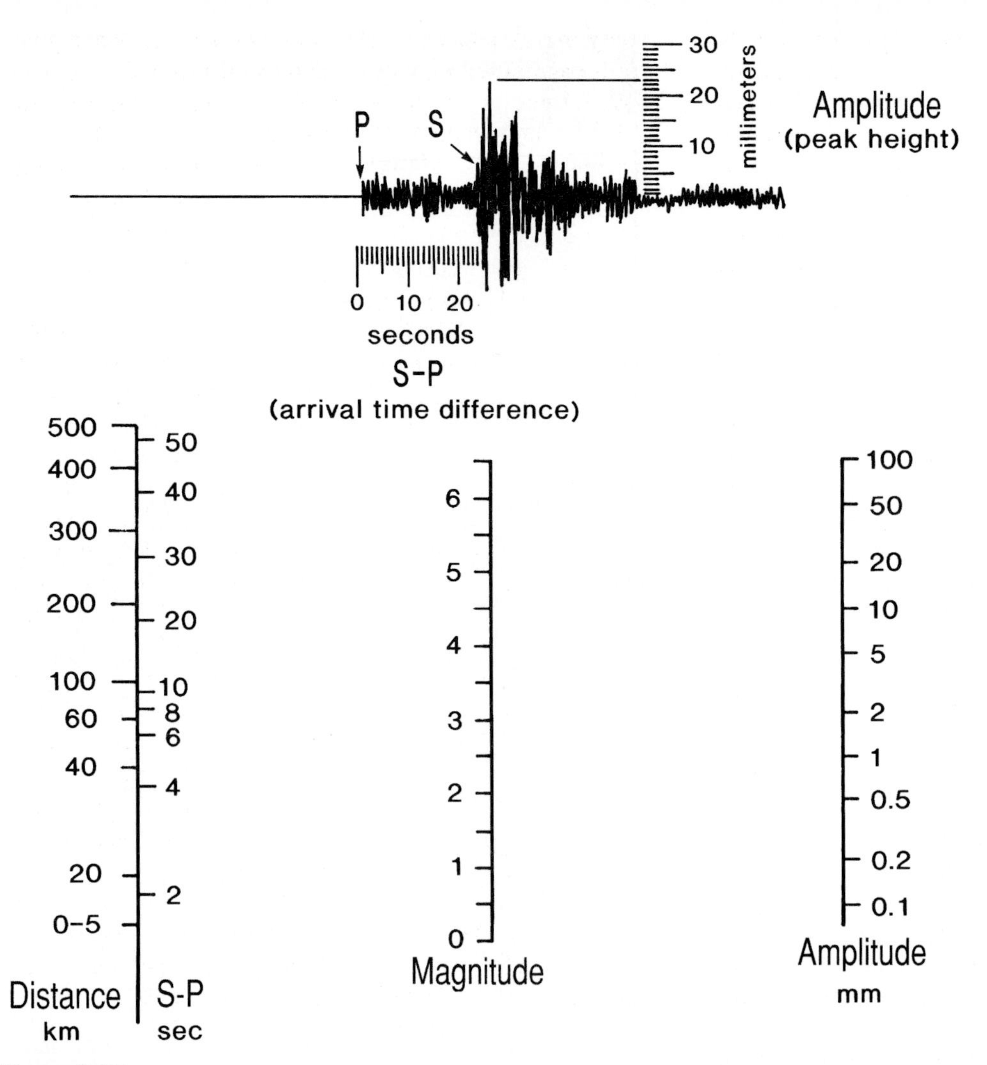

Figure 3.24 Nomograph of the Richter scale allowing earthquake magnitudes to be estimated. On the seisomogram, read the difference in arrival times of P and S waves in seconds, and plot the value on the left column of the nomograph. Next read the amplitude of the peak height of the S wave and plot this value on the right column. Draw a line between the two marked values and it will pass through the earthquake magnitude on the center column.

Table 3.3 Earthquakes in the World Each Year

Magnitude	Number of Quakes Per Year	Description
8.5 and up	0.3	
8–8.4	1	Great
7.5–7.9	3	
7–7.4	15	Major
6.6–6.9	56	
6–6.5	210	Strong (Destructive)
5–5.9	800	Moderate (Damaging)
4–4.9	6,200	Light
3–3.9	49,000	Minor
2–2.9	350,000	Very minor
0–1.9	3,000,000	

becomes larger. For example, the Richter scale assesses the 1906 San Francisco earthquake and the 1964 Alaska earthquake as both being of magnitude 8.3. However, using other scales, the San Francisco earthquake is probably equivalent to a Richter magnitude 7.8 and the Alaska seism is equivalent to a 9.2. The Alaska earthquake was at least 100 times bigger in terms of energy.

The Richter scale is now restricted to measuring only local earthquakes with moderate magnitudes (noted as M_L). Because earthquakes generate both body waves that travel through the Earth and surface waves that follow the Earth's uppermost layers, two other magnitude scales have long been used: m_b and M_s. The body wave (m_b) scale uses amplitudes of P waves with 1-to-10-second periods, whereas the surface wave scale (M_s) uses Rayleigh waves with 18-to-22-second periods. Early on, all magnitude scales were considered to be equivalent, but now we know that earthquakes generate different proportions of energy at different periods. For example, larger earthquakes with their larger fault-rupture surfaces radiate more of their energy in longer-period seismic waves. Thus, for great and major earthquakes, body wave magnitudes (m_b) will significantly underestimate the actual size of the earthquake. Even a composite of these three methods of determining earthquake magnitude (M_L, m_b, and M_s) does not necessarily yield the true size of an earthquake.

Moment Magnitude Scale Seismologists have moved on to other measures to more accurately determine earthquake size. The seismic moment (M_o) relies on the amount of movement along the fault that generated the earthquake, that is, M_o equals the shear strength of the rocks times the rupture area of the fault times the average displacement (slip) on the fault. Moment is a more reliable measure of earthquake size; it measures the amount of strain energy released by the movement along the whole rupture surface. Seismic moment has been incorporated into a new earthquake magnitude scale by Hiroo Kanamori, the moment magnitude scale (M_w), where:

$$M_w = 2/3 \log_{10} (M_o) - 6$$

The moment magnitude scale is used for big earthquakes. It is more accurate because it is tied directly to physical parameters such as fault-rupture area, fault slip, and energy release. Other earthquake scales use indirect measures such as how much a seismograph needle moves.

The two largest moment magnitudes measured to date are the 1964 Alaska earthquake (M_s of 8.3; M_w of 9.2) and the 1960 Chile earthquake (M_s of 8.5; M_w of 9.5). Both these gigantic earthquakes occurred at subduction zones (see Chapter 4). A variety of energetic events are placed on a scale for comparison in Figure 3.25.

Ground Motion during Earthquakes

Seismic waves radiate outward from a fault movement. The interactions among the various seismic waves move the ground both vertically and horizontally. Buildings usually are designed to handle the large vertical forces caused by the weight of the building and its contents. They are designed with such large factors of safety that the additional vertical forces imparted by earthquakes are typically not a problem. Usually, the biggest concern in designing buildings to withstand large earthquakes is the sideways push from the horizontal components of movement (Figure 3.26).

Acceleration

Building design in earthquake areas must account for **acceleration.** As seismic waves move the ground and buildings up and down, and back and forth, the rate of change of velocity is measured as acceleration. For analogy, when your car is moving at a velocity of 25 mph on a smooth road, you feel no force on your body. But if you stomp on the car's accelerator and rapidly speed up to 55 mph, you feel a force pushing you back against the car's seat. Following the same thought, if you hit the brakes and decelerate rapidly, you feel yourself being thrown forward. This same type of accelerative force is imparted to buildings when the ground beneath them moves during an earthquake.

Continuing the analogy further, if you hold your arm upright in front of you and wave it back and forth, you create rapid acceleration and high velocity, but no damage is done because the weight of your arm is small and the inertial forces are low. However, because force is acceleration times mass, if a building weighing thousands of tons is subjected to the same acceleration, the acceleration will produce large inertial forces that will be difficult for the building to withstand. If these forces last long enough, the building may fail.

The usual measure of acceleration is that of a free-falling body pulled by gravity; it is the same for all objects, regardless of their weight. The acceleration due to gravity is 9.8 meters per second squared (32 ft/sec squared), which is referred to as 1.0 g and is used as a comparative unit of measure. Weak buildings begin to suffer damage at horizontal accelerations of about 0.1 g. At accelerations between 0.1 to 0.2 g, people have trouble keeping their footing, similar to being in the corridor of a fast-moving train or on a small boat in high seas. A problem for building designers is that earthquake accelerations have locally been in excess of 1 g. In the hills above Tarzana, California, the 1994 Northridge earthquake generated phenomenal accelerations—1.2 g vertically and 1.8 g horizontally.

Periods of Buildings and Geologic Foundations

The concepts of period and frequency also apply to buildings and geologic foundations. Visualize the shaking or vibration of a 1-story house and a 30-story office building. Do they take the same amount of time to complete one cycle of movement, to shake back and forth one time? No. Typical periods of swaying for buildings are about 0.1 second per story of height. The 1-story house shakes back and forth quickly at about 0.1 second per cycle. The 30-story building sways much slower with a period of about 3 seconds per cycle.

The periods of buildings also are affected by their construction materials. A building of a given height and design will have a longer period if it is made of flexible materials such as wood or steel, or its period will be shorter if it was built with stiff materials such as brick or concrete.

Geologic materials also have their own natural periods. For example, hard rocks may vibrate or shake with a period of about 0.5 second, whereas a soft sediment may have a period exceeding 2 seconds.

When seismic waves of a certain period carry a lot of energy and they pass through geologic materials that have the

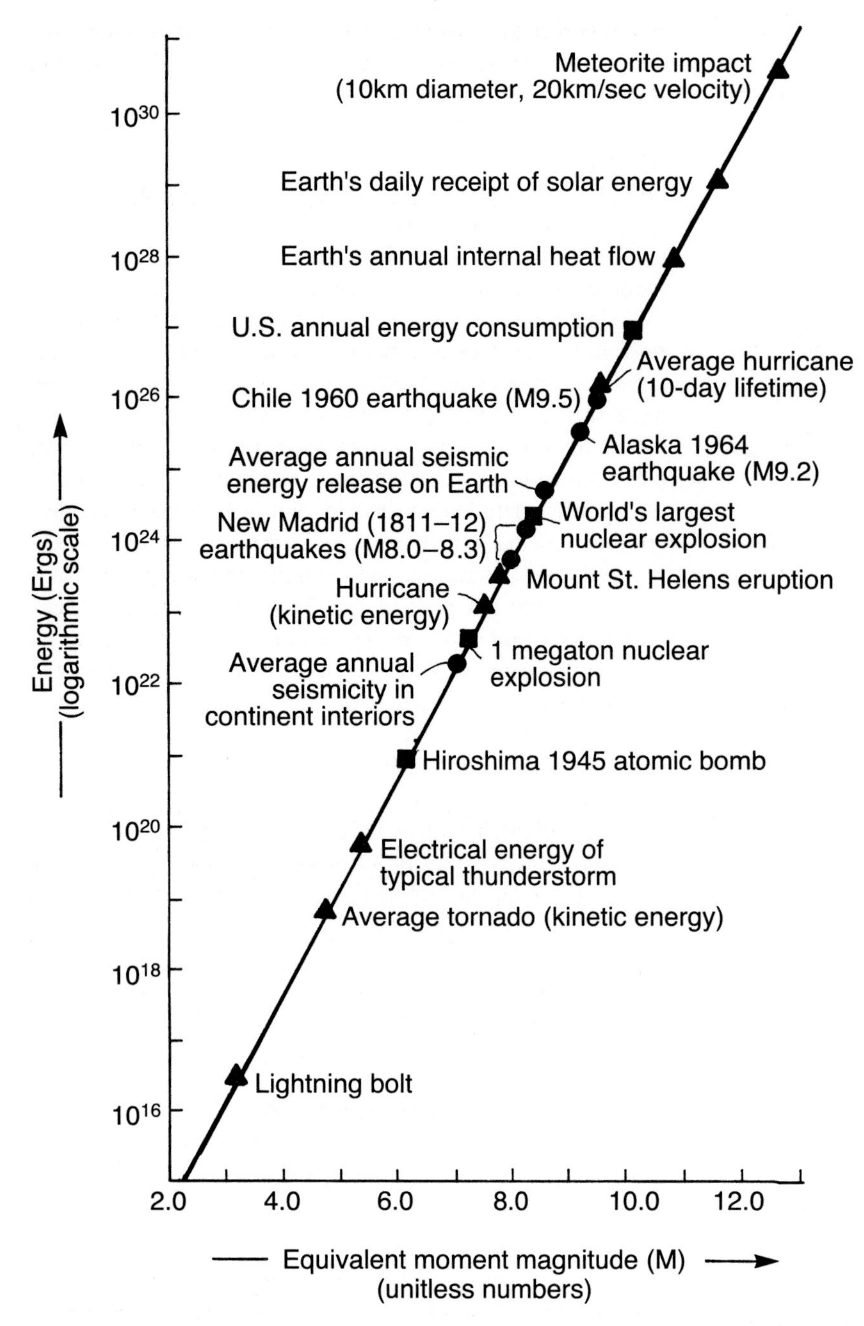

Figure 3.25 Equivalent moment magnitude of a variety of seismic (dots), human-made (squares), and other phenomena (triangles).

Source: A. C. Johnston, "An earthquake strength scale for the media and the public" in *Earthquakes and Volcanoes* 22 (no. 5): 214–16. U.S. Geological Survey.

same period, then the shaking is amplified. Even more, when the period of a building is the same as its geologic foundation and both match the periods of high-energy seismic waves, then the combined shaking or **resonance** is amplified further. The resonance created by shared periods for seismic waves, geologic foundations, and buildings is a common cause of the catastrophic failure of buildings during earthquakes.

Understanding the concept of shared periods and resonance may be furthered by visualizing a tall flagpole with a heavy metal eagle on top. First, if you shake this pole, you will quickly learn that the pole has a strong tendency to move back and forth only at a certain rate or period. If the flagpole swings a complete cycle in two seconds, it has a period of two seconds. Second, if seismic waves of a two-second period begin to shake the ground, the amount of movement of the flagpole will start to increase. The pole is now resonating, and the forces it must withstand have increased. Third, if the ground the flagpole rests on also vibrates with a period of two seconds, then the resonance is increased even more, and the greater forces created by the combined periods may cause destruction.

Figure 3.26 Inadequately braced house failed due to horizontal acceleration during the 1971 San Fernando earthquake.
Photo by Al Boost.

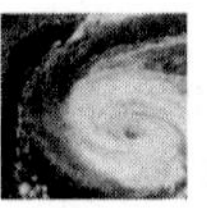

Earthquake Intensity—What We Feel during an Earthquake

During the tens of seconds that a large earthquake lasts, we feel ourselves rocked up and down and shaken from side to side. It is an emotional experience, and the drama of our personal accounts varies according to our locations during the shaking and according to our personalities. But for personal narratives to have meaning that can be passed on to succeeding generations, common threads are needed that bind the accounts together. In the late 1800s, descriptive schemes appeared that were based on the intensity of effects experienced by people and buildings. The most widely used scale came from the Italian professor Giuseppi Mercalli in 1902; it was modified by Charles Richter in 1956. The scale has 12 divisions of increasing intensity labeled by Roman numerals (Table 3.4).

Table 3.4 Modified Mercalli Scale of Earthquake Intensity

Intensity	Description
I	Not felt except by a very few people under especially favorable circumstances.
II	Felt by only a few people at rest, especially those on upper floors of buildings or those with a very sensitive nature. Delicately suspended objects may swing.
III	Felt quite noticeably indoors, especially on upper floors, but many people do not recognize it as an earthquake. Vibrations are like those from the passing of light trucks. Standing automobiles may rock slightly. Duration of shaking may be estimated.
IV	Felt indoors during the day by many people, outdoors by few. Light sleepers may be awakened. Vibrations are like those from the passing of heavy trucks or as if a heavy object struck the building. Standing automobiles rock. Windows, dishes, and doors rattle; glassware and crockery clink and clash. In the upper range of IV, wooden walls and frames creak.
V	Felt indoors by nearly everyone, outdoors by many or most. Awakens many. Frightens many; some run outdoors. Some broken dishes, glassware, and windows. Minor cracking of plaster. Moves small objects, spills liquids, rings small bells, and sways tall objects. Pendulum clocks misbehave.
VI	Felt by all; many frightened and run outdoors. Excitement is general. Dishes, glassware, and windows break in considerable quantities. Knickknacks, books, and pictures fall. Furniture moves or overturns. Weak plaster walls and some brick walls crack. Damage is slight.
VII	Frightens all; difficult to stand. Noticed by drivers of automobiles. Large bells ring. Damage negligible in buildings of good design and construction, slight to moderate in well-built ordinary buildings, considerable in badly designed or poorly built buildings, adobe houses, and old walls. Numerous windows and some chimneys break. Small landslides and caving of sand and gravel banks occur. Waves on ponds; water becomes turbid.
VIII	Fright is general and alarm approaches panic. Disturbs drivers of automobiles. Heavy furniture overturns. Damage slight in specially designed structures; considerable in ordinary substantial buildings, including partial collapses. Frame houses move off foundations if not bolted down. Most walls, chimneys, towers, and monuments fall. Spring flow and well-water levels change. Cracks appear in wet ground and on slopes.
IX	General panic. Damage considerable in masonry structures, even those built to withstand earthquakes. Well-built frame houses thrown out of plumb. Ground cracks conspicuously. Underground pipes break. In soft sediment areas, sand and mud are ejected from ground in fountains and leave craters.
X	Most masonry structures are destroyed. Some well-built wooden structures and bridges fail. Ground cracks badly with serious damage to dams and embankments. Large landslides occur on river banks and steep slopes. Railroad tracks bend slightly.
XI	Few, if any, masonry structures remain standing. Great damage to dams and embankments, commonly over great distances. Supporting piers of large bridges fail. Broad fissures, earth slumps and slips in soft, wet ground. Underground pipelines completely out of service. Railroad tracks bend greatly.
XII	Damage nearly total. Ground surfaces seen to move in waves. Lines of sight and level distort. Objects thrown up in air.

Table 3.5 Comparison of Magnitude, Intensity, and Acceleration

Richter Magnitude	Mercalli Intensity		Acceleration (% g)
2 and less	I–II	Usually not felt by people	less than 0.1–0.19
3	III	Felt indoors by some people	0.2–0.49
4	IV–V	Felt by most people	0.5–1.9
5	VI–VII	Felt by all; building damage	2–9.9
6	VII–VIII	People scared; moderate damage	10–19.9
7	IX–X	Major damage	20–99.9
8 and up	XI–XII	Damage nearly total	over 100 = over 1 g

Earthquake magnitude scales are used to assess the energy released during an earthquake; earthquake intensity scales assess the effects on people and buildings (Table 3.5). The difference between magnitude and intensity can be illuminated by comparison to a lightbulb. The wattage of a lightbulb is analogous to the magnitude of an earthquake. Wattage is a measure of the power of a lightbulb, and magnitude is a measure of the energy released during an earthquake.

A lightbulb shining in the corner of a room provides high intensity light nearby, but the intensity of light decreases toward the far side of the room. The intensity of shaking caused by a fault movement is great near the epicenter, but, in general, it decreases with distance from the epicenter. This generalization is offset to varying degrees by variations in geologic foundations and building styles.

Mercalli intensities also are crucial for assessing magnitudes of historical events before there were instrumented records, thus allowing us to assess recurrence intervals between major earthquakes.

Shake Maps Mercalli intensity maps can be generated by computer immediately following an earthquake using your input. After you feel an earthquake, go to the "Did You Feel It?" webpage at http://pasadena.wr.usgs.gov/shake, enter your ZIP code, and then answer questions such as "Did the earthquake wake you up?" and "Did objects fall off shelves?" Within minutes a computer-generated Shake Map will begin taking shape providing information so quickly that it is useful to emergency-response personnel. Within a couple of hours the computer-produced Shake Map will show Mercalli intensities for each ZIP code in the region.

Mercalli Scale Variables

The intensity of an earthquake depends on several variables: 1) earthquake magnitude; 2) distance from the hypocenter/epicenter; 3) type of rock or sediment making up the ground surface; 4) building style—design, kind of building materials, height; and 5) duration of the shaking. These factors need to be considered in assessing the earthquake threat to any region and even to each specific building.

Earthquake Magnitude The relation between magnitude and intensity is obvious—the bigger the earthquake (the more energy released), the higher the odds are for death and damage.

Distance from Hypocenter/Epicenter The relation between distance and damage also seems obvious; the closer to the hypocenter/epicenter, the greater the damage. But this is not always the case, as will be seen in Chapter 4 with the 1989 World Series (Loma Prieta) and 1985 Mexico City earthquakes.

Foundation Materials The types of rock or sediment foundation are of paramount importance. For example, hard rock foundations vibrate at high frequencies and can be excited by energetic P and S waves from a nearby epicenter; soft or water-saturated sediments vibrate at low frequencies and can be amplified by surface (L) waves from distant earthquakes; and steep slopes often fail as landslides when severely shaken.

Building Style The building style is of vital importance. What causes the deaths during earthquakes? Not the shaking of the earth but our buildings, bridges, and other structures that collapse and fall on us. **Earthquakes don't kill, buildings do.** Buildings have frequencies of vibration in the same ranges as seismic waves. High-frequency P and S waves will have their vibrations amplified by 1) rigid construction materials, such as brick or stone, and 2) short buildings. If this type of building rests on high-frequency, hard-rock foundations and is near the epicenter, beware!

Low-frequency surface waves will have their movements increased in tall buildings with low frequencies of vibration. If these tall buildings also lie on low-frequency, soft, water-saturated sand or mud and are distant from the epicenter, then disaster may strike.

Duration of the Shaking The duration of the shaking is underappreciated as a significant factor in damages suffered

Sidebar

Design of Buildings in Earthquake-Prone Areas

One of the problems in designing buildings for earthquake country is the need to eliminate the occurrence of resonance. How can this be done? 1) Change the height of the building; 2) move most of the weight to the lower floors; 3) change the shape of the building; 4) change the type of building materials; and 5) change the degree of attachment of the building to its foundation. For example, if the earth foundation is hard rock that efficiently transmits short-period (high-frequency) vibrations, then build a flexible, taller building. Or if the earth foundation is a thick mass of soft sediment with long-period shaking (low frequency), then build a stiffer, shorter building. For building materials, wood is flexible and light weight, has small mass, and is able to handle large accelerations. Concrete has great compressional strength but suffers brittle failure all too easily under tensional stress. Steel has **ductility** and great tensional strength, but steel columns fail under compressive stress.

The building components that must handle ground motion are basic. In the horizontal plane are floors and roofs plus trusses (Figure 3.27). In the vertical plane are: 1) shear walls, 2) braced frames, and 3) moment-resisting frames.

Floors, Roofs, and Trusses

Horizontal resistance is offered to seismic waves by properly constructed floors, roofs, and trusses that act to transfer seismic-induced forces among vertical resistance elements (shear walls and frames). An important component in building resistance is how securely the floors and roofs are tied or fastened to the walls so they do not separate and fail.

Shear Walls

Walls that are designed to receive horizontal forces from floors, roofs, and trusses and transmit them to the ground are called shear walls. In a building, shear walls must be strong themselves, as well as securely connected to each other and to the horizontal elements. In a simple building, seismic energy moves the ground, producing inertial forces that move the horizontal elements. This movement is resisted by the shear walls, and the forces are transmitted back to the ground.

A "house of cards" is a shear-wall structure, although each "wall" does not have much strength. The walls must be at right angles and preferably in a simple pattern (Figure 3.28). The house of cards is made enormously stronger if horizontal and vertical elements are all securely fastened; e.g., by taping them together.

A structure commonly built with insufficient shear walls is the multistory parking garage. Builders do not want the added expense of more walls, which then eliminate parking spaces and block the view of traffic inside the parking structure. These buildings are common casualties during earthquakes (Figure 3.29).

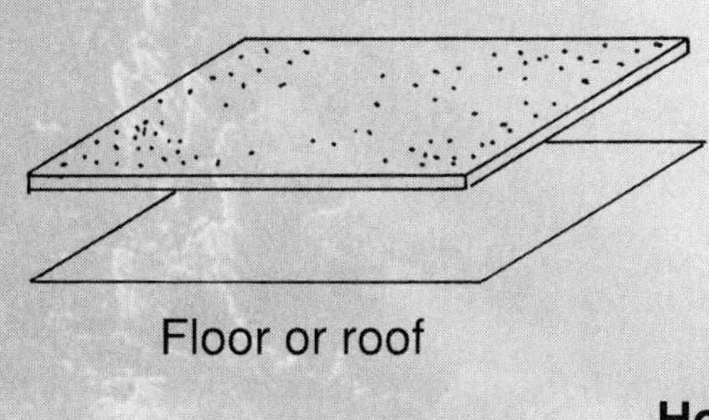

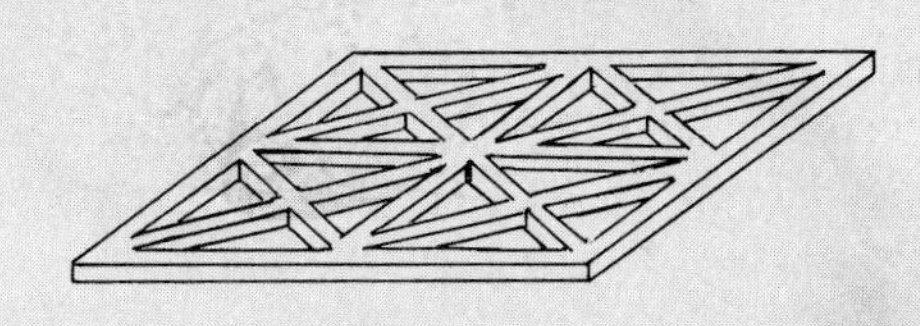

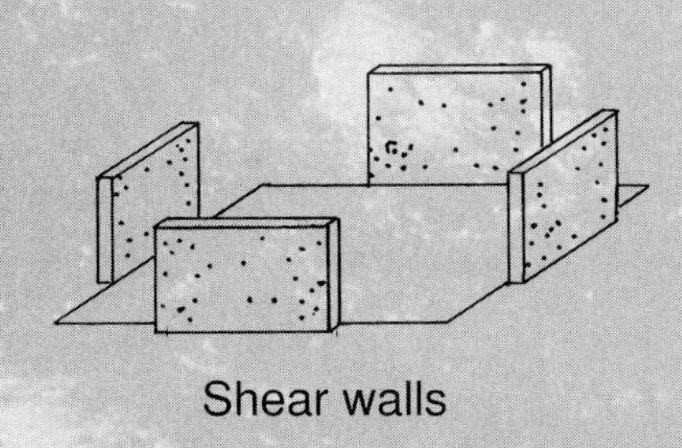

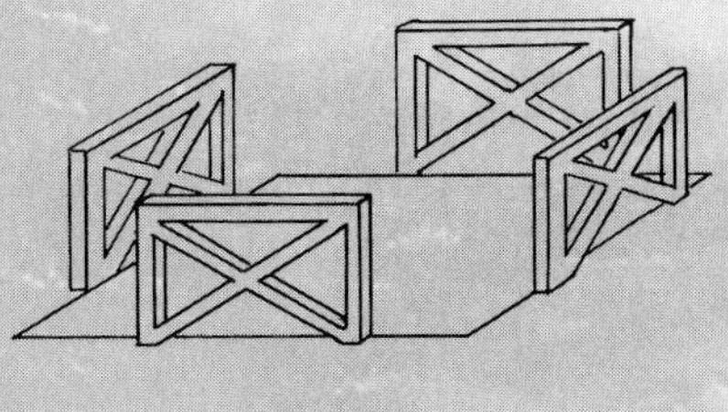

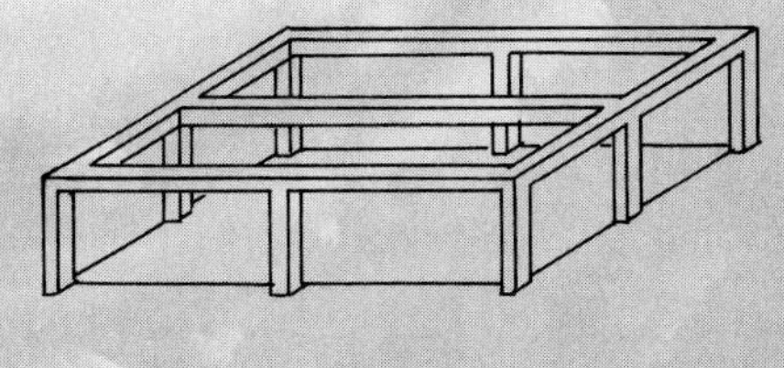

Figure 3.27 The building components of seismic resistance.

Source: "Improving Seismic Safety of New Buildings," 1986, Federal Emergency Management Agency.

Sidebar

Design of Buildings in Earthquake-Prone Areas (Continued)

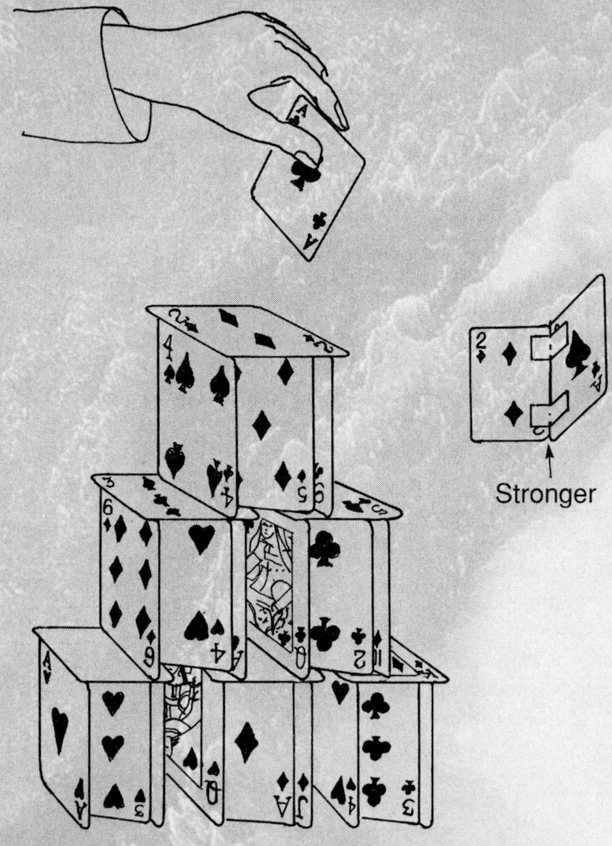

Figure 3.28 A "house of cards" is a shear-wall structure. Earthquake resistance is greatly increased by tying the "walls" together.

Source: "Improving Seismic Safety of New Buildings," 1986, Federal Emergency Management Agency.

Figure 3.29 Automobile parking structure at California State University, Northridge collapsed during 17 January 1994 earthquake. The structure had 2,500 parking spaces and was built in 1991 for $11.5 million.

Photo by Gregory A. Davis.

Figure 3.30 A six-story building with a braced frame incorporated in its design.

Photo by Pat Abbott.

Braced Frames

Bracing is another way to impart seismic resistance to a structure. The bracing gives strength to a building and offers resistance to the up, down, and sideways movements of the ground (Figure 3.30). The bracing should be made of ductile materials that have the ability to deform without rupturing.

Moment-Resisting Frames

Buildings with moment-resisting frames typically are steel structures with stiff, welded joints. The integrated framework of ductile steel offers much resistance to seismic forces. Moment-resisting frames are preferred by many architects today because they create far fewer internal obstructions than a building filled with shear walls. This allows more freedom for aesthetic design and internal space utilization.

Base Isolation

When the earth shakes, the energy is transferred to buildings. How can buildings be saved from this destructive energy? Modern designs employ **base isolation** where devices are placed on the ground or within the structure to absorb part of the earthquake energy. For example, visualize yourself standing on roller blades during an earthquake. Would you move as much as the earth? Base isolation uses wheels, ball bearings, shock absorbers, "rubber doughnuts," and other creative designs to isolate a building from the worst of the shaking (Figure 3.31). The

Sidebar

Design of Buildings in Earthquake-Prone Areas (Continued)

goal is to make the building react to shaking much like your body adjusts to accelerations and decelerations when you are standing in a moving train or bus. This concept has recently been used in building San Francisco's new airport terminal. The 115-million-pound building rests on 267 stainless steel sliders that rest in big concave dishes. When the earth shakes, the terminal will roll up to 20 inches in any direction.

Figure 3.31 The Office of Disaster Preparedness in San Diego County is housed in a 2-story, 7,000 ft^2 building sitting on top of 20 lead-impregnated rubber supports (base isolators) that each weigh 1 ton.
Photo by Pat Abbott.

Table 3.6 Magnitude versus Length of Shaking

Richter Magnitude	Duration of Strong Ground Shaking in Seconds
8–8.9	30 to 90
7–7.9	20 to 50
6–6.9	10 to 30
5–5.9	2 to 15
4–4.9	0 to 5

and lives lost. Consider the shaking times in Table 3.6. For example, if a magnitude 7 earthquake shakes vigorously for 50 seconds, rather than 20, the increase in damages and lives lost can be enormous.

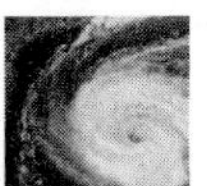

A Case History of Mercalli Variables

The San Fernando Valley, California, Earthquake of 1971

The San Fernando Valley (Sylmar) earthquake of 9 February 1971 occurred within the northwestern part of the Los Angeles megalopolis at 6 A.M. causing 67 deaths (including nine heart attacks). One of the most critical factors in determining life loss from earthquakes is the time of day of the event. In California, the best time for an earthquake for most people is when they are at home; the typical one- and two-story, wood-frame houses are usually the safest buildings to occupy.

Earthquake Magnitude The magnitude was 6.5 on the Richter scale with 35 aftershocks of magnitude 4.0 or higher occurring in the first seven minutes after the main shock. This is a lot of energy to release within an urban area.

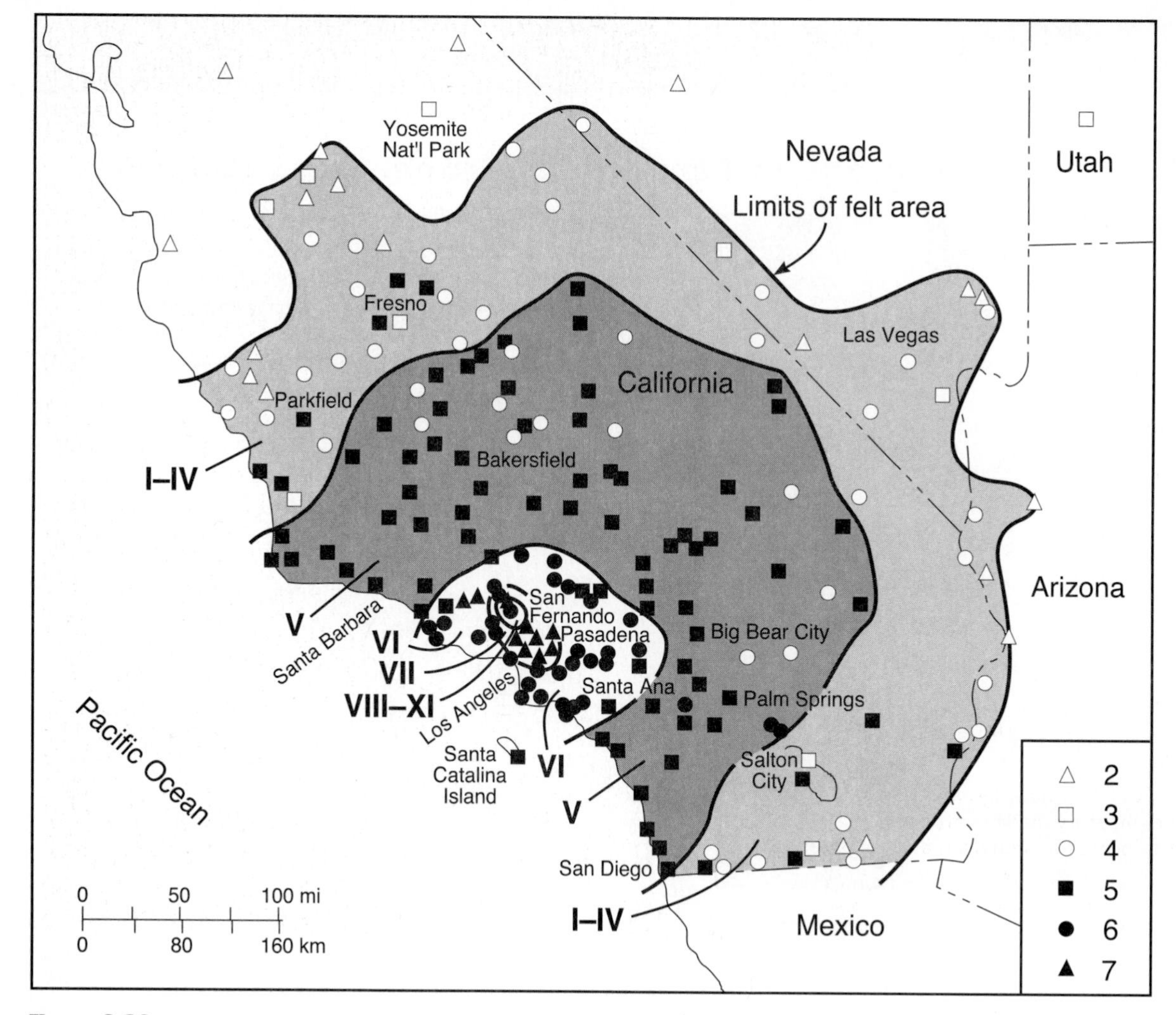

Figure 3.32 Contour map of Mercalli intensities from the San Fernando earthquake of 9 February 1971 shows an overall decrease in intensity away from the epicenter.

Distance from Epicenter The distance from the epicenter was a fairly consistent variable in this event. A rather regular bull's-eye pattern resulted from contouring the damages reported in Mercalli numerals (Figure 3.32).

Foundation Materials The types of foundation materials were not a major factor in this event.

Building Style Poorly designed buildings, bridges, and dams were the major problem. Three people died at the Olive View Hospital with the collapse of its "soft" first story featuring large plateglass windows. "Soft" first-story buildings support the heavy weight of upper floors without adequate shear walls or braced frames to withstand horizontal accelerations (Figure 3.33a). Many of these buildings still exist, despite their known high odds of failure during earthquakes (Figure 3.33b).

Another hospital failure was responsible for 47 deaths. Some of the pre-1933 buildings at the Veterans Administration Hospital used hollow, clay-tile bricks to build walls designed to carry only a vertical load. Many of the hollow-core clay bricks shattered under horizontal accelerations measured up to 1.25 times gravity.

Freeway bridges collapsed and took three lives. A freeway bridge is a heavy horizontal mass (roadbed) suspended high atop vertical columns. Swaying of these top-heavy masses, which have poor connections between their horizontal and vertical elements, resulted in collapse as support columns moved out from under elevated roadbeds (Figure 3.34). The lessons learned from these 1971 failures had not been acted upon by 1989, when the Interstate 880 elevated roadway collapsed in Oakland during the World Series earthquake and killed 42 people.

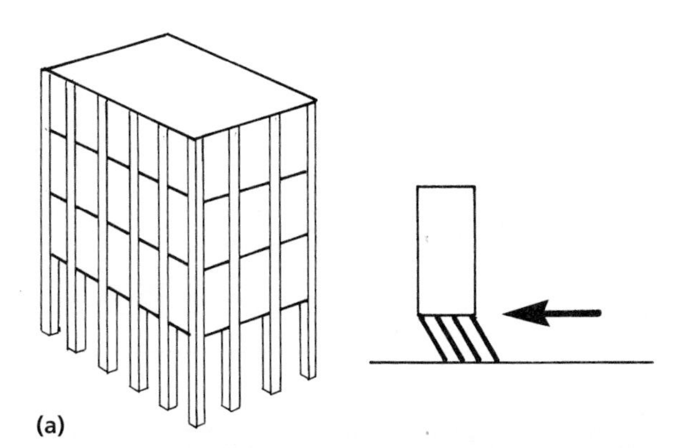

(a)

(b)

Figure 3.33 Buildings with "soft" first stories. **(a)** There is inadequate bracing on the first floor and no shear walls to transmit seismic loads to the ground. Thus, seismic stresses are concentrated at the join between the first and second floors, causing the building to move sideways and flatten the first story. **(b)** This is an eight-story medical-office building atop a "soft" first story. It is located in a California city near active faults.

(a) "Improving Seismic Safety of New Buildings," 1986, Federal Emergency Management Agency.

(b) Photo by Pat Abbott.

(a)

(b)

Figure 3.34 Bridge collapses. **(a)** Collapse of freeway bridge crushed two people in a truck. **(b)** If the earthquake had occurred later than 6 a.m., even more people would have been killed by these bridge collapses.

Photos by Al Boost.

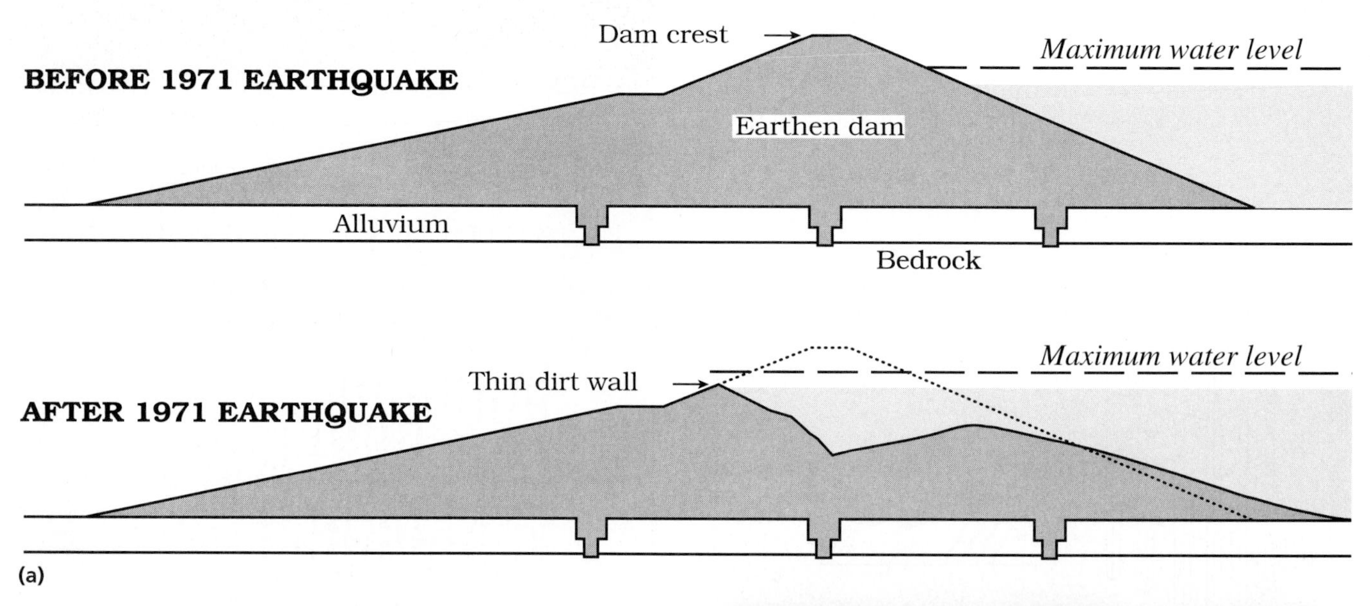

Duration of Shaking The strong ground shaking lasted 12 seconds. Earthquakes in the magnitude 6 range typically shake from 10 to 30 seconds (Table 3.6). The significance of the relatively short time of strong shaking in San Fernando Valley is enormous. The Lower Van Norman Reservoir held 11,000 acre-feet of water at the time of the quake. Its dam was begun in 1912 as a hydraulic-fill structure where sediment and water were poured into a frame to create a large mass; this is not the way to build a strong dam. During the earthquake, the dam began failing by landsliding and had lost 30 feet of its height (800,000 cubic yards of its mass) and stood only 4 ft above the water level when the shaking stopped (Figure 3.35). What if the strong shaking had lasted another five seconds? Below the dam lived 80,000 people.

Learning from the Past, Planning for the Future The 1971 San Fernando Valley earthquake unequivocally demonstrated the hazard in the region. It has been eloquently stated that "past is prologue" and that "those who do not learn the lessons of history are doomed to repeat them." How well were the lessons of 1971 learned? Another test was painfully administered on 17 January 1994, when the magnitude 6.7 Northridge earthquake struck the immediately adjacent area (see Chapter 4). This time, 61 people died and damages escalated to $30 billion. The same types of buildings again failed, and freeway bridges again fell down. The lessons from 1971 were poorly learned.

(b)

Figure 3.35 Failure of the Lower Van Norman Dam. **(a)** Landsliding lowered dam by 30 feet. **(b)** A few more seconds of strong shaking would have unleashed the deadly force of 11,000 acre-feet of water on San Fernando Valley residents below.

(a) Data Source: U.S. Geological Survey Fact Sheet 096–95, "The Los Angeles Dam Story," January 1995.

(b) Photo by Al Boost.

Tsunami

The biggest, most feared waves of all pass mostly unnoticed across the open sea, then rear up and strike the shoreline with devastating blows. The country with the most detailed history of these killer waves is Japan, and the waves are known by the Japanese word **tsunami** (*tsu* = harbor; *nami* = waves). Being a Japanese word, it is the same in both singular and plural; there are no tsunamis, just as there are no sheeps. The reference to harbor waves emphasizes the greater heights that waves reach in inlets and harbors because the narrowed topography focuses the waves into smaller spaces. For example, an 8 m (over 25 ft) high wave on the open coast may be forced to heights of 30 m (100 ft) as it crowds into a narrow harbor.

A deadly example hit Japan on 15 June 1896, a summer day when fishermen were out to sea and beaches were crowded with vacationers. An offshore earthquake swayed the sea floor; then about 20 minutes later, the sea withdrew, only to return in 45 minutes with a sound like a powerful rainstorm. Tsunami hit all the beaches hard but reached their greatest heights of 29 m (95 ft) where they crowded into narrow inlets. The tsunami destroyed more than 10,000 homes, killing over 27,000 people. The fishermen on the open ocean did not feel the earthquake or the tsunami; they learned of it when they sailed back into a bay littered with the wreckage of their houses and the bodies of their families.

In the United States, tsunami are usually called "tidal waves," but this is rather silly as tsunami have nothing to do with the tides. Nor do tsunami have anything to do with winds or storms; they are created by big "splashes" made in the deep ocean by fault movements, volcanic eruptions or caldera collapses, landslides, meteorite impacts, and such. You can approximate a tsunami by throwing a rock into a still body of water, sending off trains of concentric waves heading away from the impact point. Imagine the size of the waves formed when a really big rock drops in the water—like the caldera collapses of the volcanoes Krakatau in 1883 or Santorini around 1628 B.C.E. (see Chapter 7).

Tsunami are most commonly created during earthquakes, more specifically subsea fault movements with pronounced vertical offsets of the sea floor that disturb the deep ocean-water mass. Water is not compressible; it cannot easily absorb the fault-movement energy, therefore the water transmits the energy throughout the ocean in the waves we call tsunami. It is the vertical-fault movements at subduction zones that most commonly cause tsunami, mostly in the Pacific Ocean (Table 3.7).

The largest historic wave known occurred on 9 July 1958, when a massive rockfall dropped into Lituya Bay, Alaska. The wave (more properly called a **seiche**) sent a wall of water up the opposite slope of the bay, stripping the forest off the slopes at elevations up to 520 m (1,700 ft) above sea level. The most powerful waves of all probably occur when a several-kilometer-diameter asteroid slams into the deep ocean.

Table 3.7 Some Notable Tsunami in Recent Times

Date		Cause	Height	Site	Deaths
1 November	1755	Earthquakes	10 m	Lisbon, Portugal	30,000
21 May	1792	Landslide	10 m	Japan	>14,000
27 August	1883	Krakatau collapse	35 m	Indonesia	36,000
15 June	1896	Earthquake	29 m	Japan	27,000
2 March	1933	Earthquake	20 m	Japan	3,000
1 April	1946	Earthquake	15 m	Alaska	175
22 May	1960	Earthquake	10 m	Chile	> 1,250
27 March	1964	Earthquake	6 m	Alaska	125
1 September	1992	Earthquake	10 m	Nicaragua	170
12 December	1992	Earthquake	26 m	Indonesia	> 1,000
12 July	1993	Earthquake	31 m	Japan	239
2 June	1994	Earthquake	14 m	Indonesia	238
17 July	1998	Landslide	15 m	Papua New Guinea	>2,200

Tsunami versus Wind-Caused Waves

The typical ocean waves created by winds vary during the course of a year. Although the **periods** and **wavelengths** of wind-blown waves vary by storm and season, they are distinctly different from those of tsunami (Table 3.8).

The contrasts in velocities of wind waves versus tsunami are also great. In deep water, the velocities of the wind-caused waves in Table 3.8 are about 17 mph for a 5-second wave and about 70 mph for a 20-second wave. The velocity of tsunami may be calculated from:

$$v = \text{the square root of } g \times D$$

where v equals wave velocity, g equals acceleration due to gravity (9.8 m/sec^2 or 32 ft/sec^2), and D equals depth of ocean water. The Pacific Ocean has an average depth of 5,500 m (18,000 ft). Calculating the square root of g times D yields a tsunami velocity of 232 m/sec (518 mph).

Tsunami have such long wavelengths that they are always dragging across the ocean bottom, no matter how deep the water. The ocean basin has enough topography on its bottom to slow most tsunami down to the 420 to 480 mph range. Upon nearing a shoreline, the increased bottom friction and internal turbulence of water slow the rush of tsunami, but they still may be moving at freeway speeds. Tsunami arrive as a series of several waves separated by periods typically in the 10 to 60 minute range. The waves are

Table 3.8 Representative Wave Periods and Lengths

		Periods	Lengths	
Wind-blown Ocean Waves	short:	5 seconds	39 m	(130 ft)
	medium:	10 seconds	156 m	(510 ft)
	long:	20 seconds	624 m	(2,050 ft)
Tsunami	long:	3,600 seconds (60 minutes)	837,000 m	(2,750,000 ft) (520 mi)

(a)

(b)

Figure 3.36 **(a–b)** Tsunami rushing across the shore and heading inland east of Hilo, Hawaii, on 1 April 1946. Photos courtesy of NOAA.

typically a meter or so in height in the open ocean and 6 to 15 m (20 to 50 ft) high on reaching shallow water, except where topography, such as bays and harbors, focuses the energy to create much taller waves.

At the Shoreline What does a tsunami look like when it comes onshore? Does it resemble the animation in the Hollywood movie "Deep Impact" where a beautiful symmetrical wave curves high above the buildings of New York City? No. A tsunami arriving at the shoreline does not look like a gigantic version of the breaking waves we see every day. A typical tsunami hits the coastline like a very rapidly rising tide or white-water wave, but it does not stop on the beach, it keeps rushing inland (Figure 3.36). There is a tendency to view the destructive power of tsunami as being mostly due to the great height of their waves, but the height of tsunami commonly are not as important as the momentum of their large masses separated by ultralong wavelengths and periods (Table 3.8). Visualize a flat or gently sloping coast hit by tsunami with a 60-minute period. The tsunami can rush inland causing destruction for about 30 minutes before the water is pulled back to help form the next wave (Figure 3.37).

Figure 3.37 Tsunami damage to Hilo, Hawaii, following the magnitude 9.5 Chile earthquake, 22 May 1960. Notice the Pacific Ocean in the background and how far inland the tsunami traveled. Photo courtesy of NOAA.

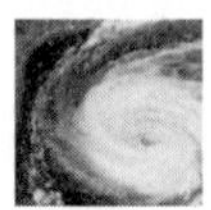

Tsunami Case Histories

Some important understanding about tsunami may be learned through case histories.

Alaska, 1 April 1946

As 1 April 1946 began in the Aleutian Islands, two large subduction movements occurred and shook the area severely. The five workers in the Scotch Gap lighthouse were wide awake and wondered what lay ahead during the dark night. The lighthouse was built of steel-reinforced concrete, and its base sat 14 m (46 ft) above mean low-water level (Figure 3.38a). Around 20 minutes after the second earthquake, a tsunami about 30 m (100 ft) high swept the lighthouse away (Figure 3.38b). This time, the first wave was the biggest; it killed all five men.

Tsunami are not just local events. The waves race across the entire Pacific Ocean at speeds approaching those of commercial jet airplanes. As tsunami enter into shallower waters, bottom friction slows them markedly. The April Fool's Day tsunami traveled about 485 mph in the deep ocean, slowing

(a)

(b)

Figure 3.38 The Scotch Gap lighthouse in the Aleutian Islands of Alaska **(a)** before and **(b)** after the tsunami unleashed by a magnitude-7.3 earthquake on 1 April 1946.
Photos courtesy of NOAA.

to about 35 mph as it neared shore in Hilo, Hawaii. Humans are no match for these massive waves (Figure 3.39); the fastest sprinter alive runs less than 25 mph. The long wavelengths also allow the crest to rush onland for long distances; there is no trough immediately behind, waiting to pull the water back to the ocean. The next tsunami may be an hour away. The 1946 tsunami killed 159 people in Hilo, Hawaii.

Sitting in the middle of the earthquake-prone Pacific Ocean basin, Hawaii receives numerous tsunami. The height of tsunami at the shoreline varies due to local topography. A tsunami hazard map has been prepared for Hawaii (Figure 3.40).

Figure 3.39 Tsunami breaking over Pier 1 in Hilo, Hawaii, on 1 April 1946. Man in foreground became one of 159 killed.
Photo courtesy of NOAA.

Chile, 22 May 1960

The most powerful earthquake ever measured occurred in Chile on 22 May 1960. Tsunami generated by this magnitude 9.5 subduction movement killed people throughout the Pacific Ocean basin. In Chile, the main seism broke loose at 3:11 P.M. on Sunday. Chileans are familiar with earthquakes, so many people headed for high ground in anticipation of tsunami. About 15 minutes after the seism, the sea rose like a rapidly rising tide, reaching 4.5 m (15 ft) above sea level. Then the sea retreated with speed and an incredible hissing and gurgling noise, dragging broken houses and boats out into the ocean. Some people took the "smooth wave" as a sign that these tsunami could be ridden out at sea, thus saving their boats. About 4:20 P.M., the second tsunami arrived

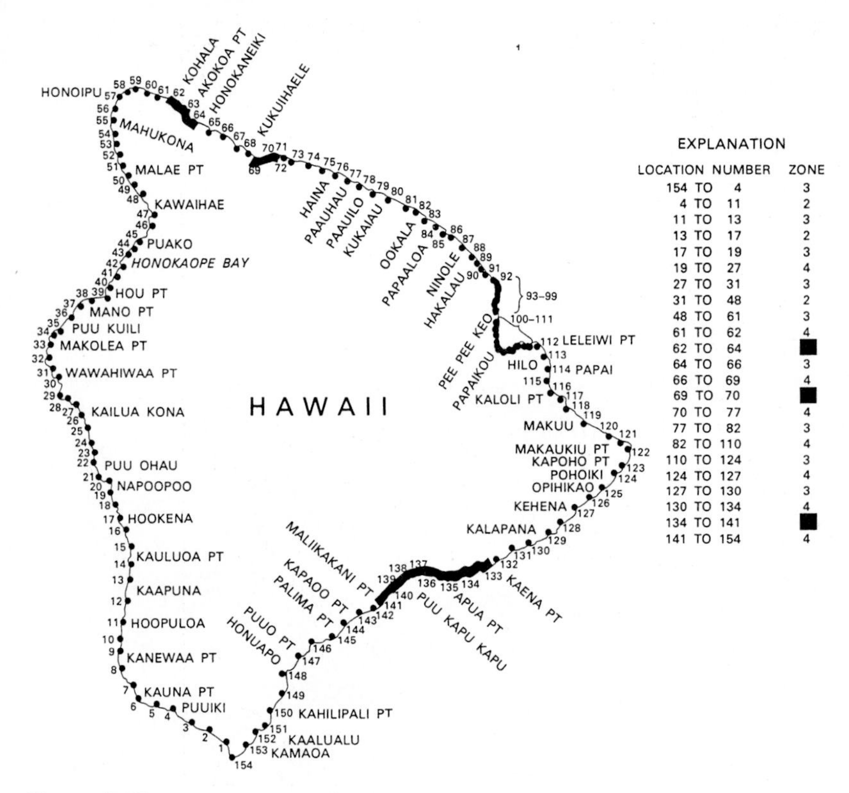

Figure 3.40 Map of tsunami hazard for the island of Hawaii. Black areas of coastline (zone 5) receive tsunami greater than 15 m (50 ft) high. Zone 4 tsunami may be greater than 9 m (30 ft) high, zone 3 tsunami greater than 4.5 m (15 ft), and zone 2 tsunami greater than 1.5 m (5 ft) high.

Source: *U.S. Geological Survey Professional Paper 1240B.*

as an 8 m (26 ft) high wave traveling at 125 mph. The wave crushed boats and their terrified passengers, as well as wrecked coastal buildings. But the third wave was the largest; it rose 11 m (35 ft) high, but it traveled at only half the speed of the second wave. Over 1,000 Chileans died in these tsunami.

Since 1960, Hawaiians are given warnings before tsunami arrive. The Pacific Tsunami Warning Center evaluates large earthquakes in the Pacific Ocean and then, using maps like Figure 3.41, provides people with hours of advance warning, including shrieking sirens. The Chilean tsunami was predicted to arrive in Hawaii at 9:57 A.M.; it arrived at 9:58 A.M. But 61 people drowned anyway, including sightseers attracted to the shore to watch the tsunami roll in (Figure 3.39). The tsunami raced on to Japan, where it killed another 185 people. The amount of energy in this set of tsunami was so great that it was recorded on Pacific Ocean tide gauges for a week as the energy pulses bounced back and forth across the entire ocean basin.

Alaska, 27 March 1964

The Good Friday earthquake in Alaska was a magnitude 9.2 monster whose tsunami ravaged the sparsely populated Alaska coastline, killing 122 people. The immense masses of land raised up and dropped down by this subduction-earthquake series (Figure 4.3) sent off large tsunami (Figure 3.41). Three hours later, Port Alberni on Vancouver Island was hit by 6+ m (21 ft) high tsunami that destroyed 58 buildings and damaged 320 others. Thanks to advance warning, the Crescent City, California, waterfront area was evacuated, and residents waited upslope while tsunami arrived and did their damage. After watching four tsunami and seeing their sizes, many people could no longer stand the suspense of not knowing the condition of their properties. Some people went down to check it out—a big mistake. In this tsunami series, the fifth wave was the biggest; it was 6.3 m (21 ft) high, and it killed 12 of the curious people. All the fatalities of this event at Crescent City were caused by the fifth tsunami.

Tsunami arrive as a series of several waves. Which wave in the series will be the biggest? It is unpredictable; in the three case histories above, the biggest wave was the first, third, and fifth, respectively.

Nicaragua, 1 September 1992

On a Tuesday evening at 6:16 P.M., an unusual earthquake occurred that was large (magnitude 7.6) but barely felt.

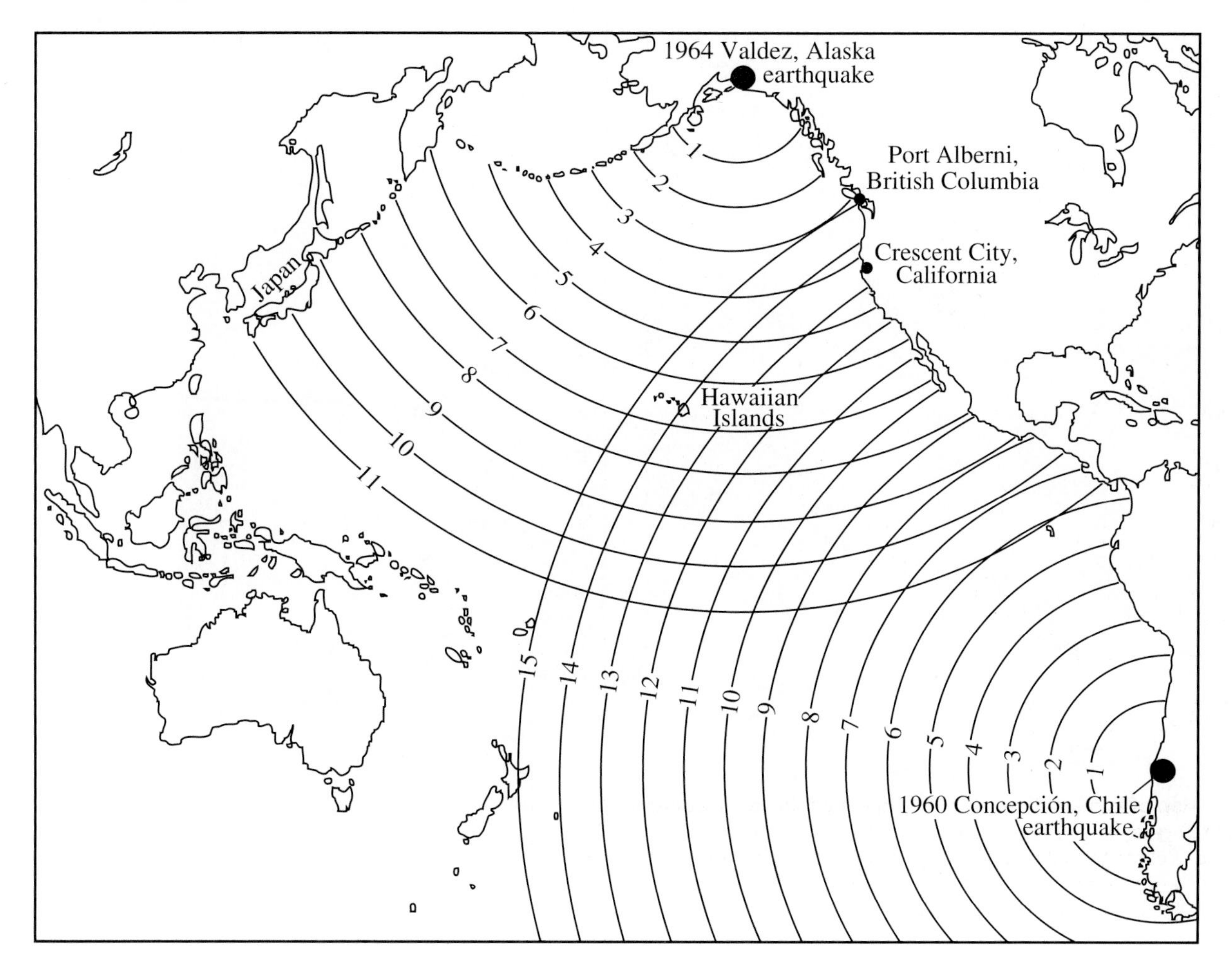

Figure 3.41 Tsunami travel times to Hawaii in hours.
Data Source: Kious, W. J. and Tilling, R. I., *The Dynamic Earth,* p. 77, U.S. Geological Survey.

About 45 minutes later, a 10 m (33 ft) high tsunami ravaged a 300 km (185 mi) long section of Nicaraguan coastline. What happened? Why did the ground shake very little yet the ocean water was agitated into large tsunami? The fault moved so slowly that the high-velocity, short-period seismic waves were relatively weak and shook the ground only slightly, about 1/100 of what is expected. However, the slow-moving fault released a lot of long-period energy into the water thus creating powerful tsunami.

The earthquake was a subduction event where a 100 km (62 mi) long segment of oceanic plate moved 1 m in two minutes. The slow-motion fault movement was especially efficient at pumping energy into water. The seawater absorbed the energy, sending tsunami onto the beach. The coastal residents were caught without warning; 13,000 homes were destroyed, and 170 people were killed, mostly sleeping children.

Papua New Guinea, 17 July 1998

At 6:49 P.M. on Friday evening, the north shore of Papua New Guinea, was rocked by a 7.1 magnitude earthquake occurring about 20 km (>12 mi) offshore. As the shaking ended, witnesses describe seeing the sea rise above the horizon and shoot spray 30 m (100 ft) high. Sounds were heard like distant thunder and then the sea slowly pulled back. About 4 to 5 minutes later, a rumbling sound was heard and a tsunami about 4 m (13 ft) high was seen approaching. But if you can see the wave coming, it is too late to escape. Many people living on the barrier beach were washed into the lagoon (Figure 3.42).

Several minutes later a second wave approached, but this one was about 14 m (45 ft) tall. A tsunami does not have the shape of a typical wave; it is more like a pancake of water. This tsunami averaged about 10 m (33 ft) high and measured 4 to 5 km (2 to 3 mi) across. Visualize this thick pancake of water pouring over the heavily populated beach at 15 miles per hour for over a minute. This tsunami event was a three-wave sequence that washed thousands of people and their homes into the lagoon. A barrier beach that hosted four villages was swept clean. The estimated 2,200 fatalities were mostly from those least able to swim—the children.

Figure 3.42 Tsunami swept villages from this sandbar into Sissano Lagoon, drowning 2,200 people on 17 July 1998. Does this sandbar remind you of coastal New Jersey, North Carolina, Balboa Island in southern California, or other sites?
Photo courtesy of NOAA.

The Papua New Guinea tsunami apparently was not caused directly by the earthquake but by a submarine landslide triggered by the shaking. This event has caused a global rethinking of the tsunami threat. Big tsunami can no longer just be considered as events caused by giant earthquakes in distant places. Tsunami can be created by smaller local faults that cause unstable sand and rock masses to slip and slide under water.

Tsunami Hazards from Landslides There are numerous sites around the world that could experience Papua New Guinea-style, landslide-generated tsunami. For example, southern California is protected from Pacific Ocean basin tsunami by its offshore system of subparallel island ridges. But these ridges have been created by active faults and their movements can trigger undersea landslides that send tsunami across the densely populated southern California coastline.

Beneath the Atlantic Ocean off the east coast of North America, new images of the sea floor show significant scars where big submarine landslides occurred. These images revived the memory of a landslide-caused tsunami that hit Newfoundland on 18 November 1929 following the magnitude 7.2 Grand Banks earthquake that occurred offshore. The tsunami arrived in three major pulses onto the Burin Peninsula in Newfoundland. Waves 3+ m (10 ft) high and a water-level runup of 13 m (43 ft) above sealevel killed 28 people and caused widespread destruction to fishing communities and the economy of Newfoundland.

Tsunami or seiche hazards exist on lake shores as well. For example, beautiful Lake Tahoe sits high in the Sierra Nevada in California and Nevada. The lake is 35 km (22 mi) long, 19 km (12 mi) wide and over 500 m (1,600 ft) deep. This broad and deep lake has been created by subparallel normal faults dropping the land between them (Figure 3.43). The faults are active. The underwater faults have a 3 to 4 percent probability of a magnitude 7 earthquake in the next 50 years. The likely fault movement could drop the lake bottom about 4 m (13 ft) and could generate 10 m (33 ft) high waves that rush over the populated shoreline. Figure 3.43 also shows debris on the lake bottom from a huge landslide that would have generated even bigger tsunami. The tsunami hazard at Lake Tahoe is another example worthy of the frequency versus magnitude discussion in Chapter 1.

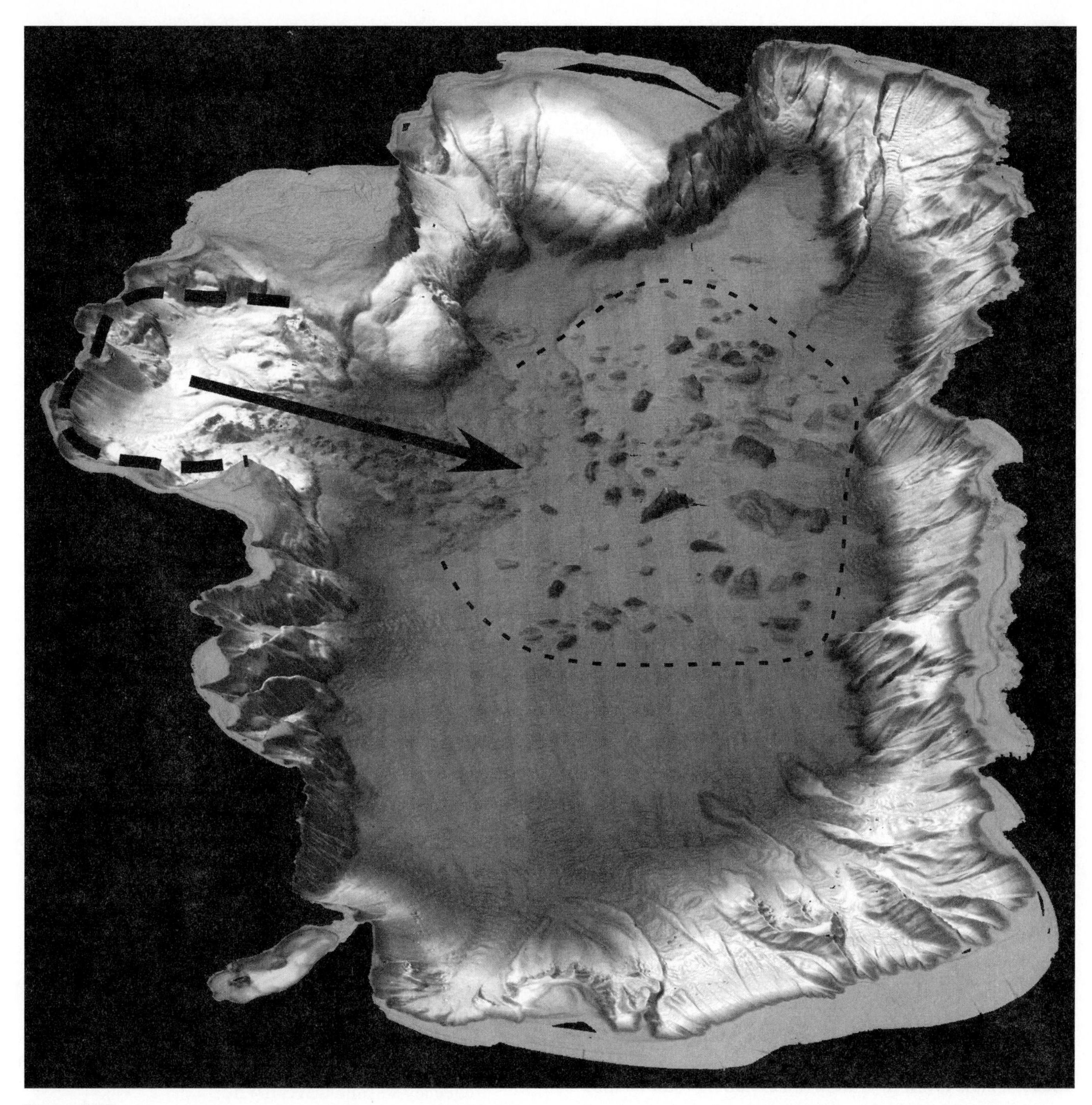

Figure 3.43 Lake Tahoe basin is created by down dropping between active faults. Notice the amphitheater (bold dashed lines) on the west side of the lake; it formed when a giant landslide pulled away and dumped debris onto the lake bottom in a tsunami-forming event. Figure from Graham Kent, Scripps Institution of Oceanography.

Summary

Earthquakes are shaking ground caused most often by sudden movements along cracks in the earth called faults. Some major faults acting for millions of years have offset rock layers by hundreds of kilometers. Sedimentary rock layers originally are continuous, horizontal, and in superpositional order (oldest on bottom, youngest on top); however, fault movements cut rocks into discontinuous masses, and in places, fault deformation has tilted rock layers and even overturned the superpositional sequence. Geologists measure the 3-D orientation of rock layers via dip (angle and direction of inclination) and strike (compass bearing of rock cutting a horizontal plane).

Dip-slip fault types have dominantly vertical movements. Normal faults are due to extensional (pull-apart) forces. Reverse faults are due to compressional (push-together) forces. Strike-slip fault types have dominantly horizontal offsets. Straddling the fault, if the right-hand side moves toward you, it is a right-lateral fault; if the left-hand side moves toward you, it is a left-lateral fault. Bends (steps) in strike-slip faults cause the land to either uplift or downdrop. Another type of fault connects offset spreading-center segments—transform faults.

Earthquakes (seisms) disperse their energy in seismic waves that radiate away from the hypocenter or point of fault rupture. The point on the surface above the fault rupture is the epicenter. Some seismic waves pass through the body of the Earth—the P waves (primary waves with a push-pull motion) and S waves (secondary waves with a shearing motion). Other seismic waves travel along the surface (Love and Rayleigh waves).

Earthquake energy is assessed by its magnitude. Different estimates of magnitude are derived from different methods based on local shaking (Richter scale), body waves (m_b), surface waves (M_s), or seismic moment (M_w). Earth has more than a million earthquakes each year, but more than 90 percent of the energy is released by the 12 to 18 largest events.

Seismic waves have different periods (time between cycles) and frequencies (number of cycles per second): period = 1/frequency. P waves commonly have from one to 20 cycles per second; surface waves commonly have one cycle every 1 to 20 seconds. Where the frequencies of seismic waves match the vibration frequencies of foundations and buildings, destruction may be great.

Earthquake effects on structures and people are assessed via the Mercalli Intensity Scale. Its variables are earthquake magnitude, distance from the hypocenter/epicenter, type of rock or sediment foundation, building style, and duration of shaking. Mercalli intensities are of more than just scientific interest because earthquakes don't kill, buildings do.

Tsunami are the biggest waves of all. Earthquakes, volcanic eruptions, landslides, and asteroid impacts disturb the deep ocean-water mass, sending off energetic waves. Earthquake-generated tsunami commonly travel almost 500 mph and may be spaced as much as 60 minutes apart. Tsunami slow down in shallow water but may still be moving at freeway speeds. Local topography, as in harbors and inlets, may focus tsunami energy, creating waves over 100 ft high that kill thousands of people.

Terms to Remember

acceleration 77
amplitude 70
base isolation 82
B.C.E. 87
body wave 70
C.E. 59
cross section 63
dip 63
dip-slip fault 63
ductility 81
fault 61
footwall 63
fracture 61
frequency 70
granite 71
hangingwall 63
hertz 70
hypocenter 65
inertia 68
joint 62
Law of Original Continuity 61
Law of Original Horizontality 61
Law of Superposition 61
left-lateral fault 64
magnitude 74
map 63
normal fault 63
period 70
permeability 63
primary or P wave 70
resonance 78
reverse fault 64
right-lateral fault 64
secondary or S wave 71
seiche 87
seism 67
seismic wave 70
seismogram 73
seismograph 68
seismology 67
seismometer 68
stress 61
strike 63
strike-slip fault 64
surface wave 70
transform fault 67
tsunami 87
wavelength 70

Questions for Review

1. Draw a cross section of a sequence of sedimentary rock layers. Label and explain the Laws of Original Continuity, Original Horizontality, and Superposition.
2. Draw cross sections of a normal fault and a reverse fault. What are the differing forces that determine which one forms?
3. Draw a map of a left-stepping, right-lateral fault. Explain what happens to the land at the step (bend) in the fault.
4. Sketch a map of a strike-slip and a transform fault. Explain their similarities and differences.
5. What do P and S seismic waves tell us about the nature of the Earth's interior?
6. How can arrival times of P and S waves be used to determine distance to the epicenter?
7. How are foreshocks distinguished from aftershocks?
8. What are the differences between earthquake magnitude and earthquake intensity?
9. Will a tall building be affected more by high- or low-frequency seismic waves? Why?
10. How fast can tsunami travel? What are typical tsunami wavelengths and periods? What causes tsunami?
11. Is the killing power of tsunami due mostly to the wave heights or to the long wavelengths that allow them to travel far inland?

Questions for Further Thought

1. Draw a cross section showing an inclined fault with a hypocenter at 15 km (9 mi) depth. Does the epicenter plot on the surface trace of the fault?

2. What is the quake potential of the Moon (moonquakes)? Does the Moon have similar numbers and magnitudes of quakes as the Earth? Why?
3. If you are in an airplane over the epicenter of a great earthquake, what will you experience?
4. How earthquake safe is your home or office? What are the nearest faults? What kind of earth materials is your home or office built upon? How will your building size, shape, and materials react to shaking? What nearby features could affect your home? What hazards exist inside your home?
5. Make a list of the similarities between snapping your fingers and the earthquake movement of a fault.
6. What are the four main causes of tsunami (see text)? Make two rank-order lists (from #1 to #4) answering these questions. Tsunami are most frequently caused by ______? Tsunami are largest when caused by _______?

Suggested Readings and References

Bernstein, Joseph. (1954, August). Tsunamis. *Scientific American,* 191, 60–64.

Bolt, B. A. (2000.) *Earthquakes* (4th ed). New York: W. H. Freeman.

Bryant, E. (2001). *Tsunami: The Underrated Hazard.* New York: Cambridge University Press.

Building Seismic Safety Council. (1986). *Improving Seismic Safety of New Buildings: A Community Handbook of Societal Implications.* Washington, D.C.: Federal Emergency Management Agency.

Driscoll, N. W., Weissel, J. K., and Goff, J. A. (2000). Potential for large-scale submarine slope failure and tsunami generation along the U.S. mid-Atlantic coast. *Geology,* 28, 407–410.

Dudley, W. C., and Lee, M. (1998). *Tsunami.* Honolulu: Univ. Hawaii Press.

Gere, J. M., and Shah, H. C. (1984). *Terra Non Firma.* New York: W. H. Freeman.

Holmes, A. (1965). *Principles of Physical Geology.* New York: Ronald Press.

Michael, A., Ross, S., and Schaff, D. (2002). The music of earthquakes. *American Scientist.*

Myles, Douglas. (1985). *The Great Waves.* London: Robert Hale.

Shelton, J. S. (1966). *Geology Illustrated.* New York: W. H. Freeman.

USGS Fact Sheet 030-01 (2001). Did you feel it? Community-made earthquake shaking maps.

Yanev, P. (1974). *Peace of Mind in Earthquake Country.* San Francisco: Chronicle Books.

Yeats, R. S., Sieh, K., and Allen, C. R. (1997). *The Geology of Earthquakes.* New York: Oxford University Press.

Videos

Raging Planet: Earthquake. (1997). Discovery Channel video (50 min.).

When the Earth Quakes. (1990). National Geographic (28 min.).

The Earth Revealed—Earthquakes. (1992). Annenberg/CPB Project (30 min.).

The Earth Revealed—Living with Earth. (1992). Annenberg/CPB Project (30 min.).

Earthquakes: Exploring Earth's Restless Crust. (1983). Encyclopedia Britannica (22 min.).

Killer Wave: Power of the Tsunami. (1997). National Geographic (60 min.).

Raging Planet: Tidal Wave (sic). (1997). Discovery Channel (50 min.).

Chapter 4

Some Earthquakes in Western North America

Nature seems to have moods in other than a poetical sense: predilections for certain deeds at certain times, without any apparent law to govern or season to account for them. She is read as a person with a curious temper; as one who does not scatter kindnesses and cruelties alternately, impartially, and in order, but heartless severities or overwhelming generosities in lawless caprice.

—Thomas Hardy, 1873, *A Pair of Blue Eyes*

Earthquakes affect the lives of people living in western North America (Figure 4.1). To understand the earthquakes that plague western North America, it is necessary to know the plate-tectonic history of the region. As the Atlantic Ocean basin opens further, both North and South America move westward into the Pacific Ocean basin, reducing its size (Figures 2.6 and 2.18). The positions of the plates and their movements through time are described with respect to mantle hot spots. The hot spots move at lesser rates than the oceanic plates and thus serve as semifixed reference points. At 30 million years ago, most of the northern portion of the Farallon plate had subducted beneath North America (Figure 4.2). At about 28 million years ago, the first segment of the Pacific spreading center made contact near today's United States–Mexico border. The remaining spreading centers to the north and south still operated as before. What connected these offset spreading-center segments? A transform fault, specifically the ancestor of the San Andreas fault.

In the last 5.5 million years, the Gulf of California has opened about 300 km (190 mi). This rifting action has torn Baja California and California west of the San Andreas fault (including San Diego, Los Angeles, and Santa Cruz) from the North American plate and piggybacked them onto the Pacific plate (Figure 4.2). The Gulf of California continues to open and is carrying the Californias on a Pacific plate ride at about 56 mm/yr (2.2 in/yr).

What is to be the fate of peninsular California? Will it break off in a giant earthquake and sink into the Pacific Ocean as some psychics predict? No! This gigantic rupture-and-sink process is impossible; it is fantasy. Remember isostasy? Continents are made of less dense rocks that float on top of denser mantle rocks. The slice of western California and Baja California will continue moving northwest toward a rendezvous with Alaska. Could the continental fragment of the Californias sink into the subduction zone along the southern edge of Alaska? No again. Isostasy prohibits the subduction of low-density continental masses into denser oceanic rocks. Instead, the Californias will plow into Alaska and become part of its southern margin. Southern California will switch from surfing beaches to ski slopes.

Today, North America has several small- to medium-sized plates subducting beneath its western edge (Figure 4.2). Much of the earthquake hazard in western North America is due to the ongoing subduction of these small plates, as well as the continuing effects of the overridden, but not forgotten, Farallon

Figure 4.1 Collapsed apartment building in Canoga Park, California on 17 January 1994. Building toppled sideways onto cars in parking lot.

Photo by Peter W. Weigand.

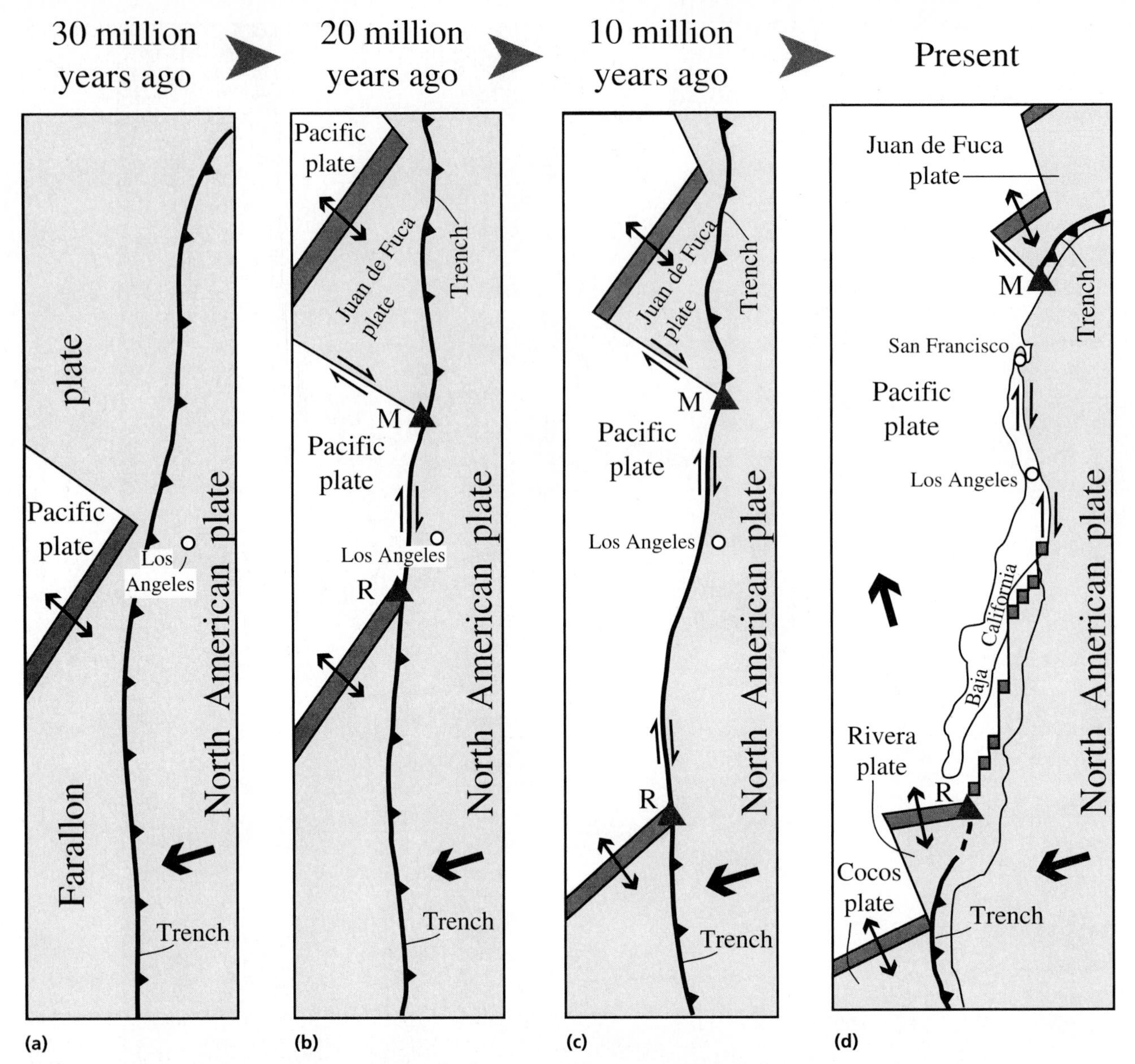

(a) (b) (c) (d)

Figure 4.2 Collision of the Pacific Ocean basin spreading center with the North American plate: **(a)** 30 million years ago—first spreading-center segment nears southern California, **(b)** 20 million years ago—growing transform fault connects remaining spreading centers, **(c)** 10 million years ago—the Mendocino (M) and Rivera (R) triple junctions continue to migrate north and south respectively, **(d)** present—the long transform fault is now known as the San Andreas fault. Interpretations based on work of Tanya Atwater.

Figure Source: Kious, W. J. and Tilling, R. I., *This Dynamic Earth:* U.S. Geological Survey, p. 77.

plate. When considering the size of the Pacific, North American, and Farallon plates, it is easy to appreciate why earthquakes affect the entirety of western North America. Consider that the Pacific plate is more than 13,000 km (8,000 mi) across and that it is grinding past the North American plate, which is over 10,000 km (6,250 mi) wide. How broad a zone is affected by these passing behemoths? Could all the stress be stored in just the rocks along the San Andreas fault system? Or must the affected zone be larger—as big as the entirety of western North America? Consideration of the scale of these gigantic plates strongly suggests that their interactions are an underlying cause of earthquakes throughout the western United States, Canada, and Mexico.

In this chapter, we will restrict our inquiry to the coastal zones to look at the seismicity associated with western North America spreading centers, transform faults, and subduction zones. In Chapter 5, we will look at earthquake history and potential for the rest of the United States and Canada.

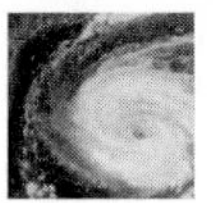

Subduction Zone Earthquakes

Most of the really large earthquakes in the world are due to subduction (Table 4.1). The insertion of an oceanic plate tens of kilometers thick into the mantle involves prodigious amounts of energy. In the United States, California is commonly called "earthquake country," but it is clear from Table 4.1 that Alaska is more deserving of this title. Over the past 5 million years, about 290 km (180 mi) of Pacific plate have been shoved under southern Alaska in the vicinity of Anchorage. This has created some shattering earthquakes.

Table 4.1 Earth's Largest Earthquakes, 1904–1984

Rank	Location	Year	M_w	Cause
1.	Chile	1960	9.5	Subduction—Nazca plate
2.	Alaska	1964	9.2	Subduction—Pacific plate
3.	Alaska	1957	9.1	Subduction—Pacific plate
4.	Kamchatka	1952	9.0	Subduction—Pacific plate
5.	Ecuador	1906	8.8	Subduction—Nazca plate
6.	Alaska	1965	8.7	Subduction—Pacific plate
7.	Assam	1950	8.6	Collision—India into Asia
8.	Banda Sea	1938	8.5	Subduction—Pacific/ Indian plate
9.	Chile	1922	8.5	Subduction—Nazca plate
10.	Kuril Island	1963	8.5	Subduction—Pacific plate

The Good Friday Earthquake, Alaska, 1964

Saint Matthew's account of the first Good Friday included: "And, behold . . . the earth did quake, and the rocks rent." His words applied again, over 1,900 years later, on Good Friday, 27 March 1964. At 5:36 P.M., in the wilderness at the head of Prince William Sound, a major subduction movement created a gigantic earthquake. This was followed in sequence by other downward thrusts at 9, 19, 28, 29, 44, and 72 seconds later as a nearly 1,000 km (more than 600 mi) long slab of 400 km (250 mi) width lurched its way deeper into the mantle. Alaska sits on the slab above the subducting Pacific plate; it was shifted horizontally (seaward) up to 19.5 m (64 ft) and was uplifted as much as 11.5 m (38 ft). This horizontal extension of the landward side resulted in another landward block being downdropped as much as 2.3 m (7.5 ft) (Figure 4.3). Some of the seafloor offsets were even greater. Over 110,000 square miles of land and sea bottom were involved in these massive movements. Hypocenter depths were from 20 to 50 km (12 to 30 mi).

The duration of strong ground shaking was a lengthy 3 to 4 minutes; it induced many avalanches, landslides, ground settlements, and tsunami. Of the 131 lives lost, 122 were due to tsunami. The town of Valdez was severely damaged by both ground deformation and a submarine landslide that caused tsunami, which destroyed the waterfront facilities. Damage was so great that the town was rebuilt at a new site. Anchorage, the largest city

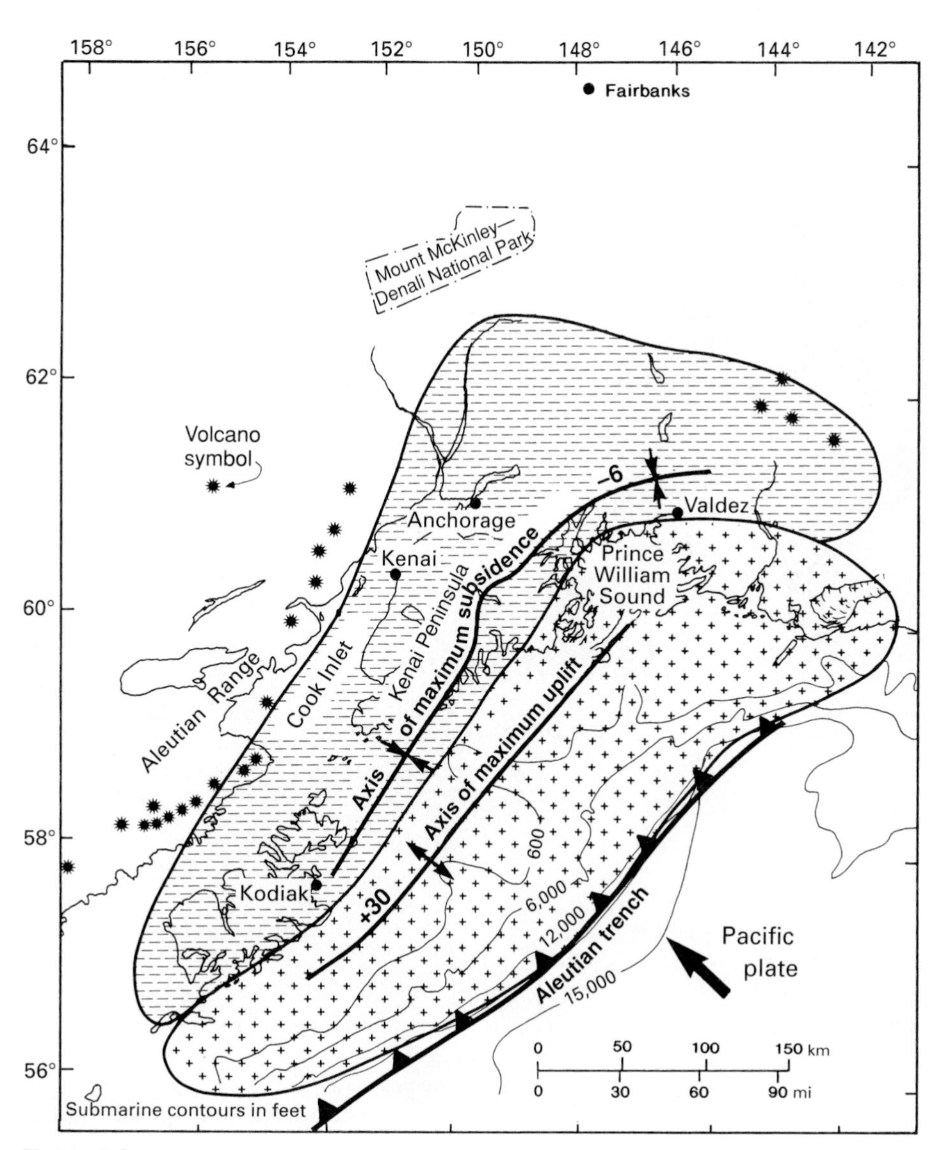

Figure 4.3 Map showing land uplifted and downdropped during the 1964 Alaska earthquake.

in Alaska, was heavily damaged by landslides. And yet, there were some elements of luck in the timing of this earthquake. It occurred late on a Friday, when few people were in heavily damaged downtown Anchorage; tides were low; it was the off-season for fishing so few people were on the docks or in the canneries; and the weather in the ensuing days was seasonally warm, thus sparing people from death-dealing cold while their homes and heating systems were out of order.

Had this earthquake occurred near a densely inhabited area, the dimensions of the human catastrophe would have been mind-boggling. As it was, the monetary damages were more than 100 times the price paid to Russia in 1867 to purchase Alaska.

Mexico City, 1985

On Thursday morning, 19 September 1985, most of the 18 million residents of Mexico City were at home, having their morning meals. At 7:17 A.M., a monstrous earthquake broke loose some 350 km (220 mi) away. Some seismic waves traveled far to deal destructive blows to many of the 6- to 16-story buildings that are heavily occupied during the working day (Figure 4.4). Building collapses killed about 8,000 people.

What caused this earthquake? The Cocos plate made one of its all-too-frequent downward movements. This time a 200 km (125 mi) long front, inclined 18° east, thrust downward and eastward about 2.3 m (7.5 ft) in two distinct jerks about 26 seconds apart (Figure 4.5). The main shock had a surface wave magnitude (M_S) of 8.1. It was followed on 21 September by a 7.5 M_S aftershock and another on 25 October of 7.3 M_S. The earthquakes were not a surprise to seismologists. Before these seisms occurred, the area was called the Michoacan seismic gap, and many instruments had been deployed in the region to measure the expected big event.

As Figure 4.5 shows, another large seismic gap waits to be filled by a major movement of the Cocos plate. The Guerrero seismic gap lies near Acapulco and is closer to Mexico City than the Michoacan epicenter.

Earthquakes Don't Kill, Buildings Do Many of the coastal towns near the epicenter received relatively small amounts of dam-

Figure 4.4 Collapsed (pancaked) seven-story building made of heavy concrete, Mexico City.
Photo courtesy of NOAA.

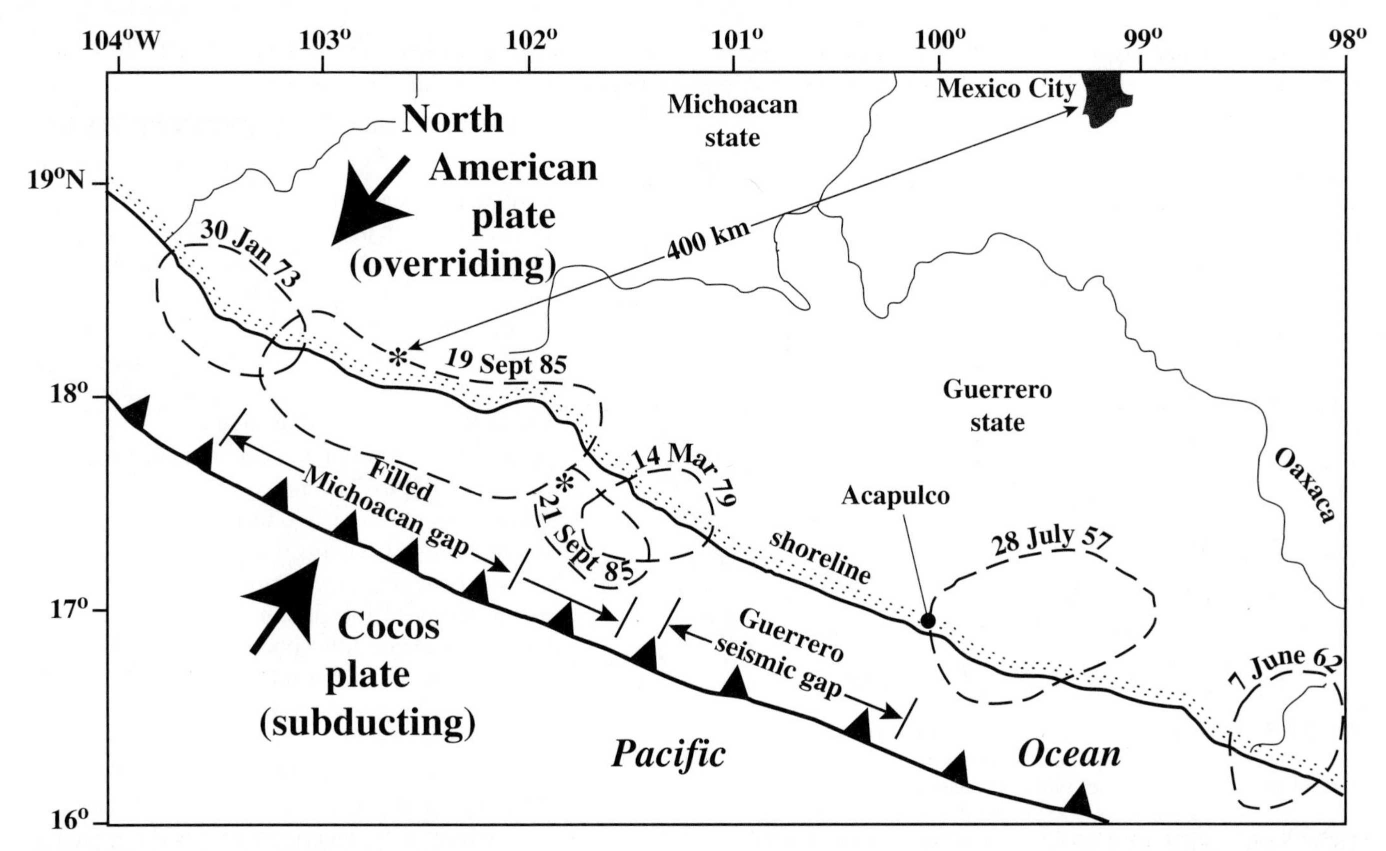

Figure 4.5 Map of coastal Mexico showing dates of earthquakes and fault areas moved (dashed lines) during Cocos plate subduction events. The Michoacan seismic gap was filled by the 1985 seisms. The Guerrero seismic gap is overdue for a major movement.

age. Yet in Mexico City, over 5,700 buildings were severely damaged, with 15 percent of them collapsing catastrophically. Why did so many buildings collapse and kill so many people when Mexico City lies 350 km (220 mi) from the epicenter? It was largely due to resonance between seismic waves, soft lake-sediment foundations, and improperly designed buildings. The duration of shaking was increased due to seismic energy being trapped within the soft sediments.

Mexico City is built atop the former Aztec capital of Tenochtitlan. The Aztecs built where they saw the favorable omen—an eagle sitting on a cactus and holding a writhing snake in its mouth. The site was Lake Texcoco, a broad lake surrounded by hard volcanic rock. Over time, the lake basin was partially filled with soft, water-saturated clays. Portions of Lake Texcoco have been drained, and large buildings have been constructed on the weak, lake-floor sediments.

Building damages were the greatest and the number of deaths the highest where three factors combined and created resonance: *Factor 1*—The earthquakes sent a tremendous amount of energy in seismic waves in the 1-to-2-second frequency band; *Factor 2*—The areas underlain by thick, soft clays vibrating at 1-to-2-second frequencies amplified the seismic waves (Figure 4.6); and *Factor 3*—Buildings of 6 to 16 stories vibrated in the 1-to-2-second frequency band. Where all three factors were in phase, disaster struck.

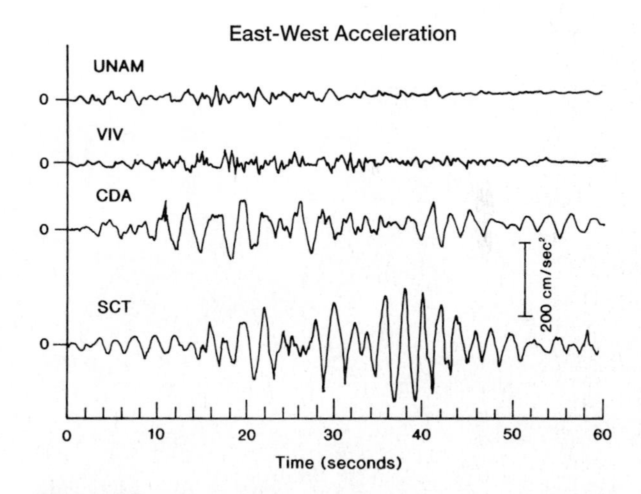

Figure 4.6 Some east-west accelerations recorded in Mexico City in 1985. The Universidad Nacional Autonoma de Mexico (UNAM) sits on a hard-rock hill and received small accelerations. The Secretaria de Comunicaciones y Transportes site (SCT) sits on soft lake sediments that amplified the seismic waves.

There were design flaws in the failed buildings (Figure 4.7) including soft first stories, poorly joined building wings, odd-shaped buildings prone to twist on their foundations, and buildings of different heights and vibration frequencies that sat close together and bumped into each other during the earthquake (Figures 4.7c and 4.8).

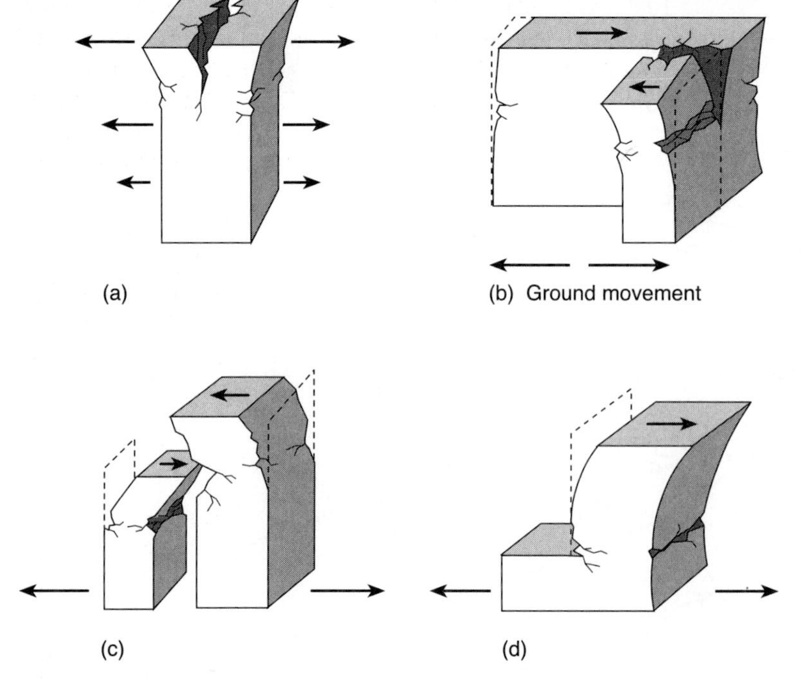

Figure 4.7 Some building-response problems during the Mexico City earthquake. **(a)** The amplitude of shaking increases up the building. **(b)** Buildings with long axes perpendicular to ground motion suffer more shaking. **(c)** Buildings with different heights sway at different frequencies and bang into each other. **(d)** A building with different heights tends to break apart.

Pacific Northwest, The Upcoming Earthquake

The 1985 Mexico City earthquake was caused by eastward subduction of a small plate beneath the North American plate. Other small plates are subducting beneath North America at the Cascadia subduction zone (Figure 4.9). No gigantic seisms have occurred yet in the Pacific Northwest in the 200 or so years since Europeans settled there. Will this area remain free of giant earthquakes? Could the Cascadia subduction zone be plugged up like a clogged drain, meaning that subduction has stopped? No. The active volcanoes above the subducting plates testify that subduction is still occurring (see Chapters 6 and 7). Could the subduction be taking place smoothly and thus eliminating the need for giant earthquakes? Probably not. The oceanic lithosphere being subducted is young, only about ten million years old. Young lithosphere is more buoyant and is best subducted when overridden by continental lithosphere. The North American continent is moving southwest at 2.5 cm/yr (1 in/yr) and colliding with the oceanic plate which is subducting along a N 68° E path at 4.3 cm/yr (1.7 in/yr). Thus, it seems certain that the subduction zone is storing elastic energy.

The Cascadia subduction zone is 1,100 km (680 mi) long. Its characteristics of youthful oceanic plate and strong coupling with the overriding plate are similar to situations in southwestern Japan and southern Chile. The stress was relieved in Japan by two earthquakes of 8.1 M_W, in 1944 and 1946. (A sequence of four to five similar seisms would cover the length of the Cascadia subduction zone.) The southern Chile stress was relieved by the world's largest measured earthquake—9.5 M_W.

Chilean earthquakes are neither rare nor small. On 20 February 1835, Charles Darwin was in Chile during his epic voyage on the HMS *Beagle* and experienced a great earthquake. His well-written descriptions of large areas of land being lifted above sea level, giant sea waves hitting the shore, and two volcanoes being shaken into action are instructive even today.

Figure 4.8 Mexico City earthquake damage due to constructing buildings with different periods of vibration next to each other. Four-story building on left repeatedly struck the taller Hotel de Carlo (middle building) causing collapse of its middle floors (see Figure 4.7c). Taller building on right also was damaged by hammering from Hotel de Carlo.

Photo courtesy of NOAA.

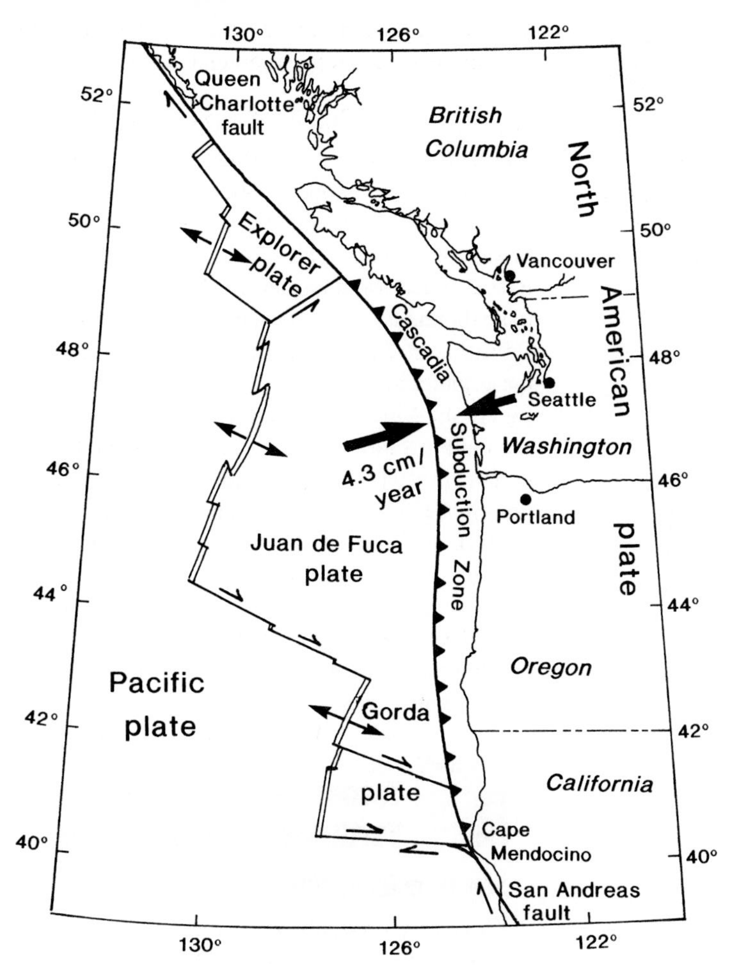

Figure 4.9 Map of small, young oceanic plates being subducted beneath the Pacific Northwest.

The Chilean events of 1960 collectively defined a downward movement of subducting Nazca oceanic plate that over a period of days involved a 1,000 km (620 mi) length and a 300 km (180 mi) width. Events of Chilean magnitude could unlock the entire Cascadia subduction zone. Figure 4.10 is a plot of the epicenters of the 1960 Chile mainshock, foreshocks, and aftershocks over a map of the Pacific Northwest to give an idea of what could happen in British Columbia, Washington, and Oregon. Could the Pacific Northwest experience a magnitude 9 earthquake? Yes. It has.

Earthquake in 1700 Recent work by Brian Atwater has shown that the last major earthquake here occurred about 9 P.M. on 26 January 1700 and was about magnitude 9. How is this known? By two converging lines of evidence. 1) Counting the annual growth rings in trees of drowned forests along the Oregon-Washington-British Columbia coast shows that the dead trees have no rings after 1699. Apparently the ground dropped during the earthquake and seawater got to the tree roots, killing them between August 1699 and May 1700, between the end of one growing season and the beginning of the next one (Figure 4.11). 2) The Japanese maintain detailed records of tsunami occurrences and sizes that they correlate to earthquake magnitudes and locations around the Pacific Ocean. Tsunami of 2 m (7 ft) height that hit Japan from midnight to dawn point to a 9 P.M. earthquake along the Washington-Oregon coast on 26 January 1700.

What does British Columbia-Washington-Oregon experience during a magnitude 9 earthquake? Three to five minutes of violent ground shaking and tsunami 10 m (33 ft) high surging onshore 15 to 40 minutes after the earthquake. Energy will be concentrated in long-period seismic waves; this will present challenges for tall buildings and long bridges.

What will the next magnitude 9 earthquake, along with its major aftershocks, do to cities like Portland, Tacoma, Seattle, Vancouver, and Victoria? When will the next magnitude 9 earthquake occur in the Pacific Northwest?

Spreading-Center Earthquakes

The Red Sea spreading centers and Dead Sea transform fault zone of the Old World have analogues in the New World with the spreading centers that are opening the Gulf of California and moving the San Andreas fault (Figure 4.12). Geologically, the Gulf of California basin does not stop where the sea does at the northern shoreline within Mexico. The opening ocean basin continues northward into the United States and includes the Salton Sea and the Coachella and Imperial Valleys at the ends of the Salton Sea. The Imperial Valley region is the only part of the United States that sits on opening ocean floor. In the geologic past, this region was flooded by the sea. However, at the present time, fault movements plus the huge volume of sediment deposited by the Colorado River hold back the waters of the Gulf of California. If the natural dam is breached, the United States will trade one of its most productive agricultural areas for a new inland sea.

The spreading-center segment at the southern end of the Salton Sea is marked by high heat flow, glassy volcanic domes, boiling mud pots, major geothermal energy reservoirs (subsurface water heated to nearly 400°C by the magma below the surface), and swarms of earthquakes associated with moving magma (Figure 4.12).

The Salton Trough is one of the most earthquake-active areas in the United States. There are seisms caused by the splitting and rifting of continental rock and swarms of earthquakes caused by forcefully moving magma. The Brawley seismic zone at the southern end of the Salton Sea commonly experiences hundreds of earthquakes in a several-day period. For example, in four days in January 1975, there were 339 seisms with magnitudes (M_L) greater than 1.5; 75 were

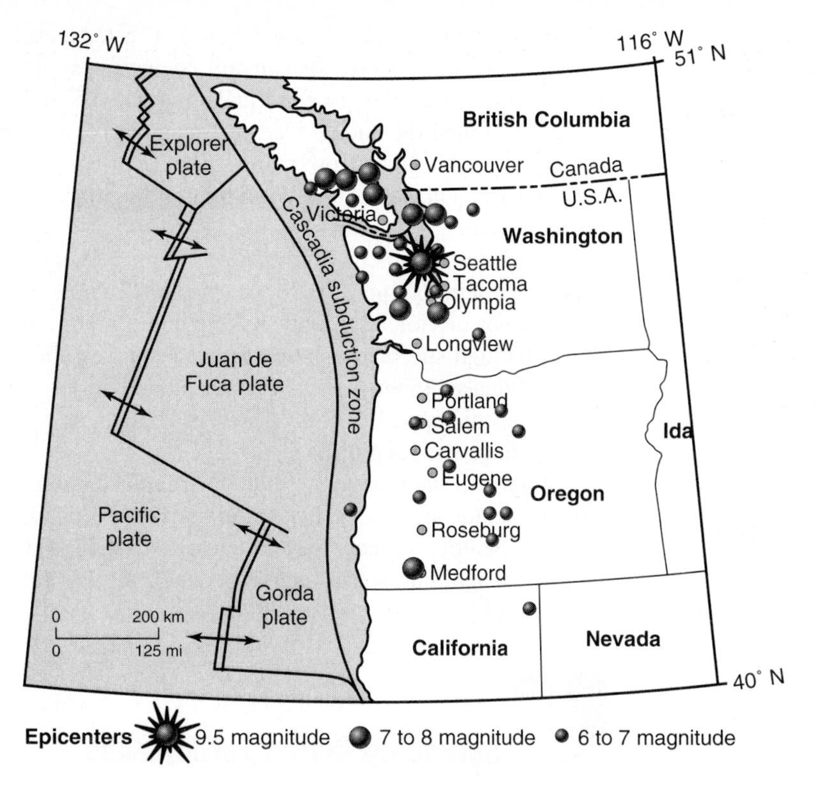

Figure 4.10 Epicenters for the 1960 Chile earthquake sequence are plotted over the Cascadia subduction zone. Earthquake magnitudes were one 9.5; nine between 7 to 8; and 28 between 6 to 7.

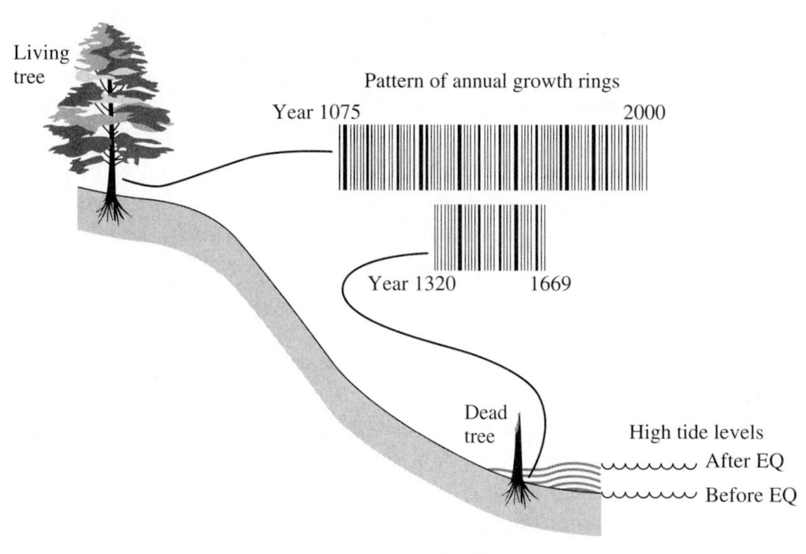

Figure 4.11 Annual growth rings in drowned trees along the Oregon–Washington–British Columbia coast tell of their deaths after the 1699 growing season. Seawater flooding occurred as land dropped during a magnitude 9 earthquake.

greater than 3 M_L, with the largest tremor at 4.7 M_L. Because hot rock does not store stress effectively, energy release takes place via many smaller quakes. The larger earthquakes in the valley are generated by ruptures in brittle continental rocks.

Notice in Figure 4.12 that the San Andreas fault ends at the southeastern end of the Salton Sea at the northern limit of the spreading center. Notice also that other major faults, such as the Imperial, San Jacinto system, Cerro Prieto, Elsinore, and Laguna Salada, also appear to be transform faults that line up with spreading-center segments. From a broad perspective, all these subparallel, right-lateral, transform faults are part of the San Andreas plate boundary fault system carrying peninsular California to the northwest. Large earthquakes on these faults in recent years in the area covered by Figure 4.12 include a 7.1-M_L quake on the Imperial fault in 1940, a 6.5-M_L event in the San Jacinto system in 1942, a 6.6-m_b quake on the Imperial fault in 1979, a 6.4-m_b seism on the Imperial fault in 1981, and a 6.6-M_S quake in the San Jacinto system in 1987. These are large earthquakes, but deaths and damages for each event typically are not high because the region is sparsely inhabited, most buildings are low, and the frequent shakes weed out inferior buildings.

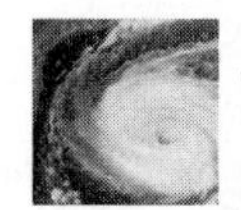

Transform Fault Earthquakes in California

San Francisco, 1906

Early in the twentieth century, San Francisco was home to about 400,000 people who enjoyed a cosmopolitan city that had grown during the economic boom times of the late nineteenth century. During the evening of 17 April 1906, many thrilled to the special appearance of Enrico Caruso, the world's greatest tenor, singing with the Metropolitan Opera Company in Bizet's *Carmen.* But several hours later, at 5:12 A.M., the initial shock waves of a mammoth earthquake arrived to begin the destruction of the city. Eyewitness accounts abound. One early riser told of seeing the earthquake approach as the street before him literally rose and fell like a series of ocean swells moving toward shore. The renowned American psychologist William James was in residence at Stanford University and reacted thusly:

> When I felt the bed begin to waggle, my first consciousness was one of gleeful recognition of the nature of the movement. "By Jove," I said to myself, "here's B's old earthquake after all." And then it went crescendo, "and a jolly good one it is, too."
>
> Sitting up involuntarily, and taking a kneeling position, I was thrown down on my face as it went fortis shaking the room exactly as a terrier shakes a rat. Then everything that was on anything else slid off to the floor, over went the bureau and chiffonier with a crash, as the fortissimo was reached; plaster

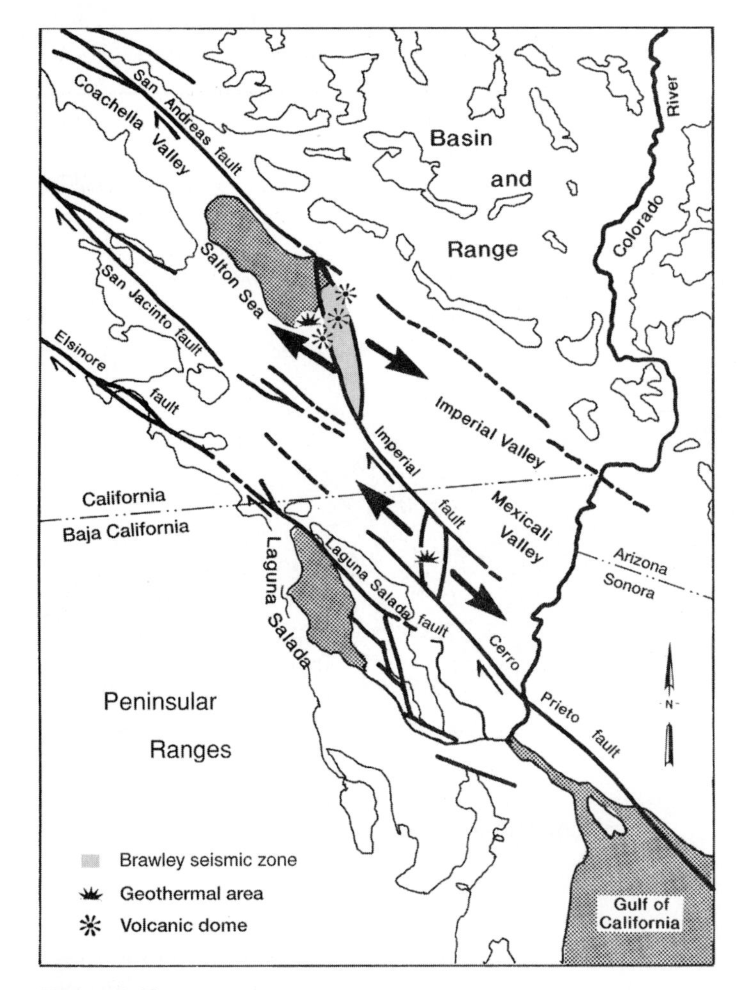

Figure 4.12 Map of northernmost Gulf of California. Note the two spreading centers (shown by large, diverging arrows) and the right-lateral (transform) faults associated with them.

> cracked, an awful roaring noise seemed to fill the outer air, and in an instant all was still again. My emotion consisted wholly of glee and admiration; glee at the vividness which such an abstract idea or verbal term as "earthquake" could put on when translated into sensible reality and verified concretely; and admiration at the way in which the frail wooden house could hold itself together in spite of the shaking. I felt no trace whatever of fear; it was pure delight and welcome. "Go it," I almost cried aloud, "and go it stronger!"

During a noisy minute, the violently pitching Earth emitted dull booming sounds joined by the crash of human-made structures. When the ground finally quieted, people went outside and gazed through a great cloud of dust to view the destruction. Unreinforced masonry buildings lay collapsed in heaps, but steel-frame buildings and wooden structures fared much better. Another factor in the building failures was the nature of the ground they were built on. Destruction was immense in those parts of the city that were constructed on top of artificial fill that had been dumped onto former bay wetlands or into stream-carved ravines to create building sites.

As repeated aftershocks startled and frightened the survivors, a danger even greater than the earthquake began to grow. Smoke arose from many sites as fires fed on the wood-filled rubble. Unfortunately, the same earthquake waves that wracked the buildings also broke most of the water lines, thus hindering attempts to stop the growing fires. From the business district and near the waterfront, fires began their relentless intrusion into the rest of the city. Desperate people tried dynamiting buildings to stop the fire's spread, but they only provided more rubble to feed the flames or even blew flaming debris as far as a block away, where it started more fires.

Meanwhile, in a residential district of side-by-side, multistory wooden homes, a housewife inadvertently began the "ham and eggs" fire when she tried to prepare breakfast without realizing her chimney had fallen away. This new fire joined the existing flames to burn for three days and two nights. The San Francisco Bulletin reported that

> The most dreadful feature of the whole panorama was the intense silence and the intense motion—the colors were neither those of day or night, but fierce, vivid, frenzied tones unimaginable outside the crater of a volcano. The background was a sulky and lurid glow like the unearthly flush on the face of a dying man.

The fires did about ten times as much damage as the earthquake itself; fire destroyed buildings covering 490 city blocks. Death and destruction were concentrated in San Francisco, where 315 people died, but the affected area was much larger. About 700 deaths occurred in a 430 km (265 mi) long belt of land running near the San Andreas fault. Towns within the high-intensity zone, such as San Jose and Santa Rosa, were heavily damaged, yet other cities to the east of the narrow zone, such as Berkeley and Sacramento, were spared significant damage. Problems continued in the months that followed as epidemics of filth-borne diseases sickened Californians; more than 150 cases of bubonic plague were reported. When all the fatalities from earthquake injuries and disease are included, the death total from the earthquake may have been as high as 5,000.

One of the intriguing aspects of disasters is the energizing effects they have on many survivors. Hard times shared with others bring out the best in many people. Shortly after this earthquake, the resilient San Franciscans were planning the Panama-Pacific International Exposition that was to impress the world and leave behind many of the beautiful buildings that tourists flock to see today. You can't keep a good city down.

San Andreas Fault Earthquakes

The San Andreas fault is part of a complex system of subparallel faults (Figure 4.13). The San Andreas fault proper is a 1,200 km (750 mi) long, right-lateral fault. In 1906, the northernmost section of the fault broke loose within a 3 km wide right-step just southwest of the city limits of San

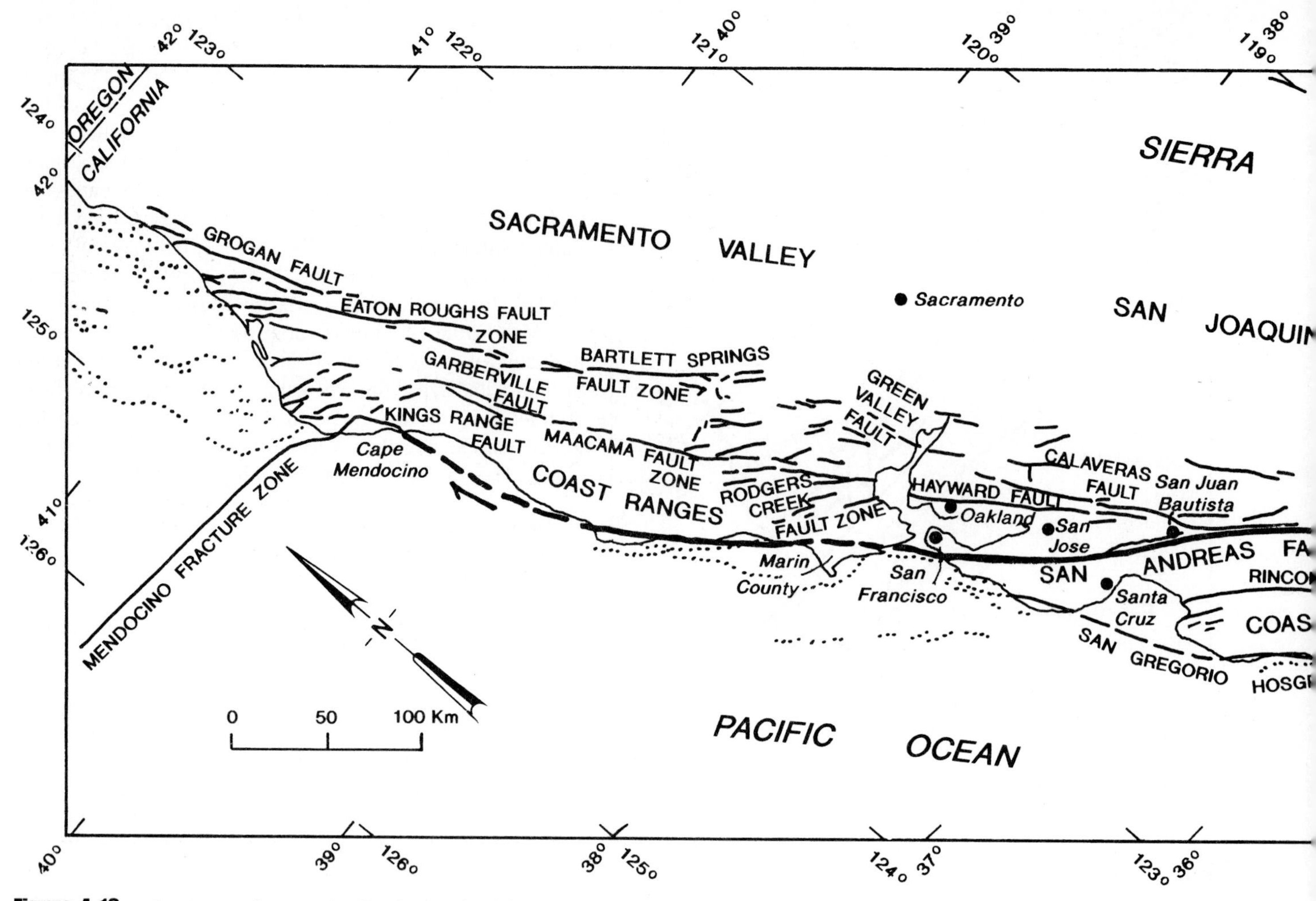

Figure 4.13 The San Andreas and other faults of California.

Source: R. E. Wallace, "General Features: The San Andreas Fault System, California" in *U.S. Geological Survey Professional Paper 1515,* 1990.

Francisco, rupturing northward and southward simultaneously (Figure 4.14). When it stopped shifting, the ground between Cape Mendocino and San Juan Bautista had been ruptured (Figure 4.13); this is a distance of 430 km (265 mi). The earthquake had a moment magnitude estimated at 7.8 resulting from 60 seconds of fault movement. When movement stopped, the western side had shifted northward a maximum of 6 m (20 ft) horizontally. In the peninsula south of San Francisco, fault movements have formed elongate topographic low areas now filled by lakes, and some of the land offset by the 1906 movements has been smoothed out and built on (Figure 4.15). In the vertical plane, the fault movement completely ruptured the 15 to 20 km thick brittle layer in the region. The amount of fault movement in 1906 died out to zero at the northern and southern ends of the rupture.

Today, the San Francisco section of the San Andreas fault has a deficit of earthquakes. Apparently this is a "locked" section of the fault (Figure 4.16). Virtually all the stress from plate tectonics is stored for many decades until the fault finally can take no more and ruptures in a big event that releases much of its stored energy in a catastrophic movement.

The San Andreas fault has different behaviors along its length. The next section to the south, between San Juan Bautista and Cholame (Figures 4.13 and 4.16), has frequent small- to moderate-sized earthquakes. This is a "creeping" section of the fault where numerous earthquakes accommodate the plate-tectonic forces before they build to high levels. The creeping movements of the fault are shown by the millimeters per year of ongoing offset of sidewalks, fences, buildings, and other features. Earthquakes in this fault segment do not seem to exceed magnitude 6. These are still significant seisms, but they are small compared to events on adjoining sections of the fault.

The San Andreas fault segment between Cholame and San Bernardino is another locked zone that is deficient in earthquake activity (Figures 4.13 and 4.16). However, on 9 January 1857, this segment of the fault broke loose at its

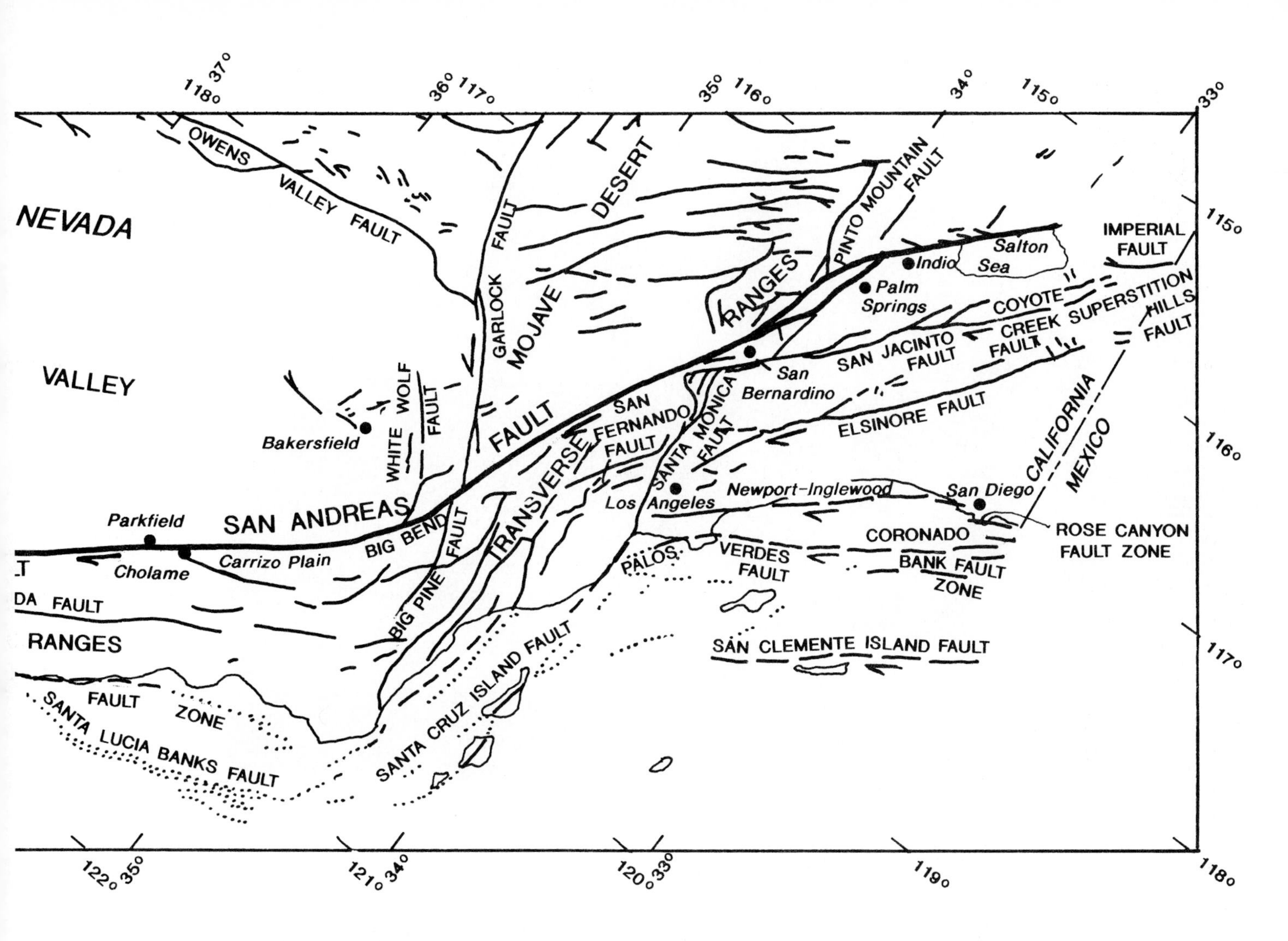

Figure 4.14 Looking south-southeast down the San Andreas fault. View is over Bodega Head and Tomales Bay toward the epicenter of the 1906 San Francisco earthquake.

Photo by John S. Shelton.

Figure 4.15 Looking southeast along the trace of the San Andreas fault. San Francisco airport and part of San Francisco Bay are in left center. Linear lakes in right center (e.g., Crystal Springs Reservoir) are in the fault zone. In bottom center, the land offset by the 1906 fault movement has been bulldozed and covered with houses!
Photo by John S. Shelton.

northwestern end, and the rupture propagated southeastward in the great Fort Tejon earthquake with a magnitude of about 8.3. Due to the one-way advance of the rupture front, the fault movement lasted almost three minutes. The ground surface was broken for at least 360 km (225 mi), and the maximum offsets in the Carrizo Plain (Figure 4.17) were a staggering 9.5 m (31 ft). One of the offset features was a circular corral for livestock that was split and shifted to an S-shape by the fault movement. In 1857, the region was sparsely settled, so the death and damage totals were small. The next time a great earthquake occurs here, the effects may be disastrous.

The southernmost segment of the San Andreas fault, from San Bernardino to the Salton Sea, is a complex zone that has not generated a truly large earthquake in historic times (Figure 4.18). It has locked zones within it.

California has a well-recorded history for only about 150 years. How can we extend our knowledge of earthquakes into the prehistoric past? Large earthquakes commonly leave evidence of their actions in sediments. Earthquake history can be read in sediments using the techniques of **neotectonics** (*neo* meaning "young") and **paleoseismology** (*paleo* meaning "ancient").

In December 1988, using paleoseismologic analyses, a group of geologists forecast earthquake sizes and probabilities for some major faults in California (Figure 4.23). They placed a 30 percent probability on a magnitude 6.5 earthquake occurring on the Loma Prieta segment of the San Andreas fault within 30 years. Ten months later, the magnitude 7.1 World Series earthquake occurred there.

In 1999, the working groups stated that there is a 70 percent probability of at least one magnitude 6.7 earthquake striking the San Francisco Bay region before 2030 and an 85 percent probability of a magnitude 7 or higher earthquake in southern California before 2024.

World Series (Loma Prieta) Earthquake, 1989

In 1989, the World Series of baseball was a Bay Area affair. It pitted the American League champion Oakland Athletics against the National League champion San Francisco Giants. Game three was scheduled in San Francisco's Candlestick Park, where the Giants hoped the home field advantage would help them win their first game. It was Tuesday, 17 October, and both teams had finished batting practice, which

Figure 4.16 Historic behavior of some California faults. The northern "locked" section of San Andreas fault ruptured for 265 miles in 1906 (magnitude 7.8). The central "creeping" section has frequent smaller earthquakes. The south-central "locked" section ruptured for 225 miles in 1857 (magnitude 8.3). The southernmost San Andreas awaits a major earthquake. The Owens Valley fault ruptured for 70 miles in 1872 (magnitude 7.7). A magnitude 7.7 seism occurred on White Wolf fault in 1952, and a magnitude 7.3 seism happened in Mojave Desert in 1992.

Source: "The San Andreas Fault," *U.S. Geological Survey.*

Figure 4.17 The San Andreas fault slashes across the Carrizo Plain. Notice the ridges and basins caused by local squeezing and pulling apart.

Photo courtesy of Pat Abbott.

Figure 4.18 Space shuttle view northward across the Elsinore, San Jacinto, and San Andreas faults.

Photo courtesy of NASA.

was watched by 60,000 fans at the park, along with a TV crowd of another 60 million fans in the United States and millions more around the world. At 5:04 P.M., 21 minutes before the game was scheduled to start, a distant rumble was heard, and a soft thunder rolled in from the southwest, shook up the fans, and stopped the game from being played. San Francisco experienced another big earthquake, and this time, it shared it with television viewers. After the earthquake, the San Franciscans at Candlestick Park broke into a cheer, while many out-of-staters were seen heading for home.

What caused this earthquake? An 83-year-long pushing match between the Pacific and North American plates resulted in a 42 km (26 mi) long rupture within the San Andreas fault zone. The southernmost section of the fault zone that moved in 1906 had broken free and moved again. There were several different aspects to the earthquake this time: 1) the fault rupture took place at depth; 2) the fault movement did not offset the ground surface; 3) there was a significant component of vertical movement; and 4) the fault rupturing lasted only 11 seconds, an unusually short time for a magnitude 7.1 event.

Sidebar

Neotectonics and Paleoseismology

The San Andreas fault slashes through the land with compressive bends causing land to uplift, and pull-apart bends causing land to drop down (Figures 4.19 and 4.20). The downdropped or fault-dammed areas within the fault zone can become sites of ponds, receiving: 1) sand washing in from heavy rains, 2) clays slowly settling from suspension in ponded water, and 3) vegetation that lives, dies, and is buried by clay and sand. These processes produce a delicate record of sediment layers that may be disturbed and offset by later fault movements. This is a record we can read.

Older (more deeply buried) layers have existed longer (Figure 4.21) and have been offset by more earthquake-generating fault movements. The amount of fault offset is proportional to an earthquake's magnitude; the greater the offset of sediment layers, the bigger the earthquake. These principles suggest a method to determine the approximate sizes of prehistoric earthquakes. Simply dig a trench through the sediment infill of a fault-created pond and read the fault offsets recorded in the sediments (Figures 4.21 and 4.22). Sediment layers in trench walls can be traced by digging a network of intersecting trenches to gain a three-dimensional view of fault offsets through time.

Figure 4.19 A close-up of the San Andreas fault at Wallace Creek in the Carrizo Plain. Notice the offset streams and the ponded depressions formed at the fault. Have the movements been right or left lateral?

Photo courtesy of Pat Abbott.

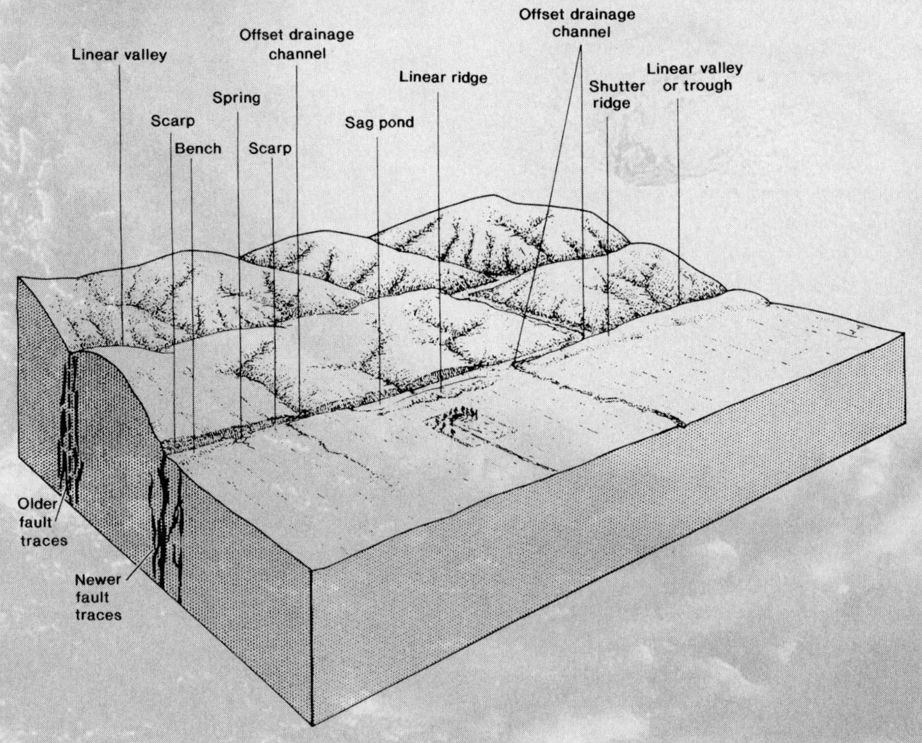

Figure 4.20 Schematic diagram of topography along the San Andreas fault in the Carrizo Plain. Notice the sag pond here and in Figure 4.19. Sediments deposited in these depressions allow the prehistoric record of earthquakes to be read.

Source: *Misc. Geol. Invest., U. S. Geological Survey.*

Sidebar

Neotectonics and Paleoseismology (Continued)

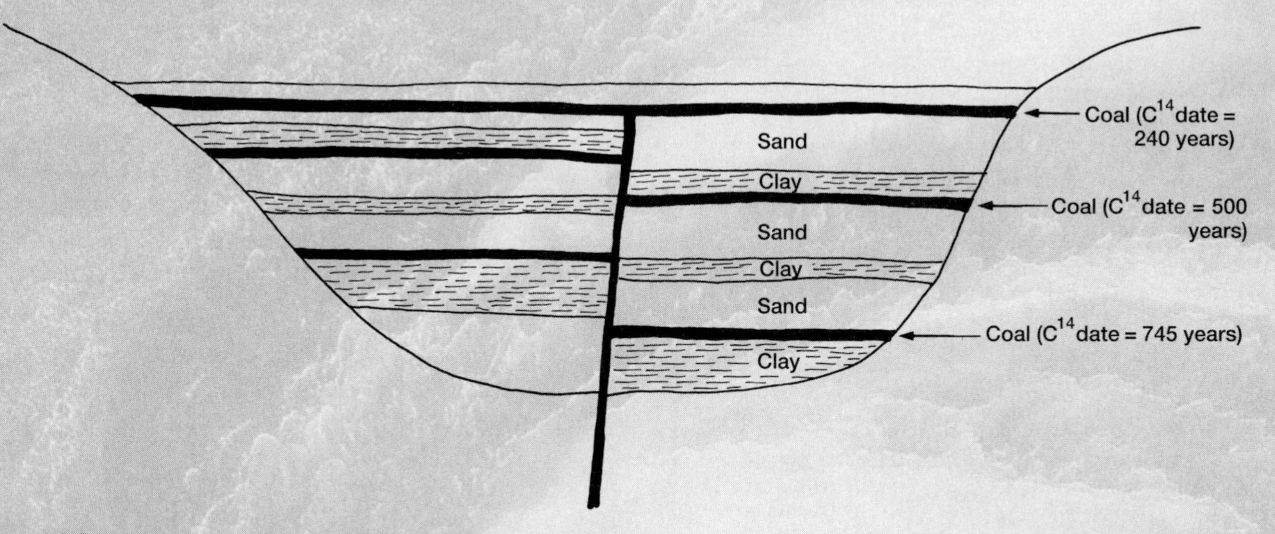

Figure 4.21 Schematic cross section of trench wall across a fault-created pond. The fault offsets the once continuous sediment layers. Notice that an upper layer of organic material formed 240 years ago is unbroken. At depth, a 500-year-old organic-rich layer has been offset. Deeper still, a 745-year-old organic-rich layer has been offset twice as much, indicating two major fault movements since it formed. What is the approximate recurrence interval between earthquakes at this site? When might the next big earthquake be expected here?

Figure 4.22 A trench wall across the San Andreas fault at Pallett Creek. Sandy layers are whitish, clayey layers are grayish, and organic-rich layers are black. Black layer in the center formed about 1500 C.E. It has been offset 1.5 m (5 ft) horizontally and 30 cm (1 ft) vertically since 1500 C.E.
Photo courtesy of Pat Abbott.

Dates of prehistoric earthquakes can come from analyzing amounts of radioactive carbon in organic material (e.g., logs, twigs, leaves, and **coal**) in the sediment layers. All life uses carbon as a fundamental building block. Most carbon occurs in the isotope C^{12}, but a small percentage is radioactive carbon (C^{14}) produced in the atmosphere by bombardment of nitrogen atoms with subatomic particles emitted from the Sun. Carbon is held in abundance in the atmosphere as carbon dioxide (CO_2). All plants and animals draw in atmospheric CO_2, and their wood, leaves, bones, shells, teeth, etc., are partly built with radioactive carbon. As long as an organism lives, it exchanges carbon dioxide with the atmosphere via **photosynthesis** or breathing. The percentage of radioactive carbon in a plant or animal remains the same as that of the atmosphere during the organism's lifetime. However, when an organism dies, it ceases taking in radioactive carbon, and the radiocarbon in its dead tissues decays with a half-life of 5,730 years. The presence of organic material allows determination of time of death and hence the age of enclosing sediments. This places actual dates into faulted sedimentary layers. The determination of real dates allows the estimation of recurrence intervals for earthquakes; i.e., how many years pass between earthquakes at a given site.

Figure 4.21 is a schematic representation of a trench-wall exposure of faulted pond sediments demonstrating how fault-rupture sizes and recurrence intervals may be determined. A real example of a faulted pile of ponded sediments is shown in Figure 4.22. Here at Pallett Creek along the San Andreas fault, Caltech geologist Kerry Sieh has determined that earthquakes with 6 m (20 ft) of horizontal offset recur about every 132 years. However, this size seism has occurred as close together as 44 years and as far apart as 330 years.

Sidebar

Neotectonics and Paleoseismology (Continued)

Earthquake Prediction—Intermediate and Long-Term

Can we predict earthquakes on intermediate to long-term scales using the paleoseismology approach? It seems to work well for some faults but not for others. San Diego State University geologist Thomas K. Rockwell classifies fault-movement timing into three groups. 1) Quasi-periodic movements. These faults have major movements at roughly equal time intervals. This regular pattern can be defined using the trenching and radiocarbon dating of paleoseismology. 2) Clustered movements. These faults have adjacent fault segments move during several decades, then they cease movement for a century or millennium until the next cluster begins. The best example of clustered movements is occurring right now on the North Anatolian fault in Turkey (Figure 2.31). 3) Random movements. These faults are inherently unpredictable; they have no definable pattern for their major movements. The San Andreas fault seems to be in this category.

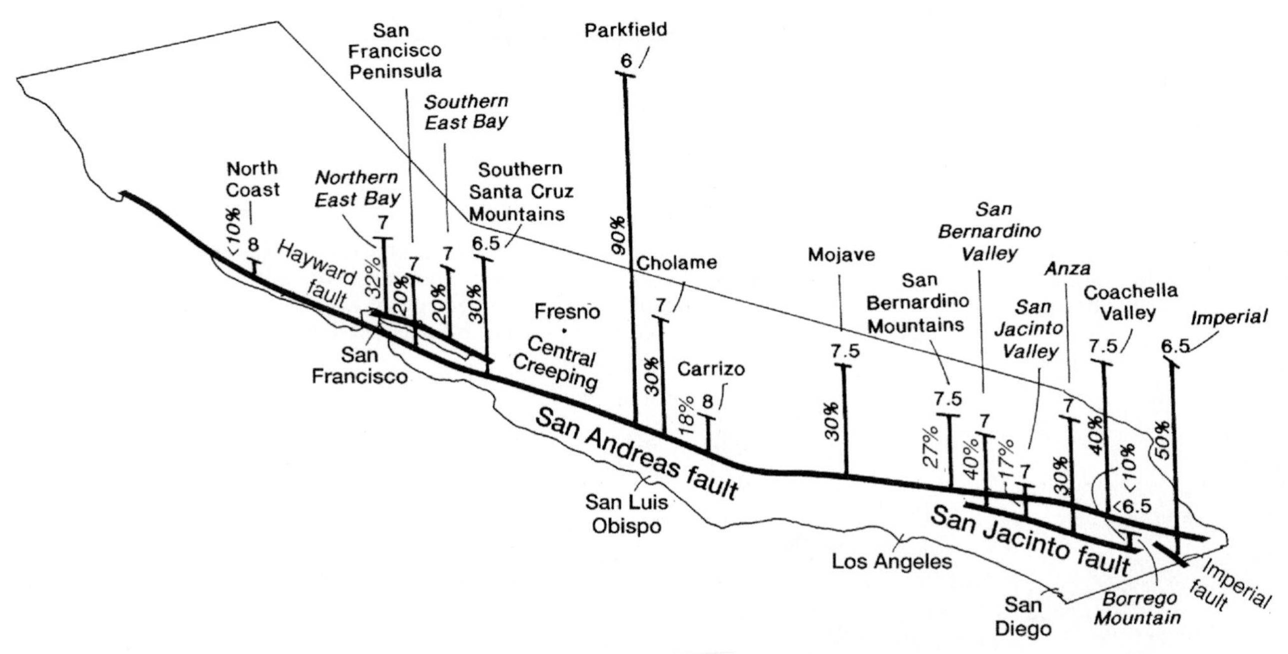

Figure 4.23 Working group analyses of expected earthquake magnitudes and their probabilities of occurring before the year 2030. Forecasts are based on historic records and trench-wall offsets of sediments dated by radiocarbon analyses and global positioning system (gps) measurements.

Movement occurred in a gently left-stepping constraining bend of the San Andreas fault zone (Figure 4.24). Long-term compressive pressures along this left step have uplifted the Santa Cruz Mountains. This step in the San Andreas fault is near where the Calaveras and Hayward faults split off and run up the east side of San Francisco Bay. The epicenter of the 1989 seism was near Loma Prieta, the highest peak in the Santa Cruz Mountains. Loma Prieta is the official name for this earthquake; it follows the rule of taking the name from the most prominent geographic feature near the epicenter. Nonetheless, it remains known to many people as the World Series earthquake.

It is difficult for a fault to move around a left-stepping bend. Constraining bends commonly "lock up"; thus, movements at a bend tend to be infrequent and large. This left step in the San Andreas also causes the fault plane to be inclined 70° to the southwest (Figure 4.25). The fault movement initiated at 11.5 mi (18.5 km) depth and slipped for 7.5 ft (2.3 m). The motion can be resolved into 6.2 ft (1.9 m) of horizontal movement (strike slip) and 4.3 ft (1.3 m) of vertical movement (reverse slip). Stated differently, the western or Pacific plate side moved 6.2 ft to the northwest, and a portion of the Santa Cruz Mountains was uplifted 14 in (36 cm). Although the fault did not rupture the surface, the uplifted area was 3 mi wide and had numerous fractures in the uplifted and stretched zone. Many of the cracked areas became the sites of landslides.

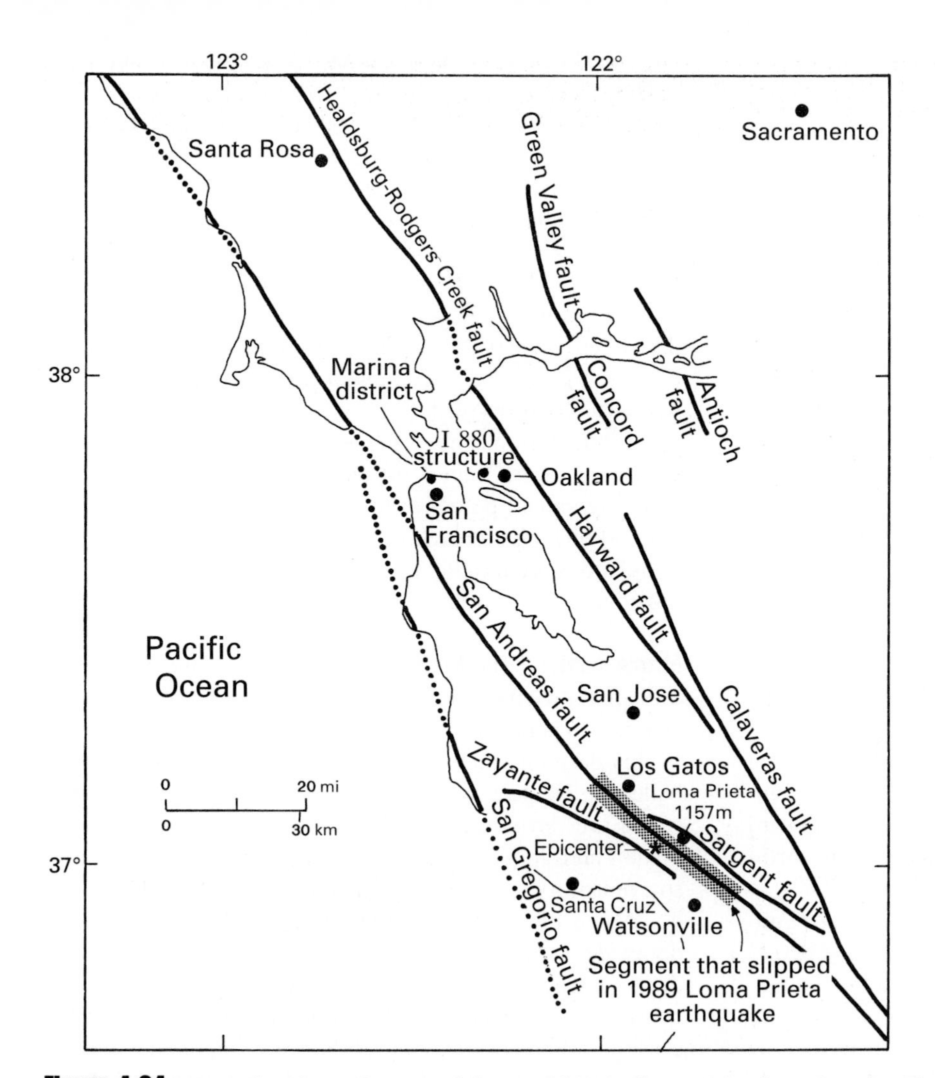

Figure 4.24 Map showing epicenter of the World Series (Loma Prieta) earthquake. The San Andreas fault takes an 8° to 10° left step in the ruptured section. The left step also is where the Calaveras and Hayward faults split off from the main San Andreas trend.

Source: "Lessons Learned from the Loma Prieta Earthquake of October 17, 1989" in *U.S. Geological Survey Circular 1045,* 1989.

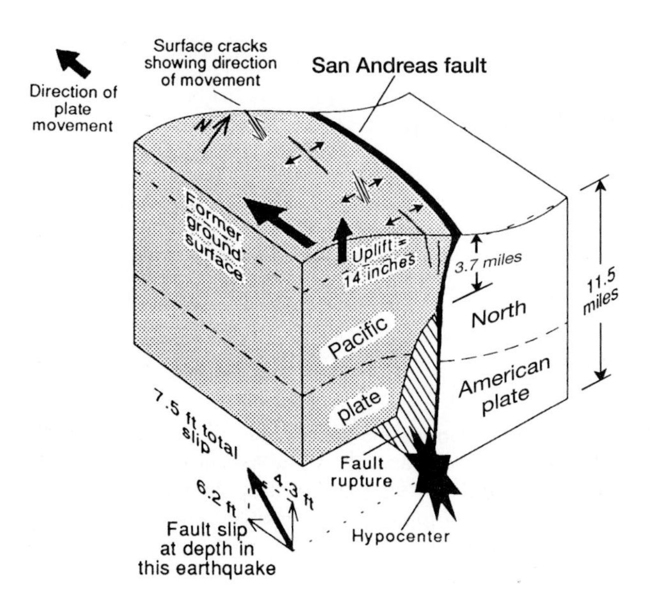

Figure 4.25 Schematic diagram of fault movement within the San Andreas zone in the World Series earthquake. The San Andreas fault dips 70° southwest because of the left-step bend. Fault movement began at 11.5 mi (18 km) depth and moved 6.2 ft (1.9 m) horizontally and 4.3 ft (1.3 m) vertically. Fault movement died out upward and did not rupture the ground, although the surface bulged upward 14 in (36 cm). Think three-dimensionally here: because of the dipping fault plane, will the epicenter plot on the ground-surface trace of the San Andreas fault? No.

Source: "Lessons Learned from the Loma Prieta Earthquake of October 17, 1989" in *U.S. Geological Survey Circular 1045,* 1989.

The main shock had a surface-wave magnitude (M_S) of 7.1 and was followed by numerous aftershocks, as is typical for large earthquakes. The Loma Prieta area had been a relatively quiet zone for earthquakes since the 1906 fault movement (Figure 4.26); before 1989, the Loma Prieta region had been a "seismic gap." As the numerous epicenters in cross section (a) in Figure 4.26 show, the San Andreas fault section to the south moves frequently, generating numerous small earthquakes. But the same plate-tectonic stresses affecting the creep zone also affect the locked or seismic-gap zone. How does a locked zone catch up with a creep zone? By infrequent but large fault movements. Notice in cross section (b) how the World Series (Loma Prieta) main shock and aftershocks filled in the seismic gap in cross section (a). This demonstrates some merit for the seismic-gap method as a forecasting tool. Figure 4.26 also shows another seismic gap, south of San Francisco in the heavily populated midpeninsula area (this is the area of elongate lakes shown in Figure 4.15). The 1989 fault movement has increased the odds by another 10 percent for a large earthquake in the Crystal Springs Reservoir area in the next 30 years (Figure 4.23).

In the World Series earthquake, the fault ruptured at greater than 2 km/sec in all directions simultaneously, upward for 13 km (8 mi) and both northward and southward for over 20 km (13 mi) each. Table 3.6 indicates that earthquakes with magnitudes of 7 to 7.9 usually rupture for 20 to 50 seconds, yet this radially spreading, 7.1-magnitude rupture lasted only 11 seconds. Had it lasted the expected 20 to 30 seconds, numerous other large buildings and the double-decker Embarcadero Freeway in San Francisco would have failed catastrophically. As it was, the event left 67 people dead or dying, 3,757 injured, and over 12,000 homeless; caused numerous landslides; disrupted transportation, utilities, and communications; and caused about $6 billion in damages.

Earthquakes Don't Kill, Buildings Do In the epicentral region, serious damage was dealt to many older buildings. The short-period P and S waves wreaked their full effects on low buildings built of rigid materials. Common reasons for failure included poor connections of houses to their foundations, buildings made of unreinforced masonry (URM) or brick-facade construction, and two-to-five-story buildings deficient in shear-bearing internal walls and supports. In Santa Cruz, four people died, and the Pacific Garden Mall, the old city center of historic brick and stone buildings that had been preserved and transformed into a tourist mecca, was virtually destroyed.

From the epicentral region, the seismic waves raced outward at more than 3 mi/sec. Some longer-period shear waves remained potent even after traveling 100 km (more than 60 mi). Upon reaching the soft muds and artificial-fill foundations around San Francisco Bay, these seismic waves had their vibrations amplified.

Marina District The Marina District is one of the most beautiful areas in San Francisco. It sits on the northern shore of the city next to parks, the Golden Gate Bridge, and the Bay itself. In this desirable and expensive district, five residents died, building collapses were extensive, and numerous building-eating fires broke out due to 1) amplified shaking, 2) deformation and liquefaction of artificial-fill foundations, and 3) soft first-story construction that led to building collapses.

Much of the Marina District is built on artificial fill dumped onto the wetlands of the Bay to create more land for development. Ironically, much of the artificial fill was the debris from the San Francisco buildings ruined by the 1906 earthquake. Seismic waves in 1989 were amplified in this artificial fill. Some fill underwent permanent deformation and settling, and some formed **slurries** as underground water and loose sediment flowed as fluids in the process of **liquefaction** (Figure 4.27a). Liquefaction in the Marina District

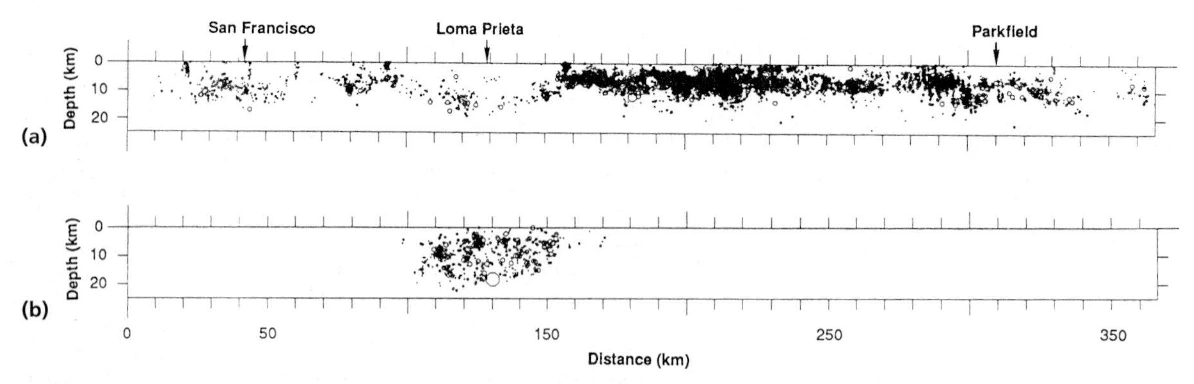

Figure 4.26 Cross sections of seismicity along San Andreas fault, 1969 to early 1989. **(a)** Notice the dense concentrations of hypocenters in the central creeping section of the fault from south of Loma Prieta to Parkfield, as well as the "seismic gap" in the Loma Prieta area. **(b)** Notice the deep hypocenter of the 1989 main shock plus the numerous aftershocks. Putting the two cross sections together fills the seismic gap. Are there other seismic gaps in cross section (a)? Yes, south of San Francisco in the Crystal Springs Reservoir area (see Figure 4.15), just west of the densely populated midpeninsula area. When will this seismic gap be filled?
Source: "Lessons Learned from the Loma Prieta Earthquake of October 17, 1989" in *U.S. Geological Survey Circular 1045,* 1989.

in 1989 brought to the surface pieces of glass, tar paper, redwood, and other debris from 1906 San Francisco.

The central cause of building failure was design flaws. Because the Marina District is home to many affluent people, they need places to park their cars. But where? The streets are already overcrowded, and basement parking garages would be below sea level and thus flooded. A common solution has been to clear obstructions from first stories of buildings to make space for car parking. Unfortunately, this creates a soft first story. Where are the internal walls, lateral supports, and bracing needed to support the upper one to four stories? They were removed or sacrificed to make room for cars; thus, buildings simply pancaked and became one story shorter (Figure 4.27b).

Interstate 880 The most stunning tragedy associated with the World Series earthquake was the crushing of 42 people during the collapse of a double-decker portion of Interstate 880 in Oakland (Figure 4.28). The elevated roadway was designed in 1951 and completed in 1957. A 1.25 mi long section collapsed: 44 slabs of concrete roadbed, each weighing 600 tons, fell onto the lower roadbed and crushed some vehicles to less than 1 ft high. The section that collapsed was built on young, soft San Francisco Bay muds. The elevated freeway structure had a natural resonance of 2 to 4 cycles per second; the bay-mud foundation produced a 5- to 8-fold amplification of shaking in that range. The seismic waves excited the muds causing the

Magnified

Magnified

(a)

(b)

Figure 4.27 **(a)** Water-saturated sediment usually rests quietly. However, when seismic waves shake, sand grains and water can form a slurry and flow as a liquid. When earth materials liquefy, building foundations may split and buildings may fail. **(b)** A typical Marina District building collapse. Three residential stories sat above a soft first story used for car parking; now, the four-story building is three-stories tall.

Photo courtesy of Dames and Moore.

Figure 4.28 The Cypress double-decker section of Interstate 880 in Oakland was completed in 1957. It failed in the 1989 seism and dropped 1.25 miles of upper roadbed onto the lower roadbed, crushing many vehicles and people.

Photo courtesy of Dames and Moore.

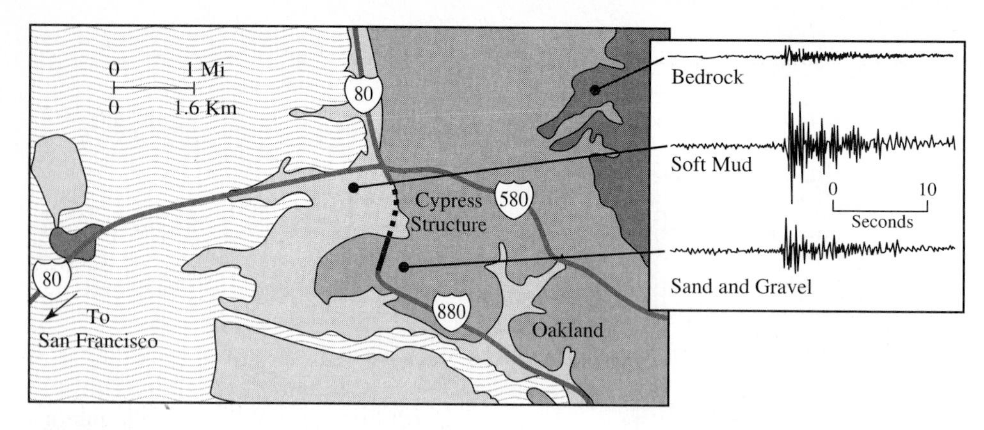

Figure 4.29 The portion of Interstate 880 elevated roadway built on top of soft bay mud collapsed (dashed black line) while the portion resting on rock still stood (solid black line). Notice how the shaking was amplified in the soft mud.

heavy structure to sway sharply (Figure 4.29). The portion of I 880 elevated roadway built on firmer sand and gravel stood intact; the portion standing on soft mud collapsed catastrophically.

The weak foundation was compounded by a flawed structural design. The joints where roadbeds were connected to concrete-support columns were not reinforced properly. Cracks initiated at the joints causing failure of supporting columns, which slid off the crushed areas of the joints and dropped the upper roadbed onto the lower level (Figure 4.30). Was this bridge failure a surprise? Not really. These lessons had been learned 18 years earlier in the 1971 San Fernando earthquake (Figure 3.34), but no one had corrected this disaster in waiting.

An ironic and deadly footnote to this disaster lay in the mode of failure. There was a delay between the initial shock and the final collapse, which allowed some people a brief time to plan. Some maneuvered their vehicles under beams next to support columns, and others got out of their cars and walked under the same supports, thinking that these were the strongest parts of the structure. Tragically, they were the weakest: this is where failure was most catastrophic, and no one survived there.

Figure 4.30 The support columns of the Interstate 880 structure failed at the joints. There were 20 #18 bars of steel in each column, but they were discontinuous at the joints and failed there.
Photo courtesy of Dames and Moore.

Bay Area Earthquakes—Past and Future

The historic record of California earthquakes is about 225 years old, and only the last 150 years or so is reasonably good. This historic record is shorter than the recurrence times for major movements on most faults. Nonetheless, the San Francisco Bay Area has enough information contained in newspaper accounts, diaries, personal letters, and similar sources to piece together a fairly accurate history of nineteenth-century earthquakes, and it is quite different from the twentieth-century record. During the nineteenth century, earthquakes with magnitudes greater than 6 were much more common (Figure 4.31). There were seven destructive seisms in the 70 years before the 1906 San Francisco earthquake, averaging a large earthquake every decade. Then came the monstrous movement of the San Andreas fault in 1906. This 265-mile-long rupture removed so much of the plate-tectonic stress stored in the rocks that several decades of the twentieth cen-

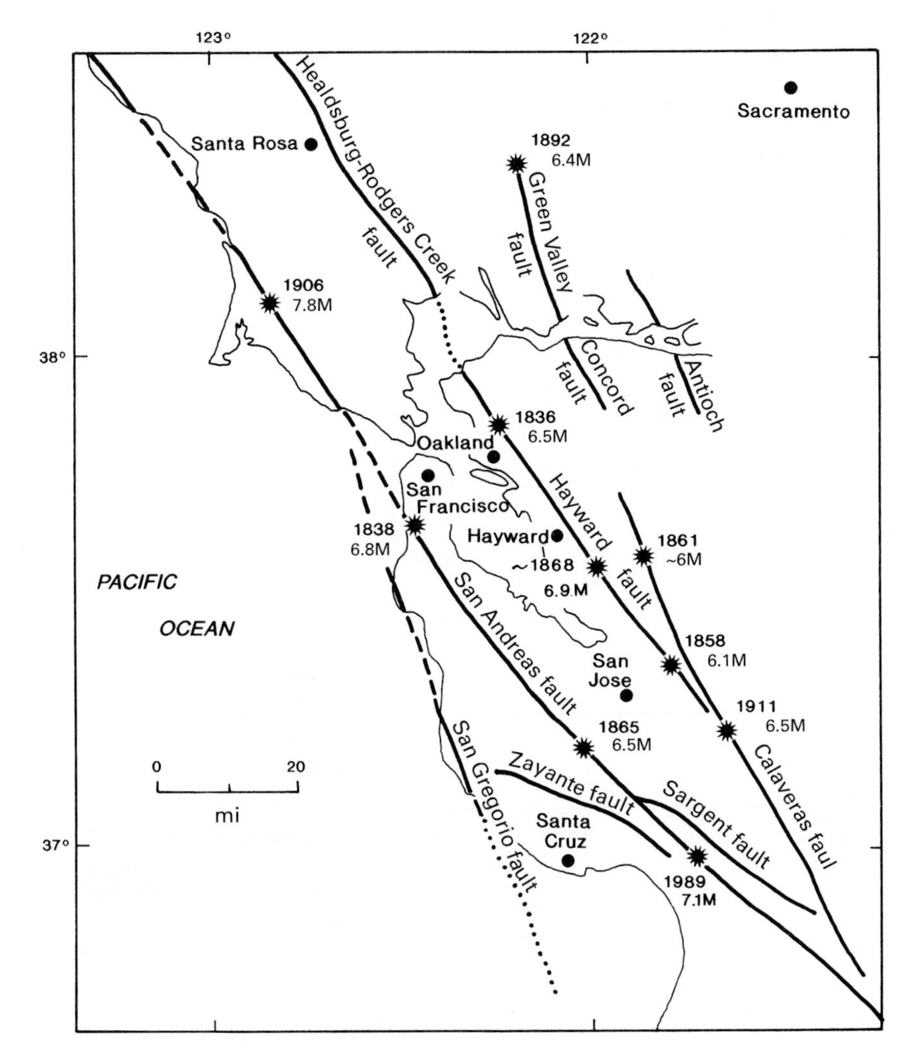

Figure 4.31 Locations and approximate sizes of some larger Bay Area earthquakes.

Source: *U.S. Geological Survey.*

tury were effectively free of large earthquakes (Figure 4.32). But large earthquakes returned to the southern part of the Bay Area beginning in the 1970s (Figure 4.33). What are the patterns in these data? Do they have any significance?

Pattern 1 Common large earthquakes versus rare giant shakes: The movement of the Pacific plate past the North American plate in the Bay Area seems to be satisfied by either a magnitude 6 to 7 earthquake roughly every decade (nineteenth century) or a magnitude 8 earthquake every century (twentieth century). Which pattern is preferable for this heavily developed and populated region? (Not that we have any choice.) Which pattern causes the least amount of death, damage, and psychological distress? Will the twenty-first century be like the nineteenth or twentieth?

Pattern 2 East-west pairings of earthquakes (Figure 4.31): In 1836, Oakland experienced a quake of about magnitude 6.5; this was followed two years later on the San Francisco Peninsula with a seism of about magnitude 6.8. In the southern Bay Area, a large shallow earthquake near Santa Cruz in 1865 was followed three years later with a shake of about magnitude 6.9 near Hayward. Will this paired occurrence of East Bay and West Bay earthquakes reoccur?

Pattern 3 Northward progression of earthquakes: The large earthquakes of 1865 and 1868 were preceded by five moderate earthquakes that moved northward up the Calaveras fault. Figure 4.33 shows four moderate to large earthquakes that have moved from south to north up the Calaveras fault. Does this repeat pattern suggest an upcoming large seism on the Hayward fault? This region today is populated by more than one million people in 10 cities. It is a built-upon fault covered by schools, hospitals, city halls, houses and the University of California. If a seism like the 1868 earthquake occurs soon, the California Division of Mines and Geology estimates that up to 7,000 people might die. The number of deaths will depend in part on the time of day that the earthquake occurs. When is the worst time for an earthquake? During the middle of the work and school day, when the maximum number of people are occupying the larger, older structures. When is the best time for an earthquake? During the night, when most people are home and asleep in their beds. In general, California houses handle earthquake shaking quite well.

Kobe, Japan, 1995 vs. Oakland, California, 20??

Kobe, Japan, 1995 Japan is also hit hard by earthquakes less mighty than great subduction events. The most expensive earthquake in history ($100 billion in property losses) hit at 5:46 A.M. on 17 January 1995 when a right-lateral, strike-slip fault movement began within a right step on the Nojima fault, rupturing simultaneously in both northeast and southwest directions, including through the city of Kobe (Figure 4.34). The 50 km (30 mi) long rupture event took 15 seconds to offset the land 1.7 m (5.6 ft) horizontally and 1 m (3.3 ft) vertically. Earthquake magnitude was 6.9 (M_W) setting some soft sediment areas of Kobe shaking strongly for 100 seconds. Kobe is a major port, the third busiest in the world. The 1.5 million residents of the city are packed into a narrow belt of land partly reclaimed from the bay with artificial fill. These weak sediments liquefied and performed poorly

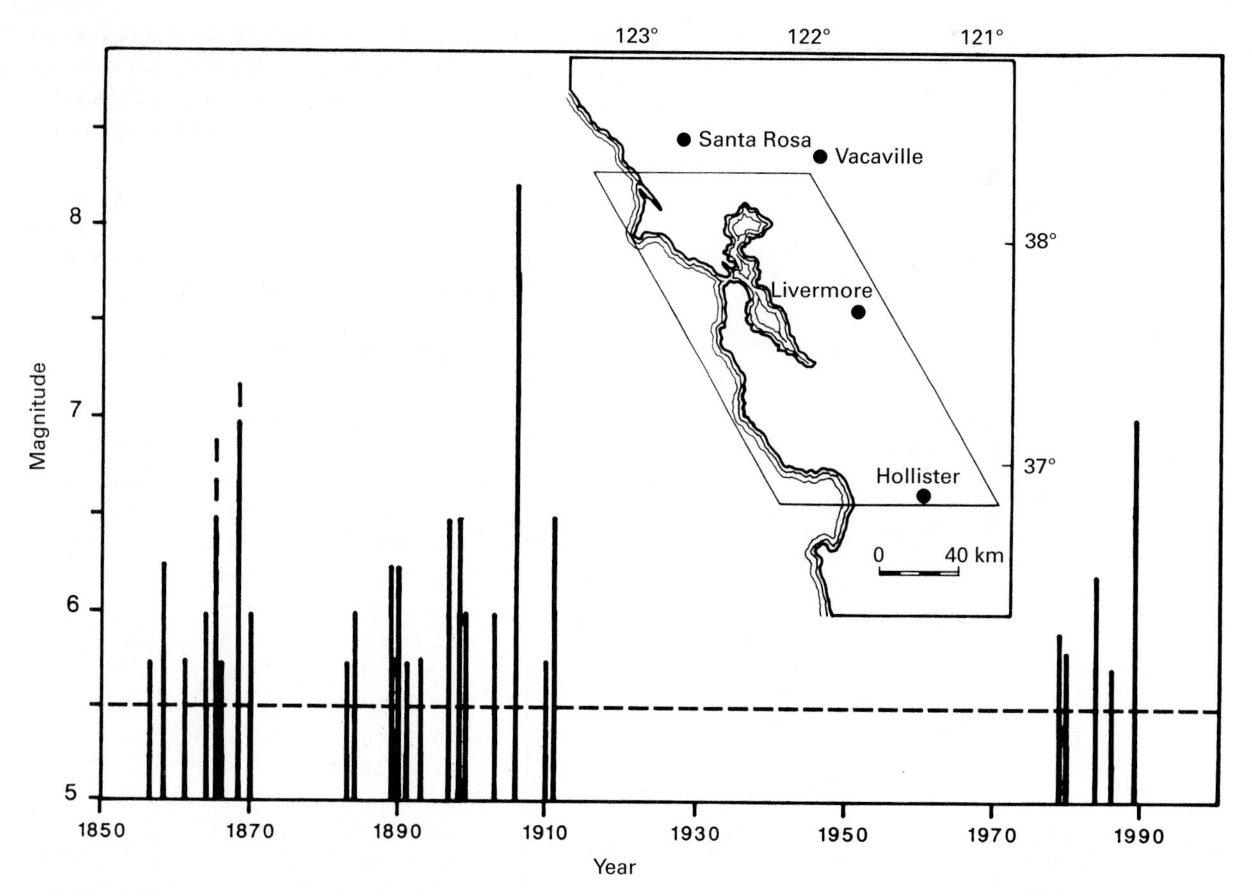

Figure 4.32 Distribution of earthquakes with magnitudes greater than 5.5 near San Francisco Bay, 1849–1990. Index map shows area of earthquake epicenters.

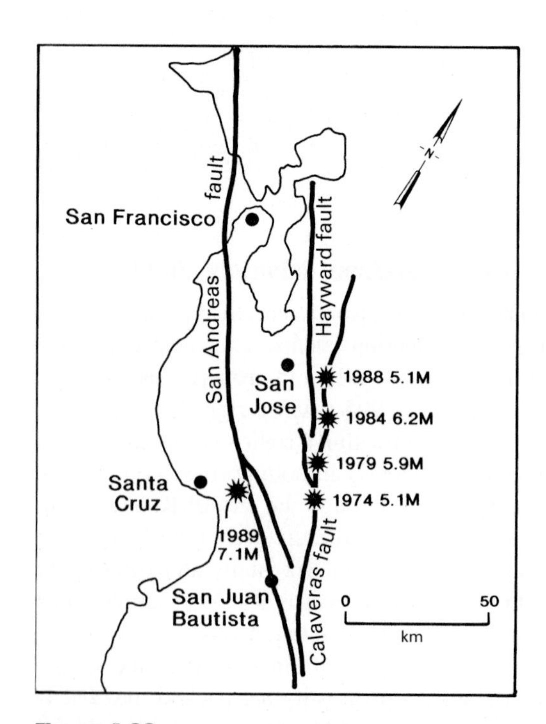

Figure 4.33 Locations and magnitudes of recent larger earthquakes in the south San Francisco Bay region.

during the seism. Despite suffering a magnitude 7 earthquake in 1596 and a magnitude 6.1 shaker in 1916, for unknown reasons, Kobe was not considered to have a strong threat of earthquakes.

In the 1995 event, many old wooden buildings with heavy tile roofs and little lateral support collapsed on sleeping residents, causing many of the 6,308 fatalities. The destroyed wooden buildings provided kindling for more than 140 fires, but luckily, the air was calm, and the lack of winds helped firefighters control the blazes. Nonetheless, 152,000 buildings were damaged or destroyed by the earthquake, and fire consumed the equivalent of 70 U.S. blocks. The infrastructure of Kobe was severely impaired as highways, railways, and port facilities were knocked out and water, sewer, gas, and electrical-power systems were severed. In addition, the recovery time for the economy takes years, and there have been increases in suicides, spousal abuse, and alcoholism.

Failed structures included massive bridges, elevated highways, and pillars supporting train tracks. The Japanese philosophy has been to build thick columns and pillars meant to stand through ground shaking using their strength, analogous to an oak tree (Figure 4.35). After these wide-

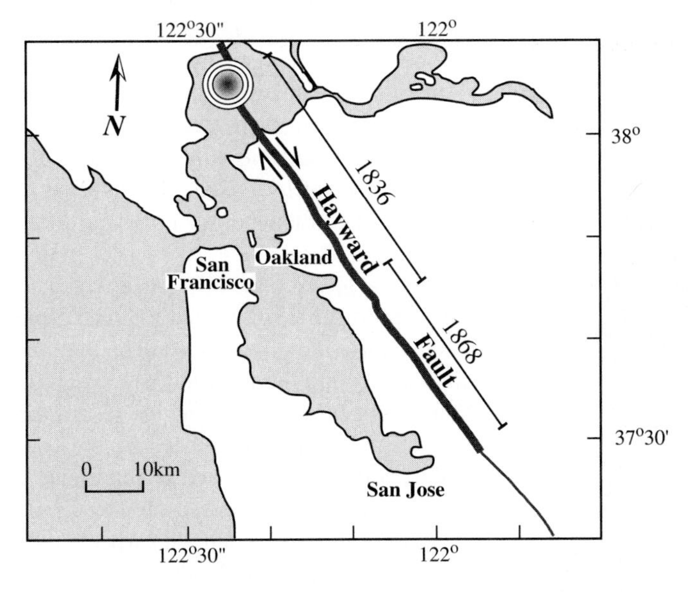

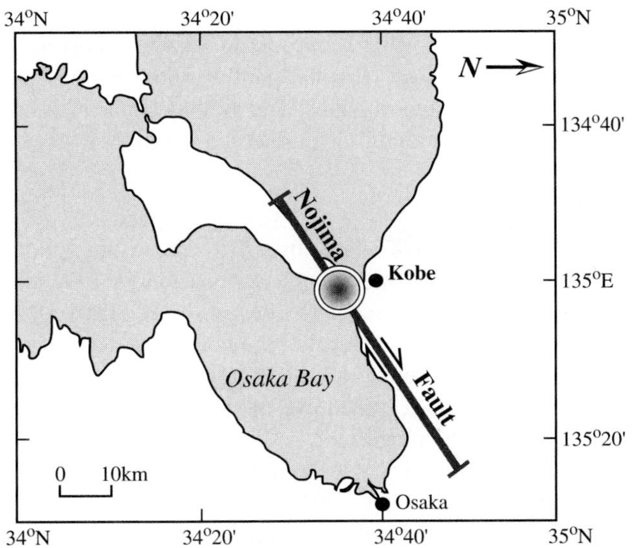

Figure 4.34 The northern section of the Hayward fault looks very similar to the Nojima fault which caused the 1995 Kobe earthquake.

spread failures, there may now be more agreement with the U.S. philosophy of designing columns that sway with the shaking, analogous to a reed.

Oakland, California, 20?? It is interesting to compare the Kobe earthquake with the upcoming seism on the northern part of the Hayward fault (Figure 4.34). Both ruptures are about 50 km long and run through heavily urbanized, major port cities with large areas of weak bay sediments and artificial fill. The closely affected East Bay area (Contra Costa and Alameda Counties) is home to 2.25 million people. At depth below 13 km, the Hayward fault is moving about 9 mm/year but the last rupture of the overlying rock along this fault section occurred in 1836. The probability of this 50 km long fault segment causing a magnitude 7 earthquake before 2020 is estimated at 28 percent. For the combined San Francisco Bay region and its 6.5 million people, there is a 70 percent (+/– 10 percent) chance that another Loma Prieta-size earthquake will occur on a fault crossing through the urban area before 2030.

For the northern section of the Hayward fault, rupture is expected to begin at the northern end and move southward for about 22 seconds with about 2 m (6 ft) of slip extending down about 13 km (8 mi). Like the 1995 Kobe earthquake, the next movement of the Hayward fault will cause tens of billions of dollars in property losses, and deaths may total in the thousands.

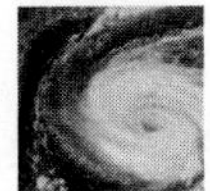

How Faults Work

Old View

The popular explanation of how faults move has been the elastic-rebound theory developed after the 1906 San Francisco earthquake. Based on surveyor's measurements of ground along the San Andreas fault, it appears that Earth stresses cause deformation and movement on both sides of a fault (Figure 4.36a and b). However, the rocks along the fault itself do not move in response to this stress because they are rough and irregular, resulting in strong interlocking bonds with **friction** that retards movement. But as the landmasses away from the fault continue to move, the buildup of stress along the fault becomes so overpowering that it finally ruptures the rocks at the fault, and both sides quickly move forward to catch up, and even pass, the rocks away from the fault (Figure 4.36c). After a fault movement, all the stress is removed from the area, and the buildup of stress begins anew. The elastic-rebound theory is somewhat analogous to snapping a rubberband or twanging a guitar string: it even accounts for aftershocks. This idea has held sway for 90 years and is described in most textbooks. It still works as a first approximation to reality, but a better understanding has emerged in the last few years.

Newer View

Movements along a fault may be better visualized as windows of opportunity. Fault movement begins at a hypocenter and then propagates outward for a certain distance and length of time. How much of the stored stress is released during an earthquake depends on the number of seconds the fault moves. For example, if there were 12 m of unreleased stress along a section of fault and the rupture event, passing by from front to end, lasted long enough for only 6 m of movement, then only half of the stress would have been released. An analogous event might be opening a locked gate to a long line of people. If the gate is held open only long enough for half the people to enter and is then closed and locked, the other people will simply have to wait until the

Figure 4.35 Despite their large size, stiff, massive support beams (note car for scale) for Kobe elevated expressway failed because of lack of ductility.

Photo courtesy of NOAA.

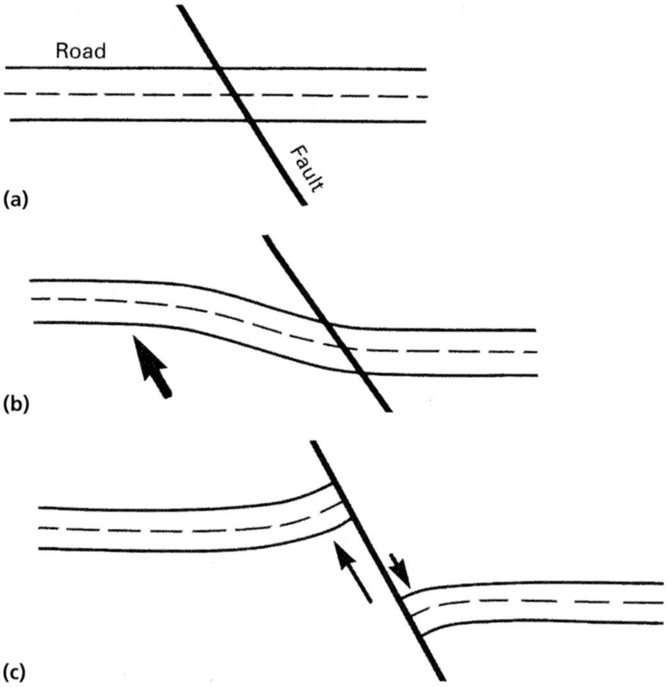

Road

Fault

(a)

(b)

(c)

Figure 4.36 Elastic-rebound theory. **(a)** An active fault with a road as a reference line. **(b)** Deformation occurs along the fault, but friction of rock masses at the fault retards movement. **(c)** Finally, the deformation is so great that the fault ruptures, and the two sides race past each other and may actually catch up with and move past the rocks farther away.

next time the gate opens. This is an important modification of elastic-rebound theory. The elastic-rebound theory has said that after a big earthquake, most of the stress is removed from the rocks and considerable time will be required for stresses to build again to a high-enough level to create another big earthquake. We no longer think this is true.

Another way to visualize this concept of fault movement is to imagine rolling out a large carpet to cover an auditorium floor. Visualize that the carpet misses covering the floor to the far wall by a foot. You try but can't pull the rug the rest of the way to the wall: it won't move because the friction is simply too great. However, if you create a large ripple in the carpet and push the ripple across the auditorium floor, then the carpet can be moved against the far wall. Faults may act the same way. A small portion of a fault may slip, creating a ripple that concentrates elastic energy at its leading edge. The farther the ripple travels, the bigger the earthquake. The moving ripple may encounter different amounts of unreleased stress in different areas of the fault.

Landers, California, 1992 New insight on how faults work was provided by the Landers area earthquakes in 1992 and 1999. This earthquake sequence began on 22 April 1992 with the right-lateral movement of the magnitude 6.1 Joshua Tree earthquake (Figure 4.37). Right-lateral movements along the fault trend resumed two months later at 4:58 A.M. on 28 June with the magnitude 7.3 Landers earthquake.

A third earthquake, triggered by the first two, broke loose a few hours later. At 8:04 A.M. on 28 June, the magnitude 6.3 Big Bear earthquake came from a left-lateral movement that ruptured northeast toward the center of the Landers ground rupture. The ruptures of the 28 June earthquakes form a triangle with the San Andreas fault as the base (Figure 4.37). These fault movements have acted to pull a triangle of crust away from the San Andreas fault, thus reducing the pressures that hold the San Andreas fault together and keep it from slipping.

Activity continued on this trend on 16 October 1999 with the right-lateral movement of the Hector Mine earthquake in a magnitude 7.1 event (Figure 4.37). Is this sequence of earthquakes finished? Probably not.

Examining the Landers earthquake records to see what happened during the 23 seconds that the faults moved 70 km (43 mi) can teach us a lot about *how faults move.*

1. Fault movements commonly are restricted to one fault and the rupture front often stops at large bends or steps in the fault, thus ending the earthquake. The Landers earthquake was different. It began right-lateral movement on the Johnson Valley fault and traveled

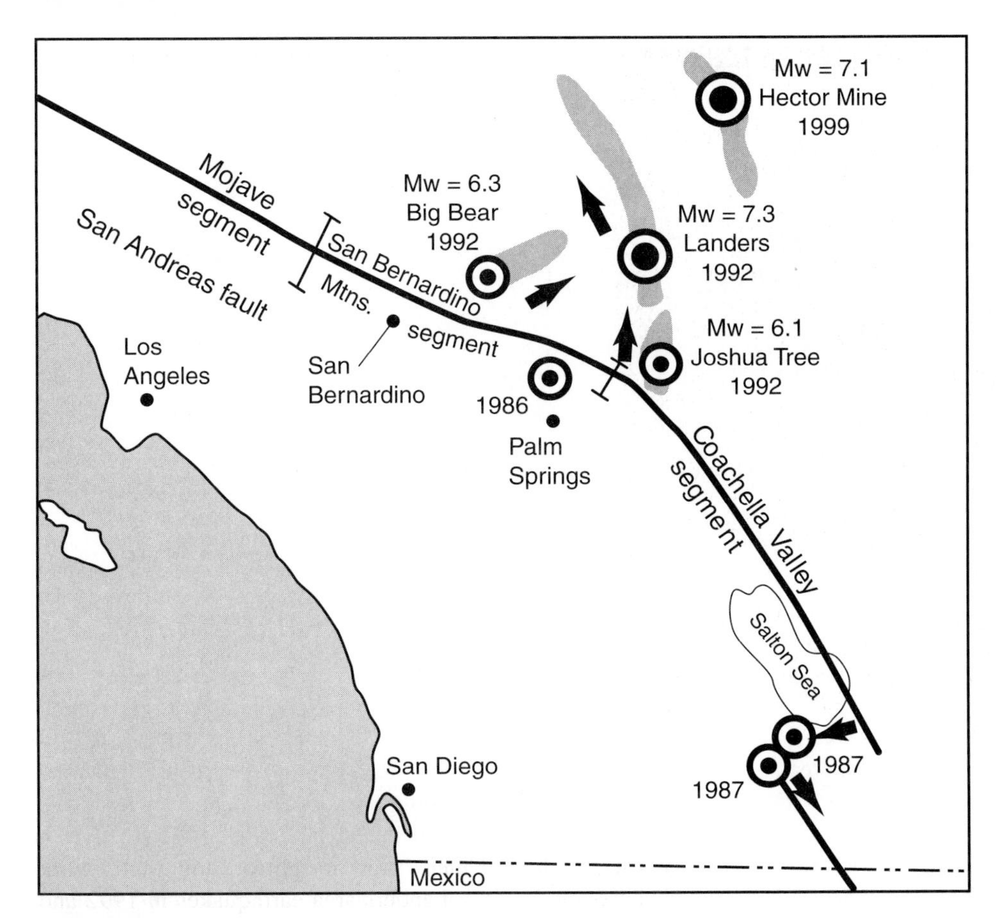

Figure 4.37 Map of major earthquakes near the northern and southern ends of the Coachella Valley segment of the San Andreas fault. The triangular block of crust near the northern end has pulled northward and away, reducing the pressures that keep this fault segment from moving.

northward about 20 km (12 mi) until reaching a right-step, pull-apart zone. The rupture front slowed, but moved through the step and continued moving northward on successive faults for another 50 km (30 mi) until finally stopping within a straight segment of the Camp Rock fault (Figure 4.38).

2. Rupture velocity on the Johnson Valley fault was 3.6 km/sec (8,000 mph), slowing almost to a stop in the right step, then continuing northward at varying speeds.
3. The amount of slip on the faults varied from centimeters to 6.3 m (21 ft) along the fault lengths and below the ground. Figure 4.39 shows the movements calculated by seismologists Dave Wald and Tom Heaton. Look at their cross section and visualize the fault in movement as the rupture front snaked its way northward, up, down, and not always involving all the fault surface.
4. Notice how the amount of fault movement at the ground surface differs from that at depth (Figure 4.39).
5. Fault movement, and shaking, at the hypocenter was modest compared to what came later.
6. As the rupture front moved northward, only a small portion of the fault was slipping at any one time. Fault

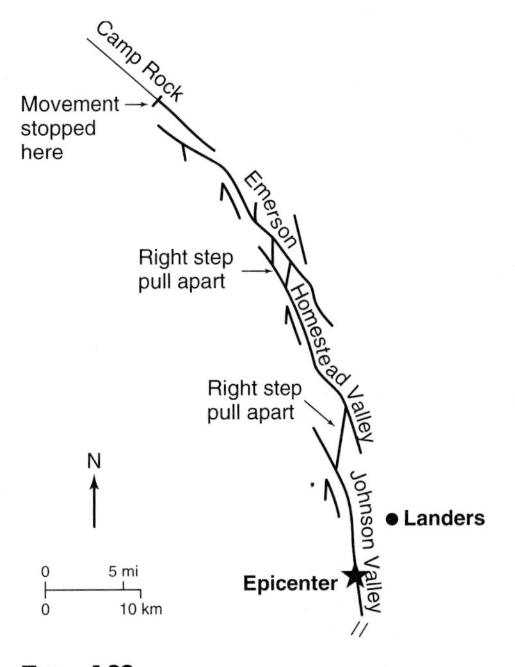

Figure 4.38 Northward-rupturing faults in 1992 Landers earthquake. Rupture front slowed at right steps then moved onto adjacent faults before stopping in the middle of a straight segment.

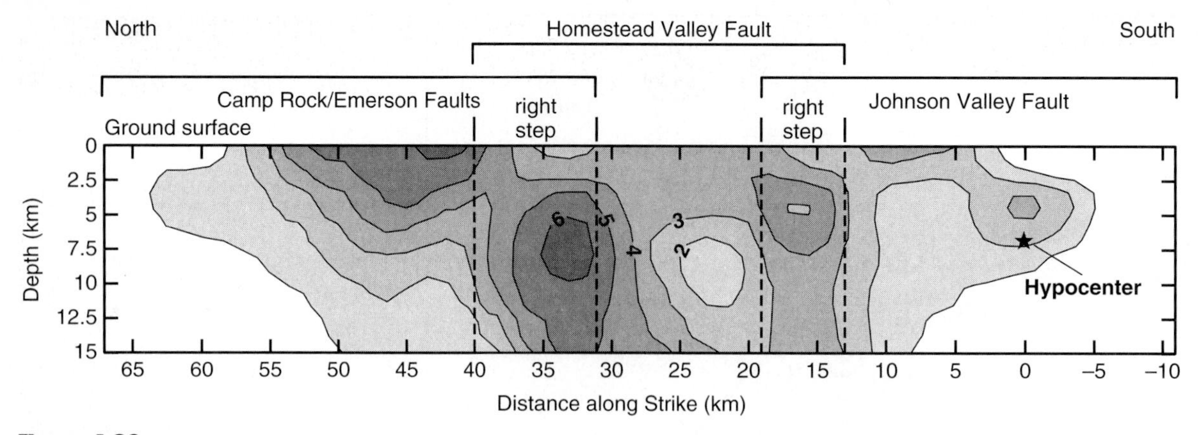

Figure 4.39 Slip on faults varied from cms to 6.3 m during movements of 1992 Landers earthquake. Contour interval of slip areas is 1 meter.

From Wald and Heaton in *Seismological Society of America Bulletin,* 84:668–691, 1994.

movement lasted 23 seconds but the longest any one fault portion moved was less than 4 seconds.

7. While the rupture front was slowed in a right step, the amount of slip behind the rupture front kept increasing until enough stress built up to cause movement through the step.
8. The earthquake triggered other earthquakes to the north in Nevada, northern California, Utah and Yellowstone Park, Wyoming. Fortunately for the Los Angeles megalopolis, the northward-moving fault directed its strongest seismic waves to the north into the sparsely inhabited desert.
9. In the 1992 Landers earthquake the faults moved from south to north; in the 1999 Hector Mine earthquake it was the opposite, as the fault moved mostly from north to south.
10. Fault patches with little or no movement on Figure 4.39 may become the origination points for future earthquakes.

Southern San Andreas Fault

Will a major earthquake occur soon on the southern San Andreas fault? The 1992 seisms occurred near the northern end of the Coachella Valley segment of the San Andreas fault. In 1988, this fault segment was assigned a 40 percent probability of having a magnitude 7.5 earthquake before 2018 (Figure 4.23). Is this probability increased by the 1992 events? Yes.

Neotectonic analyses indicate that the last four major earthquakes on the Coachella Valley segment occurred about 250 years apart, with the most recent one happening about 1680, over 320 years ago. Since 1986, seven large earthquakes have occurred at both ends of this fault segment (Figure 4.37). This is similar to the pattern in northern California before the great 1906 San Francisco earthquake.

What if the overdue movement on the Coachella Valley segment isn't just a magnitude 7.5 event that stops at its northern bend (Figure 4.37)? What if it also triggers movement of the San Bernardino Mountains segment? The combined magnitude would then be about 7.8, and the fault rupture would pass through heavily populated areas, such as San Bernardino. What if the movement continued even farther northward and the Mojave segment ruptured in turn? The combined earthquake magnitude would then exceed 8.

Is the above earthquake scenario reasonable or just an exercise in creating terror? The Coachella Valley segment last ruptured about 1680, has a repeat time of about 250 years, and has accumulated about 6 m of unreleased stress. The San Bernardino Mountains segment last ruptured in 1812 and has at least 4.3 m of unreleased stress. The Mojave segment last ruptured in 1857, has a repeat time of about 130 years, and has at least 4.7 m of unreleased stress. The rupture of all three of these fault segments in one earthquake is a reasonable assumption. The San Andreas fault in northern California ruptured through three similar segments in the 1906 San Francisco earthquake.

Thrust Fault Earthquakes in Southern California

The zone of dangerous faults in Southern California is much wider than in the San Francisco Bay area (Figure 4.13). This is largely because the major plate-bounding fault (the San Andreas) is bent so far to the west, it makes it difficult for the Pacific plate to slide along on its northwestward journey. The constriction at the "Big Bend" has caused shredding and slicing, which have split southern California into a complex maze of faults. Many of these faults trend northwest

Sidebar

Earthquake Prediction—Short-Term

Our knowledge of earthquakes is quite impressive. With plate tectonics we know *why* and *where* they occur, mostly along plate edges. With neotectonic analysis we can know *how big* and *how often* earthquakes have occurred on any fault. However, many people are not satisfied; they want the same type of short-term prediction for earthquakes that they receive daily for the weather. How close are we to being able to give short-term predictions of earthquakes? We are not close; we don't have a workable theory; and it seems quite possible that the detailed behavior of faults is too unpredictable to ever allow short-term prediction of earthquakes. There have been theories of earthquake prediction that seem logical, and they still receive coverage in textbooks, but they have all proven false. Science is a demanding thought process. Beautiful ideas may have no substance. Creative hypotheses may have no validity. The truth is elusive.

A public eager for short-term prediction of earthquakes contains many gullible people. In 1977, Charles Richter commented that "Journalists and the general public rush to any suggestion of earthquake prediction like hogs toward a full trough . . . [Prediction] provides a happy hunting ground for amateurs, cranks, and outright publicity-seeking fakers." Some people wanting prediction will grasp at almost anything.

Example 1. Much ballyhoo surrounds the rhymed prophecies published in 1555 by the French doctor Michel de Notredame (Nostradamus). Vaguely worded statements by Nostradamus are believed by some people today to predict earthquakes in our time. At the risk of being rude, the prophecies appear as truth only to undisciplined minds unable or unwilling to sort fact from fiction.

Example 2. An early 1990s prediction event occurred when a dying economist named Iben Browning decided to fill his final days with personal excitement by predicting a major earthquake in the mid-U.S. similar to the earthquakes of 1811–1812 (see Chapter 5). Scientists could readily see that his predictions were based on an old failed hypothesis, but an uncritical print and electronic media went on a binge of emotional coverage as a horde of TV crews and reporters descended on New Madrid, Missouri, eagerly awaiting the earthquake that would never come.

Example 3. A false lead was followed with fanfare as the U.S. Geological Survey predicted a magnitude 6 earthquake on the San Andreas fault in the Parkfield area (see Figure 4.13, left center of right-hand page) based on the pattern of historical seismicity. Parkfield experienced magnitude 5.5 to 6 earthquakes six times in the historical period—in 1857, 1881, 1901, 1922, 1934, and 1966. Discounting the 1934 seism, there seemed to be a pattern of an earthquake about every 22 years. U.S. Geological Survey scientists predicted the next earthquake would occur in 1988 plus or minus five years. Thus in 1984, the Parkfield Prediction Experiment was launched by deploying an unprecedented array of instruments in the field with a large team of scientists to interpret every detail of the earthquake that would come by January 1993. Then reality stepped in—the earthquake has yet to occur as of this writing.

What is our current understanding of the possibilities of short-term predictions of fault movements? First, there is no reason why the fault rupture process must occur with any regularity. Second, although it may not be hopeless to look for precursors to earthquakes, there clearly is more to earthquake triggering than can be explained simply by the steady loading of plate-tectonic stress onto faults which rupture in evenly spaced, characteristic earthquakes. The bottom line for each person is this: short-term prediction of earthquakes is not forthcoming, so plan your life accordingly. Organize your home and office to withstand the biggest earthquake possible in your area, and then don't worry about when that day will come.

with right-lateral offsets reflecting the direction of movement of the Pacific plate.

Additionally, there is another class of active faults created by southern California pushing into the "Big Bend" of the San Andreas fault. These faults are mostly east-west-oriented **thrust faults** (reverse faults) where horizontal slabs of rock (hangingwall) move up inclined fault surfaces, thus narrowing valleys and uplifting mountains. Because many of these thrust faults do not reach the ground surface, they are called **blind thrusts.** Satellite measurements of ground movement using the **global positioning system (GPS)** tell us that the Los Angeles region is experiencing a compressive shortening of 10 to 15 mm/yr. The measured deformation could generate an earthquake with a magnitude in the mid-6s every six years, plus a seism of magnitude 7 every ten years.

Northridge, California, 1994

Monday, 17 January 1994, was a holiday celebrating the birth of Martin Luther King, Jr. But at 4:31 A.M., the thoughts of most of the 12 million people in the Los Angeles area were taken over by a magnitude 6.7 (M_W) earthquake. One of the many thrust faults that underlie the San Fernando Valley, the Pico blind thrust, ruptured at 19 km (11.8 mi) depth and moved 3.5 m (11.5 ft) northward as it pushed up the south-dipping fault surface (Figure 4.40). Northridge and other cities sitting on the upward-moving

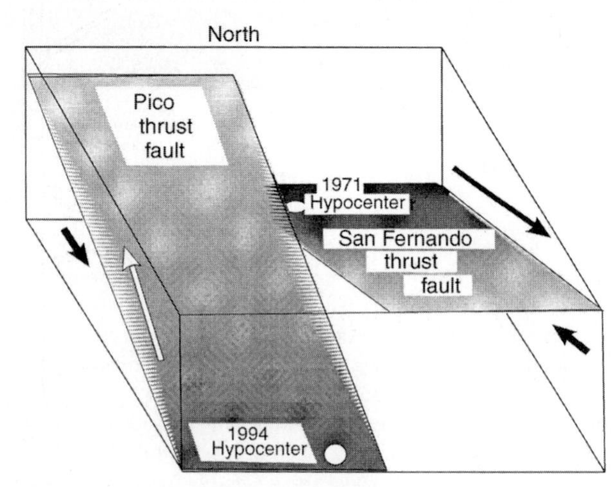

Figure 4.40 Block diagram of thrust-fault movements that created the 1994 Northridge and 1971 San Fernando earthquakes. In 1994, the Pico blind thrust fault moved 3.5 m (11.5 ft) upward from a 19 km (11.8 mi) deep hypocenter. The cities "riding piggyback" on the north-moving thrust plate experienced intense ground shaking. In 1971, a block moved southward up the north-dipping surface of the San Fernando thrust fault from a 15 km (9.3 mi) deep hypocenter.

fault slab (hangingwall) were subjected to some of the most intense ground shaking ever recorded. Ground acceleration was as high as 1.8 g (180 percent of gravity) horizontally and 1.2 g vertically. (At 1.0 g vertical acceleration, unattached objects sitting on the ground will be thrown up into the air.) This intense shaking caused the widespread failure of buildings (Figure 4.41) and bridges, which killed 61 people, injured 9,000 more, and caused $20 billion in damages. The damages included the disabling of the world's busiest freeway system, creating months of problems for drivers (Figure 4.42). In terms of deaths, this earthquake can be viewed as a near miss due to its early morning occurrence; analysis of the failed buildings leads to an estimate of 3,000 people who would have died if the seism had occurred during working hours.

The 1994 Northridge event was similar to the 1971 San Fernando earthquake, which had a magnitude of 6.5 and killed 67 people (see Chapter 3). In 1971, the movement was up a north-dipping thrust fault that abuts the blind thrust that moved in 1994 (Figure 4.40).

Measurements of Earth deformation in the Los Angeles region from global satellites (GPS) suggest that during the twentieth century, the area has been experiencing magnitude 6 earthquakes at only one-half the expected rate and magnitude 7 seisms at only one-third the long-term pattern. The 1971 and 1994 earthquakes may be omens for a more earthquake-active twenty-first century. The death and destruction from numerous magnitude 6.5 to 7 earthquakes on thrust faults within the city of Los Angeles would exceed the problems caused by a magnitude 8 or greater event on the San Andreas fault some 50 to 100 km away. However, for the 3 million people in the

Figure 4.41 Collapse of Bullocks department store in Northridge Mall. Rigid brick wall failed during ground movement.

Photo by Kerry Sieh, Caltech.

Figure 4.42 Collapse of Interstate 5 and State Highway 14 during Northridge earthquake. Vertical supports and horizontal roadbeds vibrate at different frequencies and they pulled apart.

Photo by Kerry Sieh, Caltech.

San Bernardino area, the magnitude 8 or greater event remains the biggest concern because the San Andreas fault literally runs through their backyards.

The Big One

One commonly hears talk of the Big One when discussing California earthquake hazards. Just what might the "Big One" be like? It could have more damage and deaths than are commonly imagined. Evaluations of earthquake hazards in Los Angeles typically focus on two separate fault systems. 1) The San Andreas running along the eastern and northern

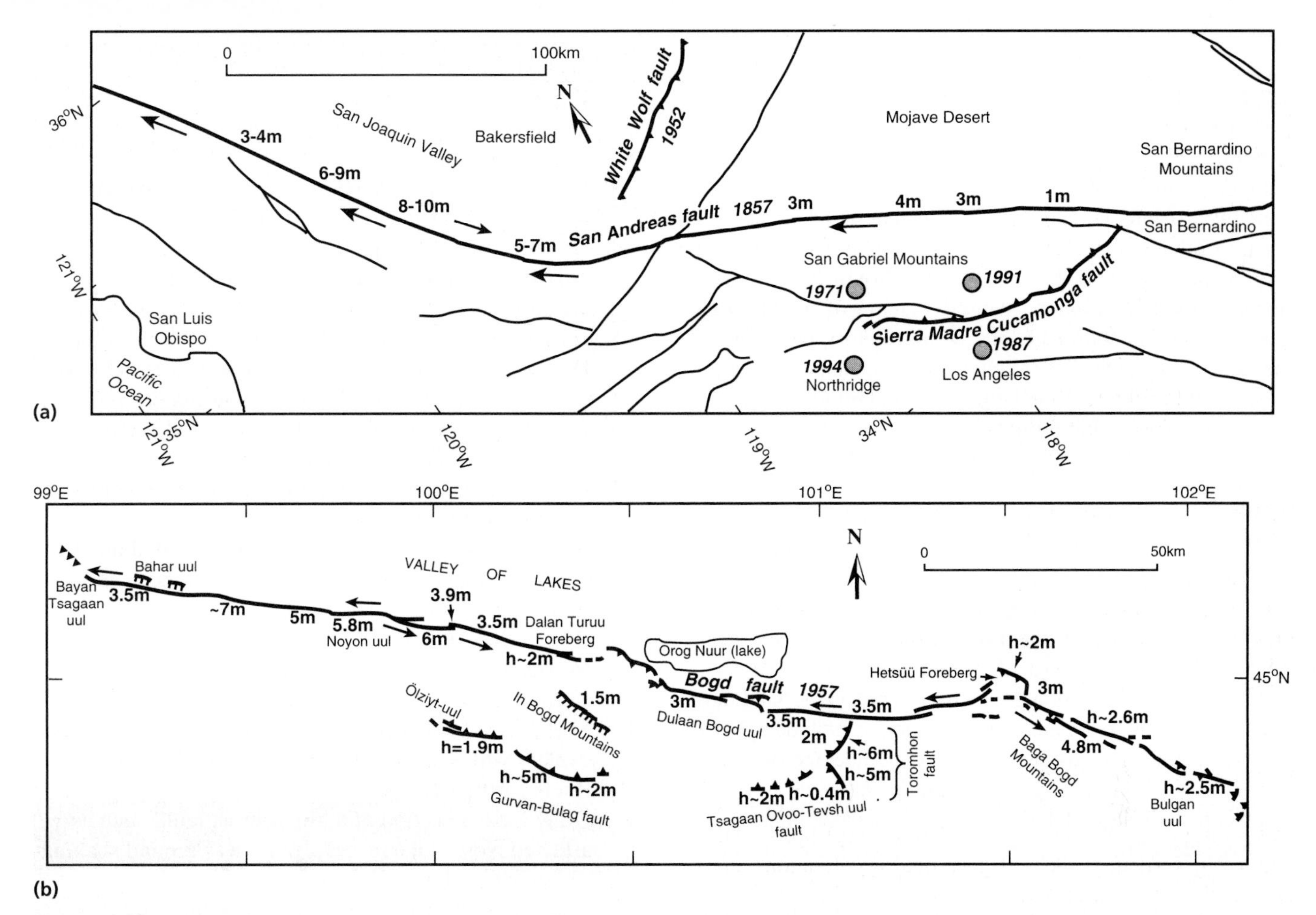

Figure 4.43 Maps of faulting in southern California **(a)** and the Gobi-Altay region of Mongolia **(b)**. **(a)** Note amounts of offset along San Andreas fault during 1857 earthquake. Dark circles are epicenters of recent earthquakes on thrust faults. **(b)** Note offsets along faults in 1957 Mongolian earthquake. The main strike-slip fault moved at the same time as the associated thrust faults.

sides of the megalopolis, which last had major movements in 1812 and 1857 (Figure 4.43a). 2) The east-west oriented thrust faults that caused the 1971 San Fernando and 1994 Northridge earthquakes. A sobering analogy from Mongolia offers a worst-case scenario for what could happen in Los Angeles. In 1957, in Mongolia, a major strike-slip fault moved with a magnitude 8 energy release at the same time that associated thrust faults moved in magnitude 7 events (Figure 4.43b). Due to the similar geometries of the fault systems in Mongolia and Los Angeles, it is reasonable to visualize the Big One as the 1857, 1971, and 1994 earthquakes all occurring simultaneously. The mind boggles when visualizing the damages associated with this Big One.

Summary

The frequent earthquakes of western North America are mostly due to plate tectonics. The westerly moving North American plate has overrun the Pacific Ocean spreading center along most of California. To the south, ongoing spreading has torn Baja California from mainland Mexico, and Baja California, San Diego, Los Angeles, and Santa Cruz are now riding on the Pacific plate toward Alaska at 5.6 cm/yr. To the north, spreading still occurs offshore from northernmost California, Oregon, Washington, and southern British Columbia. The two separated spreading centers are connected by a long transform fault—the San Andreas fault. Its earthquakes include a magnitude 8.3 caused by a 225 mi long rupture in central California in 1857, a magnitude 7.8 due to a 265 mi long rupture passing through the San Francisco Bay region in 1906, and a magnitude 7.1 unleashed by a 25 mi long rupture near Santa Cruz in 1989.

The largest earthquakes along western North America are due to subduction beneath the continent. The magnitude 9.5 Chile earthquake in 1960 was due to subduction of Nazca plate, the 9.2 Alaska earthquake in 1964 was due to subduction of the Pacific plate, and the magnitude 8.1 Mexico City event in 1985 was caused by subduction of the Cocos plate. The plates subducting beneath Oregon, Washington, and British Columbia generated a magnitude 9 earthquake on 26 January 1700, and will do so again in the future.

Major losses of life and property damage are commonly due to problems with buildings. In San Francisco in 1906, unreinforced-masonry buildings collapsed, especially those built on artificial-fill foundations. The worst damage was done by two and one-half days of fires that raged unchecked because ground shaking had broken water pipes, rendering firefighters largely helpless. In Mexico City in 1985, 1-to-2-second-period shear waves caused shaking of muddy, former lake-bottom sediments and 6- to 16-story buildings at the same 1-to-2-second frequency. The resonance of seismic waves, mud-sediment foundations, and tall buildings caused numerous catastrophic failures.

Prehistoric earthquakes may be interpreted using faulted pond sediments. The amount of offset of sediment layers is proportional to earthquake magnitude. Organic material in sediment layers can be dated by measuring the amount of radioactive carbon present. These techniques were used to forecast a 30 percent probability of a magnitude 6.5 earthquake in the Loma Prieta area by the year 2018. In 1989, a magnitude 7.1 event occurred.

Earthquake numbers and sizes have varied in the San Francisco Bay region. In the nineteenth century, magnitude 6.5 to 7 events occurred at an average of one per decade; the twentieth century was dominated by the magnitude 7.8 event in 1906. Another pattern has been the pairing of earthquakes on the west and east sides of the Bay, as in 1836 to 1838 and 1865 to 1868.

Southern California has not had its expected share of large earthquakes. The southern segment of the San Andreas fault is the only one not to have a long rupture in historic time. In prehistory, it has ruptured every 250 years on average, but the last big movement was in 1680.

The large left step in the San Andreas fault in the Los Angeles area causes compressive ruptures along east-west-oriented thrust faults as in the 1971 San Fernando and 1994 Northridge events. This type of earthquake will continue to occur.

Terms to Remember

Questions for Review

1. Sketch a plate-tectonic map along western North America from Alaska through Mexico. Label the spreading centers, subduction zones, and transform faults. Label the maximum earthquakes expected along the coastal zones.
2. If present seafloor spreading trends continue, what will happen to Baja California, San Diego, Los Angeles, and Santa Cruz?
3. Which part of the United States sits in an opening ocean basin? Evaluate the earthquake threat there.
4. Sketch a map and explain the elastic-rebound theory of faulting. How has the theory been modified in recent years?
5. How long were the surface ruptures in the 1906 San Francisco and 1857 Fort Tejon earthquakes? What was the maximum offset of the surface during each quake?
6. Evaluate the earthquake hazards in locked versus creeping segments of a fault.
7. Evaluate the seismic gap in the San Andreas fault south of San Francisco.
8. Draw a cross section and explain how faulted pond sediments can be used to tell the magnitudes and frequencies of ancient earthquakes.
9. What factors combined to cause the resonance in Mexico City that was so deadly in the 1985 earthquake?
10. Sketch a Marina District (San Francisco) dwelling and explain why so many failed during the 1989 Loma Prieta earthquake.
11. Draw a cross section of a blind-thrust fault, such as affected Northridge in 1994. Why was ground shaking so intense in this earthquake?
12. What is usually the worst time of day for a big earthquake to strike a city in the western United States?

Questions for Further Thought

1. So-called psychics are commonly quoted in the media with predictions that California will break off along the San Andreas fault and sink beneath the sea. Is this possible? Why not?
2. Which U.S. states are on the Pacific plate?
3. Which would be the better of two bad choices for an urban area: a magnitude 6.5 to 7 earthquake every 15 years or a magnitude 8 every century?
4. Why is the zone of active faults so much wider in southern than in northern California?
5. If a magnitude 9 earthquake occurred in the Cascadia subduction zone offshore from the Pacific Northwest, what might happen in Seattle, Portland, and other onshore sites?

Suggested Readings and References

Anderson, J. G., Bodin, P., Brune, J. N., Prince, J., Singh, S. K., Quaas, R., and Onate, M. (1986). Strong ground motion from the Michoacan, Mexico, earthquake. *Science,* 233, 1043–49.

Atwater, T. (1989). Plate tectonic history of the northeast Pacific and western North America. *The Geology of North America* (Vol. N, 21–72). Boulder, Colo.: Geological Society of America.

Boraiko, A. A. (1986). Earthquake in Mexico. *National Geographic,* 169, 654–75.

California Division of Mines and Geology. (1990). The Loma Prieta (Santa Cruz Mountains), California earthquake of October 17, 1989. Special Publication 104.

Canby, T. Y. (1990). Earthquake. *National Geographic,* 177, 76–105.

Earthquake Engineering Research Institute (1996) Scenario for a Magnitude 7.0 Earthquake on the Hayward Fault. EERI HF-96.

Heaton, T. H., and Hartzell, S. H. (1987). Earthquake hazards on the Cascadia subduction zone. *Science,* 236, 162–68.

Iacopi, R. (1976). *Earthquake Country.* Menlo Park, Calif.: Lane Books.

Rial, J., Saltzman, N. G., and Ling, H. (1992). Earthquake-induced resonance in sedimentary basins. *American Scientist,* 80, 566–78.

U.S. Geological Survey. (1982). The Imperial Valley, California, earthquake of October 15, 1979. Professional Paper 1254.

U.S. Geological Survey. (1989). Lessons learned from the Loma Prieta, California, earthquake of October 17, 1989. Circular 1045.

U.S. Geological Survey. (1999). Understanding earthquake hazards in the San Francisco Bay region, California. Fact Sheet 152-99.

Wallace, R. E. ed. (1990). The San Andreas fault system, California. U.S. Geological Survey Professional Paper 1515.

Videos

Killer Quake. (1994). NOVA/KCET-TV (60 min.).

The Day the Earth Shook. (1995). NOVA (55 min.).

The San Francisco Earthquake—October 17, 1989. (1989). ABC News (60 min.).

When the Bay Area Quakes. (1990). University of California–Berkeley (20 min.).

Earthquake. (1990). Nova/WGBH-Boston (55 min.).

Loma Prieta Earthquake. (1992). U.S. Geological Survey (53 min.).

Earthquake Country. (1987). United Kingdom Channel 4 (52 min.).

The American Experience: The Great San Francisco Earthquake. (1988). MOIRA Productions (60 min.).

Raging Planet—Earthquake. (1997). Discovery Channel (50 min.).

Chapter 5

More United States and Canadian Earthquakes

Eventually, everything east of the San Andreas fault will break off and fall into the Atlantic Ocean.

—Michael Grant, c. 1982, *San Diego Union*

An earthquake is deemed "significant" by the U.S. Geological Survey if it has a magnitude of 6.5 or greater, or causes deaths or considerable damage. In 1986, there were eight significant earthquakes in the United States. These included four in California associated with the transform-fault margin of the Pacific and North American plates and two in the Aleutian Islands of Alaska above the subducting Pacific plate. However, significant tremors also shook Ohio and Idaho. On 31 January 1986, a magnitude 5 earthquake near Painesville and Mentor in northeast Ohio was felt over an area of 220,000 km^2 that included 13 states and part of Canada. Seventeen people were treated for injuries. The only earthquake-caused death in the United States in 1986 happened on 12 March in the Lucky Friday mine near Mullan, Idaho. The seism was only magnitude 2, but the shaking caused highly stressed rocks in the mine to burst, killing a miner.

Awareness is growing that destructive and death-dealing earthquakes are a widespread problem, not just something that happens in California. Figure 5.1 is a map of the United States that shows the epicenters of twentieth-century earthquakes with greater than 4.5 magnitude. Seisms occur throughout the country (Table 5.1).

Figure 5.1 shows that earthquake epicenters cluster in certain areas. This chapter will examine specific earthquakes and their causes in nine regions; four in the western United States where the influence of active plate tectonics is still being felt, four in the stable (tectonically "inactive") central and eastern United States, and finally the relationship between earthquakes and volcanism in Hawaii.

Western North America: Plate Tectonic-Related Earthquakes

The immense areas and volumes of the Pacific and North American plates cause tremendous buildups and releases of energy throughout western North America (Figures 2.6 and 4.2). Remember that the North American plate is still moving southwest at 2.5 cm/yr (1 in/yr), the Pacific plate is moving northwest at 8 cm/yr (3 in/yr), and a tremendous volume of the Farallon plate has subducted beneath North America (Figure 4.2). All of this activity has lifted up the western United States and created an extensive region of high mountains and elevated plateaus.

Sidebar

Human-Triggered Earthquakes

Ohio Rocks On 25 January 2001, Ashtabula Township in Ohio was rocked by a magnitude 4.5 earthquake. The shaking damaged 50 houses and businesses as ceiling tiles fell, plaster cracked, and gas lines ruptured forcing people to evacuate. This earthquake was the biggest in a series that began on 13 July 1987 with a magnitude 3.8 event. Why did earthquakes begin and keep recurring in this industrial port city on the shores of Lake Erie? In 1986, a 1.8 km (1+ mile) deep well was drilled to inject hazardous wastes underground. For seven years, beginning in 1986, millions of gallons of waste-carrying liquids were forced down the well under pressure; earthquakes began in 1987. Was this a coincidence? No. It has already been proved that pumping fluids into or out of the ground causes earthquakes. It was shown at the Rocky Mountain arsenal in Denver in the early 1960s and at the Rangely oil field in Colorado in the early 1970s. In Ashtabula, the pressurized fluids pumped underground are flowing slowly outward from the well. Upon encountering faults at depth, the fluids add stress to underground faults causing movements big enough to do damage at the surface.

Dam Earthquakes On 1 August 1975, the town of Oroville in northern California was shaken by a magnitude 5.7 earthquake that knocked down chimneys and dumped the bottles off store shelves. Was this earthquake related to the construction of the 236 m (775 ft) high Oroville Dam and the filling of Lake Oroville in 1968? Yes. This is a common occurrence; build a dam and impound a reservoir of water, then earthquakes follow. First, impounding a reservoir causes the earth to sink isostatically. Second, and more importantly, water seeping through the floor of the reservoir flows slowly underground throughout the region pushed by the large body of reservoir water above it. The underground water moves downward and outward as an advancing front of high pressure that may reach a fault and cause it to move. For analogy, visualize what makes the water flow through the pipes in your house. In most cases, the water comes from a higher-elevation water tank or reservoir that pushes the water down through the pipes.

Bomb Blasts Underground nuclear explosions in Nevada have triggered earthquakes. Some of the atomic-bomb blasts released energy equivalent to magnitude 5 earthquakes. The bomb explosions triggered significant increases in earthquakes in their region during the 32 hours after the blast.

Anytime the level of stress or pressure is changed on rocks below the ground, earthquakes are possible. We humans can cause or trigger earthquakes.

Adjustments to this plate-tectonic activity cause earthquakes from British Columbia and Washington to New Mexico, from Montana to California, and at other points throughout the West.

Pacific Northwest: Oregon, Washington, and British Columbia

Puget Sound, Washington, 1949 and 1965 In recent years, April has brought seismic jolts to the cities in the Puget Lowlands (Figure 5.2). Two of these significant earthquakes were caused by down-to-the-east movements of the subducting Juan de Fuca plate. At 11:55 A.M. on 13 April 1949, a jolt arose from a movement 54 km (34 mi) below the Tacoma-Olympia area. The surface wave magnitude was 7.1 (M_S) and eight people lost their lives. It could have been worse, since it happened during the day and badly damaged many schools, but, luckily, it was the week of spring vacation, so the schools were largely vacant.

At 7:28 A.M. on 29 April 1965, the plate again lurched downward, this time at 60 km (37 mi) depth below the Tacoma-Seattle area. The 6.5 M_S seism killed seven people. In 1984 dollars, the destruction totaled $150 million in 1949 and $50 million in 1965. In each case, settling of soft sediments and artificial fill during the shaking caused major problems for structures built on them. There was substantial damage to older masonry buildings with inferior mortar and to buildings with inadequate ties between vertical and horizontal elements. Split-level homes suffered more than their share of damage as their different sections vibrated at different frequencies, helping tear them apart.

Puget Sound Update, 2001 It happened again. On 28 February 2001, a magnitude 6.8 earthquake radiated out from a hypocenter 52 km (32 mi) below the surface and an epicenter near the 1949 event (Figure 5.2). The seismic waves arrived at 10:54 A.M. on a Wednesday and shook the 3+ million residents of the Puget Sound for 45 seconds. In Olympia, the earthquake cracked the dome of the State Capitol, made the legislators' offices unusable and the governor's home uninhabitable. In Seattle, 30 people were caught on top of the swaying Space Needle, bricks fell from the Starbucks headquarters building onto parked cars, and Bill Gates' talk at a hotel was interrupted as overhead lights crashed to the floor and frightened people who knocked down others in their hurry to get outside. The earthquake

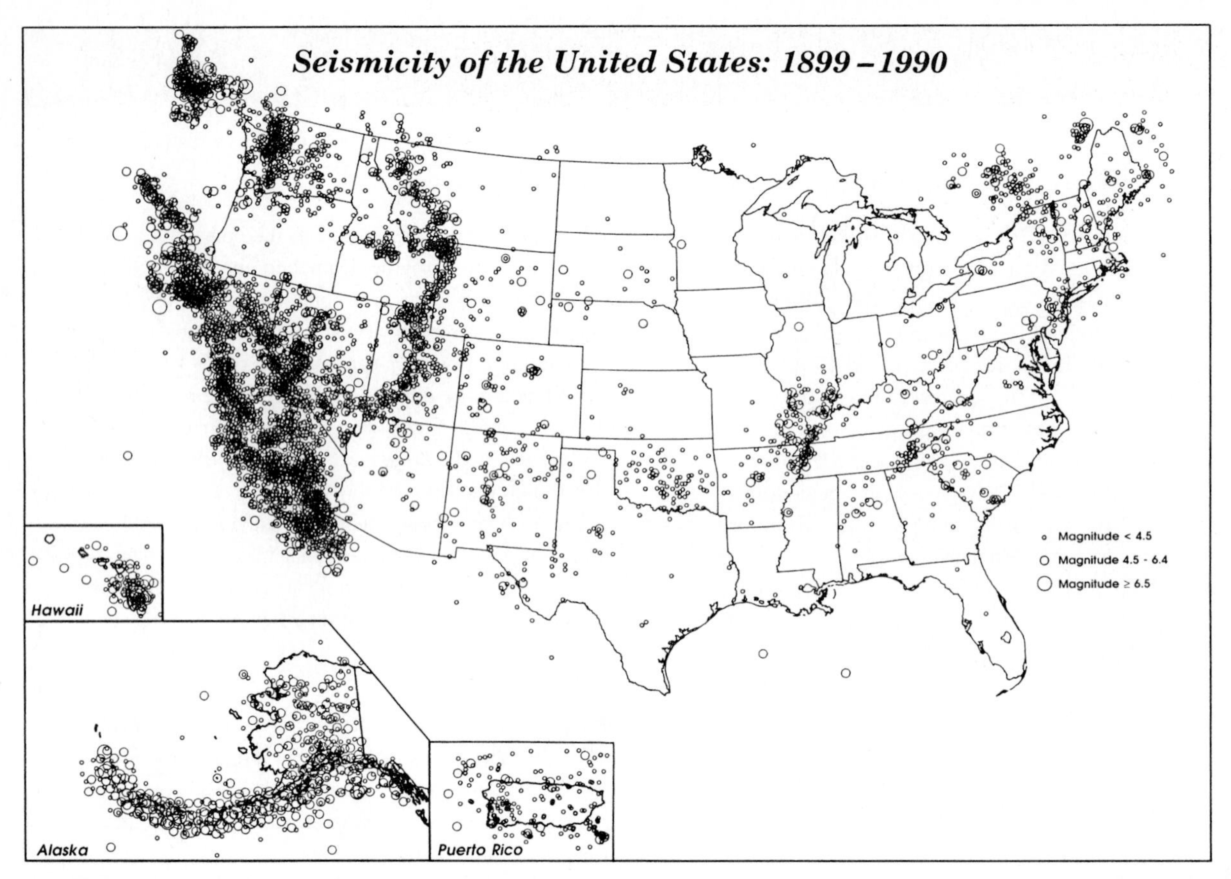

Figure 5.1 Epicenters of earthquakes in the United States, southern Canada, and northern Mexico, 1899–1990.
Source: U.S. Geological Survey National Earthquake Information Center.

killed no one, injured about 400 people, and caused about $2 billion in damages.

For all the damage the earthquake did, the damage it did *not* do is even more significant. Following the 1965 earthquake, the state of Washington improved their building codes and made many improvements, such as tying more homes to their foundations, removing water-storage tanks from the tops of school buildings, and strengthening more than 300 highway bridges. These investments were more than repaid in damages prevented and lives saved during the 2001 earthquake.

Future Earthquakes in the Washington Region Primary emphasis on earthquake hazards in the Pacific Northwest has been focused upon the subducting plates. The hypocenters in the 1949, 1965, and 2001 events were on the subducting plate at depth. When seismic waves travel upward for over 30 miles, they lose some of their energy, thus lessening the shaking of the ground surface. But a magnitude 6.5 to 7 earthquake on a near-surface fault would present a great threat, and it now appears that such seisms do occur in the region. The subducting Juan de Fuca plate is only 10 to 15 million years old and is warm and buoyant enough to couple with the North American plate. The coupling is shown by strike-slip faults on the Juan de Fuca plate that also cut the overriding North American plate.

The Seattle fault zone is oriented east-west and runs along the south side of Interstate 90 through the city of Seattle (Figure 5.2). The fault zone is 4 to 6 km (2.5 to 3.7 mi) wide and has three or more south-dipping reverse faults. A major fault movement occurred there about 1,100 years ago, as indicated by the following evidence. 1) The former shoreline at Restoration Point was uplifted about 7 m (23 ft) above the high-tide line in a single fault movement. This earthquake appears to have had a magnitude in the mid-7s, several times larger than the 1989 World Series event in the San Francisco Bay area. 2) Numerous large landslides occurred at this time, including some that carried trees in upright growth position to the bottom of Lake Washington. The age of these trees was determined by carbon-14 dating. 3) Several tsunami deposits have been recognized in the sediment layers of the area. Logs and trunks carried or buried by these large waves date to the same time period. 4) The same date appears in the ages of six major rock **avalanches** in the Olympic Mountains. The ava-

Table 5.1 **Some Significant Earthquakes in North America (excluding Alaska, California, and Mexico)**

Date	Location	Intensity	Magnitude	Fatalities
11 Jun 1638	near Plymouth, Massachusetts	IX		
5 Feb 1663	Charlevoix, Quebec	X	7.5 M	
9 Nov 1727	near Newberry, Massachusetts	VII	5+ M	
16 Sep 1732	St. Lawrence River, Ontario	IX	6.1 M	7
18 Nov 1755	offshore Cape Ann, Massachusetts	VII	6.3 M	
16 Dec 1811	New Madrid, Missouri	XI	8.2 M	
23 Jan 1812	New Madrid, Missouri	XI	8.1 M	
7 Feb 1812	New Madrid, Missouri	XI	8.3 M	
4 Jan 1843	Marked Tree, Arkansas	VIII	6.4 M	
2 Apr 1868	south tip of Hawaii	X		148
20 Oct 1870	Charlevoix, Quebec	IX	6.6 M	
14 Dec 1872	North Cascades, Washington	IX	7.0 M_s	
31 Aug 1886	Charleston, South Carolina	X	7.7 M_s	
31 Oct 1895	near Charleston, Missouri	IX	6.8 M	
31 May 1897	Giles County, Virginia	VIII	6.3 M_s	
26 May 1909	Aurora, Illinois	VIII		
2 Oct 1915	Pleasant Valley, Nevada	X	7.6	
6 Dec 1918	Vancouver Island, British Columbia	VIII	7 M_s	
28 Feb 1925	Charlevoix, Quebec	IX	7 M	
27 Jun 1925	Manhattan, Montana	IX	6.7 M	
12 Aug 1929	Attica, New York	IX	5.8 M	
18 Nov 1929	Grand Banks, offshore Canada	X	7.2 M	27
20 Apr 1931	Lake George, New York	VIII		
16 Aug 1931	near Valentine, Texas	VIII	6 M	
20 Dec 1932	Cedar Mountain, Nevada	X	7.2 M	
12 Mar 1934	Kosmo, Utah	VIII	6.6 M_s	2
18 Oct 1935	Helena, Montana	VIII	6.2 M_s	4
31 Oct 1935	Helena, Montana	VIII	6.0 M_s	
1 Nov 1935	Timiskaming, Ontario	IX	6.4 M	
9 Mar 1937	Anna, Ohio	VIII	5.3 m_b	
5 Sep 1944	near Massena, New York	IX	6 M	
23 Jun 1946	Vancouver Island, British Columbia	VIII	7.2 M_s	1
13 Apr 1949	Olympia, Washington	VIII	7.1 m_b	8
6 Jul 1954	Fallon, Nevada	IX	6.6 M	
23 Aug 1954	Fallon, Nevada	VIII	7.0 M	
16 Dec 1954	Dixie Valley, Nevada	X	7.2 M	
17 Aug 1959	Hebgen Lake, Montana	X	7.5 M_s	28
29 Apr 1965	Puget Sound, Washington	VIII	6.5 m_b	7
9 Nov 1968	southern Illinois	VII	5.5 m_b	
28 Mar 1975	Pocatello Valley, Idaho	VIII	6.1 M	
29 Nov 1975	Hawaii	VIII	7.2 M_b	2
27 Jul 1980	near Maysville, Kentucky	VII	5.1 m_b	
28 Oct 1983	Borah Peak, Idaho	IX	7.3 M_s	2
31 Jan 1986	northeast Ohio	VI	5	
12 Mar 1986	Mullan, Idaho	II	2	1
25 Mar 1993	Scotts Mill, Oregon	VII	5.6	
20 Sep 1993	Klamath Falls, Oregon	VII	5.9 & 6.0	2
3 May 1996	Seattle, Washington	VII	5.4	
24 Oct 1997	southern Alabama	VII	4.9 M_w	
25 Sep 1998	Sharon, Pennsylvania	VI	5.2	
28 Feb 2001	Puget Sound, Washington	VIII	6.8 M	
20 Apr 2002	Plattsburgh, New York	VII	5.1	
18 Jun 2002	Southwest Indiana	VII	5	

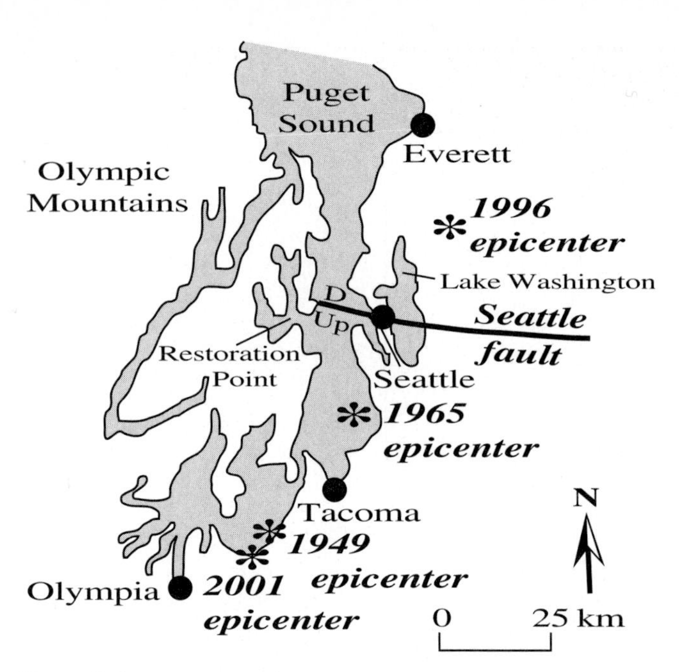

Figure 5.2 Map of Puget Sound area. U (up) and D (down) refer to the Seattle fault.

lanches apparently were shaken into action by the earthquake. 5) Coarse sediment layers on the bottom of Lake Washington were formed by downslope movement and redeposition of sediment in deeper waters. These distinctive deposits (**turbidites**) appear to have been caused by the same earthquake. Putting it all together, it appears that the Puget Sound area is subject to major fault movements that rupture the ground surface that cities are built on.

Part of Seattle sits on a 10 km (6+ mi) deep basin filled with soft sediments that shake severely during an earthquake. Seattle residents were reminded of this earthquake hazard on 3 May 1996 when a magnitude 5.4 seism struck northeast of the city. Shaking in downtown Seattle's Kingdome was intense enough in the 7th inning of a baseball game between the Seattle Mariners and Cleveland Indians to cause postponement of the game. The owner of the Seattle Mariners then tried to use the earthquake as justification for breaking his lease with the Kingdome. When the next major earthquake (> magnitude 6.5) occurs on the Seattle fault, it may cause stunning levels of death and destruction.

Western Great Basin: Eastern California, Western Nevada

Owens Valley, California, 1872 The famous naturalist John Muir was in his cabin in Yosemite Valley when:

> At half past two o'clock of a moon-lit morning in March, I was awakened by a tremendous earthquake, and though I had never before enjoyed a storm of this sort, the strange thrilling motion could not be mistaken, and I ran out of my cabin, both glad and frightened, shouting, "A noble earthquake!" feeling sure I was going to learn something. The shocks were so violent and varied, and succeeded one another so closely, that I had to balance myself carefully in walking as if on the deck of a ship among waves, and it seemed impossible that the high cliffs of the Valley could escape being shattered. In particular, I feared that the sheer-fronted Sentinel Rock, towering above my cabin, would be shaken down, and I took shelter back of a large yellow pine, hoping that it might protect me from at least the smaller outbounding boulders. For a minute or two the shocks became more and more violent—flashing horizontal thrusts mixed with a few twists and battering, explosive, upheaving jolts—as if Nature were wrecking her Yosemite temple, and getting ready to build a still better one.

What happened on 26 March 1872? The fault zone on the western side of the Owens Valley broke loose along a length of 160 km (100 mi). This is the third longest fault rupture in California history (Figure 5.3). Today, Highway 395 runs in a north-south direction, right along the faults. The 1872 faulted zone is up to 15 km (10 mi) wide, with vertical drops (normal faulting) of as much as 7 m (23 ft) and horizontal offsets (right lateral) up to 5 m (16 ft). The epicenter was near the town of Lone Pine, where 27 people, about 10 percent of the residents, were crushed to death in the collapse of their adobe (dried mud blocks) and stone houses. The seism is estimated to have had a magnitude of 7.8 to 8. So, big earthquakes do happen far away from the coastal zone and the San Andreas fault.

The Western Great Basin Seismic Trend This earthquake belt runs through eastern California and western Nevada and has a recognizable line of epicenters (Figure 5.1) and faults (Figure 5.4). In historic time, Nevada has averaged one earthquake with a magnitude in the 6s per decade and one with a magnitude in the 7s every 27 years. Why so many earthquakes? In the last 30 million years, the region between the eastern Sierra Nevada in California and the Wasatch Mountain front in central Utah has expanded in an east-west direction, opening up by several hundred kilometers (Figures 5.5 and 5.6). This extended area is known as the Great Basin or Basin and Range province (Figure 4.18). Nevada, in the heart of the extended province, has about doubled in width. As much as 20 percent of the relative motion between the Pacific and North American plates may be accommodated in the Basin and Range province. Extensional, pull-apart tectonics stretch the area, leaving numerous north-south-oriented, back-tilted mountain ranges separated by downdropped, sediment-filled basins (Figure 5.5). The extension is accomplished with normal faulting, so vertical separation dominates over horizontal slippage.

Some major earthquakes of historic times have occurred in the western part of the Basin and Range province (Figure 5.4). 1) On 2 October 1915, a large earthquake occurred south of Winnemucca in Pleasant Valley, Nevada. This magnitude 7.7 event ruptured the surface for 59 km (37 mi). The slip was dominantly vertical (normal) with displacements up to 5.8 m (19 ft) (Figure 5.7). Some fault strands had right-lateral components of offset up to 2 m (6.5

Figure 5.3 View north in Owens Valley. Fault offsets are subparallel and to left of Highway 395; note the town of Lone Pine (in center of photo) is downdropped. Lake is in right-stepping pull-apart between 2 fault segments. Alabama Hills are in left center and the Sierra Nevada in upper left.

Photo by John S. Shelton.

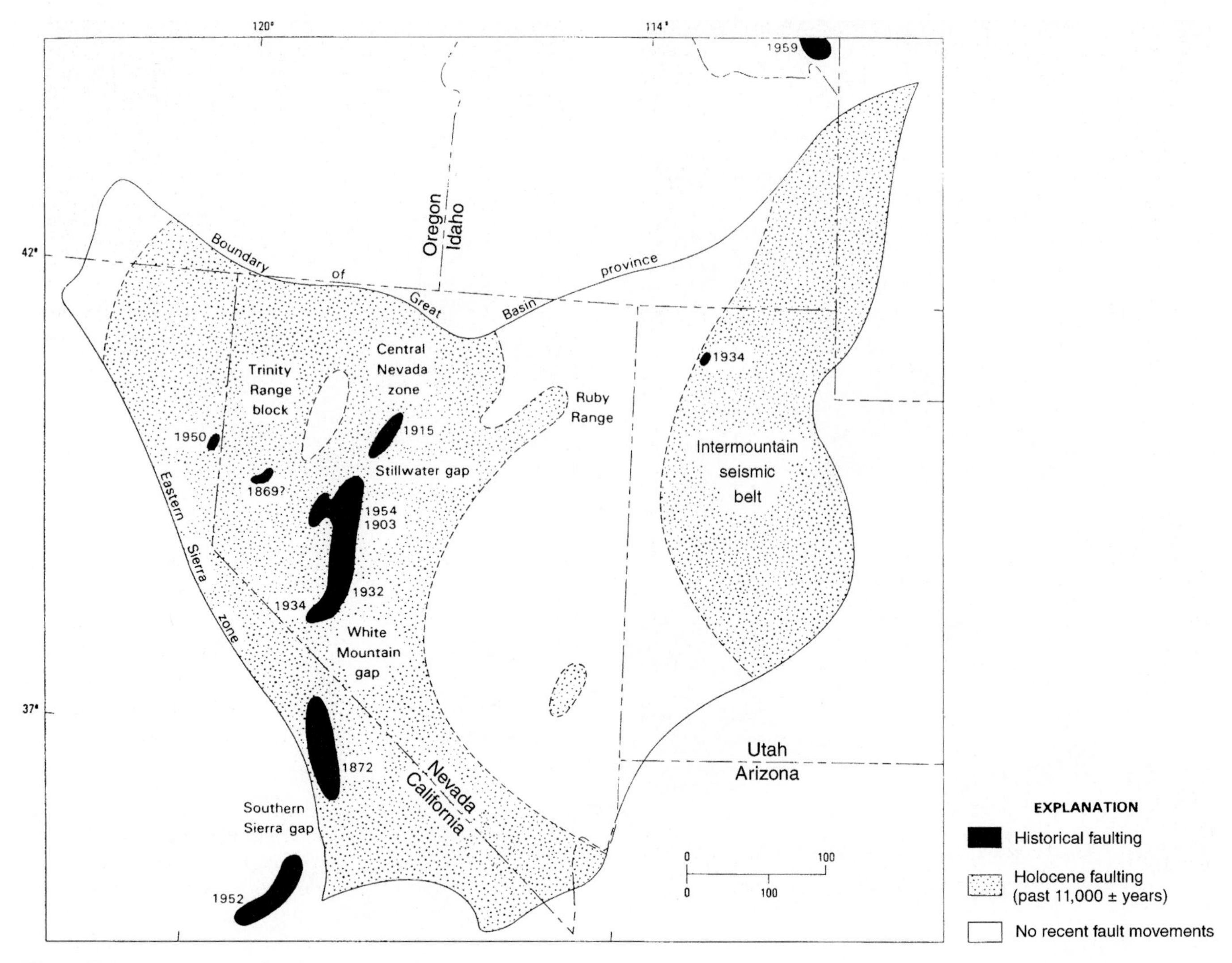

Figure 5.4 Generalized map of historic faulting in the western Great Basin.

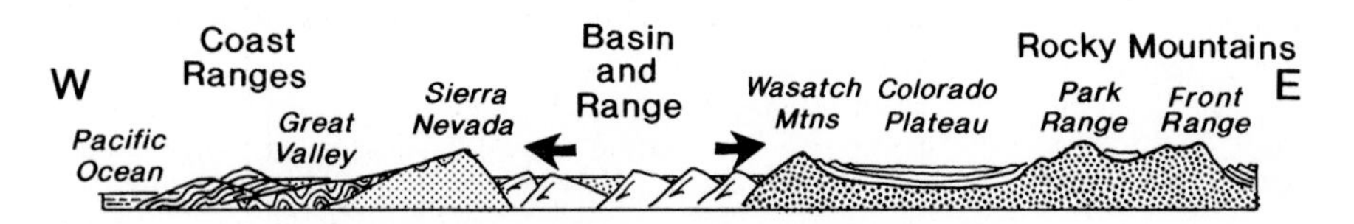

Figure 5.5 Schematic cross section oriented west-east across the western United States. Basin and Range province has stretched to double its initial width. This extension has created normal faults that generate earthquakes.

ft). 2) On 21 December 1932, a magnitude 7.3 event occurred near Cedar Mountain, Nevada, rupturing the ground for 61 km (38 mi). 3) The year 1954 was a big one for earthquakes in Nevada. Events included a magnitude 6.6 on 6 July and a 6.9 on 24 August near Fallon, and two shocks of 7.3 and 6.9 rocked Dixie Valley on 16 December. Figure 5.4 shows several gaps in the belt of historic seismicity. Residents in these seismic gaps may be in for some surprises.

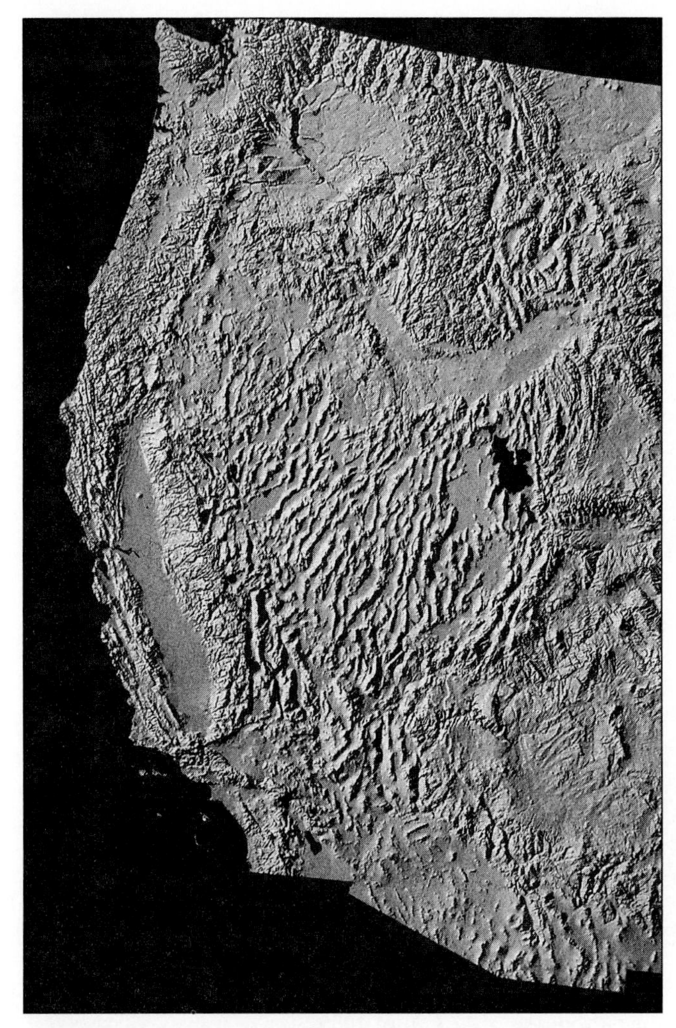

Figure 5.6 Computer-generated image of topography in the western United States. Notice in the center the Basin and Range topography of north-south-oriented mountains separated by linear valleys.
Photo courtesy of *U.S. Geological Survey.*

Figure 5.7 A portion of the fault scarp created during the 1915 earthquake in Pleasant Valley, Nevada.
Photo courtesy of Pat Abbott.

Intermountain Belt: Utah, Idaho, Wyoming, Montana

Hebgen Lake, Montana, 1959 The Rocky Mountains in the summertime are a beautiful place to be. On the moonlit evening of 17 August 1959, campers were settled into their spots at the Rock Creek Campground at the foot of the high walls of the Madison River Canyon. But at 11:37 P.M., an odd wind blew briefly down the canyon at high velocity. The wind was created by the push of an enormous rock slide. The south wall of the canyon dropped 43 million cubic yards of rock that slid down the steep slope, across the Madison River, and moved about 150 m (500 ft) up the north wall (Figure 5.8). It entombed 26 campers. The gigantic landslide buried the canyon to depths of 67 m (220 ft) and created a natural dam that began trapping a large body of water—Earthquake Lake.

What caused this life-ending landslide? At Hebgen Lake, directly west of Yellowstone National Park, two sub-parallel faults (Hebgen and Red Canyon) moved within five seconds of each other with 6.3 m_b and 7.5 M_S events (Figure 5.9). These two normal faults had their southwestern sides drop 7 and 7.8 m (23 and 26 ft) down fault surfaces inclined 45 to 50° to the southwest, also dropping the northern end of Hebgen Lake. The foreman at the Hebgen Lake Dam was awakened by the earthquake, dressed, and went outside for a look. In foreman Hungerford's own words:

> The dust was so intense you could hardly see. You could hardly breathe, or anything. It obscured the moon. We went to the river gage. . . . Just as we got to it, we heard a roar and we saw this wall of water coming down the river. . . . We thought the dam had broken. . . . Then we went up to the dam. When we got there we couldn't see much, but I walked over to the edge of the dam and all we could see was blackness. There was no water. No water above the dam at all, and I couldn't imagine what had become of it. By that time the dust had started to clear, and the moon had come out a little. And then here came the water. It had all been up at the other end of the lake. . . .We rushed back when we heard the water coming. We could hear it before we could see it. When it came over the dam, it was a wall of water about three to four feet high completely across that dam, and it flowed like that for what seemed to me to be 20 minutes, but possibly it could have been 5 or 10. I have no idea of time. It flowed for a while, and then it started to subside. Then it all cleared away, and no water again. The lake was completely dry as far as we could see. All we could see down the dam was darkness again. It seemed like a period of maybe 10 to 15 minutes, and the water came back, and then it repeated the same thing over again.

Hungerford was eyewitness to a spectacular **seiche** event where lake water sloshed back and forth for 11.5 hours. A seiche is analogous to what happens to your bathwater when you stand up quickly from a full tub of water—it sloshes back and forth. Over 50 mi^2 of land on the northern side of Hebgen Lake dropped down over 10 ft. The warping of the lake floor set off a huge series of seiches.

Figure 5.8 Madison Canyon landslide and resulting lake, caused by the earthquake of 17 August 1959. Photo taken 24 days later by John S. Shelton.

Borah Peak, Idaho, 1983 Just after 7 A.M. on 28 October 1983, the Lost River fault broke free at 16 km (10 mi) below the surface and ruptured northwestward 0.45 m (1.5 ft) horizontally and 2.7 m (9 ft) vertically for a 7.3 M_S event (Figures 5.9 and 5.10). When the fault finished moving, Borah Peak, Idaho's highest point, was 0.3 m (1 ft) higher, and the floor of Thousand Springs Valley was several feet lower. The ground shaking caused Thousand Springs Valley to live up to its name as underground water squeezed out by the subterranean pressures spouted fountains 3 to 6 m (10 to 20 ft) high. The escaping water erupted in volcano-like fashion, throwing out sandy sediment to build low cones and filling their craters with water.

The Intermountain Seismic Belt The Intermountain seismic belt is a northerly trending zone at least 1,500 km (930 mi) long and about 100 to 200 km (60 to 125 mi) wide (Figure 5.9). The belt extends in a curved pattern from southern Nevada and northern Arizona into northwestern Montana. The seismic activity here closely follows the Great Basin's boundary with the Colorado Plateau and the middle and northern Rocky Mountains (Figure 5.4). In effect, the seismic belt is the eastern boundary of the extending Basin and Range province. The bounding faults on the eastern side of the Great Basin are down-to-the-west, whereas the bounding faults on the western side (in eastern California and western Nevada) are down-to-the-east. The earthquakes reaffirm that this part of the world is being stretched and pulled apart.

In historic times, large seisms have occurred in eastern California and western Nevada on the west and in Montana and Idaho on the east but not on long sections of faults in Utah. About 75 percent of Utah's population lives within sight of the **scarps** of the 370 km (230 mi) long Wasatch Front, the zone of normal faults separating the mountains from the Great Basin (Figure 5.11). No large earthquakes have been reported along the Wasatch Front faults since the arrival of Brigham Young in 1847. In 1883, the famous geologist G. K. Gilbert warned the people of Utah of their

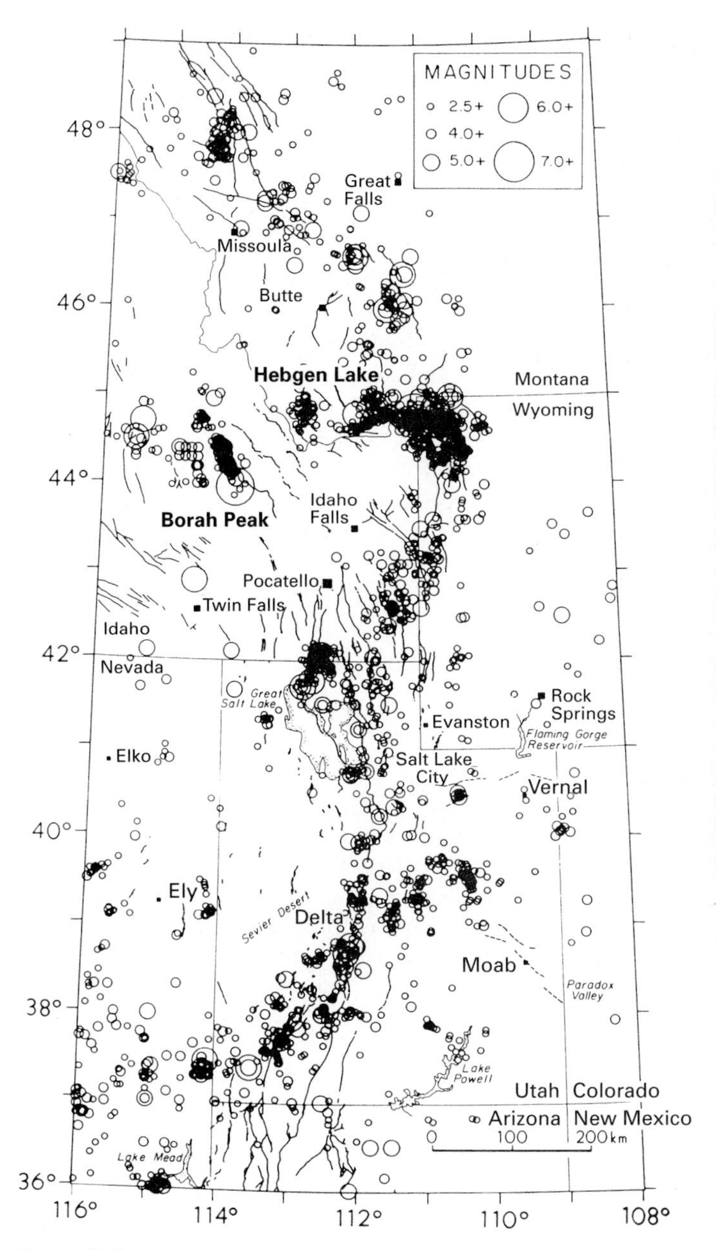

Figure 5.9 Earthquake epicenters in the intermountain seismic belt, 1900–1985.

Figure 5.10 The 26 October 1983 Borah Peak earthquake was a 7.3 M_S event with 8.2 ft (2.5 m) of vertical offset. Notice there also was some left-lateral offset. Mt. Borah (in background) was uplifted slightly by this event.

Photo courtesy of NOAA.

earthquake threat and the danger for their towns in an article in the *Salt Lake Tribune.* The sharply defined faults show obvious potential for earthquakes (Figure 5.12). The fault segments shown in Figure 5.11 are each capable of events like those of Hebgen Lake and Borah Peak. In the last 6,000 years, a magnitude 6.5 or stronger earthquake has occurred about once every 350 years on one of the Wasatch system faults. Parts of Salt Lake City, Provo, and Ogden lie on soft lake sediments that will shake violently during a large seism. The fault segment near Brigham City has not moved in the last 2,400 years and is a likely candidate for a major event.

Rio Grande Rift: New Mexico, Colorado, Westernmost Texas, Mexico

The Rio Grande rift is one of the major continental rifts in the world. It is a series of interconnected, asymmetrical, fault-block valleys that extend for more than 1,000 km (620 mi) (Figure 5.13). Here, the continental crust is being heated and stretched. The crust responds by thinning, which involves extensional (normal) faulting. In the last 26 million years, there has been about 8 km (5 mi) of crustal extension near Albuquerque, New Mexico, a rate of about 0.3 mm/yr. The dominant motion on the faults is vertical, and the offset totals 9 km (5.5 mi). The rift basin is strikingly deep in places, yet most of the vertical relief created by fault offsets has been lessened by the copious quantities of volcanic materials and sediments that have poured into the rift over millions of years.

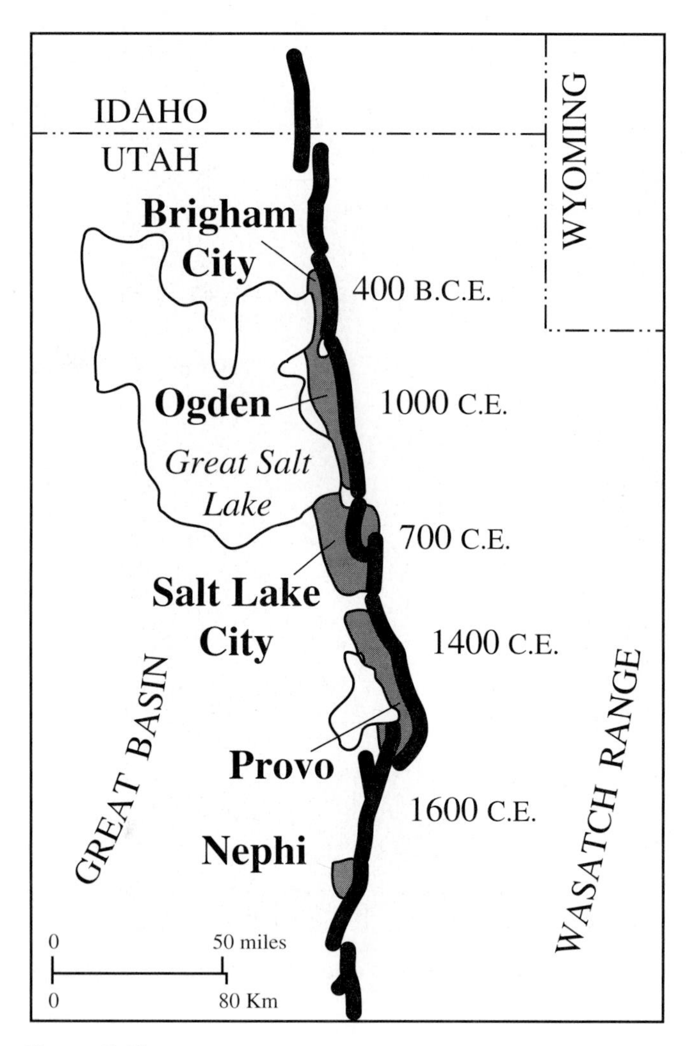

Figure 5.11 Map of faults along the Wasatch Front, Utah. The Wasatch fault has several segments. Dates of the most recent magnitude 6.5 or greater seisms are shown.

The topographic trough of the rift valley has attracted a major river (Rio Grande), which in turn has enticed human settlers in need of water. Today's settlements include Albuquerque, Socorro, and Las Cruces in New Mexico, El Paso in Texas, and Ciudad Juarez in Mexico. Historic earthquakes have had only small to moderate magnitudes, but the continental lithosphere continues to extend, thus presenting a real hazard for large earthquakes.

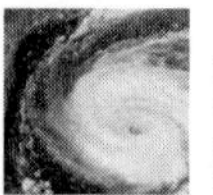

Intraplate Earthquakes: "Stable" Central United States

The map of earthquake epicenters in the United States (Figure 5.1) shows that the western third of the country has an elevated level of seismic activity. But there are clusters of epicenters in the "stable" central and eastern United States, the intraplate regions away from the active plate edges. There are not as many epicenters, but some individual earthquakes are just as big.

New Madrid, Missouri, 1811–1812

A succession of earthquakes rocked the sparsely settled central part of the Mississippi River Valley at the time of the War of 1812. Between 16 December 1811 and 15 March 1812, Jared Brooks, an amateur seismologist in Louisville, Kentucky, recorded 1,874 earthquakes. He classified eight of them as violent and another ten as very severe. The four largest events occurred on 16 December 1811 (two), 23 January 1812, and 7 February 1812. The hypocenters were located below the thick pile of sediments where the Mississippi and Ohio Rivers come together, at the upper end of the great Mississippi River **embayment** (Figure 5.14). These major seisms are called the New Madrid earthquakes, taking their name from a town of 1,000 people. Although few people were killed, the destruction of ground and buildings at New Madrid tolled the end of its importance as "The Gateway to the West."

The following is excerpted from an eyewitness account of a New Madrid earthquake.

> Accompanying the noise, the whole land was moved and waved like waves of the sea, violently enough to throw persons off their feet, the waves attaining a height of several feet, and at the highest point would burst, throwing up large volumes of sand, water, and in some cases a black bituminous shale, these being thrown to a considerable height, the extreme statements being forty feet, and to the top of the trees. With the explosions and bursting of the ground there were flashes, such as result from the explosion of gas, or from the passage of the electric fluid from one cloud to another, but no burning flames; there were also sulphuretted gases, which made the water unfit for use, and darkened the heavens, giving some the impression of its being steam, and so dense that no sunbeam could find its way through. With the bursting of the waves, large fissures were formed, some of which closed again immediately, while others were of various widths, as much as thirty feet, and of various lengths. These fissures were generally parallel to each other, nearly north and south, but not all. In some cases instead of fissures extending for a considerable distance there were circular chasms, from five to thirty feet in diameter, around which were left sand and bituminous shale, which latter would burn with a disagreeable sulphorous smell. . . .

The region is composed of thick deposits of water-soaked, unconsolidated sands and muds dropped by the Mississippi River. These loose materials intensified the shaking of the earthquakes, and the weak sediments flowed like water, erupted as sand volcanoes, and in places quivered like Jell-O. Several long-lasting effects of the New Madrid earthquakes can still be seen in the topography. A 240 km (150 mi) long area alongside the Mississippi River sank into a broadly depressed area, forming two new lakes: Lake St.

Figure 5.12 Aerial view eastward over Salt Lake City to the Wasatch fault running along the base of the mountains.
Photo by John S. Shelton.

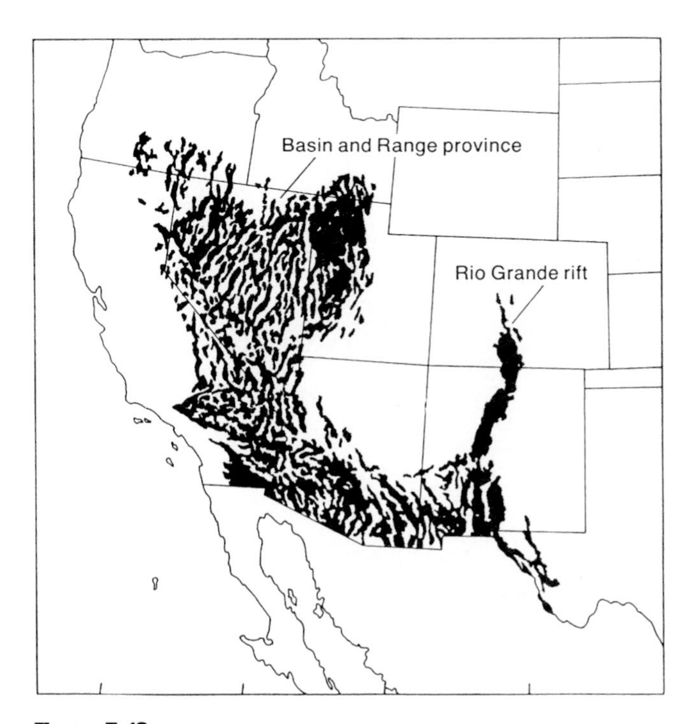

Figure 5.13 Extended topography of Basin and Range province and Rio Grande rift. The sediment-filled basins are colored dark.

Francis, 60 km (37 mi) long and 1 km (0.62 mi) wide, and Reelfoot Lake in Tennessee, 30 km (19 mi) long, 11 km (7 mi) wide, and up to 7 m (23 ft) deep. Reelfoot Lake, now a bird sanctuary, hosts the gray trunks of cypress trees drowned over 185 years ago; they still stand as silent testimony to the area's earth-wrenching events (Figure 5.15). Other topographic features created by the seisms include: 1) long, low cliffs across the countryside and streams with new waterfalls up to 2 m (7 ft) high; 2) domes squeezed up as high as 6 m (20 ft) and with lengths as great as 24 km (15 mi); and 3) former swamplands uplifted and transformed into aerated soils.

The New Madrid earthquakes have never been equaled in the history of the United States for the number of closely spaced, large seisms and for the size of the felt area (Figure 5.16). The earthquakes were felt from Canada to the Gulf of Mexico and from the Rocky Mountains to the Atlantic seaboard, where clocks stopped, bells rang, and plaster cracked. These mammoth earthquakes were not a freak occurrence. The oral history of the local American Indians tells of earlier dramatic events. Neotectonic analyses of sediment and wood indicate major earthquakes also occurred roughly around the years 500, 900, 1300, and 1600.

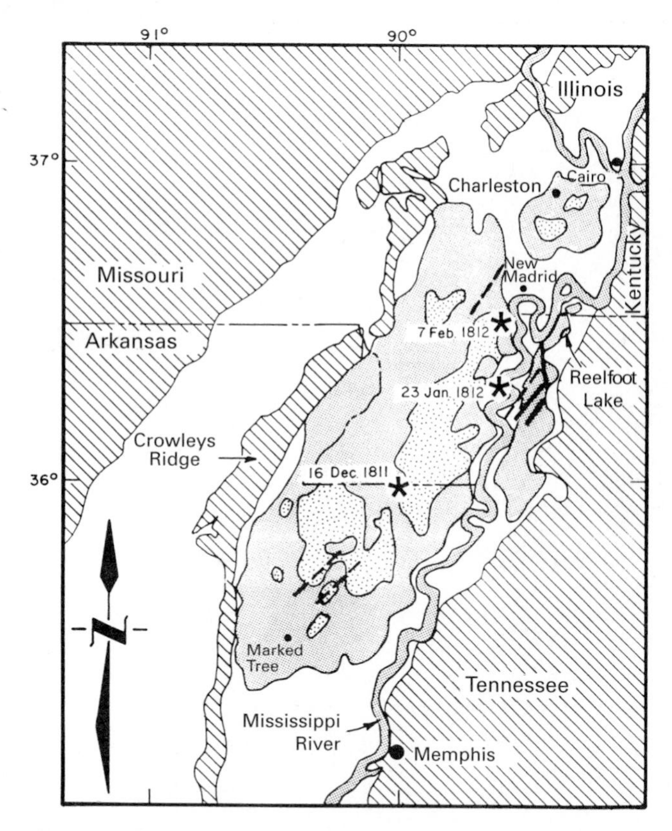

Figure 5.14 Map of New Madrid region showing epicenters of large earthquakes and ground effects of the 1811–1812 events. Marked Tree, Arkansas, is the site of a magnitude 6.4 event in 1843; Charleston, Missouri, had a magnitude 6.8 seism in 1895.

Are the sizes of the felt areas in Figure 5.16 a good indicator of earthquake magnitude? Were the New Madrid seisms many times bigger than the 1906 San Francisco earthquake? Not necessarily. The size of the felt area is related to the types of rocks being vibrated. The New Madrid seisms shook the rigid basement rocks (more than one billion years old) of the continental interior. They rang like a bell, and the seismic energy was transmitted efficiently and far. The San Francisco earthquake took place in younger, tectonically fractured rocks that quickly damped out the seismic energy, thus confining the shaking to a smaller area. The New Madrid and San Francisco earthquakes were of similar magnitudes.

The four largest New Madrid seisms had estimated body wave magnitudes (m_b) of 7.2, 7.0, 7.1, and 7.4 and estimated moment magnitudes (M_w) of 8.2, 7.8, 8.1, and 8.3. When large earthquakes return again to the upper Mississippi River region, the potential for death and destruction is staggering (Figure 5.17). 1) The area has a large population (e.g., St. Louis, Memphis); 2) most buildings were not designed to withstand large seisms; 3) the wide extent and great thickness of soft sediments will amplify seismic vibrations (remember the 1985 Mexico City and 1989 World Series events); and 4) a very large area will be subjected to strong shaking. The effects of a magnitude 8 earthquake could include deaths in the thousands and damages in the tens of billions of dollars. If there is good news, it is the low frequency of occurrence for these extra large earthquakes. The lessons of history here must be learned and all new con-

Figure 5.15 Trunks of cypress trees drowned in Reelfoot Lake after water was dammed by the New Madrid earthquakes.
Photo courtesy of *U.S. Geological Survey.*

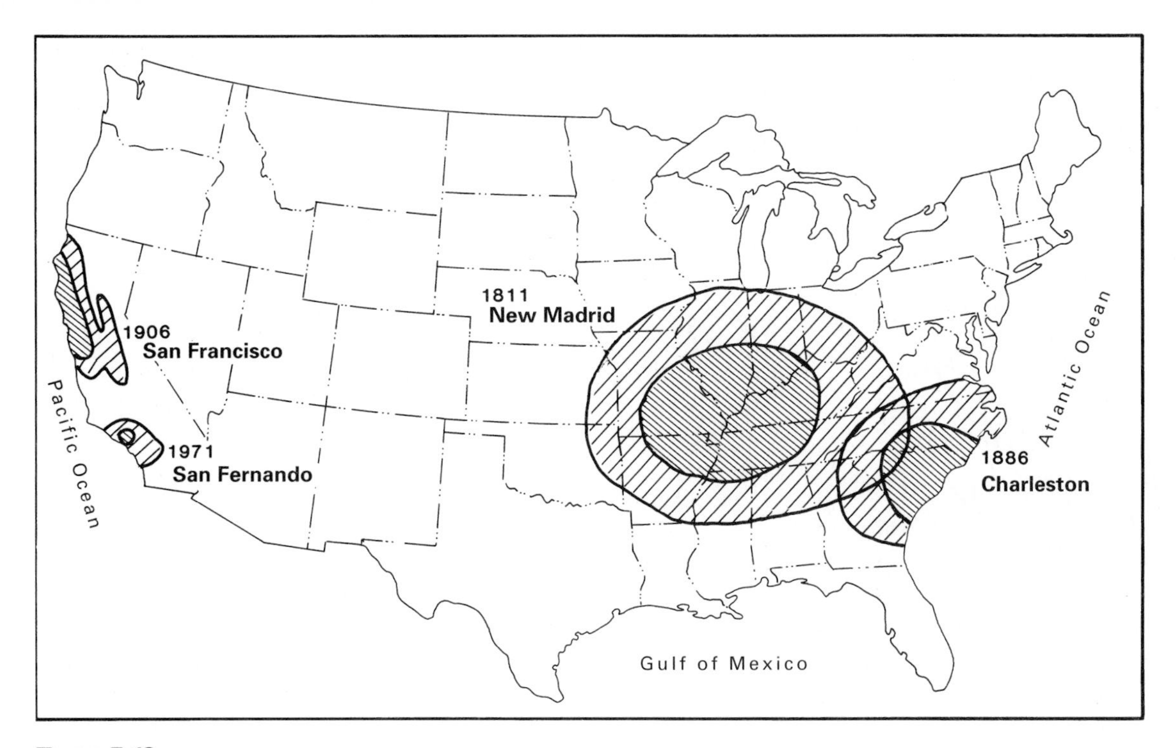

Figure 5.16 Felt areas of some large earthquakes in the United States. Inner-ruled areas are Mercalli intensities greater than VII; outer-ruled areas are intensities VI to VII.

struction in the region must be built to withstand major earthquakes.

The earthquake threat in this region also includes lesser events. Paleoseismologic analysis of trenches cut across faults and folds has led the U.S. Geological Survey to forecast a 90 percent chance of a magnitude 6 to 7 earthquake here within the next 50 years. Remember the 1994 Northridge earthquake was a magnitude 6.7, and it killed 61 people and caused $40 billion in damages. Although earthquakes in the central United States have low probability, they have high impact.

Another pertinent question concerns the cause of these seisms. Since New Madrid sits in the continental interior, away from the active plate edges, why would such earthquakes occur here? The answer is unresolved, but a look at the geologic history of the region provides some understanding.

Reelfoot Rift: Missouri, Arkansas, Tennessee, Kentucky, Illinois

Figure 5.1 shows that earthquake epicenters in the New Madrid area line up in an elongate pattern. The close-up view in Figure 5.14 shows that the epicenters of the large earthquakes also line up along the Mississippi River Valley. Figure 5.18 is a map of the southern and eastern United States depicting the distribution of coastal-plain sediments—the sands and muds dropped by rivers eroding the North American landmass. The Mississippi embayment stands out as a prominent feature. Why are sediments here deposited so much farther into continental North America? Why does the sediment distribution parallel the epicenters? Is it a coincidence that this same linear pattern keeps reappearing? No. Is it random chance that the Mississippi River flows along the course that it does? No.

The results of studies of seismic waves, gravity, and magnetism define a linear structural feature in the basement rocks underlying the New Madrid region (Figure 5.19). There is a northeast-trending depression at depth that is more than 300 km (190 mi) long and about 70 km (43 mi) wide. It is linear, has nearly parallel sides, and is about 2 km (1.2 mi) deeper than the surrounding basement rocks. In short, it is an ancient rift valley, known as the Reelfoot rift, formed about 550 million years ago. Similar features today include the Rio Grande rift in New Mexico (Figure 5.13) and the East African Rift Valley (Figures 2.23 and 2.26). The ancient Reelfoot rift has been filled with sedimentary rocks, and later the whole region was covered by extensive, younger sediments (Figure 5.19). Today, the opening Atlantic Ocean basin pushes North America to the west-southwest, and some of the ancient faults of the Reelfoot rift possibly are being reactivated to produce the region's earthquakes. Are these earthquakes unique to the New Madrid area? Or does the United States have other failed rifts at depth that could pop forth with major earthquakes?

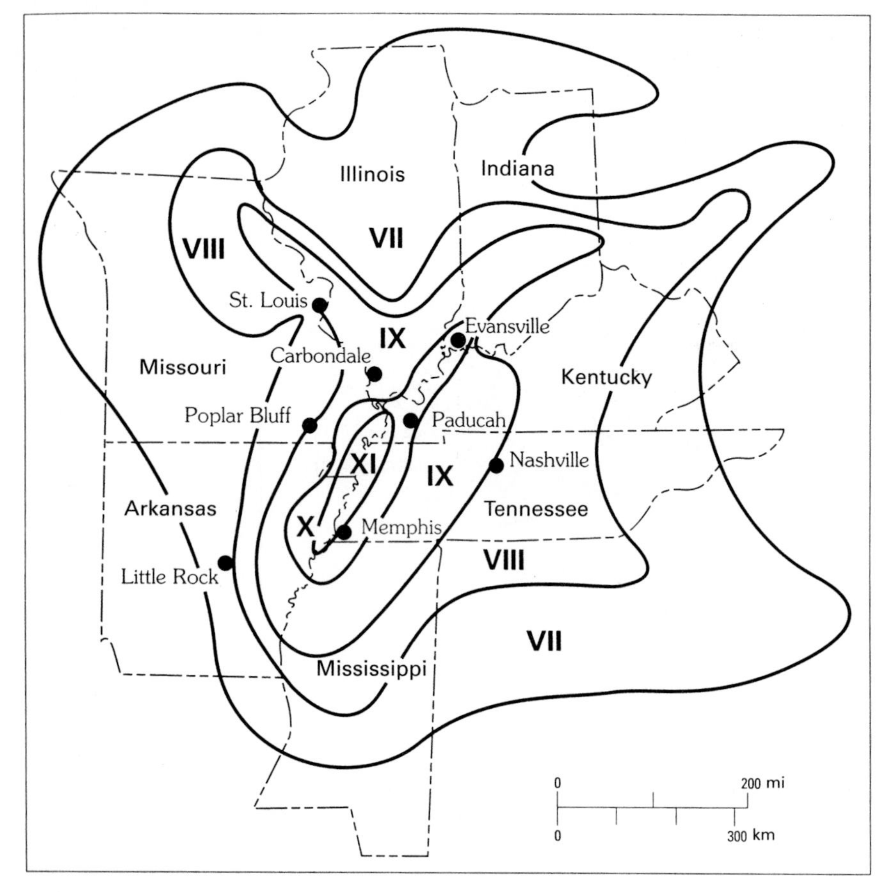

Figure 5.17 Map showing estimated Mercalli intensities expected from a recurrence of an 1811–1812 New Madrid earthquake. Intensity VIII and above indicates heavy structural damage.

Source: R. M. Hamilton and A. C. Johnston, "Tecumseh's prophecy: Preparing for the next New Madrid earthquake" in *U.S. Geological Survey Circular 1066.*

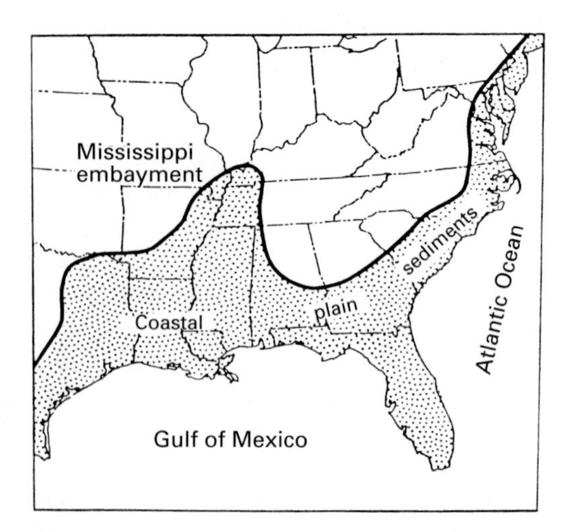

Figure 5.18 Map of coastal-plain sediments deposited by rivers eroding North America. Note that the Mississippi River embayment juts northward well into North America. Why? There is a failed rift at depth. A smaller, failed rift heads off into southern Oklahoma.

Ancient Rifts in the Central United States

It is the fate of all continents to be ripped apart from below. Continents are rifted, then drifted and reassembled in different patterns. Sometimes the rifting process stops before separating a continent. Figure 5.20 shows some rift arms developed 220 to 180 million years ago as Pangaea was torn apart. Some rift arms succeeded, combining to create today's Atlantic Ocean basin. Other rift arms failed and left behind weakened zones within continents. The Reelfoot rift, now occupied by the Mississippi River, is a prominent **failed rift;** it encompasses the epicenters of large earthquakes. Other failed rifts, from even older plate-tectonic histories, also exist beneath the surface in North America (Figure 5.21).

Failed rifts remain as zones of weakness that may be reactivated by later plate-tectonic stresses to once again generate earthquakes. Because failed rifts are deeply buried, they are difficult to study. Yet they raise significant questions. What are the frequencies of their major earthquakes? In general, the recurrence intervals for major earthquakes appear to be from a few hundred to more than a thousand years. How great an earthquake might be produced at each rift? The New Madrid earthquake series offers a sobering

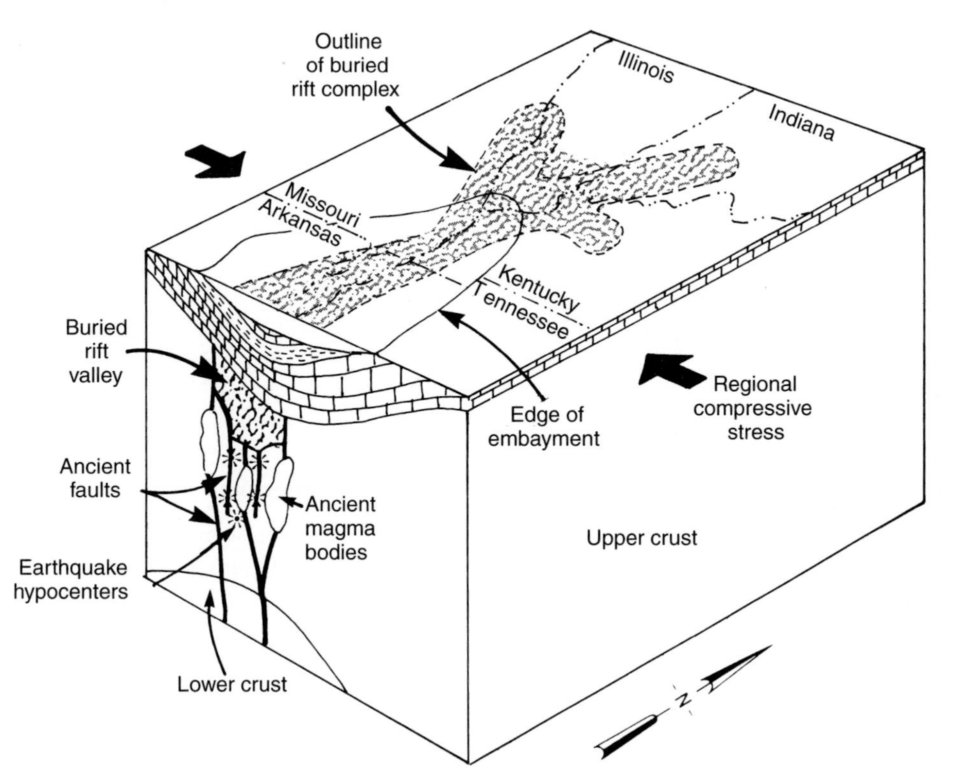

Figure 5.19 Schematic block diagram of the Reelfoot rift, the ancient failed rift valley beneath the upper Mississippi River embayment. Large earthquakes are likely caused by present tectonic stresses triggering failures on ancient faults.
Source: *U.S. Geological Survey Professional Paper 1236L.*

benchmark. Approximately 83 percent of the large earthquakes recorded in the central United States are at or near the sites of ancient rifts.

The buried, ancient rifts in Figure 5.21 correlate with active fault zones at the surface. There are several examples. 1) The St. Louis arm corresponds to the Ste. Genevieve fault zone. 2) The Rough Creek rift is expressed as the Rough Creek fault zone, appearing to continue eastward as the Kentucky River fault zone. Trenches dug across the Rough Creek fault have exposed the sedimentary records of reverse-fault movements with 1.1 m (more than 3.5 ft) of offset. 3) The southern Oklahoma rift corresponds with the frontal-fault system of the Wichita Mountains. Although this zone does not generate earthquakes at present, the land surface testifies to major earthquakes. The Meers fault is dramatic enough to make any Californian proud, but this fault strikes N63°W across southwestern Oklahoma. Its fault scarp is 5 m (16 ft) high and 27 km (17 mi) long, and has left-lateral offset ranging from three to five times greater than vertical offset. At least two major fault ruptures have occurred there in geologically recent time. 4) The southern Indiana arm is overlain by the Wabash Valley fault zone, which appears to connect with the New Madrid zone. Prehistoric earthquakes read in the sedimentary record suggest seisms with m_b equal to 6.3 and more than 7. Damaging earthquakes occur in the area about once a decade. Examples include a magnitude 5.5 in November 1968, a magnitude 5.2 in June 1987, and a magnitude 5 in southwest Indiana on 18 June 2002.

Intraplate Earthquakes: Eastern United States

New England Earthquakes

New England has a long record of significant earthquakes. On 11 June 1638, just 18 years after the Pilgrims landed in Plymouth, Massachusetts, a sizable earthquake rocked them. It rattled dishes, shook buildings, and in general frightened the Europeans, who were unfamiliar with earthquakes. Due to the limited number of settlements, it is difficult to pinpoint the location of the fault movement that generated this earthquake. However, a suggested epicentral site lies offshore from Cape Ann; estimates of Mercalli intensity range all the way up to IX.

On 9 November 1727, an earthquake rattled the East Coast from Maine to Delaware. The epicenter was near Newbury, Massachusetts (Figure 5.22), and the shaking caused chimneys and stone walls to fall and cellar walls to collapse. Some uplands were dropped down to become wet lowlands, and some wet lowlands were uplifted and became

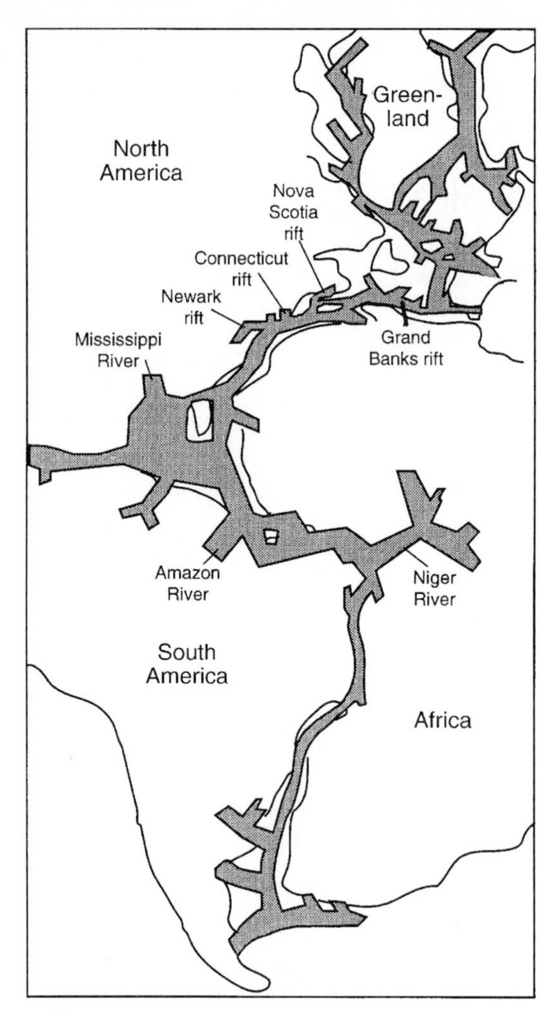

Figure 5.20 Schematic map of rifts that tore at Pangaea about 220 million years ago. Successful rifts combined to open the Atlantic Ocean basin.

dry enough to support grasses. **Quicksand** conditions were common during the earthquake.

Shortly before dawn on a frigid 18 November 1755, the entire eastern seaboard from Nova Scotia to South Carolina was shaken with an earthquake that began offshore from Cape Ann, Massachusetts. In Boston, so many chimneys were reported to have toppled that some streets were made impassable by the debris. The seism is estimated to have had a magnitude of about 6.3, but the shaking was so severe that residents reported seeing the land rolling with waves like the surface of the sea. This earthquake occurred just 17 days after the epic earthquakes in Lisbon, Portugal, and it fired up the doom-and-gloom preachers who saw the seism as just punishment for the sins of New Englanders.

Many of these earthquakes may be related to the faults that bound former rift valleys (Figure 5.20). The ancient faults may be reactivating and failing due to current stresses. Do all the rift-bounding faults have the potential for future seismic activity? The historic record is not long enough to properly answer this question. But if the answer is yes, then virtually the length of the Atlantic Coastal province could receive a significant shake sometime.

When the next magnitude 6 or greater earthquake strikes the eastern United States, the resultant destruction is likely to be proportionately greater than for a similar seism in the western part of the country. In the east, earthquake ground motion is transmitted more effectively in the older, more solid rocks, so damages may be experienced over a wider area. Also consider 1) the population density of the East, 2) the large number of older buildings not designed to withstand earthquake shaking, and 3) the concentration of industrial and power-generating facilities, including nuclear reactors.

St. Lawrence River Valley Earthquakes

The St. Lawrence is another river whose present path results from occupying an ancient tectonic structure. During late Precambrian–early Paleozoic time, some 600 to 500 million years ago, a major rift valley extended through the region (Figure 5.22). This now-buried rift coincides with most of the significant earthquakes in southeastern Canada. Seisms within the rift valley commonly reach magnitude 7, yet in unrifted continent nearby, the largest earthquakes are usually only in the magnitude 5 range. The most active area along the St. Lawrence River Valley is an 80 km (50 mi) by 35 km (22 mi) zone near Charlevoix, northeast of Quebec City. Here, earthquakes of magnitudes 6 to 7 occurred in 1534, 1663, 1791, 1860, 1870, and 1925. Why the concentration of large seisms in this one relatively small area? Charlevoix was the site of a meteorite impact some 350 million years ago. The impact caused intensive fracturing of the area, including arcuate faults. These impact-caused fractures may be being reactivated today under the stresses generated by the opening Atlantic Ocean basin.

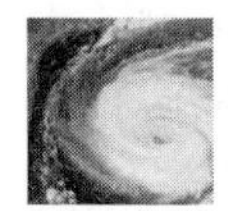

Fracture-Zone Hypothesis of Major Earthquakes

The difficulty of explaining the causes of earthquakes in eastern North America has led to a speculative hypothesis based on the transform faults that offset the mid-Atlantic Ocean spreading center. The transform faults continue northwestward as great **fracture zones** that seem to line up with some onland seismic zones (Figure 5.23). If the fracture zones continue beneath the North American continent, they would be linear trends of weakness. The ongoing westward spreading of the North Atlantic plate could cause earthquake-producing movements on the fracture zones beneath eastern North America.

The Newfoundland fracture zone lines up with numerous earthquakes, including the 18 November 1929 Grand Banks event. This earthquake occurred offshore, causing a

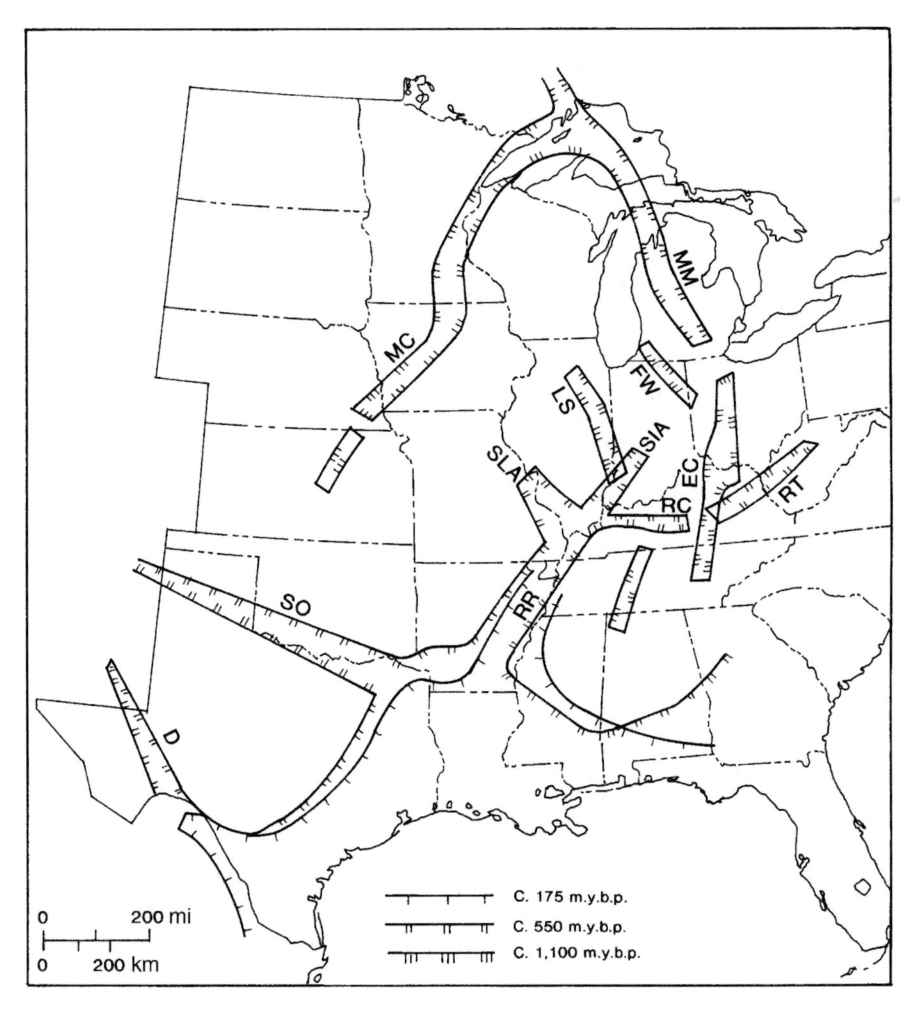

Figure 5.21 Map showing approximate locations of buried, ancient rifts in the central United States. Rifting occurred during three principal times—around 220 to 175 million years ago, 600 to 500 million years ago, and 1,100 to 1,000 million years ago. Some older rifts were apparently rifted again under later plate-tectonic regimes. Rifts are: D, Delaware; EC, East Continent; FW, Fort Wayne; LS, La Salle; MC, Mid-Continent; MM, Mid-Michigan; RC, Rough Creek; RR, Reelfoot rift; RT, Rome Trough; SIA, Southern Indiana Arm; SLA, St. Louis Arm; and SO, Southern Oklahoma.

Source: D. W. Gordon, *U.S. Geological Survey Professional Paper 1364*.

mammoth submarine landslide that in turn generated tsunami that swept away and drowned 27 people from Newfoundland's south coast.

The New England **seamounts** are a chain of submerged, extinct volcanoes extending under New England and lining up with the Cape Ann epicenters and the northwest-trending belt of seisms extending through Boston and into Ottawa, Canada (Figure 5.1).

Notice that the Blake fracture zone extends through Charleston and the South Carolina seismic belt (Figure 5.23).

Charleston, South Carolina, 1886

Charleston sits alongside a beautiful bay, a charming city with distinguished buildings, wide boulevards, and inviting gardens. The presence of the port helped the city develop as a wealthy trading center. Yet Charleston has another side. In the mid-1800s, it was a hotbed of secessionist fervor; the first shots of the Civil War were fired over its harbor at Fort Sumter on 12 April 1861. After the war ended four years later, Charleston was

> a city of ruins, of desolation, of vacant houses, of widowed women, of rotting wharves, of deserted warehouses, of acres of pitiful and voiceless barrenness.

Yet by the mid-1880s, Charleston had been restored as a center of wealth, aesthetic buildings, and cultural achievement. Even the damages wrought by an 1885 hurricane were not enough to slow the city. Then came 31 August 1886, a typical sultry summer day. At 9:50 P.M., the quiet, breezeless evening was shattered by the largest earthquake to occur east of the Appalachian Mountains in historic time. Sixty seconds of shaking left 60 people dead, and once again, the remaining citizens would have to put their city back together.

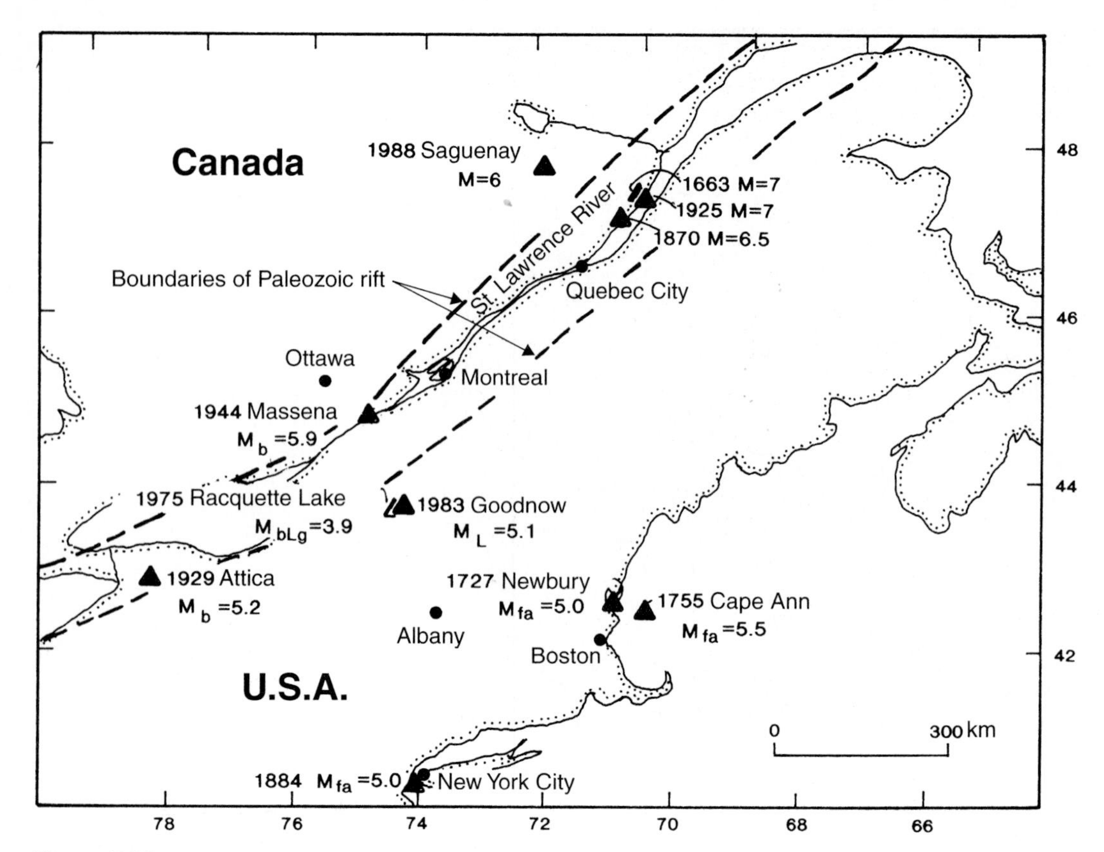

Figure 5.22 Some earthquake locations in the St. Lawrence River valley area and the approximate location of the 600 to 500 million-year-old rift valley. M_{fa} equals magnitude estimated from felt area. Large earthquakes northeast of Quebec City lie in circular Charlevoix seismic zone.

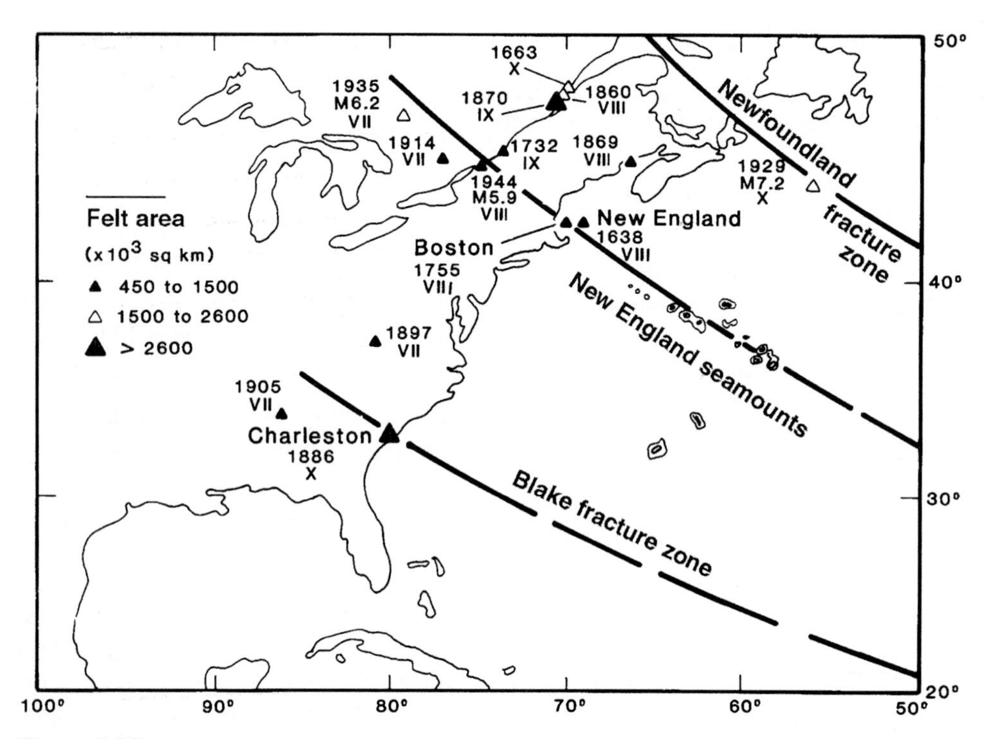

Figure 5.23 Location of earthquake epicenters in the eastern United States and Canada and fracture-zone extensions of transform faults on the mid-Atlantic spreading center.

Figure 5.24 Some damage from the 1886 earthquake in Charleston, South Carolina.

Photo by J. K. Hillers, *U.S. Geological Survey.*

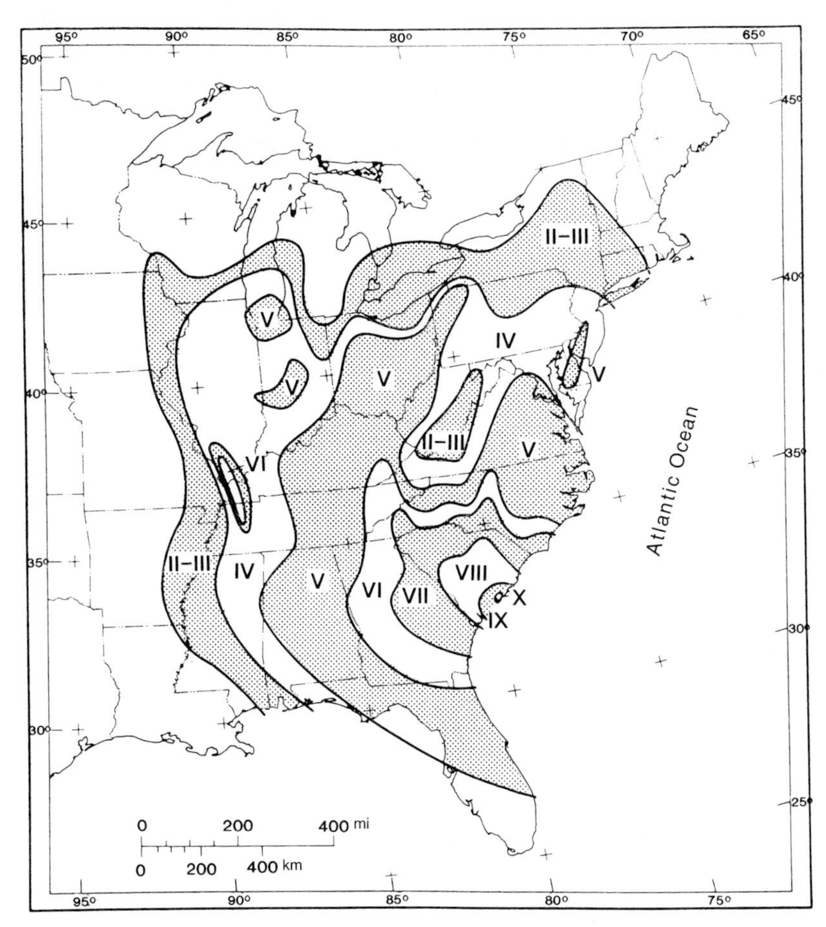

Figure 5.25 Mercalli intensity map for the 31 August 1886 earthquake near Charleston, South Carolina.

Source: G. A. Bollinger, *U.S. Geological Survey Professional Paper 1028,* 1977.

About 90 percent of the buildings were damaged or destroyed (Figure 5.24).

The earthquake apparently had a body-wave magnitude (m_b) of 6.7 and a surface-wave magnitude (M_s) of 7.7. The event produced no surface faulting, so the fault movement may have occurred below 20 km (12 mi) depth. The large magnitude corresponds to a rupture length of about 30 km (19 mi) and a rupture width on the fault surface of about 19 km (12 mi). The seism was felt over a very large area, and damages were widespread as well (Figure 5.25). The large felt area is typical of the eastern United States and is largely attributable to the persistence of longer-period seismic waves (e.g., one second), which simply do not die down as quickly as they do in the western United States. The continental rocks at depth are geologically old and rigid, causing the region to "ring like a bell" and transmit seismic waves far and wide.

How rare is an earthquake of this size for Charleston? Sediments exposed in trench walls, augmented by radiocarbon dates, tell of at least three other similar-sized earthquakes in the area in the last 3,000 to 3,600 years. Thus, large seisms may be expected about every 1,000 years.

Earthquakes and Volcanism in Hawaii

When one thinks about natural hazards in Hawaii, it is volcanism that comes to mind. But the movement of magma can cause earthquakes, including large ones (Table 5.2). When rock liquefies, its volume expands, and neighboring brittle rock must fracture and move out of the way. The sudden breaks and slips of brittle rock are fault movements that produce earthquakes. When magma is on the move at shallow depths, it commonly generates a nearly continuous swarm of relatively small earthquakes referred to as **harmonic tremors.** Figure 5.26 shows that the earthquakes below Kilauea volcano are dominantly near-surface events.

Magma movements also cause larger-scale topographic features and larger earthquakes with magnitudes in the 6s and 7s. The land surface is commonly uplifted due to the injection of magma below the ground surface. But the land surface is also commonly downdropped due to withdrawal of magma. Figure 5.27 shows some downdropped valleys on Kilauea; valley walls are normal faults.

Kilauea is "supported" on the northwest by the gigantic Mauna Loa volcano and the mass of the Big Island of Hawaii. However, on its southeastern side, there is less support; Kilauea drops off into the Pacific Ocean. The effects of subsurface magma movement, both compressive during injection and extensional during removal, combine with gravitational pull to cause large movements on normal faults.

On 29 November 1975, one of the seaward-inclined normal faults moved suddenly in a 7.2 M_S seism. It happened at 4:48 A.M., when a large mass slipped for 14 seconds with a movement of about 6 m (20 ft) seaward and 3.5 m (11.5 ft) downward. The movement of this mass into the sea caused tsunami up to 12 m (40 ft) high. Campers sleeping on the beach were rudely awakened by shaking ground; those who didn't immediately hustle to higher ground were subjected to crashing waves. Two people drowned. This fault movement had an effect on subsurface magma analogous to shaking a bottle of soda pop—gases escaping from magma unleashed an 18-hour eruption featuring magma fountains up to 50 m (165 ft) high.

Table 5.2 Some Large Earthquakes in Hawaii

Date	Location	Intensity	Magnitude
2 Apr 1868	Southeast Hawaii	X	~7.4
5 Oct 1929	Honualoa, Hawaii	VII	6.5 M_s
22 Jan 1938	North of Maui	VIII	6.7 M_s
25 Sep 1941	Mauna Loa, Hawaii	VII	6.0 M_s
22 Apr 1951	Kilauea, Hawaii	VII	6.5
21 Aug 1951	Kona, Hawaii	IX	6.9
30 Mar 1954	Kalapana, Hawaii	VII	6.5
26 Apr 1973	Southeast Hawaii	VIII	6.3
29 Nov 1975	Southeast Hawaii	VIII	7.2 M_s
16 Nov 1983	Mauna Loa, Hawaii	VII	6.6 M_s
25 Jun 1989	Kalapana, Hawaii	VIII	6.5

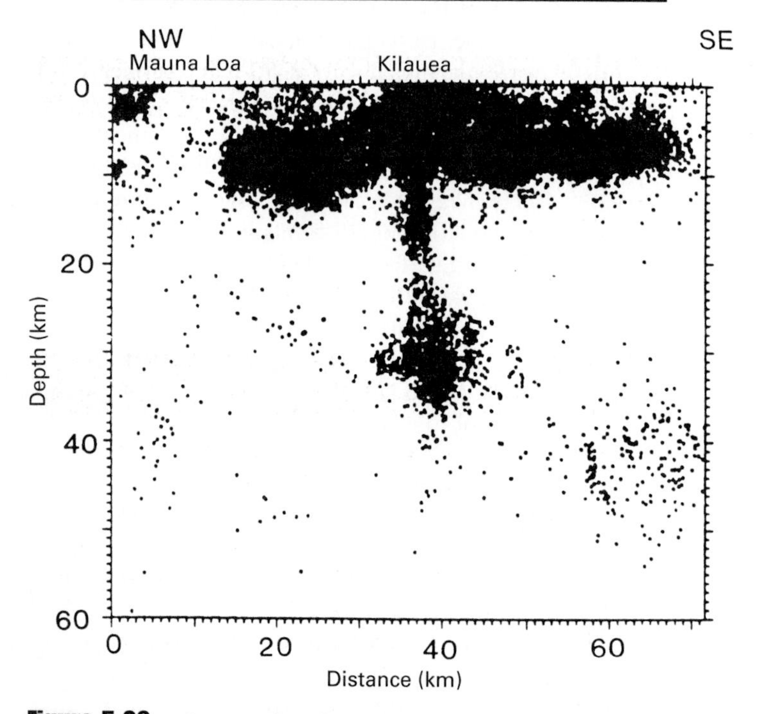

Figure 5.26 Cross section showing hypocenters beneath Kilauea volcano on flank of larger Mauna Loa volcano, southeastern Hawaii, 1970–1983.

From F. W. Klein and R. Y. Koyanagi, "The Seismicity and Tectonics of Hawaii" in *Geological Society of America, Decade of North American Geology,* Vol. N. Copyright © Geological Society of America. Reprinted by permission of the author.

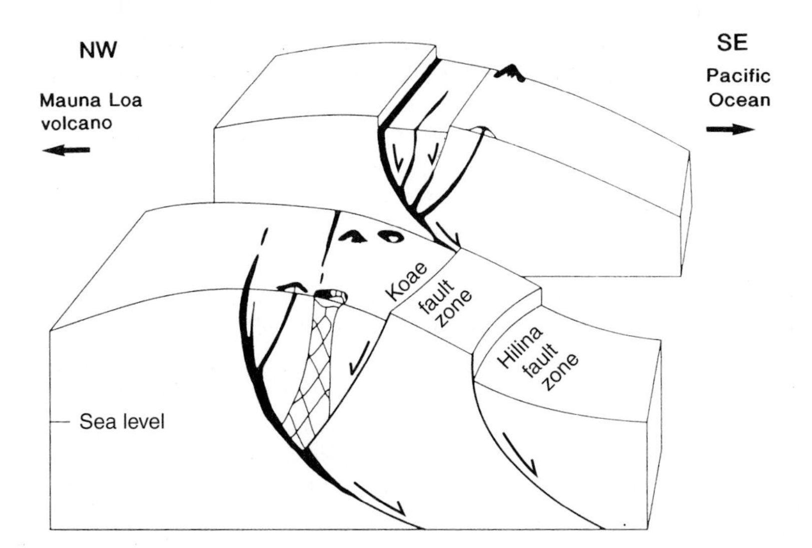

Figure 5.27 Schematic block diagrams of southeastern flank of Kilauea volcano. Intruding magma (black) forces brittle rock to break and move, generating earthquakes. Gravity-aided sliding down normal faults causes more earthquakes as rock masses slide southeastward into the ocean.

Summary

Earthquakes occur throughout North America. Most are in the West along the edges of the active plates, but the central and eastern regions also have earthquakes—not as many but just as large.

In the Pacific Northwest, stresses are transmitted upward from the subducting plates, forming strike-slip faults that rupture the surface, as in Seattle.

The Basin and Range province between eastern California and central Utah is an actively extending area. For example, Nevada has about doubled in west-east width in the last 30 million years. Normal faults accommodate most of the extension, unleashing earthquakes up to magnitude 8.

In the central United States, ancient rift valleys remain from failed spreading centers. The ancient rifts today are zones of weakness whose faults can be reactivated due to long-distance effects of Atlantic plate spreading and Pacific plate collision. The Reelfoot rift, occupied today by the Mississippi River, had earthquakes in 1811 and 1812 with moment magnitudes (M_w) of 8.2, 7.8, 8.1, and 8.3. Other rift valleys are associated with earthquakes throughout North America.

Charlevoix, Quebec, has frequent earthquakes up to magnitude 7. The region was intensely fractured by the impact of an ancient asteroid, and the fractured rocks apparently move under stresses within the moving plates.

Linear trends of earthquakes along eastern North America seem to line up with transform faults of today's mid-Atlantic ridge. Away from the spreading centers, the transform faults act as fracture zones. Reactivation of these zones of weakness may be responsible for major earthquakes in Charleston, South Carolina, and in New England.

The underground movement of magma in Hawaii generates earthquakes by forcefully rupturing brittle rocks. Land is uplifted as magma is injected and dropped down when magma is removed; these land movements may be sudden, earthquake-generating events.

Humans have triggered earthquakes by pumping water underground under pressure, by building dams and impounding water which seeps underground under pressure, and by underground explosions of atomic bombs.

Terms to Remember

avalanche 130
embayment 138
failed rift 142
fracture zone 144
harmonic tremor 147
quicksand 144
scarp 136
seamount 145
seiche 135
turbidite 132

Questions for Review

1. Explain three ways that humans have caused or triggered earthquakes.
2. How does increasing depth to hypocenter affect surface shaking?
3. What types of evidence indicate major movements on surface faults in Washington?
4. What tectonic process has affected the Basin and Range province, from eastern California to Utah, during the last 30 million years?
5. What type of fault movement best characterizes the Basin and Range province? What are the highest magnitude earthquakes generated there this century?

6. Explain what a seiche is. How could you create a small one?
7. The "stable" central United States has earthquakes clustered in distinct areas. What is a likely control on these earthquake locations?
8. When and how did the Reelfoot rift form? Explain its history of earthquakes. What does the future hold?
9. Sketch a map and explain how the spreading center/transform faults of the mid-Atlantic ridge might cause earthquakes in the Charleston, South Carolina and Boston areas.
10. Explain some Hawaiian volcano processes that cause earthquakes.

Questions for Further Thought

1. Some people suggest that earthquakes usually occur at certain times of day. Does this make sense? Is there a pattern to the times of the earthquakes discussed in Chapters 2, 3, 4, and 5?
2. Assess the earthquake hazard in Salt Lake City.
3. What controls the course of the Rio Grande in New Mexico? Is there an earthquake threat also?
4. Compare the ability to withstand earthquake shaking of downtown buildings and bridges in West Coast cities versus those in the mid-continent or East Coast of the United States.
5. Humans can trigger earthquakes, therefore should we? Can we set off medium-sized earthquakes in a controlled fashion that will prevent a large earthquake in an area? Make a list of pros and cons for earthquake control.

Suggested Readings and References

Active Tectonics. (1986). Washington, D.C.: National Academy Press.

Adams, J. (1992). Paleoseismology: A search for ancient earthquakes in Puget Sound. *Science,* 258, 1592–93.

Baldridge, W. S., and Olsen, K. H. (1989). The Rio Grande rift. *American Scientist,* 77, 240–47.

Blakely, R. J., Wells, R. E., Weaver, C. S., and Johnson, S. Y. (2002). Location, structure, and seismicity of the Seattle fault zone. *Geological Society of America Bulletin,* 114:169–77.

Earthquakes in Missouri. (Undated). Rolla, Mo.: Missouri Department of Natural Resources.

Gordon, D. W. (1988). Revised instrumental hypocenters and correlation of earthquake locations and tectonics in the central United States. U.S. Geological Survey Professional Paper 1364.

Hamilton, R. C., and Johnston, A. C., eds. (1990). Tecumseh's prophecy: Preparing for the next New Madrid earthquake. U.S. Geological Survey Circular 1066.

Hansen, M. C. (2000). Earthquakes in Ohio. *Ohio Geological Survey Leaflet 9.*

Johnston, A. C., and Kanter, L. R. (1990, March). Earthquakes in stable continental crust. *Scientific American,* 68–75.

Klein, F. W., and Koyanagi, R. Y. (1989). The seismicity and tectonics of Hawaii. *Decade of North American Geology.* (Vol. N, 238–52). Boulder, Co.: Geological Society of America.

Noson, L. L., Qamar, A., and Thorsen, G. W. (1988). Washington State earthquakes. Washington Division Geology and Earth Resources, Information Circular 85.

Nuttli, O. W., et al. (1986). The 1886 Charleston, South Carolina earthquake—A 1986 perspective. U.S. Geological Survey Circular 985.

Slemmons, D. B., Engdahl, E. R., Zoback, M. D., and Blackwell, D. D. eds. (1991). *Neotectonics of North America.* (Decade map volume). Boulder, Co.: Geological Society of America.

Stein, R. S., and Buckman, R. C. (1986, June). Quake replay in the Great Basin. *Natural History,* 29–35.

Videos

Hidden Fury: The New Madrid Earthquake Zone. (1993). Bullfrog Films (27 min.).

Earthquake Awareness and Risk Reduction in Utah. (1991). Utah Geological Society (25 min.).

Earthquake Risk in the Central United States. (1988). Federal Emergency Management Agency (9 min.).

Surviving the Big One. (1990). KCET-TV (58 min.).

Subject to Change. (1988). Pacific Bell (17 min.).

Chapter 6

Volcanic Eruptions: Plate Tectonics and Magmas

It is useful to be assured that the heavings of the Earth are not the work of angry deities. These phenomena have causes all their own.

—Lucius Annaeus Seneca, c. 63 C.E., *Naturales Quaestiones*

The dangers of volcanic eruption are obvious, but the quiet spells between active volcanism are seductive. Some people are lured to volcanoes like moths to a flame, even those who should know better. On 14 January 1993, volcanologists attending a workshop in Colombia, as part of the international decade of natural disaster reduction, hiked into the summit crater of Galeras Volcano to sample gases and measure gravity. They were looking for ways to predict imminent eruptions. The volcano had been quiet since July 1992, but during their visit, an unexpected, gas-powered secondary eruption killed six in the scientific party—four Colombians, a Russian, and an Englishman. Their deaths were not an unusual event (Table 6.1). They serve as a small-scale example of the larger drama played out when a volcano suddenly buries an entire city. During long periods of volcanic quiescence, people tend to build cities near volcanoes. For example, 400,000 people live on the flanks of Galeras Volcano, defying the inevitability of a large, life-snuffing eruption.

An individual volcano may be active for millions of years, but its eruptive phases are commonly separated by centuries of inactivity, lulling some into a false sense of security. Around 410 B.C.E., Thucydides wrote, "History repeats itself." We know well that those who do not learn the lessons of history are

Table 6.1 Volcanologists Killed by Eruptions

Year	Volcano	Total Deaths	Dead Volcanologists
1951	Kelut, Indonesia	7	3
1952	Myojin-sho, Japan	31	9
1979	Karkar, New Guinea	2	2
1980	St. Helens, United States	62	2
1991	Unzen, Japan	43	3
1991	Lokon-Umpong, Indonesia	1	1
1993	Galeras, Colombia	9	6
1993	Guagua Pichincha, Ecuador	2	2
2000	Semeru, Indonesia	2	2

Figure 6.1 Feet of a child killed by an eruption from Vesuvius in 79 C.E. The flexed toes and feet were an involuntary contraction when surrounded by 930°F (500°C) volcanic material. Death was instantaneous.

Photo from Mastrolorenzo, G., Petrone, P., Pagano, M., Incoronato, A., Baxter, P., Canzanella, A., Fattore, L., in *Nature*, 410:769, 2001.

doomed to repeat them. Every year, a large number of people inadvertently sacrifice their lives to volcanic eruptions.

Volcanoes deal out overwhelming doses of energy no human can survive. The skeletons of people killed by the eruption of Vesuvius in 79 C.E. testify to the lethal energy they experienced. In the coastal city of Herculaneum, 300 skeletons were found in life-like positions in boat chambers at the beach. These people had not been battered or suffocated; they did not display any voluntary self-protection reactions or agony contortions. In other words, their vital organs stopped functioning in less than a second, in less time than they could consciously react. The types of bone fractures, tooth cracks, and bone coloration indicate the victims were covered by volcanic material at about 500°C (930°F). At this temperature their soft tissues vaporized; their feet flexed in an instantaneous muscle contraction (Figure 6.1).

How We Understand Volcanic Eruptions

Two primary building blocks of knowledge are paramount to understanding volcanic eruptions.

1. Plate tectonics give us great insight into understanding earthquakes, now it will help us understand volcanoes.
2. Magmas vary in their chemical composition, their ability to flow easily, their gas content, and their volume. These variations govern whether eruptions are peaceful or explosive.

First, we will take a brief look at plate tectonics and volcanism. Second, we will examine magma variations and how they control eruptive style. Third, we will apply this knowledge globally to understand why volcanoes occur where they do, why only some volcanoes explode violently, and in what ways volcanoes can kill.

Plate-Tectonic Setting of Volcanoes

Over 90 percent of volcanism is associated with the edges of tectonic plates (Figure 6.2). Most other volcanism occurs above hot spots (Figure 2.14). Over 80 percent of Earth's magma extruded through volcanism takes place at the oceanic spreading centers. Solid, but hot and ductile, mantle rock rises upward into regions of lower pressure, where up to 30 to 40 percent of the rock melts and flows easily as magma on the surface (Figure 6.3). The worldwide rifting process releases enough magma to create 20 km^3 of new oceanic crust each year. Virtually all this volcanic activity takes place below sea level and is thus difficult to view.

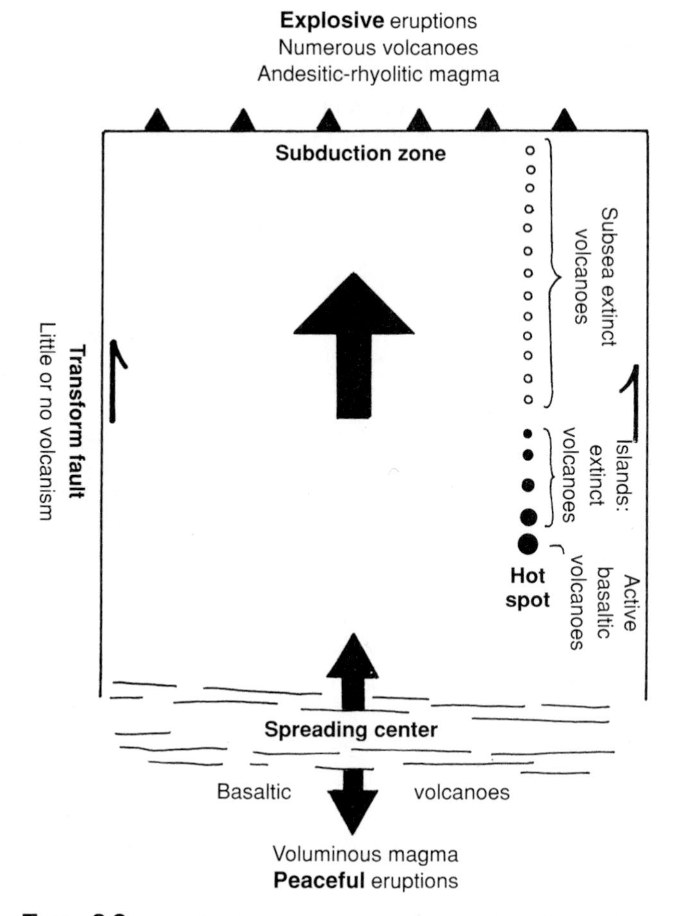

Figure 6.2 Idealized oceanic plate showing styles of volcanism.

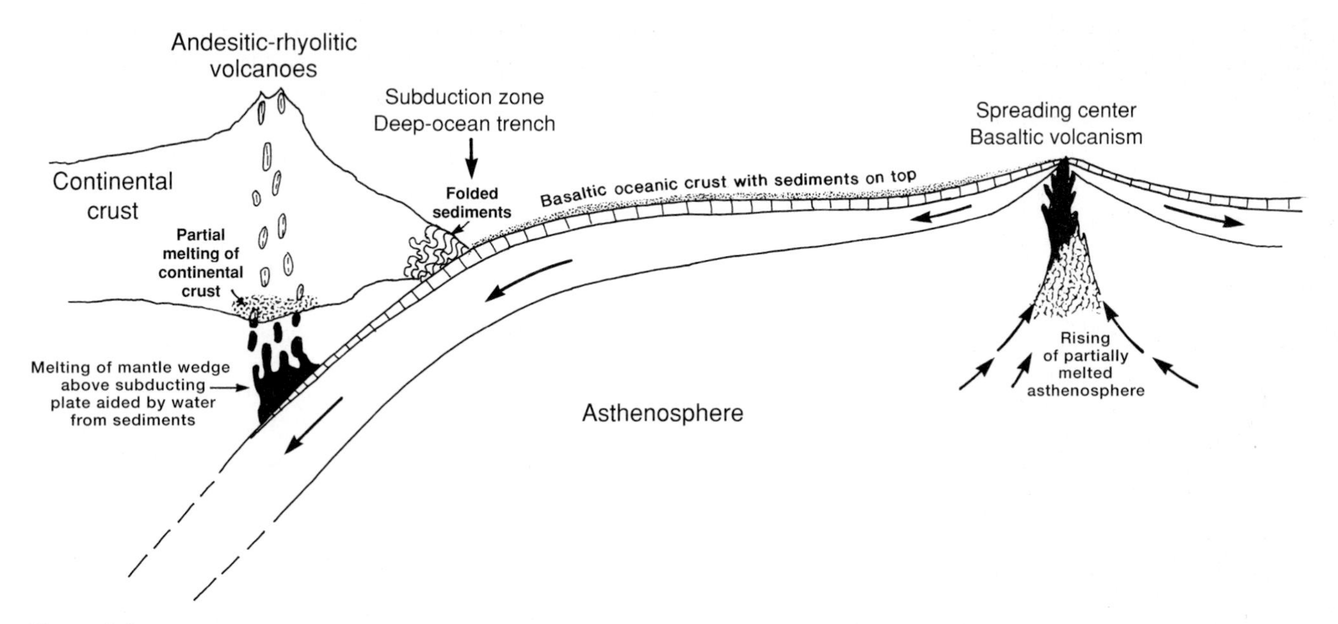

Figure 6.3 Idealized cross section showing production of basaltic magma at spreading centers. Plates pull apart, and some asthenosphere liquefies and rises to fill the gap. Andesitic-rhyolitic magmas are created above subduction zones where rising magma partially melts the continental crust on its way up, thus altering the melt by increasing SiO_2 content and viscosity.

Subduction zones cause the tall and beautiful volcanic mountains we see at the edges of the continents, but the volume of magma released at subduction zones is trivial in comparison to that of spreading centers. Subduction zones account for the eruption of 7 to 13 percent of all magma. The down-going plate carries oceanic-plate rock covered with water-saturated sediments into much hotter zones (Figure 6.3). The presence of water lowers the melting point of rock. Rising magma partially melts some of the continental crust it passes through. This adds some new melt of different composition to rising plumes of magma. Each rising plume has its own unique chemical composition.

Transform faults have little or no associated volcanism. Thinking three-dimensionally, this is understandable. At a transform fault, the two plates simply slide past each other in a horizontal sense and at all times keep a quite effective "lid" on the hot asthenosphere below.

From a volcanic disaster perspective, the differences are clear. Oceanic volcanoes are relatively peaceful, whereas subduction-zone volcanoes are explosive and dangerous. Ironically, humans tend to congregate at the seaward edges of the continents, where the most dangerous volcanoes operate.

People commonly speculate upon whether an individual volcano is active, dormant, or extinct. Because of the strong hope that a volcano is extinct and the nearby land is thus available for use, many dormant volcanoes are misclassified. But consider this: A subduction zone commonly lasts for tens of millions of years, and its province of volcanoes is active for the entire time. An individual volcano may be active for hundreds of thousands to several million years, despite "slumbers" of centuries between eruptions. As a rule of thumb, if a volcano has a well-formed and aesthetic conical shape, it is active. A pretty shape is dangerous.

Why do spreading-center volcanoes have relatively peaceful eruptions? And why do subduction-zone volcanoes explode violently? The answers to these questions are found in knowing how different magmas behave.

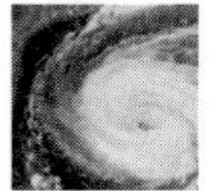

Chemical and Mineral Composition of Magmas

Although there are 92 naturally occurring elements, a mere eight make up more than 98 percent of the Earth's crust (Table 6.2). The next four most abundant elements add another 1.2 percent to the crust, bringing the weight percent contributed by these 12 elements to 99.23 percent. The remaining 0.77 percent includes gold, silver, copper, carbon, sulfur, tin, and many other familiar elements.

Oxygen and silicon are so abundant that their percentages dwarf those of all other elements. Oxygen atoms carry negative charges (–2), while silicon atoms are positively charged (+4). As magma begins to cool to a solid, its silicon and oxygen atoms will be the first to bond. Silicon and oxygen link up with four oxygen atoms ($4 \times -2 = -8$) surrounding a central silicon atom (+4) to form the silicon-oxygen tetrahedron (SiO_4)(Figure 6.4). The SiO_4 tetrahedron presents a –4 charge on its exterior that attracts and ties up positively

Table 6.2 Common Elements of the Earth's Crust (Weight Percent)

Element	Weight %		Element	Weight %
Oxygen (O^{-2})	45.20%			
Silicon (Si^{+4})	27.20			
Aluminum (Al^{+3})	8.00			
Iron ($Fe^{+2,+3}$)	5.80	plus	Titanium ($Ti^{+3,+4}$)	0.86%
Calcium (Ca^{+2})	5.06		Hydrogen (H^{+1})	0.14
Magnesium (Mg^{+2})	2.77		Phosphorus (P^{+5})	0.10
Sodium (Na^{+1})	2.32		Manganese ($Mn^{+2,+3,+4}$)	0.10
Potassium (K^{+1})	1.68		Total	99.23%
Total	98.03%			

Table 6.3 Crustal Elements in Weight-Percent Oxides

Continental Crust		Oceanic Crust	
SiO_2	60.2%	SiO_2	48.7%
Al_2O_3	15.2	Al_2O_3	16.5
Fe_2O_3	2.5	Fe_2O_3	2.3
FeO	3.8	FeO	6.2
CaO	5.5	CaO	12.3
MgO	3.1	MgO	6.8
Na_2O	3.0	Na_2O	2.6
K_2O	2.9	K_2O	0.4

Figure 6.4 A silicon atom with +4 charge is linked to four oxygen atoms each with a −2 charge.

charged atoms. After negatively charged oxygen, the 11 elements of greatest abundance are all positively charged (Table 6.2); they are attracted to, and bound up by, oxygen. This process is so common and voluminous that elemental abundances in the crust are usually listed in combination with oxygen (as oxides). The weight percentages of elements are quite different for continental versus oceanic crust (Table 6.3). Marked differences in oxide percentages produce magmas of variable composition and behavior.

The eight most common elements bond in different configurations to make up hundreds of different **minerals.** The process of mineral formation in a cooling magma is called **crystallization.** Just as with elements, a degree of simplicity occurs in a crystallizing magma because the overwhelming majority of the Earth's crust is composed of just eight common rock-forming minerals. Laboratory experiments and microscopic examination of rock-forming minerals have shown the order in which these minerals crystallize from a cooling magma (Figure 6.5).

Magmas at the surface with temperatures around 1,000 to 1,200°C (1,830 to 2,190°F) have two separate lines of mineral growth:

1. Iron and magnesium will link up with aluminum and the silicon-oxygen tetrahedron as magma temperature decreases to sequentially form four distinct and discontinuous families of minerals—olivine, pyroxene, amphibole, and biotite mica.
2. Calcium will combine with Al and SiO_4 to begin forming the plagioclase feldspar family, a continuous and gradational series of minerals. As temperature decreases, progressively more sodium (and less calcium) is locked within the plagioclase crystal structure. By the time magma has cooled down to the 800 to 1,000°C (1,470 to 1,830°F) range, it is largely depleted in Fe, Mg, and Ca. Now, potassium crystallizes within muscovite mica and potassium-rich feldspar minerals, and excess Si and O combine without other elements to make the mineral quartz.

Just as elements combine to make minerals, so minerals aggregate to make **rocks** (Figure 1.23). Magmas have a broad range of compositions resulting in many different types of igneous rocks which generations of geologists have classified into a dizzying array of rock names. Nonetheless, a working understanding can be gained while considering only three magma types and the three clans of igneous rocks that form from them. The rock types are based on their silicon and oxygen (SiO_2) percentages (Table 6.4). If the magma cools and solidifies below the surface, it crystallizes as **plutonic rocks,** named for Pluto, the Greek god of the underworld. If the magma reaches the surface, it forms **volcanic rocks,** named for Vulcan, the Roman god of fire. The left side of Figure 6.5 shows three main types of volcanic rocks next to their respective mineral compositions. These magmas behave differently due to their varying temperatures, water contents, and viscosities.

Viscosity, Temperature and Water Content of Magmas

The fluidity of a liquid is measured by its **viscosity,** its internal resistance to flow. The lower the viscosity, the more fluid is the behavior. For example, tilt a glass of water and

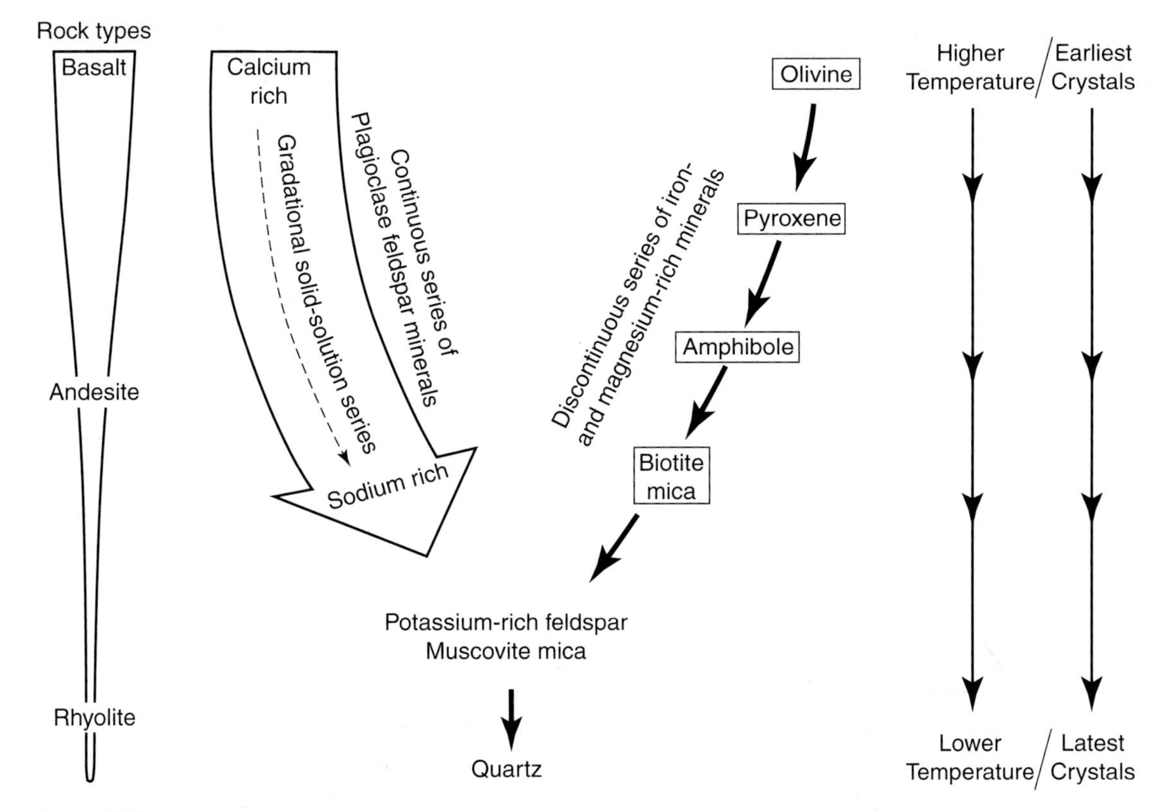

Figure 6.5 Order of crystallization of minerals from a cooling magma.

Table 6.4 Igneous Rock Types

Magma Type	Plutonic Rock	Volcanic Rock
$SiO_2 < 55\%$	Gabbro	Basalt
$SiO_2 = 55–65\%$	Diorite	Andesite
$SiO_2 > 65\%$	Granite	Rhyolite

watch it flow quickly; water has low viscosity. Now tilt the same glass but filled with honey and watch the slower flow; honey has higher viscosity. Low-viscosity magma flows somewhat like ice cream on a hot day. High-viscosity magma barely flows. The viscosity of magma is changed by various means.

1. Higher temperature lowers viscosity; it causes atoms to spread farther apart and vibrate more vigorously, thus atomic bonds break and deform more, resulting in increasing fluidity. Consider the great effect of temperature on rhyolite magma (Table 6.5). At 600°C (1,100°F), rhyolite viscosity is five orders of magnitude (100,000 times) more viscous than at 900°C (1,650°F).
2. Silicon and oxygen (SiO_2) increase the viscosity of magma because they form abundant silicon-oxygen tetrahedra (Figure 6.4) that link up in chains, sheets, and networks, creating more joins and bonds between atoms, which in turn make flow more difficult.
3. Increasing content of minerals increases viscosity. Magma is a mixture of liquid and minerals that have crystallized from it. Mineral content in magma varies from none to being the majority of the mass.

Magma contains dissolved gases held as **volatiles;** their solubility increases as pressure increases and as temperature decreases. You can visualize the pressure—temperature relations with a bottle of soda pop as an analogy. Carbon dioxide (CO_2) gas is dissolved in the soda pop and kept under pressure by the bottle cap. Pop the cap off the bottle, reducing pressure, and some volatiles escape. As the uncapped bottle warms, more volatiles are lost.

A good grasp of volcanic behavior can be gained by considering the properties of three types of magma and the rocks they become—**basalt, andesite,** and **rhyolite** (Table 6.5). Notice that the highest temperatures and lowest SiO_2 contents are in basaltic magma, giving it the lowest viscosity and easiest fluid flow. The lowest temperatures and highest SiO_2 contents occur in rhyolitic magma, material so viscous that it commonly does not flow. Table 6.5 also states that about 80 percent of the magma reaching the Earth's surface is basaltic, with only about 10 percent andesitic and 10 percent rhyolitic. Why the difference? Basaltic magma has

Table 6.5 Comparison of Three Types of Magma

Volcanic Rock Types	**Basalt**	**Andesite**	**Rhyolite**
Rock Description	Black to dark gray; contains Ca-plagioclase, pyroxene, olivine	Medium to dark gray; contains amphibole, pyroxene, intermediate Ca-Na-plagioclase	Light-colored; contains quartz, K-feldspar, biotite, Na-plagioclase
Volume at Earth's Surface	80%	10%	10%
SiO_2 Content	45–55%	55–65%	65–75%
	increasing →		
Temperature of Magma	1,000–1,250°C	800–1,000°C	600–900°C
	decreasing →		
Viscosity	Low (melted ice cream)		High (toothpaste)
	increasing →		
Water Dissolved in Magma	~0.1 – 1 wt. %	~2 – 3 wt. %	~4 – 6 wt. %
	increasing →		
Gas Escape from Magma	Easy		Difficult
	increasing difficulty →		
Eruptive Style	Peaceful		Explosive
	increasing explosiveness →		

the lowest viscosity, so more of it reaches the surface. On the other hand, the more viscous rhyolitic magmas are so sluggish that they tend to be trapped deep below the surface where they cool, solidify, and grow into the larger mineral crystals of plutonic rocks, such as granite.

In magma, water is the most abundant dissolved gas. As magma rises toward the surface and pressure decreases, water dissolved in the hot magma becomes gas and forms steam bubbles. In low-viscosity basalt magma, steam escapes easily and magma is low in dissolved water content (Table 6.5). As high-viscosity rhyolite magma rises and steam bubbles form, they have difficulty escaping from the gooey magma and have to burst their way out.

When the basaltic volcanoes of Hawaii begin to erupt, it is a tourist event: "Come see the red-hot magma flow." Although such an eruption makes a thrilling show, it is a relatively peaceful happening. Why is it safe? Because the dissolved gases escape from the low-viscosity magma with relative ease. Compare this behavior with the eruption of a rhyolitic magma of lower temperature, higher percentage of SiO_2, and very high viscosity. When rhyolitic magma oozes out onto the ground surface, the pressure within the magma is reduced and the dissolved gases expand in volume. But how do gases escape from their entrapment in sticky magma? By exploding. Spectators at the eruption of rhyolitic magma frequently die. When it comes to volcanic hazards, the greatest problem is how easily the dissolved gases can escape from the magma. As Frank Perret stated: "Gas is the active agent, and magma is its vehicle."

Plate-Tectonic Setting of Volcanoes Revisited

Knowing about magma viscosity and volatile content allows us to revisit our earlier questions about plate tectonics and volcanism. Why does the vast majority of Earth's magma pour out at spreading centers, and in relatively peaceful eruptions? And why does the magma above subduction zones commonly explode violently?

Spreading centers are ideal locations for volcanism because: 1) they sit above the high-temperature asthenosphere; 2) the asthenosphere rock has low percentages of SiO_2; and 3) the oceanic plates pull apart causing hot asthenosphere rock to rise, experience lower pressure, and change to magma that continues to rise. This magma is high-temperature, low SiO_2, low-volatile content, low-viscosity basalt allowing easy escape of gases (Figures 6.2 and 6.3). Spreading centers combine all the factors that promote the peaceful eruption of magma.

When a subducting oceanic plate reaches a depth of about 100 km (over 60 mi), magma is generated and rises to-

Sidebar

Volcanoes and the Origin of the Ocean, Atmosphere, and Life

The elements in volcanic gases are dominated by hydrogen (H), oxygen (O), carbon (C), sulfur (S), chlorine (Cl), and nitrogen (N). These gaseous elements combine at the Earth's surface to make water (H_2O), carbon dioxide (CO_2), sulfur dioxide (SO_2), hydrogen sulfide (H_2S) with its rotten egg smell, carbon monoxide (CO), nitrogen (N_2), hydrogen (H_2), hydrochloric acid (HCl), methane (CH_4), and numerous other gases. The dominant volcanic gas is water vapor; it is commonly greater than 90 percent of total gases.

Notice how the elements of volcanic gases (C, H, O, N, S, Cl) differ from the elements of volcanic rocks (O, Si, Al, Fe, Ca, Mg, Na, K). The elements of volcanic gases make up the oceans, the atmosphere, and life on Earth, but they are rare in rocks. The 4.5 billion years of heat flow from the Earth's interior has "sweated" out many light-weight elements and brought them to the surface via volcanism. Billions of years of volcanism on Earth goes a long way to explaining the origin of the continents, the oceans, the present atmosphere, and the surface concentration of the CHON elements (carbon, hydrogen, oxygen, nitrogen) of which all life on Earth is composed and on which it depends.

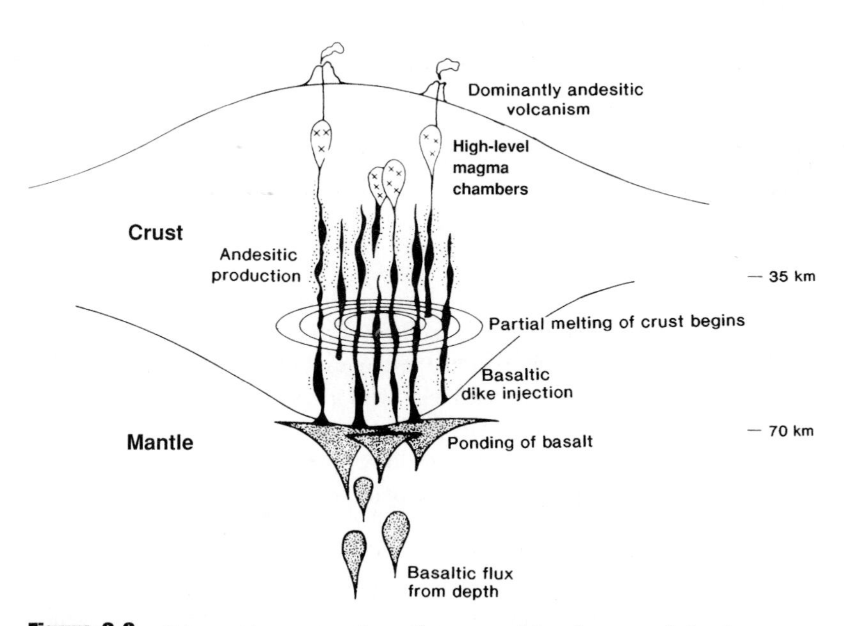

Figure 6.6 Schematic cross section of magma rising from a subduction zone and being contaminated en route.

ward the surface (Figure 6.3). The subducting plate stirs up the mantle causing the hotter rock at depth to rise and melt as pressure decreases. A significant reason that magma forms here is because the subducting plate carries a cover of sediments, water, and hydrated minerals down with it. Water, even in slight amounts, promotes partial melting by lowering the temperature for rock to melt. The partial melting process affects only those minerals with lower melting temperatures. As this partial melt rises upward, it in turn melts part of the overlying crust to produce magmas of highly variable compositions (Figure 6.6). Magma compositions depend upon the amount of crustal rock melted and incorporated into the rising magma. In general, magma temperature decreases while SiO_2, water content, and viscosity increase. All these changes in magma add to its explosive potential.

How a Volcano Erupts

The Earth's internal energy flows outward as heat (see Chapter 1). The eruptions of volcanoes are rapid means for the Earth to expel some of its internal heat.

Figure 6.7 Ash blows out of Soufriere Hills Volcano on Montserrat in 1997. All the houses and vegetation shown were incinerated several weeks later.

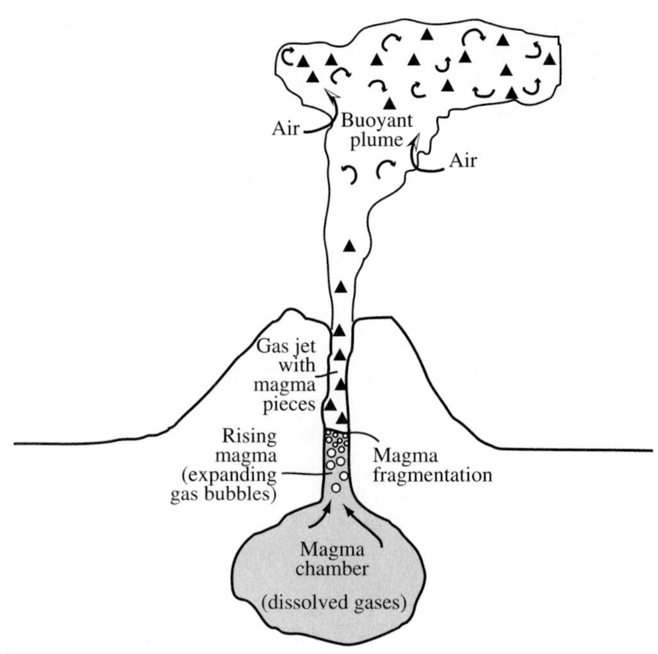

Figure 6.8 Anatomy of an eruption. As magma rises to levels of lower pressure, gas comes out of solution, forming bubbles which overwhelm magma and create a gas jet leading to a buoyant plume.

A volcanic eruption begins with heat at depth. Superheated rock will rise to levels with lower pressure, and some solid rock may change phase to liquid magma, resulting in volume expansion leading step by step to eruption (Figure 6.7).

Magma is generated by the melting of existing rock. Rock may melt by: 1) lowering the pressure on it, 2) raising its temperature, or 3) increasing its water content. Most magma is generated by decreasing the pressure on hot rock. For example, as the solid, but mobile, hot rock of the mantle rises upward it experiences progressively less pressure and spontaneously melts, without the addition of more heat. Melting caused simply by a decrease in pressure is called **decompression melting.** The process of decompression melting is so important it is worth restating—most of the rock that melts to form magma does so because the pressure on it decreases, not because more heat is added.

The largest nearby reservoir of superhot, ready-to-melt rock exists in the nearly molten asthenosphere; this rock, hot enough to flow without being liquid, is the main source of magma. As this superheated rock rises, the pressure on it decreases allowing some rock to melt. The hot, rising rock/magma mixture also raises the temperature of rock it passes through, thus melting portions of the overlying rocks.

If pressure in the asthenosphere or lithosphere is decreased, some rock melts with a resultant increase in volume that causes overlying rocks to fracture. The fractures allow more material to rise to lower pressure levels, causing more rock to liquefy. For example, at a depth of 32 km (20 mi), basaltic rock melts at 1,430°C (2,600°F) but this same rock will melt at only 1,250°C (2,280°F) at the Earth's surface. Since upward-moving rock/magma reaches ever-lower pressures, rising rock can liquefy and magma can increase in fluidity, which in turn causes more superheated rock to become magma.

Magma at depth does not contain gas bubbles because the high pressure at depth keeps volatiles dissolved in solution. But as magma rises toward the surface, pressure continually decreases, and gases begin to come out of solution, forming bubbles that expand with decreasing pressure (Figure 6.8). The added lift of the growing volume of gas bubbles helps propel magma upward through fractures or pipes toward an eruption. Gas bubbles continue increasing in number and volume as magma keeps rising upward to lower pressures. When gas-bubble volume reaches around 75 percent, gas can overwhelm magma, fragmenting the magma into pieces which are carried up and out by a powerful gas jet (Figure 6.9). Upon escape from the volcano, the gas jet draws in air that adds to buoyancy in the turbulent, rising plume (Figure 6.8).

Figure 6.9 Remarkable view into the crater of Mount Pinatubo just as a major explosive eruption was beginning, 1 August 1991.

Sidebar

How a Geyser Erupts

The eruption of water superheated by magma is called a **geyser.** The name is drawn from the Icelandic word *geysir,* meaning "to gush or rage." Areas of geyser activity include Iceland, Chile, Yellowstone Park in the United States, North Island of New Zealand, and Kamchatka Peninsula of Russia. All of these sites share common characteristics: subsurface water is present and heat is abundant. Water from snow, rain, streams, and lakes is pulled below the ground surface by gravity, where it slowly moves through the network of voids presented by fractures and cavities or **pores** in rocks. The downward-circulating water encounters heat from a near-surface body of magma, absorbs some of that heat, and then erupts (Figure 6.10). For a geyser to form, heat must be abundant. For example, the heat flow at Upper Geyser Basin in Yellowstone Park is more than 800 times higher than average.

This simple description ignores the complex interplay of temperature and pressure that combine to set off an eruption. Water boils at 100°C (212°F) at sea level. At the 2,150 m (more than 7,000 ft) elevation of Yellowstone Park, with the reduced pressure of a thinner atmosphere above it, water will boil upon reaching 93°C (193°F). However, water encountered while drilling 332 m (1,088 ft) below the surface at Norris Geyser Basin in Yellowstone Park was still liquid at 241°C (465°F). How can water still be liquid at those temperatures? Water remains liquid because the pressure at that depth is too great for it to change to steam. What does it take to create steam at great depths below the surface? Either higher temperature or lower pressure will do the job.

Water circulating at thousands of feet below the surface can be heated to temperatures far above 100°C (212°F) without boiling because the pressure of the overlying groundwater body is so great (Figure 6.11). However, when some of this superheated water does boil, its volume expands as liquid changes to steam, helping lift surrounding water upward to lower pressure levels. There, at lesser depths and lower pressures, some of the water flashes to steam; this helps lift more superheated water to lower pressure levels, etc. Thus, superheated water rises up into successively lower pressure levels making geysers erupt.

What triggers the gushing of a geyser? Reduction of pressure on superheated water causes it to change from liquid to gas triggering the geyser eruption, analogous to the reduction in pressure that causes hot rock to change from solid to liquid triggering a volcano eruption. A geyser eruption usually follows this sequence of events: 1) at depth, superheated water flows out of tiny pressurized cracks into geyser reservoirs of larger volume;

Sidebar

How a Geyser Erupts (Continued)

2) as water temperature rises, some water will flash to steam; 3) the steam bubbles rise to lower pressure levels, expanding continuously; 4) steam and bubbles become so abundant that they overwhelm the water, carrying it upward to levels of lower pressure, causing continual conversion to steam along the upward route; and 5) finally—the spectacular eruption.

Figure 6.10 Bursting eruption of Seismic geyser, Yellowstone Park, Wyoming.
Photo by D. E. White, *U.S. Geological Survey,* August 1968.

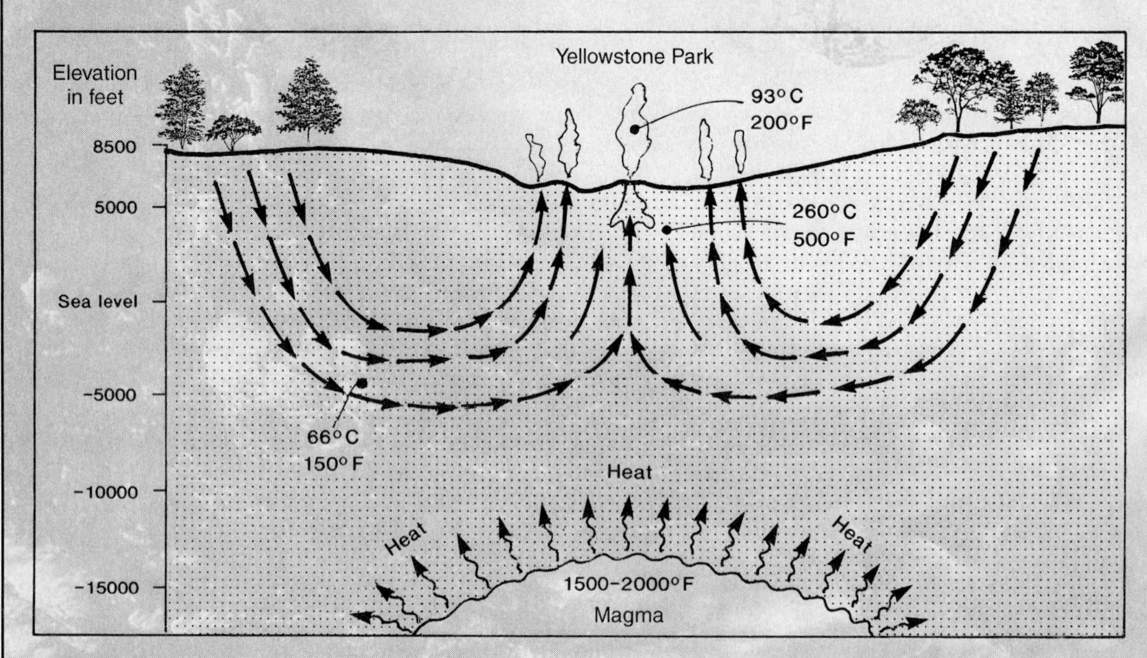

Figure 6.11 Surface water is pulled below ground by gravity, flows through holes in rocks, is superheated by magma, and through complex reductions in pressure, erupts at the surface as a geyser.

Table 6.6 Volcanic Materials

Lava	Aa	Rough, blocky surface
	Pahoehoe	Smooth, ropy surface
	Pillow	Ellipsoidal masses formed in water
Pyroclastic	Air-fall fragments	
	Fine ash (dust)	Flour-size material
	Coarse ash	Sand size
	Cinders	Marble- to golf-ball size
	Blocks	Big fragments, solid while airborne
	Bombs	Big fragments, liquid while airborne
	Volcanic tuff	Rock made of smaller fragments, e.g., deposit of a hot, gas-charged flow
	Volcanic breccia	Rock made of coarse, angular fragments, e.g., deposit of a water-charged debris flow
Glass	Obsidian	Nonporous glass
	Pumice	Porous froth

Figure 6.12 Small-scale pahoehoe near Halemaumau, Hawaii. Photo by Pat Abbott.

Some Volcanic Materials

Magmas vary in their viscosity and dissolved-gas (volatile) content. Low-viscosity magma that reaches the surface typically moves as lava flows with easy gas escape yielding nonexplosive eruptions. High-viscosity magma holds its volatiles making gas escape difficult. Gas is forced to burst out of the magma yielding explosive eruptions. Gas blasting into the atmosphere takes along chunks of magma known as **pyroclastic** debris (pyro = fire; clastic = fragments).

Nonexplosive Eruptions Lava flows are especially typical of basaltic magma and they exhibit a variety of textures (Table 6.6). Highly liquid lava may cool with a smooth, ropy surface called **pahoehoe** (Figure 6.12). Slower flowing, more viscous lava commonly has a rough, blocky texture called **aa.** If the lava reaches the sea or a lake, it cools rapidly into ovoid or toothpaste-looking forms called **pillow lava.**

Explosive Eruptions Gaseous explosions break rock and tear apart magma into pyroclastic debris with a wide range of sizes from dust to huge blocks and bombs (Table 6.6; Figure 6.13). Airborne pyroclasts have their coarsest grains fall from the atmosphere first, closest to the volcano, followed by progressively finer material at greater distances away (Figure 6.14). An air-fall deposit can be recognized by its layering and the sorting of pyroclasts into layers of different sizes. Pyroclastic debris also can be blasted out over the ground surface as high-speed, gas-charged flows that dump material quickly, producing indistinct layering and little or no sorting of the various-size particles.

Magma reaching the surface can solidify so quickly that crystallization cannot take place because there is no time for atoms to arrange themselves into the ordered atomic structures of minerals. When magma cools this fast it produces glass (Table 6.6). Cooled volcanic glass is known as **obsidian.** When gas escapes quickly and violently from lava, it may produce a frothy glass full of holes left by former gas bubbles; this porous material, known as **pumice,** contains so many holes it can float on water.

Volcanic Landforms and Eruption Styles

Rising magma creates many volcanic landforms on the Earth's surface as well as producing subsurface plutonic-rock bodies (Figure 6.15). The various volcanic landforms are created by different styles of eruption (Figure 6.16). Non-explosive eruptions can be subdivided into *Icelandic* (see beginning of Chapter 7), *Hawaiian,* and *Strombolian* types, and explosive eruptions can described as *Vulcanian, Plinian,* and *Caldera* types. Applying what we have learned about magmas allows us to see the linkages between eruptive behaviors and landforms. We can forecast eruption styles anywhere in the world using *The 3 Vs of Volcanology: Viscosity, Volatiles, Volume. Viscosity* may be

Figure 6.13 (a) Large blob of magma cooled while airborne and fell as volcanic bomb, Irazu Volcano, Costa Rica.
Photo by Pat Abbott.

Figure 6.13 (b) Pyroclastic bombs kill people every year.
Drawing by Jacobe Washburn.

low, medium, or high; and it controls whether magma flows away or piles up. *Volatile* abundance may be low, medium, or high; and volatiles may ooze out harmlessly or blast out explosively. *Volume* of magma may be small, large, or very large. Volume correlates fairly well with eruption intensity; the greater the volume, the more intense the eruption. By mixing and matching the values among the 3 Vs, you can define volcanic landforms (Table 6.7) and eruptive styles. The different types of eruptions that occur by varying the 3 Vs can be seen as animations on an interactive CD created by the Smithsonian Institution.

Shield Volcanoes: Low Viscosity, Low Volatiles, Large Volume

The rocks of **shield volcanoes** are formed mostly from the solidification of lava flows of basalt. These lava flows are low viscosity, contain less than one weight-percent volatiles, and are so fluid that they travel for great distances somewhat analogous to pouring pancake batter on a griddle. Each basaltic flow cools to form a gently dipping, relatively thin volcanic rock layer. Many thousands of these lava flows must cool on top of each other over a long time in order to build a big volcano. A shield volcano, such as Mauna Loa in Hawaii, has a great width compared to its height, whereas a volcano built of high-viscosity magma, such as Mount Rainier in Washington, has a great height compared to its width (Figure 6.17).

Hawaiian-type Eruptions As with virtually all volcanic eruptions, Hawaiian-type eruptions commonly are preceded by a series of earthquakes as rock fractures and moves out of the way of swelling magma. When these fractures split the ground surface, they suddenly reduce pressure, allowing gas to escape from the top of the magma body. This can create a beautiful "curtain of fire" where escaping jets of gases form lines of lava fountains up to 300 m (1,000 ft) high. Also common in the Hawaiian eruption is formation of a low cone with high fountains of magma. After the initial venting of gas, great floods of basaltic

Table 6.7 Volcanism Control by the 3Vs (Viscosity, Volatiles, Volume)

Viscosity +	Volatiles +	Volume =	Volcanic Landforms
low	low	large	shield volcanoes
low	low	very large	flood basalts
low/medium	medium/high	small	scoria cones
medium/high	medium/high	large	stratovolcanoes
high	low	small	lava domes
high	high	very large	calderas

Figure 6.14 Volcanic ash covers a house near Mount Pinatubo in the Philippines, June 1991.
Photo by R. P. Hoblitt, *U.S. Geological Survey.*

lava spill out of the fissures and flow down the mountain slopes as red-hot rivers (Figure 6.18). These eruptions may last from a few days to a year or more. Although few lives are lost to Hawaiian volcanism, the ubiquitous lava flows engulf and incinerate buildings, bury highways, cause drops in property value of homes near the latest flow, and cause some homeowners to lose their peace of mind.

Hawaiian volcanoes capable of eruption include Haleakala on the island of Maui, the five volcanoes that make up the island of Hawaii, plus the growing but still subsea volcano of Loihi. In the last 200 years, eruptions have occurred only on the three southernmost volcanoes on the island of Hawaii and below sea level on Loihi. The island-to-be (Loihi) is located about 30 km (19 mi) off the southeastern shore of Hawaii. Loihi's peak is about 969 m (3,175 ft) below sea level, and the weight of the overlying ocean water suppresses the explosiveness of the eruptions for now, but the volcano is building upward impressively.

In general, the volcanism on Hawaii is relatively peaceful and acts as a magnet attracting tourists to witness nature's spectacle. But there are exceptions to this statement.

Killer Event of 1790 Although less than 0.5 percent of Hawaiian magma is blown out as pyroclastic material, there are the rare killer events. In 1790, traveling parties from King Keoua's army were caught and many people were killed by a blast from Kilauea Volcano. The army was passing through the area but was stopped by eruptions. After three days of

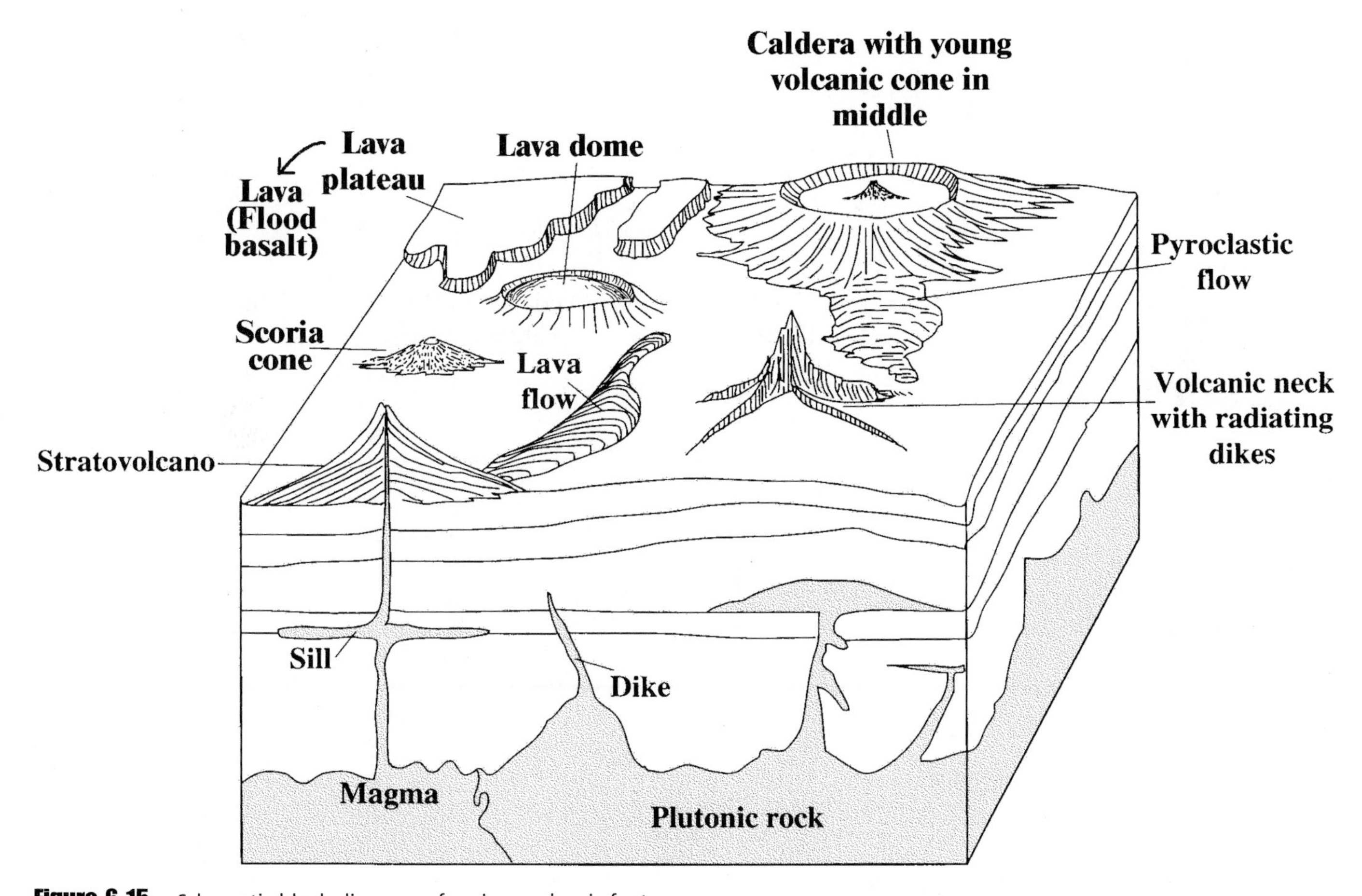

Figure 6.15 Schematic block diagram of various volcanic features.
Data Source: Schmidt, R. G., and Shaw, H. R., *Atlas of Volcanic Phenomena, U.S. Geological Survey,* 1971.

Eruption Type	Explosivity	Composition	Volcanic Landform
ICELANDIC TYPE 	Nonexplosive (VEI = 0–1)	Basalt (low viscosity)	Volcanic plateau and small shield volcanoes
HAWAIIAN TYPE 	Nonexplosive (VEI = 0.1)	Basalt (low viscosity)	Large shield volcanoes
STROMBOLIAN TYPE 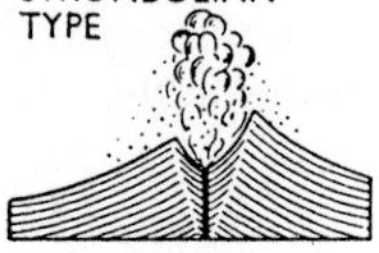	Low (VEI = 1–3)	Basalt to Andesite (low-to-moderate viscosity)	Scoria cones
VULCANIAN TYPE 	Moderate (VEI = 2–5)	Basalt to Rhyolite (moderate-to-high viscosity)	Scoria cones and Stratovolcanoes
PLINIAN TYPE 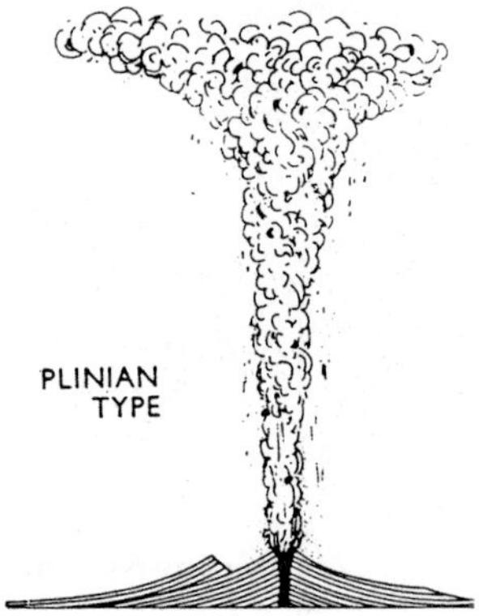	High (VEI = 3–8)	Andesite to Rhyolite (high viscosity)	Stratovolcanoes
CALDERA TYPE 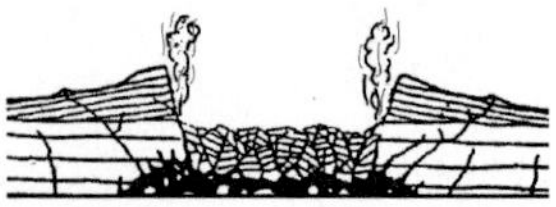			Calderas

Figure 6.16 Types of volcanic eruptions.

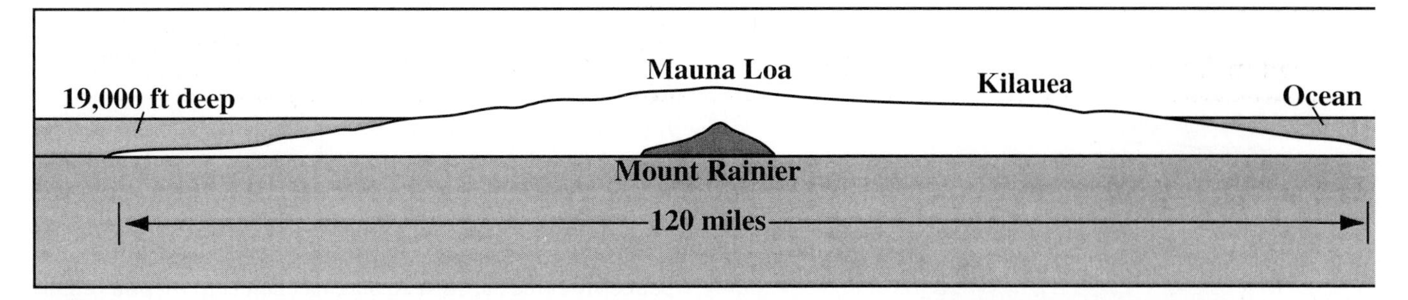

Figure 6.17 A shield volcano, such as Mauna Loa in Hawaii, has a great width compared to its height. A stratovolcano, such as Mount Rainier in Washington, has a great height compared to its width.

Data Source: Tilling, R. I., et al., *Eruptions of Hawaiian Volcanoes, U.S. Geological Survey,* 1987.

waiting, it split into three parties of about 80 people each. As the parties marched southwest down the trail from Kilauea, disaster struck. An explosion column burst upward, with a **base surge** sweeping outward as a dense, basal cloud. Base surges can travel at hurricane speeds as masses of ground-hugging hot water and gases with or without magma fragments. The base surge in 1790 overtook King Keoua's middle party, killing them all. The victims huddled together, grasping each other to withstand the hurricane-force blast, but the hot gases seared their lungs and the intense heat scorched their skin. The base surge caught up with the lead party but it had weakened, allowing most people to survive. The trailing party was alongside the blast and suffered no deaths or injuries. Although low-viscosity magma is not likely to explode, this case history shows how it can heat groundwater to cause an eruption including a base surge of superheated steam. The 1790 event is worthy of remembrance by today's watchers of Hawaiian eruptions—at least seek the high ground during your viewing.

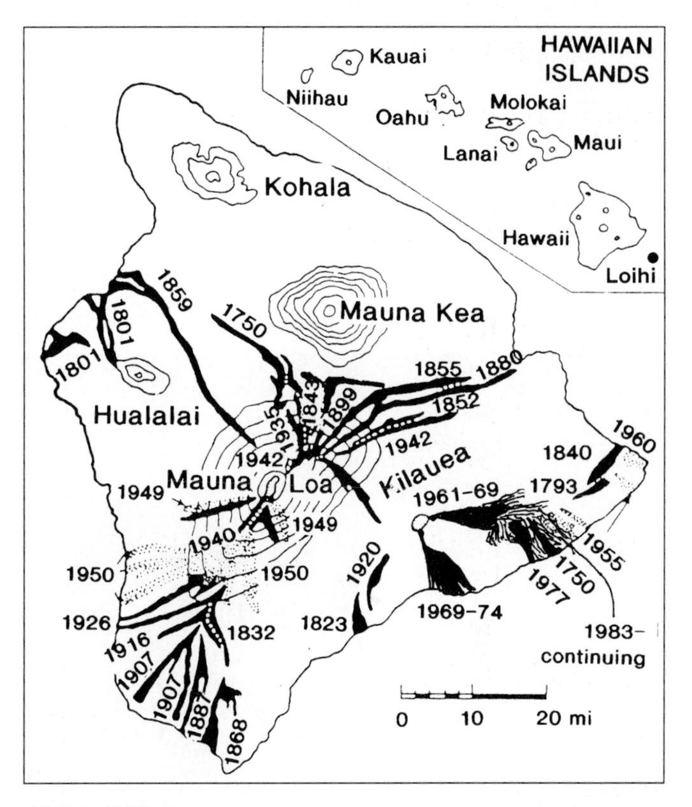

Figure 6.18 Map of Hawaii showing some historic lava flows.

Flood Basalts: Low Viscosity, Low Volatiles, Very Large Volume

Flood basalts are the largest volcanic events known on Earth. Two important descriptive facts are: 1) the immense amounts of mass and energy they pour onto Earth's surface, and 2) in the geologically brief time of 1 to 3 million years. Eruptions from individual volcanoes transfer a lot of heat from the Earth's interior to the surface, but the most impressive movements of heat occur with flood basalts. The numbers that describe the volumes of magma they erupt and the surface areas they bury with lava are so large they are hard to visualize. For example, 250 million years ago, up to 3 million km^3 (800,000 mi^3) of basalt flowed out and covered almost 4 million km^2 (1.5 million mi^2) of Siberia. Visualize basalt flows covering an area about 1,200 miles by 1,200 miles with lavas tens of meters thick. How does this area compare with the area of your state or province? Visualize your entire state or province buried beneath lava tens of meters thick.

Flood basalts occur on all continents and on all ocean floors, but none have occurred in historic times. A flood basalt obviously devastates a region, but can they have global effects? Yes, not from the lavas directly, but from the climate-modifying volatiles such as carbon dioxide (CO_2) or sulfur dioxide (SO_2) they release into the atmosphere. Is it a coincidence that the greatest mass extinction known occurred during the time the Siberian flood basalt was being erupted? Probably not.

Another flood-basalt episode occurred 65 million years ago when about 1.5 million km^3 (360,000 mi^3) of basalt

Figure 6.19 Paricutin Volcano erupting in 1943. This 400 m (1,300 ft) high scoria cone is in the State of Michoacan, Mexico.
Photo by *U.S. Geological Survey.*

Figure 6.20 Ship Rock is a volcanic neck in northwestern New Mexico. It is made of resistant rock formed from magma that cooled inside the central conduit of a volcano now eroded and gone.
Photo by *U.S. Geological Survey.*

flowed out and covered about 1.5 million km^2 (580,000 mi^2) of the Deccan region of India. This also coincides with a mass extinction, the famous one that includes the dinosaurs. Flood basalts will be mentioned again in Chapter 14 on the great dyings (mass extinctions).

Scoria Cones: Medium Viscosity, Medium Volatiles, Small Volume

Scoria cones are conical hills, typically of low height, formed of basaltic to andesitic pyroclastic debris piled up next to a volcanic vent. Scoria cones commonly are produced during a single eruptive interval lasting from a few hours to several years. When that eruption ceases, they usually do not erupt again. After the excess gas has been expelled from the subsurface magma chamber, a scoria or **cinder cone** may have lava pour out of its **crater,** the basin on top of the cone that usually is less than 2 km diameter.

Strombolian-type Eruptions Scoria cones are built mainly by Strombolian-type eruptions (Figure 6.16). The volcano Stromboli, offshore from southwestern Italy, has had almost daily eruptions for millennia. Its central lava lake is topped by a cooled crust. Even the tidal cycle disrupts the lava-lake crust, thus triggering eruptions. Gas pressure builds quickly beneath the crust, and eruptions occur as distinct and separate bursts up to a few times per hour. Each eruption tosses pyroclasts tens to hundreds of meters into the air. For many centuries, tourists have climbed Stromboli to thrill at the explosive blasts, but usually every year a few of those tourists die when hit by large pyroclastic bombs. Strombolian eruptions are not strong enough to break the volcanic cone.

On 20 February 1943, a new volcano was born as eruptions blasted up through a farm field near the village of Paricutin in the state of Michoacan, Mexico (Figure 6.19). The volcano erupted for nine years, building a distinctive scoria cone. Pyroclastic debris and lava flows buried about 100 square miles of land and destroyed the towns of Paricutin and San Juan de Parangaricutiro.

After eruption ceases, a volcano may have its central conduit plugged with hard, solidified magma. If the volcano is composed mainly of weak pyroclastic debris, such as in a scoria cone, then erosion can strip away the softer material of the volcanic cone leaving the resistant rock of the volcanic conduit standing tall as a **volcanic neck** (Figure 6.20).

Stratovolcanoes: High Viscosity, High Volatiles, Large Volume

Stratovolcanoes, or **composite volcanoes,** commonly are steep-sided, symmetrical volcanic peaks built of alternating layers of pyroclastic debris successively capped by high-viscosity andesitic to rhyolitic lava flows that solidify to form protective caps. Stratovolcanoes may show marked variations in their magma compositions from eruption to

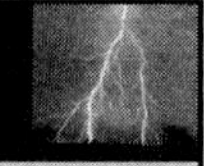

Sidebar

Volcanic Explosivity Index (VEI)

How often do big volcanic eruptions occur? On average, about once every three years according to the volcanic explosivity index (VEI). Combining the historic record with the geologic information stored in the rock record, the major volcanic eruptions occurring between the years 1500 to 1980 were evaluated for their size. Factors evaluated include: 1) volume of material erupted, 2) how high the eruption column reached, and 3) how long the major eruptive burst lasted (Table 6.8). During the 481-year interval studied, 126 major eruptions occurred with the number increasing in modern times. The increase in big eruptions in the nineteenth and twentieth centuries is certainly due to better reporting of events, rather than an actual increase in major eruptions.

The VEI ranges from 0 to 8. The biggest event since 1500 C.E. was the VEI 7 eruption of Tambora in 1815 in Indonesia. This eruption caused a cooling of the world climate during the following year (see Chapter 9). Four VEI 6 events occurred in the 481-year period including the 1883 eruption of Krakatau, also in Indonesia (described later in this chapter). Volcanic events with high VEI values are those of Vulcanian and Plinian-type eruptions. A fifth VEI 6 event occurred in 1991 when Mt. Pinatubo erupted.

Table 6.8 Volcanic Explosivity Index (VEI)

VEI	**0**	**1**	**2**	**3**	**4**	**5**	**6**	**7**	**8**
Volume of Ejecta (m^3)	$<10^4$	10^4–10^6	10^6–10^7	10^7–10^8	10^8–10^9	10^9–10^{10}	10^{10}–10^{11}	10^{11}–10^{12}	$>10^{12}$
Eruption Column Height (km)	<0.1	0.1–1	1–5	3–15	10–25	>25			
Eruptive Style	<-------Hawaiian------->			<-------Vulcanian------->					
		<---Strombolian--->				<-------------Plinian------------->			
Duration of Continuous Blast (hours)	<----------- <1 ----------->			<--- 1–6 --->		<----------- >12----------->			
					<------ 6–12 ------>				

After Newhall and Self (1982).

Figure 6.21 Shishaldin Volcano, a symmetrical stratovolcano rising 2,857 m (9,372 ft) above sea level, Aleutian Islands, Alaska. Photo by *U.S. Geological Survey.*

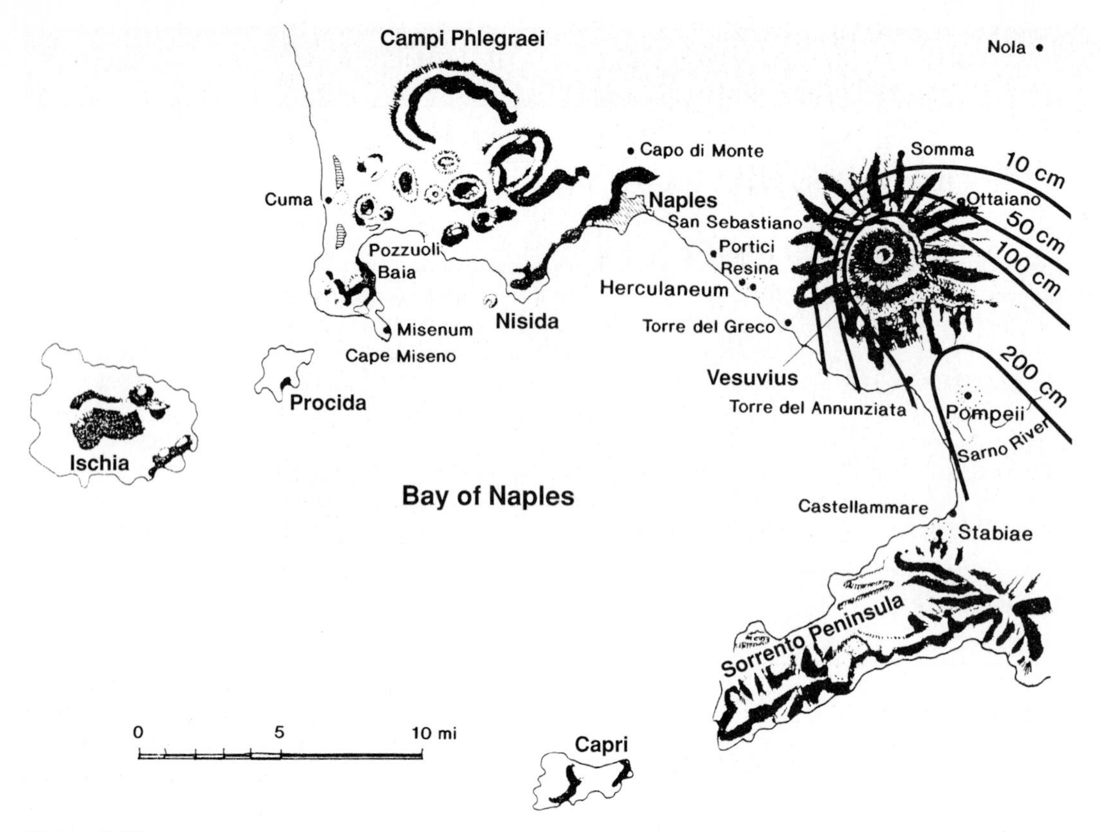

Figure 6.22 Map of Bay of Naples area showing location of Mount Vesuvius. Pumice fallout from 79 C.E. eruption is contoured in centimeters. Pompeii and Stabiae were buried by pumice; Herculaneum by lahars.

eruption, and their eruptive styles include Vulcanian and Plinian. Some of Earth's most beautiful mountains are stratovolcanoes, e.g., Mount Fuji in Japan, Mount Kilimanjaro in Tanzania, Mount Shasta in California, Mount Rainier in Washington, and Shishaldin Volcano in Alaska (Figure 6.21).

Vulcanian-type Eruptions All volcanoes take their name from Vulcan, the Roman god of fire and blacksmith for the gods. The prototypical volcano is one of the Aeolian Islands in the Tyrrhenian Sea north of Sicily. The fire and smoke emitted from the top of the mountain reminded observers of the chimney of Vulcan's forge, so the mountain was named Vulcano. Vulcanian-type eruptions alternate between thick, highly viscous lavas and masses of pyroclastic material blown out of the volcano. Some Vulcanian-type eruptions are more violent blasts of high-viscosity magma loaded with trapped gases. The material blown out during eruptions covers wide areas. Vulcanian-type eruptions commonly are the early phase in the eruptions of other volcanoes as they "clear their throats" before emitting larger eruptions.

Plinian-type Eruptions **Plinian eruptions** are named after the 17-year old Pliny the Younger in honor of his detailed written observations of the 79 C.E. eruptions of Vesuvius that claimed the life of his well-known uncle Pliny the Elder. In Plinian eruptions, the volcano "throat is now clear," and incredible gas-powered vertical eruption columns carry pyroclastic debris, including lots of pumice, up to 30 miles (50 km) into the atmosphere. The Plinian eruption is a common final phase in a major eruptive sequence. About 2 to 3 Plinian eruptions occur each century.

Vesuvius, 79 C.E. The most famous of all volcanoes probably is Vesuvius, and the most famous of all eruptions must be the Vulcanian and Plinian events of 79 C.E. It was then that the cities of Pompeii and Herculaneum were buried and forgotten for over 1,500 years. Vesuvius began as a submarine volcano in the Bay of Naples. It grew greatly in size, and its rocky debris filled in the waters that once separated it from mainland Italy (Figure 6.22). What is the cause of the volcanism at Vesuvius and the neighboring volcanoes of Stromboli, Vulcano, Etna, and others? The subduction of Mediterranean sea floor beneath Europe to make room for the northward charge of Africa.

A reminder of the natural hazards in Italy arrived on 5 February 62 C.E. when a major earthquake destroyed much of Pompeii and caused serious damage in Herculaneum and Neapolis (Naples). Earthquakes, although not as large, were a common occurrence for the next 17 years. Pompeii had been a center of commerce for centuries. In 79 C.E., the city had a population of about 20,000 people, 8,000 of whom were slaves. Robert Etienne described it as:

> An average city inhabited by average people, Pompeii would have achieved a comfortable mediocrity and passed peacefully

Figure 6.23 Body cast of man killed by a pyroclastic flow from Vesuvius in late August in the year 79.

Figure 6.24 First big explosive blast from Mount Pinatubo, 15 June 1991. A pyroclastic flow is moving downhill to the left. Photo by R. S. Culbreth, *U.S. Geological Survey.*

into the silence of history, had the sudden catastrophe of the volcanic eruption not wiped it from the world of the living.

The 24th of August 79 C.E. was a warm summer day, but then Vesuvius began erupting and the day became even hotter. Vesuvius blew out 4 km^3 of pyroclastic materials. About half of the old volcanic cone was destroyed. Pompeii lay downwind and was buried by pumice fragments accumulating up to 3 m (10 ft) deep (Figure 6.22). Death was not always quick. Some bodies were found inside houses on top of thick layers of pumice, giving evidence of hours of struggle by people fighting to stay alive. Their hands held cloths over their mouths as they tried to avoid asphyxiation from gases seeping out of the pumice (Figure 6.23).

Many other people were found near the sea (Figure 6.1). There was time for them to flee the falling pumice, but ground-hugging **pyroclastic flows,** full of hot gases, finished them off (for modern example of a major Vulcanian-type eruption, see Figure 6.24). About 4,000 people died. The more-distant town of Stabiae was also mostly destroyed. It was here that Pliny the Elder died; the weak heart of the overweight man failed at age 56 under the stress of the farthest-reaching gas-rich flow.

Following the Vulcanian-type eruption, the volcano entered a second phase, the Plinian phase, where it blew immense volumes of pyroclasts up to 32 km (20 mi) high in the atmosphere. The height of the eruption column varied as the volcanic energy waxed and waned. During weakened intervals, the great vertical column of ashes would temporarily collapse, sending surges and pyroclastic flows down the volcano slopes. Pompeii was buried under an additional 6 to 7 feet of pyroclastic debris. These were the flows that finished off the surviving Pompeiians.

The fine ashes settling out from great heights affected a much larger region. Pliny the Younger was at Misenum and wrote:

> And now came the ashes, but at first sparsely. I turned around. Behind us, an ominous thick smoke, spreading over the earth like a flood, followed us. "Let's go into the fields while we can still see the way," I told my mother—for I was afraid that we might be crushed by the mob on the road in the midst of darkness. We had scarcely agreed when we were enveloped in night—not a moonless night or one dimmed by cloud, but the darkness of a sealed room without lights. To be heard were only the shrill cries of women, the wailing of children, the shouting of men. Some were calling to their parents, others to their children, others to their wives—knowing one another

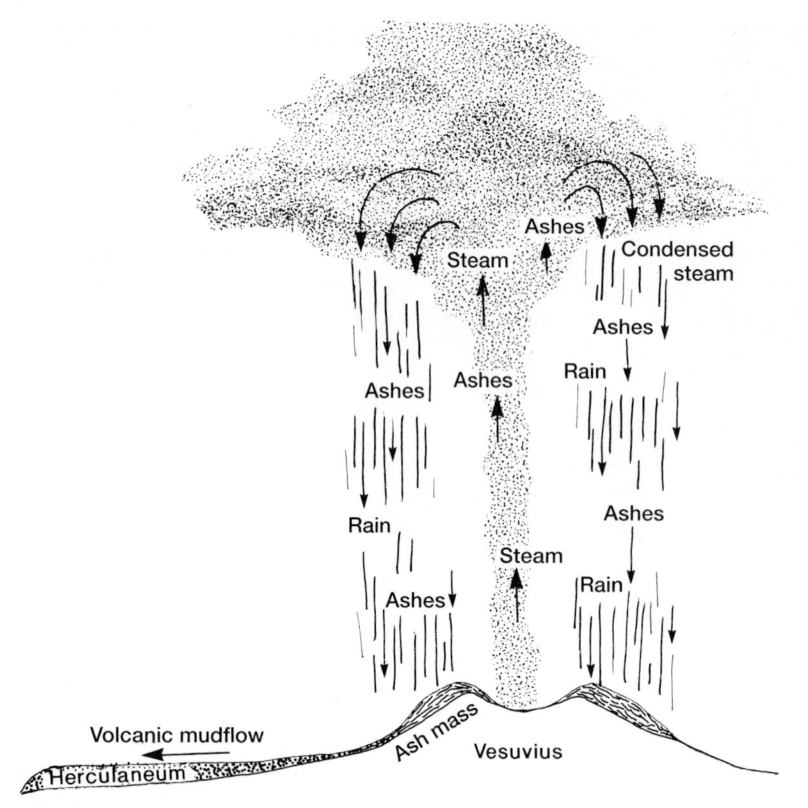

Figure 6.25 "Volcano weather" and formation of lahars. Prolonged vertical eruption leads to accumulation of debris on steep slopes of the volcano. Steam blown upward into cold, high altitudes condenses and falls back as rain. The stage is set: steep slopes + loose volcanic debris + heavy rain = lahars.

> only by voice. Some wept for themselves, others for their relations. There were those who, in their very fear of death, invoked it. Many lifted up their hands to the gods, but a great number believed there were no gods, and that this was to be the world's last eternal night.

A Plinian eruption not only blows ashes to great heights but also volcanic gases. Water, as abundant steam, can be blown high into the atmosphere, cooling and condensing, then falling back down as rain—heavy rain. Some volcanic eruptions create their own "weather." Rain falling on thick piles of pyroclastic debris, sitting unstably on the steep slopes of Vesuvius, set off thick volcanic mudflows (Figure 6.25). These gravity-pulled mass movements of muddy volcanic debris are known as **lahars,** an Indonesian word (for modern example, see Figure 6.26). Lahars buried the city of Herculaneum up to 20 m (65 ft) deep in pumice, ashes, and volcanic rock fragments jumbled together in a confused mass. However, this was during the second phase of the eruption and most people had used the day or two before to clear out of the area, so the loss of life was not nearly as great as at Pompeii. Today, the town of Ercolano lies on top of the mudflows burying Herculaneum. The lessons of history have not been well learned here.

The timing of major eruptions of Vesuvius offers an interesting lesson. Apparently Vesuvius did not have a major eruption from the seventh century B.C.E. until 79 C.E. People had at least 700 years to lose their fears and yield to the allure of the rich agricultural soils on Vesuvius. After 79 C.E., large eruptions occurred more often: in 203, 472 (ashes blown over much of Europe), 512, 685, 993, 1036 (first lava flows in historic time), 1049, and 1138–1139. Then nearly 500 years passed. Time to forget the past and recolonize the mountain. But in 1631, Vesuvius poured out large volumes of lava that destroyed six towns; mudflows ruined another nine towns, and about 4,000 people perished. The long periods of volcanic quiescence in the last 2,700 years, one about 700 years long and another of nearly 500 years, seem like long times to short-living, land-hungry humans, but this is the time schedule of an active volcano. A lack of appreciation for the time involved between eruptions leads to many active volcanoes being falsely regarded as extinct.

Since 1631, the eruptions have been mostly of Strombolian type and thus not as dangerous. There were 18 eruption cycles between 1631 and 1944; each lasted from 2 to 37 years with quiet intervals ranging from 0.5 to 6.8 years. Since 1944, Vesuvius has been quiet. Is this interval of calm setting the stage for a large Vulcanian/Plinian-type event? It is not known for sure, but consider that almost three million people live within reach of Vesuvius today, including about one million people living on the slopes of the volcano.

Lava Domes: High Viscosity, Low Volatiles, Small Volume

Lava domes form when high-viscosity magma cools quickly, producing a hardened dome or plug a few meters to several kilometers wide, and a few meters to one kilometer high. Lava domes can form as quickly as a few hours, or they may continue to grow for decades. The formation of lava domes can be visualized as part of a larger eruptive process. When a large volume of hot rock/magma rises and undergoes decompression melting, dissolved volatiles are freed. Much of the freed gases rise and accumulate at or near the top of the magma mass. When a major eruption occurs, these gases power the initial Vulcanian-type blast and then power the succeeding Plinian-type eruption that lasts until the excess volatiles have escaped. What type of magma remains? Often it is a low-volatile, high-viscosity paste that oozes upward slowly and cools quickly, forming a plug in the throat of the volcano. Figure 6.27 shows the lava dome emplaced in Mt. Katmai in southern Alaska following its 1912 eruption, the biggest eruption of the twentieth century.

Lava domes can provide spectacular sights. After the 1902 eruptions of Mont Pelee in the Caribbean killed more than 30,000 people (see Chapter 7), a lava dome formed as a great spine that grew about 10 m/day and rose above the top of the volcano. The spine of hardened magma was forced upward by the pressure of magma below until it stood over 300 m (>1,000 ft) higher than the mountaintop, like a giant cork rising out of a bottle.

Do lava domes present hazards? Yes, in the 1990s they were responsible for 129 deaths: 19 from Soufriere Hills Volcano on Montserrat in 1997 (see color photo in Chapter 1), 66 from Mt. Merapi in Indonesia in 1994, and 44 from Mt. Unzen in Japan in 1991. The hardened, brittle lava dome rock can fail in a gravity-pulled landslide from the mountain, or magma trapped below the brittle lava dome can break out in a violent eruption.

Figure 6.26 Lahars from Vulcanian-type eruption of Mount Pinatubo destroyed bridges to Angeles City, Philippines 12 August 1991.

Photo by T. Casadevall, *U.S. Geological Survey.*

Calderas: High Viscosity, High Volatiles, Very Large Volume

Caldera-forming eruptions are the largest of the violent, explosive volcanic behaviors. Calderas are large depressions, over 2 km in diameter, formed by roof collapse into partially emptied magma reservoirs. Some involve the collapse of stratovolcano peaks into their magma chambers, such as Crater Lake in Oregon, Krakatau in Indonesia, and Santorini in the eastern Mediterranean (Figure 6.28). Collapses usually follow a sustained Plinian eruption that opens void space inside the stratovolcano. The piston-like collapse of the volcano peak down into the magma chamber can force out large volumes of magma that flow outward as pumice-rich sheets.

Even larger caldera eruptions occur that are bigger than stratovolcanoes; they leave holes in the ground. The calderas of these truly large eruptions typically are "negative volcanoes," broad and deep depressions such as Lake Yellowstone in Wyoming, or Long Valley in California (see Chapter 7).

Figure 6.27 Novarupta lava dome formed as hardened magma plugged the central magma pipe of the 1912 eruption of Katmai Volcano in southern Alaska. The dome is 800 feet across and 200 feet high.

Photo courtesy of *U.S. Geological Survey.*

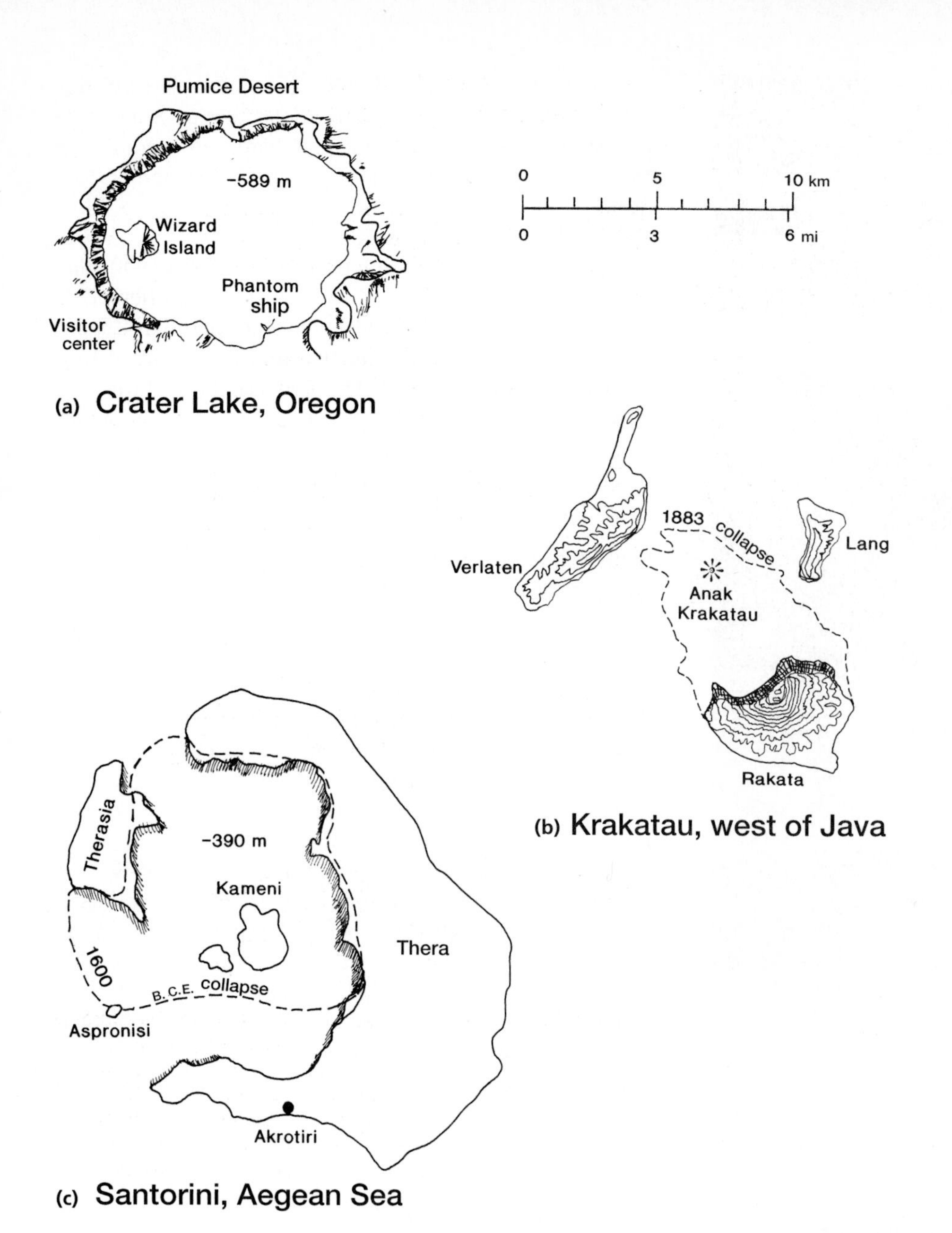

Figure 6.28 Map of some collapse calderas. **(a)** The nearly circular caldera of Crater Lake, Oregon, formed about 5,677 B.C.E. **(b)** Nearly circular outline of old volcano Krakatau; crudely ovoid shape of the 1883 collapse; and the new and growing volcano, Anak Krakatau. **(c)** Caldera in Volcano Santorini that collapsed into the Aegean Sea about 1628 B.C.E.

The most recent example of this type of eruption occurred 74,000 years ago at Toba, on the island of Sumatra in Indonesia. The caldera at Toba is 30 km (20 mi) by 100 km (60 mi) long and has a central raised area inside it that is over 1 km high. The raised area formed during the millennia following the giant eruption; this resurgent topography inside the caldera gives these features their name—**resurgent calderas.**

Crater Lake (Mount Mazama), Oregon Crater Lake is one of the jewels in the U.S. National Park system. Its intense blue waters are pure and lie cradled in a high-rimmed, nearly circular basin. Crater Lake is about 9.5 km (6 mi) across and as deep as 589 m (1,932 ft)(Figure 6.29).

Several thousand years ago, the stratovolcano Mount Mazama stood about 3,660 m (12,000 ft) high as one of the Cascade Range volcanoes above the Cascadia subduction zone (Figure 6.30a). Over 7,600 years ago, a major eruption began blowing sticky magma out of the mountain as glassy, gas bubble-filled pumice and ashes (Figure 6.30b). The magma had too high a viscosity to flow as a liquid, so it erupted as pyroclastic flows and Plinian columns. As the erupted material grew in volume, its debris covered much of the Pacific Northwestern United States and part of Canada with a thick, dis-

Figure 6.29 Crater Lake, Oregon, fills the caldera of Mount Mazama which collapsed in the year 5677 B.C.E. Wizard Island is in lower center of photo.

Photo by T. Casadevall, *U.S. Geological Survey.*

tinctive ash layer that is easily recognizable. Mazama ash is found in the Greenland glacier within the ice layer formed during the snowfall season of the year 5677 B.C.E. About 40 km^3 of magma was ejected. Evacuation of this immense volume of magma left so tremendous a void below the surface that the weakened mountain peak collapsed and moved down in piston-like fashion into the emptied magma chamber (Figure 6.30c). The collapse produced a caldera, about 10 km (6 mi) across that has collected the water for Crater Lake and hosted the growth of a 1,000-year-old successor volcanic cone called Wizard Island (Figures 6.28a and 6.29).

The eruption of Mount Mazama affected American Indians as evidenced by moccasin tracks and artifacts found beneath the distinctive ash layer. What have caldera-collapse events wrought elsewhere?

Krakatau, Indonesia, 1883 Today, Krakatau (Krakatoa) is a group of Indonesian islands in the Sunda Strait between Sumatra and Java. It is part of the grand arc of volcanoes built above the subducting Australia-India plate. Krakatau is a big stratovolcano that builds up out of the ocean and then collapses. Its larger outline is still distinguishable (Figure 6.28b).

From the ruins of an earlier collapse, magmatic activity built Krakatau upward through the seventeenth century. After two centuries of quiescence, volcanic activity resumed on 20 May 1883. By August 1883, moderate-sized Vulcanian eruptions were occurring from about a dozen vents. At 2 P.M. on 26 August, a large blast shot volcanic ashes and pumice 28 km (17 mi) high as one of the cones collapsed

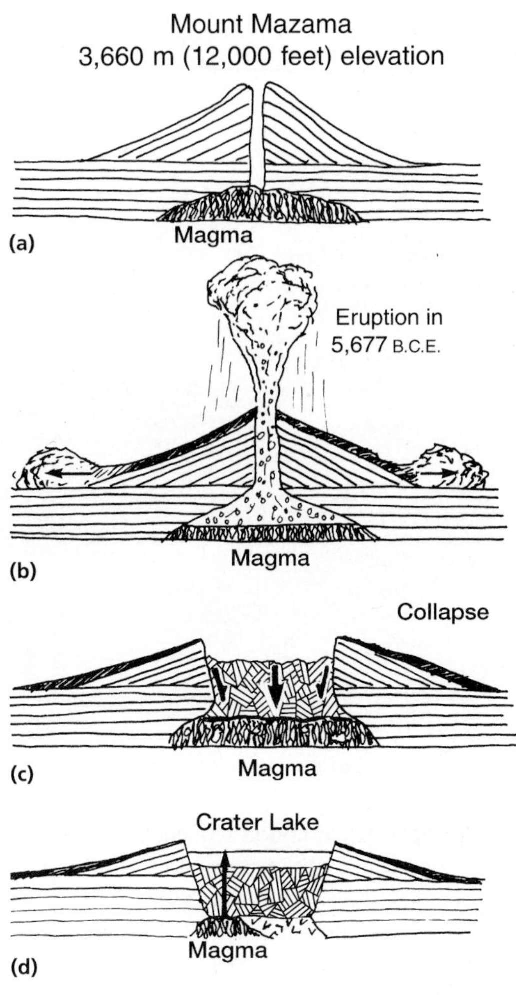

Figure 6.30 How Crater Lake formed. **(a)** Mount Mazama volcano stood high. **(b)** Gaseous eruption in the year 5677 B.C.E. emptied a huge volume of viscous magma. **(c)** The gigantic eruption left a void inside the weakened mountain, and the unsupported top fell down into the emptied magma chamber. **(d)** The waters of Crater Lake now fill the caldera, and a small new volcanic cone (Wizard Island) has built above lake level.

into the sea, setting off huge tsunami. Eruptions were so noisy that night that sleep was not possible in western Java, including the capital city of Djakarta (then called Batavia). The early morning hours of 27 August were rocked by more ear-hammering eruptions, and further volcanic collapses sent more giant tsunami to wrack the coastal villages. Day was turned to night-like darkness as heavy clouds of volcanic ashes blocked the sunlight. At 10 A.M., a stupendous blast rocketed a glowing cloud of incandescent pumice and ashes 80 km (50 mi) into the atmosphere. This blast was distinctly heard 5,000 km (3,000 mi) away.

The 10 A.M. volcano collapse sent tsunami higher than 35 m (115 ft) sweeping into bays along the low coastlines of Java and Sumatra. The volcanic eruptions caused tsunami that destroyed 295 towns and smashed or drowned an

Figure 6.31 Aesthetic shot of the inside of Santorini crater with the classic white-washed Greek homes on the cliff edge.

estimated 36,000 people. The eruption sequence blew out 18 km^3 of material (95 percent fresh magma and 5 percent pulverized older rock), creating a subterranean hole into which 23 km^2 of land collapsed (Figure 6.28b). Where islands with elevations of 450 m (1,476 ft) had stood, there now was a hole in the sea floor 275 m (900 ft) deep.

The amount of magma erupted at Krakatau in 1883 was less than half that of the Mount Mazama eruption. But Krakatau collapsed into the sea, sending off big waves.

In 1927, Krakatau came back as a new volcanic cone sprang into action. Called Anak Krakatau—"child of Krakatau"—it is still growing (Figure 6.28b). We will hear more from Krakatau.

Santorini and the Lost Continent of Atlantis As the Mediterranean oceanic plate subducts beneath Europe, it causes numerous volcanoes. One of the biggest is the stratovolcano Santorini in the Aegean Sea. Today, Thera is the largest island in a circular group marking the sunken remains of Santorini (Figure 6.28c). Thera is one of the most popular tourist sites in the Greek islands (Figure 6.31), but around 1628 B.C.E., Santorini underwent an explosive series of eruptions that buried the Bronze Age city of Akrotiri on Thera to depths of 70 m (230 ft) in four distinct phases.

The eruption phases were determined by Floyd McCoy and Grant Heiken. Phase 1 volcanic activity is represented by a 6 m (20 ft) thick layer of air-settled pumice that gently entombed and preserved the buildings and artwork in Akrotiri (Figure 6.32). Pumice is produced when there is no

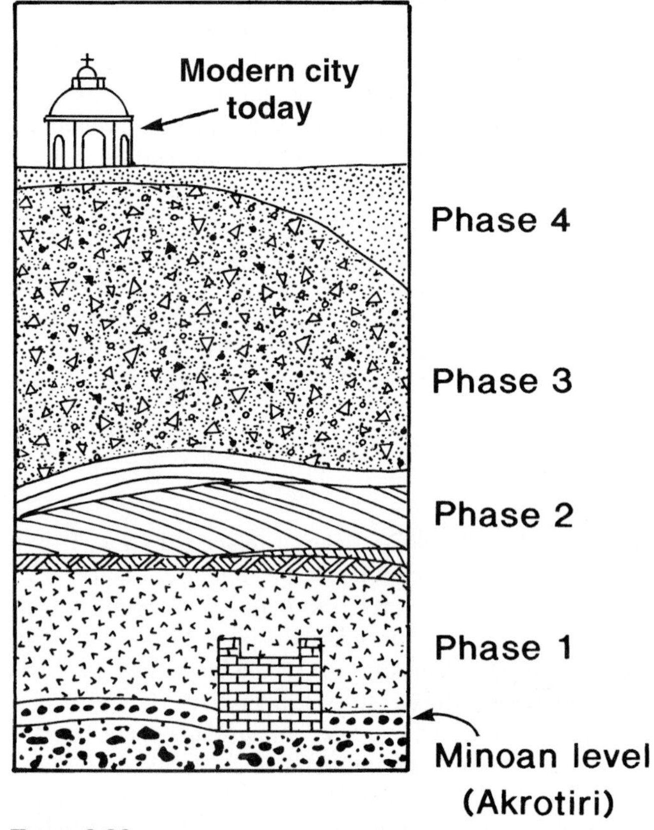

Figure 6.32 The Minoan City of Akrotiri on the Volcano Santorini was buried about 70 m deep by a four-phase eruption about 1628 B.C.E.

Source: McCoy and Heiken (1990).

Sidebar

Hot Spots

Hot spots are shallow hot rock masses/magmas or plumes of slowly rising mantle rock that create volcanism on the Earth's surface. The temperature of the rising rock is hotter than the surrounding rock by about 300°C (570°F) in the plume center and only 100°C (212°F) along the outer margin of the plume head. But this temperature difference lowers viscosity enough to start the rise toward the surface. Most hot spots are visualized as rising plumes that operate for about 100 million years.

Hot spots do not move as much as tectonic plates and are used as reference points to help chart plate movements (Figure 2.15). They occur under the oceans and under the continents (e.g., Yellowstone), in the center of plates (e.g., Hawaii), and as part of spreading centers (e.g., Iceland). In the 1970s, a survey was made of hot spots that create elevated volcanic domes with diameters greater than 200 km (125 mi). The survey counted 122 hot spots active in the last ten million years (Figure 6.33), 53 under ocean basins and 69 under continents.

The largest number of hot spots lies beneath the African plate. The drifting of Africa has been slowed by its collision with Eurasia during the last 30 million years. The slowed African plate may be acting like a thermal blanket concentrating the mantle heat beneath it. With Africa effectively stopped from making large horizontal movements, the westward movement of South America has doubled, and the mid-Atlantic Ocean spreading center is moving westward also, leaving some hot spots behind, as at Tristan de Cunha and St. Helena (Figure 6.33).

The explosiveness of volcanic eruptions above hot spots varies. They are relatively peaceful above oceanic hot spots, such as Hawaii, where low-volatile, low-viscosity, large-volume magma flows easily, analogous to spreading-center volcanism, and builds shield volcanoes (Figure 6.17). The Hawaiian hot spot is about 80 km (50 mi) in diameter, as defined by earthquake hypocenters at 60 km (37 mi) depth.

A hot spot below a spreading center means a much greater volume of basaltic magma can erupt. For example, at Iceland, the asthenosphere magma of the spreading process is augmented by deeper mantle magma to create an immense volume of basaltic rock. The combined magmas are basalt and the eruptions are peaceful enough for the citizens of Iceland to live prosperously (see Chapter 7). The mantle plume beneath Iceland is the most vigorous hot spot on Earth today. The rising plume has created crust beneath Iceland that is 4 to 5 times thicker than average.

Above continental hot spots, such as at Yellowstone National Park, the eruptions may be incredibly explosive because the rising magma breaks off and absorbs so much continental rock it creates a volatile-rich, high-viscosity, very large volume magma. The mention of a big volcanic eruption may bring to mind a tall mountain emitting a powerful explosion, but the really big eruptions emit so much magma that they leave a hole bigger than a mountain, a giant caldera that can be 100 km (>60 mi) long.

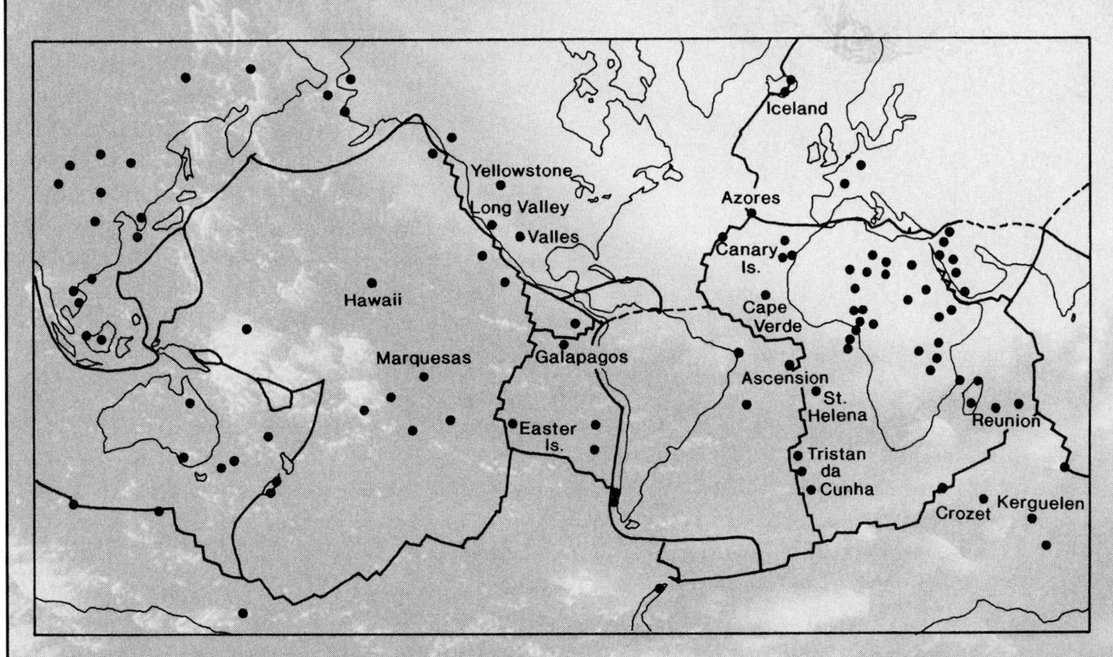

Figure 6.33 Hot spots active in the last 10 million years. Antartica is not shown but lies above 11 hot spots, raising questions about the effects of melting massive volumes of ice.

water reacting with magma, so the eruption must have been from above sea level. The settling of pumice is not necessarily fatal and usually provides time for the wise to heed the warning and evacuate. When the eruptions began, most Akrotirians picked up their most prized possessions and left. All the "brave" people who were not scared away and remained with their homes and valuables died in the second phase.

In phase 2, several meter-thick deposits formed from rapidly flowing hot water that scoured the area. This occurred when seawater reached the exposed magma chamber, escaping as destructive steam blasts.

Phase 3 deposits are a pyroclastic mass of jumbled ash, pumice, and large rock fragments that fell from the air, accumulating up to 56 m (180 ft) thick. This massive mess was produced by collapse of volcanic cones in the center of Santorini.

Phase 4 deposits are layers of ash and small rock fragments formed during the final degassing of the magma body as it spit out ground-hugging, hot gaseous clouds. Where there had been a large island made of several volcanic cones, there now existed a huge caldera with depths of 390 m (1,280 ft) below sea level. What were the effects of this eruption on the Mediterranean world?

Akrotiri was an important city, a part of the Minoan civilization based in Crete. The Minoans created an advanced civilization. In 1628 B.C.E., Akrotiri had three-story houses; paved streets with stone-lined sewers beneath them; advanced ceramic and jewelry work; regular trade with the Minoans' less-advanced neighbors in Cyprus, Syria, Egypt, and Greece; and colorful wall frescoes that depicted their wealthy and comfortable life. In short, these Minoans had a higher standard of living than many people in this part of the world today, over 3,600 years later.

The dramatic collapse of this piece of the Minoan civilization must have made an indelible impression on the people of that time. In fact, this may be the event passed down to us by Plato as the disappearance of the island empire of Atlantis, which after violent earthquakes and great floods "in a single day and night disappeared beneath the sea." Plato lived in Greece from 427 to 347 B.C.E. He told the tale in the dialogues of Critias, the historian, who recounted the visit of Solon to Egypt, where he learned the account of Atlantis from the Egyptian priests in their oral histories. About 1,200 years after the event, Plato wrote a reasonably good description of a caldera collapse with attendant earthquakes, floods (steam surges or tsunami), and a landmass sinking below the sea in a day and a night.

The eruption and caldera collapse into the sea at Santorini seem similar to the events at Krakatau 3,500 years later, except the Santorini event was bigger. The Santorini eruption is estimated to have blown out 30 km^3 of rhyolitic magma; Krakatau blew out 18 km^3. Krakatau sent out ocean waves 35 m (115 ft) high; Santorini must have done as much. Look at the map of the Aegean Sea region; it is one of the most island-rich areas on Earth. Tsunami in this region must have had a devastating effect on coastal towns and people, as well as leaving profound impressions on survivors who passed these memories down to succeeding generations. The tales of Plato, the excavations by archaeologists, and the reconstructions by volcanologists all point to a remarkably consistent story.

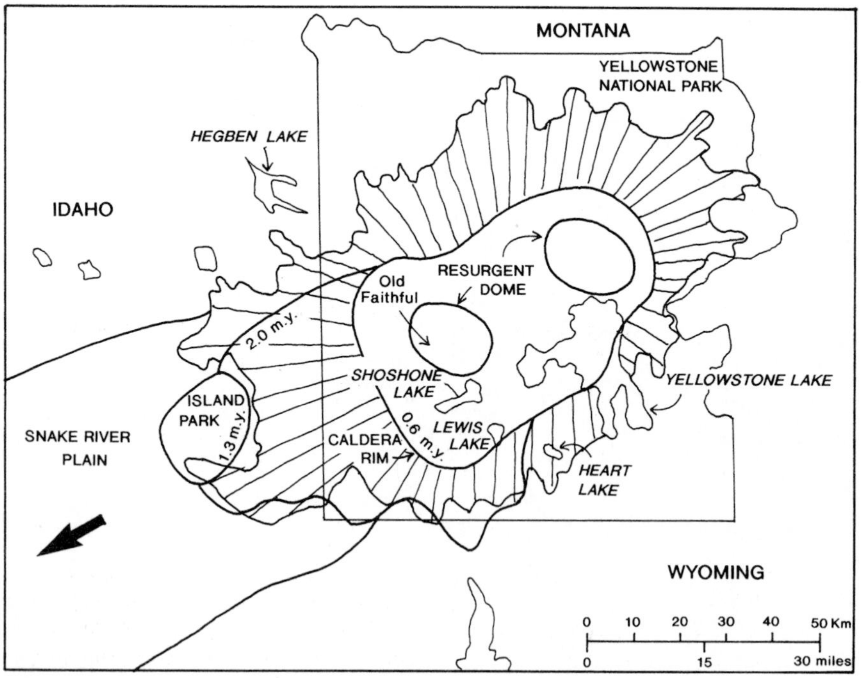

Resurgent Calderas In the United States, there are three resurgent calderas known to have had gigantic eruptions in the last million years: 1) Valles caldera in New Mexico about 1 million years ago, 2) Long Valley, California about 760,000 years ago (see end of Chapter 7), and 3) Yellowstone National Park, Wyoming about 600,000 years ago. The big depressions of resurgent calderas have a central dome or hills where part of the caldera floor has been uplifted or resurgent, in the years following the big

Figure 6.34 The Yellowstone hot spot area. North American plate is moving southwest, thus the hot-spot magma plume erupts progressively farther northeast with time. Three giant calderas have erupted in the last 2 million years—at 2, 1.3, and 0.6 million years ago. Cross-hatched area was covered by hot, killing pyroclastic flows during the eruption of 600,000 years ago.

eruption. The three U.S. resurgent calderas have different geologic settings. Valles is associated with the Rio Grande rift (Figure 5.13), Long Valley lies just east of the Sierra Nevada, and Yellowstone sits above a hot spot.

These mega-eruptive centers occur where large volumes of basaltic magma intrude upward to shallow depths. While rising upward they encounter continental rocks that melt at lower temperatures. The resultant mixture of melts creates magmas with lower temperature, higher percentages of SiO_2, high viscosity, and high content of volatiles. The buoyant, sticky magma accumulates as very large masses a few kilometers below the surface.

Yellowstone National Park A resurgent caldera exists in Yellowstone above a hot spot, a long-lived mantle plume that the North American continent is drifting across. The hot spot occupies a relatively fixed position above which the North American plate moves southwestward about 2 to 4 cm/yr. Plate movement over the hot spot during the last 15 million years is recorded by a trail of surface volcanism cut across the Snake River plain in Idaho and on into Wyoming (Figure 6.34). At present, Yellowstone National Park sits above the hot spot and a large body of rhyolitic magma lies about 5 to 10 km (3 to 6 mi) beneath it.

In the last 2 million years, three catastrophic eruptions have occurred at Yellowstone at 2 million, 1.3 million, and 600,000 years ago (Figure 6.34). Such mega-eruptions do not come often, but in a few short weeks, they pour forth virtually unimaginable volumes of rhyolitic magma, mostly as pyroclastic flows. The oldest event erupted 2,500 km^3 of magma, the middle one emptied 280 km^3, and the youngest dumped out 1,000 km^3. (Compare these magma volumes to the 1980 eruption of Mount St. Helens, which totaled 1 km^3.) An eruption of 1,000 km^3 of rhyolitic pyroclastic flows would cover a surrounding area of 30,000 km^2 with a mass of pyroclastic debris ranging from a few to more than 100 m thickness (Figure 6.34). The weight of volcanic material would cause a 500 km^2 area to sink isostatically.

The Yellowstone mega-eruption of 600,000 years ago created a giant caldera that is 75 km (47 mi) long and 45 km (28 mi) wide. Look again at Figure 6.34 and consider the size of the giant caldera and the extent of its emitted pyroclastic flows—in a matter of days, all life in the area would have died and been deeply buried.

Eruptive Sequence of a Resurgent Caldera The eruptions of these giant calderas go through a characteristic sequence (Figure 6.35). They begin when a very large volume of rhyolitic magma rises to a few kilometers below the surface, bowing the ground upward (Figure 6.35a). The magma body accumulates a cap rich in volatiles and low-density components such as SiO_2.

After a few hundred thousand years, a mega-eruption begins with a spectacular circular ring of fire as Plinian columns jet up from circular-to-ovoid fractures surrounding

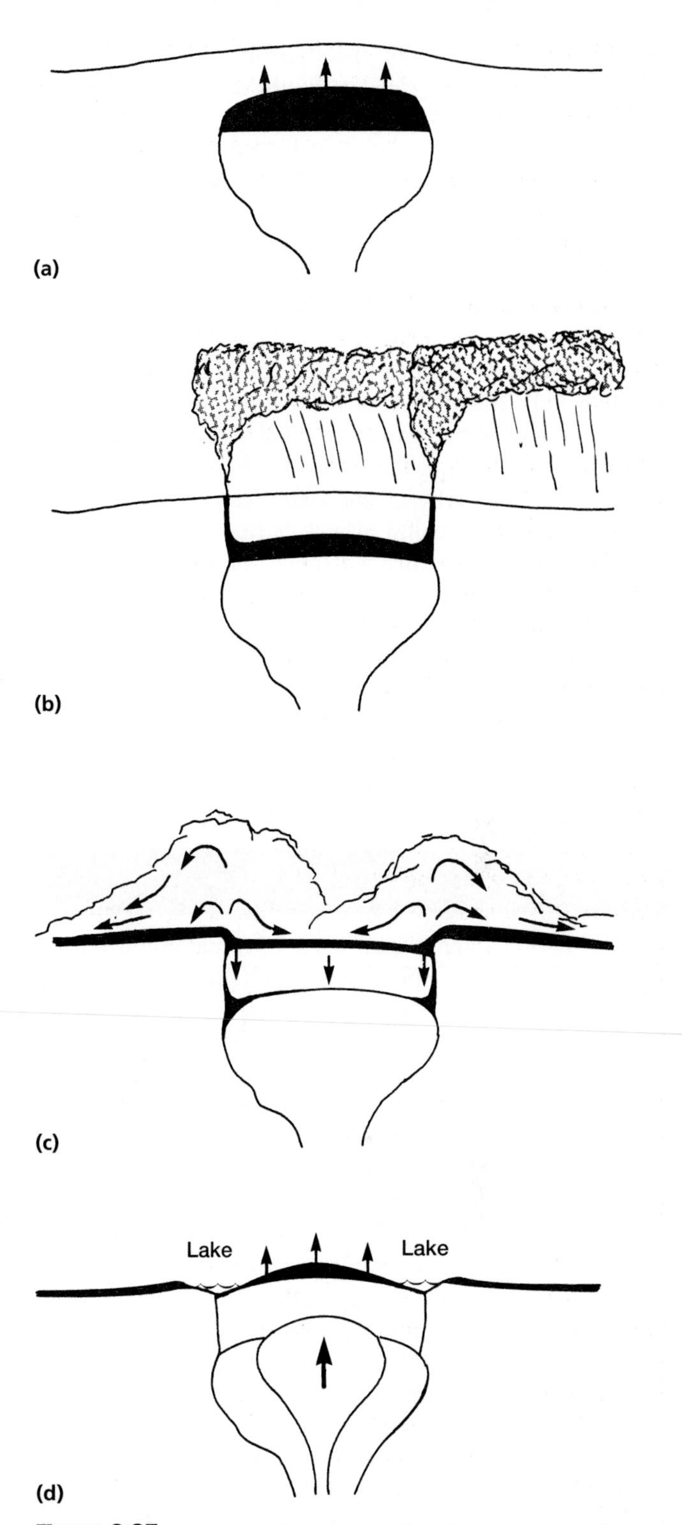

Figure 6.35 Stages in formation of a giant continental caldera. **(a)** Rising mass of magma forms low-density cap rich in SiO_2 and gases, bulging the ground surface upward. **(b)** Plinian eruptions begin from circular fractures surrounding the bulge. **(c)** Magma pours out in pyroclastic flows of tremendous volume, causing the ground surface to sink into a giant caldera. **(d)** Removal of magma decreases the crustal pressure, causing new magma to bulge up the caldera floor.

the magma body (Figure 6.35b). The escaping magma erodes the fractures, thus increasing the size of the eruptive vents so more and more magma escapes.

As greater volumes of gas "feel" the lessening pressure, the magma begins gushing out of the fractures in mind-boggling volumes (Figure 6.35c). The outrushing magma is too voluminous to all go airborne, so most just pours away from the vents as pyroclastic flows, the fastest way to remove gas-laden, sticky magma.

As the subsurface magma body shrinks, the land surface sinks as well, like a piston in a cylinder, creating a giant caldera (Figure 6.35d). The removal of 1,000 km^3 of magma creates a void, an isostatic imbalance, that is filled by a new mass of rising magma that bows up the caldera floor to create **resurgent domes** (Figure 6.34). Resurgent domes may be viewed as the reloading process whereby magma begins accumulating toward the critical volume that will trigger the next eruption.

The areas alongside resurgent domes are commonly occupied by lakes (Figure 6.34). Imagine driving the many miles from Yellowstone Lake to Old Faithful Geyser, all the time staying within the gigantic collapse caldera of the 600,000-year-old eruption. When will the next mega-eruption occur?

Summary

Some of Earth's internal heat melts rock that erupts on the surface as volcanoes. Spreading centers provide such ideal settings for volcanism that 80 percent of all extruded magma occurs there. Plates pull apart and magma rises up the fractures with relatively peaceful eruptions. Subduction-zone eruptions involve magma contaminated by incorporated crustal rock yielding high-viscosity, gas-rich magma that erupts explosively. Transform faults have little or no volcanism associated with them.

Hot rock at depth will rise buoyantly. This rock may melt and become magma due to increased temperature, decreased pressure, and/or increased water content. Most magma is produced as pressure is lowered on rising hot rock via decompression melting. When magma nears the surface, gases come out of solution and help cause volcanic eruption. Whether magma erupts peacefully or explosively depends on magma types. Eruption styles and volcanic landforms can be understood via the 3 Vs of volcanology—viscosity, volatiles, volume. Beneath the ocean basins, magmas are basaltic in composition with low contents of SiO_2 and high temperature producing low viscosity, easy escape of volatiles (gases) and peaceful eruptions. Beneath continents, rising basaltic magmas are contaminated by melting continental-crust rocks thus altering magma compositions. The resultant andesitic-to-rhyolitic magmas have high contents of SiO_2 and relatively low temperatures producing high viscosity, difficult escape for volatiles and explosive eruptions.

When magma reaches the surface and gas escapes easily, lava flows result. Low-viscosity lava flows may build shield volcanoes much wider than they are tall, e.g., Hawaii. If gas is trapped in magma then explosions result blasting pyroclastic debris into the air. A scoria cone may be built around a volcanic vent by the settling of pyroclastic debris, e.g., Paricutin. Tall symmetrical volcanic peaks are usually stratovolcanoes built of alternations of lava and pyroclastic material, e.g., Vesuvius.

The volcanic explosivity index (VEI) measures size of volcanic eruptions on a scale of 0 to 8. Between 1500 to 2002, one VEI 7 eruption occurred (Tambora, 1815) along with five VEI 6 events (e.g., Krakatau, 1883; Pinatubo, 1991).

Calderas form when their roofs collapse into partially emptied magma chambers. This occurs when a stratovolcano is too weak to stand and its peak collapses downward (e.g., Crater Lake, Oregon). If the peak falls into the ocean, major tsunami can result (e.g., Santorini, 1628 B.C.E.; Krakatau, 1883). The biggest explosive eruptions occur on continents where collapses may be bigger than mountains at resurgent calderas (e.g., Yellowstone).

Terms to Remember

Questions for Review

1. Sketch a map of an idealized tectonic plate and evaluate the volcanic hazards along each type of plate edge.
2. Diagram and explain the sequence of events leading to a geyser eruption. Include temperature and pressure changes in your answer.
3. What changes in temperature, pressure, and water content will cause hot rock to melt?

4. What common elements combine to form most igneous rocks?
5. What minerals combine to form most igneous rocks?
6. Contrast the differences between basaltic versus rhyolitic magma in terms of: SiO_2 percentage, temperature, viscosity, and mode of gas escape.
7. What determines whether volcanic activity will be a lava flow or a pyroclastic eruption?
8. Play the 3 Vs game. Pick various low, medium, and high values for viscosity, volatiles, and volume, then describe the resultant eruption styles and volcanic land forms.
9. Draw a cross section showing a scoria cone with a volcanic neck and dikes.
10. Draw a cross section showing the difference between a shield volcano versus a stratovolcano.
11. Draw a cross section illustrating a Plinian eruption.
12. Explain the factors controlling the volcanic explosivity index (VEI).
13. Draw a cross section and describe the collapse of an oceanic volcano, such as Krakatau. What usually is the biggest killer in this process?
14. Explain the eruptive behavior of a hot-spot-fed volcano on a continent.

Questions for Further Thought

1. Why do people keep returning to a volcano, such as Vesuvius, and building new cities?
2. List the beneficial aspects of volcanoes and volcanism.

Suggested Readings and References

Bullard, F. M. (1984). *Volcanoes of the Earth.* Austin: University of Texas Press.

Carey, S., and Sigurdsson, H. (1987). Temporal variations in column height and magma discharge rate during the 79 A.D. eruption of Vesuvius. *Geological Society America Bulletin,* 99, 303–14.

Decker, R. W., and Decker, B. B. (1991). *Mountains of Fire.* San Francisco: Cambridge University Press.

Etienne, R. (1992). *Pompeii, The Day a City Died.* New York: Harry N. Abrams.

Francis, P., and Self, S. (1983, November). The eruption of Krakatau. *Scientific American,* 172–87.

Heliker, C. (1992). *Volcanic and Seismic Hazards on the Island of Hawaii.* U.S. Geological Survey Publication.

McCoy, F. W., and Heiken, G. (1990). Anatomy of an Eruption. *Archaeology* (May/June), 42–49.

Newhall, C. G., and Self, S. (1982). The volcanic explosivity index (VEI). *Journal of Geophysical Research,* 1231–38.

Schmidt, R. G., and Shar, H. R. (1971). *Atlas of Volcanic Phenomena.* U.S. Geological Survey.

Smithsonian Institution (1998). *Journey Through Geology,* an interactive CD. Dubuque, Iowa: WCB/McGraw-Hill.

Tilling, R. I. (1982). *Volcanoes.* U.S. Geological Survey Publication.

White, D. E. (1984). *Geysers.* U.S. Geological Survey Publication.

Wright, T. L., and Pierson, T. C. (1992). *Living with Volcanoes.* U.S. Geological Survey Circular 1073.

Videos

Hawaii: Born of Fire. (1995). NOVA/WGBH Boston (60 min.).

Volcano. (1992). National Geographic (60 min.).

In the Path of a Killer Volcano—Mount Pinatubo. (1993). NOVA/WGBH Boston (60 min.).

The Volcano Watchers. (1987). BBC-TV, (60 min.).

The Earth Revealed—Volcanism. (1992). Annenberg/CPB (30 min.).

Kilauea: Close-up of an Active Volcano. (1994). Ka'lo Productions Hawaii (30 min.).

Deadly Shadow of Vesuvius. (1998). NOVA/WGBH (55 min.).

Chapter 7

Volcanic Eruptions Continue

Past civilizations are buried in the graveyards of their own mistakes.

—Lord Ritchie-Calder, 1970
Mortgaging the Old Homestead

Figure 7.1 A mature forest is now fallen trees pointing in the direction traveled by the volcanic blast from Mount St. Helens on 18 May 1980.

Chapter 7 directly continues the topics begun in Chapter 6. We first return to spreading-center volcanism at Iceland, examine subduction-caused explosive volcanism in the Cascade Range along the Pacific Coast of the United States (Figure 7.1) and Canada, and then examine the historic record of volcano-related fatalities to understand the specific processes that kill people. Lastly, we will look at failure and success in volcano monitoring and warning.

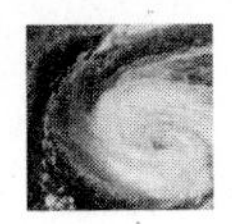

Volcanism at Spreading Centers

Most of the volcanism on Earth takes place along the oceanic ridge systems where seafloor spreading occurs. Solid, but hot and ductile, mantle rock rises upward into regions of lower pressure, where up to 30 to 40 percent of the rock melts and flows as basaltic magma. The worldwide rifting process releases enough magma to create 20 km^3 (<5 mi^3) of new basaltic oceanic crust each year. Virtually all of this volcanic activity takes place below sea level and is thus difficult to view. We see and are impressed by the tall and beautiful volcanic mountains on the edges of the continents, but the volume of magma they release is small in comparison to that of spreading centers.

Iceland

Iceland is a volcanic plateau built of basaltic lava erupted from a hot spot below the mid-Atlantic Ocean spreading center (Figure 2.21). The country is a little bit bigger than the state of Virginia; about 13 percent of its surface is covered by glaciers, and one-third consists of active volcanoes. During the nearly 1,000 years of human records, volcanic eruptions have occurred about every five years, on average. Most Icelandic eruptions do not cause deaths, but exceptions do occur.

Icelandic-type Eruptions The different volcanic behaviors have been classified by the eruptive styles of individual volcanoes (Figure 6.16). This classification is just for general purposes; each volcano varies in its eruptive behavior over time.

The most peaceful eruptions are the Icelandic-type. They are referred to as **fissure** eruptions because lava pours out of linear vents or long fractures (cracks) up to 25 km (16 mi) long. Understand Icelandic eruptions by visualizing the linear spreading center (Figure 2.22) that controls the rise of magma as it is fed upward through fractures as dikes (Figure 6.15). Eruptions can be beautiful to watch as an elongate "curtain of fire" shoots upward with varying intensity and height. Icelandic eruptions are peaceful with low-viscosity, low-volatile lava flows with easy escape of gases. Some flows are described as almost waterlike. Over time the lava flows build up and create wide plateaus of nearly horizontal volcanic rock layers.

Hawaiian-type eruptions are similar to Icelandic. They also have peaceful, basaltic flows where low-viscosity lava dominates, but over time, they build very broad shield volcanoes with high, domed mountain peaks, such as Mauna Loa and Mauna Kea.

Lava Flows of 1973 The recent story of Iceland shows that humans can make enough adjustments and accommodations to live profitably and happily next to active basaltic volcanism. The 1973 eruptions on the small island of Heimaey on the southern coast of Iceland illustrate the "peaceful" nature of these eruptions. The town of Vestmannaeyjar is built next to the premiere fishing port in Iceland. The safe harbor is itself a gift of volcanism; it was formed between ancient lava flows. On 23 January 1973, a fissure opened up only 1 km (3,300 ft) from the town of 5,300 people (Figure 7.2). By early July, the eruption had emitted 230 million m^3 of lava (Figure 7.3), and 26 million m^3 of pyroclastic material. The lava flows increased the size of the island by 20 percent. Gases vented during the eruptive sequence, other than water vapor, were dominantly CO_2 with lesser amounts of H_2, CO, and CH_4. The only fatality was one person asphyxiated in a gas-filled building.

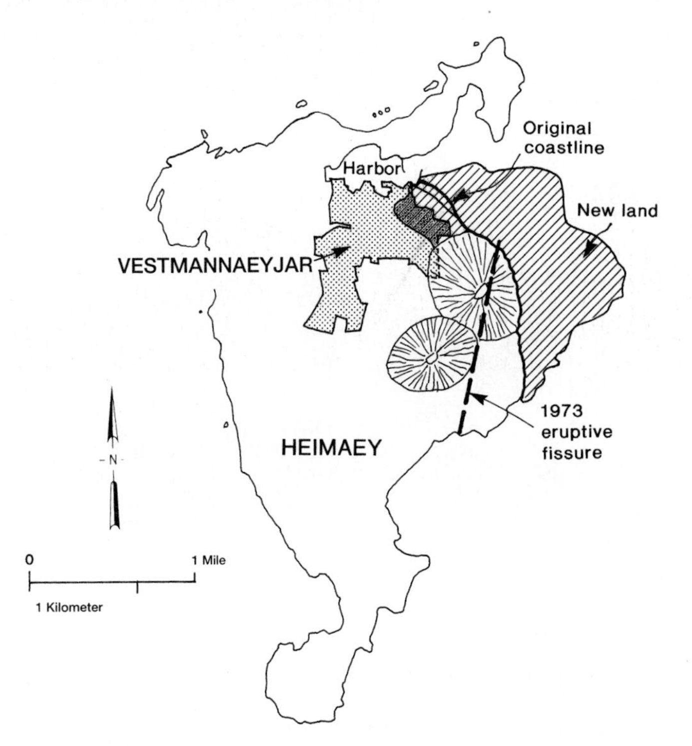

Figure 7.2 The island of Heimaey with the old coastline shown as a solid line. Striped area is new land formed by 1973 lava flow. Note that the new harbor is bigger and better protected.

Data Source: Williams, R. S., Jr., and Moore, J. G., "Man Against Volcano: the Eruption on Heimaey, Vestmannaeyjar, Iceland," *U.S. Geological Survey,* 1983.

The early lava flows on Heimaey began filling in the harbor and destroying about 300 buildings; pyroclastic fallout buried another 70 buildings. But then the Icelanders took over. Pyroclastic material was bulldozed to create barriers that diverted and controlled the flow of later lavas and even controlled the flow paths of the dense volcanic gases. To save their harbor and economic livelihood, the Icelanders sprayed seawater on the lava flows, causing rapid cooling and hardening into wall-like features that forced the lava to

Figure 7.3 An aa lava flow stopped against and between two fish-factory buildings in Vestmannaeyjar, 23 July 1973.
Photo courtesy of *U.S. Geological Survey.*

Figure 7.4 Seawater is being sprayed on the lava front to cool, harden, and stop it from closing off Vestmannaeyjar harbor, 4 May 1973.
Photo courtesy of *U.S. Geological Survey.*

flow off in another direction (Figure 7.4). This action prevented the harbor from being filled and closed. Now, with its new shape and larger size, the harbor is better than before the 1973 eruption (Figure 7.2).

When the eruptions stopped, the people set up a pipe system that poured water into the 100 m (325 ft) thick mass of slowly cooling lava. Return pumps were installed to bring the water, which had been heated to 196°F, back to the surface and into town, where it was used to heat buildings. Basaltic eruptions do not have to be killers. Humans and volcanoes can coexist in harmony, with some interruptions.

Jokulhlaup of 1996 Reading the words *glacier* and *magma* in the same sentence seems a bit odd. They are two substances that don't seem to belong together. When magma does move into contact with ice, unusual things happen.

An earthquake episode began in southeastern Iceland on 30 September 1996 as fissuring opened along the mid-Atlantic spreading center beneath the thick Vatnajokull glacier which covers 10 percent of Iceland. In two days, magma rising through the fissure melted through the 600 m (2,000 ft) thick ice cap sending steam and other gases several kilometers into the atmosphere. Meltwater flowed under the glacier to the ice-covered Grimsvotn Volcano, where it accumulated in the volcano caldera. The meltwater reached an estimated 4 km^3, but on 5 November the rising water lifted up the glacier and poured forth as an enormous **jokulhlaup,** a flood flowing at 45,000 m^3/sec (1,600,000 ft^3/sec). For two days the water flowed as the second largest river in the world, behind only the Amazon River, and then it died out, stranding blocks of ice up to 1,000 tons on the bed of the short-lived stream. The jokulhlaup water and ice blocks destroyed Iceland's longest bridge, key telephone lines, and a road.

Volcanism at Subduction Zones

Through newspapers and television, we learn of death-dealing volcanic eruptions at Galeras Volcano in Colombia, Mount Unzen in Japan, Mounts Pinatubo and Mayon in the Philippines, Mount St. Helens in Washington, Popocatepetl in Mexico, and Soufriere Hills on Montserrat. These are all subduction-zone volcanoes and they have the biggest impact on humans. Many of the regions around subduction-zone volcanoes are heavily populated and feel the wrath of the eruptions. Also, because these volcanoes erupt directly into the atmosphere, they can affect climate worldwide (see Chapter 9).

The variable behaviors exhibited by these magmas when they reach the Earth's surface are best illustrated by looking at specific volcano events and their consequences.

Cascade Range, Pacific Coast United States and Canada

Explosive eruptions are frequent happenings at the numerous volcanoes in the Pacific Northwest region of the United States and in British Columbia (Figure 7.5). The plate-tectonic process responsible for these volcanoes is identical to the cause of the region's great earthquakes—subduction. In fact, the frequent eruptions from the Cascade Range volcanoes provide clear evidence for active subduction. The melting of part of the mantle (asthenosphere) wedge above

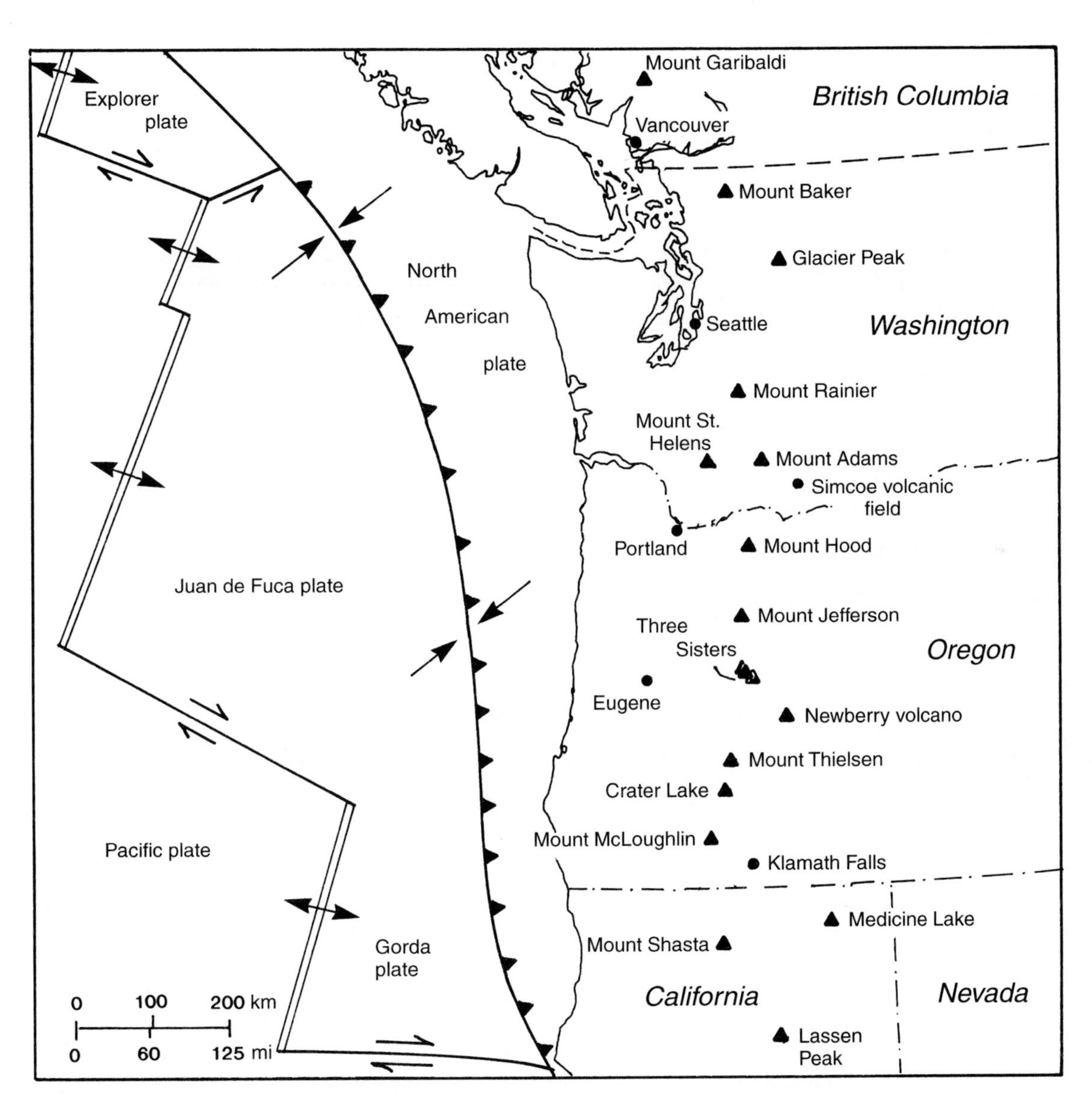

Figure 7.5 Plate-tectonic map of Cascade Range volcanoes. Volcanoes are subparallel to the subduction zone and are spaced somewhat regularly.

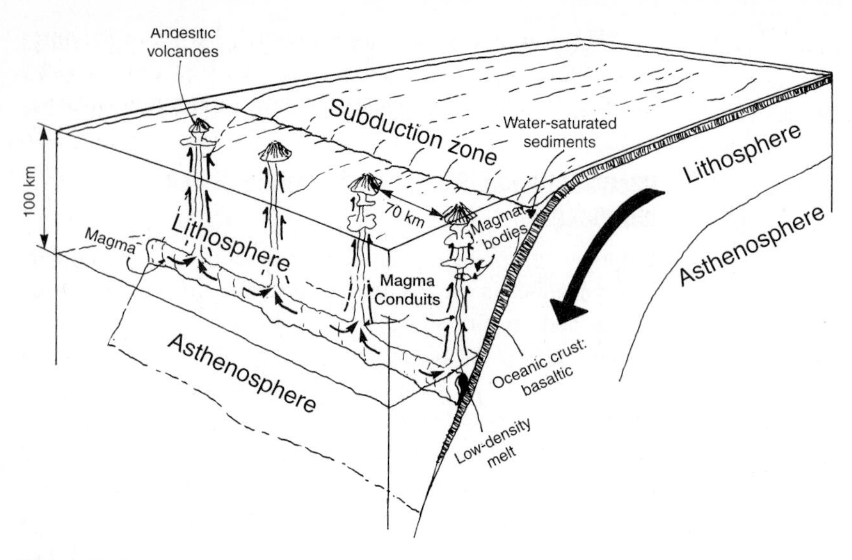

Figure 7.6 Block diagram of subducting oceanic plate carrying water-saturated sediments. Melting occurs at about 100 km (60 mi) below the surface. Magma slowly rises; some collects in subsurface "pools," and some erupts at the surface.

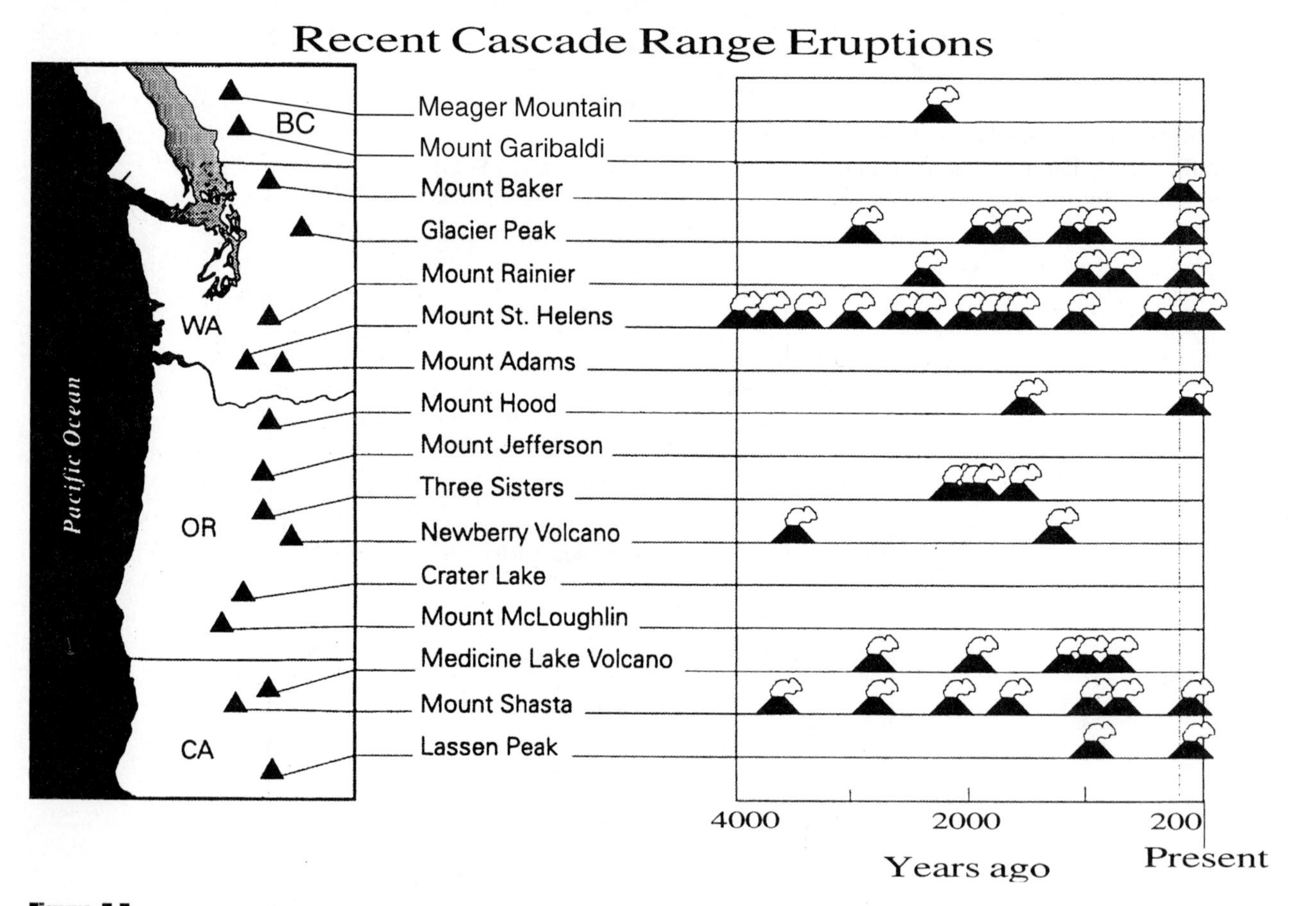

Figure 7.7 Eruption histories of Cascade Range volcanoes.

the subducting plate is aided by water released from sediments on top of the subducting plate. The magma partially melts overlying crustal rock as well, sending andesitic magma upward (Figure 7.6). Some collects in great pods and cools underground forming plutonic rocks, but some erupts explosively at the surface.

How often do major eruptions occur? An example was documented in a 1975 study of Mount St. Helens by Dwight

Crandell and colleagues. Their report stated that the large, visible volcano had formed in the last 2,500 years. Since then, Mount St. Helens has experienced major eruptions every century or two and has never been free from major volcanism for longer than 500 years (Figure 7.7). How were these prehistoric events documented? The process is the same as that used to work out dates of prehistoric earthquakes. The slopes near a volcano reveal the remains of trees knocked down by volcanic blasts (Figure 7.1). These trees may be buried by volcanic ash, incorporated in lahars, or otherwise preserved. Radiocarbon determinations of the dates when trees died also tell the dates of the volcanic eruptions that killed them (Figure 7.8). The 1975 report stated, "Although dormant since 1857, St. Helens will erupt again, perhaps before the end of this century." The geologic analysis was prophetic (Figure 7.9).

Mount St. Helens, Washington, 1980 In late March 1980, Mount St. Helens awoke from a 123-year-long slumber. Dozens of magnitude 3 earthquakes occurred each day as magma pushed its way toward the surface. On 27 March, small explosions began as groundwater and magma came in contact. The spectacle of an erupting volcano was a tremendous lure for sightseers. People flocked to Mount St. Helens. The weekend traffic was so jammed that it reminded folks of rush hour in big cities. But this was an explosive giant just warming up its act, and all nearby life was in grave danger. The governor of Washington ordered blockades placed across the roads to Mount St. Helens to keep people away. Her action was unpopular. Then, at 8:32 A.M. on 18 May 1980, the volcano blew off the top 400 m (1,313 ft) of its cone during a spectacular blast that generated about 100 times the power of all U.S. electric-power plants combined. Most of the 62 people killed had found ways around the barricades to get a better view of an erupting volcano. A look at the eruptive sequence provides a good example of how an explosive volcano does its thing (Figure 7.10).

First, Mount St. Helens achieved its beautiful conical shape during the mid-1800s (Figure 7.10a). In 1843, a SiO_2-rich lava dome grew at the volcano peak. In 1857, andesitic lava flows cooled high on the slopes. But these events also set up discontinuities, or weaknesses, within the volcanic cone.

Second, in 1980, rising magma began changing the shape of the volcano (Figure 7.10b). Earthquake hypocenters were abundant at 1 to 3 km (0.6 to 1.9 mi) depth. The seisms were recording the injection and pooling of magma. With magma forcing its way upward, the northern side of the volcano began rising. The increasing volume of magma also caused the groundwater body to expand its volume. The effect on the volcano was dramatic. By 12 April, a 2 km^2 (1.2 mi^2) area on the north flank had risen upward and outward by 100 m (325 ft). This unstable situation grew worse as the "megablister" kept growing about 1.5 m (5 ft) per day.

Third, at 8:32 A.M. on 18 May 1980, a magnitude 5.1 earthquake rocked the volcano. It triggered a gigantic landslide/avalanche as 2.5 km^3 of the north side of the mountain fell away at speeds up to 250 km/hr (150 mph) (Figure 7.10c). The avalanche was a roiling mass of fragmented rock that once was the mountaintop and side, combined with ice blocks, snow, magma, soil, and broken trees; the internal temperature of the mass was about 100°C (212°F). Part of the avalanche slammed into Spirit Lake, causing waves 200 m (650 ft) high. Another part overrode a 360 m (1,180 ft) high ridge that lay 8 km (5 mi) to the north; then it turned and moved 23 km (14 mi) down the north fork of the Toutle River (Figure 7.11). The resulting deposit was a chaotic mixture of broken rocks and loose debris that averaged 45 m (150 ft) in thickness and had a hummocky surface relief of 20 m (65 ft). Only a short time earlier, this material had been the top of the mountain. At the same time as the avalanche occurred, lahars were forming and flowing down the river valleys as rock particles mixed with water derived from melting snow and ice, from Spirit Lake, and from within the

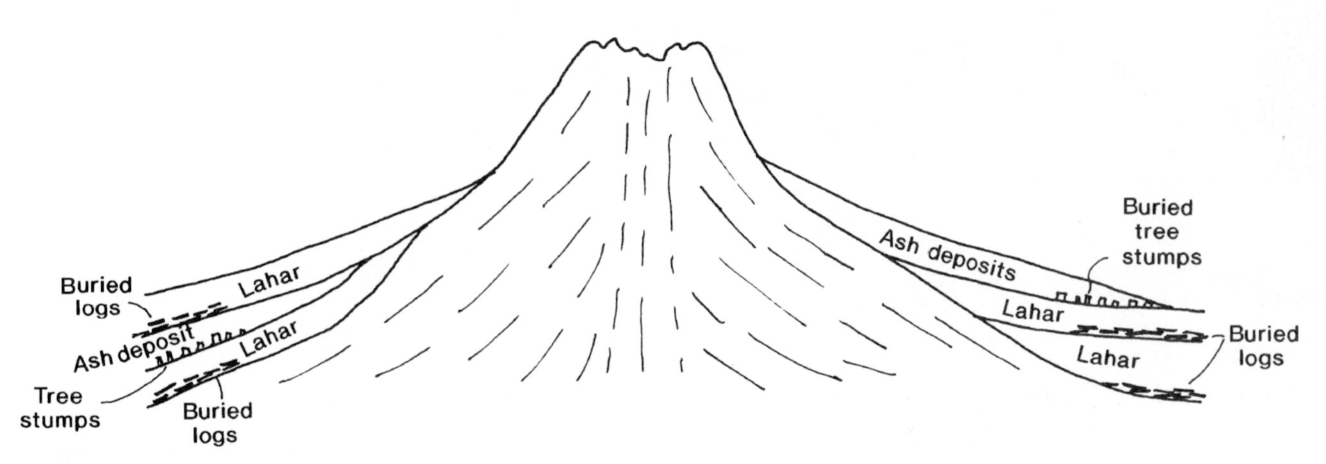

Figure 7.8 Schematic cross section of a volcano and some of its eruptive deposits. Radiocarbon dates on buried wood tell when trees died, i.e., when the volcano erupted.

(a)

(b)

Figure 7.9 Mount St. Helens, Washington. **(a)** *Before:* View to the northeast of the beautiful cone of Mount St. Helens on 25 August 1974. Mount Rainier is in the distance. **(b)** *After:* Same view on 24 August 1980, after the volcano had blown off its top 1,313 feet.
Photos by John S. Shelton.

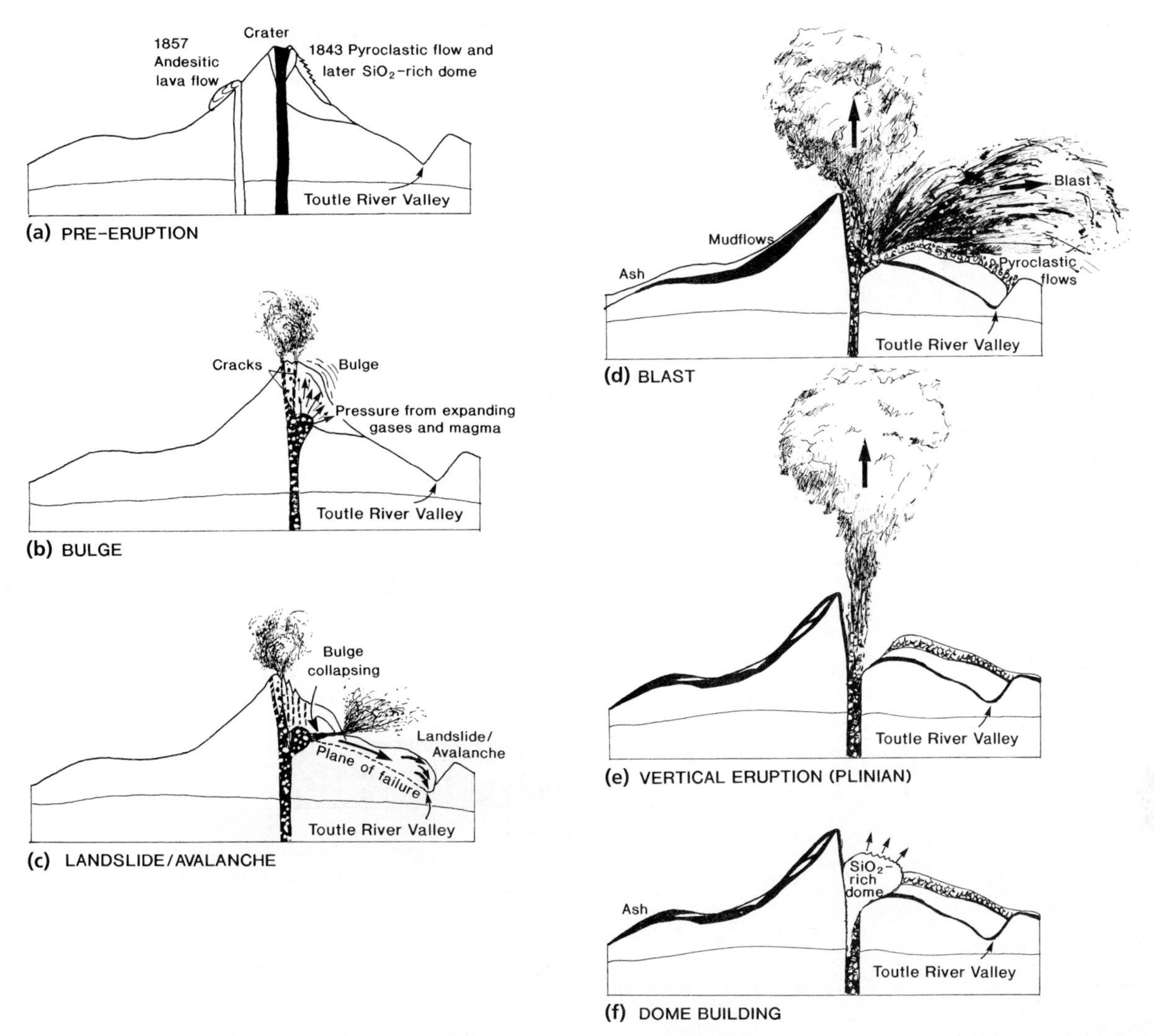

Figure 7.10 Eruptive sequence (VEI = 5) of Mount St. Helens in 1980. **(a)** The symmetrical volcanic cone was shaped in 1843 and 1857. **(b)** In late March, rising magma and expanding gases caused a growing bulge on the northern side. **(c)** At 8:32 A.M. on 18 May 1980, a magnitude 5.1 earthquake caused the bulge to fail in a massive landslide/avalanche. **(d)** The landslide released pressure on the near-surface body of magma, causing an instantaneous blast of fragmented rock and magma. **(e)** The "throat" of the volcano was now clear, and the vertical eruption of gases and small blobs of magma shot up to heights of more than 20 km (12 mi) for nine hours. **(f)** Today, the mountain is slowly rebuilding with a volcanic dome of SiO_2-rich magma.

avalanche. These slurries continued to form and flow for many hours after the eruption began. Lahars moved long distances (Figure 7.11) at speeds up to 40 km/hr (25 mph), carrying huge boulders and flowing with a consistency like wet cement.

Fourth, as the landslide began to pull away, the dramatic drop in pressure on the gaseous magma and superheated groundwater caused a stupendous blast (Figure 7.10d). The blast and surge roared outward at speeds up to 400 km/hr (250 mph). The blast overtook and passed the fast-moving avalanche, racing over four major ridges and scorching an area of 550 km^2 with 0.18 km^3 of volcanic rock fragments and swirling gases at about 300°C (572°F) (Figure 7.11). The blast was a pyroclastic flow. It was denser than air, flowing along the ground as a dark cloud with turbulent volcanic gases keeping solid rock fragments, magma bits, and splintered trees in suspension; it behaved as a very low-viscosity fluid.

Fifth, the big blast opened up the throat of the volcano, exposing an effervescing magma body. Rapidly escaping gases blew upward, carrying small pieces of magma to heights greater than 20 km (12 mi) during the Plinian phase, which lasted about nine hours (Figure 7.10e). The boiling gases carried about 1 km^3 of ashes up and away. About 0.25 km^3 of ash was blown across the United States at different heights by various wind systems. Another 0.25 km^3 formed pyroclastic flows by either spilling out of the volcano or falling down from the eruption cloud. These pyroclastic flows had temperatures of 300° to 370°C (570° to 700°F) and moved at speeds up to 100 km/hr (more than 60 mph).

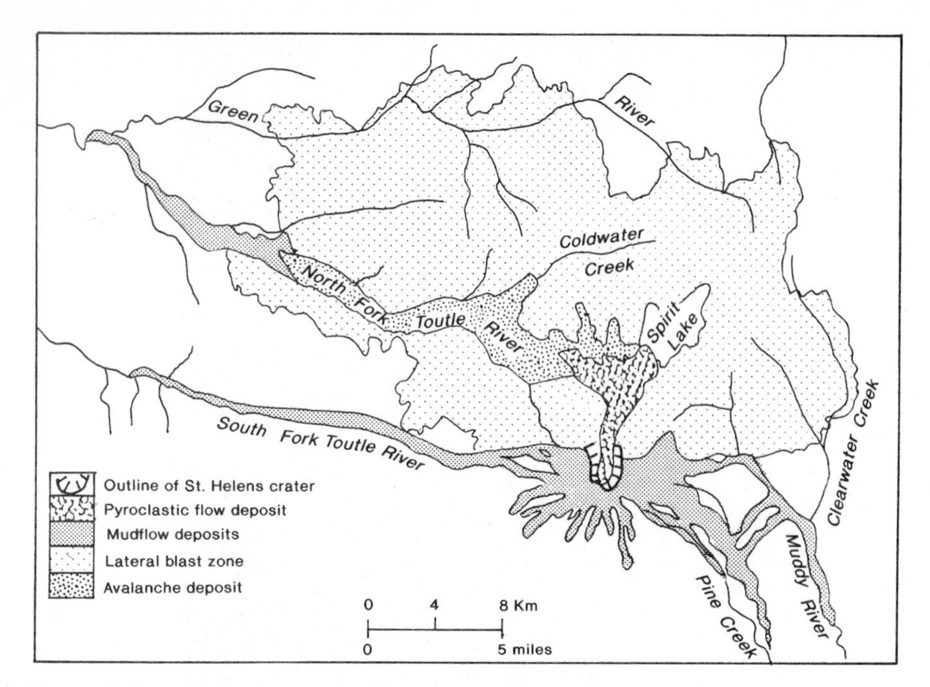

Figure 7.11 Map of materials dumped in 18 May 1980 eruption of Mount St. Helens. Avalanche deposit was from the initial landslide. Lateral blast followed immediately. Then pyroclastic flows spilled out of the exposed magma body. Throughout it all, the superheated groundwater, plus melting snow and ice, fluidized sediments on the steep slopes as lahars that ran down the valleys.

Source: R. Tilling, "Eruptions of Mount St. Helens: Past, Present and Future," 1984, *U.S. Geological Survey.*

Figure 7.12 View of the devastated northeastern side of Mount St. Helens, 20 August 1980. Mount Adams is in background.

Photo by John S. Shelton.

Figure 7.13 The Toutle River, choked with eruption debris.
Photo by John S. Shelton.

Sixth, the volcano now slowly repairs the damage done to its once-symmetrical cone as it builds an SiO_2-rich lava dome (Figure 7.10f). The magma building the lava dome has not erupted explosively probably because it lost most of its volatiles during the big eruption on 18 May 1980. The growth of this lava dome continues in the twenty-first century.

Mount St. Helens looks very different these days (Figure 7.12). Gone are the mountaintop, snowfields, forests, and lakes. The once tree-lined river valleys are now clogged with volcanic debris (Figure 7.13). But recovery is progressing well. Bacteria have eaten sludge from dirty lakes, leaving pure water that has been stocked with trout. Plants have sprouted anew in devastated ground, and animals have returned to feed on them and each other. Life is erasing the effects of the volcanic events.

Were the explosive events at Mount St. Helens a rare event? Are similar events likely at other Cascade Range volcanoes in our lifetimes?

Lassen Peak, California, 1914–17 Lassen Peak is not the typical volcano; rather, it is an unusually large (about 1 mi^3) lava dome of SiO_2-rich volcanic rock analogous to that growing in Mount St. Helens today (Figure 7.10f). Lassen Peak is built within the ruins of Mount Tehama, once a grand volcano (Figure 7.14).

Lava domes form when magma is too viscous to flow away so instead it oozes upward as a conduit-plugging mass (Figure 6.27). A lava dome grows largely from below as magma solidifies on the base of the dome. The lava dome may rise as magma builds up beneath it, and subside as magma squirts off to the side through fractures. Lava domes may be blasted to bits (Mount St. Helens, 1980); they may be lifted above the crater rim, shedding pyroclastic flows down the sides of the volcano (Mount Unzen, 1991); or they can congeal and act as a cork to help conclude an eruptive phase. Mount Lassen is one of the largest lava domes known.

Lassen Peak awakened in May 1914 with numerous eruptions, culminating on 18 July 1914 with a major episode that sent an ash cloud up over 3,350 m (11,000 ft) high. Small-scale volcanic activity continued, but large events did not return until May 1915. On 16–18 May 1915, the 1,000 ft (300 m) wide crater overfilled with sticky magma that stood

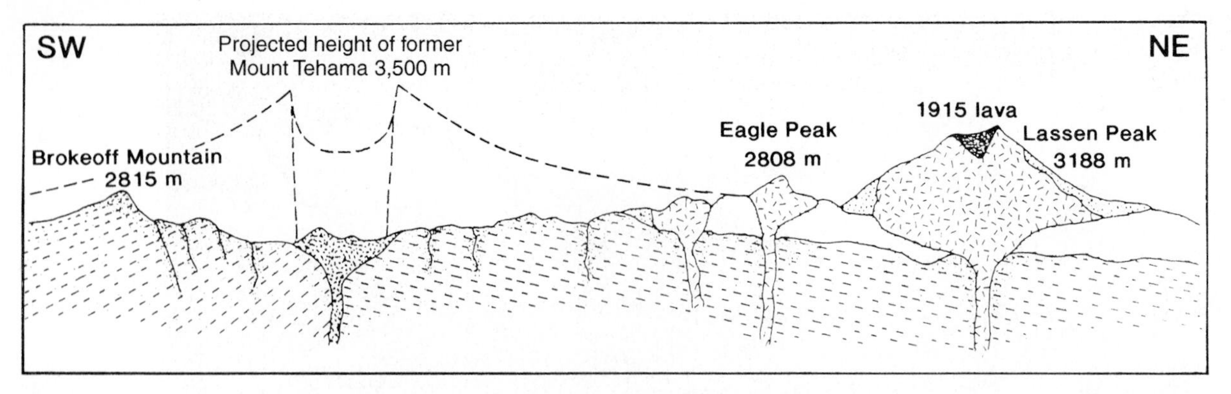

Figure 7.14 Cross section through Lassen Peak—Mount Tehama region, California.

Figure 7.15 View of north side of Lassen Peak, devastated by the 19 May 1915 eruption.
Photo by John S. Shelton.

higher than the rim. The magma was too viscous to flow over the lip, so instead, red-hot blocks broke off and rolled downslope. On 19 May, a 300 m (1,000 ft) long ribbon of lava spilled part way down the hillside to the west. Meanwhile to the east, the melting snow combined with rocky debris to set in motion a massive lahar that flowed outward 50 km (30 mi). On the same day, on the north slope, the side of Lassen Peak split, and a pyroclastic flow blasted forth as a mixture of superhot gases, fragmental rock debris, trees, and water, devastating a triangular-shaped area 6.5 km (4 mi) long and 1.6 km (1 mi) wide (Figure 7.15). Volcanic activity continued with more lahars and pyroclastic flows, and on 22 May, a broad

Figure 7.16 View from the north to Mount Shasta and Shastina. Note the network of roads being used to develop towns on top of lava flows, lahars, and avalanche deposits.
Photo by John S. Shelton.

mushroom cloud of ash was blasted 8 km (5 mi) high. Lassen remained relatively peaceful through 1916, but May and June 1917 brought renewed activity.

In three of four years, the month of May saw the start of extensive volcanic activity. Was this a coincidence? Maybe, but it is possible that as water from the melting snow sank and was heated underground, its volume expansion helped fracture Lassen Peak and reduce internal pressure enough to begin the eruptions. In the nonvolcanic year of 1916, Lassen Peak was too hot for snow to accumulate.

In the twentieth century, two Cascade Range volcanoes underwent similar eruptions with sideward-directed blasts (pyroclastic flows), far-reaching volcanic mudflows (lahars), and great vertical eruptions of ash (Plinian phase). Luckily, each of these eruption sequences took place in sparsely inhabited areas. What are the prospects for similar eruptions near towns and cities?

Mount Shasta, California Mount Shasta, at 4,318-m (14,162-ft) elevation, is the second tallest of the Cascade Range volcanoes (Figure 7.16). The third highest is Shastina (3,759 m or 12,330 ft), perched on its shoulder. The combined mountain mass is particularly impressive, standing over 3,000 m (10,000 ft) higher than its surroundings and visible from more than 160 km (100 mi) away. It is an active volcano, erupting 11 times in the last 3,400 years, including at least three times in the last 750 years. Its last eruption probably was in 1786.

The Mount Shasta area is a beautiful place to live, and the towns along the volcano base are growing. But how wise is this? The lower slopes of Mount Shasta are broad and smooth, allowing pyroclastic flows to spread widely as they move down the volcano flank (Figure 7.17). Lahars are more prone to flow through valleys, and towns lie there (Figure 7.18). The rock record gives further reason to pause and consider whether or not to build here. Figure 7.19 shows the distribution of a 300,000-year-old avalanche deposit that extends 43 km (27 mi) out from the volcano base. This catastrophic event deposited eight times more debris than Mount St. Helens did in 1980. This jumbled mass near Mount Shasta is the foundation for three towns and one large reservoir.

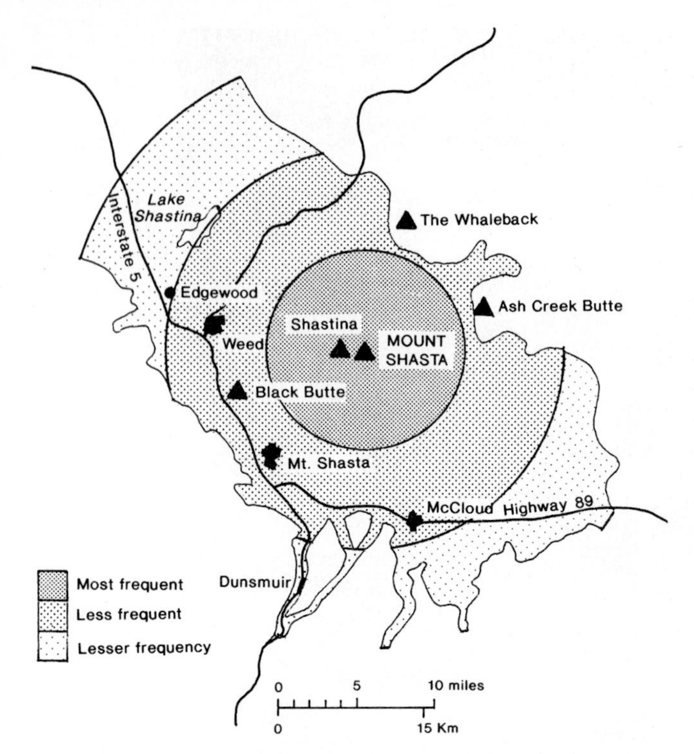

Figure 7.17 Map of Mount Shasta–Shastina region showing areas most susceptible to lateral blasts and pyroclastic flows. Note the growing towns within the danger zones.

Source: D. R. Crandell and D. R. Nichols, "Volcanic Hazards at Mt. Shasta," 1989, *U.S. Geological Survey.*

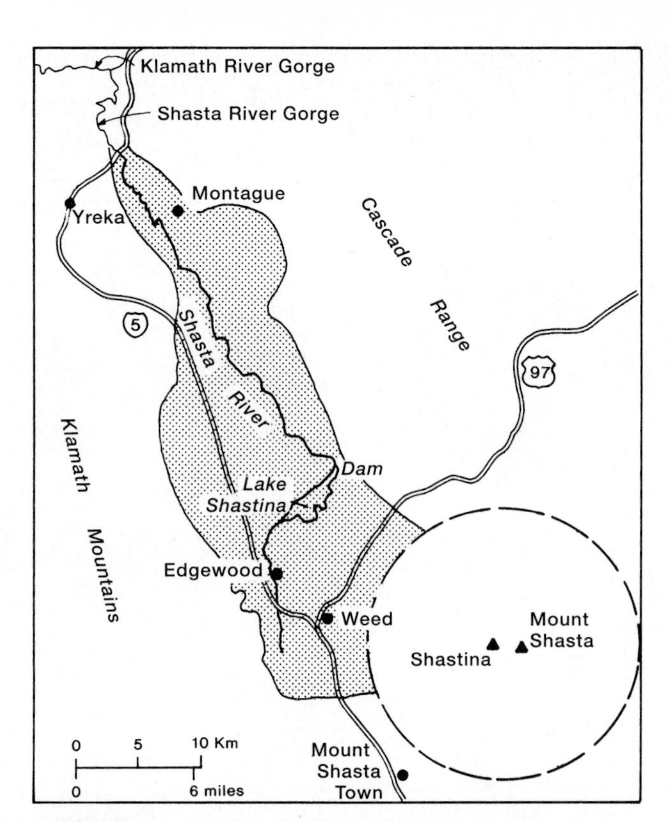

Figure 7.19 Map of 300,000-year-old avalanche deposit at base of Mount Shasta. Amount of material is eight times greater than erupted at Mount St. Helens in 1980. It forms the foundation for three towns and one reservoir.

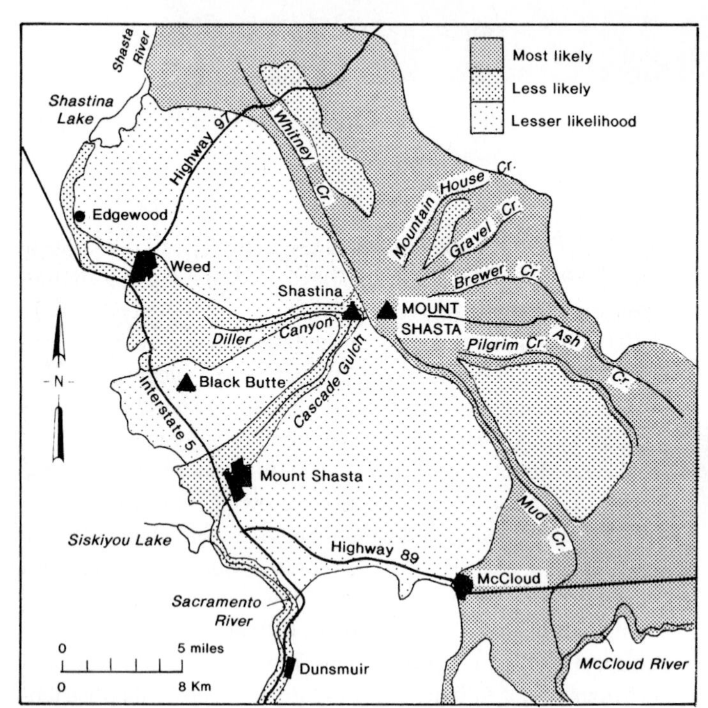

Figure 7.18 Map of Mount Shasta–Shastina area showing most likely paths for lahars. These volcanic mudflows tend to occupy the same river bottom flatlands where towns are built.

Source: D. R. Crandell and D. R. Nichols, "Volcanic Hazards at Mt. Shasta," 1989, *U.S. Geological Survey.*

Would it be advisable to draw park boundaries around the hazardous Cascade Range volcanoes and not allow towns to be built there? Volcanologists Dwight Crandell, Donal Mullineaux, and Meyer Rubin point out that

> The potential risk from future eruptions may be low in relation to the lifetime of a person or to the life expectancy of a specific building or other structure. But when dwelling places and other land uses are established, they tend to persist for centuries or even millennia.

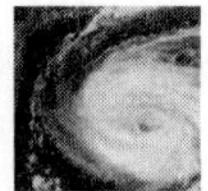

Killer Events and Processes

Volcanoes have numerous ways to kill you (Figure 7.20). They can burn you with a pyroclastic flow, slam and suffocate you with a lahar, batter and drown you with a tsunami, poison you with gas, hit you with a pyroclastic bomb, fry you with a lava flow, and more.

The Historic Record of Volcano Fatalities

Volcanoes operate all around the world. How many people do they kill? Which volcanic processes claim the most lives? The lack of written records for some time intervals and in

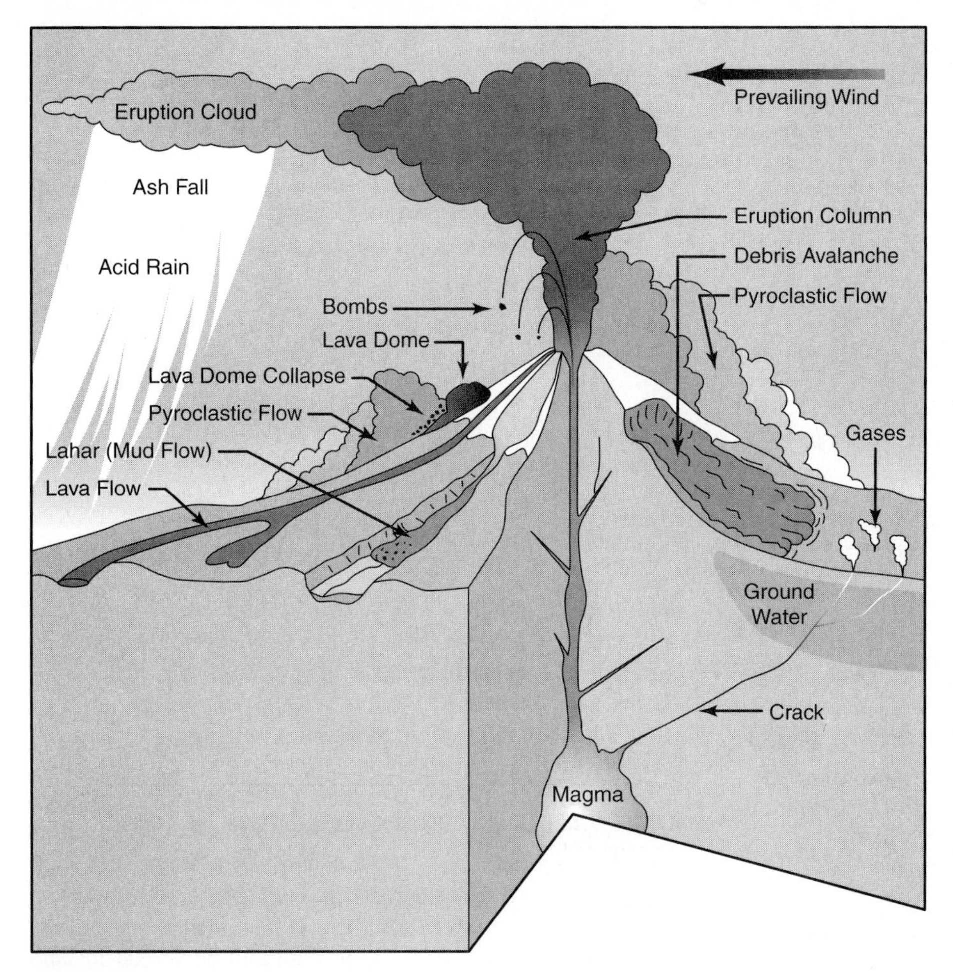

Figure 7.20 Volcanoes operate many life-threatening natural hazards.
Source: *U.S. Geological Survey Fact Sheet 002–97* (1997).

some parts of the world make these questions difficult to answer. Volcanologists Tom Simkin, Lee Siebert, and Russell Blong have studied the questions and made approximate answers. About 275,000 people have been killed by volcanic action during the last 500 years (Figure 7.21). A dozen or so volcanic processes have done the killing (Table 7.1). We will now individually examine the killer processes, and in so doing cover each multi-thousand death event named in Figure 7.21.

Pyroclastic Flows

Few events on Earth are as frightening as having a superhot, turbulent cloud of ash, gas, and air come rolling toward you at high speed. History records numerous instances of these pyroclastic flows killing thousands of people at each event. Pyroclastic flows begin in a variety of ways (Figure 7.22). Case histories teach us more about pyroclastic flows.

Mount Mayon, Philippines, 1968 Since 1616, more than 1,500 people have been killed during 40 recorded deadly eruptions

Table 7.1 Volcanic Causes of Deaths

	275,000 Deaths	530 Volcanic Events
Pyroclastic flow	29%	15%
Tsunami	21%	5%
Lahar	15%	17%
Indirect (Famine)	23%	5%
Gas	1%	4%
Lava flow	<1%	4%
Pyroclastic fall (bombs)	2%	21%
Debris avalanche	2%	3%
Flood	1%	2%
Earthquake	<1%	2%
Lightning	<1%	1%
Unknown	7%	20%

Data source: Simkin, T., Siebert, L., and Blong, R., "Volcano fatalities" in *Science*, 291:255, 2001.

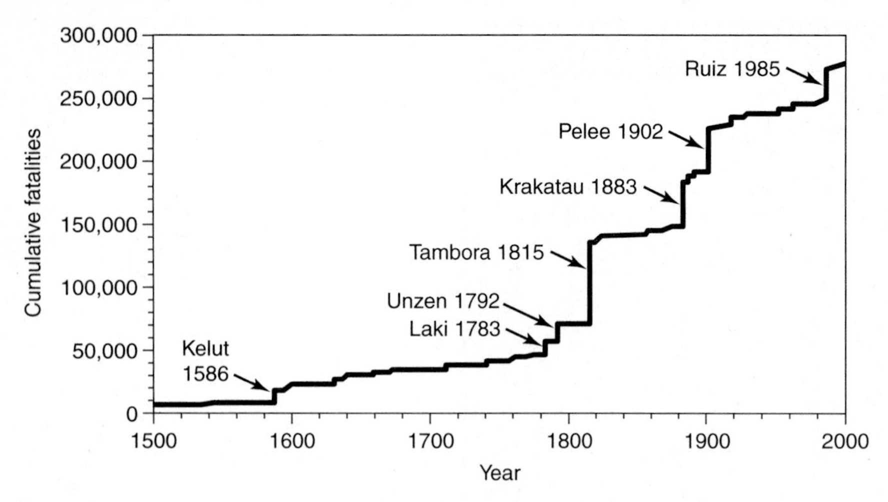

Figure 7.21 Cumulative fatalities from volcanoes during the last 500 years.
Source: T. Simkin, L. Siebert, R. Blong, *Science*, 291:255 (2001).

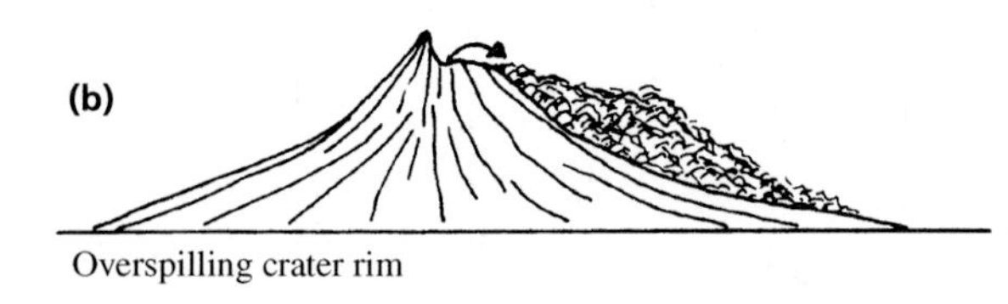

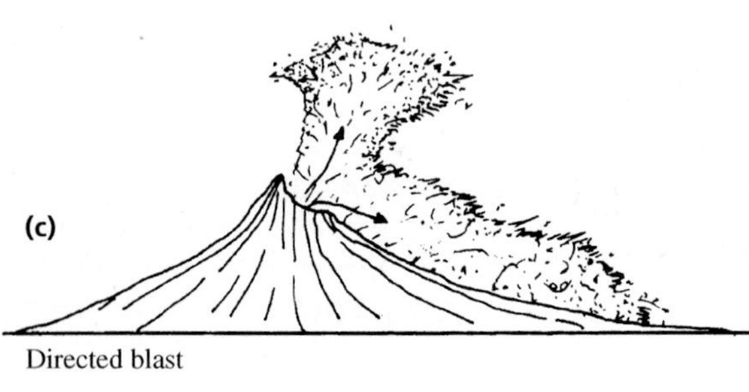

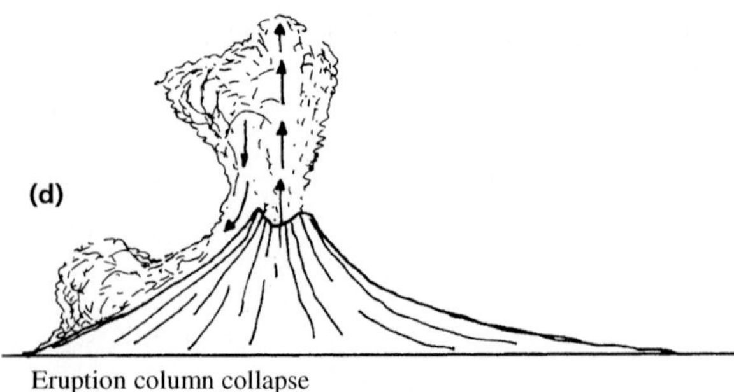

Figure 7.22 Ways of generating pyroclastic flows. **(a)** Dome collapse as at Mount Unzen, 1991; **(b)** Overspilling of crater rim as at Mont Pelée, 1902–1903; **(c)** Directed blast as at Mount St. Helens, 1980 and Mount Pinatubo, 1991; **(d)** Eruption column collapse as at Mount Mayon, 1968.

of the subduction-caused stratovolcano Mount Mayon. In 1968, a series of Vulcanian eruptions sent eruption clouds 10 km (6 mi) up into the atmosphere several times. Partial collapses of the eruption columns sent pyroclastic flows rolling down the mountain slope at velocities ranging from 50 to 100 km/hr (30 to 60 mph)(Figures 7.23 and 7.22d).

Is Mount Mayon at its deadliest when it is erupting most energetically? No, it is deadliest in weak intervals during the eruption. When the volcano is sending its eruption column upward with the most heat, hot pyroclastic material, hot gas, and intermixed air, there is time for the heat to dissipate and for the pyroclasts to cool and be spread far and wide. The most destructive phase of the eruption occurs in those moments when less energy is fed into the eruption column and the column begins to collapse, sending clouds of hot gases, ash, and pumice flowing as ground-hugging deadly pyroclastic flows.

El Chichón, Mexico, 1982 Can a pyroclastic flow travel down all sides of a volcano simultaneously? Yes, especially if they are the variety known as **pyroclastic surges,** ring-shaped base surges that occur when more steam and less pyroclastic material combine to produce a more dilute, less dense, high-velocity flow. The deadliest pyroclastic surge in modern time occurred in Mexico on 4 April 1982.

El Chichón Volcano sits in a remote part of Chiapas, the southernmost state in Mexico (Figure 7.24a). The volcano had been dormant for at least 550 years and was not considered an imminent hazard. March 1982 was a month of numerous earthquakes leading up to 29 March when an unexpected 6 hour-long Plinian eruption blasted 1.4 km^3 of rock and magma into the atmosphere. The volcano had changed (Figure 7.24b). The eruption was surprising and the pyroclastic debris settling from the atmosphere was uncomfortable, but the Plinian event was not enough to drive the rural farmers and villagers from their land. The next five days were calming for the residents as only minor volcanic activity occurred. But suddenly on 4 April, a pyroclastic surge flowed radially

Figure 7.23 Formation of a pyroclastic flow as a vertical eruption column collapses and flows downhill, Mayon Volcano, Philippines, 1968.
Photo courtesy of *U.S. Geological Survey.*

outward for 8 km (5 mi), overrunning 9 villages and killing 2,000 people. Everyone within 5 miles of the volcano, in any direction, was killed by the base surge. Following the surge, a Plinian column shot up 20 km (12 mi) high. On the same day, there were two more base surges and Plinian columns, but the last two base surges did not matter, everyone was already dead. However, the Plinian columns injected sulfur dioxide (SO_2) into the upper atmosphere and the whole world felt the effect as global climate changed (see Chapter 9).

Mount Unzen, Japan, 1991 Mount Unzen has a growing lava dome that provides a unique combination of steady magma supply and the upward lift of unstable, overhanging topography. Big hunks of lava dome frequently break off and create pyroclastic flows (Figure 7.22a). How frequent are the pyroclastic flows at Unzen? Between 1991 and 1994, more than 7,000 were recorded.

Does being the world's most active creator of pyroclastic flows cause people to avoid building and living in the area? No, cities and towns lie along the coastline near the volcano, and farming villages are built on its lower slopes. In May 1991, the lava dome in Mount Unzen began a growth spurt that attracted international attention. As the unstable lava dome grew and towered 90 m (300 ft) above the crater rim, thousands of residents were evacuated from villages and tea plantations around the base of Mount Unzen. As residents

(a)

(b)

Figure 7.24 El Chichón, Chiapas, Mexico. **(a)** *Before:* In September 1981, the lava-dome plugged volcano was not considered a big hazard. **(b)** *After:* During one week in 1982, the lava dome was destroyed leaving a 1 km diameter crater.
Photo **(a)** by René Canul D. and photo **(b)** by Robert I. Tilling, *U.S. Geological Survey.*

left, journalists and volcanologists arrived to record the numerous collapses of 200 to 300 ft high masses from the lava dome and watch them run downslope as glowing pyroclastic flows. At 4:09 P.M., on 3 June 1991, a much larger than usual mass fell off the lava dome and rolled downslope at about 60 mph, killing 43 observers including the famed French volcano photographers, Maurice and Katya Krafft.

Mont Pelée, 1902–1903, 1929–1932 The Caribbean island of Martinique in the West Indies was colonized by the French in 1635. The tropical climate was superb for growing sugar cane to help satisfy the world's growing appetite for the sweetener. On the north end of Martinique is a 1,350 m (4,430 ft) high volcano with a pronounced peak. The French called the volcano Pelée, meaning "peeled" or "bald," to describe the bare area where volcanism had destroyed all plant life during the eruptions of 1792 and 1851.

Mont Pelée spewed numerous pyroclastic flows in 1902–1903 and 1929–1932. Many of these pyroclastic flows began as hot ash and gas overspilled the crater rim in a fashion analogous to a boiling pot spilling over and onto the stove (Figure 7.22b). But Mont Pelée is best remembered for its pyroclastic flows generated by directed blasts (Figure 7.22c). This type of flow occurred at Mount St. Helens in 1980 and Mount Pinatubo in 1991 (Figure 6.24), but the deadliest example happened at Mont Pelée in 1902. When magma is highly viscous and flow is so difficult that the throat of the volcano is plugged by its own sticky magma, gas pressure builds until it is relieved by blowing out the top or the side of the volcano.

Mont Pelée, Martinique, 1902 In early spring of 1902, Vulcanian activity began. The crater began filling with extremely viscous magma, displacing boiling lake waters through a V-shaped notch (Figure 7.25). The extraordinarily

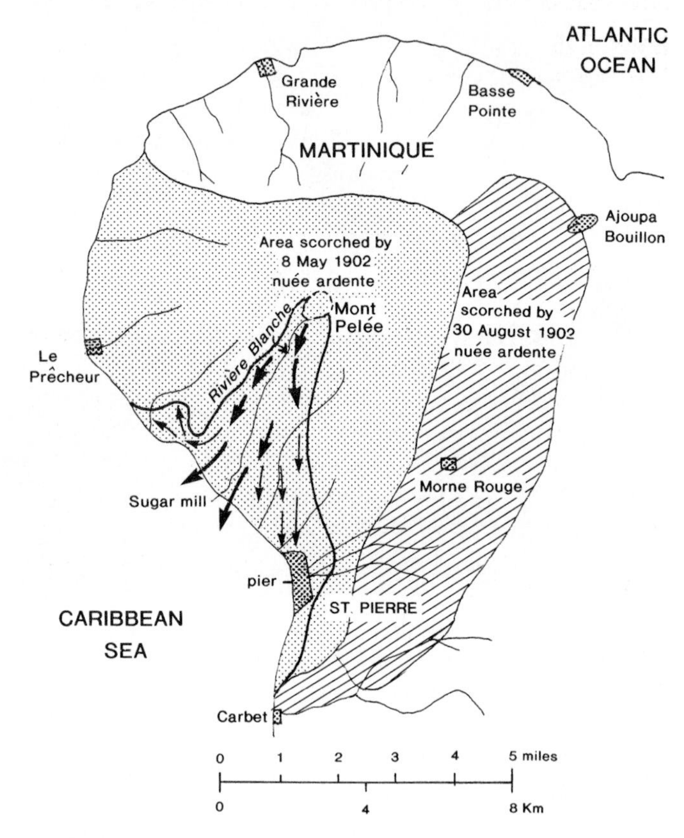

Figure 7.25 Map of Mont Pelée showing areas scorched by the largest nuée ardentes of 1902.

sticky magma kept plugging the crater. At times, superhot pyroclastic flows would gush out of the crater; at other times, they would blast out. By late April, it was obvious to most people that this trouble might get bigger. About 700 rural folks were migrating each day into St. Pierre, a city of picturesque, early seventeenth-century buildings that nor-

Figure 7.26 The nuée ardente-charred remains of St. Pierre, May 1902. Mont Pelée is in background.
Photo by Underwood and Underwood, courtesy of the Library of Congress.

mally was home to 25,000 residents. Another 300 people a day were leaving St. Pierre, which lay only 10 km (6 mi) from Mont Pelée. At a little past noon on 5 May, a large pyroclastic flow sped down the Riviére Blanche, destroying the sugar mill and 40 people. This further increased the anxiety level in St. Pierre. But there was an election coming up on 10 May, and the governor did not want everyone scattered from the island's largest city. It was likely to change the election results. Governor Mouttet and his wife went to St. Pierre and used the militia to preserve order and halt the exodus of fleeing people. Bad decision. There was no election on 10 May anyway; all the voters, including the governor, died on 8 May (Figure 7.26).

On the morning of 8 May 1902, a massive volume of gas-charged, ultrasticky magma had risen to the top of the crater. At about 7:50 A.M., there were sharp blasts that sounded like thousands of cannons being fired as trapped gas bubbles exploded and shattered magma into fine pieces. This spectacular pyroclastic flow moved as a red-hot avalanche of incandescent gases and glowing volcanic fragments, later named a **nuée ardente** (French for "glowing cloud"). The mass moved as solid particles of magma suspended in gas. It absorbed energy from: 1) the initial blast, 2) gravity, 3) gas continuing to escape from the pieces of airborne magma, creating a "popcorn" effect, 4) internal turbulence, and 5) the heating of air mixed into the flow as it moved outward and downward. The temperature at the crater is estimated to have been about 1,200°C (2,200°F), and the glowing cloud was still hotter than 700°C (1,300°F) when it hit St. Pierre. The coarsest and heaviest part of the nuée ardente flowed down the Riviére Blanche. The associated gas-ash clouds expanded in width and overwhelmed St. Pierre (Figure 7.25).

What happened to St. Pierre? The nuée ardente moved with hurricane speeds of about 190 km/hr (115 mph), but it was much denser than a hurricane because of its contained ash. The nuée ardente lifted roofs, knocked down most walls perpendicular to its path, twisted metal bars, and wrapped sheets of metal roofing around the scorched trunks of trees. Within the space of a couple of minutes, St. Pierre turned from a verdant tropical city to burned-out ruins covered by a foot of grey ash and with muddy ash plastered on those walls and tree trunks that were still standing.

What killed the people? Death was quick and came from one of three causes: 1) physical impact, 2) inhaling superhot gases, or 3) burns. The refugee-swollen population of St. Pierre was more than 30,000; two people survived. One was Auguste Ciparis, a 25-year-old murderer locked in a stone-hut jail without windows and with only a small barred grating in his door. When hot gases entered his cell, he fell to the floor, suffering severe burns on his back and legs. Four days later, he was rescued; he then spent the rest of his life showing his scarred body at circus sideshows as "the prisoner of St. Pierre."

Was it safe to be on a boat in the harbor? No. The fiery hot cloud did not stop when it hit the water. Of 18 boats in the harbor, only the British steamship Roddam survived, though it was badly burned and two-thirds of its crew were dead.

Nuée ardentes continued rolling out of Mont Pelée. St. Pierre was overwhelmed again on 20 May, but it no longer mattered. On 30 August, a nuée ardente flowed toward the southeast and scorched Morne Rouge and four other towns, killing another 2,000 people. Despite these tragic events, at present the area is fully settled once again.

Krakatau, Indonesia 1883 Can a pyroclastic flow travel across a body of water to kill you? Or does the water absorb heat from the hot gas-rich cloud quickly enough to eliminate its ability to kill? A body of water does not eliminate the hazard. The 1883 eruption of Krakatau was described in Chapter 6 but it can teach us another lesson here. During the eruptions leading up to the volcano collapse, one remarkable blast on 27 August sent out a hot gaseous pyroclastic flow that raced across the sea surface of the Sunda Straits for 25 mi (40 km) to reach the coastal Province of Katimbang on Sumatra. It flowed onshore with enough heat to fatally burn more than 2,000 people.

Tsunami

The Krakatau eruption and caldera collapse in 1883 killed more than 36,000 people (see Chapter 6). The volcanic eruptions directly killed less than 10 percent of the people; over 90 percent of fatalities were due to volcano-caused tsunami.

Mt. Unzen, Japan 1792 Can a pyroclastic flow or lava-dome avalanche cause tsunami? Yes. On 21 May 1792, an earthquake triggered a collapse from the lava dome in Mount Unzen. The avalanche volume of 0.3 km^3 (0.07 mi^3) was not impressively large. However, after it flowed the 6.4 km (4 mi) to the sea, it hit the water with enough impact to create tsunami that killed 15,000 people in the surrounding region. The tragedy inspired the construction of the Anyoji Temple and Buddhist sanctuary in memory of those killed. In 1991, the temple served as a temporary morgue for the 43 people slammed and burned to death by the 3 June pyroclastic flow.

Lahars

Viscous lahars cause many deaths and major damages as we saw in the case history of Vesuvius in 79 C.E. Many other lahar disasters have occurred, and will continue to occur.

Kelut, Indonesia, 1586, 1919 Indonesia is a nation of volcanoes—Krakatau, Tambora, Toba are part of a lengthy rogue's gallery of serial killers. Figure 7.21 shows that one of the seven deadliest volcanic events of the last 500 years was at another Indonesian stratovolcano—Kelut in 1586. It was deadly events at Kelut that brought us the Indonesian word *lahar* to describe volcanic mudflows. Why does Kelut kill so many people? The volcano supplies fresh pyroclastic materials to its slopes and they quickly decompose under the tropical climate to produce fertile soil, and that attracts people. The slopes of Kelut are intensely cultivated and densely populated. How does Kelut kill? Kelut has a large crater lake at its summit, and Kelut erupts often—15 eruptions in the last 200 years. In 1919, a surprise eruption forced 40 million m^3 of lake water onto the slopes covered with loose pyroclastic debris. The combination of water and loose pyroclasts produced three major lahars flowing down three sides of the volcano at velocities of 65 km/hr (40 mph)—and more than 5,000 people died. The longest-traveled lahar went 38 km (24 mi). What can be done? The 1919 disaster caused Dutch engineers to dig a set of tunnels to reduce the size of the crater lake by over 95 percent. Did that solve the problem? No. A 1951 eruption deepened the crater floor allowing another large lake to form. Then a 1966 eruption dumped 20 million m^3 of water onto the pyroclast-covered slopes to again form lahars; this time they killed 282 people. Will people quit coming to this dangerous volcano to live and farm its fertile soils? No.

Nevado Del Ruiz, Colombia, 1985 Does it require a big eruption to kill a lot of people? No, just consider this story. Nevado del Ruiz rises to an elevation of 5,400 m (over 17,700 ft), up where the air is cold. A 19 km^2 area on top of the mountain is covered by an ice cap 10 to 30 m (30 to 100 ft) thick with an ice volume of about 337 million m^3. In November 1984, the volcano awoke with small-scale activity.

A year later, on 10 November 1985, continuous harmonic tremors (earthquakes) foretold a coming large eruption. At 9:37 P.M., a Plinian column rose several miles high. Hot pyroclastic debris began settling onto the ice cap, causing melting. By 10 P.M., condensing volcanic steam, ice melt, and pyroclastic debris combined to send lahars down the east slopes into Chinchina, destroying homes and killing 1,800 people.

But the worst was yet to come. Increasing eruption melted more ice, sending even larger lahars flowing down the canyons to the west, out over the **alluvial fans** at the base of the mountain, and onto the floodplain of the Rio Magdalena (Figure 7.27). At 11 P.M., the first wave of cool

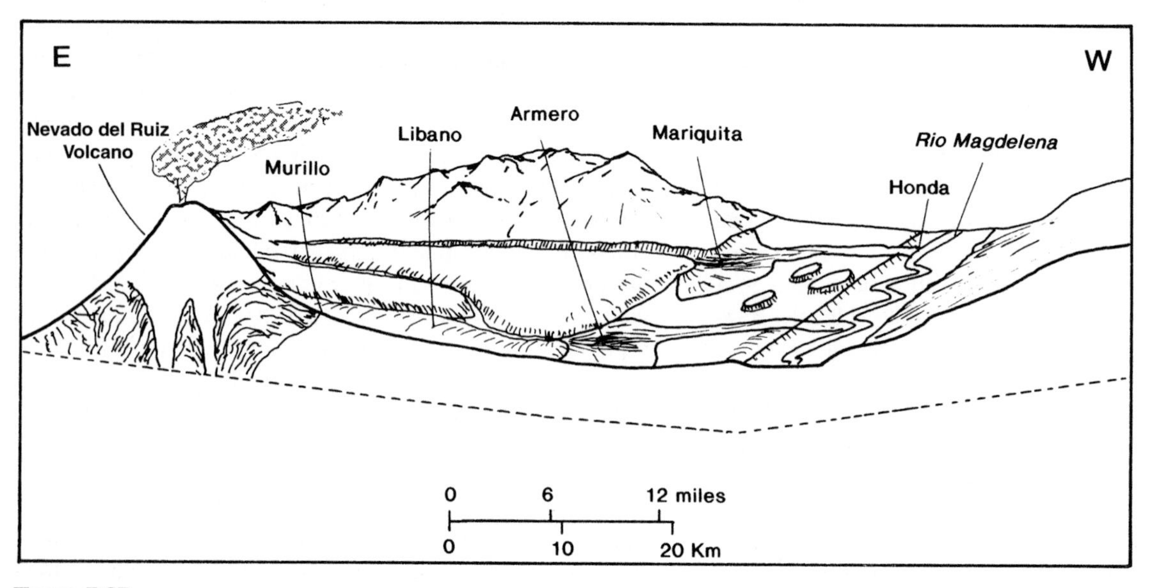

Figure 7.27 Eruption of Nevado del Ruiz in 1985 dropped hot pyroclastic debris onto glaciers, resulting in lahars.
Source: *U.S. Geological Survey.*

Figure 7.28 Most of the town of Armero, Colombia, and 22,000 of its residents lie beneath lahars up to 8 m (26 ft) thick.
Photo courtesy of the *U.S. Geological Survey.*

lahars reached the city of Armero and its 27,000 residents. These lahars had traveled 45 km (28 mi) from the mountaintop, dropping over 5,000 m (more than 16,400 ft) in elevation. In the steep-walled canyons, the lahars moved at rates up to 45 km/hr (28 mph), slowing as they flowed out onto the flatter alluvial fans.

A few minutes after 11 P.M., roaring noises announced the approach of successive waves of warm to hot lahars. Most of Armero, including 22,000 of its residents, ended up buried beneath lahars totaling 8 m (26 ft) in thickness (Figure 7.28). The 22,000 unlucky people were either crushed or suffocated by the muddy lahars.

But 5,000 people did escape. How? They were higher up the slopes. A memorable video showed a man's talking head that appeared to be resting on top of the mudflows; the man was caught by lahars and buried to his chin as he tried to escape upslope. One step slower and he would have been completely buried and suffocated. But with a bit of digging, he was freed, shaken but unharmed.

The volcanic eruption at Nevado del Ruiz was actually rather minor. Had there not been an ice cap to melt, no harm would have been done. The November 1985 lahars were a virtual rerun of the events of February 1845. The same places were buried by the same types of lahars. In 1845, the death toll was about 1,000, but because Colombia's population has grown, the dead in 1985 numbered about 24,000.

Mount Rainier, Washington—On alert Should the Seattle-Tacoma metropolitan region be concerned about lahars? Yes, Nevado del Ruiz showed that a small eruption on a glacier-capped volcano can be big trouble. Mount Rainier is the tallest of the Cascade Range volcanoes, at 4,393 m (14,410 ft). It stands 2,150 to 2,450 m (7,000 to 8,000 ft) above its adjacent areas and is a beautiful sentinel readily seen from throughout the Seattle-Tacoma urban region (Figure 7.9a). Yet Mount Rainier is number one on the danger list of many volcanologists because of its: 1) great height, 2) extensive glacial cap, 3) frequent earthquakes, and 4) active hot-water spring systems, which have weakened the mountain internally. Mount Rainier can be described as 33.6 mi^3 of structurally weak rock capped by 1 mi^3 of snow and ice; this volcanic mountain is inherently unstable. Mount Rainier is a national park and cannot be densely developed, but it nonetheless presents distinct threats to heavily populated areas. The mountain itself may fail in a massive avalanche, and/or rapidly melted ice can cause floods or lahars. Mount Rainier supports the largest glacier system of any mountain in the lower 49 states. This ice can be melted by magma moving up inside the mountain, even without active volcanism.

The rock record shows numerous far-reaching lahars in the last several thousand years (Figure 7.29). The Osceola mudflow moved about 5,600 years ago, flowing more than 120 km (75 mi) down the White River valley before spreading out onto the Puget Sound lowlands and into Puget Sound. It covers an area greater than 100 mi^2 to depths over 20 m (70 ft). The Osceola mudflow began as a water-saturated avalanche during summit eruptions of Mount Rainier. It transformed into a clay-rich lahar within two kilometers of travel as it carried 3.8 km^3 (0.9 mi^3) of material at velocities up to 45 mph out across the Puget Sound lowlands. The affected area is now home to about 100,000 people. A repeat of an Osceola-size lahar could kill thousands

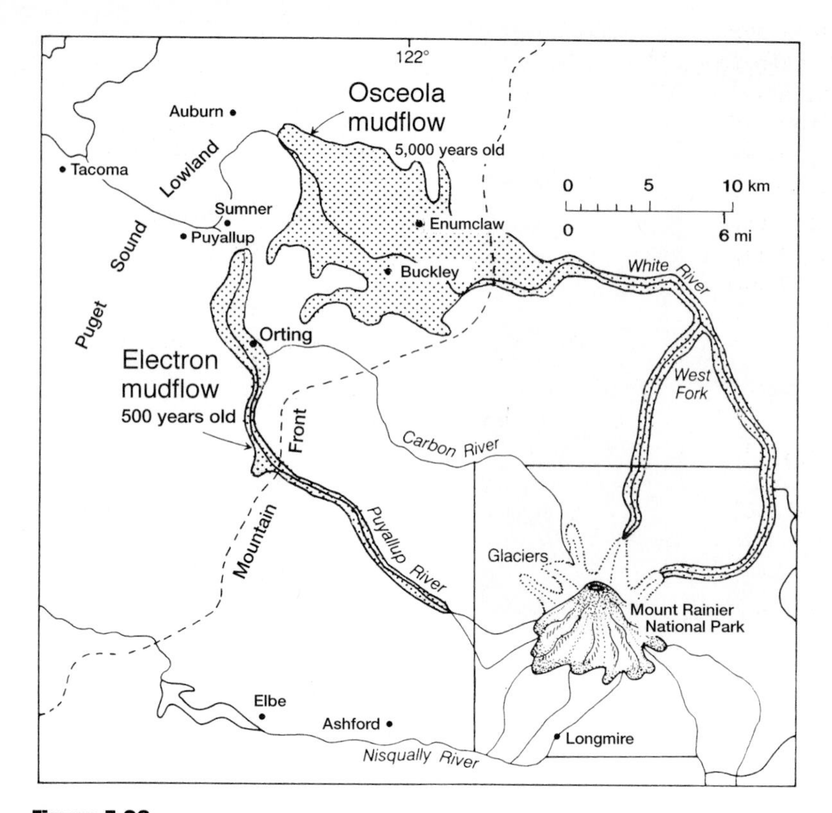

Figure 7.29 Map showing area covered by two of the many lahars that have flowed from Mount Rainier.
Source: D. R. Crandell and D. R. Mullineaux, "Volcanic Hazards at Mt. Rainier, Washington," 1967, in *U.S. Geological Survey Bulletin 1238.*

of people. To visualize what could happen, see the 1985 lahar in Armero, Colombia (Figures 7.27 and 7.28); the Osceola event was 40 times larger than the Armero lahar.

The Electron mudflow is only 500 years old; it flowed down the Puyallup River valley for 48 km (30 mi) and also out onto the Puget Sound lowlands. Today, the region is a desirable place to live, the population is growing rapidly and building homes on top of these lahar deposits. Mount Rainier's next major eruption may bring staggering property damage and deaths.

Indirect—Famine

Volcanoes not only affect humans directly, but they also get us indirectly by changing the atmosphere and climate, and by harming the plants and animals we depend upon.

Laki, Iceland Fissure Eruption of 1783 During the summer of 1783, the greatest lava eruption of historic times poured forth near Laki. After a week of earthquakes, on the morning of 8 June 1783, a 25 km (16 mi) long fissure opened and basaltic lavas gushed for 50 days. The Laki event was a textbook fissure eruption where the magma typically flowed at 5,000 m^3/sec. To better appreciate this volume of magma, consider that North America's mightiest river, the Mississippi, empties into the Gulf of Mexico at about three times this volume. When the eruption ended, an area of 565 km^2 (218 mi^2) was buried beneath 13 km^3 (3 mi^3) of basaltic lavas. The volume of ash and larger airborne fragments totaled another 0.3 km^3.

The 50 days of eruption were accompanied by the release of an enormous volume of gases that enshrouded Iceland and much of northern Europe in a "dry fog" or blue haze. This haze was rich in SO_2 (one of the visible components of today's urban smog) and an unusually large amount of fluorine. The gases slowed the growth of grasses and increased their fluorine content, helping kill 75 percent of Iceland's horses and sheep and 50 percent of the cattle. The resulting famine weakened the Icelandic people, and about 20 percent of the population (10,000 people) died. In today's world of instant communication and rapid air transport, these deaths would have been avoided.

Tambora, Indonesia, 1815 The most violent and explosive eruption of the last 200 years was another Indonesian event; it was Tambora Volcano on Sumbawa Island in April 1815. After three years of moderate activity, on 5 April, a Plinian eruption column shot up 33 km (20 mi) high and carried out 12 km^3 (2.9 mi^3) of pumice in just 2 hours. On 10 April, an even more powerful Plinian eruption blasted up to 44 km (27 mi) high for 3 hours. The magma exited with so much force that it eroded and widened the vent in the volcano thus cutting off the focused energy that drove the Plinian column. The eruption column stopped and the widened vent lay open; the volcano now had its insides exposed and it spilled its guts. On 11 April, about 50 km^3 (12 mi^3) of magma poured out of the caldera in overwhelming pyroclastic flows. The week-long eruption saw about 150 km^3 (36 mi^3) of magma erupted. Tambora once stood 4,000 m (13,000 ft) high but now its elevation was reduced to 2,650 m (8,700 ft) with a 6 km (<4 mi) wide caldera that was over 1 km deep. The volcanic explosions were audible 2,600 km (1,600 mi) away, and volcanic ash fell 1,300 km (800 mi) from Tambora. On Sumbawa Island, pyroclastic flows killed at least 10,000 people plus they destroyed the feudal kingdoms of Sanggar and Tambora including erasure of the Tambora language, the easternmost Austro-Asiatic language.

The eruption of Tambora was responsible for an estimated 117,000 deaths; about 10 percent killed by the eruption and 90 percent dying slowly at the end of a chain reaction. Pyroclastic fallout devastated crops, that led to famine and weakened people ever more susceptible to disease, and then the diseases killed them. But this was not just another Indonesian disaster, the Plinian eruptions of April 1815 so affected global climate that 1816 is known as "the year without a summer." The climatic effects of the eruption will be discussed in Chapter 9.

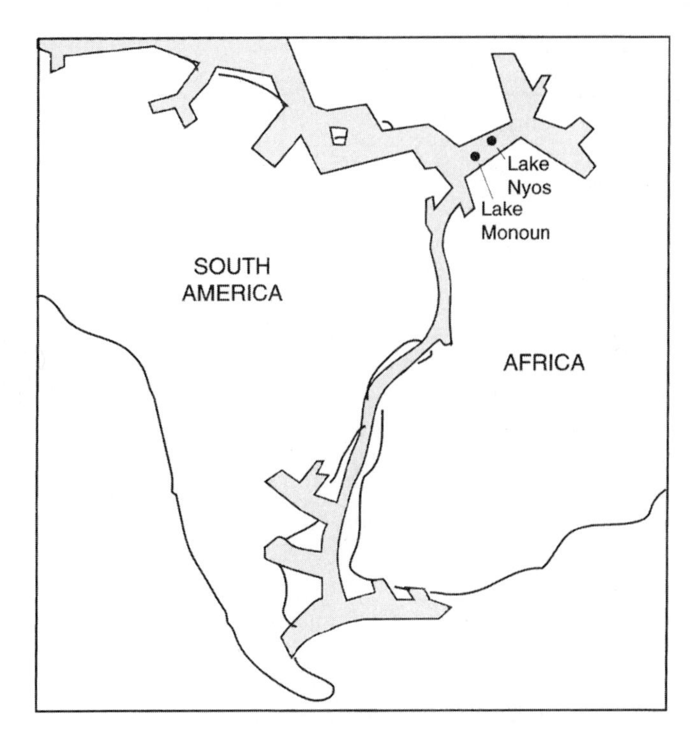

Figure 7.30 Schematic map of Africa and South America splitting apart 135 million years ago. Note the third rift that extends into Africa.

Gas

It is not just gas-powered magma that kills, gas can be deadly all by itself.

Killer Lakes of Cameroon, Africa Spreading centers commonly begin as three-armed rifts meeting at a triple junction (Figure 2.25). In northeast Africa, two rift arms have spread apart enough to create the Red Sea and the Gulf of Aden while the third arm has failed, so far, to open the East African Rift Valley into another new ocean basin (Figure 2.26). **Failed rifts** that do not open up enough to become spreading centers are common (e.g., Figure 7.30). If a rift fails to open a new ocean basin, must it stop all activity? No.

Cameroon sits near the equator in western Africa. It hosts a string of crater lakes running in a northeasterly trend. Prolific rainfalls fill the lakes and combine with the hot temperatures to cover the countryside with greenery. Lake Nyos is one of these crater lakes filled with beautiful, deep-blue water. This topographically high crater is only several hundred years old. It was blasted into the country rock by explosions of volcanic gases and is 1,925 m (6,310 ft) across at its greatest width and as deep as 208 m (680 ft).

At about 9:30 P.M. on 21 August 1986, a loud noise rumbled through the Lake Nyos region as a gigantic volume of gas belched forth from the crater lake and swept down the adjacent valleys. The dense, "smoky" rivers of gas were as much as 50 m (165 ft) thick and moving at rates up to 45 mph. The ground-hugging cloud swept outward for 25 km (16 mi). Residents of four villages overwhelmed by the gaseous cloud felt fatigue, light-headedness, warmth, and confusion before losing consciousness. After 6 to 36 hours, about half a dozen people awoke from their comas to find themselves in the midst of death: 1,700 asphyxiated people; 3,000 dead cattle; and not a bird or insect alive, or any other animal. Yet the luxuriant plants of the region were unaffected.

This shocking event raises numerous questions. What was the death-dealing gas? What was the origin of the gas? How did the gas accumulate into such an immense volume? What triggered the gas avalanche? Is this event likely to happen again?

What was the death-dealing gas? Carbon dioxide. After a lot of effort to identify some exotic lethal gas or toxic substance as the cause of the tragedy, the killer gas turned out to be simply carbon dioxide. This is the same gas we drink in sparkling spring water, soda pop, and champagne. Its toxicity at Nyos is explained by the principle set forth in 1529 by the German physician Theophrastus von Hohenheim (Paracelsus). The principle of Paracelsus states: The dose alone determines the poison. A gas does not have to be poisonous, just abundant. Life in the Nyos region was subjected to the same conditions we recreate inside the fire-extinguisher cylinders in our buildings. Fire extinguishers are loaded with carbon dioxide, which does not put out flames directly; because CO_2 is heavier than air, it deprives fire of oxygen, thus causing flames to die out. Animal life in the Nyos area was extinguished in the same fashion.

What was the origin of the gas? It had a volcanic origin, leaking upward from underlying basaltic magma. A 1,600 km (1,000 mi) long string of volcanoes, the Cameroon volcanic line, trends northeastward through several Atlantic Ocean islands, then on land through northeastern Nigeria and northwestern Cameroon. Interestingly, this is the location of the triple junction of spreading centers that ripped apart this section of Gondwanaland, helping give the distinct outlines to the Atlantic margins of South America and Africa (Figure 7.30). The two successful spreading arms are still widening the South Atlantic Ocean. The failed rift is occupied by the line of volcanism that includes the crater that forms Lake Nyos; it is not a volcanic mountain but a crater blasted through bedrock by largely gaseous explosions. The volcanic activity is not seafloor spreading per se; rather, it is a "wannabe" ocean basin that never made it but has not given up totally.

How did the gas accumulate into such an immense volume? Lakes by their nature are stratified bodies of water. Their water layers differ in density, one stacked on top of another. (This is a smaller-scale example of the density differentiation discussed for the whole Earth in Chapter 2.) Carbon dioxide, given off by basaltic magma at depth, leaks into the bottom waters of Lake Nyos, is dissolved into the heavier, lower water layer, and is held there under the pressure of the overlying water (Figure 7.31). The event of 21 August 1986 released about 1 billion m^3 of gas in about one hour. It was like a large-scale erupting champagne bottle, where removal of the cork causes a decrease in pressure, allowing CO_2 to escape in a gushing stream. As CO_2 began escaping from Lake Nyos, it churned the lake waters, thus reducing pressure and allowing about 66 percent of the dissolved gases to escape. After the event, the lake level was 1 m lower, and the water was brown from mud and dead vegetation stirred up from the bottom.

What triggered the gas avalanche? Many suggestions have been made, including volcanic eruption, landslide, earthquake, wind disturbance, or change in water temperature with resultant overturn of lake-water layers. It is interesting to note that a similar event occurred two years earlier at Lake Monoun on 15 August 1984. This was a smaller event, but it killed 37 people. Both events were in August, the time of minimum stability in Cameroon lake waters. Is this a coincidence, or is this a normal overturning of lake water during the rainy season?

Is this event likely to happen again? Definitely. The Lake Nyos gas escape left behind 33 percent of the CO_2, and more is constantly being fed through the lake bottom. In about 20 years, the lake water could again be oversaturated with CO_2. The same loss of life will occur again unless remedial actions are taken. A pump system could be installed to slowly but continuously suck up CO_2-charged bottom water and pour it into surface streams. This would prevent the CO_2 concentrations from building up to explosive levels.

Lava Flows

Lava flows are common and they are impressive, but are they big killers? No, see Table 7.1. Why don't lava flows kill more people? Usually they move too slow, but not always.

Nyiragongo, Zaire 2002 As East Africa slowly rifts away from the African Continent (Figure 2.26), magma rises to build stratovolcanoes such as Mount Kilimanjaro and Nyiragongo in the East African Rift Valley. Nyiragongo has a long-lived lava lake in its summit crater. On 17 January 2002, lava flowed rapidly down the slopes of the volcano killing more than 45 people living on the mountain. Upon reaching flatter ground the lava flows slowed but moved relentlessly toward Lake Kivu. The city of Goma lay in the path of the oncoming lava—500,000 residents plus uncounted thousands of civil war refugees from Rwanda. Lava reached the lake, but it first flowed through the heart of Goma destroying about 25 percent of the buildings, and the war refugees were forced to flee again.

How were the lava flows able to catch and kill so many people? The lava had unusually low viscosity. In 1977, Nyiragongo lava flows had exceptionally low SiO_2 content, about

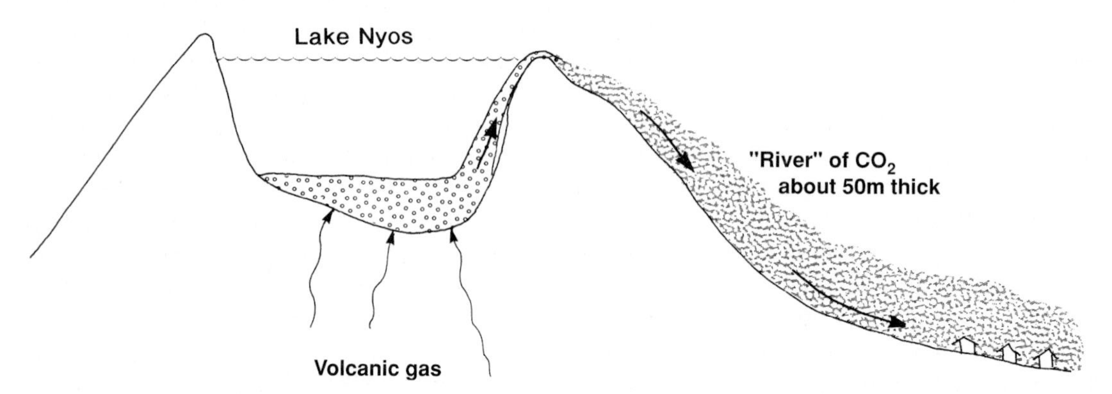

Figure 7.31 Schematic cross section of Lake Nyos. Volcanic gas is absorbed by deep-water layer. In 1986, when bottom water was disturbed, 1.2 km^3 of CO_2 gas poured down river valleys for an hour in a 50 m thick cloud. Virtually all animal life was killed; plants were unaffected.

Table 7.2	VEIs of Some Notable Volcanic Disorders
VEI (Volcanic Explosivity Index)	
8	Yellowstone 600,000 years ago
7	Tambora 1815
6	Vesuvius 79; Krakatau 1883; Pinatubo 1991
5	St. Helens 1980
4	Pelée 1902
3	Nevado del Ruiz 1985
2	—
1	—
0	Lake Nyos 1986

42 percent. (Compare this value to Table 6.5). The low-viscosity lava in 1977 flowed down the volcano slopes at about 60 km/hr (40 mph) killing an estimated 300 people.

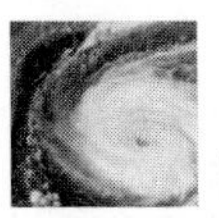

VEIs of Some Killer Eruptions

Does the total energy involved in a volcanic eruption correlate well with number of deaths? Not necessarily. The volcanic explosivity index (VEI), discussed in Chapter 6 and Table 6.8, is a semiquantitative approach to estimating the magnitude of explosive eruptions based on volume of material erupted and eruption-column height. Table 7.2 lists VEIs for some of the deadly events we have examined, and some had low VEIs; they killed with a relatively small-volume pyroclastic flow, or melted glacier ice, or gas without magma.

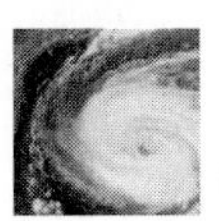

Volcano Monitoring and Warning

Can we monitor the activity of a volcano and provide advance warning before a large eruption? We have had failures and successes.

Long Valley, California, 1982 In the Long Valley–Mammoth Lakes area of California, there is abundant melting of crustal rock, although no classic hot spot exists there. About 760,000 years ago a colossal eruption blew out 150 mi^3 (>600 km^3) magma generating pyroclastic flows that covered an east-central California area greater than 1,500 km^2 with pyroclastic debris (called the Bishop Tuff) up to hundreds of meters thick. Immediately after the magma blasted out, the Earth's surface dropped nearly 2 km (>1 mi) into the void to form the Long Valley caldera (Figures 7.32 and 7.33). One pyroclastic lobe flowed 65 km (40 mi) down the Owens Valley. Before the eruption, the magma body is estimated to have had a diameter of 19 km (12 mi), with its roof 5 km (3 mi) below the surface.

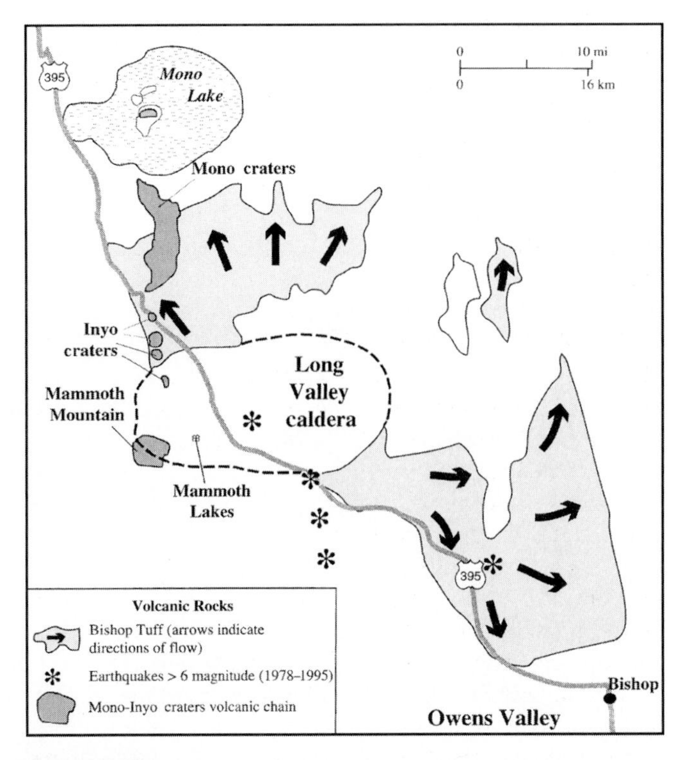

Figure 7.32 Map showing the Long Valley caldera formed by massive eruptions. Bishop Tuff is uneroded remains of pyroclastic debris from last major eruption.

Giant eruptions are rare, but these continental hot spots have fairly frequent small eruptions. There were eruptions in Long Valley 600 years ago and in Mono Lake just 150–250 years ago. Today, the main magma body is 10 km (6 mi) in diameter and 8 km (5 mi) deep (Figure 7.34).

On 25–26 May 1980, one week after the catastrophic eruption of Mount St. Helens, Long Valley was shaken by numerous earthquakes within 48 hours: four magnitude 6, tens of magnitude 4 to 5, and hundreds of smaller seisms. In the resort town of Mammoth Lakes, foundations and walls cracked, chimneys fell, while pantry and store shelves dumped their goods. Monitoring by the U.S. Geological Survey showed the resurgent dome had risen 10 inches (25 cm) in late 1979–early 1980. The dome rose another several inches by early 1982 accompanied by swarms of earthquakes. In 1980, some magma that was 8 km (5 mi) deep had risen to within 3 km (2 mi) of the surface by 1982.

Was a volcanic eruption imminent? What should be done? The affluent town of Mammoth Lakes draws most of its income from tourism; it has a year-round population of 5,500 but adds another 20,000 during winter ski season. Would issuing a formal warning of volcanic hazard do more good or harm? On 27 May 1982, the U.S. Geological Survey issued a Notice of Potential Volcanic Hazard, the lowest

Figure 7.33 View over the caldera complex of Long Valley, California.
Photo by John S. Shelton.

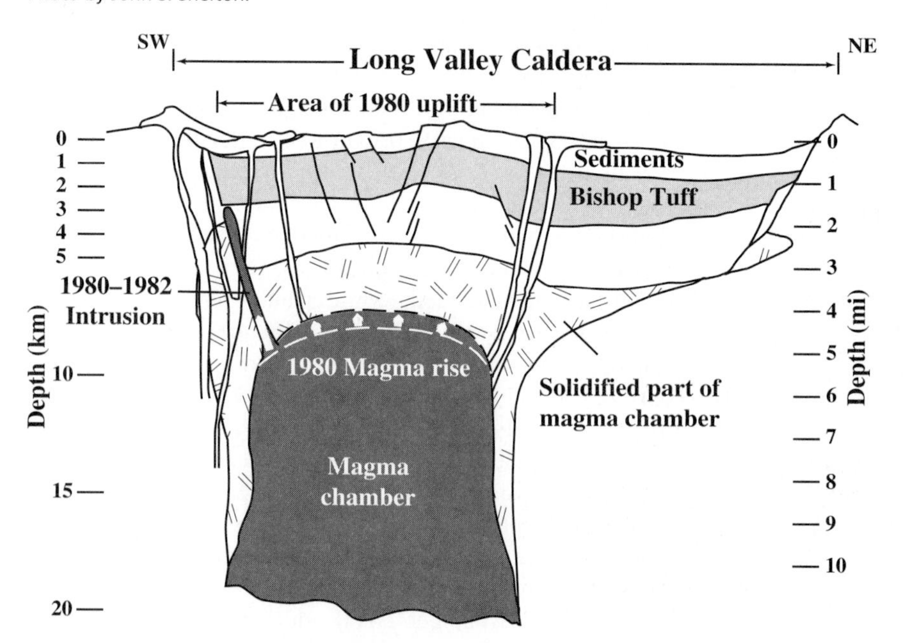

Figure 7.34 Cross section oriented northeast-southwest through the Long Valley caldera showing the size and depth to the magma body. Tongue-like intrusion moved up southern edge of the caldera in 1980–1982.

level of alert. House prices fell 40 percent overnight and tourist visits dropped dramatically. Home and business owners erupted, but the volcano did not.

In the early 1990s, trees on Mammoth Mountain began dying as large amounts of carbon dioxide (CO_2) rose up from the underlying magma and killed them. At the same time, small earthquakes resumed and the ground surface began rising. These phenomena can occur for decades or centuries; however, at large calderas they do not always mean an eruption is imminent. Many residents remain angry about the "false alarm" of 1982, while many volcanologists and emergency planners are scared to issue another volcano warning. Residents are advised to follow the motto: Prepare for the worst, but hope for the best.

To get a good view of the giant caldera that is Long Valley, look over your left shoulder as you ride up the chairlifts at the Mammoth Mountain ski resort. The big, dry valley below is the caldera (Figure 7.33).

Mount Pinatubo, Philippines, 1991 A volcano-warning success story occurred in the Philippines in 1991 before the climactic eruption of Mount Pinatubo on 15 June 1991. The volcanic eruption was the largest in the twentieth century to occur near a heavily populated area. Nearly one million people including 20,000 American military personnel and their dependents lived in the danger zone.

In March 1991, Mount Pinatubo awoke from a 500-year-long slumber as magma moved upward from a depth of 20 mi (32 km) causing thousands of small earthquakes, creating three small steam-blast craters, and emitting thousands of tons of sulfur dioxide-rich gas. U.S. and Philippine volcanologists and seismologists began an intense monitoring program to anticipate the size and date of a major eruption. On 7 June, magma reached the surface but it had lost most of its gas (like a stale glass of soda pop), so the magma simply oozed out to form a lava dome (older example in Figure 6.27). Then on 12 June (Philippine Independence Day), millions of cubic meters of gas-charged magma reached the surface causing large explosive eruptions. It was time to get out of the volcano's killing range! The message to speed up the evacuation was spread quickly and loudly. Virtually everyone, and every movable thing, left hurriedly. On 15 June the cataclysmic eruption began (Figure 7.35). It blew more than 1 mi^3 (5 km^3) of magma and rock up to 22 mi (35 km) into the atmosphere, forming an ash cloud that grew to more than 300 miles (480 km) across. The airborne ash blocked incoming sunlight and turned day into night. Pyroclastic flows of hot ash, pumice, and gas rolled down the volcano flanks (Figure 6.24) and filled valleys up to 660 ft (200 m) deep. Then, as luck would have it, a typhoon (hurricane) arrived and washed tremendous volumes of volcanic debris downslope as lahars (Figure 6.26).

How successful was the advance warning? Although almost 300 people died, it is estimated that up to 20,000 people might have died without the forceful warnings. The score card for the monitoring program from March to June 1991 shows that a monitoring expense of about $1.5 million saved 20,000 lives and $500 million in evacuated property including airplanes. What a dramatic and cost-effective success!

Figure 7.35 The 15 June 1991 Plinian-type eruption of Mount Pinatubo lasted 15 hours, sending pyroclastic flows moving downslope. VEI = 6.

Photo by Robert Lapointe, U.S. Air Force.

Summary

Spreading centers provide such ideal settings for volcanism that 80 percent of all extruded magma occurs here. Spreading centers sit on top of the asthenosphere, which yields basaltic magma that rises to fill fractures between diverging plates. Basaltic volcanoes may be successfully colonized both at spreading centers, e.g., Iceland, and oceanic hot spots, e.g., Hawaii.

Subduction-zone eruptions involve basaltic magma contaminated by crustal rock to yield highly viscous magma containing trapped gases. Their explosive eruptions make the news (e.g., St. Helens, Unzen, and Pinatubo) and the history books (e.g., Santorini, Vesuvius, and Krakatau). Transform faults have little or no volcanism associated with them.

The historic record tells of about 275,000 deaths by volcano in the last 500 years. The deadliest processes have been pyroclastic flows, tsunami, lahars, and indirectly via famine. Gas-powered pyroclastic flows can move at speeds up to 150 mph with temperatures over 1,300°F and for distances over 30 mi, e.g., Mont Pelée and El Chichón. Pyroclastic debris and water combine and flow downslope as lahars at speeds up to 30 mph and for distances up to 45 miles killing thousands at Nevado del Ruiz, and presenting a hazard to Seattle-Tacoma from Mount Rainier. Volcano-generated tsunami were mega-killers at Krakatau and Unzen. Volcanic cones can collapse, producing giant avalanches that bury entire landscapes up to 30 miles away, e.g., Mount Shasta. Giant eruptions from continental calderas can erupt more than 1,000 times as much magma as a typical volcano, e.g., Long Valley.

A volcano may be active for millions of years, but centuries may pass between individual eruptions. The time scale of an active volcano must be considered by people living nearby.

It is possible to monitor a volcano and give advance warning of a major eruption. At Mount Pinatubo in the Philippines, early warning saved up to 20,000 lives.

Terms to Remember

alluvial fan 198	jokulhlaup 182
failed rift 201	nuée ardente 197
fissure 181	pyroclastic surge 194

Questions for Review

1. How many years might one subduction zone operate? One volcano? How many years might pass between eruptions at an active volcano?
2. Explain why it is relatively safe to watch the eruption of a Hawaiian volcano but dangerous to watch a Cascade Range volcano.
3. Draw a cross section and explain how a jokulhlaup forms.
4. Draw a plate-tectonic map and explain the origin of the Cascade Range volcanoes.
5. Draw a series of cross sections and explain the sequence of events in the Mount St. Helens eruption in 1980.
6. What volcanic processes have killed the most people in the last 500 years?
7. Why do pyroclastic flows travel so fast? How do they kill?
8. Draw a cross section and explain how lahars form and move. How do they kill?
9. Explain the hazard that Mount Rainier presents to the Seattle-Tacoma region.
10. Draw a cross section and explain the sequence of events at an African killer lake, such as Nyos.
11. Can an eruption with a low VEI (low magnitude explosivity) rating kill thousands of people? How?

Questions for Further Thought

1. Is a Cascade Range volcano likely to have a major eruption during your lifetime?
2. Is it wise for towns near Mount Shasta to keep growing? What should be done about this situation?
3. Is it wise to build in river valleys below Mount Rainier, even tens of miles away?

Suggested Readings and References

Crandell, D. R., and Nichols, D. R. (1987). *Volcanic hazards at Mount Shasta, California.* U.S. Geological Survey Publication.

Crandell, D. R., Mullineaux, D. R., and Rubin, M. (1975, 7 February). Mt. St. Helens volcano: Recent and future behavior. *Science,* 187, 438–41.

Decker, R., and Decker, B. (1998). *Volcanoes.* New York: W. H. Freeman.

Francis, P. (1993) *Volcanoes: A Planetary Perspective.* Oxford: Clarendon Press.

Perret, F. A. (1937). *The Eruption of Mount Pelée 1929–1932.* Washington, D.C.: Carnegie Institute of Washington.

Sigurdsson, H. (ed.) (2000). *Encyclopedia of Volcanoes.* San Diego: Academic Press.

Simkin, T., Siebert, L., and Blong, R. (2001). Volcano fatalities—lessons from the historical record. *Science,* 291, 255.

Swanson, D. A., and Christiansen, R. L. (1973, October). Tragic base surge in 1790 at Kilauea volcano. *Geology,* 83–86.

Vallance, J. W., and Scott, K. M. (1997). The Osceola mudflow from Mt. Rainier. *Geological Society of America Bulletin,* 109, 143–63.

Volcanoes and the Earth's Interior. (1982). Readings from *Scientific American.* San Francisco: W. H. Freeman.

Videos

The 1902 Eruption of Mont Pelée. (1988). Michigan Tech University (20 min.).

Volcanoes—Our Fiery Neighbors. (1990). KGO-TV San Francisco (22 min.).

Anatomy of a Volcano—St. Helens. (1981). Time-Life (57 min.).

Volcanic Eruption in Colombia. (1988). Pan-American Health Organization (15 min.).

Volcano: Nature's Inferno. (1997). National Geographic Society (60 min).

Raging Planet: Volcano. (1998). Discovery Channel (50 min.).

Chapter 8

Mass Movements

The distinctive features of unintelligent humans are the hastiness and absoluteness of their opinions; scientists are slow to believe, and never speak without modification. The use of science and the scientific method creates an intellectual conscience which believes only up to the evidence in hand, and is always ready to concede that it may be wrong.

—Bertrand Russell, 1928, *The Outline of Philosophy*

Large volumes of material move downslope under the pull of gravity, and some do so catastrophically. These **mass movements** occur throughout the world. In the United States, no state is exempt. Many mass movements are not reported separately but are included with descriptions of the events that triggered their movements, such as earthquakes, volcanic eruptions, and major rainstorms. Mass movements in the United States annually cause about $1.5 billion in damages and 25 deaths. There are about two million mass movements each year in the United States.

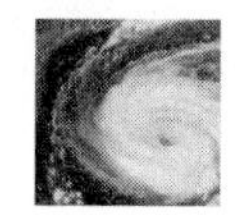

The Role of Gravity

Gravity is relentless, it operates 24 hours a day, every day of the year, and the law of gravity is strictly enforced. The constant pull of gravity is the immediate power behind the agents of erosion (Chapter 1). Rain falls, water flows, ice glides, wind blows, and waves break under the influence of gravity. But gravity can also accomplish major changes working largely by itself, without the help of any erosive agent. It is the solo work of gravity that is the subject of this chapter.

Gravity causes the downward and outward movements of landslides and the downward collapse of subsiding ground. Given enough time, gravity would pull all the land into the seas. Over the great lengths of geologic time, all slopes can fail; all slopes should be viewed as inherently unstable. Slope failures may be overpowering, catastrophic events, as when the side of a mountain breaks loose and roars downhill. Or hill slopes may just quietly deform and yield to the unrelenting tug of gravity in the very slow-moving process known as **creep.**

Gravity pulls materials downslope with a measurable force. For example, consider a 1 lb boulder resting upon a 30° slope (Figure 8.1). Gravity exerts 1 lb of pull on the boulder toward the center of the Earth; but the ground is too solid to allow the boulder to move down vertically. The trigonometric relations of a triangle allow the magnitude of the horizontal and vertical forces on the boulder to be determined. The downhill force is calculated as 1 lb × sine 30° = ½ lb. This ½ lb of force is directed downslope, toward open space; all that is needed is some initial energy to overcome inertia to begin the boulder's move-

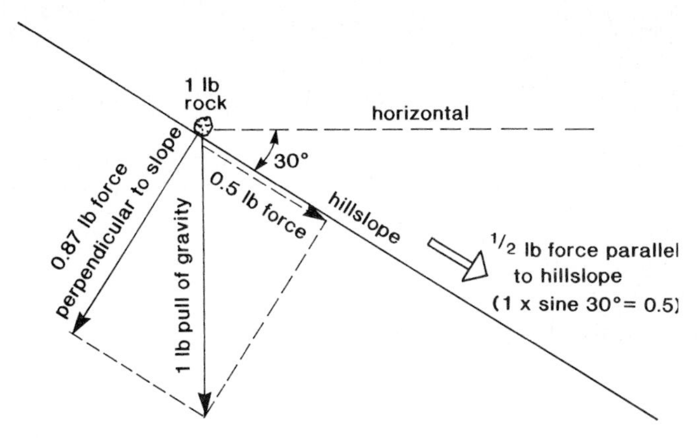

Figure 8.1 Gravitational forces acting on a 1 pound boulder sitting on a 30° slope.

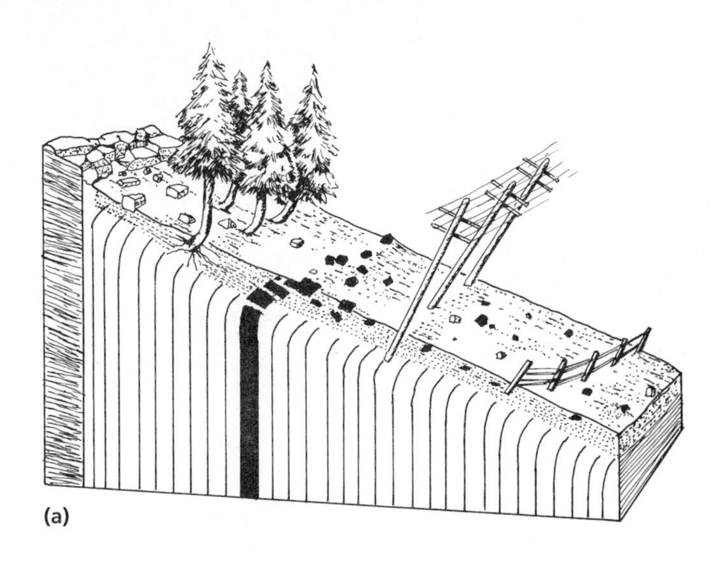

(a)

ment. Initial energy could come from an earthquake, a heavy rain, the footstep of a sheep, or. . . .

Creep

Gravity induces materials to move via creep, the slowest but most widespread form of slope failure. Creep is an almost imperceptible downslope movement of the **soil** and uppermost **bedrock** zones. Creep is most commonly seen by its effects on objects, such as telephone poles and fences, that lean downslope or by trees whose trunks have deformed due to growing upward while rooted in material that is slipping downhill (Figure 8.2a). The soil zone slips in ultraslow movements as individual particles shift and move in response to gravity; the upper bedrock zone yields to the pull by curving downslope (Figure 8.2b).

By what mechanisms do soil and rock actually move? We see that ground surfaces do not sit still; they move up and down. The volume of soil does not stay constant but instead swells and shrinks. Several processes cause swelling. 1) Soil has a high percentage of void space or **porosity.** When water filling these pores freezes, it expands in volume by 9 percent, swelling the soil volume and lifting the ground surface upward. 2) When soil rich in expandable materials, such as **clay minerals,** is wetted, it absorbs water and expands. 3) Heating by the Sun causes an increase in volume. Soil expands perpendicular to the ground surface (Figure 8.3).

Several processes shrink masses of soil, rock, and water, causing the ground surface to lower. Shrinkage occurs when soil: 1) thaws, 2) dries, or 3) cools. Lowering of the surface is influenced by the downward pull of gravity, causing a net downslope movement of particles in the soil zone (Figure 8.3). Creep is a slow downhill flow, but it can grade into faster flow types (Figure 8.4).

(b)

Figure 8.2 **(a)** Block diagram of a slope showing the effects of creep. Soil moves slowly, and bedrock deforms downhill. **(b)** Creep has deformed rock layers near Marathon, Texas.
Photo courtesy of NOAA.

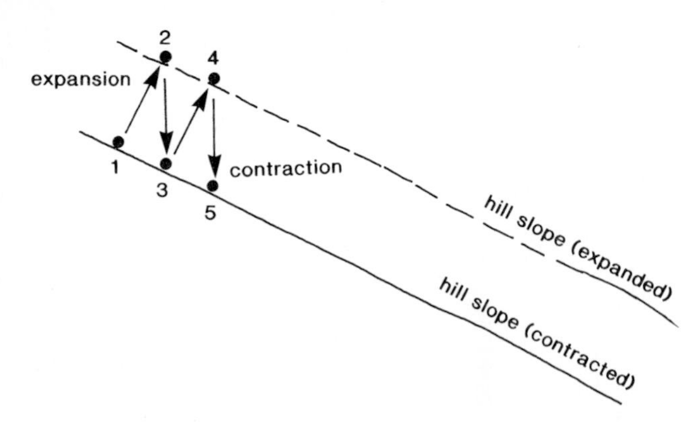

Figure 8.3 How creep works. Surface materials expand perpendicular (e.g., from point 1 to point 2) upon freezing of pore water, swelling when wetted, and heating by the Sun. Surface materials shrink (e.g., from point 2 to point 3) upon thawing, drying, and cooling. Contraction occurs under the pull of gravity and is toward the center of the Earth. The result is a net downslope movement of materials, i.e., creep.

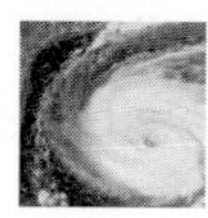

External Causes of Slope Failures

Most mass-movement fatalities are caused by the fast-moving varieties, as recorded on a tombstone in Westland, New Zealand.

> Patrick O'Brien
> Who was killed by a landslip
> on the 8th of March 1888
> Aged 37 years
> Death to him short warning gave
> Therefore be careful how you live
> Repent in time and don't delay
> For he was quickly called away

A typical landslide is a mass whose center of gravity has moved downward and outward (Figure 8.5). There is a tear-away zone upslope where material has pulled away, and a pile-up zone downslope where material has accumulated.

Figure 8.4 Large-scale creep high on hill (upper right) causes soil and rock to build up and flow downslope (lower left), central California.

Photo by John S. Shelton.

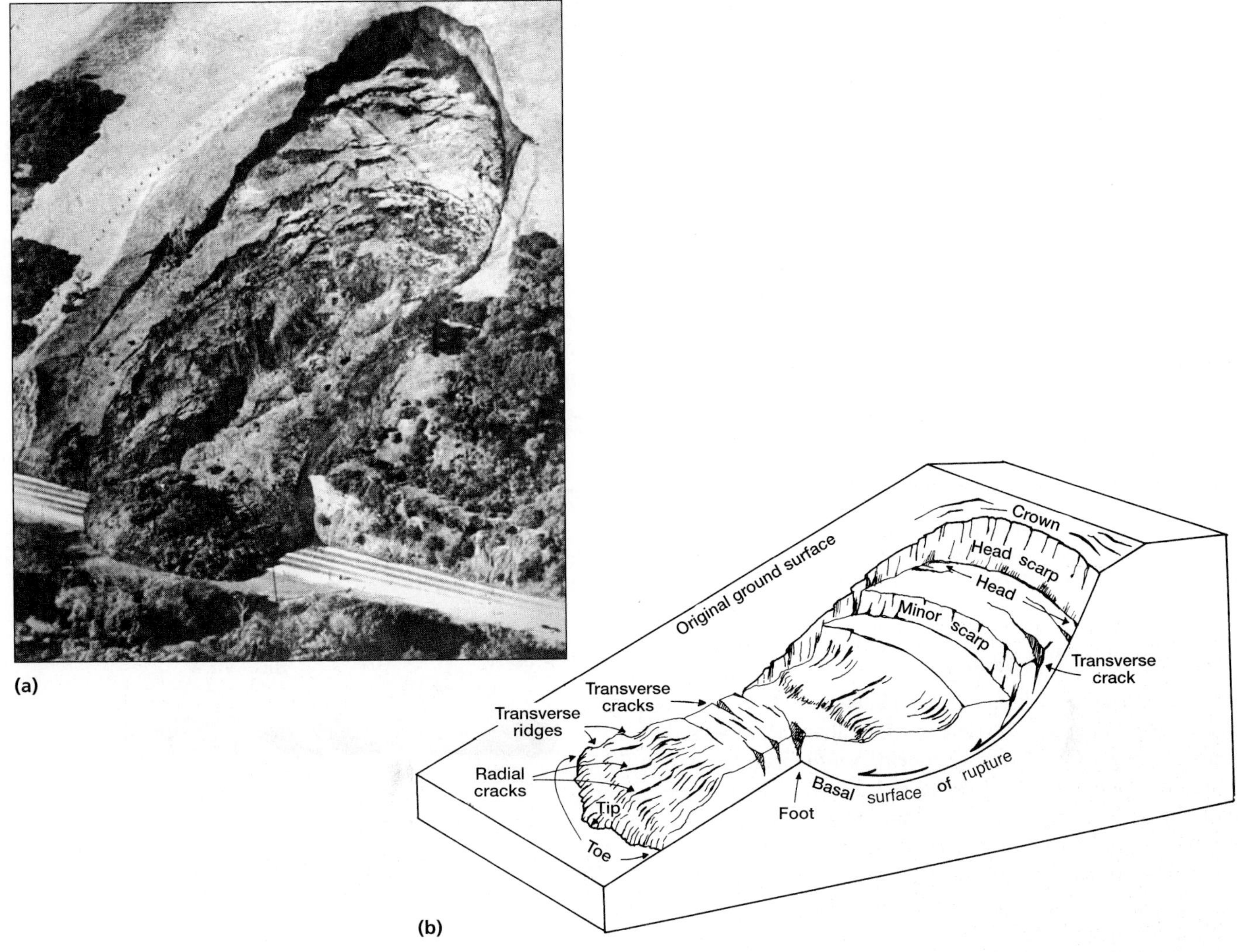

Figure 8.5 (a) Landslide showing downward-and-outward movement. (b) Some topographic features created by a landslide in its downward-and-outward movement.
Photo courtesy of the *U.S. Geological Survey.*

External processes that increase the odds of a slope failure include: 1) steepening the slope, as by fault movements, 2) removing support from low on a slope, as by stream or ocean-wave erosion, and 3) adding mass high on a slope, as in sediment deposition.

Some landslides move on top of an arcuate failure surface. Material above the failure surface can be divided into a driving mass resting inclined out of the hill and a resisting mass inclined back into the hill. In effect, the slope is an equilibrium situation where a driving mass seeks to break free and move downward and outward, but it is blocked by a resisting mass that acts as a wedge holding the driving mass and slope in place (Figure 8.6).

How can humans cause landslides to occur? We can add fill dirt high on hill slopes to make more view lots, thus adding to the driving mass (Figure 8.6). We can remove material from the base of the slope to widen or clear a road (Figure 8.5a) or to form a building lot, thus weakening the resisting mass. All too frequently, both processes are done simultaneously. Whether it is by natural or human processes, anything that steepens a slope moves it closer to failure.

The reverse is also true; the overall lessening of slope angle is one step in increasing the stability of a hill. Slope stability also depends on internal factors.

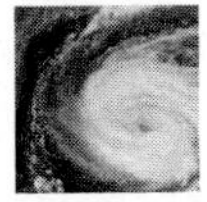

Internal Causes of Slope Failures

Beneath the surface, inside the materials underlying a slope, long-term processes are weakening the Earth and preparing it for failure. Internal causes of slope failure include: 1) inherently weak materials, 2) water in different roles, 3) decreasing cohesion, and 4) adverse geologic structures.

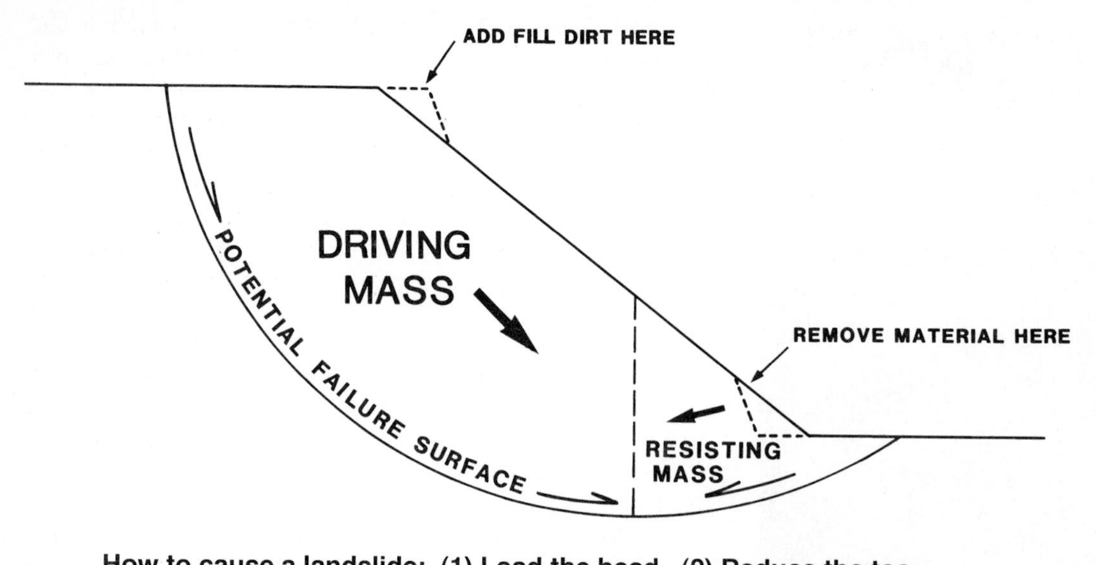

Figure 8.6 A hill slope of homogeneous materials may fail along an arcuate basal surface. The slope is in equilibrium when a driving-mass portion is kept from moving by a resisting-mass portion. Adding to the driving mass or removing from the resisting mass can cause a landslide. Does this situation bring to mind any construction practices in your area?

Inherently Weak Materials

The materials most commonly associated with earth failures are the clay minerals. Clays are the most abundant of all sediments. They form during **chemical weathering** as rocks exposed at the surface decompose and form new minerals under conditions of low temperature and pressure. The weathering occurs when acidic fluids, such as water, CO_2-charged water, and organic acids, decompose minerals. For example, a magma that cooled at a depth of 5 miles has minerals that crystallized in chemical equilibrium with high pressure many thousands of times that at the Earth's surface and temperatures of 1,100 to 1,600°C. When these minerals are exposed at the surface by long-term tectonic processes and erosion, they are out of equilibrium with the low-pressure and low-temperature surface conditions. These minerals are likely to transform into new atomic structures to achieve equilibrium. In the decomposition process, the relatively simple atomic structure of such minerals as feldspar will transform to the wildly variable structures and compositions of the clay minerals.

Although the bulk chemical compositions of feldspars and clays are similar, their internal structures are radically different. Clay crystals are very small—too small to be seen with a typical microscope. Clay minerals are built like submicroscopic books (Figure 8.7). From a top view, they are nearly equidimensional. But a side view shows a much thinner dimension that also is split into even thinner subparallel sheets, like the pages in a book. The booklike structure typically forms in the soil zone where water strips away elements, leaving many unfilled atomic positions in crystal structures. This is like building a Tinker Toy or Lego structure while leaving out a tremendous number of pieces.

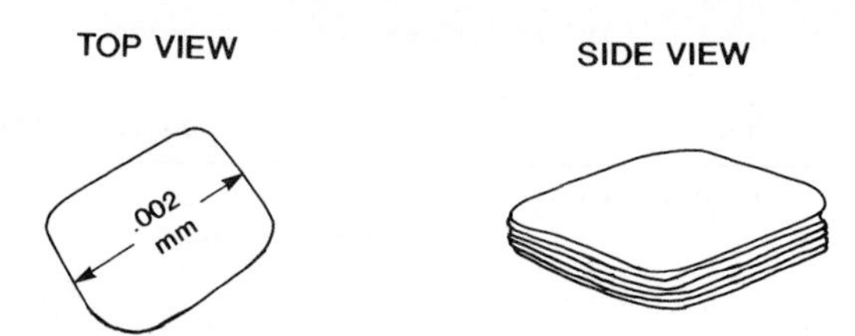

Figure 8.7 Schematic views of the size and structure of a clay mineral.

As clay minerals take in different elements and other elements are removed, they increase and decrease in strength, they expand and contract, and they may absorb water and later have it removed. The constantly changing conditions cause variations in the strength of clay minerals from month to month and year to year. Thus, there are certain times when a hill containing clay minerals is weaker, and then gravity has a better chance of provoking a slope failure.

The mechanical or strength characteristics of a rock are usually governed by the 10 to 15 percent of the rock with the finest grain size. Clay minerals may have their strength lessened by water: 1) adsorbed to the exterior of clays, thus spreading the grains apart, and 2) absorbed between the interlayer sheets, with resultant expansion.

Spectacular examples of slope failures are presented by **quick clays,** the most mobile of all deposits. Quick clays are abundant in Scandinavia and eastern Canada and occur in the northeastern United States. Quick clays begin as fine rock flour scoured off the landscape by massive glaciers and later deposited in nearby seas. The clay and silt sediments sit in a loosely packed, "house of cards" structure filled with water and some sea salts that help hold it together as a weak

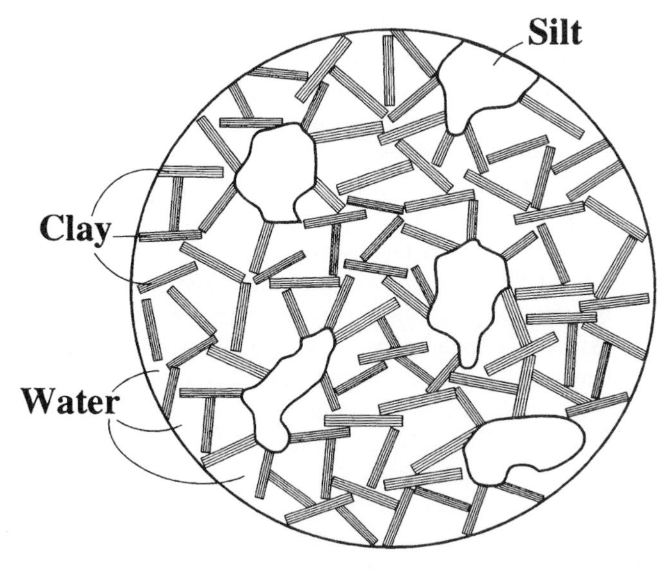

Figure 8.8 A "house of cards" structure occurs in quick clay. Platy clay minerals are stacked in an unstable configuration with silt grains and much water.

solid (Figure 8.8). When glaciers retreat, isostatic rebound lifts these clay-sediment areas above sea level (see Chapter 2). Freshwater passing through the uplifted sediments dissolves and removes much of the sea salt "glue," leaving quick clay with: 1) weak structure, 2) grains mostly less than 0.002 mm diameter, 3) water contents commonly in excess of 50 percent, and 4) a low salt content. In short, the "house of cards" structure can be collapsed by a jarring event, such as a dynamite blast or vibrations from construction equipment.

What has been solid Earth can literally turn to liquid and flow away. These words sound like an overstatement. How does land that has been plowed and farmed, built on, and occupied for many generations all of a sudden turn to fluid and flow away? The collapse of the "house of cards" with its high water content creates a muddy fluid. To really believe this, you need to see it. Check out the video entitled *Rissa Landslide, Quick Clay in Norway;* it is mind-boggling to watch solid Earth suddenly turn to fluid and flow off, carrying houses with it. Quick-clay slope failures have caused the destruction of numerous buildings and the loss of many lives. These events will recur more than once during your lifetime.

Table 8.1 Some Canadian Quick-Clay Slope Failures

Date	Site	Deaths	Damages
June 1993	Lemieux, Ontario	—	Severed County Road 16
May 1971	St. Jean Vianney, Quebec	31	40 houses destroyed
December 1962	Rivière Toulnustouc, Quebec	8	Failure initiated by workers setting off blasts
April 1908	Notre Dame de la Salette, Quebec	33	Houses destroyed
April 1895	Saint Luc de Vincennes, Quebec	5	—
1877	Sainte Geneviève de Batiscan, Quebec	5	—

Canadian Quick-Clay Slope Failures

Quick-clay hazards are common in eastern Canada. Quick-clay disasters (Table 8.1) have led to recognizing problem areas and taking preventive actions such as moving towns. The wisdom of moving people out of harms way was shown during the high rainfall year of 1993 near the former townsite of Lemieux in Ontario, Canada. During one hour in the afternoon of 20 June, a 3.5 million m^3 mass liquefied and flowed carrying forest and rangeland into the South Nation River (Figure 8.9). Eyewitnesses report seeing flowing waves of Earth carrying trees into the river and sending 3 m (10 ft) high waves both up and down river. Human development of the area is hindered. How can you live your life on ground that may turn to liquid?

Figure 8.9 A 3.5 million m^3 mass of quick clay and sand flowed into the South Nation River near Lemieux in Ontario, Canada on 20 June 1993. In the upper right is County Road 16.

Geological Survey of Canada Miscellaneous Report 56 (1993). The Lemieux landslide of June 20, 1993.

Water in Its Different Roles

Water weakens earth materials in several different ways; it does so by its: 1) weight, 2) interplay with clay minerals, 3) decreasing the cohesion of rocks, 4) subsurface erosion, and 5) pressure in pores of rocks and sediments.

1. Water is heavy. Sedimentary rocks commonly have porosities of 10 to 30 percent. When these void spaces

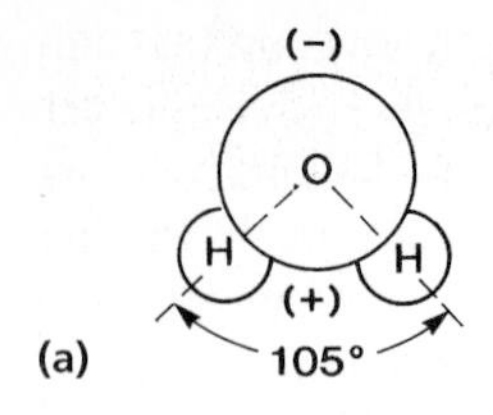

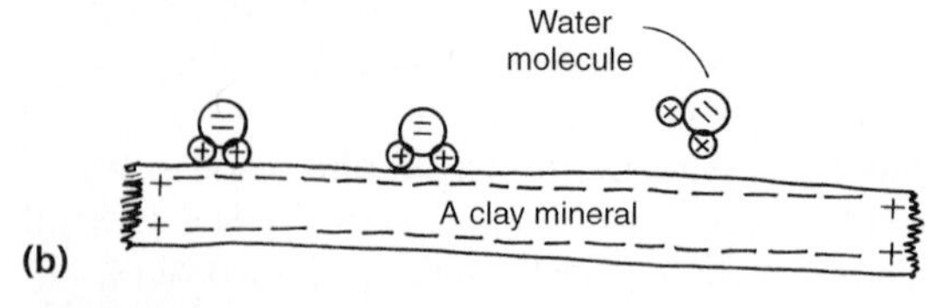

Figure 8.10 (a) Water is a bipolar molecule. Two hydrogen atoms (each H+) attach to one oxygen atom (O– –), creating an electrically neutral molecule. However, the asymmetry of the molecule makes one side slightly negative and the other side positive. (b) The positive side of water molecules attaches to the negatively charged surface of clay minerals.

are filled with water, the weights of materials are dramatically increased, thus the driving masses of slope materials are also increased and mass movements may begin.

2. Water is both absorbed (internally) and adsorbed (externally) by clay minerals with resultant decreases in strength. How does water attach so easily to clay minerals? Because of its unique interplay of charges. Water is a molecule formed by two hydrogen (H^+) atoms linking up with one oxygen (O^{--}). The two positive charges from the hydrogen atoms should cancel the two negative charges from the oxygen atom to create an uncharged or neutral molecule. But water is a bipolar molecule (Figure 8.10a) with its hydrogen atoms on one side (a positive side) and its oxygen on the other side (a negative side). Thus, water molecules can attach their positive sides against clay minerals because clay surfaces are negatively charged (Figure 8.10b).
3. Water flowing through rocks can dissolve some of the minerals that bind the rock together. The removal of cementing material decreases the cohesion of rocks and saps some of a slope's strength, preparing it for failure by mass movement.

   A death-dealing example occurred with the failure of the St. Francis dam about 73 km (45 mi) north of Los Angeles, California. The dam was a heavy concrete mass built in 1926 across San Francisquitos Canyon. Unfortunately, part of the foundation of the dam was a poorly bedded mass of gravels, sands, and muds held together by clay coatings and cemented by the mineral gypsum. In two years, lake water dissolved enough gypsum cement and weakened the clays, causing a catastrophic failure of the dam. At about midnight on 12 March 1928, the base of the dam failed, unleashing a 56 m (185 ft) high wall of water moving 29 km/hr (18 mph). The undammed water took 5 hours and 27 minutes to travel 87 km (54 mi) down the Santa Clara River to the Pacific Ocean. Traveling through the night, the raging water made permanent the sleep of about 420 people.
4. Water flowing underground can not only chemically dissolve minerals but also can physically erode loose material. Subsurface erosion (**piping**) can create extensive systems of caverns (Figure 8.11). A network of caves obviously makes a hill weaker.
5. Pressure builds up in water trapped in pores of rocks being buried deeper and deeper. As sediments pile up on the surface, their weight puts ever-more pressure on sediments and pore water at depth. Sediment grains of sand and mud pack into smaller and smaller volumes, while water, which is nearly incompressible, simply stores built-up pressure. When a pile of sediments sits on top of overpressurized pore water, the entire mass becomes less stable. The buildup of pressure within pore water has been referred to as a "hydraulic jack" that progressively "lifts up" sediments until the pull of gravity can start a massive failure. Many mass movements and slope failures are due to abnormally high **pore-water pressures.**

Quicksand occurs where sand grains are supersaturated with pressurized water. For example, if water flows upward through sands, helping lift up the sand grains, then the pull of gravity on the sand grains can be effectively canceled,

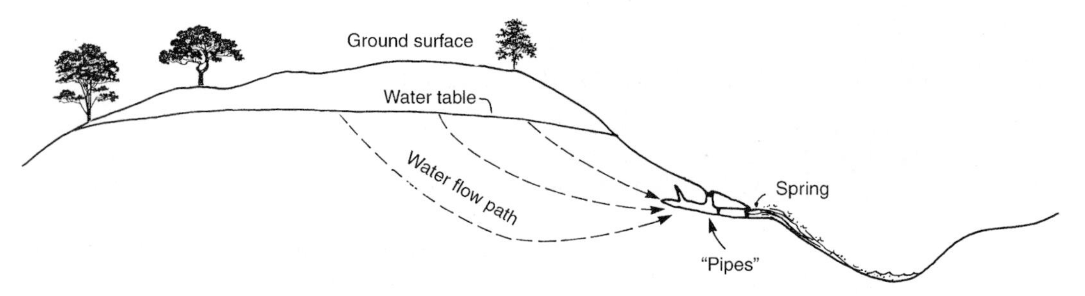

Figure 8.11 Schematic cross section of groundwater flowing through poorly consolidated rocks. Water will carry sediments to the surface at springs. This erosion creates a network of caverns that seriously weakens a hill.

Sidebar

Analysis of Slope Stability

The analysis of a slope for its resistance to failure is commonly addressed by use of the Coulomb-Terzaghi equation:

$$s = c + (p - h_w) \tan \phi$$

where s is **shear** resistance, the sum of characteristics that hold a mass in place; c is cohesion, a measure of how strongly a material is held together by interactions between particles (e.g., there is little or no cohesion between quartz sand grains, but there is much cohesion between flakes of clay); p is the weight of solids and water above a potential slide surface; h_w is the height of the water column times the unit weight of water; and *tan phi* is the tangent of the angle of internal friction—the slide surface's angle measured in degrees from horizontal.

What does this all mean? The strength of a hillside or mass of material comes from 1) its cohesion, how well it sticks together, plus 2) the weight of all its materials under the pull of gravity. These factors of stability are offset by 1) the pore-water pressure, plus 2) the angle at which a rock mass fractures, the angle of its slide surface. The failure angle is low or near horizontal for weak materials, such as clay-rich sedimentary rock, and is steep or near vertical for strong rocks, such as granites.

When pore-water pressure (h_w) equals the weight of overlying materials (p), then the only shear resistance the materials possess comes from cohesion. If these materials are low in cohesion, such as quartz sands, then the materials effectively have no strength and will fail. An example of this is quicksand.

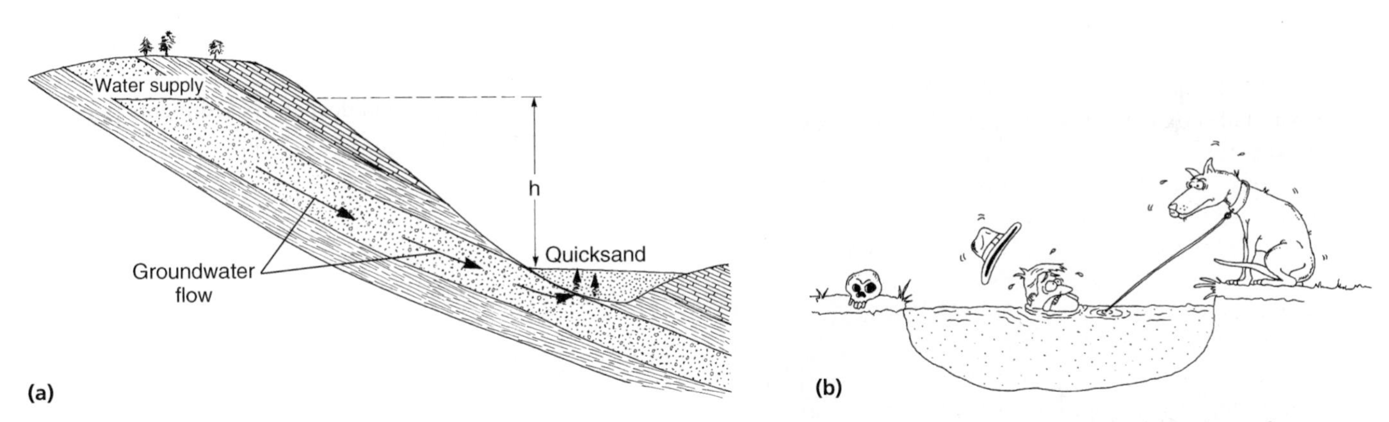

Figure 8.12 **(a)** Schematic cross section showing groundwater pushed by a high column of water (h) and reaching the surface through loose sands. The uplifting force of the escaping water equals the weight of the sand grains, thus making quicksand. **(b)** Quicksand! Drawing by Jacobe Washburn.

leaving the sand with no strength or ability to carry a load (Figure 8.12). (In effect, the pore-water pressure (h_w) equals the weight of the sands (p), leaving the cohesionless sand with no shear strength.)

If water-pressurized sand was on a slope, it would flow away; but if it sits on a flat surface, it will be quicksand. Despite what some old movies show, quicksand does not suck people or other objects down. It is rather like stepping in very thick water. Stand there long enough and you will sink below the surface (Figure 8.12b). What should you do if you get caught? If the water is not too deep, fall backward and spread your arms out; this distributes your weight broadly, like a boat on water, and you will not sink. Then call to your companions to pull you out, or if alone, slowly slide your way backward keeping your weight broadly distributed as you pull your legs from the quicksand's grasp. Quicksand holds tightly, so do not panic and flail wildly to get yourself out. Slow and easy, with a broad spreading of your body weight is the answer.

Decreases in Cohesion

When rocks are buried to depths of hundreds, thousands, and tens of thousands of feet, they are compressed into smaller volumes by the weight of the overlying materials. But this process also works in reverse. When deeply buried rocks are uncovered by erosion and exposed at the surface, the removal of great weight allows the compressed rocks to relax and expand. The expansion in volume produces fractures and increases in porosity. The stress-relaxation process reduces the strength of rocks, opening up passageways and storage places for water to attack and further weaken the rocks.

Adverse Geologic Structures

Many hill-slope masses are weak due to preexisting geologic conditions. Weaknesses exist as: 1) ancient slide surfaces, 2) rock layering dipping less than slopes, and 3) structures within the rocks, such as joints and fractures, ancient faults, thin clay seams, and contacts between soft and hard rocks.

1. Ancient slip surfaces are weaknesses that tend to be reused over time. When a mass first breaks loose and slides downslope, it tends to create a smooth, slick layer of ground-up materials beneath it (Figure 8.13). The slick layers become especially slippery when wet. It is wise to avoid building on these sites, but if building is necessary then these slick slide surfaces need to be recognized, dug up, and destroyed. Otherwise, they commonly are reactivated when wetted causing major financial losses and much heartache for building owners. Replays of this scenario are found in the news with a saddening frequency.
2. The orientation of rock layering within a hill may create either a strong or a weak condition. Where the rock layers dip at angles less than that of the hill slope, then the stage is set for slippage and mass movement (see the right-side slope in Figure 8.12a). This condition is known as **daylighted bedding** where the ends of shallow-dipping rock layers are exposed to daylight on a steeper slope. On the other side of the same hill, the same rock layers dip into the hill at a steep angle, making it difficult for a massive slide to initiate and break free. This relationship is a good one to keep in mind when you are selecting a home or building site.
3. Rocks have weaknesses that set-up slope failure where: 1) rocks are not cemented together, 2) a clay layer may provide a basal slip surface, 3) soft rock layers may slide off strong materials, 4) joints split and separate rock, or 5) an ancient fault may act as a slide surface.

Figure 8.13 A piece of slickened and grooved, basal slide surface created by the weight of an overriding slide mass. Yard watering by homeowners expanded the clay minerals, reactivating the ancient landslide and destroying seven homes in San Diego, California. Sample is about 25 cm (10 in) wide.
Photo by Michael W. Hart.

Triggers of Mass Movements

Slopes usually do not fail for just one reason. Most failures have complex causes. Over the long intervals of time a slope exists, gravity is constantly tugging and water keeps soaking in and sapping its strength. On numerous occasions, a slope almost fails. Then along comes another stress for the slope, such as heavy rains, and the slope finally fails in a massive event. Did the last stress, the saturation by heavy rains, cause the slide? Or was it just the trigger for the movement, the proverbial straw that broke the camel's back? Clearly the rains were simply the trigger, or immediate cause, for the mass movement.

It is useful to distinguish between immediate and underlying causes. The sum of all the underlying causes pushes the slope to the brink of failure, and then an immediate cause triggers the movement. Common triggers for mass movements include heavy rains, earthquakes, thawing of frozen ground, and more and more frequently, the construction projects of humans.

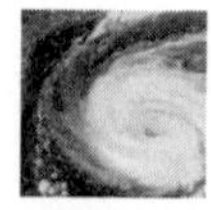

Classification of Mass Movements

Speed of movement and water content are two important factors in defining different types of mass movements (Figure 8.14). Slow-moving masses cause tremendous amounts of destruction and property damage, but rapidly moving masses not only destroy, they kill (Table 8.2).

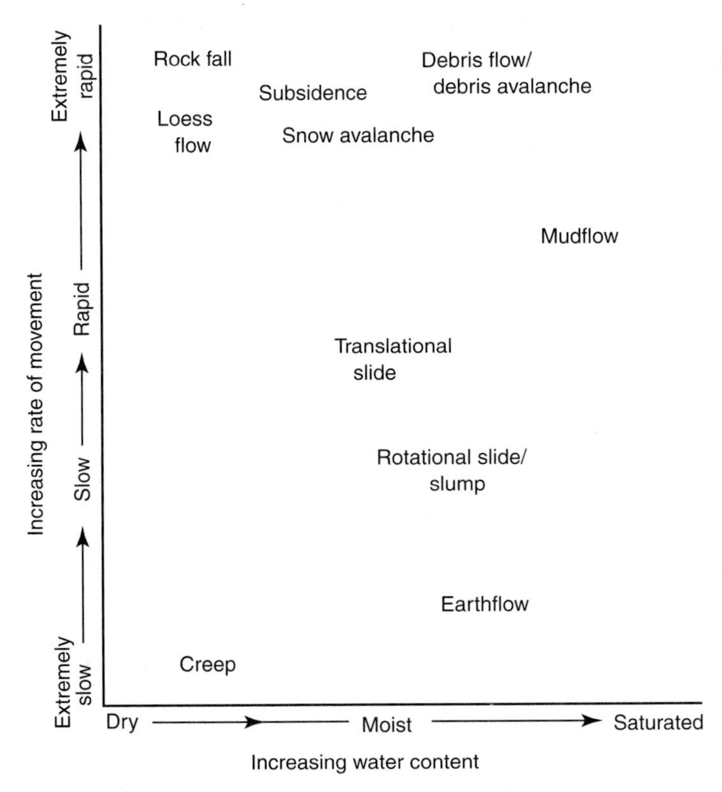

Figure 8.14 Mass-movement speed versus moisture content.

Table 8.2 Rates of Travel for Mass Movements

	Rate	
	Extremely Rapid	
10 ft/sec (6 mph)		3 m/sec
	Very Rapid	
1 ft/minute		0.3 m/min
	Rapid	
5 ft/day		1.5 m/day
	Moderate	
5 ft/month		1.5 m/month
	Slow	
5 ft/year		1.5 m/year
	Very Slow	
1 ft /5 years		0.3 m / 5 years
	Extremely Slow	

Source: D. J. Varnes (1978).

Table 8.3 Some Killer Mass Movements

Year	Place	Fatalities
1499	Kienholz, Switzerland	400
1515	Blenio Valley, Switzerland	600
1556	Shaanxi Province, China	830,000
1569	Hofgastein and Schwaz, Austria	287
1584	Yvorne, Rhone Valley, France	328
1618	Mont Conto, Switzerland	2,430
1669	Salzburg, Austria	250
1741	Pennsylvania	22
1806	Goldau, Switzerland	457
1814	Boite Valley, Italy	300
1843	Mount Ida, Troy, New York	15
1881	Elm, Switzerland	115
1892	St. Gervais, France	177
1893	Trondheim, Norway	111
1903	Frank, Alberta, Canada	70
1920	Shaanxi Province, China	200,000
1936	Nordfjord, Norway	73
1938	Kobe, Japan	505
1945	Kure, Japan	1,154
1958	Shizuoka, Japan	1,094
1959	Hebgen Lake, Montana	28
1962	Nevados Huascarán, Peru	4,000
1963	Vaiont, Italy	3,000
1964	Anchorage, Alaska (plus an earthquake)	114
1966	Aberfan, Wales	144
1966–7	Rio de Janeiro area, Brazil	2,700
1969	Nelson Co., Virginia (plus floods)	150
1970	Nevados Huascarán, Peru (plus an earthquake)	70,000
1971	St. Jean-Vianney, Quebec	31
1985	Mameyes, Puerto Rico	129
1998	Campania, Italy	180
1998	Honduras and Nicaragua	10,000
1999	Venezuela	10,000
2002	El Salvador	700

Source: After R. L. Schuster and R. W. Fleming (1986).

Rapid-moving mass movements have been big-time killers all around the world (Table 8.3).

The main types of mass movement are downward, as in falling or subsiding, or downward and outward, as in sliding and flowing (Figure 8.15). Falling is downward from a topographic high place, such as a cliff or mountain, whereas subsiding is downward via collapse of the surface. Sliding occurs where a semicoherent mass slips down and out on top of an underlying failure surface. Flowing occurs when a moving mass behaves like a viscous fluid flowing down and out over the countryside. The types of movements will be examined in a series of examples of **falls** and **flows, slides** and **subsides.**

Falls

Falls occur when elevated rock masses separate along joints, **bedding,** or other weaknesses (Figure 8.16). When a mass

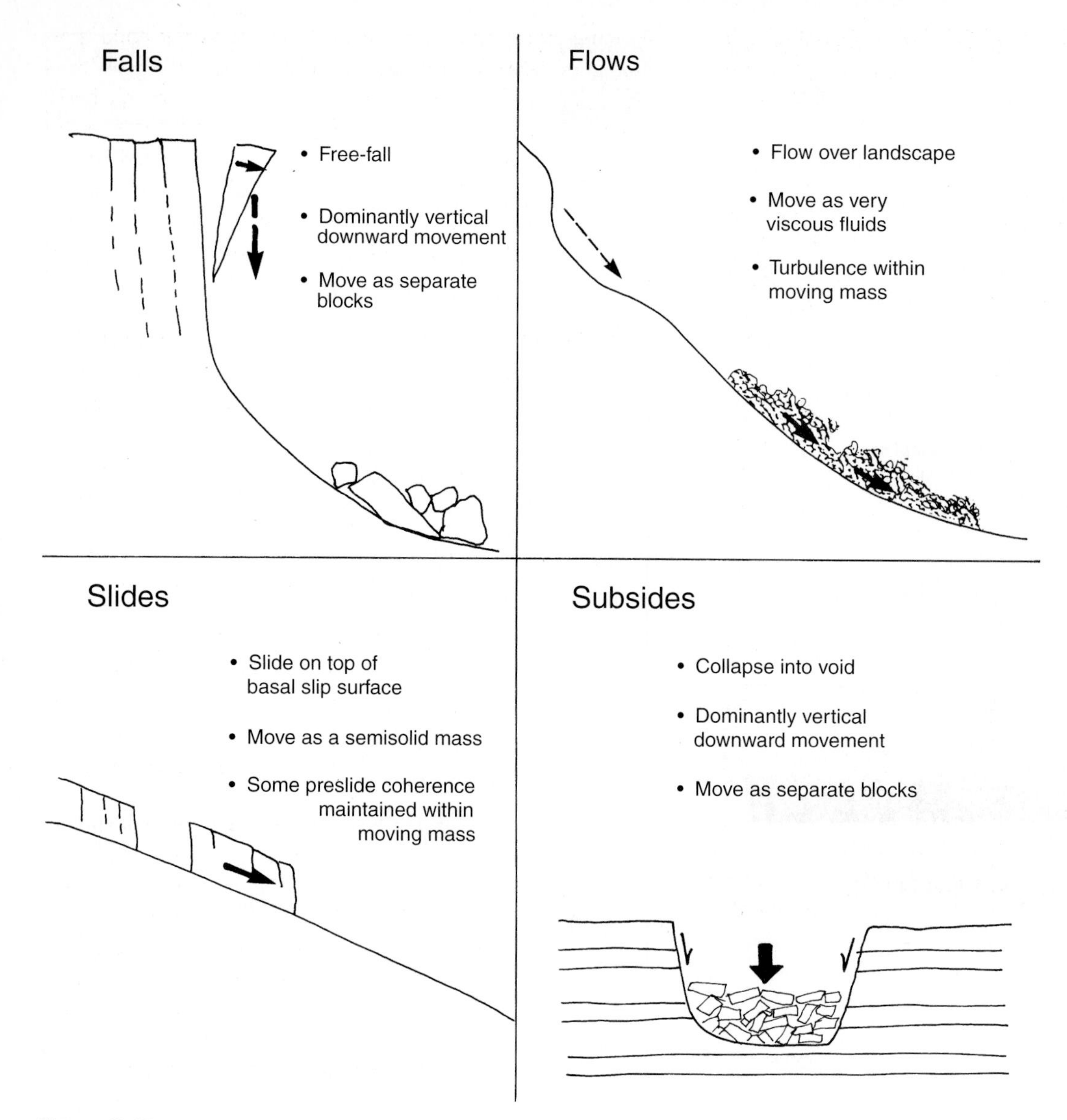
Falls
• Free-fall
• Dominantly vertical downward movement
• Move as separate blocks
Flows
• Flow over landscape
• Move as very viscous fluids
• Turbulence within moving mass
Slides
• Slide on top of basal slip surface
• Move as a semisolid mass
• Some preslide coherence maintained within moving mass
Subsides
• Collapse into void
• Dominantly vertical downward movement
• Move as separate blocks

Figure 8.15 Classification of mass movements.

detaches, it mostly falls downward through the air via free fall and then, after hitting the ground, moves by bounding and rolling. The triggers for falls may be heavy rainfall, frost wedging, earthquakes, and such. Perhaps you have been the triggering agent for small falls at one time or another?

Yosemite National Park, California During the summertime, Yosemite Valley fills with visitors awed by the steep-walled valley and its waterfalls. But steep walls are also the setting for gravity-powered rock falls. Below Glacier Point at 6:52 P.M. on 10 July 1996, a 162,000 ton mass of granitic rock pulled away from the canyon wall in two separate masses, 14 seconds apart. Each slid down 165 m (540 ft) then launched into the air for a 500 m (1,640 ft) drop in an arcing trajectory, reaching a speed of 270 mph before hitting the valley floor and shattering into a roiling cloud of pulverized rock that rolled across the valley and part way up the opposite wall before turning and flowing down canyon (see color photo of this event in Chapter 1). The falling granitic rock mass pushed air ahead of it in a gale-force blast that knocked down over 1,000 mature trees and covered 50 acres with an inch-thick blanket of pulverized rock. The fallen rock masses hit the ground creating a magnitude 3+ earthquake and a vertical column of dust 1 km high that blotted out the sun. Despite the enormous amount of energy involved in this fall-shatter-flow event, only one person was killed—a 20-year-old man crushed by a tree blown down by the airblast in front of the fallen rock mass.

Figure 8.16 A fall from a New Zealand cliff face.
Photo by Pat Abbott.

Slides

Slides, or landslides, are movements above one or more failure surfaces (Figure 8.15). Basal failure surfaces typically are either: 1) curved in a concave-upward sense or 2) nearly planar.

Rotational Slides

Rotational slides move downward and outward on top of curved slip surfaces (Figures 8.5 and 8.6). Movement is more or less rotational about an axis parallel to the slope. The center of rotation can be approximated by piercing a cross section with the point of a compass and then swinging the compass in an arc to draw a basal failure surface (Figure 8.17a). This is the Swedish Circle analysis of slope stability used to calculate driving versus resisting forces, giving a quantitative understanding of how close a slope is to failing.

When a **slump** occurs, its head moves downward and typically rotates backward (Figure 8.17b). Water falling or flowing onto the head of the slide mass ponds in the basin formed by the backward tilt. The trapped water sinks down into the slump mass, causing more instability and movement. Because the scarp at the head of a slump is nearly vertical, it is unstable, thus setting the stage for further mass movement. The toe of a slump moves upward, riding out on top of the landscape (Figure 8.17b).

Ensenada, Baja California The features of a rotational slide are well illustrated by a 1976 slump in Ensenada, Baja California. Arcuate cracks a few hundred meters long began forming on a hill slope at 90 to 120 m (295 to 400 ft) above sea level. As the cracks widened, most residents heeded this natural warning and evacuated their homes. But not everyone left; two people were asphyxiated during their sleep one night when the slow-moving slide severed the natural gas lines inside their house. The down-going slide left a pronounced head scarp (Figure 8.18). The body of the slide carried a 275 m (900 ft) long portion of Mexican Highway 1 toward the Pacific Ocean (Figure 8.19). The toe exhibited the classic upward bulge as it uplifted the sea floor to 9 m (30 ft) above sea level (Figure 8.20). Local residents lost little time in picking the abalone off the uplifted and exposed sea floor.

Although the basal failure surface of a rotational slide is concave in the direction of movement, it rarely is as perfect an arc as is drawn by the Swedish Circle analysis method (Figure 8.17a). Why? Because the failure surface usually cuts through heterogeneous materials that influence the rupture path. Discontinuities such as faults, joints, rock layering, and the presence of clay-rich seams can alter the shape of the failure surface. Slope failures beginning as rotational slides may transform downslope into translational slides.

Translational Slides

In **translational slides,** masses move down and out by sliding on surfaces of weakness, such as faults, joints, a clay-rich layer, soft rocks slipping off hard rocks, and hard rocks being spread apart by movements within underlying soft rocks. A translational slide may move as long as it sits on the downward-inclined surface and its driving mass still exists. In contrast, rotational slides move only short distances; their arcuate movements tend to restore equilibrium soon because the driving mass decreases and the resisting mass increases.

Translational slide masses behave in different fashions. 1) They may remain basically coherent as block slides. 2) The sliding mass may deform and disintegrate to form a debris slide. 3) Lateral spreading may occur where the underlying material fails and flows, thus causing the overlying coherent material to break apart and move. The various styles of translational slides may be best understood using specific examples.

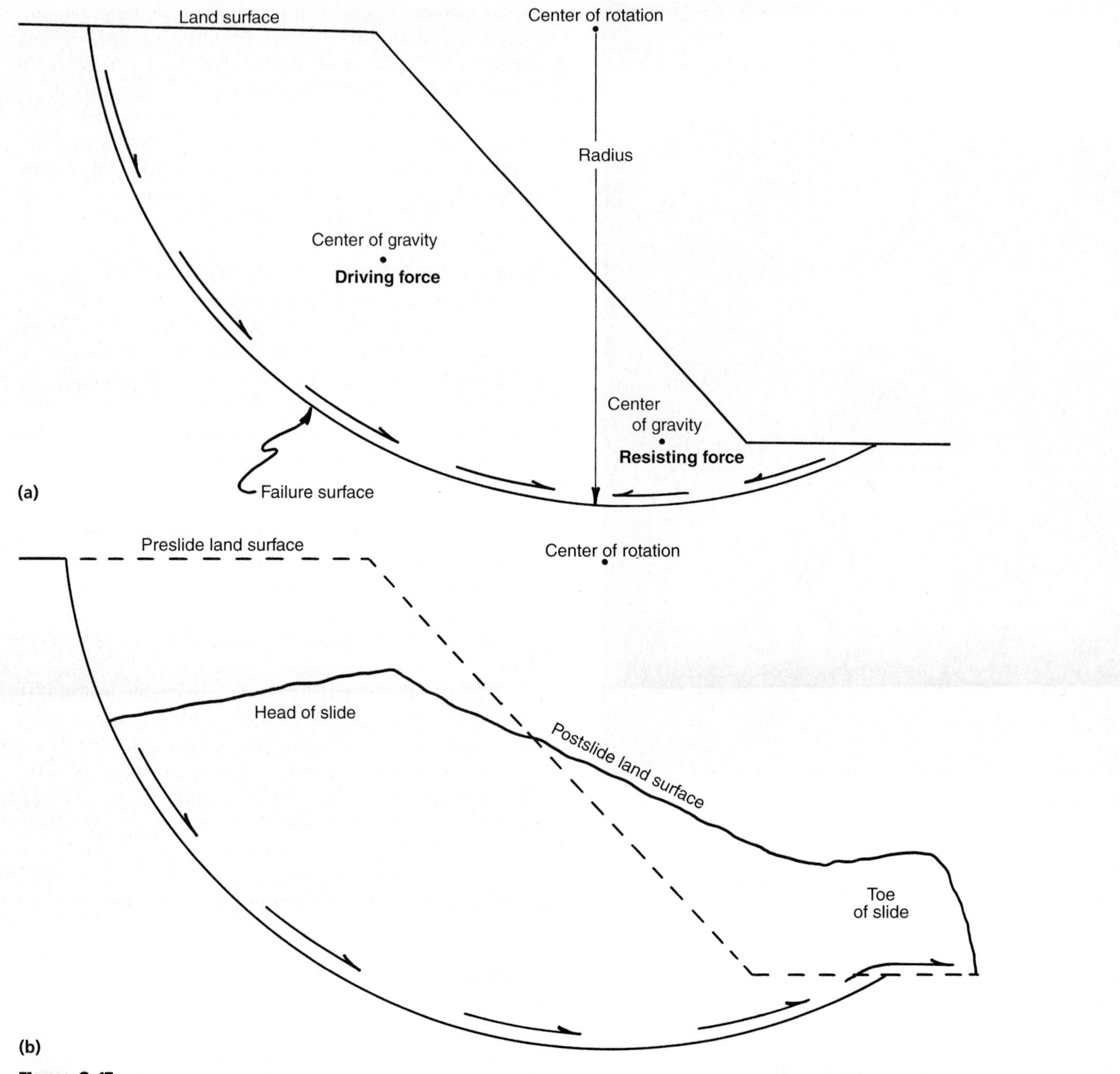

Figure 8.17 **(a)** Swedish Circle analysis of slope stability. A compass is set at the center of rotation, then an arc is swung, approximating the basal failure surface. Computations of driving and resisting forces help determine the stability of the slope. **(b)** A rotational slide with movement around a center of rotation. Notice the backward-tilted head and bulged toe.

Point Fermin, California An excellent example of a block slide lies on the Palos Verdes peninsula in Los Angeles. Just east of Point Fermin, the rocks include layers of sandstone and clay-rich mudstone inclined 10° to 22° seaward (Figure 8.21). Beginning in January 1929, a half-mile-long block slide, with 5 acres of its mass on land, began sliding slowly seaward down the inclined bedding on top of a particularly slippery clay layer. The slide surface daylights on an unsupported submarine slope, thus no resisting mass exists to hold the block in place. The slide block remains basically coherent during movement. From January 1929 to June 1930, the block shifted from 2 to 2.5 m (6.5 to 8 ft) seaward. No one was killed by this slow movement, but homes sitting atop 30 m (100 ft) high sea cliffs were slowly twisted out of shape and had to be removed (Figure 8.22). Movement apparently is triggered by excess water from yard irrigation seeping down to layers of weak clays, which expand and lose strength, and sliding begins. Slide movement rates are not constant; they accelerate after earthquakes.

Figure 8.18 A portion of the head scarp of a 1976 rotational slide north of Ensenada, Baja California. The offset shown is the total of weeks of movement. Two people died in this house when the shifting land ruptured their natural gas lines, asphyxiating them during the night.

Photo by Michael W. Hart.

Figure 8.19 View north across the side of the Ensenada slide. Note the seaward shift of Mexican Highway 1.

Photo by Michael W. Hart.

Figure 8.20 Toe of the Ensenada slide. The ocean floor was lifted above sea level by the up-bulging toe.

Photo by Michael W. Hart.

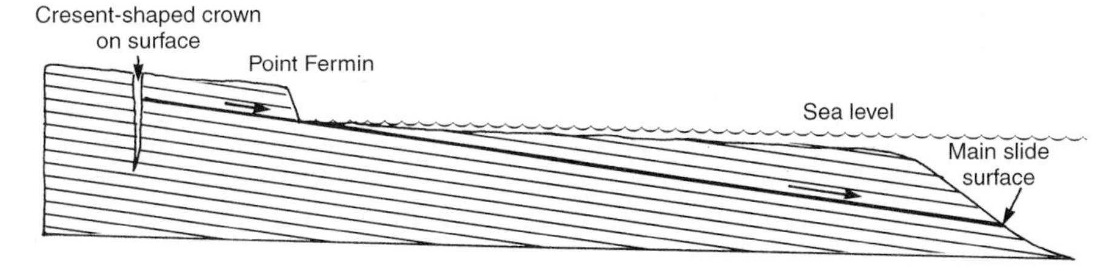

Figure 8.21 Cross section through Point Fermin showing block slide on top of an inclined slippery clay layer. Movement is toward the unsupported slope lying offshore.

Data from W. J. Miller, "The Landslide at Point Fermin, California" in *The Scientific Monthly,* 32:464–69, 1931.

Vaiont, Italy, 1963 A shocking example of the havoc that can be wreaked on inclined surfaces occurred at the Vaiont dam in Italy in 1963 with a massive debris slide. The dam was built in 1960 across a deep mountain valley in northeastern Italy near Austria and Slovenia. The reservoir impounded 150 million m^3 (316,000 acre feet) of water. Several dangerous factors exist at the site. 1) Sedimentary rock layers are folded into a troughlike configuration (**syncline**) with beds

Figure 8.22 View of the head of the Point Fermin translational slide. Note the foundation slabs of destroyed homes. San Pedro is in the background.
Photo by John S. Shelton.

on each side dipping toward the river valley (Figure 8.23). 2) Two sets of fractures split the rocks apart. Fractures formed in the last several thousand years as near-surface rocks rebounded upward due to the removal of overlying glacial ice. When the glacier was gone, the river downcut the canyon into a V-shape, and the canyon walls fractured as they expanded into the newly formed open space of the canyon. 3) Some sliding occurred in the hills long ago, leaving old slide surfaces in the rocks. 4) Some rock layers contain thin seams of weak clays. 5) Limestone in the hills has numerous **caverns** dissolved into it. 6) After dam construction, water filling the reservoir saturated the rocks at the toe of the slopes, causing elevated pore-water pressures. The adverse conditions at the dam site were joined by another: the heavy rains of August-September-October 1963. The rains added a tremendous weight of water into the unstable slopes. It was anticipated that the rains would trigger a landslide, but its size was a deadly surprise.

On the evening of 9 October 1963, most villagers were home during the heavy rains; many were still buzzing with the events of the soccer cup matches. At 10:41 P.M., the south wall of the Vaiont reservoir failed massively and rapidly, beginning the events that would kill nearly 3,000 people in the next seven minutes. The slide mass was 1.8 km (1.1 mi) long and 1.6 km (1 mi) wide, had a volume exceeding 240 million m^3, and filled part of the reservoir with rocky debris at heights up to 150 m (500 ft) above the water level (Figure 8.24). The landslide hit bottom hard, creating an earthquake felt over a wide area of Europe. The mass movements took less than 30 seconds from start to finish, moving at rates of 30 m/sec (68 mph). The slide mass displaced a huge volume of air, blowing outward with enough force to shatter windows and lift roofs from houses.

Most of the death and destruction was caused by the displaced reservoir water. A wall of water climbed over the north side of the dam at 240 m (780 ft) above water level. Subse-

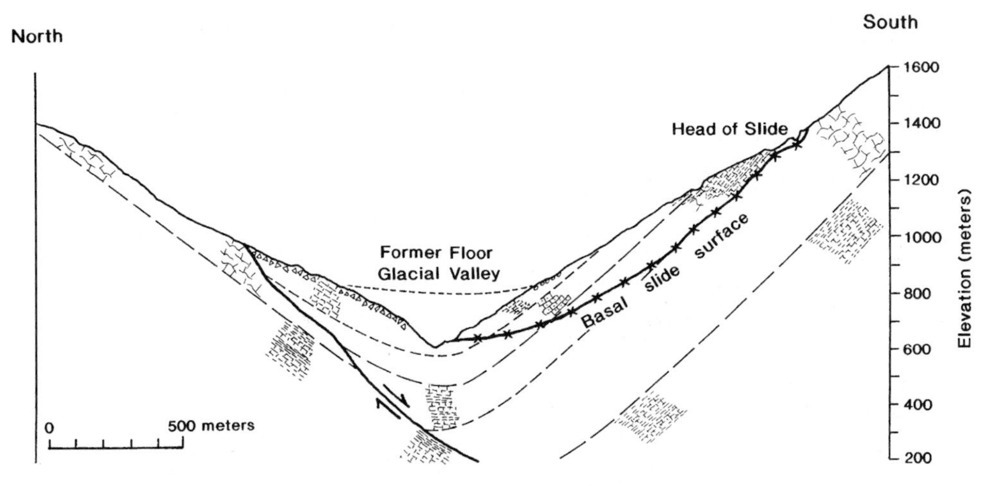

Figure 8.23 North-south cross section across Vaiont reservoir. The slide moved over fracture surfaces parallel to the former glacial valley floor. Location of this cross section shown on Figure 8.24.

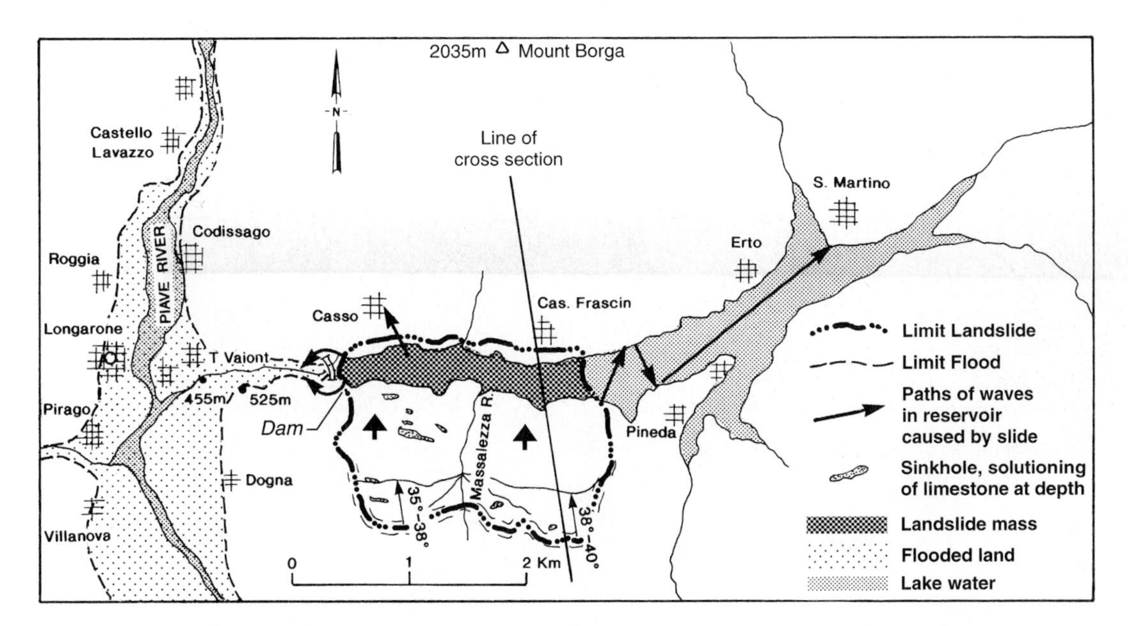

Figure 8.24 Map of Vaiont reservoir area showing positions of slide mass and towns affected by the catastrophic flood.

quent waves poured out of the reservoir, 100 m (328 ft) above the top of the dam. The combined waves emerged from Vaiont Valley as a wall of water over 70 m (230 ft) high, slamming head-on into the town of Longarone (Figure 8.24). For six minutes, Longarone was overwhelmed by raging water. When the waters passed, more than 2,000 residents lay dead.

At the upper end of the reservoir, a wave first hit the north wall of Vaiont Valley, demolishing houses. The wave bounced back to the south, hitting the Pineda peninsula, where it again rebounded. This time, the wave went northeast across the full length of the reservoir, bypassing without harm the village of Erto, before slamming with full force into San Martino (Figure 8.24).

What killed 3,000 people? A tremendous mass of rock moved down an inclined surface as a translational slide and pushed a killing wall of water out of the reservoir. The dam still stands (an engineering success) but the landslide emptied the water (a geologic failure). The event has been called the world's worst dam disaster.

Gros Ventre, Wyoming, 1925 The Gros Ventre River runs south of Yellowstone National Park as a tributary to the Snake River. Virtually the entire valley is a museum for landslides because its sedimentary rock layers are tilted 15 to 21°, daylighting into the valley (Figure 8.25). The dipping rocks are mostly beds of sandstone and limestone with some clay layers. All the rocks have been weakened by weathering, disaggregation, and solution through a long time of exposure.

In 1920, a portion of the valley wall began pulling down and away, as shown by cracks high up the north spur of

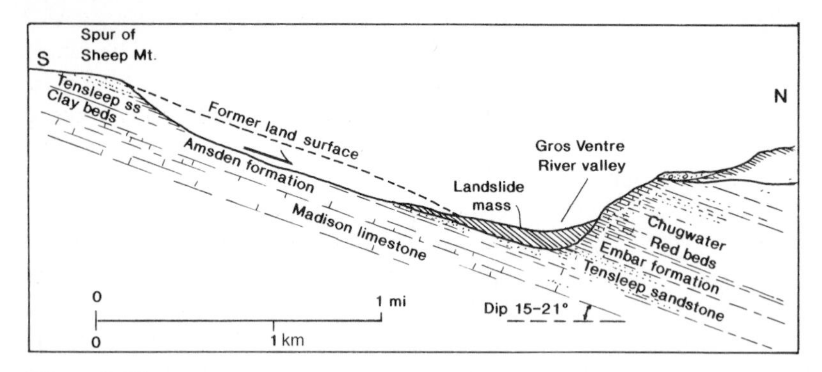

Figure 8.25 Cross section across Gros Ventre River valley showing the 1925 landslide. Notice the sandstone, limestone, and clay beds dipping to the north. Masses of daylighted rock slid translationally into the valley.

Source: W. C. Alden, "Landslide and Flood at Gros Ventre, Wyoming" in *Trans. American Institute of Mining & Metallurgical Engineers,* Vol. 76, 1928.

Sheep Mountain. One resident at the base of the slope, recognizing the event that was slowly unfolding, sold his farm five years before the slide that buried it. After years of weakening, on the afternoon of 23 June 1925, a 38.2 million-m^3 mass suddenly slid into the valley, dropping 640 m (2,100 ft) while carrying a dense pine forest intact. The debris slide ran over 100 m (350 ft) up the north wall of the valley, and then part of the slide turned back, forming a dam across the river up to 75 m (250 ft) high (Figure 8.26). At the time of the slide, the river was in flood. The large slide mass completely blocked the river, which rapidly began filling the newly formed lake basin. In just three weeks, the lake was 60 m (200 ft) deep, 0.8 km (½ mi) wide, 5 km (3 mi) long, and had a surface area of 11,000 acres.

The dam was made of broken pieces of sandstone and limestone, forming a porous barrier that water began seeping through almost immediately. By late summer, the amount of seepage through the dam was greater than the

Figure 8.26 View southwest over the Gros Ventre debris slide above Kelly, Wyoming.

Photo by John S. Shelton on 8 July 1958.

amount of river water flowing into the lake. By spring of 1927, the dam and lake had been stable for nearly two years, and many folks, including some engineers, felt a sense of security about the situation. But the winter of 1926–27 brought heavy snowfall to the Gros Ventre Mountains. In May 1927, heavy rainfall accelerated the melting of snow, causing a rapid rise in lake level. On 18 May, the dam was overtopped by rising lake water. Folks in the down-river town of Kelly (population 65) saw the rising river waters and began evacuating. Most left safely, but seven people, still carrying their precious possessions, drowned when a 5 m (16 ft) high wall of water flowed through, destroying the town. The overtopping lake water cut a channel through the dam about 90 m (300 ft) wide and 30 m (100 ft) deep. The channel allowed the rapid outflow of about 43,000 acre feet of water, dropping the lake level about 15 m (50 ft).

Turnagain Heights, Anchorage, Alaska, 1964 The magnitude 9.2 earthquake (see Chapter 4) triggered numerous mass movements in Alaska. The most destructive slide occurred in the Turnagain Heights residential section of Anchorage as a translational slide of the *lateral spreading* type. Part of Anchorage is built above a rock mass called the Bootlegger Cove Clay; it is composed of sediments that were carried away from glaciers and deposited in estuaries and nearshore marine environments. The Bootlegger Cove Clay has a zone at depth rich in clays of low strength, high water content, and high sensitivity. Above and below this weak zone are much stiffer clay units.

The Turnagain Heights homes sat above sea cliffs 21 m (70 ft) high. The slide moved toward the sea as a 2.6 km (8,500 ft) long mass extending 365 m (1,200 ft) inland. The ground surface on top of the slide broke into an irregular pattern, dropped an average of 11 m (35 ft) and moved 610 m (2,000 ft) toward the bay (Figure 8.27). Sliding did not begin until 90 seconds of violent earth shaking liquefied some weak clays at depth. Mass movement began with rotational slides at the sea cliff and extrusion of weak clays from the toe. Rotational slides cut off the weak clay layer, thus preventing further escape of liquefied clays. After the weak clay layer was sealed off and trapped at depth, the violent shaking caused it to deform internally. A large ridge sitting on top of the clay began moving coastward, causing extension of rock layers behind it (Figure 8.28). Extensional pull-apart set the stage for the widespread dropping and sliding of blocks, and oozing upward of clayey masses.

Figure 8.27 View of head of Turnagain Heights landslide in Anchorage, Alaska, triggered by the 1964 earthquake.
Photo courtesy of *U.S. Geological Survey.*

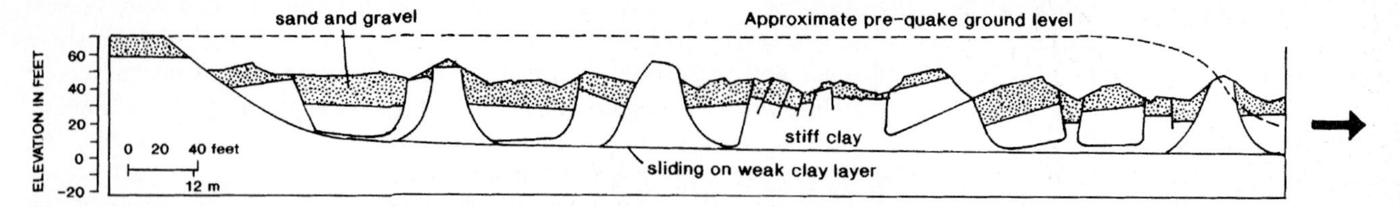

Figure 8.28 Cross section through the Turnagain Heights slide. Lateral spreading on top of weak clays caused masses to break, downdrop, and slide.

Source: *U.S. Geological Survey.*

During the last three minutes of earthquake shaking and for many seconds thereafter, the weak clay layer deformed and flowed causing lateral-spreading sliding. Failure at depth caused rupturing of the firm layer at the surface. When movements stopped, some 75 homes lay ruined.

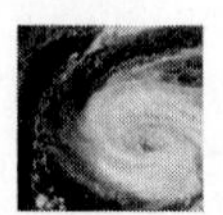

Flows

Flows are mass movements that behave like fluids. The materials of flows range from massive boulders to sand to clay to snow and ice to mixtures of them all. Water contents vary from dry to sloppy wet. Their velocities range from barely moving to observed speeds of 175 mph to calculated speeds in excess of 200 mph. Within the moving masses, internal movements dominate and slip surfaces are absent to short-lived.

There is a complete gradation from the debris slides that move on top of slip surfaces to debris flows that do not require a basal slip surface. Many names have been used to describe flows, e.g., **loess** flow, earthflow, mudflow, **debris flow,** and debris avalanche. Examples will help clarify the different types of flows.

Gansu Province, China, Loess Flow

Loose silt deposits (loess) were a major factor in a 1920 catastrophe in Gansu Province, China. The event was triggered by a large earthquake. The night of 16 December 1920 hosted a bitterly cold wind and dust storm. Around 9:30 P.M., about 30 seconds of ground shaking set in motion the events the Chinese call *shan tso-liao,* meaning "the mountains walked." In a 160 by 275 km (100 by 170 mi) area, hills of dry and weakly stable loess flowed. Entire hills moved hundreds of meters laterally, while weaker hills were shaken apart, creating fluid suspensions of silt in air that flowed down valleys, burying villages and killing 200,000 people. These deadly loess flows were rapidly moving yet dry flows of fine silt grains. Loess masses flow dry, but they move like water with **eddies, vortices,** and swirls.

One anecdote from the event comes from the "Valley of the Dead," where seven major loess flows buried almost every living thing. Yet there were three survivors who began the event on one hill slope, were rafted across the valley on the crest of a loess flow, got caught in a giant whirling vortex created where three loess flows collided, and then were spun up onto the slope of another hill. Their house, orchard, and threshing floor remained enough intact that they simply began farming their new location. Such are the vagaries of fate that they rode out the event on a firm piece of ground while thousands of their neighbors were buried alive.

Portuguese Bend, California, Earthflow

An expensive example of an ancient earthflow that is still moving exists in the Portuguese Bend area of Los Angeles. Some underlying rock layers are rich in volcanic ashes that have altered to **bentonitic** clays especially susceptible to absorbing water and swelling with resultant loss of strength.

In the Portuguese Bend area, the rocks are folded into a broad concave warp (syncline) where the rock layers dip seaward (Figure 8.29). Ocean waves eat away at the toe of the slope, thus removing support and helping keep the slope moving slowly downhill. The site is an ancient earthflow, as shown by the rolling and hummocky topography that drops about 300 m (1,000 ft) to the sea, looking like a rumpled carpet with depressions, arcuate scarps, undrained depressions ("lakes"), uptilted and downdropped blocks, highly fractured material, and crushed plastic clayey materials exhibiting flow structures (Figure 8.30). The area had been used for truck farming—a good usage for unstable ground. There is no great harm in having a plowed field slowly deform.

In the 1950s, a development of 160 fairly expensive houses was built on the former farmland on top of the active slide. The homes were not served by sewers; all fluid wastes were poured below ground via septic tanks. In 1956, about 25 percent of the area (400 acres) began moving again. It moved faster in the winter, up to an inch a day, and slower in the summer. Movement rates correlated with rainfall, with a one- to two-month lag time necessary for rain water to sink below the surface and be absorbed by clay minerals. In three years, parts of the area moved over 20 m (70 ft) seaward (Figure 8.31).

The homeowners were angry. Their beautiful homes with ocean views were breaking apart. With economic loss comes the urge to sue. Who, or what, gets the blame? 1) The seaward tilt of the rock layers; 2) the bentonitic clays that

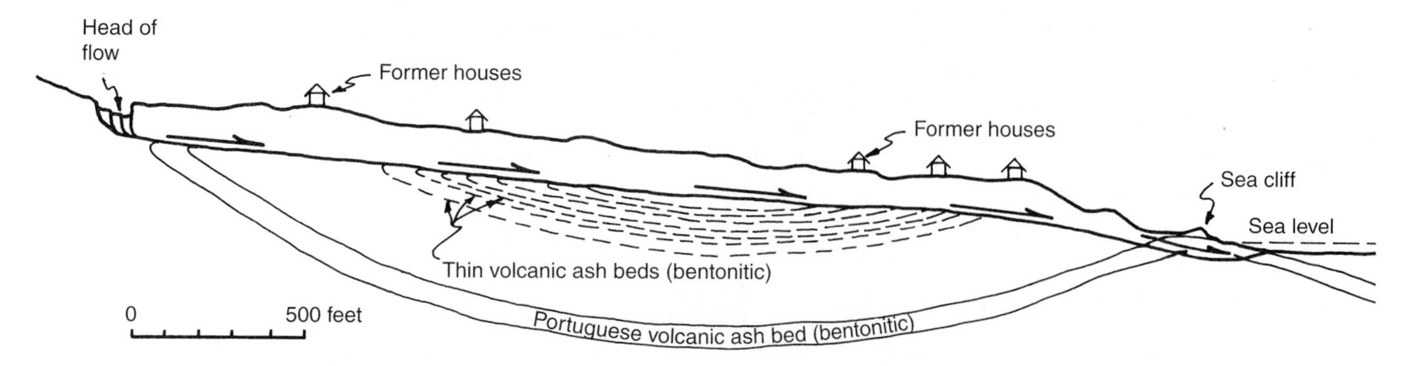

Figure 8.29 Cross section through Portuguese Bend showing seaward-dipping beds, bentonitic clay-rich rock layers, ocean waves eroding base of slope, and preexisting slide surface.

Figure 8.30 View of the Portuguese Bend area on 5 March 1958. Note the kink in the pier. Crenshaw Boulevard winds across upper right of photo.

Photo by John S. Shelton.

absorb so much water and then weaken; 3) the ocean waves that erode the toe of the slope, keeping the ocean waters muddy with material eroded from the resisting mass; or 4) the existence of ancient and still active slide surfaces with water-catching depressions? Each of these features was a good reason not to build there in the first place, but the homeowners were looking for someone with money to sue. One player with deep pockets ended up taking the blame and

Figure 8.31 View of the bulged-up toe of the Portuguese Bend flow on 24 November 1959. Note the remains of houses and roads and the damaged pier.
Photo by John S. Shelton.

paying all the bills—the County of Los Angeles. What was its crime? The county had added some fill dirt to the head of the slope while building Crenshaw Boulevard (see upper right of Figure 8.30). The fill dirt was only one small factor in an overall bad situation, but the county had money, so it was found liable and paid the homeowners' expenses.

Long-Runout Debris Flows

Many mass movements involve combinations of fall, slide, and flow at different times and places along their travel route. Common examples include slumps that change into earthflows, debris slides that become rock falls, and slumps that end as topples. But another complex movement is the most spectacular of them all. These are massive rock falls that convert into highly fluidized, rapidly moving debris flows that travel far and kill in great numbers.

Rock falls and small-volume avalanches tend to flow horizontally for distances less than twice their vertical distance of fall; these short distances of transport are due to the slowing effects of internal and external friction. However, very large rock falls, with volumes in excess of 1 million m^3, commonly travel long distances; some travel up to 25 times farther than their vertical fall. These long-runout flows, called **sturzstroms** (in German, *sturz* means "fall" and *strom* means "stream"), imply lower coefficients of internal friction. Sturzstroms have been observed moving at rates up to 280 km/hr (175 mph), even running up and over sizable hills and ridges lying in their paths.

Blackhawk Event, California A beautiful example of an ancient long-runout debris flow is the Blackhawk sturzstrom (Figure 8.32). This huge rock mass fell from the San Bernardino Mountains 17,000 years ago, flowing out onto the Mojave Desert east of Los Angeles. The Blackhawk sturzstrom has a volume of 300 million m^3 that dropped 1.2 km (0.75 mi) vertically and steeply from Blackhawk Mountain and then flowed 9 km (5.6 mi) to form a lobate tongue of rock debris from 10 to 30 m thick, 2 km wide, and 7 km long. The Blackhawk flow traveled 7.5 times farther than it

Figure 8.32 View of Blackhawk debris flow that fell from the San Bernardino Mountains 17,000 years ago and flowed out onto the Mojave Desert floor.
Photo by John S. Shelton.

fell and is estimated to have moved at about 120 km/hr (75 mph). This type of event has occurred in historic times.

Elm Event, Switzerland, 1881 With the advent of compulsory education in Europe in the nineteenth century, there arose a demand for **slate** boards to write upon in classrooms. To help satisfy this need, some Swiss farmers near Elm became amateur miners, quarrying slate from the base of a nearby mountain. By 1876, an arcuate fissure formed about 360 m (1,180 ft) above their quarry, opening about 1.5 m (5 ft) wide. By early September 1881, the quarry had become a V-shaped notch about 180 m (600 ft) long and dug 60 m (200 ft) into the slope. At this time, the upslope fissure had opened to 30 m (100 ft) wide, and falling rocks were frequent and coupled with ominous noises coming from the large overhanging rock mass. These signs caused the miners to halt work. The inhabitants assumed the rock mass would fall down; but they did not think it would also flow up a steep slope and down their mountain valley to bury 115 of them. But it did.

On 11 September 1881, the Elm event unfolded as a drama in three acts: the fall, the jump, and the surges up a slope and down the nearly flat valley floor (Figure 8.33). Act 1, the fall, was described by Mr. Wyss, the Elm village teacher, from the window of his home:

> When the rock began to fall, the forest on the falling block moved like a herd of galloping sheep; the pines swirled in confusion. Then the whole mass suddenly sank. . . .

Apparently, the formerly rigid mass of rock had already begun to disintegrate during its free fall.

In Act 2, the jump, the fallen mass hit the flat floor of the slate quarry, completely disintegrated, and then rebounded with a big jump forward, also described by teacher Wyss:

> Then I saw the rock mass jump away from the ledge. The lower part of the block was squeezed by the pressure of the rapidly falling upper part, disintegrated, and burst forth into the air. . . . The debris mass shot with unbelievable speed northward toward the hamlet of Untertal and over and above the creek, for I could see the alder forest along the creek under the stream of shooting debris.

The bottom surface of the jumping mass was sharply defined. Eyewitnesses could see trees, houses, cattle, and fleeing people under the flying debris. The upper surface was not so sharply defined; it was a cloud of rocks and dust. Residents of Untertal who saw the jumping mass coming toward them ran uphill to save themselves. This turned out to be the wrong choice, as part of the flying debris hit their hill slope, liquefied, and flowed upslope 100 m, overtaking and burying

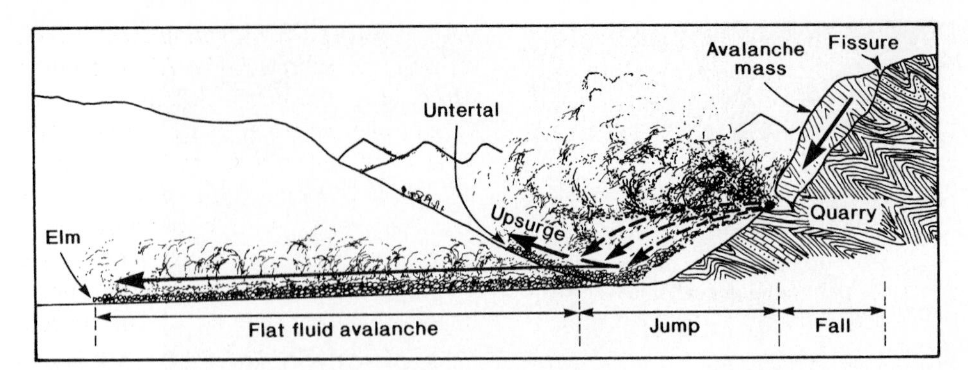

Figure 8.33 Cross section of the 1881 Elm debris avalanche in Switzerland. A drama in three acts: the fall, the jump, and the surges.

them. At the same time, some dogs and cows instinctively moved sideways, thus avoiding the debris flow.

In Act 3, the surges, the disintegrated mass of rock was now fully in contact with the ground and flowed rapidly down the valley floor. Its motion was described by Kaspar Zentner, who barely eluded the flow:

> The debris mass did not jump, did not skip, and it did not fly in the air, but was pushed rapidly along the bottom like a torrential flood. The flow was a little higher at the front than in the rear, having a round and bulgy head, and the mass moved in a wave motion. All the debris within the stream rolled confusedly as if it was boiling, and the whole mass reminded me of boiling corn stew. The smoke and rumble was terrifying. I now ran breathlessly over the bridge and bent around the corner of Rudolf Rhyner's house. Then I turned back and held myself firmly against the house. Just as I went past the corner the whole mass shot right past me at a distance less than 1 meter away. The debris flow must have been at least 4 meters high. A single step had saved me. During the last jump I noticed that small stones were whirling around my legs like leaves in the wind. The house crunched, moved and seemed to be breaking apart. I fled on hands and knees through the garden until I got to the street. I was then safe. I had no pain anywhere and no stones had hit me. I did not feel any particular air pressure.

Although the moving mass at Elm looked and behaved like a "torrential flood" and a "boiling stew," it was not a watery mass but a dry one whose internal fluid was dust and air. Visitors who later viewed the mass of deposited debris remarked how similar its appearance was to a "lava flow." The facts are these: a mass of broken rock with a volume of 10 million m^3 dropped 600 m (1,970 ft) and then flowed 2,230 m (1.4 mi) as a dry mass moving at 180 km/hr (110 mph).

Turtle Mountain, Alberta, Canada, 1903 The coal-mining town of Frank, Alberta, occupied a beautiful site in the Oldman River valley and Canadian Rocky Mountains. Its place in history changed on 29 April 1903 when residents were startled at 4:10 A.M. by the noise coming down from Turtle Mountain as a 90-million-ton mass of dipping limestone layers slid down their basal surface (daylighted bedding), dropped 3,000 ft (1 km) into the river valley, shattered, then flowed 2 mi (3 km) across the valley and climbed 400 ft (130 m) up and over the terraced valley wall on the opposite side (Figure 8.34). The whole event lasted only 100 seconds but it pulverized and buried the southern end of town, killing about 70 people. Eyewitnesses include the engineer of a train backing up to the coal mine when he heard the rock breaking high up on Turtle Mountain. He quickly switched to full speed ahead and chugged to safety as he watched miners at the loading dock sprinting for their lives be overrun and killed by the rock flow. The site is now an organized tourist attraction with the Frank Slide Interpretive Center, audiovisual program, and self-guided hiking trails.

Movement of Highly Fluidized Rock Flows (Sturzstroms) How do such large masses of debris move so far, so fast, and behave so much like a fluid? Numerous hypotheses have been proposed. 1) Some rely on water to provide lubrication and fluidlike flow, but some observed flows are masses of dry debris, as at Elm. 2) Other hypotheses invoke the generation of steam to liquefy and fluidize the moving mass, or 3) they call for frictional melting of material within the moving mass; both ideas fail because some long-runout deposits contain blocks of ice or lichen-encrusted boulders showing that no great amount of internal heat was generated nor was internal friction ever at a very high level. 4) A popular hypothesis suggests the falling mass traps a volume of air beneath it and then rides partially supported on a carpet of trapped air that enables it to travel far and fast. This idea has never been verified and seems most unlikely. For example, after its early airborne jump, the Elm sturzstrom was described as being in contact with the ground, and in fact, it dug up pipes buried a meter below the surface. Additional problems for the air-cushioned flow hypothesis are deposits with identical flow features on the ocean floor, and on the Moon and Mars where no or very little atmosphere is available.

So it appears that neither water, heat, nor a trapped cushion of air is necessary for a long-runout debris flow. So how do these large masses do it? A remarkable fact helps guide the formulation of another hypothesis. After the Elm mass fell, disintegrated, jumped, and flowed 2.23 km and the rubble had come to rest, the original layering in the bedrock of the mountainside was still recognizable. Even though the debris had flowed according to all eyewitness accounts, the hunks of debris stayed in their same relative position. In the words of German geologist Albert Heim, who studied the scene in 1881:

> When a large mass, broken into thousands of pieces, falls at the same time along the same course, the debris has to flow as

Figure 8.34 A 90-million-ton mass of rock fell, shattered, and flowed down Turtle Mountain, across the Oldman River valley, and up the opposite wall of the valley on 29 April 1903. About 70 people were killed in the town of Frank, Alberta, Canada.

Photo courtesy of NOAA.

> a single stream. The uppermost block, at the very rear of the stream, would attempt to get ahead. It hurries but strikes the block, which is in the way, slightly ahead. The kinetic energy, of which the first block has more than the second, is thus transmitted through impact. In this way the uppermost block cannot overtake the lower block and thus has to stay behind. This process is repeated a thousandfold, resulting eventually in the preservation of the original order in the debris stream. This does not mean that the energy of falling blocks from originally higher positions is lost; rather the energy is transmitted through impact. The whole body of the mass is full of kinetic energy, to which each single stone contributes its part. No stone is free to work in any other way.

Who would guess that the pieces of rubble in a rapidly moving debris stream would keep their relative positions next to their neighbors? How is this relationship to be explained? A provocative hypothesis involving acoustic energy within the moving mass has been proposed by the American geophysicist Jay Melosh. Apparently, an immense volume of falling debris produces much vibrational or acoustical energy within its mass. The jostling and bumping back and forth of fragments produces acoustic (sound) energy that propagates as internal waves. The trapped acoustic waves may act to fluidize the rock debris, allowing the rapid fall velocity to continue as rapid flow velocity in a process called **acoustic fluidization.**

Nevados Huascarán Events, Peru Nevados Huascarán is the highest peak in the Peruvian Andes. The west face of the north peak is granitic rock cut by nearly vertical joints roughly parallel to the face. At 6:13 P.M. on 10 January 1962, with no perceptible triggering event, a huge mass of rock and glacial ice fell, initiating a debris flow. The debris flowed down river valleys like bobsleds, rising higher on the outsides of valley bends and lower on the inside bends. Some debris flowed up and out of the valley at bends, but most stayed between the valley walls and issued forth as a 10 to 15 m (30 to 50 ft) high mass, spreading out as a lobe covering 3.5 km^2, including part of the town of Ranrahirca (meaning "hill of many stones"); 4,000 people died. The debris flow had a volume of 13 million m^3 and traveled at speeds of 170 km/hr (105 mph). The slide left a scar on Nevados Huascarán, including a 1 km high overhanging cliff. A final report on the tragedy stated:

> The people are adjusting to this huge scar that lies across their land and their lives. . . . But they say that Huascarán is a villain who may yet have more to say.

On 31 May 1970, a subduction-zone earthquake of magnitude 7.7 occurred beneath the Pacific Ocean 135 km (84 mi) away, with a hypocenter at 54 km (33 mi) depth. Shaking lasted 45 seconds, but before that time was up, a gigantic portion of the same west-facing slope of the north peak of Nevados Huascarán failed with a sound like a dynamite blast or sonic boom, and a cloud of dark dust obscured the mountain from view. The side of the mountain between 5,500 to 6,400 m (18,000 to 21,000 ft) elevation fell away, including a 30 m (100 ft) thick glacier. The mass was nearly 100 million m^3 of granitic rock, ice blocks, glacial sediments, and water. It moved at speeds of 280 to 335 km/hr (175 to 210 mph), devastating an area of 22.5 km^2 and killing 18,000 people (Figure 8.35).

The event began as a fall, transformed into a debris avalanche with an airborne segment, and then moved down the Rio Santa as a debris flow for over 50 km (30 mi). The sequence was:

1. A vertical fall for 400 to 900 m (1,300 to 3,000 ft) (Figure 8.36).
2. The fallen mass landed on a glacier and slid along its surface, scooping up snow.
3. The debris avalanche raced up the side of a glacial-sediment hill, launching much debris into the air.
4. For the next 4 km (2.5 mi) downslope, boulders with weights up to several tons each rained from the sky, pulverizing houses, people, and other animals. The deadly rain of megaboulders left the ground pockmarked with craters like a heavily bombed battlefield.
5. The mass recombined as a flow, reaching the 230 m (750 ft) high Cerro de Aira (*cerro* is Spanish for hill),

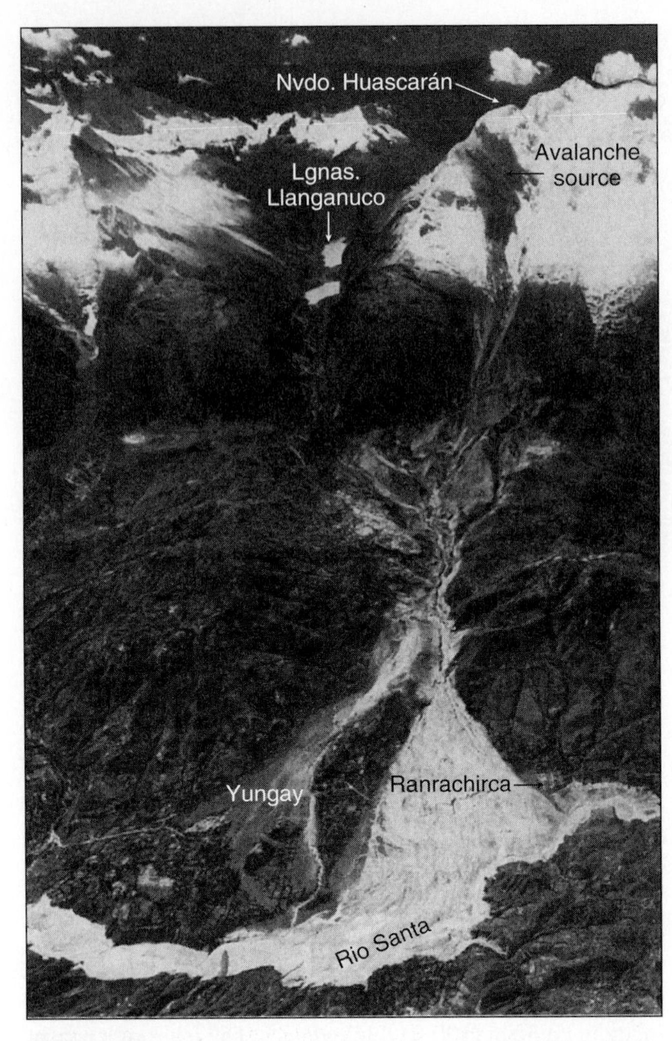

Figure 8.35 Aerial view of Nevados Huascarán and the 1970 debris avalanche that buried 18,000 people. Vertical drop from summit to Rio Santa is 4,144 m (13,592 ft), and horizontal distance is 16 km (10 mi).

Photo courtesy of the *U.S. Geological Survey.*

which had protected the city of Yungay during the 1962 event. No such luck this time. A lobe of debris overflowed the Cerro, burying Yungay and 18,000 people beneath more than 30 m (100 ft) of debris. It was especially bad timing because this was a Sunday afternoon and the population was swollen with visitors from the surrounding region. The scene was well described by survivors who witnessed the event from Yungay's highest point and the safest place in town, the cemetery.

6. Meanwhile, the main mass of material continued racing toward the Rio Santa, preceded by a strong wind pushed in front of it. The main lobe of debris swept across the Rio Santa and ran 83 m (275 ft) up the far slope, killing 60 people in the town of Matacoto, before it fell back like an ocean wave from the shore.

What does the future hold for this site? Today, the glacier above the 1962 and 1970 break-away sites has large fissures, suggesting the mountainside beneath is still fractured and unstable. The odds are high that the repopulated areas downslope will again experience another killer debris flow.

Snow Avalanches

Heavy snowfalls on steep slopes yield to the pull of gravity and fail as snow avalanches (Figure 8.37). They may be understood using the same mass-movement principles for earth and rock. Just like earthen mass movements, snow avalanches creep, fall, slide, and flow. In effect, the snowfall and wind-blown accumulations high on slopes act to "load the head" of the slope, thus setting the stage for failure. Avalanche volumes vary from small to very large. They move at rates ranging from barely advancing to measured speeds of 370 km/hr (230 mph) in Japan. Their travel distance varies from only a few meters to several kilometers.

An avalanche of about 765,000 m^3 was unleashed down the slopes of Mount Sanford in Alaska on 12 April 1981; it

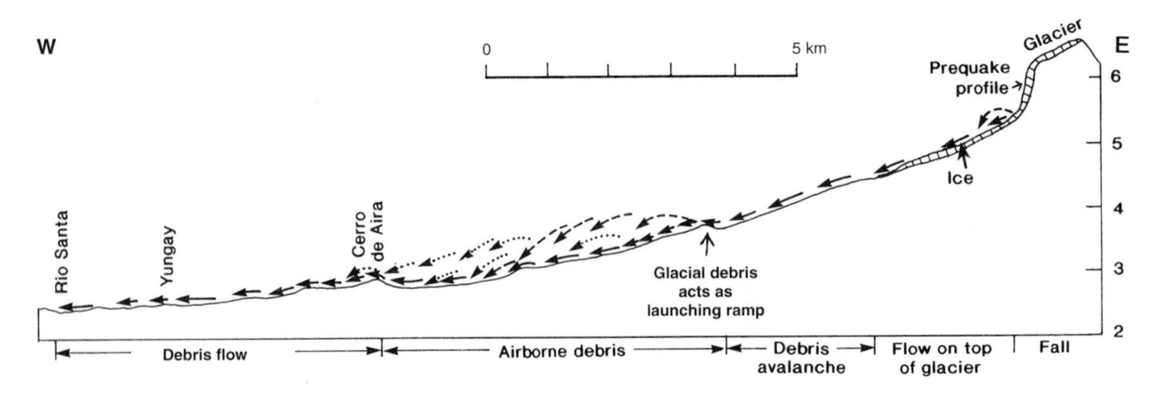

Figure 8.36 Cross section of the 1970 Nevados Huascarán debris avalanche. Key elements were the initial vertical fall, air launching over glacial-debris hills, splitting of the avalanche into two lobes, and flow as wet debris: debris avalanche (solid arrows), airborne rocks (dashed arrows), and airborne mud (dotted arrows).

Source: *U.S. Geological Survey.*

Figure 8.37 Downhill skier triggers a snow avalanche. Photo courtesy of the U.S. Forest Service/U.S. Geological Survey photo library.

dropped over 3 km (2 mi) and flowed for 13 km (8 mi), including running up and over a 900 m (3,000 ft) high ridge.

The styles of avalanches vary. Loose, powdery snow has a low amount of cohesion. This weak material may have 95 percent of its volume as pore space. It typically fails at a point source, then triggers more and more snow into moving during its downhill run. Avalanches begin where slopes are steepest (usually 30° to 45°), then move down slopes of 20° to 30°, commonly guided by topography, and finally come to a halt in the runout zone (slopes usually less than 20°).

Larger avalanches usually involve the breaking free of a slab of cohesive snow from its poorly anchored base. The failures are analogous to translational slides, where an upper mass breaks free and slides down and out on top of a layer beneath it. The slides typically turn into flows during their downslope movement. The key factor in understanding slab avalanches is recognizing that the snow mass on a mountain is made of many different layers of snow that formed at different times. Each layer of snow has its own characteristics of thickness, strength, hardness, and density. Within the pile of snow layers are melt-freeze crusts that form during times of surface exposure between snowfalls. The result is an inhomogeneous pile of separate and distinct layers that contain numerous potential failure surfaces within it.

In today's world, the major effect of avalanches is felt by people during recreational skiing, climbing, snowmobiling, and hiking. Average annual death figures for some countries are: 35 in Austria, 30 in France, 30 in Japan, 26 in Switzerland, 25 in Italy, 14 in the United States, and 6 in Canada. A "typical" avalanche may be 0.6 to 0.9 m (2 to 3 ft) deep at its head scarp, 30 to 60 m (100 to 200 ft) wide, and may drop 90 to 150 m (300 to 500 ft). If the snow is dry, avalanche speeds will be about 65 to 100 km/hr (40 to 60 mph), and if the snow is wet, speeds will be about 30 to 65 km/hr (20 to 40 mph).

An example of a larger event occurred at 3:45 P.M. on 31 March 1982 at Alpine Meadows in California. High on a slope, cracks developed about 3 m (10 ft) deep and 900 m (3,000 ft) long. The resulting avalanche of 65,000 tons of snow moved as a 10 m (30 ft) high mass traveling 130 km/hr (80 mph). The avalanche rammed into the ski resort, crushing the ski lift and buildings and killing seven people. Being in the snow-covered mountains is great sport, but while there, thinking about mass-movement principles could be worthwhile.

Submarine Mass Movements

A whole range of familiar mass movements takes place below the sea. There are rotational slumps within the sediment masses (**deltas**) deposited at the mouths of large rivers; complex failures occur within the distorted rock masses at subduction zones; and debris avalanches slide down the slopes of submarine volcanoes. The largest submarine mass movements were first recognized on the sea floor along the Hawaiian Islands volcanic chain. Slump and debris-avalanche deposits there cover more than five times the land area of the islands (Figure 8.38). Some individual debris avalanches are over 200 km (125 mi) long with volumes

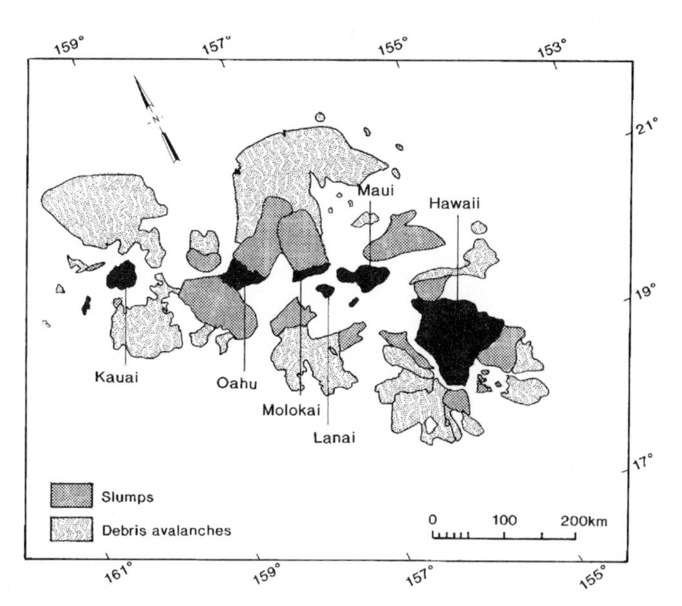

Figure 8.38 The Hawaiian Islands (black) cover less surface area than the slumps and debris avalanches that have fallen from them.

greater than 5,000 km³ (1,120 mi³), making them some of the largest on Earth. These events are not just loose debris sliding down the side of the volcano; they are catastrophic **flank collapses** where the whole side of an oceanic volcano breaks off and falls into the sea. There have been 70 flank collapses from the Hawaiian Islands in the last 20 million years.

Pause and think about this a moment—each Hawaiian Island has major structural weaknesses that lead to massive failures. The not-so-solid earth here betrays us; it can fail rapidly and massively. For example, the island of Molokai has no volcano. Where did it go? Apparently the northern part of the island fell into the ocean (Figure 8.38) leaving steep cliffs behind.

What happens to the ocean when a gigantic chunk of island drops into it and flows rapidly underwater? Tsunami. Prehistoric giant waves washed coral, marine shells, and basaltic gravels inland where they are found today as gravel layers on Lanai lying 365 m (1,120 ft) above sea level, and on Molokai over 2 km inland and >60 m (200 ft) above sea level. Tsunami of this size would not only ravage Hawaii but would cause death and destruction throughout the Pacific Ocean basin.

Where might an event like this happen next? Quite likely on the active volcano Kilauea on the southeast side of the island of Hawaii (Figure 8.39). The zones of normal faults associated with magma injection in the active volcano Kilauea also appear to be head scarps of giant mass move-

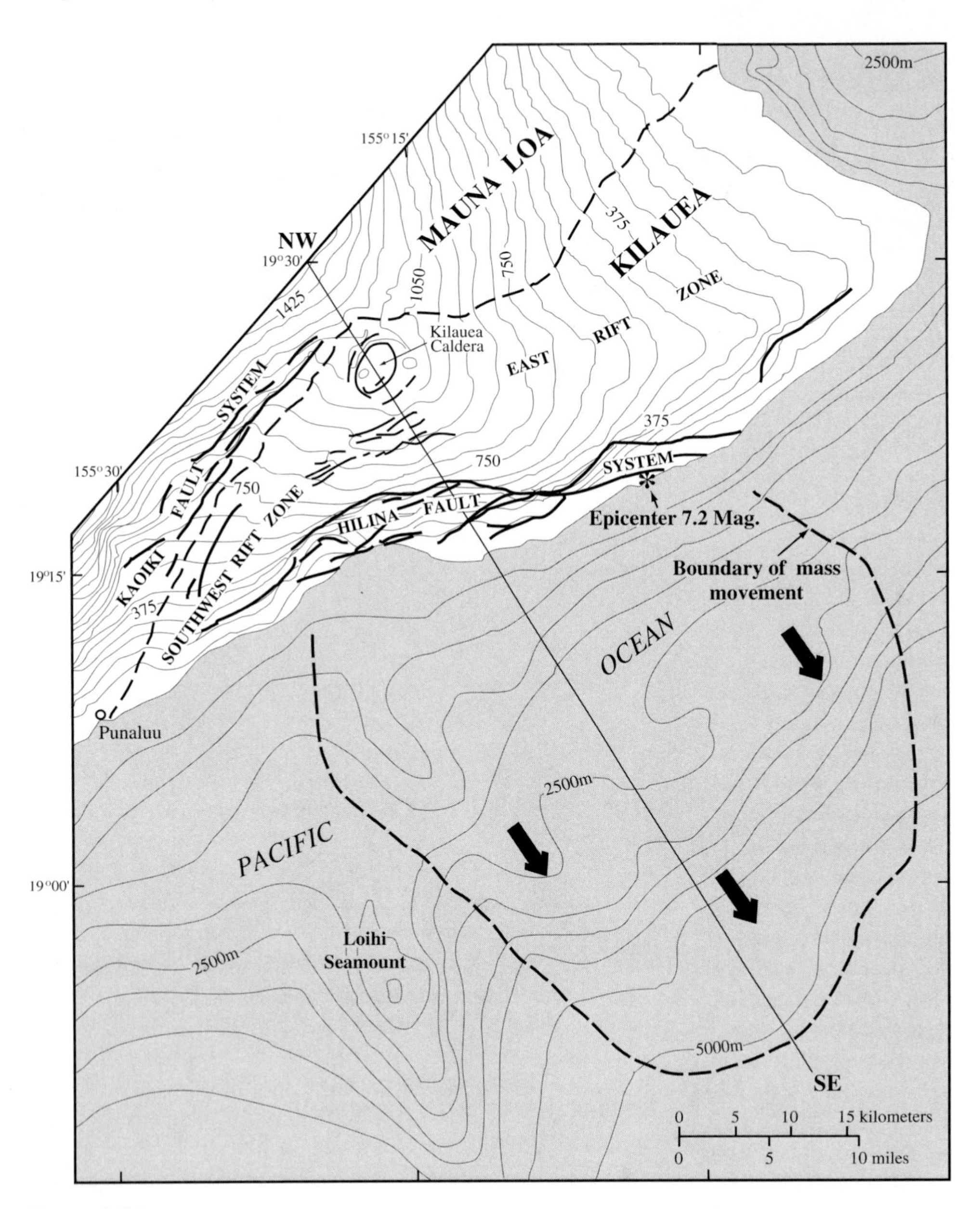

Figure 8.39 Map of Kilauea area of Hawaii showing outline of mass sliding into the Pacific Ocean.

Data Source: Lipman, P. W. et al., *U.S.G.S. Professional Paper 1276,* "Ground deformation associated with the 1975 earthquake", U.S. Geological Survey, 1985.

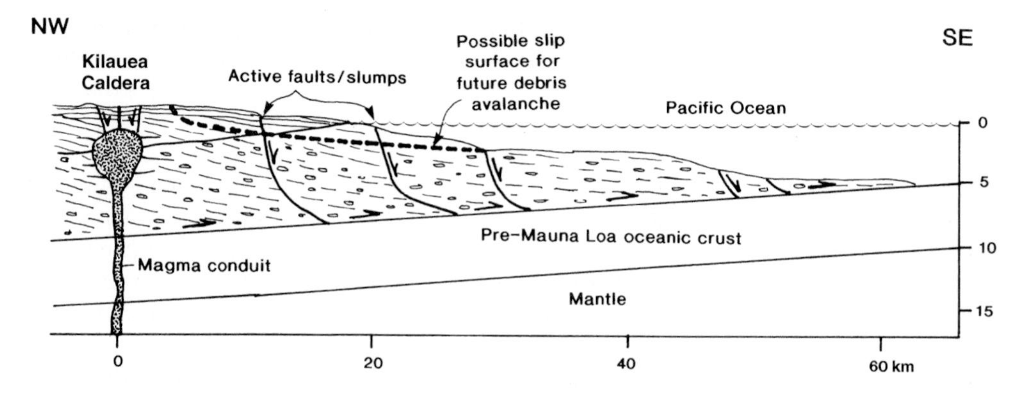

Figure 8.40 Cross section through southeastern part of the big island of Hawaii. Massive blocks are currently slumping on normal faults. Future debris avalanches may dump multisquare-mile hunks of land into the Pacific Ocean.
From James G. Moore, *U.S. Geological Survey.*

ments (Figures 5.27 and 8.40). One moving mass on southeastern Kilauea is more than 5,000 km^3 in volume and is sliding at rates up to 25 cm/yr (10 in/yr). The slide mass extends offshore about 60 km (37 mi) to depths of about 5 km (3 mi) (Figure 8.40). The area on the moving mass includes the 80 km (50 mi) long coastal area southeast of Kilauea.

During early November 2000, the Global Positioning System (GPS) measured this large block moving at an accelerated rate that peaked at 6 cm (2.3 in) per day. The days-long movement was a *silent earthquake* with an equivalent moment magnitude of 5.7. It is possible that this large hunk of Hawaii, with its more than 10,000 residents, could be plunged into the sea following a large earthquake or during a major movement of magma.

Flank collapses are not unique to Hawaii; they occur around the world. As oceanic volcanoes grow, they tend to develop sides that are too steep and thus they collapse in gigantic events. Worldwide, approximately one flank collapse occurs every 10,000 years.

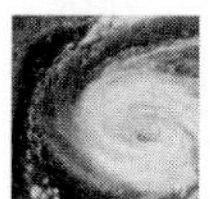

Subsidence

In subsidence, the ground moves down (Figure 8.15). The surface either sags gently or drops catastrophically as voids in rocks close. This is not the downdropping associated with tectonic plates, fault movements, or volcanism, but rather is either the slow compaction of loose, water-saturated sediments or the rapid collapse into caves.

Slow Subsidence

In many areas of the world, the ground surface is slowly sinking as fluids are removed below the surface (Table 8.4). When water or oil are squeezed out or pumped up to the surface, the removal of fluid volume and the decrease in pore-fluid pressure cause rock grains to be crowded closer together; this results in subsidence of the ground surface (Figure 8.41).

Table 8.4 Some Subsiding Coastal Cities

City	Maximum Subsidence (meters)	Area Affected (km^2)
Bangkok, Thailand	1	800
Houston, Texas	2.7	12,000
London, England	0.3	300
Long Beach, California	9	50
Mexico City, Mexico	10	3,000
Nagoya, Japan	2.4	1,300
New Orleans, Louisiana	3	175
Niigata, Japan	2.5	8,300
Osaka, Japan	3	500
San Jose, California	3.9	800
Savannah, Georgia	0.2	35
Shanghai, China	2.7	120
Taipei, Taiwan	1.9	130
Tokyo, Japan	4.5	3,000
Venice, Italy	0.3	150

The fluids within rocks help support the weight of overlying rock layers. The effect is similar to that of carrying a friend in a swimming pool where the water helps support your friend's body weight. Some examples will help illustrate the subsidence problem.

Delta Compaction, Mississippi River, Louisiana Deltas are popular sites for cities. Many resources exist side by side at deltas: fertile soils, freshwater for drinking and farming, marine foods, tempered climate, and trade and transportation possibilities where rivers from continental interiors meet the world ocean. The problem is that deltas are only loose piles of water-saturated sand and mud, and thus they compact and

(a) Sand grains (dark) and pore spaces (white) filled with air, water, or oil

(b) Clay minerals (dark) and pore spaces (white)

Figure 8.41 **(a)** Loosely packed sand has high porosity (25 to 45 percent). When grains are pressed closer together and fluids are removed, the ground surface subsides. **(b)** Loosely packed clay minerals have even higher porosity (30 to 80 percent). When clay flakes are pressed closer together and fluids are removed, the ground surface subsides.

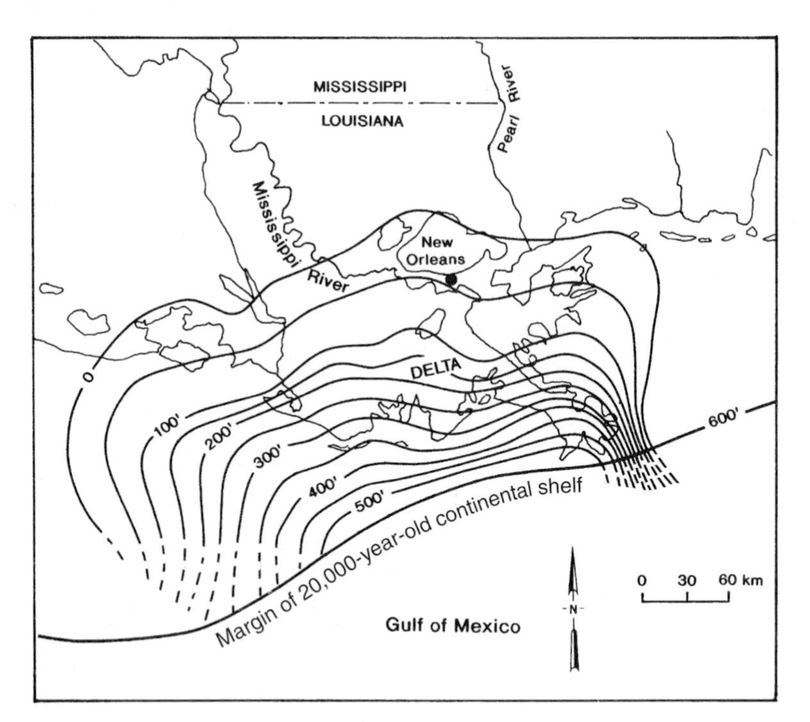

Figure 8.42 Subsidence in the Mississippi River delta area during the last 20,000 years. Contour lines are in feet.
After H. Gould, 1970.

sink. Each new layer of sediment adds weight on top of older layers, forcing the grains of sediment closer together and squeezing out more of their contained water (Figure 8.41). The Mississippi River delta at the shoreline is underlain by a 6 km (3.7 mi) thick pile of sediments deposited during the last 20 million years. In the last 20,000 years, the region has sunk markedly due to sediment compaction, dewatering, and isostatic adjustment (Figure 8.42).

New Orleans slowly subsides with the delta. The sinking of the city is accentuated in places by pumping up underground water. Parts of New Orleans have sunk about 3 m (10 ft) during the last 50 years. Today, about 45 percent of New Orleans lies below sea level, thus making the city evermore prone to damages from the high-water surges associated with hurricanes. Many other coastal cities also are subsiding due to delta compaction and groundwater withdrawal.

Oil Withdrawal, Houston-Galveston Region, Texas One of the earliest cases of fluid withdrawal causing ground subsidence occurred around the Goose Creek oil field between Houston and Galveston. Beginning in 1917, pumping brought large volumes of oil, natural gas, and associated water to the surface. As a result, the ground surface subsided enough for seawater in San Jacinto Bay to submerge the area. Because states own the mineral rights on submerged lands, the state of Texas sued to take over the oil field that now lay below seawater. Ultimately, the state lost its case.

Houston-Galveston is one of the largest metropolitan areas in the country. It has relied heavily on groundwater withdrawals to supply its needs. Thus, an area greater than 12,000 km^2 has sunk up to 2.7 m (9 ft). Another significant result of the ground settling is renewed movements on old faults (Figure 8.43). These faults act as mega-landslide surfaces for the thick pile of coastal sediments to slowly slide toward the Gulf of Mexico. The faults are active but do not produce earthquakes. However, they do break up houses, commercial buildings, roads, and other features.

Groundwater Withdrawal, Mexico City The Valley of Mexico has been a major population center for many centuries. People need water so the Aztecs, and later the Spanish, built aqueduct systems to bring water in from the surrounding mountains. In 1846, it became well recognized that a large volume of groundwater lay beneath the city. The convenience of abundant fresh water lying underfoot led to drilling wells to make large withdrawals of underground water. But the water has been withdrawn faster than the natural rate of replenishment, so the land subsides. In the center of the city the land sank about 30 feet (10 m) between 1846 to 1954. The city center now lies lower than the level of nearby Lake Texcoco.

How can land subsidence be stopped? Stop pumping out groundwater. Can the land subsidence in Mexico City be reversed? No. Groundwater withdrawal is now banned in the city center and has been moved to new wells in the north and south of the valley. Land subsidence of 1 to 3 inches (2 to 8

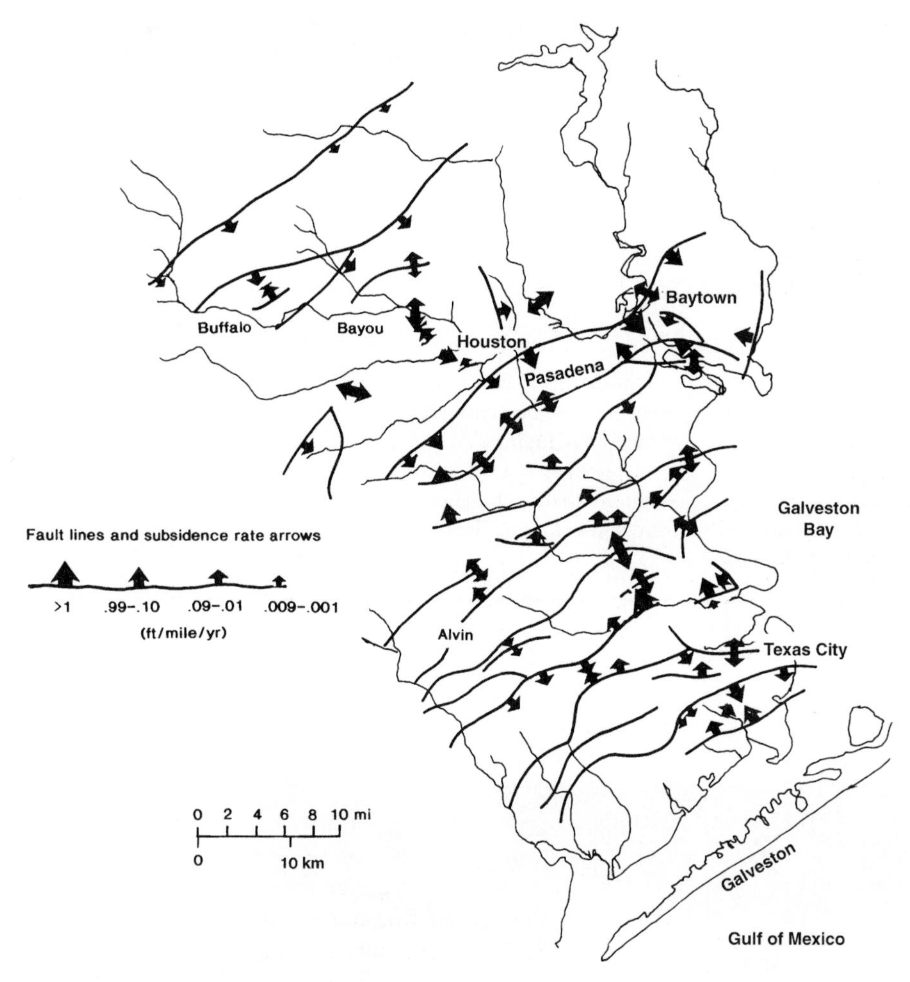

Figure 8.43 Map of some active faults in the Houston, Texas, area and relative amounts of subsidence.

cm) a year is now occurring in the new areas of groundwater withdrawal. Supplying the water needs of 20 million people is not easy in an area where the evaporation rate is greater than the precipitation rate.

Catastrophic Subsidence

Limestone Sinkholes, Southeastern United States Limestone is a different kind of rock. Most is formed in warm, shallow seas by the accumulation and disintegration of shells and skeletons of organisms that remove calcium (Ca), carbon (C), and oxygen (O) from seawater to make their skeletal material of calcium carbonate ($CaCO_3$). Some organisms build limestone directly as reefs. Other organisms die and have their shells, spines, and other mineralized remains bound together by $CaCO_3$ precipitated in pores as cement.

Limestone formed extensively in the geologic past when the central and southeastern United States were flooded by shallow, warm ocean waters. Today, the limestones in these regions have acidic freshwater flowing through them, dissolving them and forming extensive networks of caverns. When the levels of underground water drop during a drought or due to pumping of groundwater, the removal of the water lessens the internal support that helps hold up the roofs of caves. The loss of buoyant support that occurs when groundwater is drained from caves weakens some so much that their roofs collapse suddenly and catastrophically to form **sinkholes.** Every year, the print and electronic media show us examples of these sudden collapses (Figure 8.44). In Alabama alone, an estimated 4,000 sinkholes have formed since 1900.

Much of central Florida is underlain by limestone that is covered in most areas by 15 to 30 m (50 to 100 ft) of muddy sands. When the underground water body is lowered due to drought reducing the water supply or by humans making excessive withdrawals by overpumping wells, then caverns in limestone may be drained of water. Remove the water and the weakened cavern may collapse. On 8 May 1981, a small depression on the ground in Winter Park, Florida, grew to a 45 m (150 ft) diameter collapsed pit (sinkhole) within 15 hours. Before a week was up, the sinkhole was 100 m (325 ft) across and 34 m (110 ft) deep (Figure 8.45). The collapsing cavern claimed one house, several Porsches from a repair shop, and the deep end of the municipal swimming pool.

Sidebar

How to Create a Cave

Caves usually occur in limestone. The same equation that describes the formation of limestone also describes its dissolution into caves. The basic equilibrium equation is:

$$Ca^{++} + 2HCO_3^- \rightleftharpoons CaCO_3 + H_2CO_3$$

where Ca is calcium ion, HCO_3 is bicarbonate ion, $CaCO_3$ is limestone, and H_2CO_3 is **carbonic acid.** When the equation runs from left to right, limestone is precipitated. When the equation runs from right to left, limestone is dissolved. The primary variable controlling whether limestone is precipitated or dissolved is the amount of carbonic acid present. And the main variable controlling the concentration of carbonic acid is the amount of carbon dioxide (CO_2) in solution:

$$H_2O + CO_2 \rightleftharpoons H_2CO_3$$

If dissolved carbon dioxide content is high, then the water is rich in carbonic acid and limestone is dissolved. This also describes our familiar bottles of soda pop, where flavored sugar water holds large quantities of CO_2 in solution under pressure. Carbon dioxide contents are highest when water is under pressure and cold; i.e., the gas bubbles are held in solution.

Figure 8.44 View of the "December giant" sinkhole formed by roof collapse of a limestone cavern on 2 December 1972, Shelby County, Alabama. Sinkhole is 130 m (425 ft) long and 46 m (150 ft) deep.

Photo courtesy of the *U.S. Geological Survey.*

Figure 8.45 Aerial view of Winter Park, Florida, sinkhole on 10 May 1981. Sinkhole is 100 m (>320 ft) wide and 34 m (110 ft) deep. Note the failed municipal swimming pool in lower left.
Photo by UPI/George Remaine.

Summary

Gravity tugs incessantly at all landforms on Earth, commonly causing failures called mass movements. They range from the barely perceptible surface creep of hill slopes to debris flows moving in excess of 200 mph. The movement of surface materials sets the stage for creep. Soil and rock swell upward perpendicular to the surface due to freezing of pore water, wetting of clays, and volume expansion upon solar heating. They shrink straight downward under the pull of gravity upon thawing, drying, and cooling, resulting in a net movement downslope.

Slopes approach failure due to external factors, such as steepening of the slope, adding mass upslope, or removing mass low on a slope. Internal factors that make a slope weak include inherently weak materials such as clay minerals, decreasing cohesion through solution or internal erosion, weight of pore water and elevated pore-water pressure, and adverse geologic structures, such as inclined bedding, fault surfaces, and clay seams. Many factors push a slope toward failure. Movement usually is initiated by a triggering event, such as heavy rains, an earthquake, or human activities.

Major types of mass movement are downward, as in falling or subsiding, or downward and outward, as in sliding and flowing. Falls are rock masses dislodged from elevated slopes; if big enough, they may transform downslope into slides or flows.

Slides are mass movements on top of failure surfaces. Concave-upward, curved failure surfaces produce rotational slides. Here, the slide head tilts backward, the toe bulges upward, and little distance is traveled. Rotational slides are a common destroyer of property, but their slow movements rarely kill anyone. Slides on top of inclined, planar surfaces are called translational; they can travel far and fast. Translational masses may remain essentially coherent as block slides that destroy property, as at Point Fermin, California; they may disintegrate into debris slides killing thousands of people, as at Vaiont, Italy; or they may spread laterally at the surface when an underlying weak layer of rock deforms as at Anchorage, Alaska.

Flows are mass movements that behave like fluids, even when they are dry. Flows may be made of gravels, sands, muds, snow, ice, or mixtures of materials. Many names describe them, including loess flow, earthflow, mudflow, debris flow, avalanche, and sturzstrom. Loess flows in China occur when earthquakes shake loose silt into suspension in air; these flows kill hundreds of thousands of people in single events. Earthflows destroy many buildings, as in the Portuguese Bend area of Los Angeles. Sturzstroms are flows that run out especially long distances, up to 25 times longer than their vertical drop. The high fluidity of sturzstroms may be due to acoustic or vibrational energy trapped within the fallen mass, keeping grains jostling but apart. Sturzstroms can travel in excess of 200 mph and have killed thousands of people.

Subsidence occurs when the surface drops down either slowly in response to removal of subsurface water or oil, or catastrophically as in the collapse of limestone cavern roofs to form sinkholes. Extensive cavern systems in limestone occur throughout much of the central and southeastern United States.

The biggest mass movements on Earth appear to be submarine failures. The Hawaiian Islands suffer mass failures larger than 5,000 km^3. About 10,000 people live on the block that is the next most likely to fail catastrophically.

Terms to Remember

acoustic fluidization 231
bedding 217
bedrock 209
bentonite 226
carbonic acid 238
cavern 222
chemical weathering 212
clay minerals 209
creep 208
daylighted bedding 216
debris flow 226
delta 233
eddy 226
fall 217
flank collapse 234
flow 217
limestone 237
loess 226
mass movement 208
piping 214
pore-water pressure 214
porosity 209
quick clay 212
rotational slide 219
shear 215
sinkhole 237
slate 229
slide 217
slump 219
soil 209
sturzstrom 228
subside 217
syncline 221
translational slide 219
vortex 226

Questions for Review

1. Explain the mechanisms of creep. What causes soil and surface rock to swell and shrink?
2. Draw a cross section through a slope and explain external actions that are likely to cause failure as mass movements.
3. Draw a cross section through a clay mineral. Explain the physical properties that promote swelling and shrinking.
4. Draw a water molecule and explain how it links up so readily with some clay minerals.
5. Draw a cross section and explain how quicksand forms.
6. How do pore waters become pressurized? What is their role in mass movements?
7. List some adverse geologic structures inside hills that facilitate mass movements.
8. Draw a cross section and explain the Swedish Circle method of analyzing slope stability for rotational slides.
9. Draw a cross section and explain how translational slides work.
10. China has suffered hundreds of thousands of deaths in single loess-flow events. Explain how these rapid flows operate.
11. What triggering events set off rapid mass movements? For example, at Elm, Switzerland; at Vaiont, Italy; at Nevados Huascarán, Peru.
12. Compare a snow avalanche to a mass movement of soil and rock.
13. Why do coastal cities on river deltas slowly subside, e.g., New Orleans, Louisiana? Why do some cities near giant oil fields slowly subside, e.g., Houston-Galveston, Texas?
14. Why do large cavern systems form in limestone? Why do they sometimes collapse and form sinkholes?

Questions for Further Thought

1. Roadways are commonly cut into the base of slopes. When mass movements block the road, they are quickly removed. What is wrong with this whole process?
2. Draw a cross section of a hill made of inclined rock layers. On which side of the hill would you build your house?
3. Acoustic fluidization has been proposed as a mechanism for fast-moving, long-distance flows, but it is not widely accepted. What tests or analyses might be made to evaluate its reality?
4. Evaluate the mass-movement danger to coastal residents southeast of Kilauea volcano in Hawaii.
5. Can pumping up huge volumes of groundwater or oil trigger a movement on an active fault?
6. How can you protect yourself against the cavern collapses that form sinkholes?

Suggested Readings and References

Alden, W. C. (1928). Landslide and flood at Gros Ventre, Wyoming. *Transactions, American Institute of Mining and Metallurgical Engineers,* 76, 347–61.

Armstrong, B. R., and Williams, K. (1992). *The Avalanche Book.* Armstrong, Colo.: Fulcrum Publishing.

Cervelli, P., Segall, P., Johnson, K., and Miklius, A. (2002). Sudden aseismic slip on the south flank of Kilauea volcano. *Nature,* 415: 1014–1018.

Close, U., and McCormick, E. (1922, May). Where the mountains walked. *National Geographic,* 41, 445–64.

Evans, S. G., and De Graff, J. V. (Eds.) (2002). Catastrophic landslides. Geological Society of America Reviews in Engineering Geology, 15.

Hsu, K. J. (1989, January). Catastrophic debris streams (sturzstroms) generated by rockfalls. *Geological Society of America Bulletin,* 86, 129–40.

Kerr, P. F. (1963). Quick clay. *Scientific American,* 209, 132–41.

Kiersch, G. A. (1964, March). Vaiont reservoir disaster. *Civil Engineering,* 32–39.

Matthes, G. H. (1953). Quicksand. *Scientific American,* 188, 97–102.

Melosh, H. J. (1983, March–April). Acoustic fluidization. *American Scientist,* 71, 158–65.

Miller, W. J. (1931). The landslide at Point Fermin, California. *Scientific Monthly,* 32, 464–69.

Moore, J. G., Clague, D. A., Holcomb, R. T., Lipman, P. W., Normark, W. R., and Torresan, M. E. (1989). Prodigious submarine landslides on the Hawaiian ridge. *Journal of Geophysical Research,* B12, 94, 17, 465–84.

Nilsen, T. H., and Brabb, E. E. (1975). Landslides. In R. D. Borcherdt (Ed.), *Studies for seismic zonation of the San Francisco Bay region.* U.S. Geological Survey Professional Paper 941A.

Plafker, G., and Ericksen, G. E. (1978). Nevados Huascarán avalanches, Peru. Chapter 8, In B. Voight (Ed.), *Rockslides and Avalanches.* Amsterdam: Elsevier Scientific.

Schuster, R. L., and Fleming, R. W. (1986). Economic losses and fatalities due to landslides. *Bulletin of the Association of Engineering Geologists,* 23, 11–28.

Shreve, R. L. (1968). *The Blackhawk landslide.* Geological Society of America Special Paper 108.

Varnes, D. J. (1978). Slope movement types and processes. In R. L. Schuster and R. J. Krizek, (Eds.), *Landslides, Analysis and Control* (Chap. 2). Washington, D.C.: National Academy of Sciences.

Williams, G. P., and Guy, H. P. (1973). *Erosional and depositional aspects of hurricane Camille in Virginia, 1969.* U.S. Geological Survey Professional Paper 804.

Videos

The Runaway Mountain. (1995). Horizons/Pioneer Productions/BBC (50 min.).

The Rissa, Norway Landslide. (1981). Norwegian Geotechnical Institute (24 min.).

Debris-Flow Dynamics. (1984). U.S. Geological Survey (23 min.).

Landslide: Gravity Kills. (1999). Discovery Channel (52 min.).

Raging Planet: Avalanche. (1997). Discovery Channel (50 min.).

Chapter 9

Climate Change

> My kingdom is as wide as the world, and my desire has no limit. I go forward always, freeing spirits and weighing worlds, without fear, without compassion, without love, and without God. Men call me science.
>
> —Gustave Flaubert, 1874, *La Tentation de Saint-Antoine*

This chapter signals a major shift in energy sources as we move to those processes and disasters fueled by the Sun. In Chapter 1, the Sun was identified as the external source of energy that powers the hydrologic cycle and, with the help of gravity, drives the agents of erosion. Although the ultimate source of energy behind weather and climate is the Sun, the immediate force is gravity. Now we need to know how solar energy is spread about the Earth and how it helps create the short-term processes of **weather** and the long-term conditions of **climate.** The difference between climate and weather is emphasized in the humorous line—climate is what you expect, weather is what you get.

The Sun's energy heats the Earth unequally. The equatorial area faces the Sun more directly than do the polar regions. During the course of a year, the equatorial area receives about 2.4 times as much solar energy as the polar regions. The Earth's spin helps set the heat-carrying oceans and atmosphere in motion. Gravity then works to even out the unequal distribution of heat by pulling more forcefully on the colder, denser air and water masses. All of this is the province of weather and climate, a province of continual change. We experience the hour-to-hour, day-to-day, and season-to-season changes in atmospheric conditions known as weather. When weather is viewed over the longer time spans of decades, centuries, and on up to intervals of millions of years, it is called climate. Like weather, climate is marked by change and variability, and the changes occur on all time scales. This chapter is organized by the time scales of change. We begin with a look at the early Earth, then view changes operating over different lengths of time—millions of years, thousands of years, hundreds of years, and multiyears. We conclude with a look at how we humans are modifying climate.

Climate affects all life on Earth and as climate changes, all life must adapt to the new conditions. Some species thrive during and after a given change. They are joined by new species well adapted to the climate of the time. Meanwhile, other species cannot survive the changes and thus become extinct. The record of **extinctions** is covered in Chapter 14.

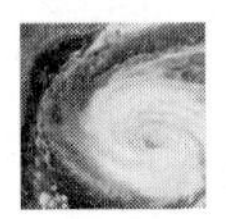

Climate

What are the factors that help determine climate? The principal ones are: 1) how much solar radiation is received by the Earth during any given time interval and

2) how much is retained in any given area. The amount of incoming solar energy commonly is treated as a constant, although it does vary over time. Conditions on Earth may act to repel or to retain solar energy, accentuating cold or warm conditions worldwide (Figure 9.1). For example, excessive cloudiness or snow and ice cover may cause more solar radiation to be reflected back to space **(albedo),** resulting in cooler conditions. Or large amounts of certain gases may accumulate in the atmosphere, allowing short-wavelength radiation from the Sun to pass through but blocking and trapping the returning long-wavelength reradiation. This phenomenon is known as the **greenhouse effect** and, with high levels of gases such as CO_2, results in a buildup of heat known as global warming.

Climate depends on long-term trends in the heating and cooling of the Earth's surface; it is largely affected by the circulation of the atmosphere and oceans. The atmosphere and oceans are in unceasing motion about the Earth and their energy systems are linked. Moisture evaporated from the ocean carries its latent heat of vaporization up into the atmosphere where it is released (see Chapter 1). Earth's atmosphere is a gigantic system of great complexity that interacts with the Sun, oceans, land, and all life to redistribute heat energy. Large-scale circulation of the winds plays the biggest role in quickly moving heat about the Earth's surface and is a major transporter of moisture (Figure 9.2). Clouds and water vapor, the uncondensed version of a cloud, carry tremendous amounts of heat around the world.

The oceans hold a much greater volume of heat due to their greater mass and higher capacity per unit volume to hold heat. However, ocean movement is much slower. Ocean surface circulation is largely created by the effects of winds blowing over the water surface. It is the deep-ocean currents that are most important in the global exchange of heat. The deep-ocean circulation is mostly driven by density differences caused by colder and/or saltier water masses sinking and flowing at depth in a global circuit (Figure 9.3). The oceanic conveyor belt of Figure 9.3 transports enormous amounts of heat around the Earth and in so doing modifies atmospheric wind patterns and moisture distribution.

We will look at the marked variability of solar heat reception, retention, and distribution in three chapters. This

Figure 9.1 An iceberg from western Greenland floating in the Labrador Current and moving toward Newfoundland. A similar iceberg sank the *Titanic.*

Photo by Anne Jennings, University of Colorado at Boulder.

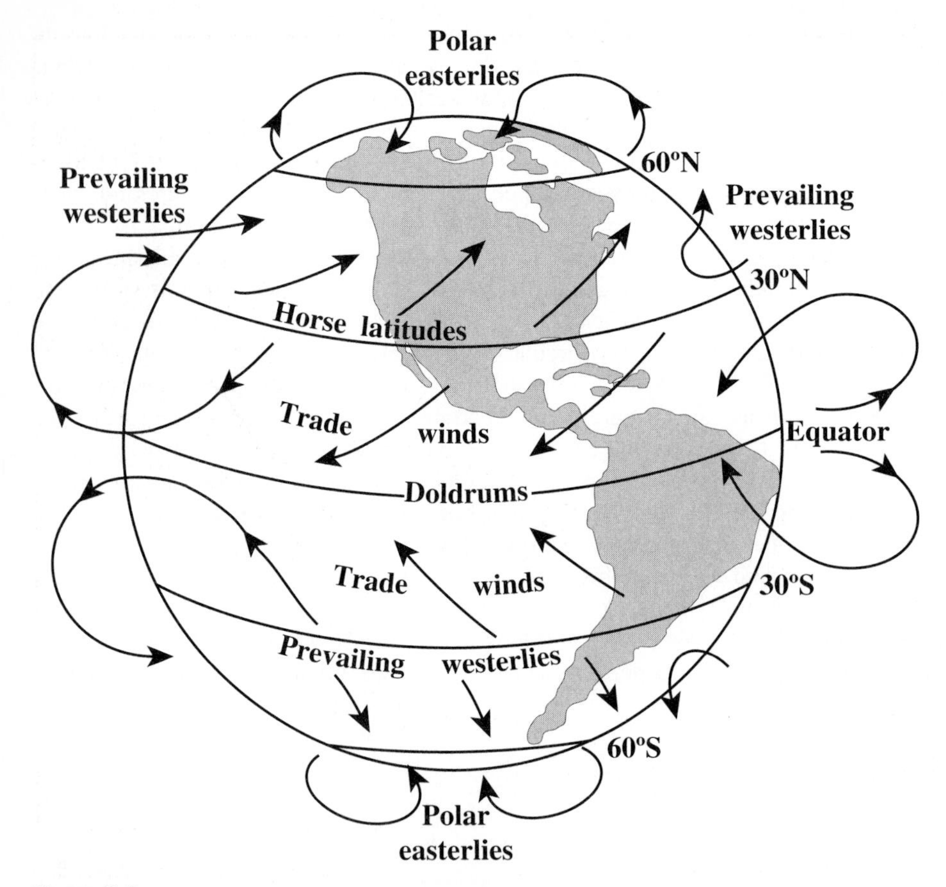

Figure 9.2 Major pattern of atmosphere circulation. Warm air rises at the equator and moves toward the poles. Cold air at the poles flows toward the equator. All moving air masses are affected by Earth's rotation.

chapter is concerned with longer intervals of time and some of the major climatic regimes that have affected Earth. As the time scales of climate become progressively shorter, our discussion will move to Chapters 10 and 11 on severe weather.

Early Earth Climate—A Runaway Greenhouse

The climatic regime of the early Earth, the third planet from the Sun, can be appreciated by looking at the atmospheric compositions of the inner planets (Table 9.1).

Venus is the second planet from the Sun and thus receives intense solar radiation. Much of that solar energy is trapped by its dense, carbon dioxide-rich atmosphere, which helps create surface temperatures of about 477°C (890°F). Life on Venus is difficult to visualize when temperatures are so high that surface rocks glow red like those in a campfire ring.

Mars is the fourth planet from the Sun, and its greater distance causes it to receive much less solar energy. However, the thin Martian atmosphere is also relatively rich in CO_2, helping hold the heat it does receive and maximizing its surface temperature. The atmospheres of Venus and Mars are thought to be little changed over more than 4 billion years, yet Earth's atmosphere has undergone a radical change from being CO_2-rich to CO_2-poor.

Why has Earth's atmosphere changed? The changes have been caused in large part by life processes. Plants remove CO_2 from the atmosphere via photosynthesis and respire O_2 as a by-product that has built up in the atmosphere. But the total amount of CO_2 locked up in plants, dead or alive, is small compared to the amounts originally in the atmosphere. Also, on the early Earth, photosynthesizing plants could not have survived in the high CO_2 levels in the atmosphere and high temperatures that existed at the beginning.

Where has all the CO_2 in the early Earth's atmosphere gone? It is stored in physical form in several ways, but about 80 percent of that CO_2 is now chemically tied up in limestone (Table 9.2). Limestone is rock composed of $CaCO_3$. Most limestone is made from the hard parts of oceanic life, such as shells, reefs, and mineralized tissue of invertebrate animals and algae. (The process by which carbon [C] is precipitated from seawater is described by the equations in the sidebar on "How to Create a Cave" in Chapter 8.) CO_2 in the atmosphere readily dissolves in water. In fact, a partial equilibrium exists between CO_2 in ocean water and the atmosphere.

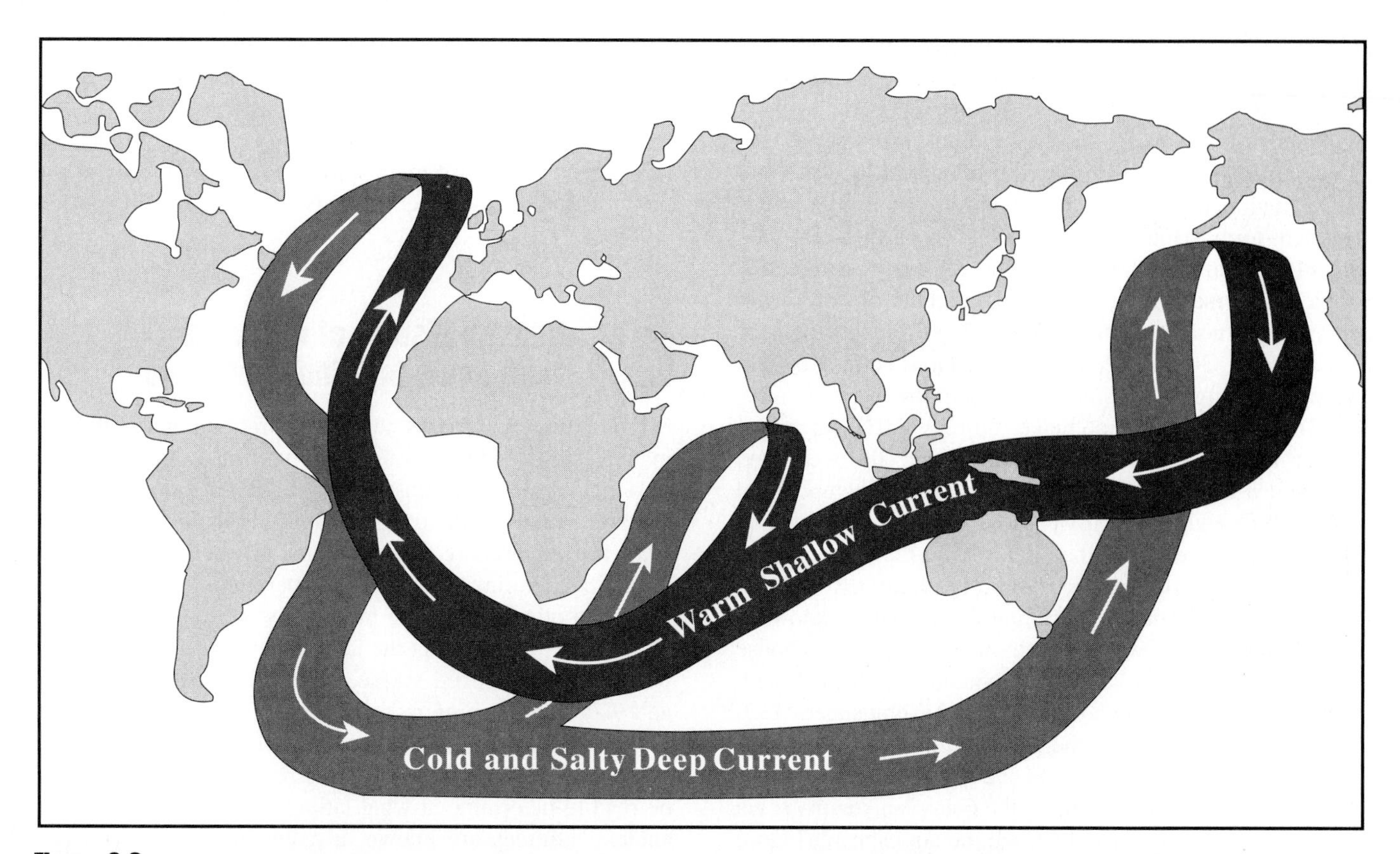

Figure 9.3 The ocean circulation system carries warm shallow water to the North Atlantic, thus keeping Europe 5 to 10°C warmer. Cooling in the Arctic increases ocean-water density causing it to sink and flow at depth southward out of the Atlantic Ocean. This ocean flow is the equivalent of 100 Amazon Rivers.

Table 9.1 Atmospheres of the Inner Planets

	Venus	Early Earth	Mars	Earth Today
CO_2	96.5%	98%	95.3%	0.037%
N_2	3.4%	1.9%	2.7%	78%
O_2	trace	trace	0.13%	21%
Ar	0.01%	0.1%	1.6%	0.93%
Temperature °C	477	290	–53	16
Pressure (bars)	92	60	0.006	1

Table 9.2 Carbon on the Earth (in gigatons = 10^9 metric tons)

Atmosphere	720
Oceans	
Total organic	1,000
Seawater layers	
Surface	670
Deep	36,730
Continents	
Living biologic mass	~800
Dead biologic mass	1,200
Fossil fuels (oil, coal, gas)	4,130
Organic matter in mudstone	15,000,000
Limestone	>60,000,000

Source: Falkowski, P., and others. (2000) The Global Carbon Cycle. *Science,* 290: 291–296.

Returning to the earliest Earth (Table 9.1), when no life was present and the atmosphere was full of CO_2, the surface temperature of the Earth would have been about 290°C (550°F). Why would Earth have been so hot? This global warming was due in part to the greenhouse effect. A glass-walled greenhouse (or your car with all the windows rolled up) admits incoming visible light, i.e., solar radiation with short wavelengths (about 0.5 micrometers). The Sun's short-wave radiation warms objects inside the greenhouse (or car). Heat builds up inside the greenhouse and is given off as infrared radiation of longer wavelengths (about 10 micrometers). However, glass is opaque to long-wavelength radiation, and thus, the outgoing radiant energy is trapped, producing a warm environment inside the greenhouse (or car).

The same greenhouse effect is produced by some of the gases in the atmosphere, such as CO_2, H_2O vapor, **methane** (CH_4), and chlorofluorocarbons. Because of its fluctuations in abundance, CO_2 is the most important of the greenhouse gases. Over geologic time, as CO_2 dissolved in water and $CaCO_3$ sediments formed, the amount of atmospheric CO_2 declined. Early photosynthesizing life on Earth pulled CO_2 out of the atmosphere causing a decline. Innumerable species of animals made skeletal material out of $CaCO_3$, reducing the amount of atmospheric CO_2 even further. Biologic use of CO_2 has lessened the greenhouse effect to yield the present congenial temperatures on Earth.

Today, carbon dioxide makes up only 0.037 percent of the Earth's atmosphere, but it helps create the weakened greenhouse effect that keeps Earth's average temperature 16°C (61°F) higher than it would be if CO_2 were absent. If CO_2 were not present in the atmosphere, the average temperature at the Earth's surface would be about −18°C (0°F), and much of life would be different from what we know.

The Earth has always been influenced by a greenhouse effect, and life has always been in dynamic equilibrium with it. However, humans are now changing the atmospheric CO_2 concentration in the atmosphere by burning tremendous volumes of plants, both living (trees and shrubs) and dead (coal, oil, and natural gas). Combining the C in plants with O_2 via fire returns large amounts of CO_2 to the atmosphere (Figure 9.4). About 6 gigatons (1 gigaton equals 10^9 metric tons) are returned to the atmosphere each year by burning fossil fuels; about 5,000 gigatons remain to be burned. The human contribution is small compared to the natural fluxes between the atmosphere and ocean, and between the atmosphere and continents, each of which exchange in excess of 100 gigatons annually. Although human changes in CO_2 and other gases are relatively small, they can be enough to trigger climate shifts that cause major problems. We will look at the modern situation at the end of this chapter.

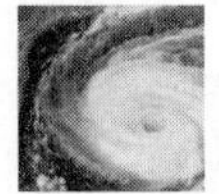

Climate History of the Earth: Time Scale in Millions of Years

Many sedimentary rocks contain information about the climate at the time they formed. Warm climates are indicated by 1) fossil reefs and most limestones; 2) the aluminum ore **bauxite,** which forms only in tropical soils; and 3) beds of evaporite minerals (e.g., **halite**) that crystallize from water bodies evaporating under high-temperature, arid climates.

Cold climates may be marked by the powerful erosion of glaciers, which sculpt the landscape (Figure 9.5), leaving polished and grooved surfaces beneath them (Figures 9.6 and 9.7) and dumping massive piles of debris (Figure 9.8). The distribution of fossil organisms tells much about ancient climates. First, the paleomagnetic record in the rocks may be read to determine at what latitude a rock formed. When ancient latitudes are known, a paleoclimatic analysis can begin. Then, for example, when fossil shells of organisms that live only in polar seas are found in abundance in rocks formed in midlatitudes, the world climate must have been colder at that time.

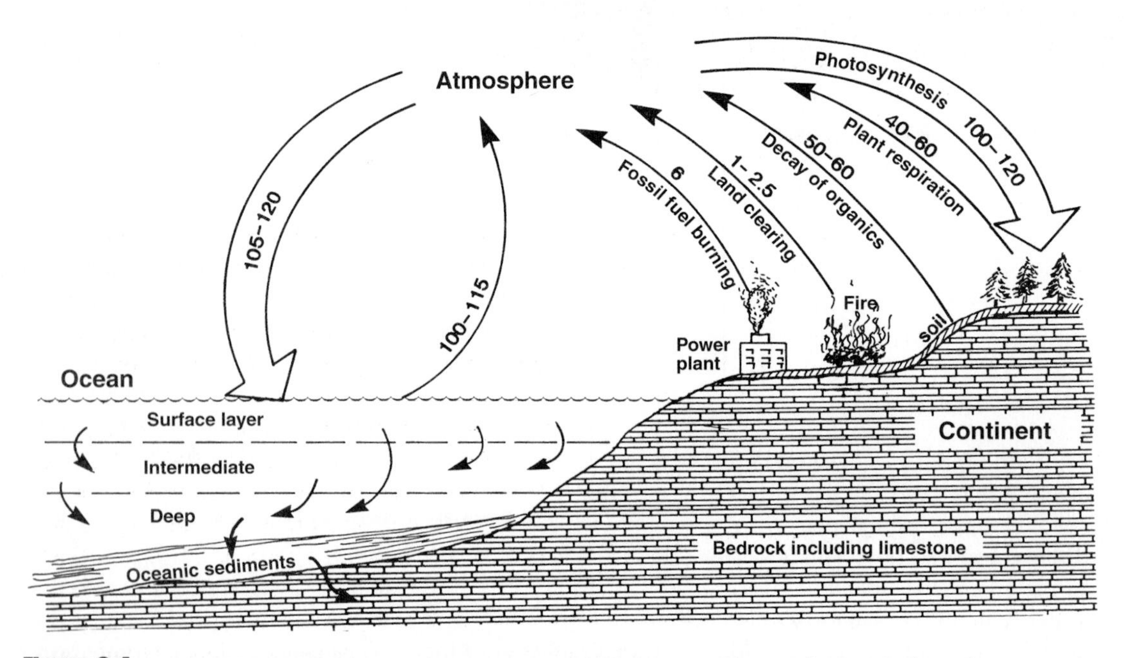

Figure 9.4 Annual cycle of carbon exchange measured in gigatons. Plants take in CO_2 from the atmosphere during photosynthesis. CO_2 is returned during plant respiration, decay after death, and by burning of forests and fossil fuels. Near equilibrium exists between CO_2 in ocean surface water and the atmosphere. The ocean "pumps" some CO_2 into deep-water storage. Organisms remove dissolved CO_2 to build shells, which end up in sediments.

Figure 9.5 Muir glacier flows into Glacier Bay, Alaska, as the present Ice Age continues.

Putting all these, and many other, kinds of information together yields an ever-improving story about the climate history of the Earth (Figure 9.9). The rocks tell of extreme variations and changes in world temperature and precipitation throughout geologic time. Not only do warm and cold intervals come and go, but they do not necessarily correlate with wet and dry periods, nor is there a pattern to the arrivals and departures of various climates. Notice some of these features:

- The Late Paleozoic glacial interval was cold and wet.
- Early Eocene time was hot and wet.
- Global cooling beginning in late Eocene time has led us into the current glacial interval.

Earth's climate depends upon the balance between incoming and outgoing heat. At any given time, the atmosphere-ocean-continent system may be gaining or losing in its overall heat budget. Global heat supply has a profound effect on water, which exists on Earth's surface at the transition between its three phases of ice (solid), liquid, and vapor (gas). Water has such a tremendous capacity to either absorb or release heat that it acts as a powerful control or **buffer** on global climate.

The surface of the Earth is divided into temperature zones of frigid, temperate, and torrid as defined by latitude (Figure 9.10). Climate seems to swing like an irregular pendulum from ages where cold temperatures of the frigid zone dominate the Earth to other times when warmth covers most of the world. During a frigid period, an Ice Age, the colder climates of the high latitudes expand in area while the area of warmer climate in the low latitudes shrinks but does not disappear (Figure 9.11). Conversely, in an era of warmth, a Torrid Age, the globe is marked by expansion of the subtropical climatic zones, while the cold-climate belts shrink

Figure 9.6 Glacially smoothed and polished surface on granite with large feldspar crystals, near Lake Tenaya, Sierra Nevada, California. Photo by John S. Shelton.

Figure 9.7 Glacial grooves (striations) carved in rocks by a former glacier. This rock is exposed west of New Haven, Connecticut. Photo by John S. Shelton.

Figure 9.8 Rocky debris dumped by a glacier (lower center of photo). Convict Lake (center of photo) is dammed behind a glacial-debris dam, Sierra Nevada, California.
Photo by John S. Shelton.

back toward the poles (Figure 9.12). Let us look at examples of times of extreme climates.

Late Paleozoic Ice Age

One of the major Ice Ages in Earth history began in Early Carboniferous time (360 million years ago) and lasted for tens of millions of years until Late Permian time (260 million years ago) (Figure 9.9). For a glacial interval to last so long, a broad-scale and long-lasting situation is required. The major factors appear to be changes in the shapes, sizes, and orientations of the continents and oceans.

1. An initial, absolute requirement for an Ice Age is having one or more large continental masses near the poles. A polar landmass is necessary to collect the snowfall that allows the buildup of immense 3 km thick continental ice sheets. Massive glaciers cannot be built on top of ocean water. In Late Paleozoic time, the continents were largely united as the single landmass Pangaea. The southern portion of the Pangaea supercontinent is known as Gondwanaland; it was composed of present-day South America, Africa, Antarctica, Australia and India. In Late Paleozoic time, as Gondwanaland moved across the south polar region, major ice sheets were always present near the South Pole. South America-Africa probably first supported the great ice sheet, then Antarctica, and finally Australia (Figure 9.13).
2. Another important consideration is ocean-water circulation. No matter how much the Sun's brightness varies, equatorial waters will receive more solar energy than polar waters. Without continents present to block ocean-water flow, the warm equatorial waters simply will circulate latitudinally (east-west) due to the spin of the Earth.

What does warm water have to do with building massive ice sheets? In the hydrologic cycle, water must first be evaporated from the ocean before clouds can form and move over cold landmasses to drop snow. Cold water is extremely difficult to evaporate; warmer water is a great help in promoting evaporation. The geologic record shows that Ice Ages are favored when oceanic circulation is more longitudinal (north-south) than latitudinal (east-west). When the continents are aligned in a north-south direction, they act to block latitudinal circulation of ocean water, thus sending warmer equatorial waters toward the poles, where evaporation can form the clouds that yield the snowfall that builds

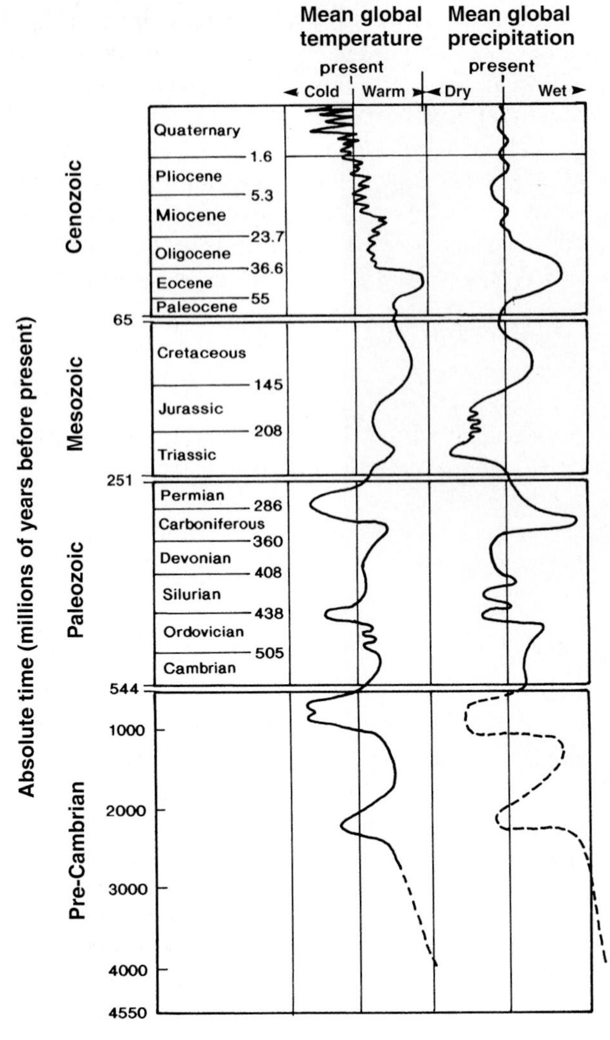

Figure 9.9 Temperature and precipitation history of the Earth.

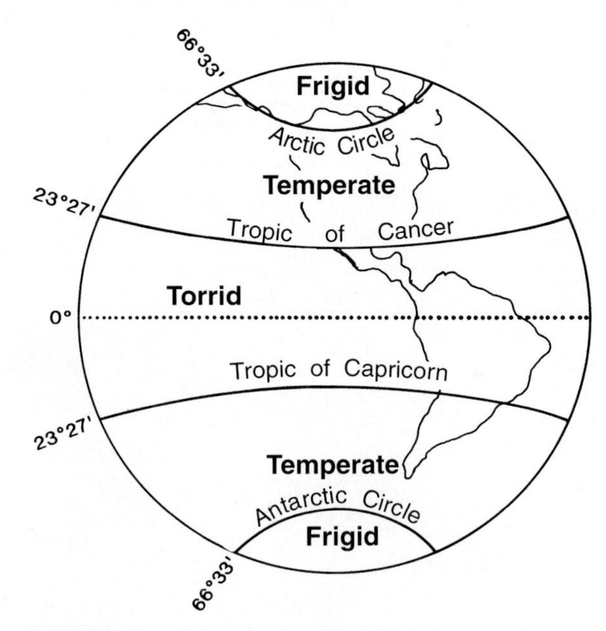

Figure 9.10 Five great divisions of the Earth's surface with respect to temperature and latitude.

up on polar landmasses as glaciers. The continents had a north-south alignment during Late Paleozoic time, as they do today (Figure 9.14) in the current Ice Age.

Why did the Late Paleozoic Ice Age end? Possibly because Gondwanaland began to break up and disperse. As the continents moved apart, ocean-circulation patterns around the world were changed. Warm waters stayed near the equator, and cold waters encircled the poles, thus drastically reducing the moisture supplied to polar landmasses. Additionally, when continents move away from the poles, no platform exists for the accumulation of snow and the building of glaciers.

Late Paleocene Torrid Age

The world was warming during Paleocene time (65–55 million years ago). There was more heat in the Paleocene oceans and atmosphere than at any time since. The equatorial zones had tropical temperatures and rainfalls higher than, but similar to, what they enjoy today; however, more poleward latitudes were markedly warmer. Sea-surface temperatures in the southern oceans near Antarctica were 10° to 15°C (18° to 27°F) warmer than today (Figure 9.15).

Ancient temperatures are determined from the ratio of stable isotopes of oxygen in the $CaCO_3$ shells (fossils) of single-celled sea life. An atom of oxygen may have either 16, 17, or 18 protons and neutrons in its nucleus. Water evaporated from oceans removes more of the lighter common oxygen (^{16}O) and less of the heavier ^{18}O. This ^{18}O-depleted water is locked up on land as ice and snow, leaving

Figure 9.11 During an Ice Age, the colder high-latitude climate expands into lower latitudes.
Drawing by Jacobe Washburn.

Figure 9.12 During a Torrid Age, the hotter low-latitude climate expands into higher latitudes.
Drawing by Jacobe Washburn.

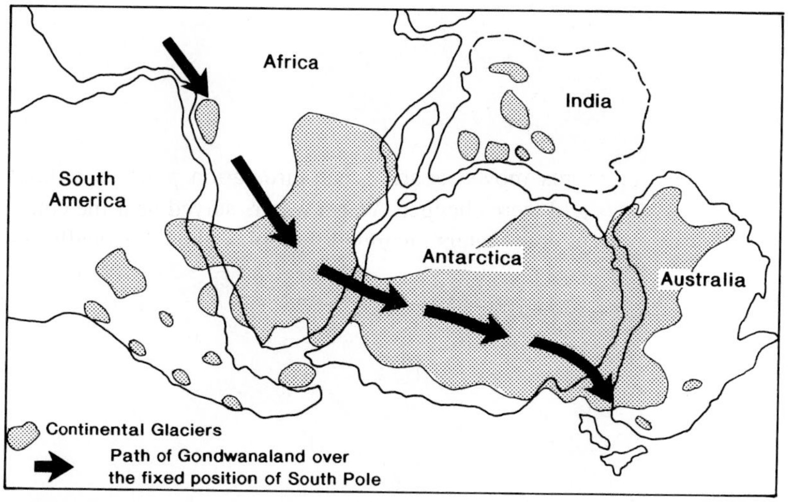

Figure 9.13 Late Paleozoic ice masses on Gondwanaland. Not all of these continental glaciers existed at the same time. Arrows approximate path of drifting Gondwanaland over the fixed position of the South Pole.

the ocean with ^{18}O enriched water. Shells constructed from seawater incorporate the $^{18}O/^{16}O$ ratio of the seawater during their lifetime within their $CaCO_3$ shell walls. Thus, measurement of the $^{18}O/^{16}O$ ratio in shells acts as a paleothermometer that is used to estimate the temperatures of ancient seas. Heavier seawater (^{18}O enriched) corresponds to cooler climates and lighter seawater (^{18}O depleted) corresponds to warmer climates.

What was the world like during the Late Paleocene Torrid Age? There was less difference in temperature between tropical and polar waters; an absence of cold, dense, sinking water at the poles; and less difference in temperature between surface and deep-ocean waters, meaning that the pull of gravity would have been less effective and ocean circulation would have been more sluggish.

Temperature differences in the atmosphere would also have decreased, resulting in more peaceful weather worldwide. There was an absence of strong seasons, weather was more constant, and rainfall was more evenly distributed throughout the year. Most of the world was wetter and warmer.

The conterminous United States was covered by either tropical or subtropical climates. Along the coastal zones, subtropical conditions existed above the Arctic circle as is shown by fossil crocodiles and palm trees. Continental ice sheets apparently did not exist anywhere in the world. Evergreen (coniferous) and warm deciduous forests covered much of the land. Hot deserts and arctic tundra covered smaller percentages of the ground. The world climate lacked extremes.

How did Earth's climate become so dominated by warmth? Several factors apparently combined to turn up the heat. 1) The equatorial zones were largely covered by oceans, allowing more absorption of solar heat. 2) As oceans warmed, areas covered by snow and ice decreased, thus exposing more land. Snow and ice reflect the Sun's rays; land absorbs heat. 3) Enormous outpourings of lavas from the opening North Atlantic Ocean are likely to have released large volumes of gases to the atmosphere, which may have increased global warming via the greenhouse effect. 4) The oceans changed their style of density differentiation. At present, cold Antarctic and Arctic waters are the densest of all waters; they sink and flow along the ocean deeps. By Eocene time, the polar water became so much warmer that the heaviest waters might have been tropical waters that had become saltier due to evaporation. The warm, oxygen-deficient, salty waters apparently sank, flowing through the ocean deeps and warming up the oceans from surface to bottom. Warm, salty water masses moving along the ocean bottoms would have affected deep-ocean life. Organisms used to living in cold, oxygen-rich bottom waters had the sudden shock of their environment becoming warm and oxygen-poor. At 55.5 million years ago, the massive change in deep-sea water temperature reached a peak, causing up to 50 percent of unicellular deep-sea animal species to become extinct—a natural disaster.

What was responsible for the final increase in warmth? The warming of ocean bottom waters about 8°C (14°F) caused melting of icy **methane hydrates** on the sea floor thus releasing methane gas to the atmosphere. What are methane hydrates? Bacteria living on the deep-ocean floor release methane as part of their life process but the overlying water is

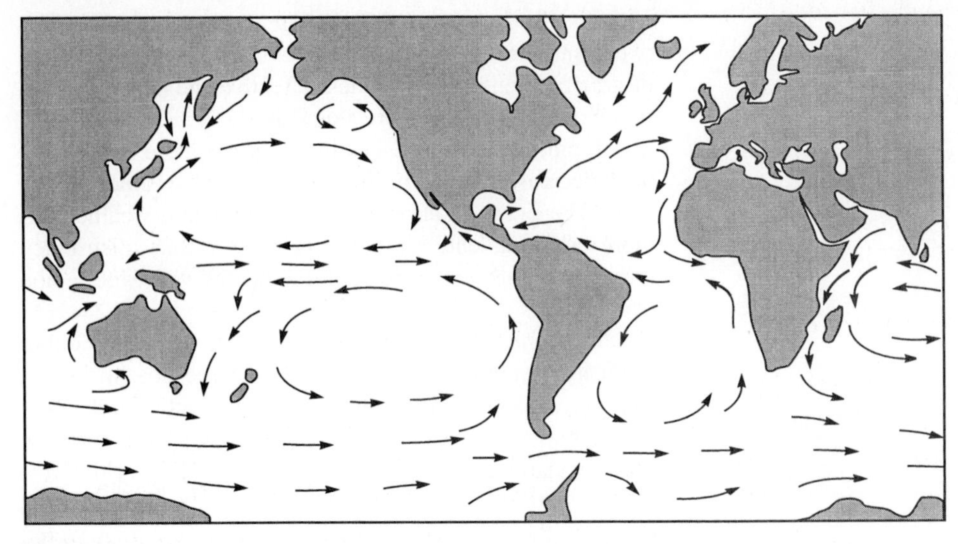

Figure 9.14 Circulation of the wind-blown surface waters of the oceans. Notice how the equatorial waters are deflected both northward and southward by the continents, thus sending warmer waters toward the poles. Also note that the only latitude not blocked by continents (60°S) has a latitudinal flow; it encircles Antarctica.

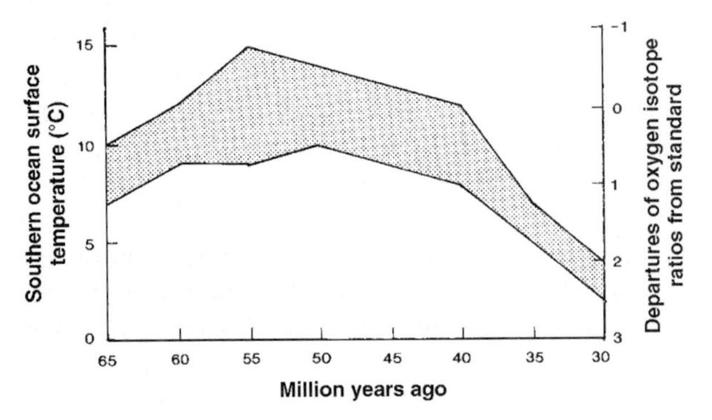

Figure 9.15 Surface temperature of the southern ocean over time, based on oxygen isotope measurements.

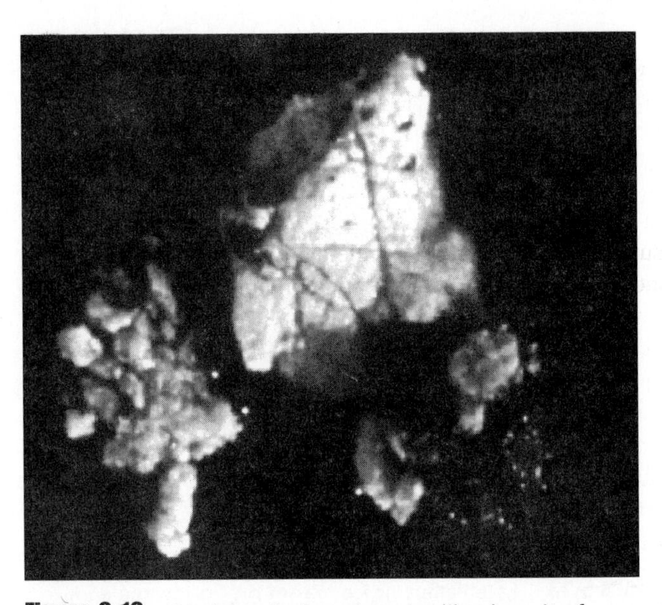

Figure 9.16 Methane hydrate is an ice-like deposit of methane trapped within near-freezing water on and under the ocean floor. Field of view is 6 in (15 cm) across.

so cold and the pressure from the weight of the overlying water is so great that the methane is locked up inside linked near-freezing water molecules to form an ice-like deposit (Figure 9.16). Methane hydrate holds more energy than all of Earth's oil, coal, and natural gas combined. It becomes unstable if temperature rises a few degrees above freezing or if pressure is less than that of 500 m (1640 ft) of overlying ocean. Today, about 15 trillion tons exist on the sea floor; melt it and the world would see a sharp greenhouse increase in temperature. Recent analyses of carbon isotopes in Late Paleocene sedimentary rocks suggest that a major release of methane occurred about 55.5 million years ago. The methane release occurred during a 10,000-year-long interval. This is a very short time for the atmosphere to receive such a large volume of a powerful greenhouse gas (methane has about a 21-times stronger capacity to trap heat than carbon dioxide). During about 250,000 years of excess methane in the atmosphere, the warmer climate caused changes in life on Earth. For example, mass extinctions on the sea floor coincided with major migrations and new appearances of many large terrestrial mammal species.

The world changed from a Late Paleozoic ice house to a Late Paleocene hot house. But climatic change is the way of the world.

Late Cenozoic Ice Age

Beginning from the temperature peak at 55.5 million years ago, the Earth began the long-term cooling trend that has carried us into our current Ice Age (Figures 9.9 and 9.15). The sequence of events has included:

- At 40 million years ago, Antarctica was surrounded by cold water.
- At 36 million years ago, small glaciers were widespread in Antarctica.
- At 14 million years ago, a continental ice sheet existed on Antarctica and mountain glaciers were in the Northern Hemisphere.
- At 5 million years ago, the Antarctic ice sheet had expanded.
- At 3 million years ago, continental ice sheets existed in the Northern Hemisphere.

Why have these changes occurred? There is no single answer. Several variables have interacted in a complex fashion to bring about the climatic cooling of the last 55.5 million years. The relative importance of these variables is the focus of current study for many scientists around the world. The main factors appear to be related to plate-tectonic changes, i.e., the changing positions of continents and oceans. 1) The climatic change is associated with the ongoing breakup of Pangaea into separate continents (Figures 2.16 and 2.17). 2) As continents drifted, seaways opened and closed, thus altering the circulation patterns within the oceans and the distribution of heat about the globe. 3) Continental masses have moved into polar latitudes, with Antarctica centering upon and rotating about the South Pole while North America and Eurasia have moved to encircle the North Pole region. 4) As snow and ice began accumulating on polar landmasses, they reflected more sunlight (increased albedo), and thus heat, back to space. 5) Circulation of the ocean water around the equator was restricted at about 23 million years ago with the closure of the eastern Mediterranean Sea and ended at 3 million years ago when volcanism completed building the Isthmus of Panama as a north-south barrier that blocked east-west ocean-water flow. 6) The area of shallow oceans has been reduced, so less water surface is available to absorb sunlight. 7) The uplifts of the Tibetan Plateau/Himalaya Mountains in Asia and the Colorado Plateau in the western United States have deflected west-to-east atmospheric circulation in the midlatitudes with resultant air flows to the north and return flows to the south.

The Last 3 Million Years The ice sheet on Antarctica is older and more stable than ice in the Arctic. The cold ocean water circulating around Antarctica (Figure 9.14) helps isolate the continent from major changes. The ice sheets on North America and Eurasia have a greater effect on global climate change because they expand and shrink in more dynamic fashion. Their initial growth as continental glaciers occurred between 3 and 2.7 million years ago and coincided with the formation of the Isthmus of Panama. What is the cause and effect here? Once Central America formed a continuous link between North and South America it blocked westward-flowing ocean water and began diverting the warm water of the Caribbean Sea and Gulf of Mexico and forcing it to flow northward along the western Atlantic Ocean (Figure 9.14). The warm water delivered to Canada and Europe caused greater evaporation and formation of water vapor, which resulted in greater snowfall that accumulated to build glaciers.

Once continental ice sheets existed in the Northern Hemisphere, they underwent complex cycles of glacial advance and retreat. The cycles appear to have been present during earlier Ice Ages and are linked strongly to regular variations in the Earth's orbit and rotation.

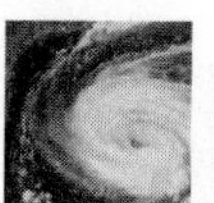

Glacial Advance and Retreat: Time Scale in Thousands of Years

During the last 2 million years, Earth has hosted about 20 Ice Age glacial advances. As an Ice Age begins, ocean surface water evaporates, and some precipitates on the continents as snow. Snow accumulates and burial pressure converts it into ice (Figure 9.17). Continental glaciers reach thicknesses of about 3 km (2 mi), deeply burying the land. The immense volumes of ice deform internally under their own weight and slowly flow out over the countryside like megabulldozers, scarring and reshaping the land (Figure 9.18). The record of glacially deposited sediments tells of numerous glacial advances and retreats. Starting in the 1970s, our knowledge of the advance-retreat history has been leaping ahead, thanks to cores of sediments taken from the ocean floor and cores of ice removed from the Greenland glacier. Each core holds the cumulative record of the annual deposits of sediment or snow (ice) that may be read like the pages of a history book using techniques such as the ratios of oxygen isotopes.

The emerging story for the last 1 million years is of worldwide glacial advances that last about 100,000 years followed by retreats that take place more rapidly—withdrawing over periods from decades to a few thousand years (Figure 9.19). What causes the cycles of slow buildup and advance of glaciers followed by rapid shrinkage and retreat? The answer lies in the cyclic peculiarities of the Earth's spin and its orbit around the Sun, each of which affects the amount of solar energy received by Earth. Verification of the importance of orbit and rotation cycles came in the 1980s, when computer analyses of data from sediment and ice cores were shown to match the theoretical astronomical framework erected by the Serbian geophysicist Milutin Milankovitch in the 1920s and 1930s. The glacial advance and retreat cycles, including primary, secondary, and tertiary fluctuations, are largely explained by three astronomical peculiarities (Figure 9.20).

1. Eccentricity of the Earth's orbit around the Sun. The shape of the orbit varies every 100,000 years from nearly circular to an eccentric ellipse. The eccentricity

Figure 9.17 Space shuttle view of the Malaspina glacier and Seward ice field in the Gulf of Alaska, 10 August 1989.
Photo courtesy of National Aeronautics and Space Administration.

Figure 9.18 Southern limit of the last glacial advance is shown by the dumped glacial debris (irregular, hilly land on left) on the Waterville Plateau, Washington.
Photo by John S. Shelton.

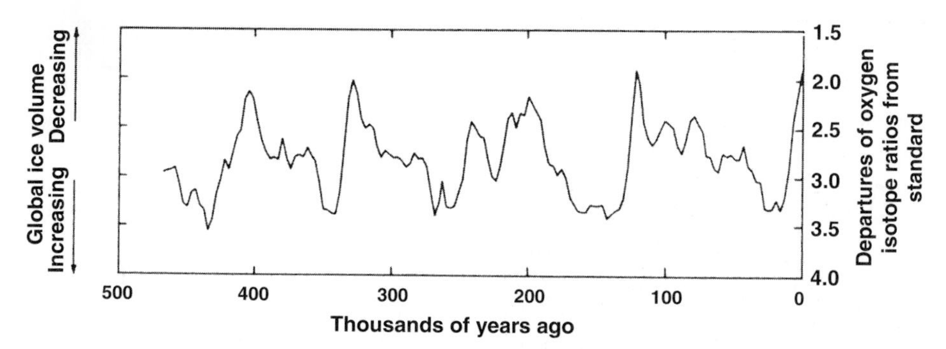

Figure 9.19 World ice-mass volumes of the last 470,000 years, based on oxygen isotope measurements.
Data from J. D. Hayes, et al., 1976.

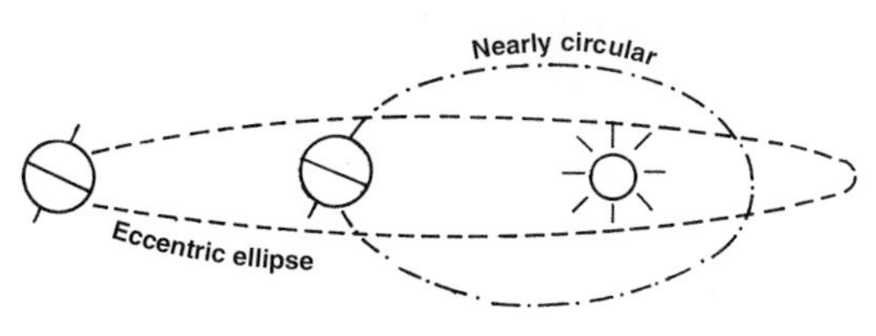

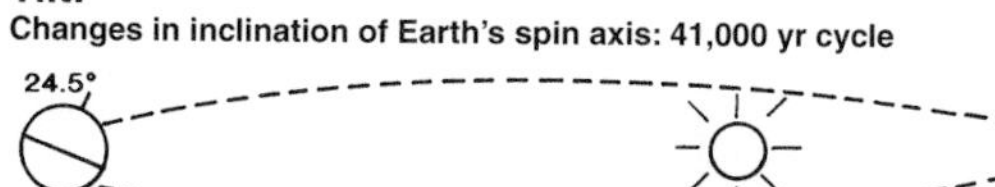

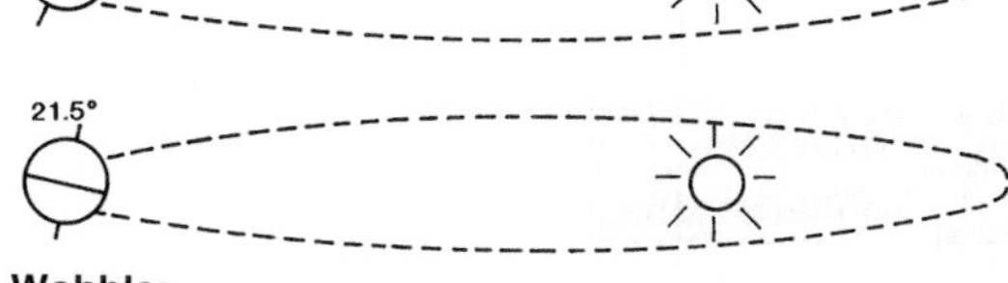

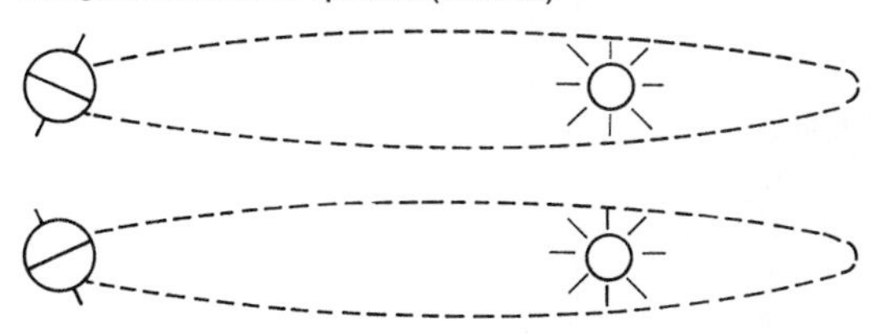

Figure 9.20 Astronomical peculiarities that affect the amount of solar energy received by the Earth. The ellipticity of orbit is exaggerated.

From *Planet Earth* by V. A. Schmit. Copyright © 1986, 1994, 1998 by Kendall-Hunt Publishing Company. Used with permission.

time cycle is similar to the broad-scale, primary length of time for each glacial advance and retreat pairing, suggesting that eccentricity sets the fundamental frequency of the cycles.

2. Tilt of the Earth's axis. The spin axis of the Earth tilts away from the orbital plane in a 41,000-year-long cycle where tilt varies from 21.5° to 24.5°. Greater tilt angles cause more increased seasonal extremes. At present, the tilt is about 23.5°.
3. Precession of the equinoxes where the direction of the tilt changes even though the angle stays the same. The effect is a wobble roughly analogous to what you see in the spin of a toy top. The wobble has a double cycle with periodicities of 23,000 and 19,000 years. Currently, the wobble places the Earth closest to the Sun during the Northern Hemisphere winter, giving it milder winters and summers than the Southern Hemisphere. The changes over time of the eccentricity, tilt, and wobble cycles have been calculated for the past and into the future (Figure 9.21). At present, the eccentricity and tilt each contribute to cooling while the wobble (precession) works to warm the climate.

The sediment- and ice-core records show that glacial advances and retreats are synchronous in both the Northern and Southern Hemispheres. How are ice masses around the opposing poles affected simultaneously by astronomical tilts and wobbles? Probably by heat transfer within the world ocean and atmosphere. Any increased heat received in one hemisphere is shared with the other. For example, the Greenland ice-core record shows significant changes. About 14,000 years ago, the Earth began to warm according to changes in the annual ice layers—the ^{18}O, CO_2, and methane contents increased. The increase in ^{18}O means more heat was available to evaporate the heavier oxygen isotope; the increase in CO_2 added to global warming via the greenhouse effect; and the increase in methane is due to an increase in swamps. These measures are symptomatic of the global warming of our current interglacial interval, where ice masses have retreated rapidly.

What was the Earth like 18,000 to 20,000 years ago when glacial ice masses were at peak extent? The continental ice sheets contained about 70 million km^3 of ice, and the ice masses had spread out to cover about 27 percent of today's land, including virtually all of Canada and part of the northeastern United States (Figure 9.22). Each ice sheet had its own cell of atmospheric high pressure that displaced midlatitude storm systems to the south. The displacement of storm systems increased midlatitude rain, turned the desert basins of the southwestern United States into a series of lakes and produced much heavier rainfall over the Mediterranean region. The high pressure over Asia kept away its monsoonal rains, thus increasing the aridity of the Indian subcontinent.

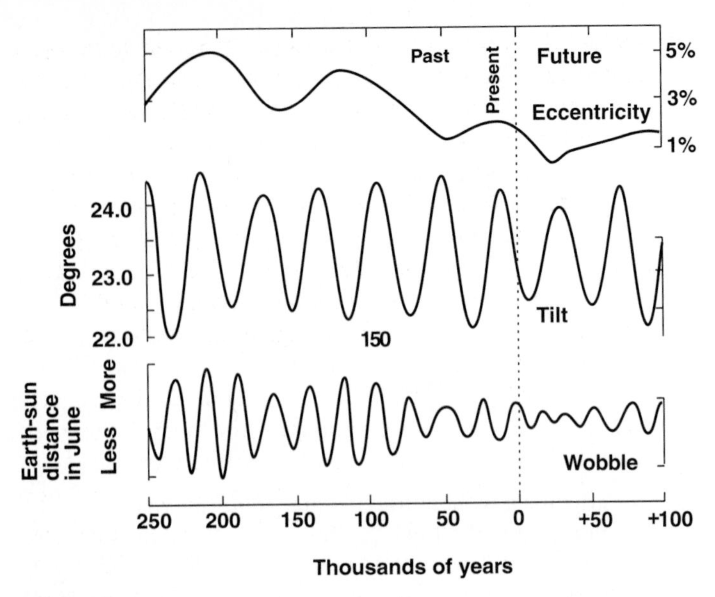

Figure 9.21 Patterns of eccentricity, tilt, and wobble for the past, present, and future.
Data from A. Berger.

The amount of seawater required to build the glaciers resulted in sea level 130 m (425 ft) lower than today (Figure 9.23). But extensive ice masses carry some of the seeds of their own destruction: shrinking ocean surface area plus colder ocean water mean less water evaporation with less snowfall. Cutting down on evaporation reduces the supply of snow necessary to maintain glaciers, thus they shrink.

Our current Ice Age is not over. We live during one of the colder intervals in Earth history, despite the current glacial retreat. About 10 percent of the continents today remain buried beneath about 25 million km^3 of ice, primarily on Antarctica and Greenland. If this ice were to melt, with or without human help, sea level would rise about 65 m (210 ft), and many of the world's major cities would be submerged. The changes associated with marked climatic fluctuations are stressful for plants and animals; they test the resilience of species and help drive ones with low tolerance for environmental change toward extinction.

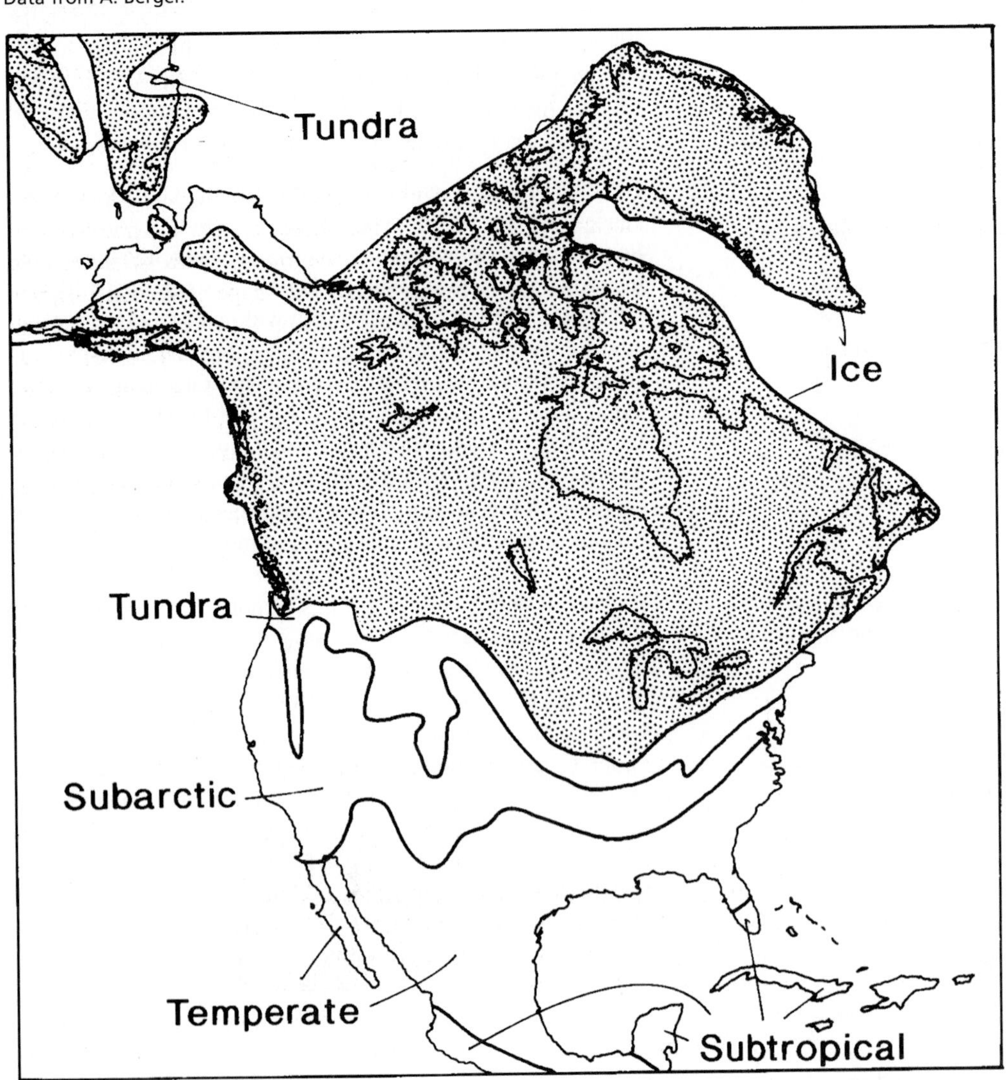

Figure 9.22 The maximum extent of glacial ice during the present Ice Age.

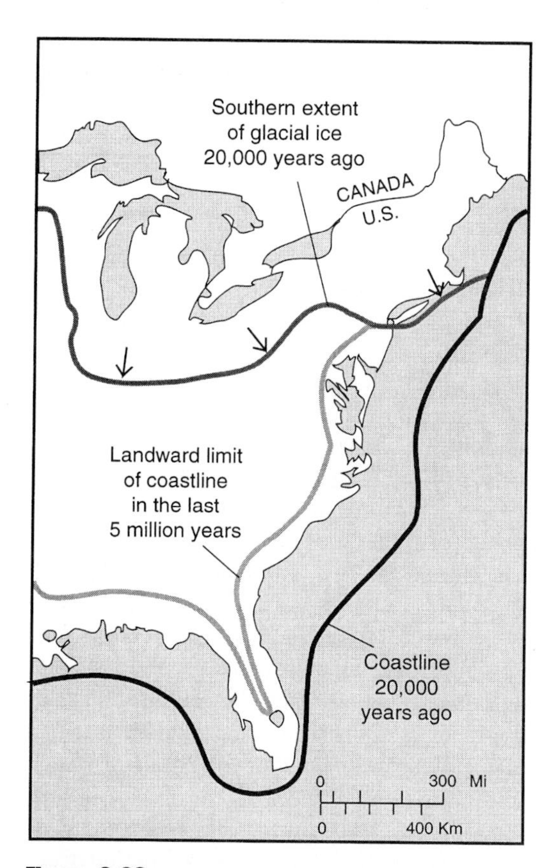

Figure 9.23 Some past positions of coastlines.
Source: S. J. Williams, et al., "Coasts in Crisis." *U.S. Geological Survey Circular 1075,* 1990.

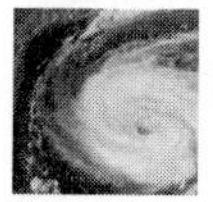

Climate Variations: Time Scale in Hundreds of Years

The air temperature over the Greenland ice cap has significant warm stages followed by colder intervals, as recorded by oxygen-isotope ratios in ice layers (Figure 9.24). Even during the heart of the latest Ice Age glacial advance, from 20,000 years ago and earlier, there were spikes of warmer temperatures. Look at the temperature conditions in Figure 9.24 following 20,000 years ago: 1) conditions began to warm; 2) the warming was interrupted by the Older Dryas cold stage; 3) the cold interval was suddenly replaced by the elevated temperatures of the Bølling period; 4) the higher temperatures deteriorated through the Allerød interval; 5) then temperatures plunged back into the depths of the Ice Age during the Younger Dryas stage from 12,900 to 11,600 years ago; and 6) last came the current interglacial period.

Look again at Figure 9.24 and note the sharp rises and falls in average annual temperature. How much did temperature rise or fall in a brief time? Temperature changes of 3° to 5° C (5° to 9° F) occurred in a few years. Rates of temperature change used to be viewed as occurring gradually analogous to using a dimmer switch to lower the lights. The rapid temperature changes recorded in Greenland ice show us that the best analogy may not be the dimmer, but the on-off switch. If one of these rapid temperature changes occurred today, life as we know it would change markedly. Rain fall patterns would change as some wet areas became dry and some dry areas became wet. Crop-growing lands of the world would change with some countries gaining and some losing.

Why the sudden jumps or drops in temperature? One suggested cause relies on changes in the North Atlantic Ocean. As the massive ice sheets on the continents were melting, there were formed enormous lakes of pure, cold water held back by ice dams. The shape of the land surface (see last section of Chapter 12) and the sediment record tell of enormous floods produced by the failure of the ice dams. Floods of cold, glacial meltwater would flow from the Mississippi, St. Lawrence, Columbia, and other rivers out on top of seawater, creating a surface layer of cold, nonsaline water. In the North Atlantic Ocean, this cold surface-water layer would alter the ocean-circulation pattern shown in Figure 9.3

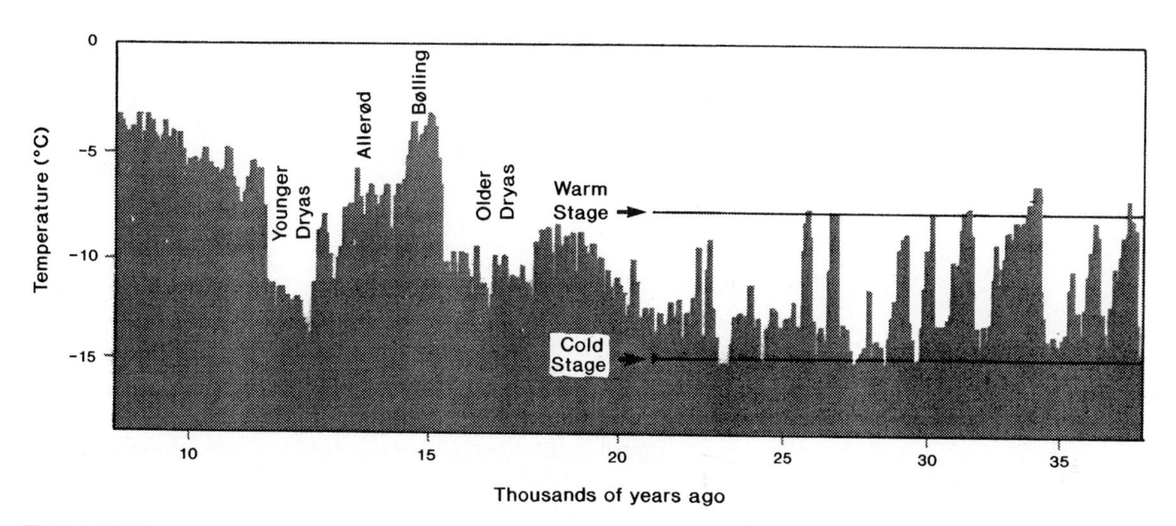

Figure 9.24 Air temperature over the Greenland continental glacier as recorded by oxygen-isotope compositions of ice at the glacier summit.

by stopping Arctic seawater from sinking and by blocking the northward inflow of Gulf Stream warm water.

As long as North Atlantic cold, salty seawater sinks and flows away as deep-ocean currents, it is replaced by warm surface water flowing up from the Gulf of Mexico and Caribbean Sea. With warm water in northern latitudes the winds pick up tremendous amounts of heat from warmer surface water and warm the adjacent lands of Greenland and Europe. But when Arctic surface water is fresh and cold, then its low density prevents sinking and its cold temperature results in colder air temperatures. This is apparently what happened during the Younger Dryas (Figure 9.24). Glacial meltwater floods from 12,900 to 12,700 years ago put cold freshwater on top of the North Atlantic Ocean shutting down the circulation system of Figure 9.3. It took another 1,100 years for solar energy to return the ocean surface to its warmer, saltier condition and the present circulation system.

Remember that sea level was 130 m (425 ft) lower at the peak of continental glaciation and that the removed water was stored on land as glacial ice. When the glaciers retreated, sea level rose by the inflow of cold, freshwater freed by the melting of glacial ice. The return of this massive volume of meltwater affected the oceanic distribution of heat; this is a climate-modifying process. The last melting of the ice sheets is recorded by the sea-level rise curve (Figure 9.25).

At about 7,000 years ago, average global temperatures were warmer and rainfall totals had risen. At this time, known as the "climatic optimum," even North Africa had enough rainfall to support civilizations. Since then, there has been a 7,000-year-long lowering of global average temperature totaling about 2°C. However, the cooling trend has had several smaller cycles of glacial expansion and contraction superimposed on it (Figure 9.26). It seems that climatic cycles can be found at any time scale we choose to use.

The Last Thousand Years

During the last thousand years, the combined effects of Earth's orbital patterns of eccentricity, tilt, and wobble caused a cooling trend but climate records show numerous variations testifying to other processes also at work (Figure 9.27). The climate variations seen in Figure 9.27 are actively being studied to learn more about: 1) the extent of the temperature fluctuations, 2) whether they were regional or occurred simultaneously around the world, and 3) what were the causes of the changes. Answers to these questions are being sought using scientific data such as oxygen isotopes in glacial ice layers and in the annual growth rings of corals, and in tree-ring widths and densities. The last thousand years finds us in the realm of human history with ever-improving observational records. Historic records studied to learn about past climates include: 1) tax records of grain and grape crops, 2) advances and retreats of mountain glaciers, 3) paintings of winter scenes showing frozen lakes, rivers, and ports, and 4) numbers of weeks per year of sea ice around Iceland.

Looking at Figure 9.27 reveals a warm period from about 1000 to 1300 C.E. referred to as the **Medieval Maximum.** During this time, northern Europeans emigrated to Iceland where the almost ice-free coast helped fishermen build thriving industries. The coastal plains of Greenland also were settled by Europeans who farmed the land. In England, wine grapes were grown and harvested.

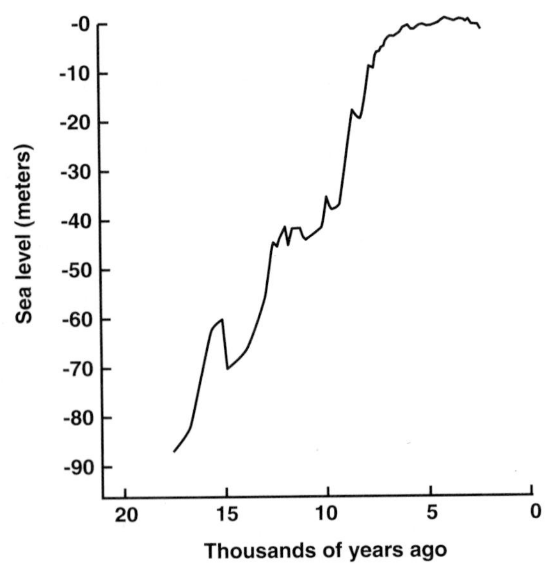

Figure 9.25 Rise in sea level in northwest Europe during the last 18,000 years.

After Morner, 1971.

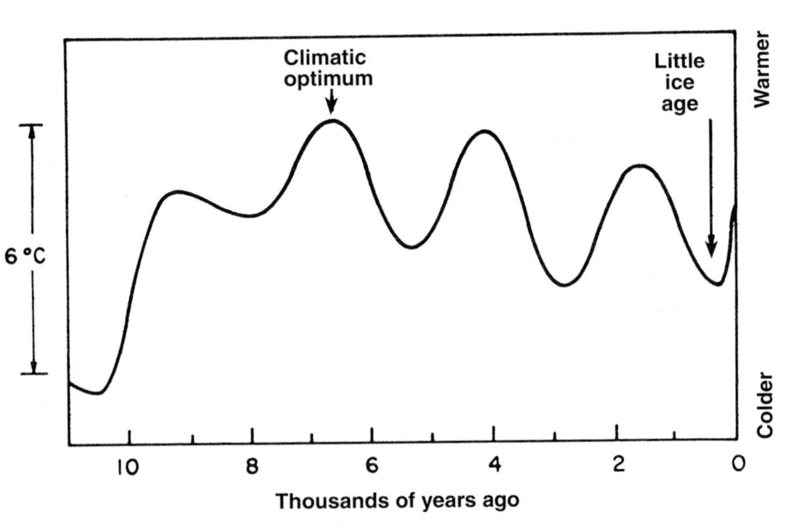

Figure 9.26 Generalized trends in global temperature for the last 11,000 years.

Data from J. Imbrie and K. P. Imbrie, *Ice Ages.* Cambridge, Mass.: Harvard University Press.

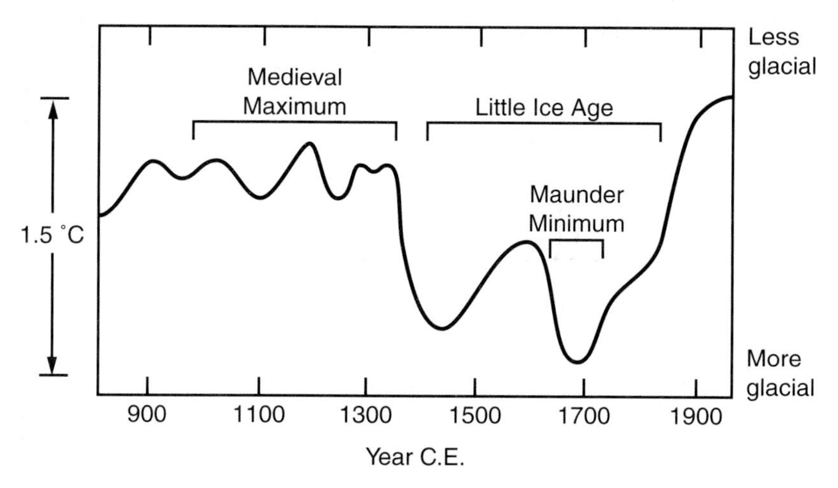

Figure 9.27 Climate of the last 1,000 years, based on European winters.

Data from J. Imbrie, and K. P. Imbrie, *Ice Ages*. Cambridge, MA: Harvard University Press.

The **Little Ice Age** affected Europe from about 1400 to 1900 C.E. It was originally defined by Francois Matthes in 1939 as an "epoch of renewed but moderate glaciation." Late in the Little Ice Age, part of northeastern Canada had accumulated permanent snowfields and the beginning of an ice sheet. Cold winters in Europe led to shorter growing seasons with reduced crop yields leading to local famine. Mountain glaciers advanced throughout Europe. The fishing industry in Iceland was slowed by many weeks of sea ice each year.

Climatic conditions during the Little Ice Age were far from constant as smaller-scale warmings and coolings occurred (Figure 9.27). One colder interval between 1645 to 1715 C.E. is known as the **Maunder Minimum.** During this time, minimal sunspot activity was noted by astronomers and it is estimated that the Sun may have been 0.25 percent weaker.

What processes were involved in the climate changes of the last thousand years? 1) Changes in Earth's orbital patterns caused cooling. 2) A lessening of solar-energy production caused cooling. 3) Volcanism caused changes (see later in chapter). 4) There probably were interactions between the ocean, the atmosphere, and the ice sheets that are yet to be understood.

Look at Figure 9.27 again, at arm's length, and notice the thousand-year pattern—a warm upswing for the first few hundred years, followed by a downswing until the twentieth century. Is this pattern part of a millennium cycle? If so, it suggests that the twentieth century was the first of a few warm centuries. Human activities added to the twentieth-century warming combined to produce a century of warmth that is unprecedented in both amount and rate in the last 1,100 years. We will discuss present and future climate trends and causes at the end of this chapter.

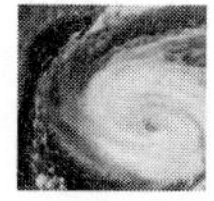

Shorter-Term Climatic Changes: Time Scale in Multiyears

To complicate the picture further, several processes change climate on time scales of one to a few years. Let us look at examples.

El Niño

The high heat capacity of water gives it the ability to absorb and store tremendous volumes of heat upon warming and to release copious quantities of heat upon cooling. The heat supply in the world ocean has a major effect on the atmosphere and on world climate. An example of ocean-atmosphere coupling is the phenomenon commonly marked in South America by the arrival of warm ocean water to Peru and Ecuador near Christmastime, where it is known as **El Niño** (Spanish for "the child").

Typical conditions in the central Pacific Ocean find high atmospheric pressure over the eastern Pacific Ocean resulting in trade winds that blow toward the equator from the north and south. The trade winds push Pacific Ocean surface waters to the west within the equatorial zone, where they absorb solar energy (Figure 9.28a). The winds push so hard that sea level is not level; it is about 1.5 ft higher on the western side of the ocean. The warm water piled up on the west side forms a pool of warm water that evaporates readily, helping produce heavy rainfalls for the tropical jungles of Indonesia and Southeast Asia and providing the environment for the Great Barrier Reef of Australia. Meanwhile, on the eastern side of the Pacific Ocean, the warm surface water blown west is replaced by cold waters rising from depth (Figure 9.28b) and from the polar regions. The colder waters along the coast evaporate less readily, and thus deserts are common along the coasts of Ecuador, Peru, Baja California, and California because of the shortage of cloud-producing water vapor.

Every two to seven years, the typical ocean-atmosphere pattern breaks down. The trade winds weaken, the atmospheric

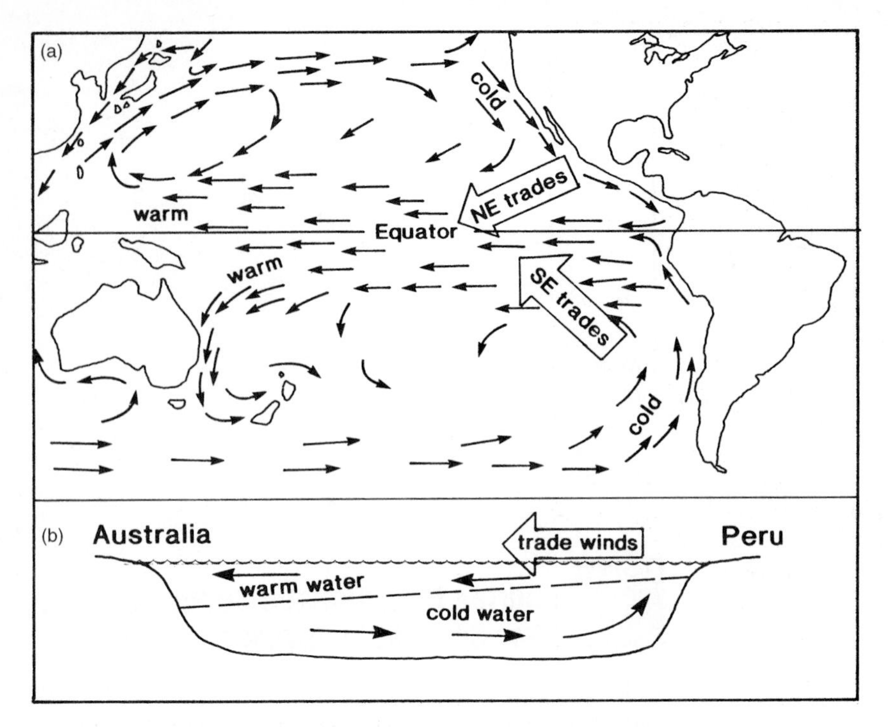

Figure 9.28 Pacific Ocean circulation. **(a)** Map. The northeast and southeast trade winds combine to push warming surface water westward across the ocean in the equatorial belt. After circling near the poles, the return water flowing along the North and South America coasts is cold. **(b)** Schematic cross section. The trade winds stack up warm surface water on the western side of the Pacific Ocean.

low pressure over Indonesia moves out over the central Pacific Ocean, and winds blow into the Pacific basin from the west (Figure 9.29). The warm surface waters then flow "downhill" toward South and Central America. Some surface currents are reversed as some winds from the west blow surface water to the east. The reversal places a huge mass of warm water against the Americas which evaporates more readily, producing more clouds and lowering the atmospheric pressure. As the pressure declines, warm moist air will flow eastward off the ocean into the Americas leading to heavier rains than the coastal deserts can handle. The high-level atmospheric winds (**jet stream**) are affected also, including a southern branch that flows eastward bringing higher rainfall to the southeastern United States and helping break apart Atlantic and Caribbean storms resulting in fewer hurricanes.

The El Niño of 1982–83 was especially strong. On the eastern side of the Pacific Ocean, the cold-water fisheries off Peru and Ecuador collapsed as ocean water warmed up as much as 14°F above normal. The warm water led to greater than normal evaporation, which fed torrential rainfalls. The rains caused overwhelming floods and mass movements from steep hillsides that killed 600 in Peru and Ecuador and severely punished the economies of those nations. The warm coastal waters also promoted heavy rainstorms in the western United States. For example, California suffered $300 million in damages, 10,000 people were evacuated due to flooding and landsliding, and 12 people were killed. Meanwhile, out in the ocean, the tropical rain belt shifted to the central Pacific Ocean helping form hurricanes which hit Hawaii and Tahiti. On the western side of the Pacific Ocean, Australia and Indonesia were covered by high-pressure air and below-average rainfalls. Australia suffered its worst drought of the century, and out-of-control bushfires whipped by high winds killed 75 people and many domestic and wild animals and caused $2.5 billion in damages (see Chapter 13).

The El Niño effects in the Pacific Ocean basin are impressive, but the ocean-atmosphere system is linked on a larger scale. In the Indian Ocean, the shifting of weather patterns is known as the Southern Oscillation; it occurs when the usual low-pressure atmosphere is replaced by high-pressure air, as measured at Darwin on the north coast of Australia (Figure 9.30). For example, in late 1996, strong weather systems in the Indian Ocean migrated into the western Pacific Ocean triggering the strong El Niño of 1997–98.

As our knowledge grows, it is increasingly evident that unusually high rains in one area and drought in another are not isolated events; rather they are both parts of a globally connected weather system. A change that occurs in one area can trigger other changes around the world, somewhat like a falling line of dominoes—knock over one and it starts the process whereby they all fall down. Every several years the tropical atmosphere goes through changes that link up around the world. The tropical Indian Ocean warms and its weather pattern blows into the western Pacific Ocean setting off an El Niño. After the El Niño wind shifts cross the Pacific Ocean and South America, the tropical Atlantic Ocean

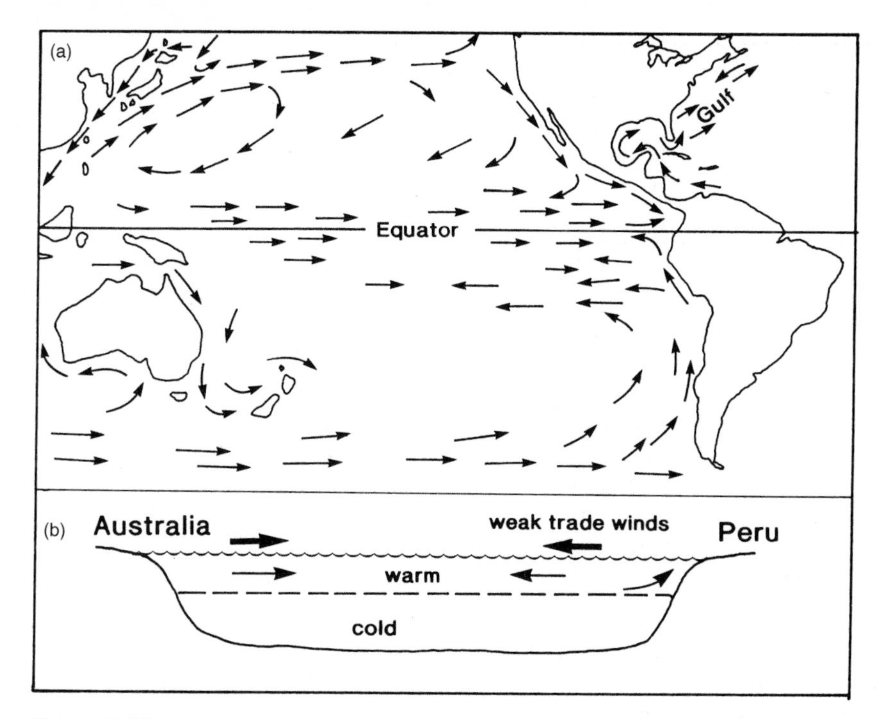

Figure 9.29 Pacific Ocean circulation during El Niño. **(a)** Map. Equatorial winds blow toward the center of the ocean from both sides. **(b)** Schematic cross section. Weakened trade winds plus winds from the west cause warm water to accumulate along the equatorial Americas.

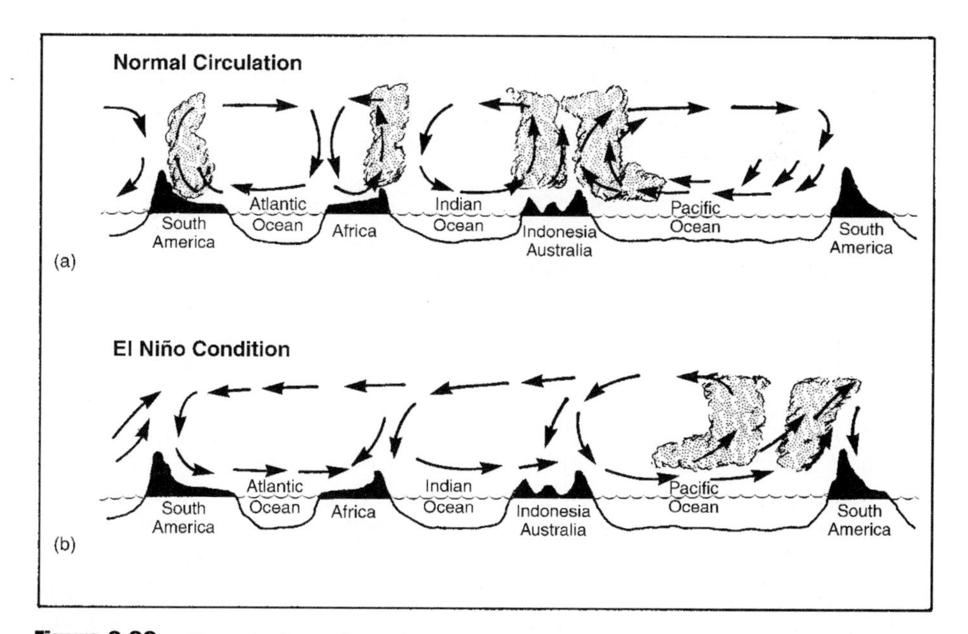

Figure 9.30 Cross sections of Southern Hemisphere atmospheric circulation. (a) With "normal circulation," moist air rises over the landmasses, condenses, and falls as rain on eastern South America, eastern Africa, and Indonesia-Australia. (b) With El Niño conditions, eastern Africa and Indonesia-Australia do not get their customary rain, while western America receives heavy rainfall.

begins to warm. This global circuit takes about 4 years to move around the Earth.

La Niña

The surface waters of the ocean are a mosaic of warmer, intermediate, and cooler water masses that exert strong controls upon regional weather. The Pacific Ocean surface-water masses of different temperature are measured by the Topex-Poseidon satellite and are shown at http://topex-www.jpl.nasa.gov. When warmer than normal surface waters extend along the west coast of the Americas during an El Niño they usually bring high rainfalls and accompanying floods.

But El Niño has a sister called **La Niña** (the girl) and she has a different personality.

La Niña occurs when cooler water moves into the equatorial Pacific Ocean and wind systems change their normal paths bringing different weather patterns across North America. A typical La Niña winter brings cold air with high rainfall to the northwestern United States and western Canada, but causes warming and below average rainfall in most of North America. The winter of 1999–2000 was typical with heavy rainfalls in the northwest, and below average rainfalls common elsewhere accompanied by numerous wildfires in the west.

There are hazards associated with the La Niña cooling in the Pacific Ocean. La Niña allows the growth of hurricanes in the Atlantic Ocean, spelling trouble for the eastern United States and Gulf of Mexico coastal areas. La Niña leads to decreased rainfall in the American southwest helping dry out the El Niño-fed vegetation, leading to wildfires as in the summers of 2000 and 2002.

El Niño and La Niña are the extreme conditions where warm or cold water masses strongly influence the distribution of rainfall. Extreme ocean conditions make weather prediction easier. There are times when the tropical Pacific Ocean is neither excessively warm nor markedly cool but instead is neutral. The weaker signal from the ocean makes weather more difficult to predict. NASA oceanographer William Patzert has suggested the term **La Nada** for this neutral condition.

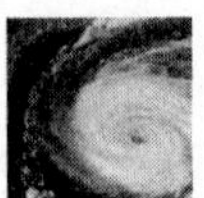

Volcanism and Climate

Benjamin Franklin recognized in 1784 that volcanism could affect the weather. He suggested that the haze and cold weather in Europe during 1783–84 were due to the massive outpourings of lava and gas at Laki, Iceland (see Chapter 7). How else can volcanism affect the climate? Large explosive Plinian eruptions can blast fine ash and gas high enough to be above the normal zone of weather. Free from the cleansing effects of rainfall at these heights, the volcanic products can float about for years and interfere with incoming sunlight.

Most of the moisture and "weather" occurs in the lowest layer of the atmosphere, the **troposphere** (Figure 9.31). The troposphere ranges from about 8 km (5 mi) thick at the poles to 18 km (11 mi) at the equator. The troposphere is warmer at its base and colder above, thus creating a basic instability as lower-level warm air rises and upper-level cold and dense air sinks. The density contrasts set off a constant mixing of tropospheric air, that is part of our changing weather pattern.

The top of the troposphere, the **tropopause,** is a significant boundary where the cooling-upward trend reverses and the air begins to warm upward through the **stratosphere** (Figure 9.31). The temperature inversion acts as a barrier or

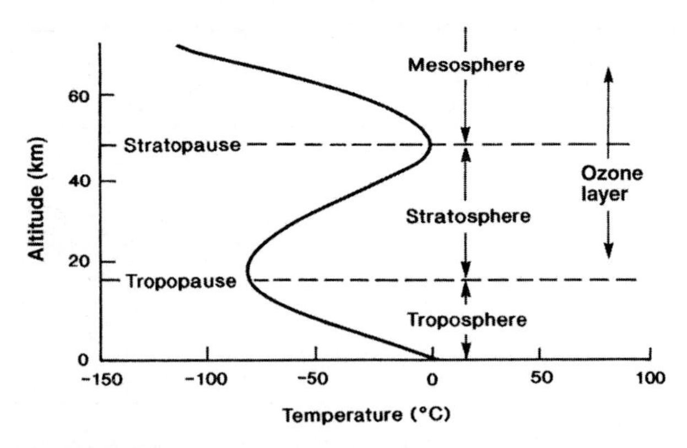

Figure 9.31 Cross section through Earth's lower atmosphere. Ozone captures solar energy. The temperature inversion at the tropopause acts as a lid holding moisture and "weather" in the troposphere.

lid that confines weather to the troposphere below. Above the tropopause lies the stratosphere, which draws its name from its stratified condition where warmer air sits on top of cooler air; this is a stable configuration. The stratosphere is home to cold, dry, and ever sunny conditions. Without rainfall and with little vertical mixing, the stratosphere allows the finest volcanic ash (0.001 mm) to stay suspended for years. Most gases blown into the stratosphere disappear into space, but sulfur dioxide (SO_2) picks up oxygen and water to form an aerosol of sulfuric acid (H_2SO_4) that may stay aloft for years. The combined ash and sulfuric acid produce **haze,** reducing the amount of sunshine that reaches the troposphere and the ground surface; thus, climatic cooling results.

El Chichón, 1982 Located in the state of Chiapas in southern Mexico is the relatively small volcano called El Chichón (which translates loosely as "bump"). Four big Plinian eruptions from El Chichón on 28 March and 4 April 1982 blew out about 0.6 km^3 of material, leaving a 1 km diameter crater and killing 2,000 people (see Chapter 7). Although the eruptions were not as big as the Mount St. Helens event in 1980, over 100 times the volume of SO_2 gases was pumped into the stratosphere along with ashes. The cloud of stratospheric gases took 23 days to circle the globe (Figure 9.32). The SO_2 gas combined with O_2 and water vapor, converting to sulfuric acid (H_2SO_4) aerosol. Sunsets were spectacular for months, beginning with a purple glow high over the horizon, changing gradually to surreal yellows and oranges as the Sun set and, finally, a red afterglow took over when the sky normally would have been dark. World temperature from this event was lowered 0.2°C.

The injection of SO_2 gases into the stratosphere by El Chichón in spring 1982 was followed by the strong El Niño phenomenon of 1982–83. Was there a relationship between El Chichón and El Niño? Or was it a coincidence? Did the aerosols from El Chichón reduce the inflow of solar energy to the Earth's surface, thus weakening the trade winds? If strong trade winds do not blow warm surface water to the

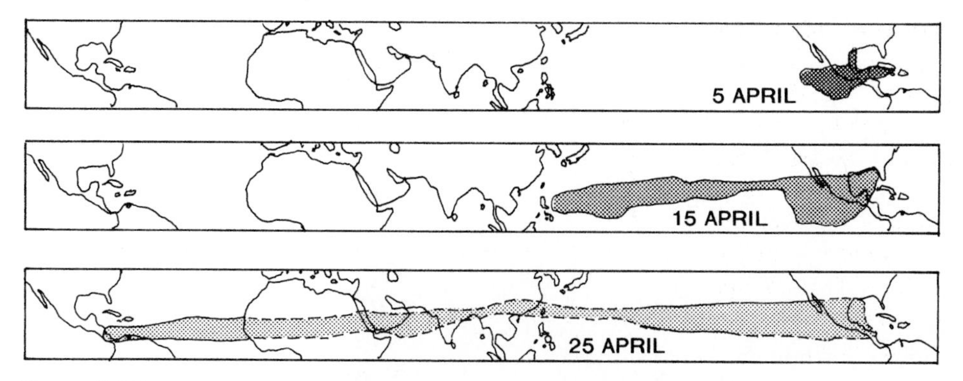

Figure 9.32 The El Chichón gaseous cloud moving west in the stratosphere.

From Robock, Alan, and Michael Matson, 1983, "Circumglobal transport of the El Chichón volcanic dust cloud" in *Science,* 221:195–97. Reprinted by permission of the author.

west across the Pacific Ocean, then an El Niño condition results with stacking of warm water against South and North America.

Mount Pinatubo, 1991 After a slumber of 635 years, Mount Pinatubo awoke to disrupt life on the Philippine island of Luzon in the spring of 1991. The 1,745 m (5,724 ft) high summit was blasted to bits and replaced by a 2 km (1.25 mi) wide caldera as up to 5 km^3 (1.2 mi^3) of dense magma was blown out as pyroclastic debris. Despite ample warnings before the explosive events, over 300 people died in pyroclastic flows and lahars. Adding to the tremendous destruction of property was a major storm that poured torrential rains on the loose pyroclastic debris, setting in motion numerous large-volume lahars.

Of climatic importance were the 20 million tons of SO_2 gas blasted into the stratosphere (Figure 9.33)—triple the volume of SO_2 released by El Chichón. The H_2SO_4 aerosols reflected 2 to 4 percent of incoming short-wavelength solar radiation back to space, causing a 20 to 30 percent decline in solar radiation directly reaching the ground. Mean global temperatures at the ground surface dropped 0.5°C. The greatest cooling occurred in the midlatitudes of the Northern Hemisphere, including the United States, where temperatures declined by 1°C. The volcanically induced cooling from SO_2 in the stratosphere more than offset the greenhouse warming that was expected in 1991–92 due to the CO_2 added to the troposphere by humans burning wood, oil, coal, and natural gas.

Figure 9.33 Vertical eruptions from Mount Pinatubo in June 1991 injected 20 million tons of sulfur dioxide (SO_2) into the stratosphere.

Photo by K. Jackson, U.S. Air Force.

Tambora, 1815 In the early 1800s, Mount Tambora, on the island of Sumbawa in Indonesia, stood 4,000 m (13,000 ft) tall. After the explosive eruptions of 10–11 April 1815, Tambora was only 2,650 m (8,700 ft) high and had a caldera 6 km (4 mi) wide and 650 m (0.4 mi) deep (Figure 9.34). About 150 km^3 (36 mi^3) of rock and magma were blasted out during the eruption, producing 175 km^3 of ashes and other pyroclastic debris. The eruption has been called the greatest in historic times, killing about 10,000 people outright by pyroclastic flows and another 117,000 indirectly through famine and disease.

The ashes and aerosols, especially sulfur dioxide (SO_2), blown into the stratosphere blocked enough sunshine to make 1816 known as "the year without a summer." Agricultural

Figure 9.34 Space shuttle view of Mount Tambora on Sumbawa Island, Indonesia. The caldera formed by the 1815 eruption is 7 km wide and 650 m deep.
Photo courtesy of National Aeronautics and Space Administration.

production was down throughout the world as global temperatures were lowered another 0.3°C during an already cold series of years. Lord Byron spent a cold and darkened summer of 1816 on the shores of Lake Geneva and described it in his poem "Darkness."

> The bright Sun was extinguish'd, and the stars
> Did wander darkling in the eternal space
> Rayless and pathless, and the icy earth
> Swung blind and blackening in the moonless air;
> Morn came and went—and came,
> and brought no day . . .

On the other side of the world in the northeastern United States, snow or frost occurred in every month of the year. It is estimated that 8 percent of the residents of Vermont were forced to leave the state due to agricultural failures. It is suggested that the Tambora-cooled climate induced famine in India, weakening the population enough to trigger a cholera epidemic. The disease then slowly migrated around the world, killing people who lived under the harshest, least sanitary conditions.

Toba, Indonesia, ~ 74,000 years ago Tambora erupted an impressive 150 km^3 of material, but if we go back 74,000 years ago, the eruption of Toba on Sumatra expelled about 2,000 km^3 of material. The Toba event is the youngest known resurgent caldera eruption. It is estimated that the Toba ash and H_2SO_4 aerosols formed a dense cloud in the stratosphere lasting for up to six years. Global cooling may have been 3 to 5ºC (5 to 9ºF) for several years. A *volcanic winter* of this magnitude may have triggered additional climate responses that prolonged the cold climate and increased the severe drought, ecological disasters, and famine. It even has been speculated that the Toba eruption effects drove down the global population of humans to just thousands of people.

Volcanic Climate Effects

The eruptions of El Chichón, Pinatubo, Tambora, and Toba give an idea of the climatic effects of volcanism. Gas-rich eruptions can decrease the amount of incoming solar radiation and thus cause agricultural production to decline, which in turn can lead to famine, disease, and death. Yet each Plinian eruption affects the climate significantly for only a year or two. The rarer resurgent caldera eruptions (e.g., Toba) may cause longer-lasting climate changes.

What if eruptions from different volcanoes kept occurring at closely spaced intervals for a large number of years? It has been reported that the Greenland ice-core record is dominated by ice with high acid content for the years of the Little Ice Age. Is the acidic ice a record of abundant SO_2-rich volcanic eruptions that helped cause the Little Ice Age? More work must be done to answer this question. Conversely, the record of global warming from 1912 to 1952 has been interpreted as being partly due to the absence of major SO_2-rich volcanic eruptions.

What are the main variables that come into play when volcanism affects climate? Principal factors include:

1. The size and rate of eruptions.
2. The heights of eruption columns.
3. The types of gases and the atmospheric level where they are placed. Sulfur dioxide in the stratosphere reflects sunlight and cools the climate below. Halogens, usually chlorine, react with ozone, destroying it and thus weakening its capacity to shield organisms from the ravages of ultraviolet rays. Carbon dioxide in the atmosphere creates a greenhouse effect.
4. Low-latitude eruptions spread atmospheric debris across more of the world and have greater global effects.

What is the worst-case scenario for volcanic effects on climate? Probably the hypothesis that blames flood-basalt volcanism for the great die-off of species, including the dinosaurs, that occurred 65 million years ago. It is suggested that the extinctions resulted from the climatic effects of the voluminous flood basalts erupted to form the Deccan Plateau in India. The proposal is based on the fact that about 2.6 million km^3 (625,000 mi^3) of basaltic lavas poured forth in as short a time as 500,000 years. What might the climatic effects have been? Proponents of this hypothesis estimate that atmospheric CO_2 increased 10 to 25 percent, with consequent global warming that raised average world temperature as much as 10°C (18°F). The elevated surface temperatures, more acidic ocean waters (carbonic, sulfuric, and nitric acids), and possibly a depleted ozone layer all combined to deal punishing blows to life on land and in the uppermost layer of the ocean. Although the processes described above are all reasonable, most quantitative calculations suggest that the effects are overstated and that they were not nearly large enough to cause the widespread deaths and extinctions.

Drought and Famine

Drought does not equal desert. Drought describes times of abnormal dryness in a region when the usual rains do not appear, and all life must adjust to the unexpected shortage of water. The lack, or reduction, of moisture can cause agricultural collapse or shortfalls bringing famine and disease and causing deaths and mass migrations to wetter areas. Famine is the slowest moving of all disasters. Earthquakes, volcanic eruptions, tornadoes, and the like all hit suddenly and with great force, then quickly are gone. But famine is slow. First, the expected rains do not arrive, then vegetation begins to wither, food supplies shrink, and finally famine sets in.

Unlike other natural disasters, drought tends to drive people apart rather than bring them closer together. The shortages of food and water lead to conflicts as people, communities, and governments battle each other for the means to survive. After an earthquake or flood, people are commonly at their best as they aid their neighbors and strangers in need; during a drought, people are typically at their worst as they fight for survival.

In the *early stage* of a famine, food is still available, but there is not enough. Healthy people can lose up to 10 percent of their body weight and still remain mentally alert and physically vigorous. In the *advanced stage,* body weight decreases by around 20 percent, and the body reacts to preserve life itself. Body cells lower their activity levels, reducing the energy needed to keep vital functions going. People sink into apathy. In the *near-death stage,* when 30 percent or more of body weight has been lost, people become indifferent to their surroundings and to the sufferings of others, and death approaches.

U.S. Dust Bowl, 1930s

One of the greatest weather disasters in U.S. history occurred during the 1930s, when several years of drought turned grain-growing areas in the center of the nation into the "Dust Bowl." Failed crops and malnutrition caused abandonment of thousands of farms and the broad-scale migration of displaced people, mostly to California and other western states. This human drama was captured in many articles and books, including *The Grapes of Wrath* by John Steinbeck.

What happened to cause the drought? Recurrent large-scale meanders in the upper-air flow created ridges of high pressure with clockwise flows resulting in descending air. The upper-level high-pressure air was already dry, but as it sank, it became warmer, thus reaching the ground hot, dry, and thirsty. As the winds blew across the ground surface, they sucked up moisture, killing plants and exposing bare soil to erosion. Wind-blown clouds of dust built into towering masses of turbulent air and dust called rollers (Figure 9.35). When they rolled across an area, the Sun was darkened, and

Figure 9.35 A typical dust storm rolling through the 1930s Dust Bowl.

Photo courtesy of the USDA Soil Conservation Service.

dust invaded every possible opening on a human body and came through every crack in a home. Dust even blew as visible masses across East Coast cities and blanketed ships at sea.

The drought began in 1930, a particularly bad time. Only months before, in October 1929, the U.S. stock market crashed, and the nation's economy began sinking into the Great Depression. By 1931, farmers were becoming desperate. For example, a group of 500 armed farmers went to the town of England in Arkansas to seek food from the Red Cross. They were denied aid, so they went to the town's stores and took the food they needed. The event drew worldwide attention. Here were farmers from the U.S. heartland who helped feed the world during the early twentieth-century and through the ravages of World War I; now they could not even feed themselves.

The dust storms became even worse in 1934 and 1936. Some of the blame for the Dust Bowl was heaped onto the farmers for plowing deeply through drought-tolerant native grasses and exposing bare soil to the winds. The plowed lands were sowed with seeds of plants that could not handle drought and thus died, exposing more soil. The farming practices were not the best, but they did not cause the drought: they just accentuated its effects. Evidence showing that droughts are common is found in the archaeological record. For example, droughts in the past led to the downfall of the Anasazi civilization of the southwestern United States and the migration of its people to areas with more dependable water supplies.

The Dust Bowl drought affected more than just local agriculture. The drought, combined with the stock market crash, caused fundamental changes in the economic, social, and political systems of the United States.

Sub-Sahelian Africa, 1968–1975

Sahel is an Arabic word for "shore" or "boundary"; the shore here is the southern margin of the great Sahara Desert. South of the Sahel, between roughly 14° to 18°N latitude, lies dry, shrubby land with grasses similar to the American Southwest. In Africa, approximately 25 million people live in this zone as nomadic herders in the north and subsistence farmers in the south. All of these people depend upon the 14 to 23 inches of annual rain that falls in downpours, typically in June and July. It is in these months that the desert air heats and rises, and is replaced by moister air from the Atlantic Ocean to the southwest: the Atlantic air yields the rainfall.

In 1968, the rainfall was only about half of normal. This was the beginning of a drought that lasted until 1975. By the time the rains finally resumed, about 200,000 people had died due to famine, millions of herd animals were dead, and many thousands of children were brain-damaged from inadequate nutrition. The economy of the region collapsed, affecting several countries, including Mauritania, Senegal, Mali, Niger, Chad, Sudan, Ethiopia, and Somalia. The drought caused massive changes in life-style, large shifts in population, and a social upheaval that continues today.

Why did the rainfall drop off so devastatingly? Droughts in the tropical zone are commonly due to the position of the intertropical convergence zone where the trade winds from each hemisphere meet (Figure 9.2). The trade winds begin around 30° latitude where air descends and moves toward the equator as dry winds picking up moisture from the lands they cross. When these winds meet near the equator, they rise and cool, dropping their voluminous moisture on the tropics. In much of 1968–1975, the convergence zone moved about 2° farther south, leaving sub-Sahelian Africa exposed to drying winds that picked up moisture to be dropped farther south.

The geological and archaeological records show that periodic droughts and arid conditions along the Sahara Desert's southern margin are common. The recurrence of droughts through time apparently reflects earlier shifts of the convergence zone. In the future, the convergence zone will shift again, leaving drought behind it. The human situation will be even more difficult due to deforestation, formation of hardgrounds, and increased population.

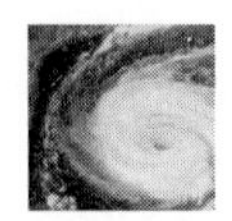

The Twentieth and Twenty-First Centuries

We frequently read and hear that global warming is in progress. To appreciate the twentieth-century warming, it is necessary to examine the climate history of the preceding 1,000 years. Figure 9.27 shows average temperatures fluctuating about 1.2°C (2°F) between the late 800s and the late 1800s. But Figure 9.27 also shows temperature rising at the close of the nineteenth century to begin the twentieth century at as high a temperature as existed in the preceding 1,000 years.

How much did global temperature rise in the twentieth century? Average global surface temperature rose 0.6ºC (+/– 0.2ºC), or 1ºF. This increase elevated global temperature to its highest level in over 1,000 years.

Could a person feel the climate warming of the twentieth century? No, because the climate warming is small compared to the day-to-day temperature fluctuations of weather. Although the human body cannot feel the warming, the human eye and brain can record many examples of climate warming (Table 9.3).

Did it warm continuously through the twentieth century? No, most of the warming occurred in two time intervals: 1910 to 1945 and since 1976. It appears that the early warming was largely due to a hotter Sun and a lack of global volcanism. The present warming is about double that of 1910 to 1945, and it is mostly due to increases in greenhouse gases in the atmosphere.

How much of the twentieth century warming was due to natural processes and how much was due to human activities? Natural processes causing a net increase in temperature of about 0.2º C with changes in Earth's orbital patterns causing a cooling of –0.02ºC that was offset by a hotter Sun netting a +0.2ºC warming. Human activities were responsible for the remaining 0.4ºC increase in global temperature.

How do humans cause the global climate to warm? We add large volumes of greenhouse gases to the atmosphere each year, and the amounts we add increase each year. At the beginning of this chapter we looked at the early Earth greenhouse, but now we need to revisit the greenhouse effect in its modern condition and see what we are doing to increase it.

The Greenhouse Effect Today

The greenhouse effect is not something new. Earth has always had an atmosphere and always had its surface climate warmed by the greenhouse effect, but the strength of the greenhouse has varied. Remember that early Earth surface temperatures were about 290ºC (570ºF). Our present surface temperatures are radically lower and have varied little during the past few centuries when the human race made great advances in many areas of life. Many people question whether it is wise for us to release huge volumes of greenhouse gases and change a climate system we have thrived under, but we are changing it anyway.

What greenhouse gases are we adding to the atmosphere now? Carbon dioxide, methane, nitrous oxide, ozone, and several industrially produced gases including the chlorine- and fluorine-bearing chlorofluorocarbons. Table 9.4 lists these greenhouse gases, states how responsible each has been for global warming, and assesses the relative ability of each gas to trap heat via the gas's **global warming potential (GWP)** compared to carbon dioxide.

Remember what these gases do. Chapter 1 describes the solar energy reaching Earth as having short wavelengths in or near the visible spectrum. Earth's atmosphere reflects back about 30 percent of incoming solar radiation, but 23 percent passes through to power the hydrologic cycle while the remaining 47 percent is absorbed by air, sea, and land. As absorbed heat builds up, some is reradiated outward in the longer, infrared wavelengths, but the greenhouse gases prevent their escape from Earth. The greater the volume of greenhouse gases in the atmosphere, the greater the amount

Table 9.3 Some Observed and Measured Effects of Global Warming in Recent Decades

Top 3 km (2 mi) of ocean water absorbed ~90% of added greenhouse warmth; other 10% melted ice and is held in atmosphere.

Satellites show 10% decrease in snow and ice cover since 1960s.

Freeze-free periods are lengthening in mid- and high-latitudes.

Asymmetrical warming in many regions. Daily low temperatures are increasing at twice the rate of daily high temperatures.

Longer growing seasons.

Earlier arrival of spring climate in Europe and North America.

1. Earlier breeding of birds.
2. Earlier arrival of migrant birds.
3. Earlier appearance of butterflies.
4. Earlier spawning of amphibians.
5. Earlier flowering of plants.

Population shifts in latitude and altitude.

1. Europe and New Zealand: tree line climbing to higher altitudes.
2. Alaska: expansion of shrub-covered area.
3. North Atlantic Ocean and offshore California: increasing abundance of warm-water species.
4. Europe and North America: 39 butterfly species extended their ranges northward up to 200 km (125 mi).
5. Costa Rica: lowland birds extended their ranges to higher elevations.
6. Britain: 12 bird species extended their ranges northward an average of 19 km (12 mi).
7. Canada: red foxes extended their range northward while arctic fox range retreated.

Table 9.4 Greenhouse Gases

Gas	Relative Percent Responsible for Greenhouse Warming	Ability to Trap Heat (compared to CO_2 = 1)
Carbon dioxide (CO_2)	60	1
Methane (CH_4)	16	21
Nitrous oxide (N_2O)	5	310
Ozone (O_3)	8	2,000
Chlorofluorocarbons (CFCs)	11	~12,000

of reradiated heat that is absorbed and prevented from escaping into space. The trapped heat is held in Earth's climate system. Global warming is so important that we need to know something about each of these gases.

Carbon Dioxide (CO_2) About 60 percent of the greenhouse warming caused by humans comes from releasing CO_2 into the atmosphere (Table 9.4). How does carbon cycle through Earth's surface environments? The element carbon is a major building block of life on Earth. CO_2 is removed from the atmosphere by plants during photosynthesis to build their tissue. Upon death much of the organic tissue is oxidized, and CO_2 both returns to the atmosphere and dissolves in water. Humans have disturbed the carbon cycle by decomposing plants at ever-increasing rates and thus causing CO_2 to increase in the atmosphere and water. In the year 1800, the CO_2 concentration in the atmosphere was about 280 parts per million (ppm), but it had increased to about 370 ppm in 2000. The increase is greater than 30 percent, and about 70 percent of that increase has occurred since 1950.

How do we decompose plants to release CO_2? In two main ways: burning wood and burning fossil fuels. About 20 percent of the increased atmospheric CO_2 is due to humans burning wood to clear land for agriculture, to heat homes, and to make charcoal for furnaces. Fossil fuels are coal, oil, and natural gas. They form during transformation of dead plant material in swamps, river deltas, and other organic-rich environments after burial beneath sediments. Over 80 percent of the energy that powers our global societies is generated by burning fossil fuels, and these energy-producing processes have added about 80 percent of the excess CO_2 now in the atmosphere.

How is CO_2 removed from the atmosphere? The scorecard for the latter part of the twentieth century shows about 20 percent is removed by plants during photosynthesis, about 25 percent dissolves in ocean water, and the remaining 55 percent stays in the atmosphere and traps reradiated heat.

Methane (CH_4) About 16 percent of modern greenhouse warming has come from adding methane to the atmosphere. Notice in Table 9.4 that the global warming potential or heat-trapping ability of methane is 21 times greater than that of carbon dioxide. Imagine what global warming would be like if methane were released each day in the same volume as carbon dioxide.

Air trapped in ice tells us that methane concentrations in the year 1750 were about 700 parts per billion (ppb), but they have risen more than 150 percent since then. The increase in atmospheric methane was slow in the nineteenth century and rapid in the twentieth century.

How is methane released to the atmosphere? It is released during decomposition of vegetation in oxygen-poor environments such as swamps, rice paddies, and cattle digestive systems. Bacteria remove carbon (C) from dead vegetation, and if oxygen is absent, the carbon combines with hydrogen (H) to make methane (CH_4).

How much methane is released by natural processes and how much by human activities? About 30 percent of methane release occurs by natural decomposition, mostly in wetlands and secondarily via termites. About 70 percent is given off by human activities, listed in order of importance: burning fossil fuels, growing rice, maintaining livestock, and lesser amounts from landfills, burning wood, and rotting of animal waste and human sewage.

Remember from earlier in this chapter that the hottest climate in the last 65 million years occurred when deep-ocean water warmed enough to melt icy methane hydrates on the sea floor, thus releasing a huge volume of methane gas into the atmosphere. If the slow warming of the deep oceans occurring today continues for enough decades to melt methane hydrates, then a warm climate could become a torrid climate.

Nitrous Oxide (N_2O) Nitrous oxide is another contributor to the greenhouse effect (Table 9.4). N_2O is produced naturally by bacteria removing nitrogen from organic matter, especially within soils. Humans cause the release of nitrous oxide via our agricultural activities, including use of chemical fertilizers. The second important way humans release N_2O is by combustion burning of fuels in car and truck engines.

Ozone (O_3) **Ozone** (O_3) is a greenhouse gas in both the stratosphere and the troposphere (Table 9.4). It is a gaseous molecule composed of three atoms of oxygen rather than the usual two-atom molecule (O_2). Ozone in the stratosphere acts as a greenhouse gas, helping warm the stratosphere. It is this heat that places the "lid" on the troposphere (Figure 9.31).

Ozone is effective at absorbing ultraviolet (UV) radiation emitted by the Sun, thus shielding life from dangerous rays. The UV rays that do pass through the atmosphere and make it to the ground surface cause sunburn and skin cancer.

Ozone is also a principal component of the smog that chokes urban atmospheres. Our automobiles and industries emit gases, some of which react with sunlight to produce the ozone that makes our eyes water and lungs ache. The ozone story is well described by the saying that pollutants are merely resources that are in the wrong place. Ozone in the stratosphere shields us from killing UV rays, but ozone in the air we breathe weakens and shortens our lives.

Chlorofluorocarbons (CFCs) Chlorofluorocarbons do not occur naturally. They are examples of gases produced solely by humans. CFCs are used as coolants in refrigerators and air conditioners, foam insulation in buildings, solvents, and other uses. Chlorofluorocarbons are not only greenhouse gases (Table 9.4), they also aid in the destruction of the ozone in the stratosphere that helps shield life from damaging ultraviolet (UV) rays. CFCs may remain in the atmosphere for a century, causing so many problems that international treaties have been signed restricting their usage.

Twentieth Century Greenhouse Gas Increases Why did we release such great volumes of greenhouse gases in the twentieth century? The gases were a byproduct of many praiseworthy activities such as providing energy for industries, homes, and personal automobiles, and from growing rice and raising livestock for human consumption. It took a long time to recognize how much the climate could be changed by these activities that raised the standard of human existence.

Another significant factor in the increase of greenhouse gases was the twentieth-century growth of the human population. In 1900 the world population of humans was about 1.5 billion, but the population exploded to 6 billion in 1999. Think about that: the human population doubled twice in the twentieth century from 1.5 billion in 1900 to 3 billion in 1960, and then to 6 billion in 1999. Even the most conservative estimate for twenty-first-century population growth forecasts another doubling to 12 billion people. Most people desire the affluent lifestyle of the industrialized world; this means the billions alive today who don't have that lifestyle plus the billions of people yet to be born. The greenhouse-gas-caused global warming will be a growing political issue throughout your lifetime.

The Twenty-First Century

How will global climate change in the twenty-first century? This is a complex question because its answer involves predicting many variables, such as greenhouse gas content of the atmosphere, temperatures around the planet, ocean warmth and circulation patterns, and wind strengths and positions. The question is addressed by constructing **global climate models (GCMs)** involving complex computer simulations. One GCM covering the twenty-first century forecasts a surface temperature increase of 2.5°C (4.5°F) from today (Figure 9.36). GCMs emphasizing CO_2 increases in the atmosphere forecast temperature increases of 1.5° to 4.5°C (2.7° to 8°F) in the next 50 to 100 years due to doubling CO_2 from its preindustrial value of 280 parts per million (ppm) to 560 ppm.

What global changes are likely in the twenty-first century? Significant melting of: 1) mountain glaciers, 2) the edges of the Greenland ice sheet, and 3) the West Antarctic ice sheet. In the Arctic, significant melting of sea ice and permafrost will occur and forests will advance north inside the Arctic Circle. As ice melts and seawater warms and expands, global sea level will rise from 1 to 3 feet (30 cm to 1

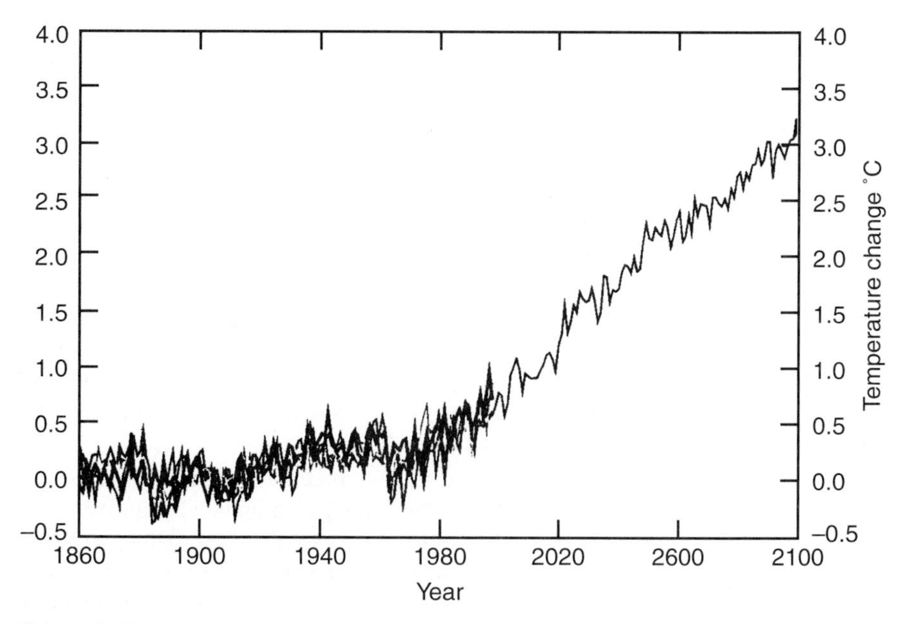

Figure 9.36 Projected change in average global temperature at Earth's surface relative to 1881–1920.
From Stott et al., 2000.

m). Some regions on Earth will become hotter and drier, others will become cooler and wetter. Agricultural productivity will increase in some areas and decrease in others. For example, recurrent, severe summer droughts are expected in the grain-growing middle of the United States.

A major climatic shift will occur if the present deep-ocean circulation system is altered (Figure 9.3). At present, the circulation involves sinking of salty water in the North Atlantic Ocean, but this could stop due to inflow of fresh water from melting glaciers.

We know that releasing greenhouse gases into the atmosphere causes climate warming, but we do not know what natural changes may occur. Will the Sun emit less (or more) energy in the twenty-first century? Will global volcanism increase and add a cooling effect, or will it be virtually absent? The future is difficult to predict. But look how life has changed. For thousands of years humans lived at the mercy of climate changes, but now *we* change the climate, for better or worse.

Summary

Change occurs at nearly every historical and geological time scale used, from hours or days to millions of years. Many processes affect climate, each with its own operating principles and time table.

The amount of solar energy received by the Earth varies over time and with geographic position. The cold water and air of the polar areas and the warm water and air of the equatorial belt are in motion due to the Earth's spin and the pull of gravity. Short-term changes in atmospheric conditions are known as weather, while longer-term changes are referred to as climate.

Early in Earth's history, the atmosphere was rich in carbon dioxide, and surface temperatures were about 290°C. Now, over 99 percent of CO_2 is locked up in limestone, and CO_2 is only 0.037 percent of the atmosphere. The presence of atmospheric gases, such as CO_2, water vapor, and methane, create a greenhouse effect where incoming, short-wavelength solar radiation passes through the atmosphere, but heat reradiated by the Earth is in longer wavelengths that are unable to pass back through the atmosphere. Thus, Earth's surface temperature rises with the increasing abundance of greenhouse gases. At present, humans are burning wood, coal, natural gas, and oil, returning large amounts of CO_2 to the atmosphere. This leads to global warming. Global surface temperature rose 0.6°C (1°F) in the twentieth century to reach the highest temperature in over a thousand years.

Warm climates of the past are deduced from evidence such as fossil reefs, tropical soils, evaporite mineral bodies, and widespread fossils of tropical and subtropical organisms. Ancient cold climates are interpreted from features such as glacially deposited debris, ice-polished and grooved rock surfaces, and wide distribution of fossils of cold-water organisms.

Torrid Ages are times when tropical and subtropical conditions cover much of the Earth. They commonly involve buildup of greenhouse gases and extensive equatorial seas that absorb solar energy.

Ice Ages have occurred several times in Earth's history and they last for millions of years. They require: 1) large continents at the poles to support 3 km thick glaciers and 2) continents aligned to deflect warm ocean water toward the poles, where it can evaporate, then fall as snow on land to build glaciers.

Advances and retreats of glaciers during an Ice Age occur on a time scale of thousands of years. They are largely due to changes in the orbit and rotation of Earth, affecting the amount of solar energy received. Earth's orbit is eccentric, moving closer, then farther away from the Sun in a 100,000-year cycle. The angle of Earth's spin axis varies in its tilt toward the Sun on a 41,000-year cycle. The tilt direction changes over a double cycle with periodicities of 23,000 and 19,000 years. The effects of variations in orbit, tilt, and wobble may reinforce or cancel each other. At present, orbit and tilt contribute to cooling while wobble helps warming.

Climate change occurs on hundred-year time scales also. For example, sea level can rise and fall by 200 m (650 ft) as glaciers wax and wane. When glaciers retreat rapidly, massive volumes of cold, fresh meltwater cover the ocean surface, causing big changes in ocean circulation, weather, and life.

Climate also changes on the scale of years. In the El Niño condition, warm water in the Pacific Ocean shifts positions. For example, the trade winds usually push warm water to the east in the southern Pacific Ocean, producing tropical conditions and the Great Barrier Reef on the western side of the ocean but leaving coastal deserts on the eastern side. When the system shifts, it results in cooler water off Australia, yielding less rainfall and leading to massive bushfires. At the same time, warm water off the west coast of the Americas may yield heavy rains.

Volcanism has major effects on climate. For example, when ash and SO_2 are blasted through the troposphere into the stratosphere, they block some incoming solar radiation, leading to cooling. The Mount Pinatubo eruption in 1991 produced so much SO_2 that its cooling effect more than offset the greenhouse warming caused by humans burning fossil fuels and releasing CO_2 for two years.

The rock record tells us much about past climates and former life. Climate often changes dramatically, and over 99 percent of all species that have ever lived are extinct. Yet collectively, life is firmly interwoven into Earth processes and is almost impossible to extinguish. However, the hold on existence for individuals and species is fragile and subject to a sudden end.

Terms to Remember

albedo 243
bauxite 246
buffer 247
climate 242
El Niño 259
extinction 242
global climate model (GCM) 269
global warming potential (GWP) 267
greenhouse effect 243
halite 246
haze 262
jet stream 260
La Nada 262
La Niña 262
Little Ice Age 259
Maunder Minimum 259
Medieval Maximum 258
methane 246
methane hydrate 251
ozone 268
stratosphere 262
tropopause 262
troposphere 262
weather 242

Questions for Review

1. What are the differences between climate and weather?
2. Explain the greenhouse effect in some detail.
3. The earliest Earth had an atmosphere loaded with CO_2 in a runaway greenhouse climate. Explain where that atmospheric CO_2 has gone.
4. Draw a cross section through the atmosphere that defines troposphere, tropopause, stratosphere, and ozone layer.
5. Climate is related to amount of solar radiation received on Earth. How is incoming solar radiation affected by: 1) continental ice sheets? 2) volcanic ash in the stratosphere? 3) elevated levels of atmospheric CO_2? 4) SO_2 blown into the stratosphere by volcanic eruption?
6. What information about ancient climates is suggested by: 1) fossil reefs? 2) bauxite, the aluminum ore? 3) bodies of sea salt? 4) the area covered by a fossil species? 5) polished and grooved surfaces in rocks?
7. How can oxygen-isotope ratios be used as an ancient thermometer?
8. What causes glacial advances and retreats during an Ice Age?
9. How much can sea level drop during an Ice Age? What effect does this have on ocean-water temperature and evaporation rates?
10. When massive continental ice sheets melt, what happens to: 1) sea level? 2) deep-ocean-water circulation? 3) salinity of sea-surface water? 4) organisms living near the sea surface?
11. Describe what happens to produce the El Niño phenomenon. What changes does it bring to Australia? To California?
12. How much did global surface temperature rise in the twentieth century? How much is temperature projected to rise in the twenty-first century?

Questions for Further Thought

1. What is the relationship between continental drift and the existence of an Ice Age?
2. What would be the global effect of melting the West Antarctic ice sheet? What would happen to New Orleans, Miami, and other coastal cities?
3. Considering the climatic history of the Earth, are you alive at a typical time?
4. If major volcanic eruptions occurred nearly every year for a century, what might happen to global climate?
5. Humans are causing global warming by burning wood, coal, natural gas, and oil and thus returning CO_2 to the atmosphere. What global changes may result?

Suggested Readings and References

Bryant, E. A. (1991). *Natural Hazards.* Cambridge: Cambridge University Press.

Climate in Earth History. (1982). Washington, D.C.: National Academy Press.

Crowley, T. J., and North, G. R. (1996). *Paleoclimatology.* New York: Oxford University Press.

Dawson, A. G. (1992). *Ice Age Earth.* London: Routledge.

Frakes, L. A. (1979). *Climates Throughout Geologic Time.* Amsterdam: Elsevier.

Imbrie, J., and Imbrie, K. P. (1979). *Ice Ages, Solving the Mystery.* Cambridge: Harvard University Press.

Lamb, H. H. (1982). *Climate, History and the Modern World.* London: Methuen.

Post, W. M. (1990, July–August). The global carbon cycle. *American Scientist,* 78, 310–26.

Rampino, M. R., and Self, S. (1984, January). The atmospheric effects of El Chichón. *Scientific American,* 48–57.

Ruddiman, W. F. (2001). *Earth's Climate: Past and Future.* New York: W. H. Freeman.

Self, S., Rampino, M. R., Newton, M. S., and Wolff, J. A. (1984, November). Volcanological study of the great Tambora eruption of 1815. *Geology,* 12, 659–63.

Stommel, H., and Stommel, E. (1983). *Volcano Weather, the Story of 1816, the Year without a Summer.* Newport, R.I.: Seven Seas Press.

Stott, P. A. et al. (2000). External control of 20th Century temperature by natural and anthropogenic forcings. *Science,* 290, 2133–37.

Suess, E., Bohrmann, G., Greinert, J., and Lausch, E. (1999, November). Flammable Ice. *Scientific American,* 76–83.

Walther, G. R. et al. (2002). Ecological responses to recent climate change. *Nature,* 416, 389–95.

Williams, S. J. et al. (1990). *Coasts in Crisis.* U.S. Geological Survey Circular 1075.

Videos

The Miracle Planet—Riddles of Sand and Ice. (1994). Nova/KCST (60 min.).

Planet Earth—The Blue Planet. (1986). Annenberg/CPB (60 min.).

Planet Earth—The Climate Puzzle. (1986). Annenberg/CPB (60 min.).

Time Lapse Observations of Columbia Glacier. (1980). U.S. Geological Survey (5 min.).

Hubbard Glacier—Russell Fiord. (1986). U.S. Geological Survey (25 min.).

Evidence for the Ice Age. (1961). Encyclopedia Britannica (19 min.).

Severe Weather

The evening and night winds here were . . . charged with a something that did not burden them elsewhere. They brought it up from that sinister Bay to the west, whose movement she and he were hearing now. It was a presence—an imaginary shape or essence from the human multitude lying below: those who had gone down in vessels of war, East Indiamen, barges, brigs, and ships of the Armada—select people, common, and debased, whose interests and hopes had been as wide asunder as the poles, but who had rolled each other to oneness on that restless sea-bed. There could almost be felt the brush of their huge composite ghost as it ran a shapeless figure over the isle, shrieking for some good god who would disunite it again.

—Thomas Hardy, 1897, *The Well-Beloved*

Weather kills. People drown in floods, are struck down by random bolts of lightning, are battered and drowned in hurricanes, and are chased and tossed by tornadoes (Figure 10.1). Severe weather causes about 75 percent of the yearly deaths and damages from natural disasters. The destruction wrought by storms and associated phenomena kills hundreds of people in the United States each year (Table 10.1); more than are killed by earthquakes, volcanoes, and mass movements combined.

Before different types of severe weather are examined, some basic principles need to be understood.

Figure 10.1 A tornado with wind speeds over 200 mph destroys houses in Pampa, Texas on 8 June 1995.

Table 10.1 Deaths Due to Severe Weather in the United States, 1940–2001

Event	Approximate Yearly Deaths
Lightning	143
Flood	108
Tornado	101
Hurricane	35
	387 fatalities per average year

Source: National Weather Service. (2002) Hazstats.

Table 10.2 Thermal Properties of Selected Materials

	Density × (gm/cm³)	Specific Heat = (cal/gm/°C)	Heat Capacity (cal/cm³/°C)
Air	0.0013	0.24	0.00031
Quartz sand	1.65	0.19	0.31
Granite	2.7	0.19	0.51
Water	1.0	1.0	1.0

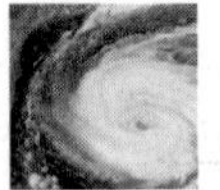

Weather Principles

Water and Heat

Water has a remarkable ability to absorb heat (see sidebar in Chapter 1). It has the highest **heat capacity** of all solids and liquids, except liquid ammonia (NH_3) (Table 10.2). Sand and rock have smaller **specific heats** and heat capacities, so even though land heats to higher temperatures, it does so only to shallow depths, and the resultant heat held is small compared to that absorbed by the same volume of water. Not only can water absorb more heat per unit volume, but solar radiation penetrates to depths of several hundred meters where it is absorbed and carried away by moving water masses. The transmission of heat in flowing water (or air) is the process of **convection.** But heat transfer by convection is not possible for rock because it is solid and does not readily flow. Rock also does not transfer heat well by **conduction,** so the daily fluctuation of heat energy reaches down only a meter or so.

As an everyday illustration of the differences in heat capacity, convection and conduction, remember how unbearably hot beach sand can be on your bare feet at the same time that ocean or lake water is much cooler. At night, the situation reverses as the sand, with its low heat capacity, releases its heat and becomes uncomfortably cool to bare feet, while the water temperature has changed but little.

Water vapor in the atmosphere ranges from near 0 to 4 percent by volume, but its importance in determining weather is great. The amount of water vapor in the air is measured as **humidity.** The higher the temperature of the air, the more water vapor it can hold. As air cools, its ability to hold water vapor decreases. At any air temperature, the water vapor content can become great enough to saturate the air and achieve a relative humidity of 100 percent. When relative humidity reaches 100 percent, then excess water vapor condenses and forms liquid water. The temperature at which an air mass reaches 100 percent humidity is its **dew point temperature.**

Water absorbs, stores, and releases tremendous amounts of solar energy (see Chapter 1) in all three states: solid, liquid, and gas. Ice melts to water when supplied with about 80 calories of heat per gram (80 cal/g) of ice; this energy is stored as **latent heat** in the water (Figure 10.2). Water evaporates to water vapor when it absorbs about 600 calories per gram of water; this energy is stored in water vapor as the **latent heat of vaporization.** Ice can change to water vapor directly without passing through a liquid state, but it requires both the 80 cal/g for melting and the 600 cal/g for evaporation. The process of changing directly from solid to gas is called **sublimation.** You probably have seen sublimation when dry ice (frozen CO_2) changes to vapor.

The latent heat carried in water vapor is released upon condensation to liquid water. The **latent heat of condensation** gives up the same number of calories as were absorbed during evaporation (Figure 10.2). When water freezes to ice, it releases stored energy as the **latent heat of fusion;** it gives up the same number of calories as were absorbed during melting. Water vapor can change directly to ice in deposition; it gives up both the latent heats of melting and evaporation. You probably have seen deposition of water vapor to ice in the build up of "frost" in the freezer compartment of a refrigerator.

Water and water vapor are most important for absorbing solar radiation, transporting heat about our planet, and releasing that heat. All of this heat transport helps prevent extreme ranges in temperature on Earth.

Atmospheric Heating

Solar **radiation** is absorbed in massive amounts in the equatorial belt facing the Sun and extending from about 32°N to 34°S latitudes. Polar latitudes receive far less of the Sun's energy and, in fact, show a net cooling because the heat reradiated back to space is greater than the amount locally gained from the Sun. Some of the excess heat of the low-latitude equatorial zone is transported to the high-latitude polar regions (Figure 10.3). The midlatitudes (like the continental United States) are zones of energy transfer. Cold air flows equatorward and hot winds move poleward, transferring much heat, especially carried in water vapor. This transport of energy in moving air masses is often experienced as severe storms.

When a cool air mass aloft sinks, it compresses and becomes denser but warmer as it sinks, thus creating higher pressure. This is an **adiabatic process** where temperature changes without gaining heat. The amount of heat in the

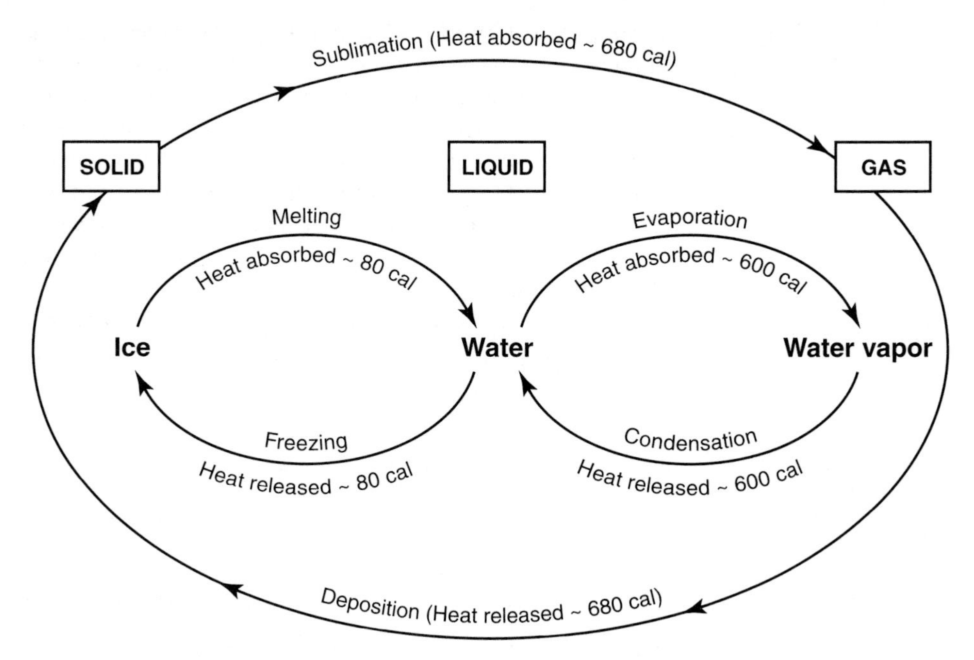

Figure 10.2 Water changing state from solid to liquid to gas absorbs heat. Water changing state from gas to liquid to solid releases heat.

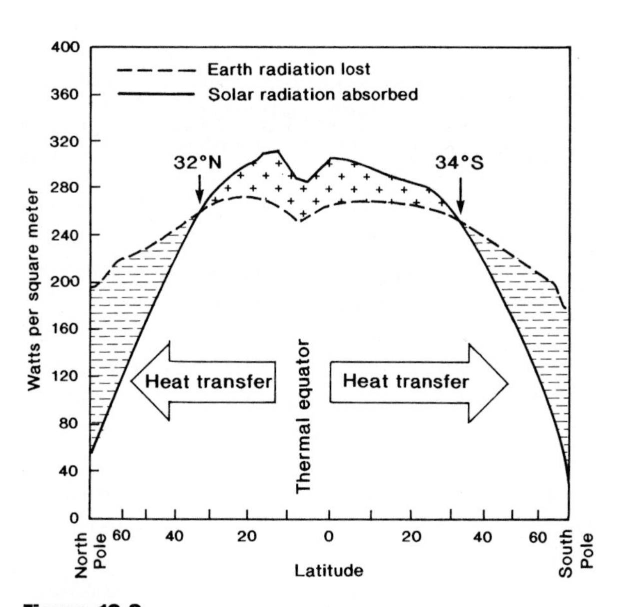

Figure 10.3 Energy radiated from inside Earth and energy absorbed from solar radiation are plotted against latitude. Poleward from latitudes 32°N and 34°S, the energy loss deficit increases. Heat is transported poleward from tropics via ocean and atmosphere, tending toward energy equilibrium.

Source: NOAA Meteorological Satellite Laboratory, Washington, D.C.

sinking air mass does not change, but because it compresses into a smaller volume, the temperature rises.

A heated air mass rises and expands, becoming a less dense mass that has lower pressure. The process of expanding causes cooling, an adiabatic process where temperature changes without loss of heat. Although the total heat remains the same, it is spread through an expanded, larger volume of air, thus the temperature drops. Adiabatic temperature changes produce cooling when a mass of air rises and expands in volume, and warming when an air mass descends and compresses into a smaller volume.

A rising mass of air moves upward through ever-decreasing pressure. As atmospheric pressure decreases, the rising air mass expands and cools adiabatically about 10ºC per km of rise. As the air cools, it has less ability to hold water vapor; thus, its relative humidity increases. When the rising air reaches 100 percent humidity, excess water vapor will condense and form clouds. The altitude where 100 percent humidity is reached is known as the **lifting condensation level.** When water vapor condenses, it releases the latent heat it absorbed when evaporated. The released latent heat slows the rate of upward cooling to about 5ºC per km of rise.

The atmosphere responds readily to the Sun's radiation; warm air rises and cool air sinks. Air masses flow horizontally from high pressure toward low pressure, seeking a balance or equilibrium condition. But the vertical movement of air is small compared to its horizontal motion (i.e., wind). The flow of air is greatly affected by the Earth's rotation and by the Coriolis effect.

Coriolis Effect

If the Earth's surface were as smooth as a billiard ball and solar energy were received equally over its surface, then the atmosphere would be rather still. But the Earth rotates rapidly and sets cold and warm air masses into motions that are altered by topography. The velocity of rotation on Earth's surface varies by latitude from 1,037 mph (464 km/sec) at the equator to 0 mph at the poles (Figure 10.4). Because there are different velocities at different latitudes, bodies moving across latitudes will follow curved paths. This is the

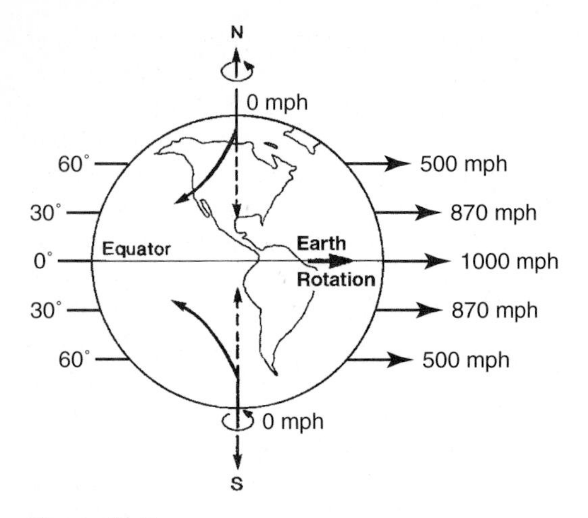

Figure 10.4 The Coriolis effect describes how large air and ocean masses tend to follow curving paths because of the rotating Earth. Looking down the direction of movement (dashed lines), paths veer toward the right (solid line) in the Northern Hemisphere and toward the left (solid line) in the Southern Hemisphere.

Coriolis effect, named for the French mathematician Gaspard Coriolis who described it in 1835.

In the Northern Hemisphere, all large moving masses will sidle off to their right-hand side when viewed down the movement direction; in the Southern Hemisphere, large moving bodies will veer toward the left (Figure 10.4). The magnitude of the Coriolis effect on a moving body increases as: 1) its path lengthens, 2) its time of motion lengthens, 3) the number of latitudes crossed increases, 4) the farther north or south it travels, and 5) its size increases. The Coriolis effect is important in determining the movement paths of ocean currents, large wind systems, and hurricanes; possibly important for a large thunderstorm; probably not important for individual tornadoes; and certainly not important for water draining down kitchen sinks or toilets.

The Coriolis effect is greatest near the poles and least at the equator. Away from the equator (above 20°N and S latitudes), the Coriolis effect causes winds to veer into arcuate paths. For example, watch the course hurricanes travel and see how prominently their paths curve with distance away from the equator (Figures 11.2, 11.6, 11.8).

The Merry-Go-Round Analogy A good way to visualize the Coriolis effect is to go with friends to a local playground that has a merry-go-round, a large circular wheel that rotates horizontally around a pole. When the merry-go-round is spinning counterclockwise, visualize being above it and looking down onto an analogy for the Northern Hemisphere. As the merry-go-round whirls rapidly, which person is moving faster, one in the center (North Pole) or one on the outside (equator)? The outside (equator) rotates much faster (compare to Figure 10.4). If the person riding in the center aims a squirt gun directly at the rider on the outer rim, will the water hit the targeted person? Probably not, the person will have rotated away. When viewed from above, the path of the squirted water will appear to curve to the right.

Now spin the merry-go-round in the opposite direction, in a clockwise pattern analogous to the Southern Hemisphere viewed from above the South Pole. All other factors are the same, but the water squirted from the center to the outside now appears to curve to the left (compare to Figure 10.4).

Global Wind Pattern

Consider the wind pattern on an idealized Earth with no landmasses, a perfectly spherical shape, and no tilt or wobble to its spin, but with the Coriolis effect intact (Figure 10.5). The intense sunshine received in the equatorial belt causes air to expand and rise into the upper troposphere, where it cools and drops its condensed moisture as rain on the tropics.

The cool, dry, upper-elevation air then spreads both north and south, becoming compressed as it flows toward higher latitudes with their smaller areas. Around 30°N and S latitudes, the now-denser air sinks at the Subtropical High Pressure Zone, warming adiabatically as it descends and returning to the surface as a warm and dry air mass. Some of the descending air flows poleward as westerly winds (westerlies), and some flows equatorward as the trade winds (Figure 10.4). The ascending air at the equator and descending air around 30°N and S latitudes create semicircular air circulation routes known as **Hadley cells** (Figure 10.6).

The warm air flowing from the Subtropical High as the trade winds has low moisture content; thus, precipitation is scarce. The warm, dry winds of the subtropical belts between 30° and 20°N and S latitudes are responsible for many of the world's great deserts, such as the Sahara and Kalahari of Africa, the Sonora of North America, and the Great Australia.

Cold air flows over the land from both poles (Figure 10.5). As air masses move equatorward, they flow across ever-wider latitudinal spans. Around 60°N and S latitudes, they collide with the westerly wind masses, and both rise at the subpolar low.

The global wind pattern described above is modified by the interference of continental masses, including mountain ranges, and by seasonal warming and cooling affecting both the Northern and Southern Hemispheres. The equatorial region Hadley cells and the polar regions operate pretty much as described. However, the midlatitudes are a much more turbulent zone where competing polar and tropical air masses transfer their energies back and forth as the seasons and location of the polar-front jet stream vary, commonly creating severe weather conditions (Figure 10.6).

Jet Streams

Jet streams are relatively narrow bands of high-velocity winds that flow from west to east at high altitudes. There are two main jet streams in both the Northern and Southern Hemispheres, a polar jet and a subtropical jet. They occur high in the atmosphere where the major air circulation systems meet (Figure 10.6).

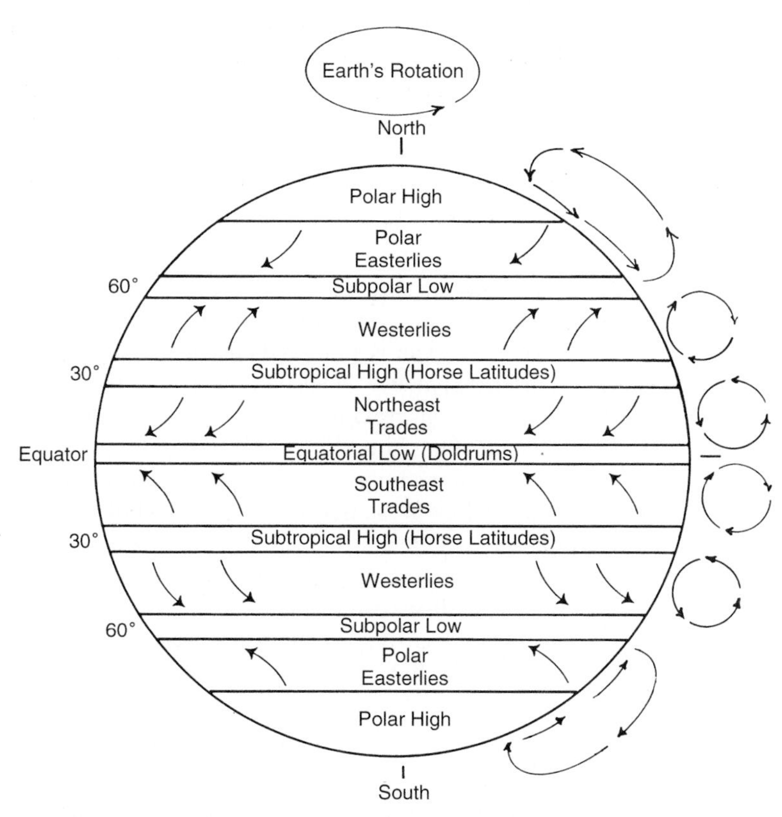

Figure 10.5 Idealized wind pattern on an Earth without landmasses or tilt or wobble in its spin.

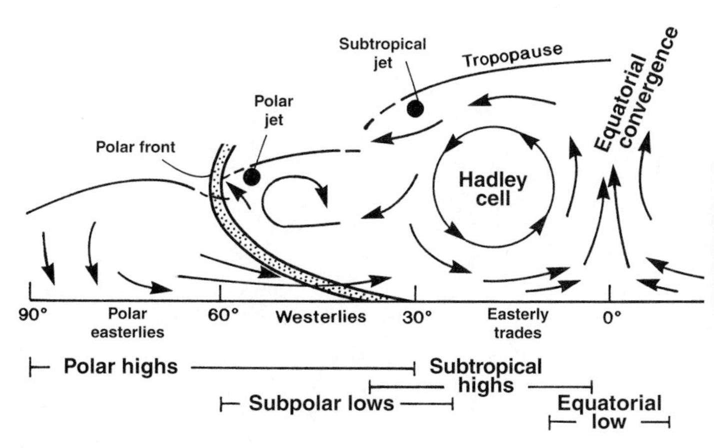

Figure 10.6 Cross section of air circulation between the equator and a pole.

The subtropical jet stream runs over the tropics around 25° latitude about 13 km (8 mi) above the ground (Figure 10.7). The tropical atmosphere absorbs heat; the air expands, becomes less dense, and rises to begin circulating in a Hadley cell (Figure 10.6). The warm air in the equatorial region forms a thick mass of air with molecules in energetic motion causing atmospheric pressure. By contrast, cooler air of the regions is denser, the molecules move less, and most of the mid-latitude air lies nearer the ground. Traveling up through a column of warm air, there is a slower rate of pressure decrease than moving upward through a column of cold air. Expanded warm air creates a thicker tropical mass with higher pressure aloft sitting next to a thinner, cooler, midlatitude air mass with lesser pressure in its upper part. The result aloft is that higher-pressure tropical air flows both "downhill" and also down the upper atmosphere pressure gradient toward the poles, creating strong poleward flows of air. But these north to south, or south to north, air flows occur on a rapidly spinning Earth which converts them to belts of high-speed jet stream winds. The subtropical jets do not play as dominant a role in weather as do the polar jets.

The most powerful and variable jet stream is the polar jet, which races along the boundary or front between the polar air cell (above the polar easterly winds) and the mid-latitude air cell (above the westerly winds) (Figure 10.8). The greater temperature contrast between these cells results in higher velocity winds. Polar jet streams flow from west to east over the midlatitudes at elevations of about 10 to 14 km (32,000 to 46,000 ft). A polar jet is a belt of winds about 1,000 km wide (over 600 mi), flowing as fast as 600 km/hr (370 mph) in its central "core" or "tube." A polar-front jet stream's path is ever-changing, like that of a meandering river. Meanders in the flow can bend so much that, locally, jet-stream flow directions may be to the

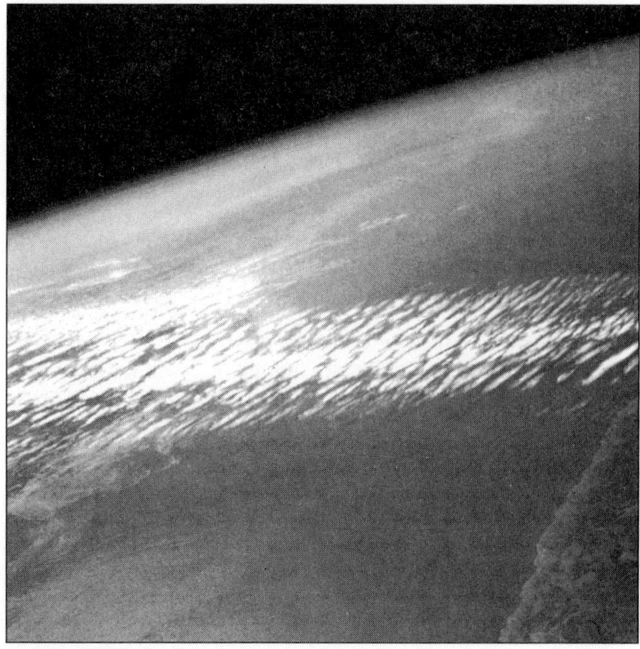

Figure 10.7 The subtropical jet stream is marked in this Space Shuttle photograph by the elongate band of clouds crossing the Red Sea, 5 April 1984.
Photo courtesy of NASA.

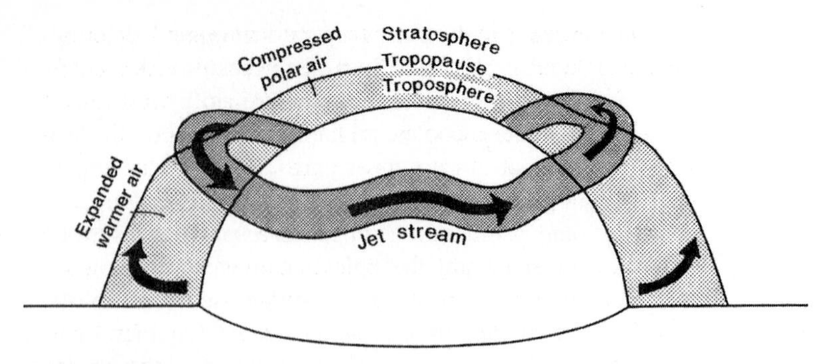

Figure 10.8 The polar-front jet stream flows at high altitudes in a meandering path from west to east. It exists where the expanded volume of warm air slopes down toward the compressed volume of cold air.

From J. Eagleman, *Severe and Unusual Weather.* Copyright © 1983 Van Nostrand Reinhold, New York. Reprinted by permission. All Rights Reserved.

north, south, or west. As the flow path twists and turns, it may cut off and abandon some flow loops, temporarily achieving a straighter west-to-east flow.

Polar-front jet streams also change position with the seasons. In the Northern Hemisphere, it flows over Canada during the summer, when the warm air volume is greatest; during the winter, it migrates southward over the United States as the volume of Northern Hemisphere cold air increases to its maximum. During each hemisphere's winter, the atmospheric temperature contrasts between pole and equator are greatest, and each polar jet races its fastest. When polar jets reach speeds around 200 km/hr (120 mph), they have significant effects in moving heat and air masses, as well as in provoking storms.

The polar-front jet stream results from temperature differences, but its existence in turn influences the movement and behavior of warm and cold air masses. The polar jet stream flows from west to east, under the influence of the Earth's rotation, in both the Northern and Southern Hemispheres.

Air Masses

The air masses that move across North America come mainly from several large source areas (Figure 10.9). The polar air masses are cool to cold, while the tropical air bodies are warm to hot. Air masses that gather over land are dry, whereas those that form over water are moist. The dominant direction of air-body movement is from west to east under the influence of the Earth's rotation. Thus, air masses that build over the northern Pacific Ocean have a much greater chance of affecting the United States than those that form over the North Atlantic Ocean.

Fronts

Different air masses do not readily mix; they are separated along boundaries called **fronts.** (The term "front" came out of World War I, from the battlefronts where armies clashed.) Many of the clouds and much precipitation are associated with fronts. A front is a sloping surface separating air masses that differ in temperature and moisture content. Weather fronts are commonly associated with severe weather and violent storms. The largest frontal system in the Northern Hemisphere separates cold polar air from warm tropical air along the polar front (Figure 10.6).

The advance of a cold front can produce a wedge, lifting warm air up to higher altitudes. When rising warm air is moist and unstable, it often forms tall clouds that may produce thunderstorms (Figure 10.10a). In Figure 10.11, an advancing cold front is seen pushing into and bending the rain falling from a warm air mass.

You can create and observe your own weather front at home. Open the door to the freezer compartment of your refrigerator and watch the cold air mass flow out. The cold air body moves into the warmer, moister air of the room, causing clouds and small-scale precipitation.

A warm front leads the advance of a warm air mass in a flatter wedge. The lighter-weight warm air flows up and over a cooler air mass along a gentle slope. The warm air cools as it rises along the broad and gentle front, commonly producing widespread clouds (Figures 10.10b and c).

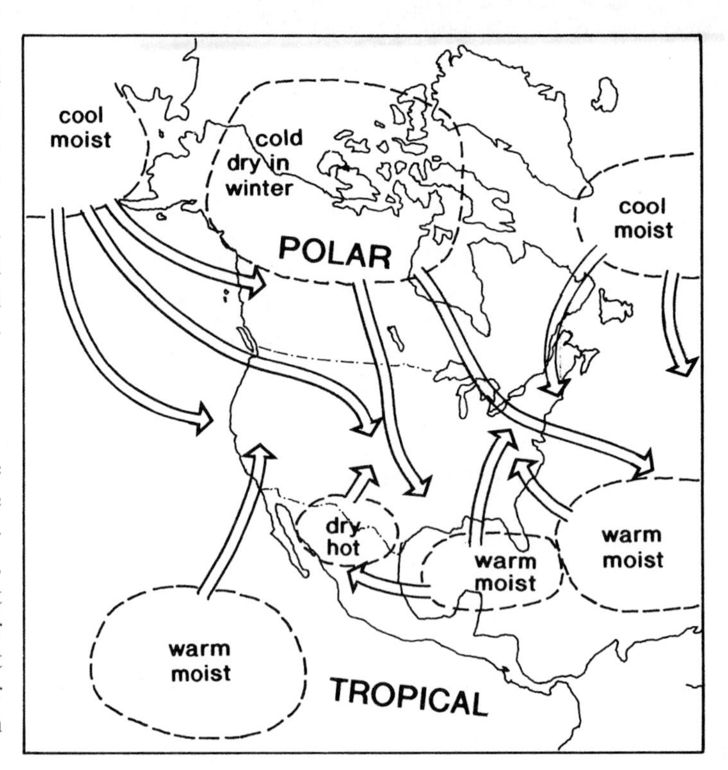

Figure 10.9 Map showing areas where large air masses acquire their temperature and moisture characteristics before moving about North America.

From J. Eagleman, *Severe and Unusual Weather.* Copyright © 1983 Van Nostrand Reinhold, New York. Reprinted by permission. All Rights Reserved.

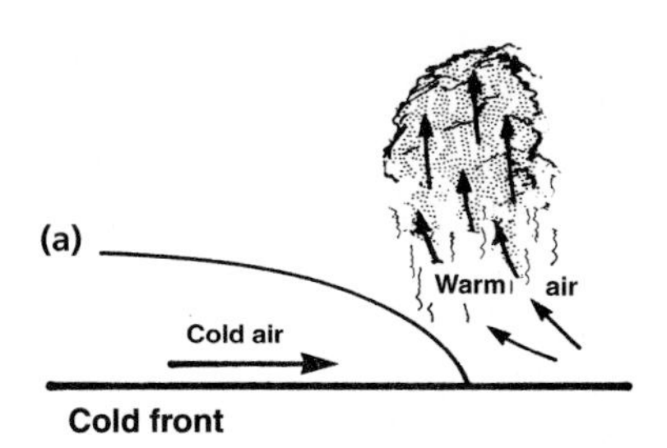

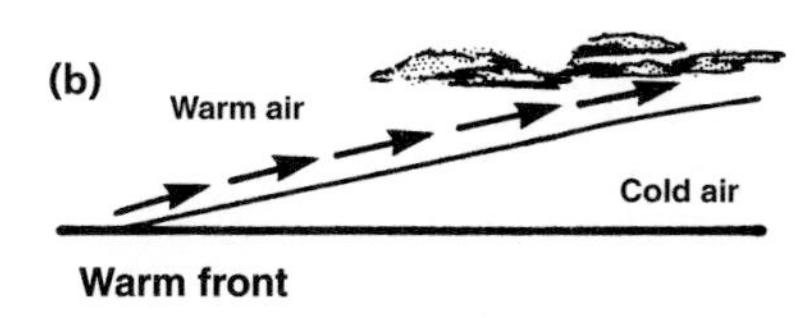

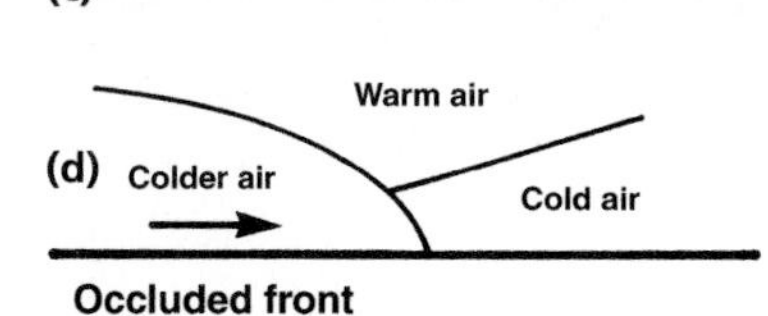

Figure 10.10 Schematic cross sections of fronts and air masses. **(a)** A cold front wedges under warm air, forcing it upward. **(b)** A warm front runs up and on top of a cold air mass. **(c)** A warm, moist air mass moving north (to the right) runs up a shallow sloping cold front producing clouds over the north shore of Lake Erie, Ontario, Canada. **(d)** An occluded front occurs where three different air masses meet.
Photo by John S. Shelton.

When warm, cool, and cold air masses come together, an occluded front is formed with air mass boundaries above the ground as well as upon it (Figure 10.10d).

Rotating Air Bodies

Rising warm air in the Northern Hemisphere creates a low-pressure zone that is fed a surface inflow of air that moves counterclockwise (Figures 10.12 and 10.13). Descending air in a Northern Hemisphere high-pressure zone flows out over the ground surface as clockwise-blowing winds.

The meanders in the polar-front jet stream can help create rotating air bodies; they rotate horizontally about vertical cores. In the Northern Hemisphere, the bends in the west-to-east flow of the polar jet create areas of diverging air at southward-bending *troughs* of lower pressure and regions of converging air at northward-extending *ridges* of higher pressure (Figure 10.14). A trough in the jet stream refers to a bend that is concave northward whereas a ridge is a bend that is convex northward.

The lower-pressure zone at a trough forms the core of a cyclonic circulation, a counterclockwise flow. In a **cyclone,** the winds include a surface inflow of winds toward the low-pressure core, feeding a large updraft of vertically rising air that cools, forming clouds and rainy weather, as well as producing an upper-level outflow of air.

The above process is reversed at a ridge in the jet stream (Figure 10.14). Here, the upper-level air flow imparts a clockwise rotation about a high-pressure zone; this is an **anticyclone,** moving clockwise. At an anticyclone, air converges in the upper atmosphere and descends, adding to the high-pressure center, then flows outward over the Earth's surface. The descending air warms and usually creates dry and windy conditions on the ground.

Midlatitude Cyclones

Much of the midlatitude severe weather in the Northern Hemisphere occurs via cyclones: air masses rotating counterclockwise about a low-pressure core. This nontropical cyclonic activity occurs at a number of scales. The largest-scale cyclones are linked to jet-stream troughs (Figure 10.14). A trough helps form a large frontal cyclone with a cold front drawn in from the north and, frequently, a warm front extending toward the east. A cold front may wedge under a warm, moist air mass, sending it upward to form thick clouds, and, if the air is unstable enough, possibly a line of **thunderstorms.** A warm front may move eastward, flowing up onto a cooler air body and producing widespread low clouds and precipitation (Figures 10.10 and 10.15).

In the northeastern United States, these cyclones can create potent storms known as *nor'easters.* When a low-pressure system moves up the northeastern United States coastline, its counterclockwise circulation on its western or landward side draws cold air down from the north. Meanwhile, its eastern or seaward side picks up moisture from the Atlantic Ocean to feed into the cyclone. The resultant nor'easters can be disasters, as described in the sections "White Hurricane," "Blizzard," and "Ice Storms."

Cyclonic air flow also characterizes individual thunderstorms within a large frontal cyclone, but the radius of a thunderstorm is much smaller (e.g., 10 km versus 1,000 km). Although the thunderstorm radius is only about 1 percent that of a frontal cyclone, its momentum is greater and its wind speeds are higher. Within a cyclonic thunderstorm,

Figure 10.11 A cold front moving south bends the rain falling from a warm air mass near Sioux Falls, South Dakota.
Photo by John S. Shelton.

Figure 10.12 Space shuttle photo in the Northern Hemisphere of a giant, low-pressure air mass rotating counterclockwise.
Photo by NASA.

an even smaller-radius cyclone, a **tornado,** may spring forth. The radius of a tornado funnel cloud may be only a few percent that of a thunderstorm, but again, the smaller radius of the spinning tornado brings its mass closer to the axis of rotation thus causing even higher wind speeds.

The Eastern U.S. "White Hurricane" of 1993

Shortly before spring began in 1993, an immense cyclone moved in and covered the eastern United States from Florida to Maine and from the Appalachian Mountains to the Atlantic Ocean. Between 12 to 15 March, normal life was put on hold for most people living beneath the cyclone. Winds gusted to over 100 mph, driving snow and sleet into the eastern seaboard and sending shallow seas surging into coastal communities. When the cyclone passed, 238 people had died—from Cuba to the Maritime Provinces of Canada, 48 sailors were lost, and damages exceeded $800 million. The hardest-hit area was Florida, where over 50 tornadoes spun out of the storm and killed more people than Hurricane Andrew did in 1992.

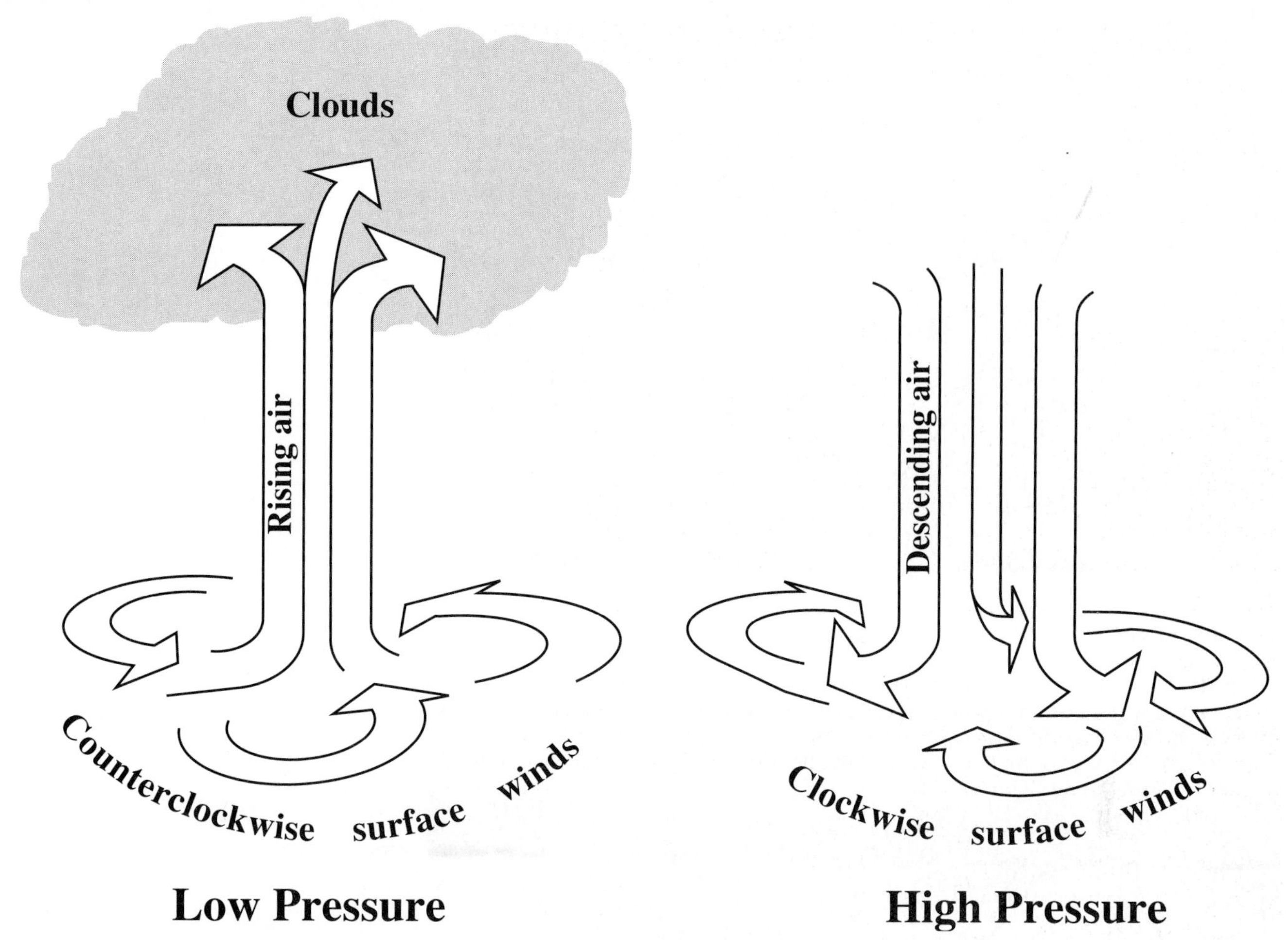

Figure 10.13 Air rises at a low-pressure zone in the Northern Hemisphere being fed by counterclockwise surface winds. Descending air at a high-pressure zone flows over the ground surface as clockwise winds.

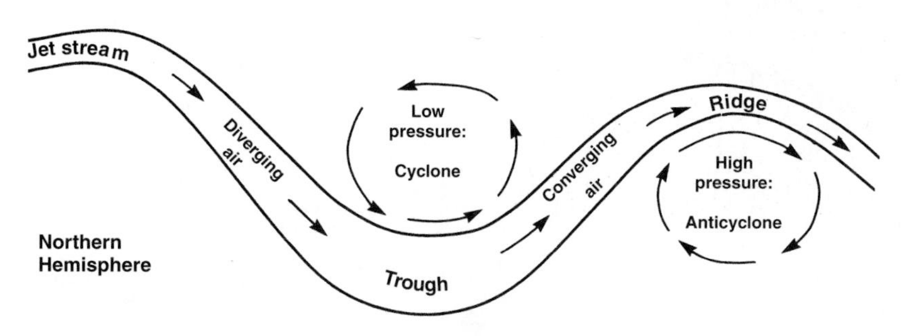

Figure 10.14 Influence of the polar-front jet stream in creating counterclockwise cyclonic flow about a low-pressure zone and clockwise anticyclonic winds around a high-pressure core.

Why did such a huge winter storm hit such an unusually large area so late in the season? How did it combine some of the worst aspects of both a blizzard and a hurricane? The U.S. weather map for 12 March 1993 shows a large trough in the jet stream and three fronts migrating toward it (Figure 10.16). The collision of two of these fronts would have made a significant storm, but the conflict between all three fronts created a "storm of the century."

The scene was set with unusually low pressure (Figure 10.17). The low-pressure zone in the Gulf of Mexico caused big trouble with its warm moist air and line of violent thunderstorms (Figure 10.16). Then a trough in the jet stream, with its associated very low pressure, drew in a fast-moving front of frigid arctic air from the north as well as a rainy and snowy east-moving front off the Pacific Ocean. The collision among the three air masses began in Florida. The low-pressure zone and three colliding fronts rode up the eastern seaboard with the jet stream, savaging everything in their path.

This late-winter cyclone was more intense than some hurricanes and was even bigger than the legendary blizzard that struck the northeastern United States on the same days of March in 1888. The 1888 blizzard brought wind gusts up to 85 mph and snow drifts 20 ft deep. In New York City, the heavy snow immobilized the city, leaving about 400 people dead.

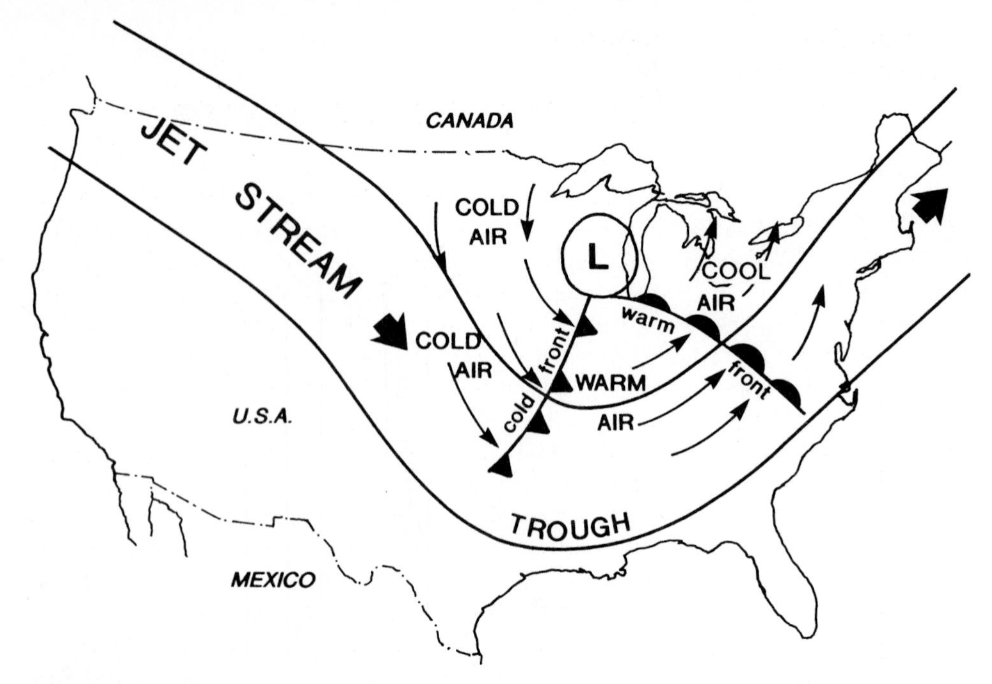

Figure 10.15 A trough in the jet stream helps cause a large-scale frontal cyclone formed of horizontally rotating winds around a low-pressure core. The cold front sweeping down from the north wedges beneath warm air, lifting it to form a line of thunderstorms. The warm front coming from the southwest flows up a gentle slope on top of the cooler air mass to the east to form widespread clouds and rain.

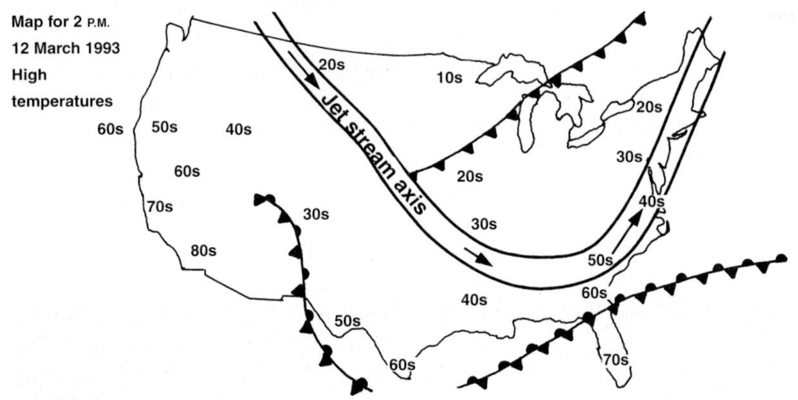

Figure 10.16 Weather map for 2 P.M., 12 March 1993 showing the day's high temperatures, a large trough in the jet stream, and three significant fronts. The very low-pressure southeastern front moved up the Atlantic coast, attracting the Pacific and Arctic fronts to a three-way collision.

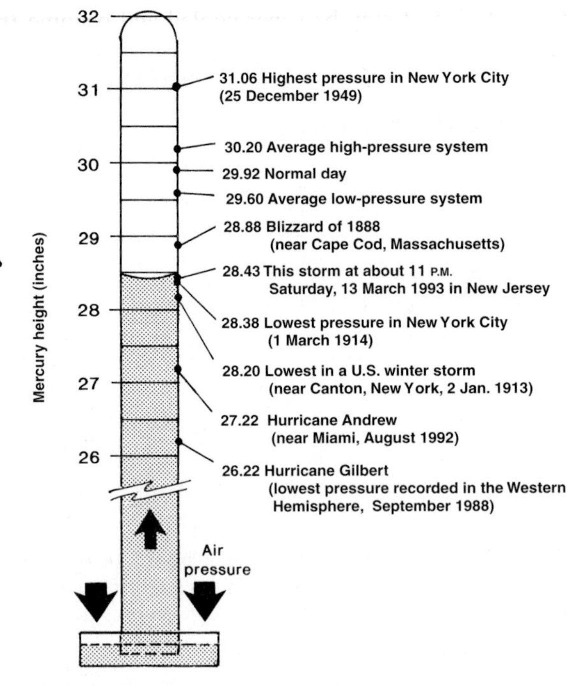

Figure 10.17 Air pressures recorded as mercury height (in inches) in a barometer. The barometer is a tube with one end closed, filled with mercury, and turned upside down in a dish of mercury. The mercury in the tube rises and falls depending on the amount of air pressure on the mercury in the dish. Commonly, high pressure indicates clear weather, and low pressure means clouds and rain.

Blizzards

One of winters most unpleasant events is the **blizzard.** In an average year, as many people are killed by blizzards as die during tornadoes. Blizzards occur when a long cyclone brings cold winds blowing at least 60 km/hr (37 mph), below freezing temperatures (less than 20°F), and much falling/blowing snow. The winter cyclone may travel slowly although its winds blow rapidly. A blizzard can occur without snowfall; on freezing days, strong winds can pick up and blow snow dropped by an earlier storm. Blizzard times are tough times; they test your ability to live. Fatalities occur: 1) by heart attacks while shoveling snow and pushing stuck cars, 2) when automobiles slide and collide, 3) when people slip on ice and fall, or 4) when folks get lost and freeze.

Northeastern United States, 6–8 January 1996 A powerful Canadian blizzard blew into the northeastern United States dumping record snowfalls in Ohio, Pennsylvania, West Virginia, and New Jersey. Wind speeds commonly exceeding 50 mph blew blinding attacks of snow. The blizzard killed 154 people; 80 of those fatalities occurred in Pennsylvania. In this case, the bitter conditions of the blizzard were immediately followed by warm weather with moderate to heavy rains which melted the record snows, unleashing destructive floods. In a few days, the region was pushed from one extreme to another, resulting in 187 deaths and $3 billion in damages.

Ice Storms

When it is cold, clouds may drop precipitation as snow flakes or ice particles. Falling snow and ice commonly pass downward through air warm enough to melt the snow/ice and turn it into rain. If the falling rain then enters a below freezing air layer near the ground, the rain may re-freeze into tiny (~1/16 in) ice particles called **sleet.**

If falling rain is not in sub-freezing air long enough to freeze to ice, it may be super cooled and become **freezing rain.** When freezing rain collides with the ground or other solid objects it turns to ice. Coatings of ice a few inches thick add so much weight to tree limbs, power lines, and even house roofs that they may snap and collapse.

Canadian Ice Storm, 5–9 January 1998 Ice storm 98 was the most expensive natural disaster in the history of Canada. During 80 hours of freezing rain, at least 25 people died of hypothermia and $7 billion in damages occurred in Ontario, Quebec, New Brunswick, and Newfoundland as the power system collapsed beneath the weight of the clinging ice. Ice storm 98 also severely affected Maine, New Hampshire, Vermont, and upstate New York. Millions of trees were split and downed by the ice as well. The destruction included 130 major power transmission towers, 30,000 wooden utility poles, and 120,000 km (74,000 mi) of power lines and telephone cables. In the dead of winter, hundreds of thousands of people were without power for up to 4 weeks. The damage to the Canadian life-support systems was so great that repair was impossible; they had to be rebuilt. Due to our dependence on fragile power lines, the disruption of lives was greater in 1998 than if the same storm had occurred in 1898.

How a Thunderstorm Works

Air temperature normally decreases upward from the ground surface up through the troposphere. When denser cool layers of air overlie less dense warmer air, the situation is unstable. The degree of atmospheric instability increases as the temperature differences increase between warm bottom air and overlying cool air. Warm low-altitude air is less dense and it wants to rise upward. Once vertical lifting begins, the warm air mass will continue to rise as long as it is less dense than the surrounding air.

If heat builds up in dry air near the surface, the warm air will rise and cool adiabatically after a moderate ascent, and no thunderstorms will form. However, if the air near the ground is both warm and moist, the warm air may rise high enough to pass through the lifting condensation level, allowing condensation of water vapor to begin. Once condensation is occurring, the rising cloud also is fueled by a large and important energy source—release of the latent heat that was absorbed during evaporation. This latent heat provides the fuel to help form thunderstorms, tornadoes, hurricanes, and other severe weather.

Thunderstorms occur where warm, moist air has absorbed enough heat and moisture to be significantly less dense than the surrounding air. Warm, buoyant air rises high, condensing into towering clouds (Figure 10.18). This process of vertical transfer of heat in a rising air mass is known as convection. As water vapor in the rising air mass condenses, it releases latent heat, adding to the warmth of the rising air and helping it climb even higher.

Most individual thunderstorms form on sunny days in the late afternoon or early evening, when temperatures of the

Figure 10.18 Warm buoyant air rises high forming a "torchere lamp" cloud in San Diego at 7:20 P.M., 29 August 1999.
Photo by Pat Abbott.

ground surface and lower air are the highest. A thundercloud begins with an initial updraft of warm, moist air, maybe aided by wind pushing up a hillslope or by surface-wind collision. The early stage of thunderstorm development requires a continuous supply of rising, warm, moist air to keep the updraft and cloud mass growing (Figure 10.19a).

When the amount of ice crystals and water drops becomes too heavy for the updrafts to support, some upper-level precipitation begins. Falling rain causes downward drag, developing downdrafts and pulling in cooler, dryer air surrounding the tall cloud mass. In the thunderstorm's mature stage, the cloud-mass top commonly spreads out as an icy cap (Figures 10.19b and 10.20). Updrafts and downdrafts operate side-by-side as warm, moist air is rising high at the same time that cool, dry air is descending rapidly. This is the most violent stage of the thunderstorm, with gusty winds pummeling the ground and tossing about any airplane daring to fly through the storm. Rain is heavy, **thunder** and **lightning** are powerful, and ice crystals may pummel the ground as **hail.**

The dissipating stage is reached when downdrafts drag in so much cool, dry air that it damps out the updrafts of warm, moist air necessary to fuel the thunderstorm (Figure 10.19c). Without new moisture, the tall thundercloud mass evaporates in the dry air of the downdrafts.

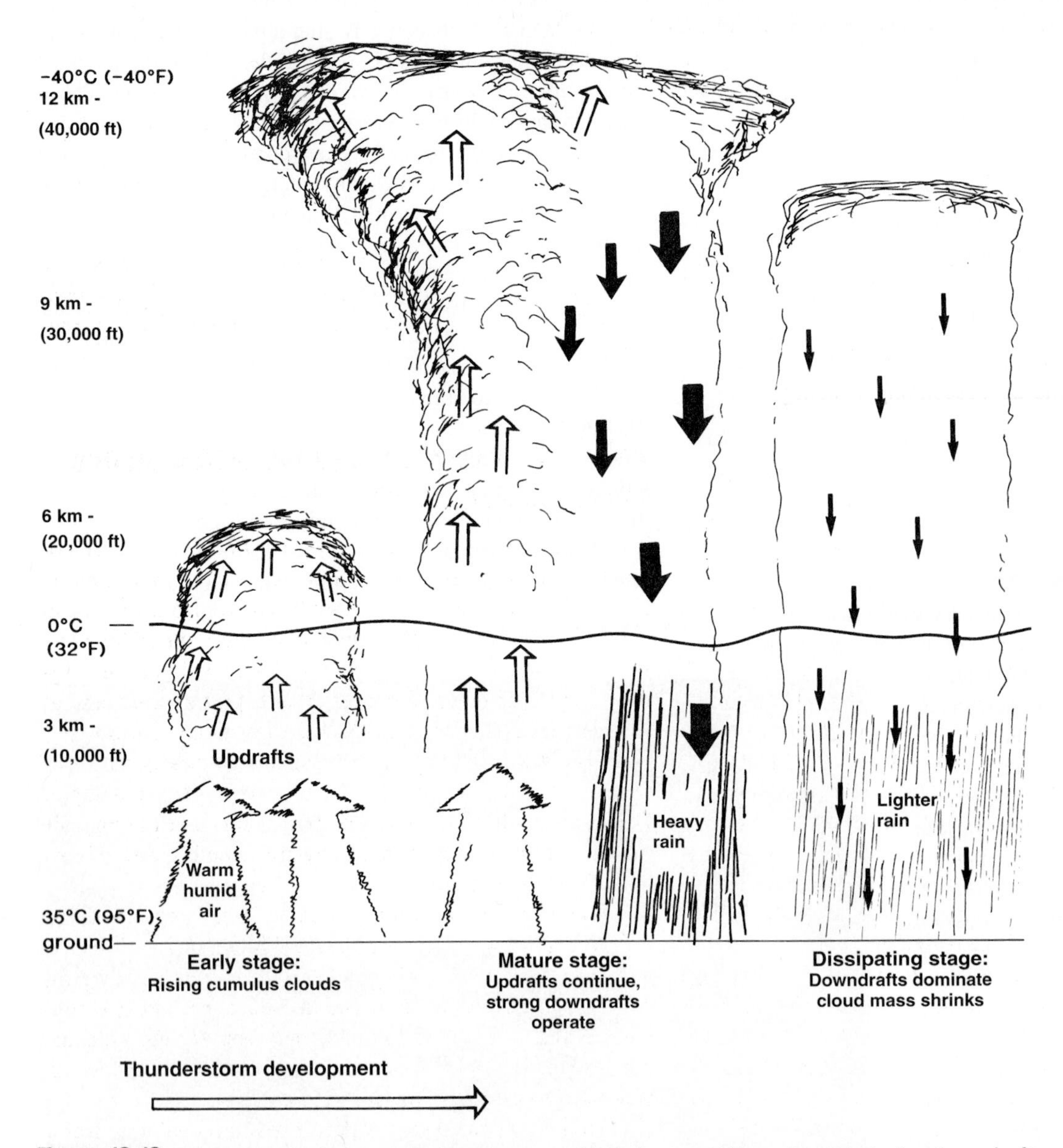

Figure 10.19 Stages in the history of a thunderstorm. **(a)** *Early stage:* Warm, humid air rises in updrafts; cooling causes condensation to form clouds; air keeps rising as long as it is warmer than surrounding air. **(b)** *Mature stage:* Ice crystals and large rain drops become too heavy for updrafts to maintain in suspension; ice and rain fall, forcing air to move in downdrafts, while updrafts still pump warm, moist air into the thunderstorm. **(c)** *Dissipating stage:* Downdrafts dominate, warm and moist updrafts cease, and rain becomes lighter as clouds begin to evaporate.

Figure 10.20 Giant symmetrical convective cloud on a hot summer afternoon near the Wyoming-Montana border.
Photo by John S. Shelton.

Microbursts: An Airplane's Enemy

In the mature stage of a thunderstorm, violent downdrafts of air may occur (Figure 10.19b). These sudden, strong downrushes of wind are known as **microbursts** and are composed of a mass of cold air descending violently along with rain and maybe hail. These convective downdrafts are commonly tens of meters to one kilometer in diameter and they may occur in grouped sequences. When a wet microburst hits the ground it does so with enough violence to leave a splattered pattern of damage that looks like the work of a tornado.

A crude analogy can be made to dropping a water-filled balloon from the roof of a house and watching it splatter with force on the ground. What if the "water balloon" is a heavy ball of wind with rainwater descending at 168 mph? The danger to airplanes is obvious. Microbursts are especially hazardous to airplanes during takeoff and landing. The airplane is so close to the ground that the unexpected downdraft of a microburst can push the plane into the ground before the pilot has a chance to react.

Shortly after a 4:10 P.M. takeoff from New Orleans International Airport on 9 July 1982, Pan American flight 759 was hit by a wet microburst from a mature thunderstorm. The microburst bounced the plane down onto a residential neighborhood, killing 8 people on the ground and 145 persons on board the aircraft.

While approaching Dallas–Fort Worth International Airport for a landing during a thunderstorm at 5:06 P.M. on 2 August 1985, Delta flight 191 was just coming out of a wall of extremely heavy rain and lightning when a rounded bulge was seen dropping down from the cloud. This was a wet microburst that shoved the plane onto the ground, killing 133 of 163 passengers. The 30 survivors were seated at the plane's tail.

Many other microburst incidents have been reported. They present a continuing threat to aircraft and there probably will be more of these late afternoon downings of low-flying planes.

Thunderstorms in the Conterminous United States

Thunderstorms develop both in isolation and as parts of larger weather systems. All of them require that warm, moist air be elevated; this air commonly comes from the warm waters of the Gulf of Mexico. The distribution of thunderstorms in the United States shows an unequal pattern (Figure 10.21). Florida has the most thunderstorms; west-central Florida averages more than 100 days of thunderstorms per year. The low-lying Florida peninsula is a meeting place for weather systems from the warm waters of the Atlantic Ocean and Gulf of Mexico. The converging warm, moist air masses are commonly forced upward, creating the ubiquitous thunderstorms.

Frequent thunderstorms affect a broad region in the central and southern United States (Figure 10.21). Most of these events occur in spring and early summer, when warm, moist air from the Gulf of Mexico advances up over the United States. The Gulf air masses meet strong, cold air masses moving down from the north, which underride the warm air masses, helping lift them upward.

Thunderstorms also develop in the mountainous west-central United States, where warm, moist Gulf air comes in contact with warm, dry, desert air masses that force the less-dense, moist air upward, sometimes with the help of the locally mountainous topography.

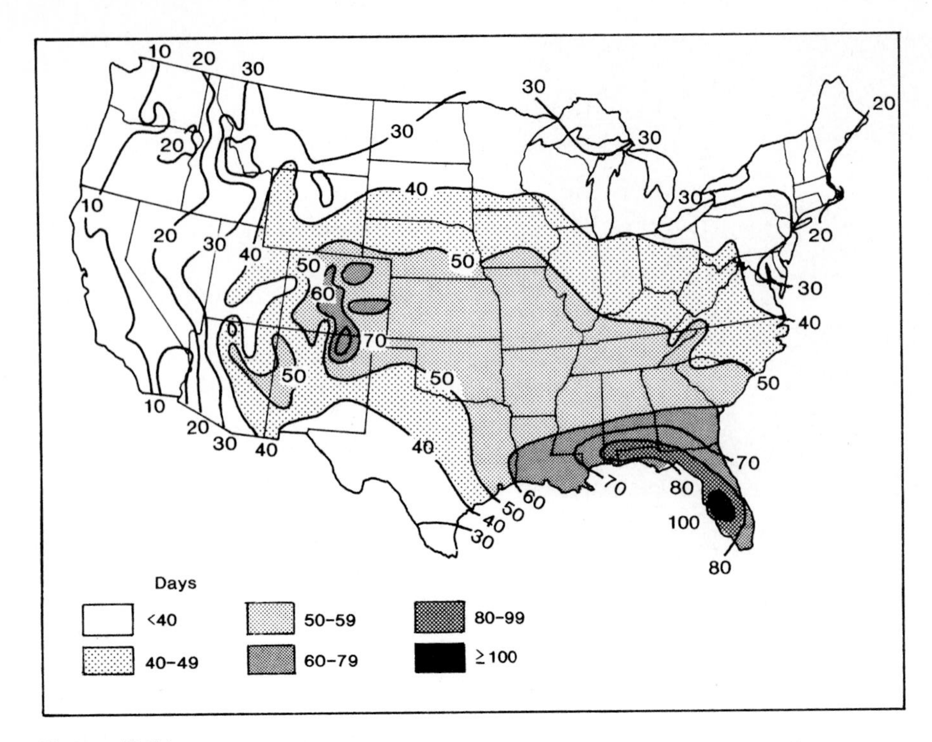

Figure 10.21 Number of days with thunderstorms each year in the United States.
From J. Eagleman, *Severe and Unusual Weather.* Copyright © 1983 Van Nostrand Reinhold, New York. Reprinted by permission. All Rights Reserved.

Thunderstorms wreak havoc with their 1) heavy rains, 2) flash floods, 3) hail, 4) lightning-caused deaths and fires, and 5) high-speed winds, which can be either straight-line blasts or rotating tornadoes.

Heavy Rains and Flash Floods

Thunderstorms are major suppliers of water to many areas of the United States. The abundance of water they deliver makes them positive events, yet there is a price to be paid. Thunderstorms are high-intensity, localized, short-term events that can also bring hail, lightning, and flash floods. A prime example of thunderstorm activity is found in central Texas.

Central Texas Some of the most intense precipitation in the world occurs in short lengths of time over small areas of central Texas (Figure 10.22). Many of the U.S. and world record rainfalls in the 1-to-24-hour range were established here. For example, a May 1935 thunderstorm dumped 22 inches (56 cm) of rain on D'Hanis in just 2¾ hours. In September 1921, the region around Thrall was drenched with 32 inches (81 cm) of rain in only 12 hours; flash flooding from this deluge drowned 215 people and caused $19 million in damages.

Why do such heavy rainfalls occur here? First, there is a ready source of precipitation in the warm, moist winds that flow in from the Gulf of Mexico. Second, the air flow over the gently sloping coastal plain may be interrupted and turned upward along the Escarpment formed by the Balcones fault zone. The updraft of warm, moist air creates an ideal situation for the formation of thunderstorms (Figure 10.19). The Balcones Escarpment runs for 545 km (340 mi) from Del Rio on the Mexican border through San Antonio and Austin and north to Waco. It stands roughly at right angles to the dominant winds that blow in from the Gulf of Mexico. The Escarpment was formed by down-to-the-coast normal fault movements primarily between 19 and 16 million years ago. Today, the height of the eroded fault scarp ranges from 100 to 500 ft (30 to 150 m) and this is high enough to influence the weather.

The Balcones Escarpment is also the dividing line between two grand physiographic divisions of North America, the Great Plains to the west and the Coastal Plain to the east. The geologic fault is also a "cultural fault" separating the cotton economy of the Old South on the fertile soils of the Coastal Plain from the cattle economy of the Old West on the deeply eroded and dissected limestone of the plateaus. The faults themselves act as conduits for spring water. The springs enticed the original settlers into starting the towns that have grown into the large cities of today.

The growing urban areas along the Balcones Escarpment are randomly visited by torrential downpours from thunderstorms. For example, on 11 May 1972, a thermally unstable air mass lifted upward to form a localized thunderstorm near New Braunfels, Texas. The storm dumped 30 cm (12 in) of rain in only one hour; up to 41 cm (16 in) fell in only four hours (Figure 10.22). The Guadalupe River rose from 0.9 m (3 ft) to 9.5 m (31 ft) in just two hours. Flash

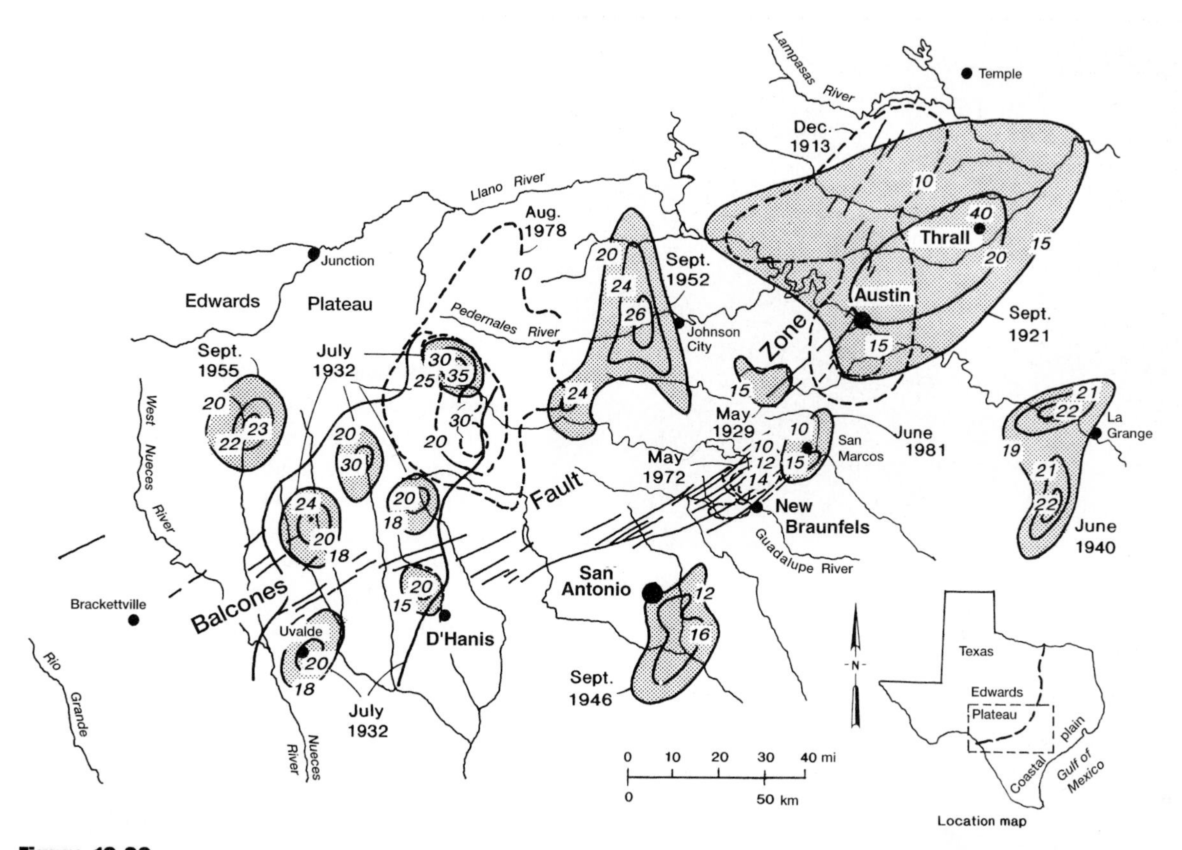

Figure 10.22 Map showing localized areas of high rainfall for some Balcones Escarpment thunderstorms in Texas. Rainfall depths are contoured in inches.

Data summarized from Breeding and Montgomery, 1954; Colwick and others, 1973; U.S. Army Corps of Engineers, 1964; U.S. Soil Conservation Service, 1954, 1958; and Williams and Lowry, 1929.

flooding claimed 17 lives in New Braunfels and caused $18 million in damages. The large Canyon Reservoir with its flood-control potential sits just 17 km (11 mi) upstream from New Braunfels; yet the localized thunderstorm dumped its heavy rains on the small, unprotected area below Canyon Dam, wreaking havoc.

Hail

Hailstones are semispherical, layered ice balls dropped from some thunderstorms (Figure 10.23). In North America, hail kills humans only occasionally, but it takes a tremendous toll on agriculture, as well as roofs and automobiles. Some estimates of United States crop damages due to hail range up to 2 percent of total crop value.

Thunderstorms that drop large, damaging hailstones are irregularly distributed in the United States (Figure 10.24). Important requirements for hail are 1) large thunderstorms with buoyant hot air rising from heated ground and 2) upper-level cold air creating maximum temperature contrasts, resulting in 3) the strong updrafts needed to keep hailstones suspended aloft while adding coatings of ice onto ever-growing cores. Comparison of thunderstorm abundance in Figure 10.21 with abundance of large hailstorms in Figure 10.24 reveals a marked difference. Thunderstorms are common in Florida, but the cold air necessary for hail formation is uncommon. Destructive, large hail abounds in the colder midcontinent.

A cross section through a hailstone (Figure 10.23) reveals accretionary layers of ice with an onionlike appearance. The layering indicates that both updrafts and downdrafts are needed to suspend and circulate a growing ice ball while it adds layers. A hailstone in Kansas had a diameter of 44 cm (over 17 in); updrafts must be very powerful to keep hailstones this large and heavy suspended in the air.

Hail is most common in the late spring and summer. The timing of hailstorms is related to the position of the polar-front jet stream. In April, Texas and Oklahoma receive much hail, but by June, most of the hailstorms have migrated northward into Montana, Wyoming, and South Dakota. In July, much of the hailstorm activity has moved into Canada. For example, in July of 1996 and 1998, four hailstorms hit Calgary, Alberta, causing total damages in excess of $450 million.

What can be done to protect against hail damage? An American Society of Civil Engineers study calls for development of economical materials for roofs and exterior walls that are resistant to impacts by hailstones.

Figure 10.23 A broken hailstone shows the layers of its growth, west Texas.

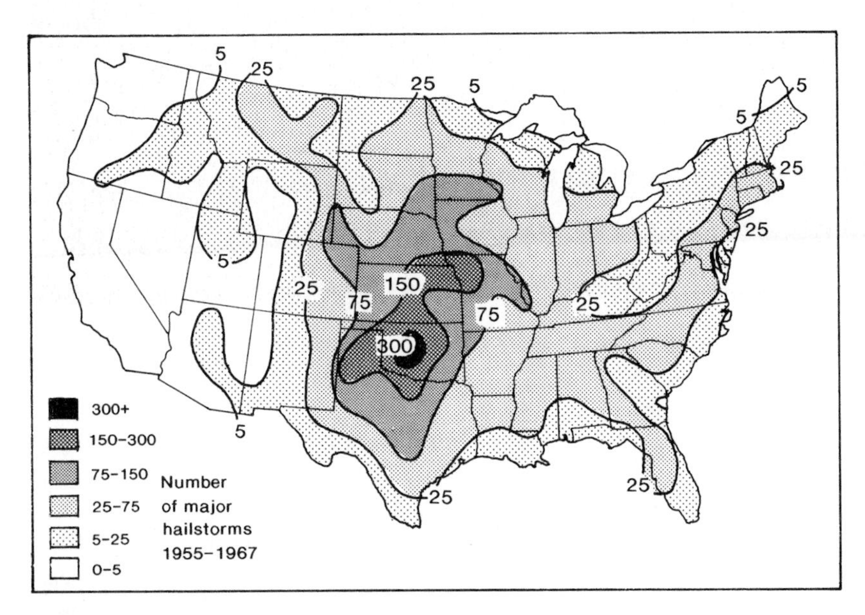

Figure 10.24 Major hailstorm frequency, 1955–1967. Contours of number of storms producing hail with 3/4-inch or larger diameter.

Data from Pautz, 1969.

Lightning

Thunderstorms generate lightning (Figure 10.25); thunder is caused by lightning. Lightning is the leading cause of forest fires and kills many people in the United States. Hurricanes, tornadoes, and floods are dramatic events that take many human lives in brief dramatic episodes, whereas lightning usually kills people one at a time.

Where does lightning occur? Its distribution is the same as thunderstorms (Figure 10.21). More thunderstorms develop in central Florida than anywhere else in the United States, and most lightning deaths happen in Florida (Figure 10.26). When does lightning claim its victims? In late spring and summer, when thunderstorms occur after moist air and sun-heated ground combine to build tall, unstable clouds. Lightning deaths plotted by month show thunderstorms are most active in summertime (Figure 10.27). Where does lightning kill people? Mostly outdoors (Table 10.3).

How Lightning Works You can make your own "lightning." Drag your feet across a carpet and become a negatively charged "thundercloud." Now touch a metal door handle and feel the "lightning" as the negative charges bolt from you to the positive charges on the metal.

The lightning of thunderstorms involves a similar flow of electric current as areas with excess positive charges seek a balance with places having excess negative charges. During the buildup of tall clouds, charged particles separate, creating an abundance of positive charges up top and an excess of negative charges down low (Figure 10.28).

The charge imbalance apparently comes about as the freezing and shattering of supercooled water drops initiates charge separations that are distributed by updrafts and downdrafts within the thundercloud. The charge separations occur during the cloud buildup of the early stage (Figure 10.19a), then lightning bolts forth during the mature stage (Figure 10.19b).

A thundercloud interacts electrically with the ground. The abundance of negative charges in the basal part of the cloud induces a buildup of positive charges on the ground surface. This sets in play the principle of physics that opposites attract each other and, in this case, produce lightning. Lightning can move from cloud to earth, earth to cloud, or cloud to cloud.

Lightning moves at speeds over 6,000 miles per second and typically includes several strokes, all occurring within one-half second. Thanks to high-speed photography, it is now possible to explain the basic sequence within a lightning flash. 1) Static electricity builds up within the lower thundercloud and induces opposite charges on the ground. 2) Discharge begins within the cloud and initiates a dimly visible, negatively charged stream of electrons propagating downward (Figure 10.29a). 3) The conductive stream moves earthward in 50 m (165 ft) jumps as a stepped leader (Figure 10.29b). 4) As the stepped leader nears the ground, the electric field at the surface increases greatly, shooting

Figure 10.25 Lightning from a late afternoon thunderstorm lights up the cloud and connects with the ground in Tucson, Arizona.

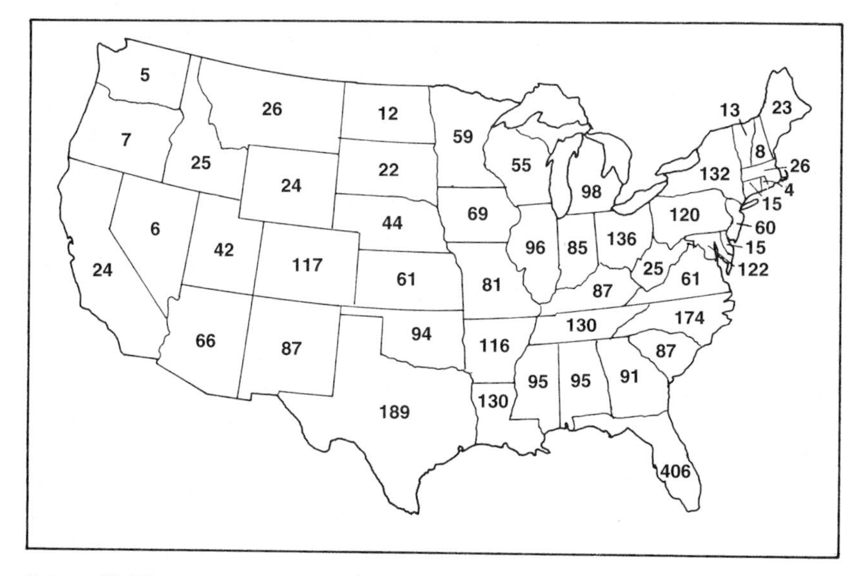

Figure 10.26 Lightning deaths in the United States, 1959–2001. In the 43 years of data, only two states escaped lightning deaths—Alaska and Hawaii.

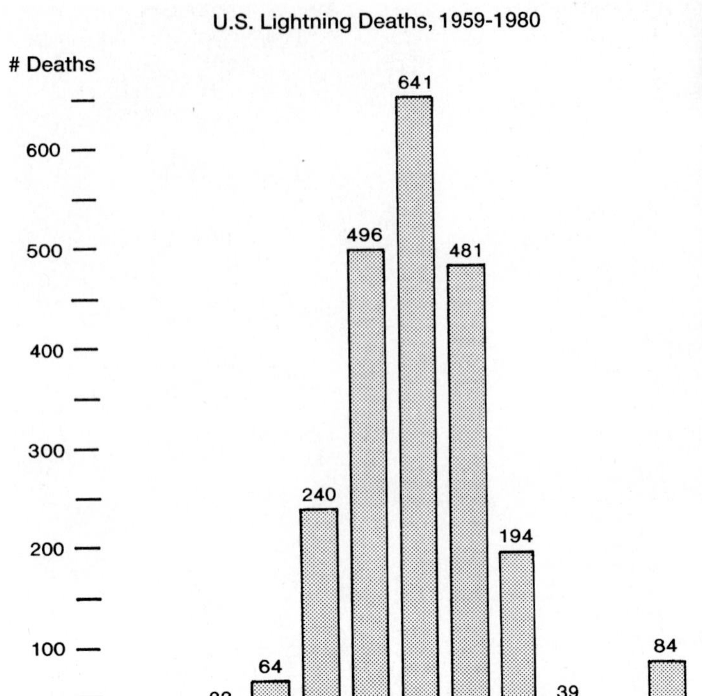

Figure 10.27 Lightning deaths in the United States by month, 1959–1980. Note: The 84 deaths in December include 81 killed in Maryland on 8 December 1963, when lightning caused the crash of a Boeing 707 jetliner.

Table 10.3 Location of U.S. Lightning Deaths, 1959–1980

Locations	
Open fields, ball fields	26%
Under trees	15%
On boats and in water-related activities	12%
On tractors and heavy road equipment	6%
Golf courses	5%
Via telephone	1%
Unspecified	35%
	Total 100%

Source: H. N. Vigansky. (1982). *General Summary of Lightning*. Asheville, N.C.: National Climatic Center.

streamers of positive sparks upward and connecting with the stepped leader about 50 m (165 ft) above ground (Figure 10.29c). 5) The connection initiates the return stroke, sending positive charges up to the cloud with a brilliant flash (Figure 10.29d). 6) More lightning strokes occur as charges flow between the cloud and the earth.

Several different strokes all occur within the split-second event we call a lightning bolt. If you have seen a lightning bolt appear to flicker, you have witnessed the several different up-and-down strokes that comprise a given "bolt." The electrical discharge of lightning can briefly create temperatures as high as 55,000°F. The high temperatures of lightning flash heat the surrounding air, causing it to expand explosively; this expansion of air produces the sound waves we call thunder.

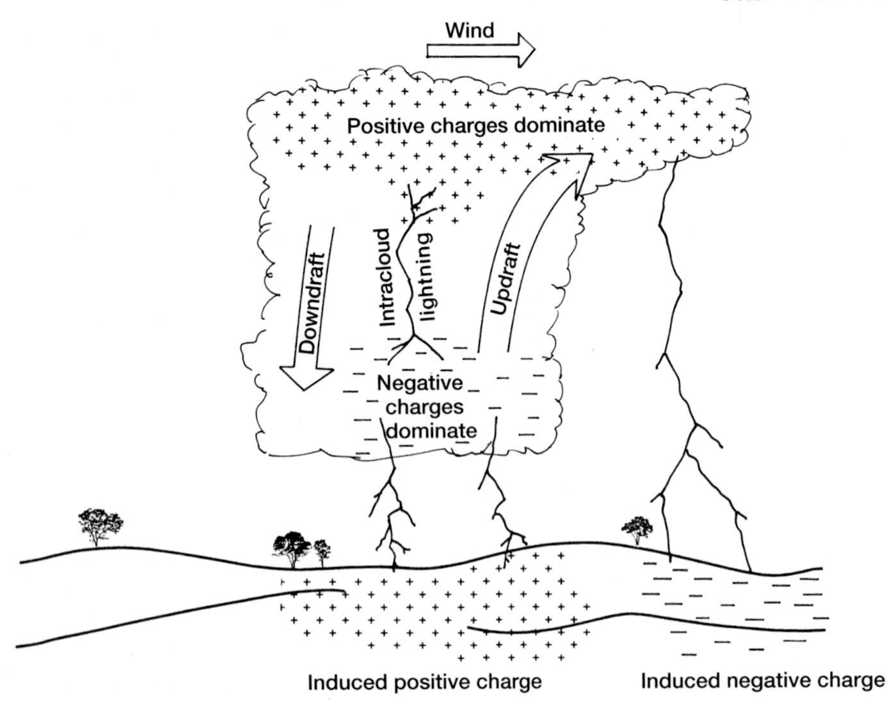

Figure 10.28 Schematic view of charge separation within a thundercloud and the induced charges on the earth beneath. When electric energy is great enough, opposite charges will attract and connect to form lightning.

Don't Get Struck Mark Twain said "Thunder is good, thunder is impressive; but it is lightning that does the work." Lightning bolts can strike ten miles away from the thundercloud. If you hear thunder, you are in an area of risk. How can you avoid being struck by lightning? 1) Get inside the house, but don't touch anything. Lightning can flow through plumbing, electrical, or telephone wires. People have been killed while talking on the phone. 2) Get inside a car or truck, but don't touch anything. Lightning usually travels along the metal surface of the vehicle then jumps to the ground through the air, a wet tire surface, or through the tire, causing it to blow out. 3) If you are caught outside, move to a low place.

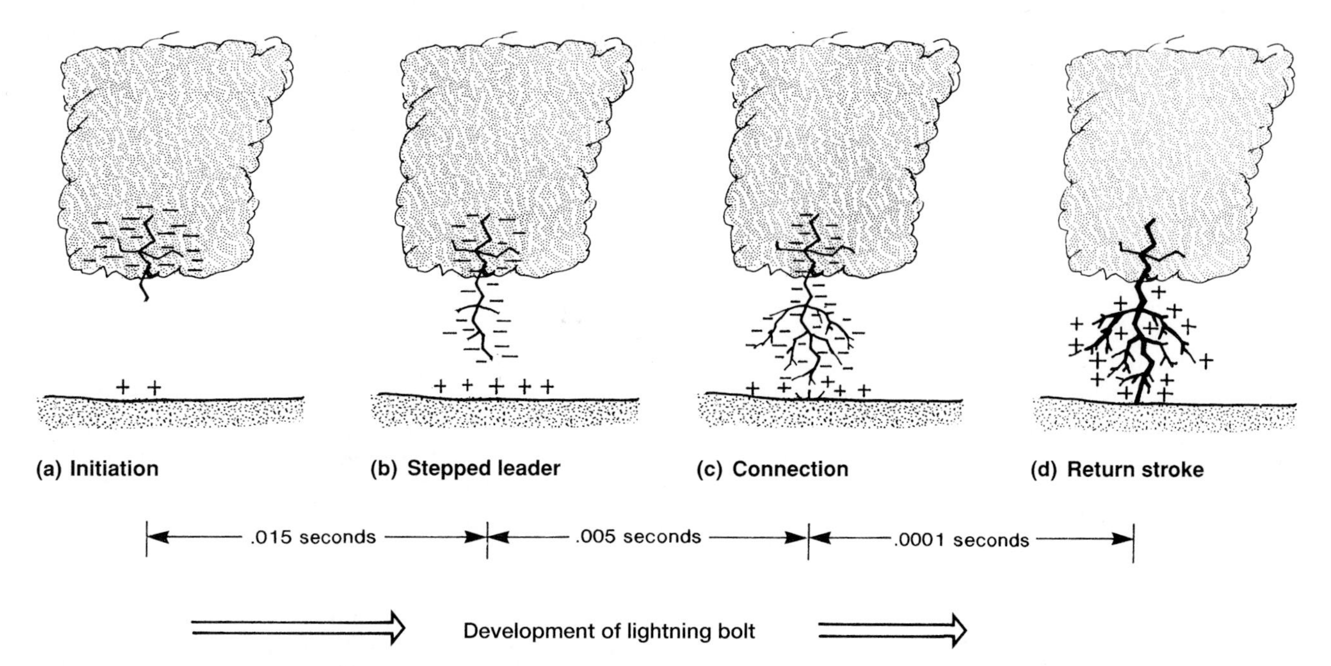

Figure 10.29 Steps in creating a lightning bolt. **(a)** *Initiation:* Charge separation in cloud builds up static electricity. **(b)** *Stepped leader:* Negative charges move in dimly visible stream downward in intermittent steps. **(c)** *Connection:* When leader nears ground, a positive discharge leaps up, completing the attachment. **(d)** *Return stroke:* The connected path flashes bright as charges exchange between cloud and ground in several events all totaling about one-half second.

Avoid tall structures, including trees. Don't lie down; lightning can flow through the earth. Assume the lightning position: crouch down on the balls of your feet with your hands over your ears.

Winds

Winds can arrive with tremendous power; they may be hot or cold, and they may blow either straight-line or rotating. The rotating winds of tornadoes capture our imaginations; they are the most feared offspring of a thunderstorm. Yet straight-line winds can be as damaging as a small tornado; they can kill people, extensively damage buildings, and cause tragic incidents with airplanes.

Straight-line Winds A widespread, powerful wind storm is referred to as a **derecho** (Spanish for straight ahead). Thunderstorms advancing on a line can have their individual energies combined to form a line of ferocious winds with hurricane-force gusts. An individual thunderstorm can send out bursts of wind at 60 mph for 10- to 15-second durations. But an organized line of storms in a region can generate a derecho lasting 10 to 15 minutes.

Ontario to New York Derecho, 15 July 1995 A 160 km (100 mi)-wide mass of thunderstorms traveling 80 mph in Ontario, Canada, moved south into New York at about 5:30 A.M. During the next two hours, it rushed through the Adirondack Mountains, across the Albany area, before losing strength in New York City. Wind gusts measured at 106 mph blew down millions of trees, five of which killed people.

Severe thunderstorms feed off solar energy, thus they usually occur during the hottest part of the day, in the late afternoon or early evening. Why did this derecho occur in the early morning hours? In mid-July 1995, a heat wave was in progress. Late evening temperatures on 14 July were still in the mid-80s as was the humidity. The hot and humid air supplied energy to the mass of thunderstorms traveling through the night. As the cloud mass raced across New York, it drew the hot, humid surface air upward until it condensed, releasing its contained energy. Derecho winds flowed down from the fast-moving cloud mass, causing extensive damage.

Tornadoes

A tornado is a rapidly rotating column of air usually descending from a large thunderstorm. Tornadoes are the most violent storms on the surface of the Earth; they have the highest wind speeds of any weather phenomenon. The strongest tornadoes are more intense than the biggest hurricanes, but they affect smaller areas. The Great Plains region of the central United States plays host to about 70 percent of the tornadoes that occur on Earth. U.S. tornadoes typically move from southwest to northeast at speeds up to 100 km/hr (62 mph), with rotating wind speeds sometimes in excess of 500 km/hr (310 mph).

Only slightly more than 1 percent of U.S. tornadoes have wind speeds in excess of 200 mph, but they are responsible for over 70 percent of deaths. The core of the whirling vortex is usually less than 1 km wide and acts like a giant vacuum cleaner sucking up air and objects (Figure 10.30).

Funnel clouds initially form thousands of feet up in the atmosphere, and many never touch the ground. Tornadoes may touch ground only briefly, or they may stay in contact for many kilometers, moving along an irregular path toward the northeast, but with abrupt changes in direction.

Tri-State Tornado, 18 March 1925

A little before 1 P.M. on 18 March 1925, the residents of Annapolis, Missouri, heard a sound "like a thousand freight trains" heading their way. It was an unusually broad tornado moving at about 60 mph along a northeasterly path. After traveling down Main Street and leveling the town, the tornado spun on toward Murphysboro, Illinois, and its 11,000 residents. The extrawide tornado destroyed everything in a nearly one-mile-wide path. Some of the wrecked buildings caught on fire, and just like in earthquakes, the event destroyed the water-supply system. Firefighters watched helplessly as flames devoured some people trapped in wrecked buildings. In Murphysboro, 210 people were killed either by the tornado force or the ensuing fires. On a lighter note, the business cards of the Reverend H. W. Abbot were plucked out of his ruined home and delivered to Palestine, Illinois, some 210 miles away.

Along its path, the tornado never seemed to lose energy. Residents of DeSoto, Illinois, watched the advancing tornado with its suspended load of house and auto parts and uprooted trees, seemingly defying the law of gravity. The tornado spun toward the town school, lifting off its roof and knocking down the walls on 125 students and teachers, 88 of whom died. The dead students were laid out on the school lawn, but many parents were unable to claim their bodies—they too were dead.

A railroad engineer faced the tornado coming straight down the tracks toward him. He responded by accelerating

Figure 10.30 Powerful tornado moves into Oakfield, Wisconsin, causing $50 million in damages and injuring 19 people on 18 July 1996. This is a good place to view a tornado—through a car window.

as fast as possible and trying to blast through the tornado. He succeeded, but the roofs were stripped off his train cars like lids off sardine cans.

This "Tri-State tornado" is the largest known. It traveled on a N 21°E path for 353 km (219 mi) across parts of Missouri, Illinois, and Indiana, devastating 23 cities and towns in the widest swath ever recorded—up to 1 mi wide. This single tornado claimed 689 lives.

How a Tornado Works

What is necessary to turn an ordinary thunderstorm into a tornado-spinning monster? In the central United States, several conditions typically occur simultaneously: 1) a northerly flow of marine tropical air from the Gulf of Mexico that is humid and has temperatures at the ground in excess of 75°F; 2) a cold, dry air mass moving down from Canada or out from the Rocky Mountains at speeds in excess of 50 mph; and 3) jet-stream winds racing east at speeds in excess of 150 mph. These three air masses, all moving different directions, set up shearing conditions, imparting spin to a thundercloud (Figure 10.31).

The warm, moist Gulf air lifts vertically, releasing its latent heat and forming a strong updraft that is sheared and spun at midlevels by the fast-moving polar air and then twisted in another direction at its upper levels by the jet stream. The corkscrew motion is enhanced by vertical movements of air: warm air rising on the leading side, with cool air descending on the trailing side. The rotation of the winds is achieved without involving the Coriolis effect; tornadoes are too small for their rotation to be caused by Coriolis forces.

But most large thunderclouds do not spin off tornadoes. What is different about those clouds that do? The question cannot be answered exactly, but we are beginning to see patterns. In an ordinary *single-cell thunderstorm* (Figure 10.32a), warm air rises by convection to build a nearly vertical cloud mass. In the cold upper air, water vapor condenses and rain falls back down through the thundercloud, cooling the lower air and slowing the upward flow of energy-carrying warm air.

Sometimes wind shear tilts the thundercloud mass and it may grow into a *super-cell thunderstorm* (Figure 10.32b). The tilt allows the warm air to rise on the advancing or leading side

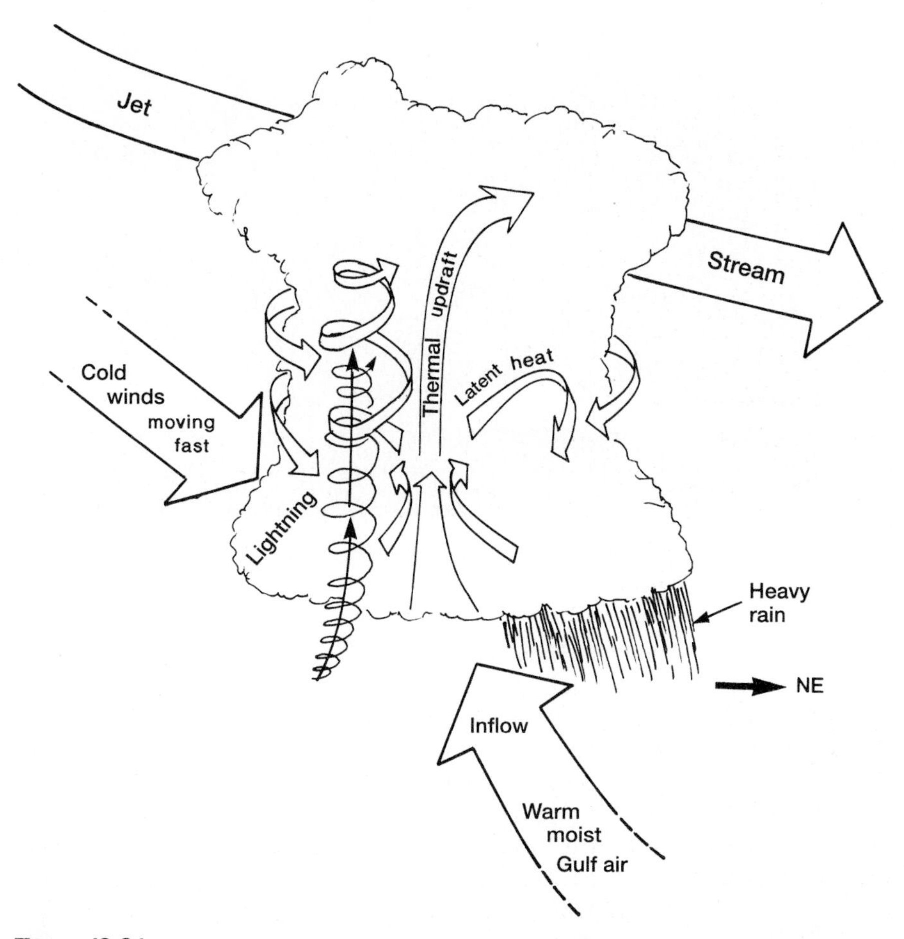

Figure 10.31 Components of a tornado. A warm, moist mass of Gulf of Mexico air collides with a fast-moving mass of polar air, causing updrafts that build up into the polar-front jet stream. Three fast-moving air masses, all going different directions, impart shears, causing rotation. A tornado receives additional energy via lightning and latent heat released by rainfall.

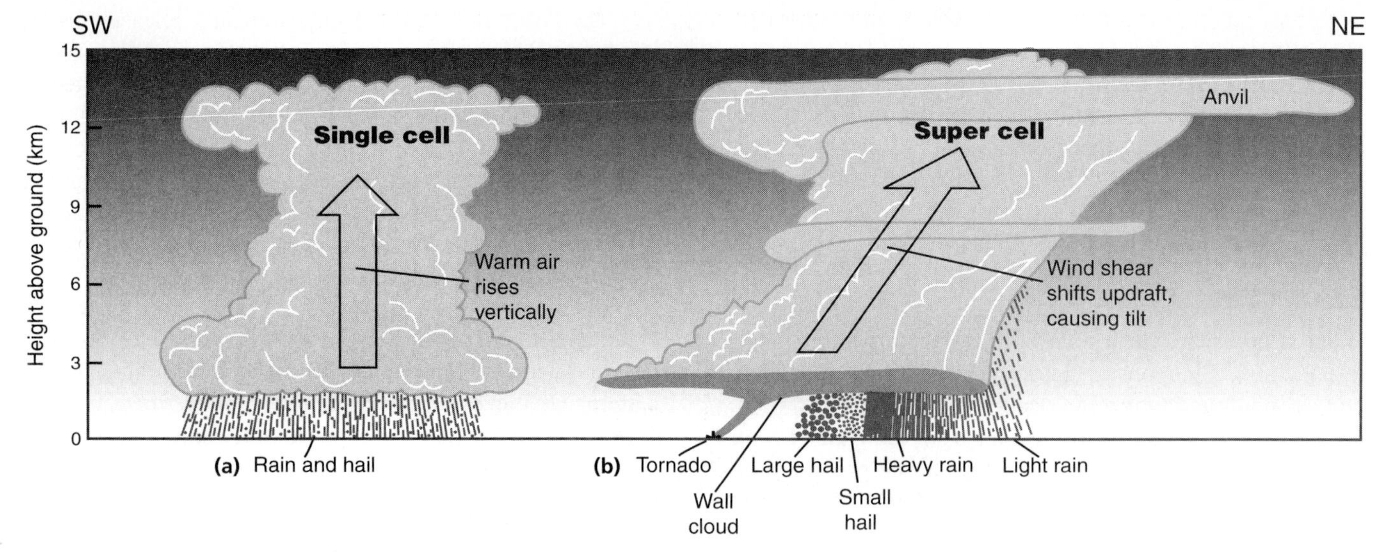

Figure 10.32 Types of thunderstorms. **(a)** *Single-cell thunderstorm:* Warm moist air rises vertically, forms rain which falls down through cloud cooling it down. **(b)** *Super-cell thunderstorm:* Tilted thunderstorm has rain and hail on leading side with tornadoes on trailing side.

of the cloud and this is where most of the rain falls. On the rear or trailing side, downdrafts of cool air exist and it is here that tornadoes usually form. Rotation may develop in a wide zone in the thunderstorm; then, as the rotating core pulls into a tighter spiral, its speed increases dramatically, its angular momentum is preserved. This principle is analogous to ice skaters spinning with arms outstretched who, pulling their arms toward their bodies, spin faster. In other words, the smaller the diameter of a rotating mass, the faster it spins.

In those instances when a mobile team of tornado chasers is able to get a Doppler radar unit close enough to a tornado, the radar data tell some interesting tales. At least some tornadoes have downward-moving air in the center surrounded by a cone- or cylinder-shaped funnel that rapidly spirals upward. This indicates that as the tornado is sucking up huge volumes of air plus debris from the ground, some of the air supplied to this mega-vacuum cleaner comes from a central downdraft.

The rotating wind speeds of a tornado are highest a few hundred feet above the ground. This is most likely due to the winds at ground level being slowed by the frictional resistance of earth, trees, buildings, cars, and such. Tornadoes range in orientation from vertical to horizontal (Figure 10.33). Once formed, tornadoes derive additional energy from electrical discharges of lightning within the clouds and from latent heat released by heavy rainfall at the cloud's front.

Tornadoes in the United States and Canada

The interior of the United States is the tornado capital of the world. It is over these gently sloping lands that air masses collide, spinning out tornadoes that usually head northeast under the influence of the polar jet stream (Figure 10.34). Tornadoes can occur at any time of the year but are most common in late spring and early summer (Figure 10.35). Tornado deaths in the United States are most common in the spring (Figure 10.36). The furious energy of these twisters generates wind speeds that stretch one's imagination. Wind damages are gauged by the Fujita scale (Table 10.4). Tornadoes with wind speeds exceeding 260 mph (115 m/sec) cause "incredible" damages.

Figure 10.33 A wide tornado moves across the plains toward the aptly named Last Chance, Colorado on 21 July 1993.

Tornadoes have three main destructive actions. 1) High-wind speeds blow away buildings and trees. 2) Winds rushing up the funnel have a lifting force in the updraft. 3) An explosive situation exists because of the differences in air pressure

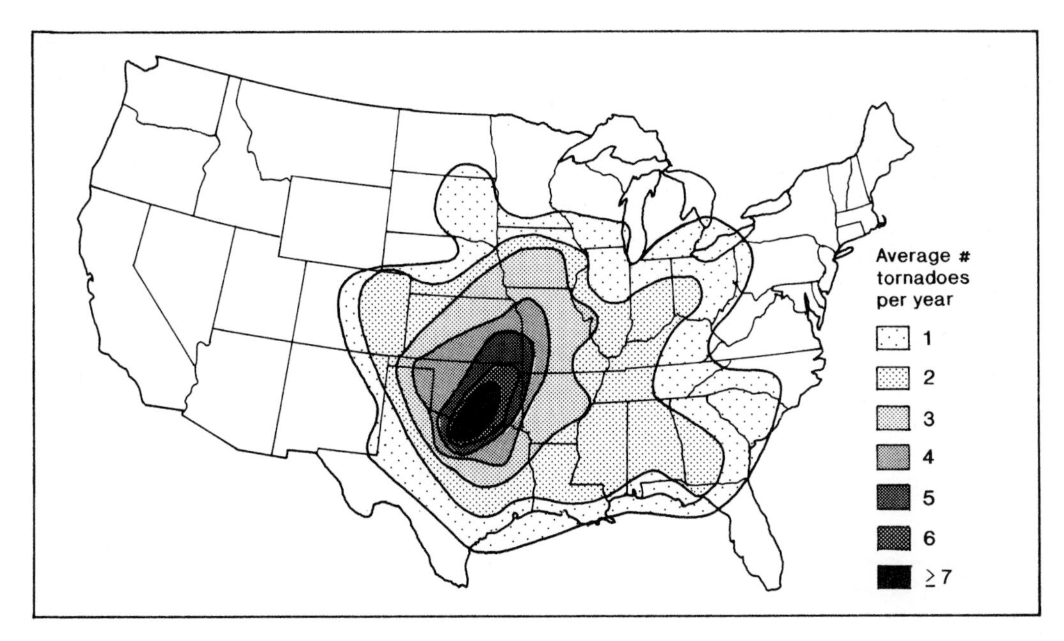

Figure 10.34 Average annual occurrence of major tornadoes. General direction of travel is toward the northeast.

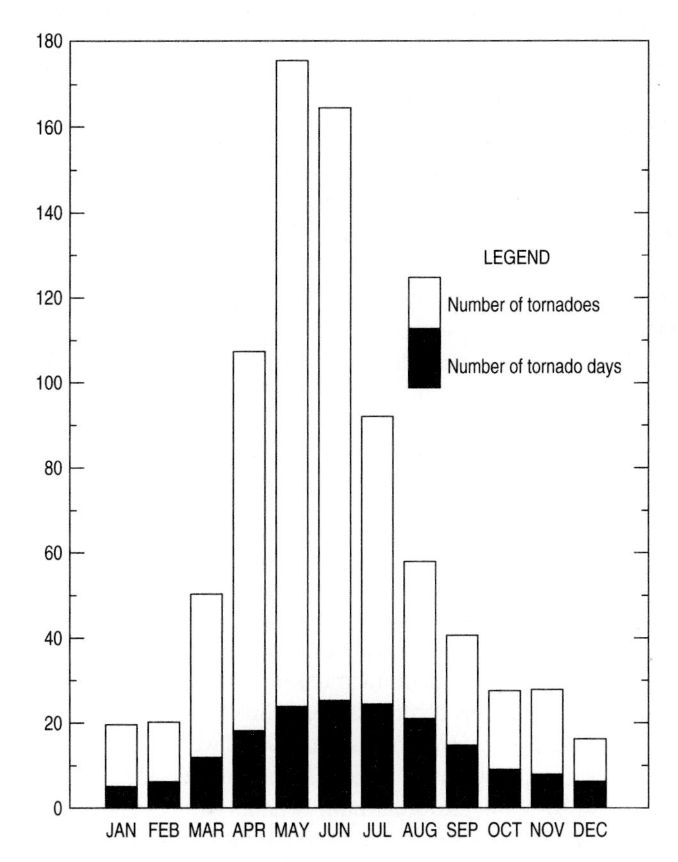

Figure 10.35 Average number of tornadoes each month in the United States, 1950 to 1999.

Based on data from NOAA.

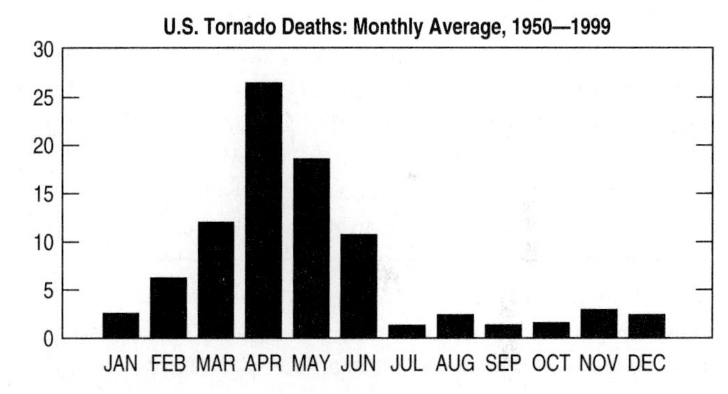

Figure 10.36 Average number of tornado deaths each month in the United States, 1950 to 1999.

Based on data from NOAA.

Table 10.4 Fujita Wind Damage Scale

Fujita Wind Damage Scale				U.S. Tornadoes, 1992	
				Number	Percentage
F-0	under 72 mph	(32 m/sec)	light	696	54
F-1	73–112	(33–50)	moderate	411	32
F-2	113–157	(51–70)	considerable	129	10
F-3	158–206	(71–92)	severe	43	3
F-4	207–260	(93–116)	devastating	13	1
F-5	261–318	(117–142)	incredible	1	<1
F-6	>318	>142	unexpected	0	0
				1293 total	100

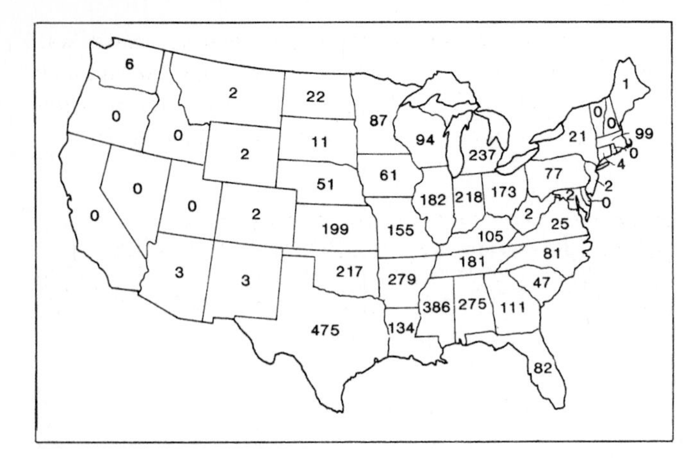

Figure 10.37 Distribution of U.S. tornado deaths, 1950–1994.

Table 10.5 Deadly Dozen States for Tornado Deaths

Total Deaths	Deaths per Capita
1. Texas	1. Mississippi
2. Mississippi	2. Arkansas
3. Arkansas	3. Kansas
4. Alabama	4. Oklahoma
5. Michigan	5. Alabama
6. Indiana	6. Indiana
7. Oklahoma	7. Tennessee
8. Kansas	8. Nebraska
9. Illinois	9. Louisiana
10. Tennessee	10. Missouri
11. Ohio	11. Kentucky
12. Missouri	12. Texas

between the very low pressure inside a tornado funnel and the higher pressures outside it. For example, a closed-up building contains air under higher pressure; the passage of a low-pressure tornado funnel can cause an unventilated building to explode outward due to the differences in air pressure.

Tornado deaths in the United States are dealt out unequally (Figure 10.37). A belt of states running from the Gulf of Mexico into Canada gets most of the carnage, while the coastal states are largely spared the agony. Note that 11 states suffered no tornado fatalities between 1950 and 1994. The deadly dozen states are listed in Table 10.5. But if the number of tornado deaths by state are divided into the population of each state, it yields a per capita ranking of deaths that rearranges the deadly dozen states (Table 10.5). The 11 states where the highest percentage of residents die by tornado are all low-population states. On a per-capita basis, Texas drops from deadliest down to number 12.

The litany of tornado deaths skips no years. There is some encouragement to be found in the declining numbers of tornado deaths during the last eight decades (Table 10.6). Tornadoes are an omnipresent threat that must be faced. Either we play Russian roulette and hope they don't hit buildings, or else buildings need to be designed and built to withstand tornado forces. High winds are a difficult test for any structure, but when a tornado passes over a building, there can be a 10 to 20 percent drop in air pressure, causing tightly closed buildings to explode. Some of the exploded debris is sucked up with the updrafting air, which may be rising at over 100 mph (45 m/sec). Design with nature, or run the risk of dying by nature.

Table 10.6 U.S. Tornado Deaths by Decade, 1920–1999

Decade	Recorded Deaths
1920–1929	3,169
1930–1939	1,944
1940–1949	1,788
1950–1959	1,409
1960–1969	935
1970–1979	987
1980–1989	522
1990–1999	580

Tornado Deaths Average for 86 years from 1916 to 2001 = 145 deaths per year

Source: Climatological Data, National Climatic Center.

Who dies during tornadoes? Tornadoes preferentially kill: 1) old people, 2) mobile-home residents, 3) occupants of exterior rooms with windows, and 4) those unaware of broadcast tornado alerts. Residents of frame houses can run into interior rooms to gain some protection. But where do mobile-home dwellers run to hide? There are no interior rooms and who wants to run outside into a severe thunderstorm with heavy rain, lightning, hail, and flying debris? Almost half of the Americans killed by tornadoes die inside their disintegrating mobile homes (Table 10.7).

Are you safer in a car or a mobile home? Many mobile homes can be tipped over by 80 mph (35 m/sec) winds whereas many modern cars have low centers of gravity and streamlined shapes that require wind speeds of about 120 mph (55 m/sec) to tip over. Also most cars are built to provide some protection during a rollover or collision. Many mobile-home dwellers facing a tornado threat would do better to run outside and sit inside their cars.

It is an interesting commentary on human nature that so many people in the central United States are not worried about the dangers posed by tornadoes yet wonder how Californians can live with the threat of earthquakes. Meanwhile, most Californians are quite comfortable with earthquakes, but many do not want the stress of living with tornadoes. Many people seem to be most comfortable with the events of their home area.

Table 10.7 U.S. Tornado Fatalities by Location, 1985–2001

Mobile home	40%
Permanent home	30%
Vehicle	10%
School/Church	7%
Outdoors	5%
Business	5%
Unknown	3%

Data from NOAA.

The Super Outbreak, 3–4 April 1974

Scanning the death totals in Table 10.6 makes it apparent that not as many people in this generation are dying from tornadoes as in past generations. Earlier warnings are now broadcast, and people are better prepared. However, no amount of preparation could protect everyone from the events of April 1974.

The weather scene on 2 April 1974 included: 1) a cold front spreading snow in the Rocky Mountains, 2) a low-pressure system moving east, 3) increasingly humid air over the 75°F water of the Gulf of Mexico, 4) a strong polar jet stream with a bend flowing from Texas to New England, and 5) a dry air mass coming from the southwest attracted to the low-pressure system. As the dry, desert air mass moved toward the Mississippi River, it overrode the moist Gulf air, forming an **inversion layer** that trapped unstable moist air below.

On 3 April, all the weather systems came together. The unstable, moist air from the Gulf of Mexico began bursting up through the inversion layer, forming huge anvil-shaped thunderclouds that were set spinning by the other converging air masses. At about 1 P.M., there began the greatest tornado assault ever: in 16 hours, 147 tornadoes touched ground in 13 states (Figure 10.38). The barrage included six tornadoes of F-5 force: two in Ohio, two in Alabama, and one each in Indiana and Kentucky. The mighty six were a decade's worth, all in a few hours; each touched ground for over 50 km (30 mi), and two stayed down for more than 160 km (100 mi). Most of the tornadoes touched down during the warm hours between 4 to 9 P.M.; these are typical hours for tornado touchdown.

The area beneath the tornado-generating fronts was about 1,270,000 km^2 (490,000 mi^2), but only an unlucky 1,550 km^2 (600 mi^2) was hit hard. The destruction wrought by the super outbreak was overwhelming (Table 10.8).

Table 10.8 Damages from the Super Outbreak Tornadoes

335	people killed
1,200	people hospitalized
over 7,500	houses destroyed
over 6,000	houses severely damaged
2,100	mobile homes destroyed
over 4,000	farm buildings destroyed
1,500	small businesses destroyed or severely damaged
27,600	families suffered significant losses

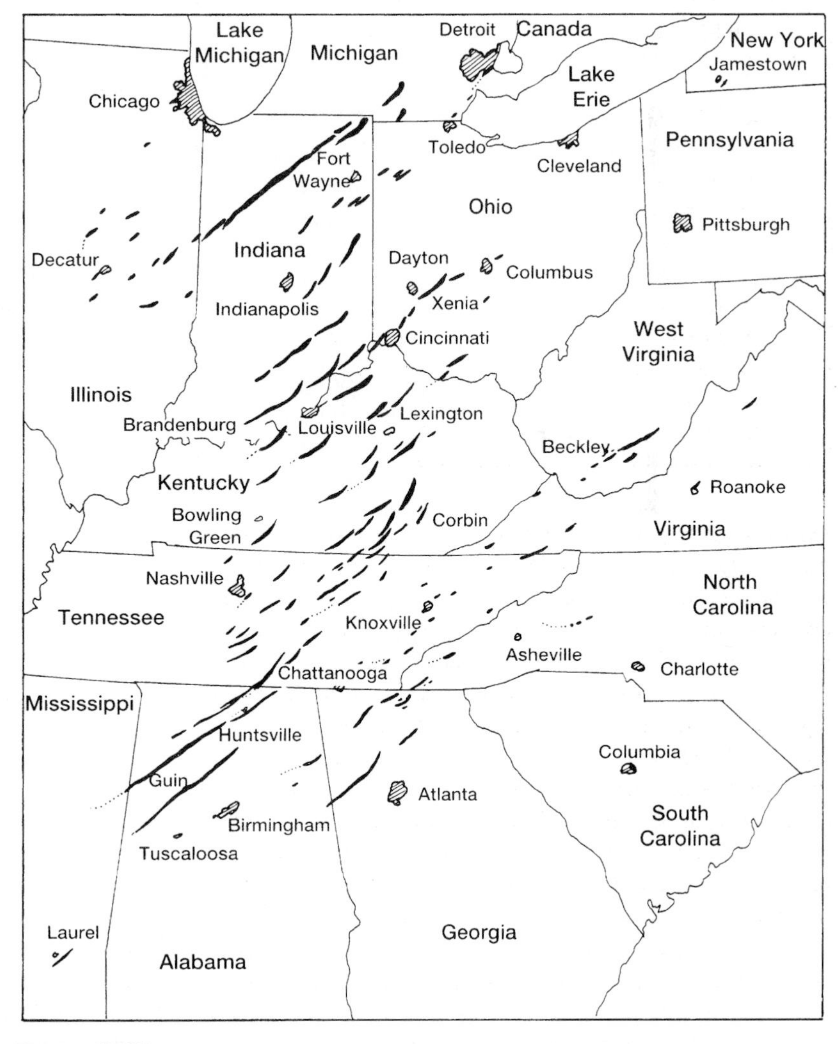

Figure 10.38 Paths etched across the ground by 147 tornadoes on 3–4 April 1974. The northeasterly trend is typical.

Based on a map prepared by T. T. Fujita at the University of Chicago.

Table 10.9 Cities Struck by Tornadoes, May 1997–March 2000			
Date	**City**	**Fujita Scale Intensity**	**Deaths**
12 May 97	Miami, Florida	F1	—
8 Apr 98	Birmingham, Alabama	F5	33
16 Apr 98	Nashville, Tennessee	F3	1
21 Jan 99	Little Rock, Arkansas	F3	3
9 Apr 99	Cincinnati, Ohio	F4	7
3 May 99	Oklahoma City, Oklahoma	F5	42
3 May 99	Wichita, Kansas	F4	6
11 Aug 99	Salt Lake City, Utah	F2	1
8 Mar 00	Milwaukee, Wisconsin	F2	—
28 Mar 00	Fort Worth, Texas	F2	4

Tornadoes and Cities

Cities create their own weather. In the 1800s it was recognized that Berlin, London, and Paris were warmer than the surrounding countryside. A study of Atlanta shows it to be an urban heat island with night-time temperatures up to 10°C (18°F) warmer than the adjoining areas. Atlanta has developed by cutting down 380,000 acres of forest and replacing it with buildings and roads. The urban concrete, asphalt, and stone absorb heat during the day and radiate heat at night. The warm air rising above a city creates its own low-pressure cell, a convecting plume of heat that can rise, cool, condense, and form thunderstorms.

There is a commonly heard tale that tornadoes never strike big cities. It is a myth. The United States averages about 1,200 tornadoes per year. Most of these tornadoes occur in the central states where there are few big cities available to strike. Cities cover very little ground in a large region, thus they are small targets. Yet in the three years from Spring 1997 to Spring 2000, approximately 3,600 tornadoes occurred and 10 of them struck large metropolitan areas (Table 10.9).

The recent tornado strikes on cities include assaults on downtowns and their skyscraper buildings. The Nashville tornado sent pedestrians scurrying into high-rise buildings for protection. In Salt Lake City, the tornado ripped across the Delta Center arena and through the Salt Palace Convention Center in the early afternoon. Downtown Fort Worth was battered by wind and hail at 6:30 P.M., that knocked the windows out of tall buildings and destroyed the offices and businesses inside. The Oklahoma City tornado was noteworthy for its record-high wind speeds of 318 mph measured 65 feet above ground using Doppler radar. If there is a capital city for tornadoes in the United States it would have to be Oklahoma City. Sitting in the heart of Tornado Alley, Oklahoma City has felt the wrath of at least 103 tornadoes between 1893 and 1999.

Safe Rooms The traditional protection against dying in a tornado has been to go underground, into a cellar. But more and more houses are being built without cellars. What can a homeowner do?The Oklahoma City tornado showed the value of a new type of shelter, the safe room. In the interiors of houses, closets or bathrooms are being built as safe rooms with 12-inch-thick concrete walls, a steel door, and a concrete roof. These safe rooms are reminiscent of bank vaults. Heeding the Oklahoma City tornado warning, some residents went into their safe rooms and emerged later to find their homes blown away. Although their houses were destroyed, they suffered no deaths or even injuries.

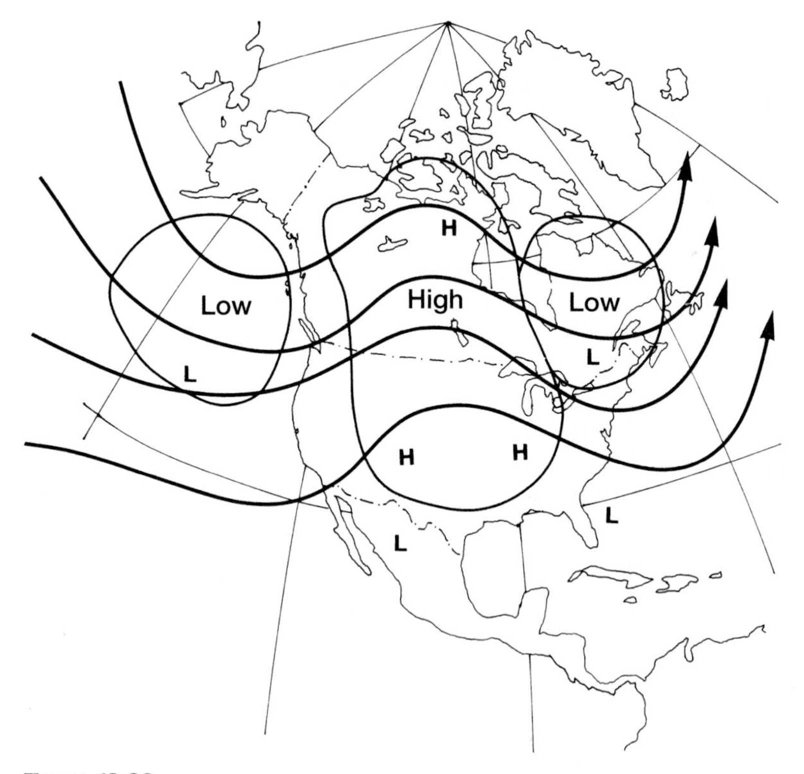

Figure 10.39 Dry conditions in the central United States are commonly caused by a long-lasting, high-pressure ridge in the upper troposphere. The ridge in the upper-level wind-flow path over the central parts of Canada and the United States causes anticyclones (clockwise rotations) where warm air aloft descends, warming further and lowering its humidity, and then sucks up moisture from the land below. The midcontinent high also blocks the northward flow of moist Gulf of Mexico air.

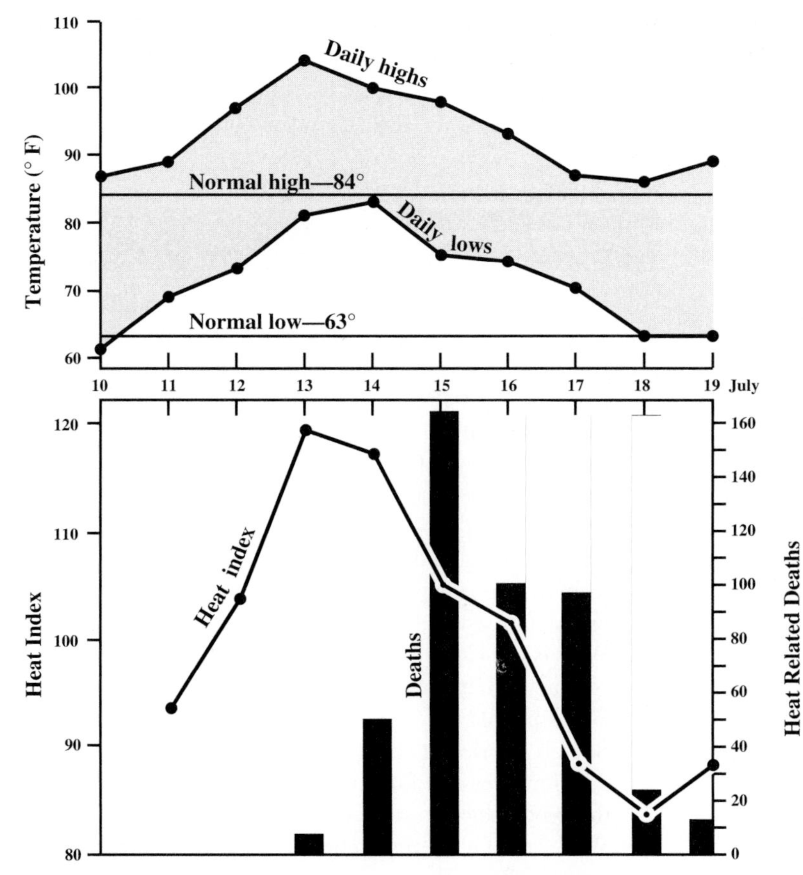

Figure 10.40 Daily maximum and minimum temperatures in Chicago, Illinois, during 10 to 19 July 1995, and the numbers of heat-related deaths. The heat index involves both temperature and humidity.

Data Source: NOAA/U.S. Department of Commerce, "Heat Wave Natural Disaster Survey Report," pg. 23, figs 9, 10, U.S. Dept. Commerce, July 1995.

Heat Wave in Chicago, July 1995 Dry and hot weather in the central United States and Canada are commonly associated with high-pressure atmospheric conditions (Figure 10.39). In July of 1995, a strong, upper-level ridge of high pressure sat on top of a slow-moving, hot and humid air mass on the surface. During the 3-day period from 13 to 15 July, heat records were broken at numerous locations in the central and northern Great Plains. What made this heat wave especially difficult was its combination of both high maximum and high minimum temperatures (Figure 10.40). The surface air mass did not cool much at night because its high humidity (high water content) held so much heat.

The health effects of the heat wave became apparent as bodies were recovered from overheated dwellings. When human body temperature reaches 105°F, hyperthermia sets in and death becomes a possibility. Notice in Figure 10.40 that most deaths do not occur at the peak of the heat wave, but rather they happen for days afterward as bodies weaken and fail. From 13 to 27 July, 465 people in Chicago died of heat-related causes. By way of comparison, there were no heat-related fatalities from 4 to 10 July.

Who died during the heat wave? The most affected were people without access to air conditioning and especially elderly folks; the greater their ages, the higher their death percentages. An interesting statistic is that more than half the deceased lived on the top floor of their buildings, where heat buildup was greatest.

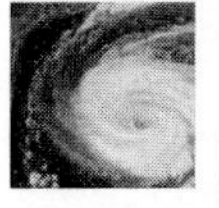

Extreme Heat

Dry weather may occur on long- or short-time scales. As continents drift and ocean basins open and close, large regions may be cut off from moisture supplies and kept in long-term desert conditions. On a shorter time scale, changes in the position of the jet stream or atmospheric convergence zones may keep moisture-bearing air away from large areas for many years, bringing on drought conditions. On an even shorter time scale are heat waves.

Heat Wave

The days of high temperatures during a heat wave may be one of the least appreciated of the weather disasters. A heat wave is not visible; it does not leave piles of wreckage. The heat wave is an invisible, silent killer and it can kill in large numbers.

Summary

Most of the death and destruction from natural disasters comes via severe weather.

Solar radiation is received in abundance between 32°N and 34°S latitudes. Much of this heat is transferred as latent heat in water vapor by winds moving aloft toward the poles. Cold polar air is dense and flows equatorward. The midlatitudes are the transfer zone between the equatorial and polar air masses; they have the most severe weather.

Where tropical air in the heat-expanded troposphere meets the cold, compressed polar air, a west-to-east high-level air flow exists—the fast-moving, polar-front jet stream. The position of the polar jet migrates across the United States with the seasons. In the summer, it is over Canada; in the winter, it is near the Gulf of Mexico. The polar jet stream plays a large role in moving heat and air masses and is involved in many severe weather situations.

The paths of large, moving air and water masses are deflected by the force of the Earth's rotation. In the Northern Hemisphere, moving masses veer to their right, in the direction of movement; this is the Coriolis effect. In the Southern Hemisphere, moving objects veer to their left.

Water has a remarkable ability to absorb and release heat. Energy is absorbed in water vapor during evaporation (latent heat of vaporization) and released during precipitation. Energy is absorbed in liquid water during melting (latent heat of fusion) and released during precipitation as snow.

Air masses vary in their temperature and water-vapor content. Different air masses do not readily mix; they are separated along boundaries called fronts. Much severe weather occurs along fronts.

Rotating air bodies create some of the most severe weather via thunderstorms, tornadoes, and hurricanes. In the Northern Hemisphere, rotation is counterclockwise as cyclonic circulation. Cyclones have a low-pressure core, so surface winds flow inward toward the core, feeding a large updraft of rising air that cools to form clouds and sometimes rain. Many of the largest cyclonic circulations are linked to troughs (large bends concave toward the North Pole) in the polar jet stream. Intense cyclones a few miles across can be thunderclouds commonly producing heavy rain, lightning, thunder, and hail, and sometimes spinning off even smaller-radius cyclones—tornadoes. The smaller the radius of a rotating air mass, the faster its wind speeds. Tornado winds can exceed 300 mph.

Heat waves are silent killers that prey on the elderly. Droughts are years of rainfall shortage. In the central United States, droughts have resulted from high-pressure ridges (large bends convex toward the North Pole) in the polar-front jet stream that foster anticyclonic circulation. An anticyclone rotates clockwise with dry air descending down its core, warming further, and evaporating moisture from the lands below.

Terms to Remember

Questions for Review

1. What is latent heat? Is it absorbed or released during: a) melting, b) freezing, c) evaporation, d) condensation?
2. Draw a cross section and explain the cause of the polar-front jet stream. Why does its position vary across the United States during the course of a year?
3. What are the relationships between high- and low-pressure zones and between cyclones and anticyclones?
4. Draw a map and explain the relationships between troughs and ridges in the polar-front jet stream and air circulation as cyclones and anticyclones.
5. Why is the Coriolis effect always to the right in the Northern Hemisphere and to the left in the Southern Hemisphere?
6. Sketch a series of cross sections showing the stages of development of a late-afternoon thundercloud. Label the processes occurring in the cloud during each stage.
7. What is the relationship between thunder and lightning?
8. How does hail form? Where does most of the large hail fall in the United States?
9. Why do higher wind speeds develop in a tornado than in a hurricane?
10. Draw a map showing the polar jet stream conditions associated with drought in the central United States.

Questions for Further Thought

1. Why are air temperatures near the seashore cooler during the day and warmer during the night than in inland areas?
2. Is it more dangerous to live in earthquake or tornado country?
3. Do tornadoes have major impacts on large cities?
4. What is the difference between a drought and a desert?

Suggested Readings and References

Abbey, Robert F., Jr., and Fujita, T. Theodore (1981). Tornadoes: The tornado outbreak of 3–4 April 1974. In *The Thunderstorm in Human Affairs* (37–66). Norman: University of Oklahoma Press.

American Society of Civil Engineers (2001). *The Ten Most Wanted.* Institute for Business and Home Safety.

Baker, Victor R. (1977). Stream-channel response to floods. *Geological Society of America Bulletin,* 88, 1057–71.

Battan, Louis J. (1974). *Weather.* Englewood Cliffs, N.J.: Prentice-Hall.

Bluestein, Howard B. (1999). *Tornado Alley.* New York: Oxford University Press.

Bomar, George W. (1983). *Texas Weather.* Austin: University of Texas Press.

Bryant, Edward A. (1991). *Natural Hazards.* New York: Cambridge University Press.

Davis, Robert E., and Dolan, Robert. (1993, September–October). Nor'easters. *American Scientist,* 81, 428–39.

Eagleman, Joe R. (1983). *Severe and Unusual Weather.* New York: Van Nostrand Reinhold.

Krider, E. Philip. (1983). Lightning damage and lightning protection. In *The Thunderstorm in Human Affairs* (111–24). Norman: University of Oklahoma Press.

Miller, Peter. (1987, June). Tracking tornadoes. *National Geographic,* 171, 690–715.

Parker, Sybil P., ed. (1988). *Meteorology Sourcebook.* New York: McGraw-Hill.

Smith, Howard E., Jr. (1982). *Killer Weather.* New York: Dodd, Mead.

Williams, Jack. (1992). *The Weather Book.* New York: Vintage Books.

Videos

Tornado! (1985). NOVA/WGBH (60 min.).

Lightning. (1995). NOVA/WGBH (60 min.).

Tornado Video Classics, vols. 1 and 2. (1994). The Tornado Project (180 min. total).

Clouds. (1988). Pinnacle Productions (57 min.). Uses time-lapse photography and music.

Tornadoes!! (1993). Norman Beerger Productions (60 min.). Forty-six tornadoes set to music.

Raging Planet:Blizzard. (1998). Discovery Channel (50 min.).

Raging Planet:Lightning. (1997). Discovery Channel (50 min.).

May's Fury. (1999). Oklahoma News Channel 4. The 3 May 1999 Oklahoma City tornado.

Chapter 11

Hurricanes and the Coastline

Miracles appear to be so, according to our ignorance of nature, and not according to the essence of nature.

—Michael de Montaigne, 1580, *Essays I*

Hurricanes

A **hurricane** is the only natural disaster that is given a human name. Andrew, Camille, Hugo, Iniki, Mitch, Pauline, and their kin share family traits, but each has its own personality. Each hurricane "lives" for enough days that we get to know its individual characteristics. Hurricanes are large **tropical cyclones.** They are heat engines that convert the heat energy of the tropical ocean into winds and waves. They are huge storms that can generate winds over 150 mph (240 km/hr) (Figure 11.1). Hurricanes can push massive mounds of seawater onshore as **surges** that temporarily raise sea level over 20 feet (6 m); and their heavy rains can cause dangerous floods, killing people well away from the coastline.

Tropical Storm Allison poured rain onto Houston, Texas, on 7 June 2001, then left the city only to return unexpectedly on 9–10 June. On its return, Allison dumped 14 in (36 cm) of rain onto the already saturated ground. Flooding drowned 2,500 animals involved in medical research at the Texas Medical Center and claimed the lives of 24 people in the region. One poignant story occurred in the Bank of America building after employees were warned that their cars in the underground parking garage needed to be moved before they were flooded. A 42-year-old law clerk took an elevator down to retrieve her car from the fourth underground level. Upon reaching the third level, flood water shut off power to the elevator. The slowing rising water drowned the unfortunate woman trapped in her elevator jail.

Andrew, August 1992

Andrew was born in Africa. On 13 August, it had developed into thunderstorms over West Africa. It then moved out over the Atlantic Ocean as rainy, low-pressure wind waves that converged at low angles to form a rotating air mass. By 17 August, the central circulation had intensified into a **tropical storm,** but high-level winds disrupted the upward growth of the rotating core of clouds. Weak and disorganized, Andrew drifted west across the Atlantic Ocean. By Friday, 21 August, Andrew had moved to 1,000 mi off Florida when the upper-level winds died down, allowing growth in cloud height and wind strength. At the same time, a high-pressure zone built to the north, forcing Andrew to travel westward over energy-yielding warmer waters (Figure 11.2). On Saturday, its

Figure 11.1 Space shuttle view of a hurricane (cyclone) over the northern Indian Ocean, 13 November 1984. Note the well-developed eye. Wind speeds of 160 km/hr (100 mph) pushed a 3 m (10 ft)-high surge onshore, killing 430 people in India.
Photo courtesy of NASA.

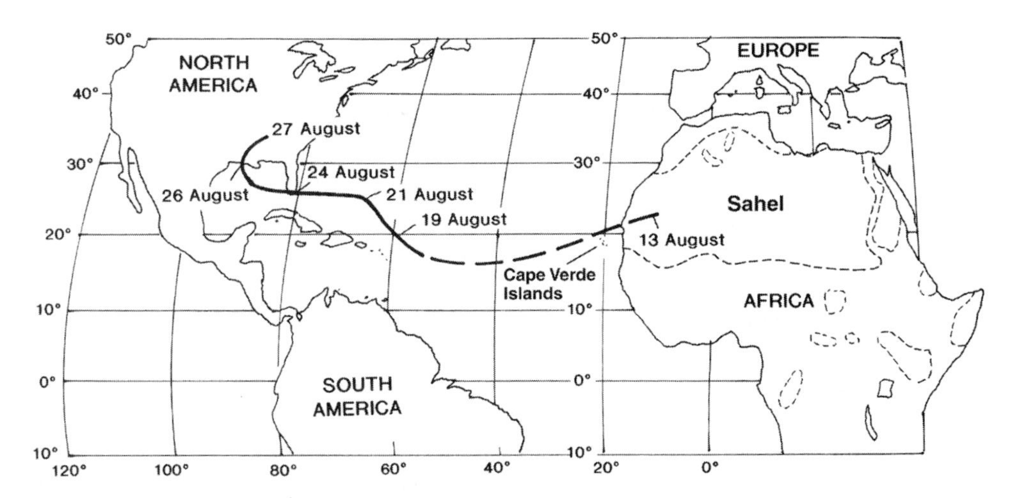

Figure 11.2 The odyssey of Hurricane Andrew, August 1992.

wind speeds blew above 74 mph (119 km/hr); Andrew had grown to hurricane strength. On Sunday, 23 August, the revitalized Andrew moved through the northern Bahamas with wind speeds of 150 mph (240 km/hr), killing four people. But the worst was yet to come.

After 3 A.M. on Monday, 24 August, Andrew crossed southern Florida with a vengeance. In Andrew's 25-mile-wide path of destruction lay the residences of over 350,000 people. Mobile homes and abundant poorly constructed houses were no match for Andrew's winds. By this time, sustained wind speeds were 155 mph (250 km/hr) with gusts up to 175 mph (282 km/hr), and there were tornado-like "spin-up vortices" with wind speeds around 200 mph (320 km/hr). Typically, most death and destruction from hurricanes come from the surging seas that travel with them. Andrew had a surge of 5.2 m (16.9 ft) height, but this time it was winds that did the most damage. The hurricane left 33 people dead, 80,000 buildings destroyed or severely damaged and another 55,000 heavily damaged but still habitable, thousands of cars demolished, and most trees downed or stripped leafless. Animals fared poorly also; even the Dade County Metro Zoo was faced with the escape of many animals. But more fear was spread by another escape, that of AIDS-infected monkeys from a research lab.

Andrew lost a lot of strength laying waste to southern Florida; the landfall cut off its warm-water energy supply. But after crossing the Florida peninsula, Andrew moved onto the warm waters of the Gulf of Mexico and recouped enough energy to attack the Louisiana coastline with 120 mph (190 km/hr) winds early on Wednesday, 26 August. Andrew tore through an area of marshland, sugar cane farms, and small rural towns, killing another 15 people and stirring up muddy, organic-rich bay and lake waters, exhausting the water's oxygen supply and suffocating hundreds of millions of fish. Andrew spent most of its remaining energy as heavy rains dropped on Mississippi on 27 August, and then it dissipated.

Andrew was the most destructive hurricane in United States history, with $30 billion in damages; it was the third strongest hurricane in the twentieth century. The only stronger hurricanes were one on Labor Day 1935 that crossed the Florida Keys, killing 405 people, and Camille, which slammed into Mississippi in mid-August 1969 with a 7.3 m (24 ft)-high sea surge, leaving 256 dead.

As with earthquakes, most hurricane-related deaths are due to poorly constructed buildings, not the natural event itself. Mobile homes are a dangerous place to be during a hurricane. Many houses in southern Florida have roofs covered with shingles that are only stapled down, frameworks that are only weakly supported, and no storm shutters to cover windows. Once a window is broken, hurricane winds enter a house, tear up its inside and lift off its roof. A study of Andrew's damages concluded that up to 40 percent of the losses could have been avoided if buildings had been constructed to meet the wind-resistance standards of the South Florida Building Code. Poor construction and lax enforcement of building codes caused much unnecessary suffering.

How a Hurricane Works

A hurricane is a storm of the tropics. Heat builds up in the tropics during long, hot summers, and hurricanes are one means of exporting excess tropical heat to the midlatitudes. Before a hurricane develops, several requirements should be met: 1) seawater should be at least 80°F (27°C) in the upper 200 ft (60 m) of the ocean; 2) air must be unstable, warm, and humid; and 3) upper-level winds should be weak and preferably blowing in the same direction the developing storm is moving.

The development of a hurricane begins with a low-pressure zone that draws in a poorly organized cluster of thunderstorms with weak surface winds; this is a **tropical disturbance.** As surface winds strengthen and flow more efficiently around and into the center of the growing storm, it becomes a **tropical depression** and receives an identifying number. The storm rotates in a counterclockwise (cyclonic) fashion around a central core in the Northern Hemisphere. The converging surface winds meet at the central core, which acts like a chimney, sending warm, moist air flowing rapidly upward toward the stratosphere (Figure 11.3). The rising

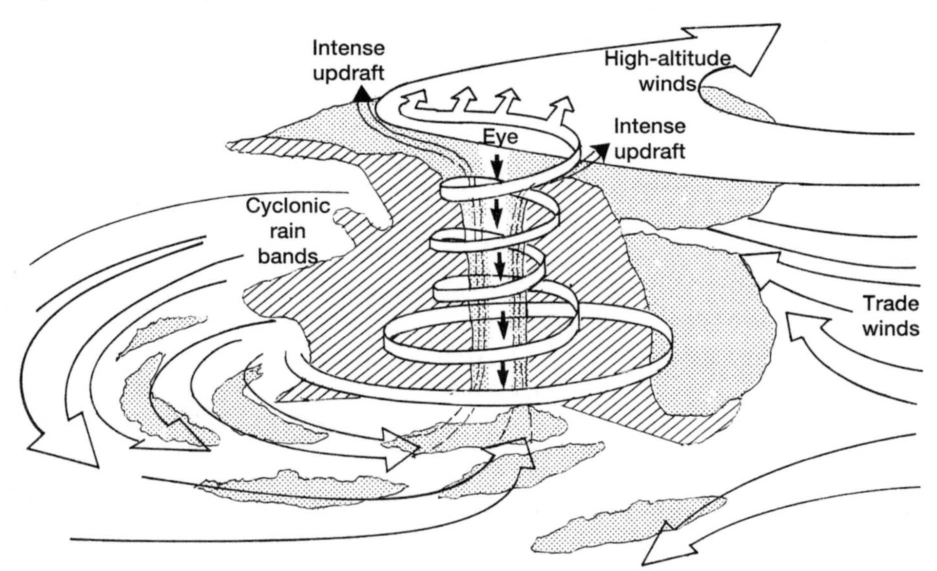

Figure 11.3 Schematic drawing through a hurricane. Low-altitude trade winds feed moisture and heat to the eye. Updrafts rise rapidly up the core (eye) wall and are helped away by high-altitude winds.

Source: U.S. Department of Commerce, 1971.

moist air cools, reaches its dew point temperature where water vapor condenses, thus releasing prodigious quantities of latent heat. The released heat warms the surrounding air, causing stronger updrafts which, in turn, increase the rate of upward flow of warm, moist air from below.

The converging winds continue to spiral up the core wall at ever-increasing speeds as the cyclonic system grows in strength. When sustained surface-wind speeds exceed 39 mph, it has become a tropical storm (surface winds from 39 to 74 mph). It matures to hurricane status when the surface winds consistently exceed 74 mph. The strength of a hurricane depends on the speed that surface winds can flow into the central core, race up its sides, and easily flow out and away in the upper atmosphere. As the central core or column becomes a more efficient "chimney," the hurricane grows stronger.

Figure 11.4 Hurricane Fran passes Florida on its way to landfall in North Carolina, September 1996.

Hurricane Energy Release

A hurricane acts as a heat engine transferring heat from the warm, moist air above tropical seas into the core of the hurricane. As air rises into the hurricane, latent heat is released in staggering quantities. The average hurricane generates energy at a rate 200 times greater than our worldwide capacity to generate electricity. The kinetic energy of winds in a typical hurricane is about half our global electrical capacity. Summing up, the energy released in a hurricane by forming clouds and rain is 400 times greater than the energy of its winds.

The Eye

As increasing amounts of wind blow faster into the center of a tropical storm, it becomes difficult for all winds to reach the center. The result is a spiraling upward cylindrical wind mass in the center of the storm. When surface-wind speeds reach about 74 mph, none of the wind reaches the center of the storm, resulting in the calmer clear area known as the eye (Figures 11.3 and 11.4). Because the distinctive eye forms at wind speeds of about 74 mph (119 km/hr), this wind speed defines the threshold where a tropical storm has grown strong enough to be called hurricane.

Inside the eye, air sinks. The cool air aloft sinks into the center of the core on top. As this air descends, it warms and absorbs moisture, leaving the core clear and cloud-free to form the "eye" of the hurricane (Figure 11.3).

The eye wall is the cylinder-shaped area of spiraling-upward winds that surround the eye. The eye wall of a hurricane has the strongest winds; they surround the relatively calm eye. Hurricane Andrew allowed recognition of a new twist in hurricanes. Embedded in Andrew's eye wall were small twisting vortices analogous to eddies in a river. The twisting eddies were about 500 feet in diameter. When they drifted into the intense updrafts of the eye wall, they were stretched vertically, decreasing their diameters and increasing their speeds to about 80 mph. Consider the effects upon the ground. On the side of a spin-up vortex moving in the same direction as the hurricane's rotation, the two speeds are additive, e.g., 130 mph plus 80 mph equals 210 mph. On the other side of a spin-up vortex, the winds oppose each other, e.g., 130 mph minus 80 mph equals 50 mph. This phenomenon helps explain why houses on one side of a Florida street were demolished while houses on the other side suffered only minor damages.

Not all coastal residents are hit by the same wind velocities; they vary along the coastline (Figure 11.5). If you are on the "right-hand side" of the tropical cyclone, you experience the speed of the storm body *plus* the wind speeds. If you are on the "left-hand side," you feel the wind speed *minus* the storm motion (Figure 11.5).

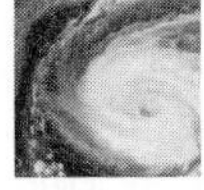

Hurricane Origins

Hurricanes are storms from low latitudes, i.e., from the tropics. They differ significantly from storms formed in higher latitudes. Hurricanes have unique aspects. 1) Heat released by condensation of water vapor inside a hurricane is its main energy source. 2) Hurricanes that move onto land weaken rapidly. 3) Fronts are not associated with hurricanes. 4) The

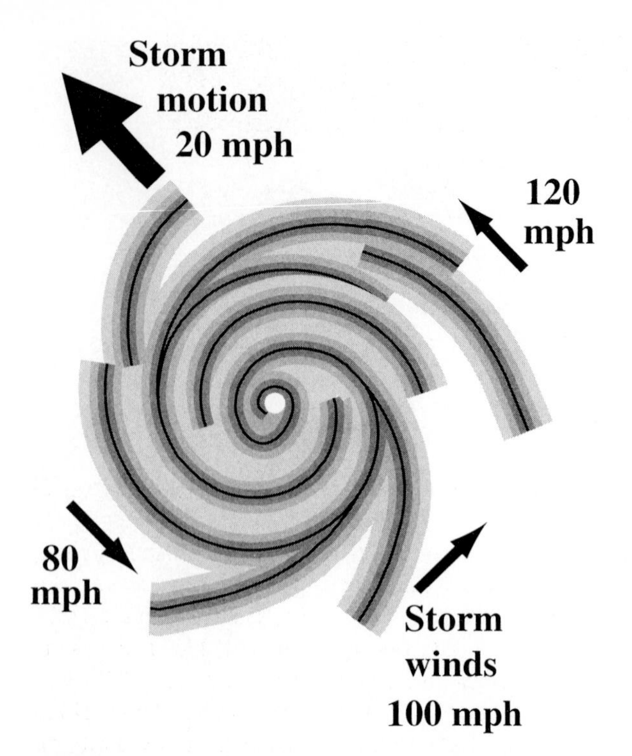

Figure 11.5 Tropical cyclones hit the coastline with different wind speeds. Storm motion and wind velocity may combine or subtract.

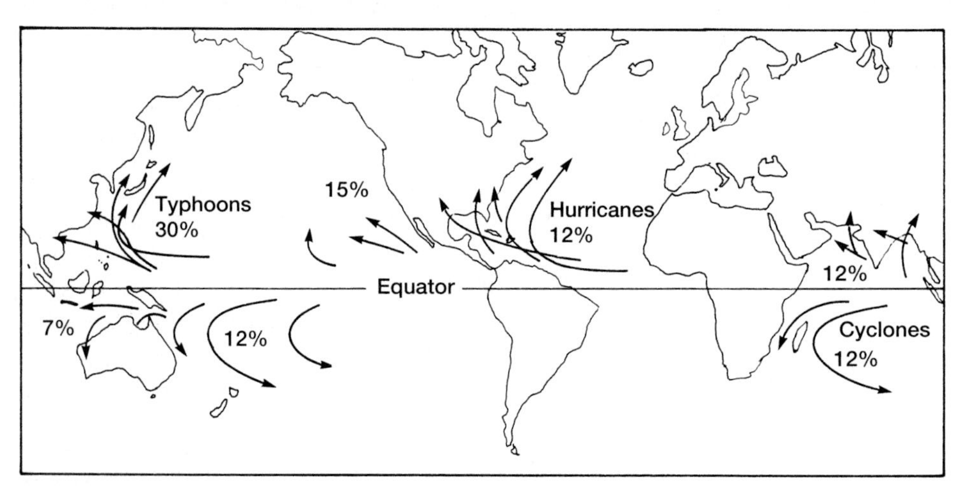

Figure 11.6 Map of common areas where hurricanes form, typical paths they travel, and annual percentage of the Earth's large cyclones occurring in each region. Note that they are called cyclones in the Indian and South Pacific Oceans and typhoons in the west Pacific Ocean.

weaker the high-altitude winds, the stronger a hurricane can become. 5) Hurricane centers are warmer than their surroundings. 6) Hurricane winds weaken with height. 7) Air in the center of the eye sinks downward.

In the United States we know them as hurricanes, but they go by different names in different parts of the world. In the Indian Ocean, they are cyclones, and in the western Pacific Ocean, they are **typhoons** (Figure 11.6). All are rotating, low-pressure weather systems with warm cores that generally form over warm seawater between 5° to 20° latitude and then travel off to deliver their heat to higher latitudes. Hurricanes do not form along the equator because there the Coriolis effect is zero. The Coriolis effect is so weak within 5 degrees north or south of the equator that there is not enough rotation to build hurricanes. Even an already formed hurricane could not cross the equator because without the Coriolis effect it would lose its rotation.

Each year about 84 tropical cyclones (hurricanes, typhoons, cyclones) form on Earth. United States residents think the North Atlantic Ocean–Caribbean Sea–Gulf of Mexico area is where the action is, but on a global scale it accounts for only 10 of the 84 events. The typhoons of the northwest Pacific Ocean hit Japan-China-the Philippines about three times as often and the storms can be larger.

The strength of tropical cyclones and the damages they inflict are assessed by the Saffir-Simpson scale (Table 11.1). In Category 1, winds damage trees and unanchored mobile homes. Category 2 winds blow some trees down and do major damage to mobile homes and some roofs. In Category 3, winds blow down large trees and strip foliage, destroy mobile homes, and cause structural damage to small buildings. In Category 4, all signs are blown down, damages are heavy to windows, doors, and roofs, flooding extends miles inland, and coastal buildings suffer major damage. In Category 5, damages are severe to windows, doors, and roofs, small buildings are overturned and blown away, and damages are major to all buildings less than 5 m (15 ft) above sea level and within 500 m (1,640 ft) of the shoreline.

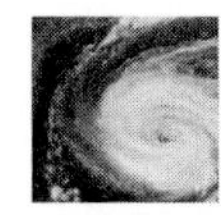

North Atlantic Ocean Hurricanes

Hurricanes in the North Atlantic Ocean are large, mobile and long-lasting (Table 11.2). Each year witnesses from 4 to 21 tropical storms and hurricanes in the North Atlantic–Caribbean Sea–Gulf of Mexico region and the coastline of the United States is frequently crossed by landfalling tropical cyclones (Table 11.3).

The arrival of rotating tropical-weather systems is an annual event in the United States. In the 93 years from 1900 to 1992, the U.S. Gulf and Atlantic coastlines were hit by 153 tropical cyclones of hurricane strength, averaging 1.7 per year. Of these 153 hurricanes, only three were category 5 in strength and fourteen were category 4. These figures exclude the additional large number of destructive hurricanes that hit the Caribbean islands and Central America.

Table 11.1 Saffir-Simpson Hurricane Damage Potential Scale

	Barometric Pressure		Wind Speed		Storm Surge		
	(millibars)	(inches)	(mph)	(km/hr)	(feet)	(meters)	Damages
Category 1	≥980	over 28.92	74–95	119–154	4–5	1.2–1.5	Minimal
Category 2	965–979	28.50–28.91	96–110	155–178	6–8	1.8–2.4	Moderate
Category 3	945–964	27.91–28.49	111–130	179–210	9–12	2.7–3.7	Extensive
Category 4	920–944	27.17–27.90	131–155	211–250	13–18	4–5.5	Extreme
Category 5	<920	less than 27.17	over 155	>250	over 18	>5.5	Catastrophic

Table 11.2 General Characteristics of North Atlantic Hurricanes

Storm diameter	200 to 1300 km (125 to 800 mi)
Eye diameter	16 to 70 km (10 to 44 mi)
Surface wind speed	≥74 mph (>120 km/hr)
Direction of motion	westward then northward
Life span	1 to 30 days

Table 11.3 North Atlantic Tropical Storms and Hurricanes, 1899–1999

	Frequency (Year)	
Category	Maximum	Minimum
Tropical Storms and Hurricanes	21 (1933)	4 (1983)
Hurricanes	12 (1969)	2 (1982)
Major Hurricanes (Winds >110 mph)	7 (1950)	0 Many times (e.g. 1994)
U.S.A. Landfalling Tropical Storms & Hurricanes	8 (1916)	1 Many times (e.g. 1997)
U.S.A. Landfalling Hurricanes	6 (1916, 1985)	0 Many times (e.g. 1994)
U.S.A. Landfalling Major Hurricanes	3 (1909, 1933, 1954)	0 Many times (e.g. 1998)

Data from www.nhc.noaa.gov.

Hurricanes form when sea surface temperatures are warmest, and in the North Atlantic Ocean this occurs in late summer (Figure 11.7). The warmest weather occurs earlier in the summer, but sea surface temperatures are highest at the end of summer because the ocean water with its high heat capacity keeps absorbing solar energy all summer long. Looking at the arrival times of the deadliest hurricanes in the twentieth-century United States also reveals an abundance in September (Table 11.4).

Cape Verde-type Hurricanes

Hurricanes such as Andrew begin as storms in the western Sahel region of Africa (Figure 11.2) that lies below the Sahara Desert. These storms are low-pressure systems that travel westward as **tropical waves** within the trade wind belt. Upon reaching the warm water of the subtropical Atlantic Ocean, some of these storms strengthen rapidly and may even reach tropical storm status near the Cape Verde Islands (Figure 11.2). These Cape Verde-type tropical cyclones are blown across the Atlantic Ocean by the trade winds between 5° and 20°N latitude picking up heat from warm ocean water. Approaching the Western Hemisphere they commonly move north on clockwise-curving paths due to the Coriolis effect. As a hurricane moves farther north the Coriolis effect strengthens.

Hurricane Paths The paths followed by hurricanes are difficult to predict in detail because they adjust to other high- and low-pressure atmospheric systems they encounter. But in a broad sense there are a few main influences on tropical cyclone paths. 1) Trade winds blow the tropical cyclone toward the west (Figure 10.5). 2) The Coriolis effect adds a curve to the right that progressively increases in strength with increasing distance from the equator (Figure 10.4). 3) An extensive high-pressure zone called the Bermuda High commonly sits above the North Atlantic Ocean. Hurricane paths vary depending on the size and position of the Bermuda High (Figure 11.8). When the Bermuda High is small and to the north, hurricanes may curve northward around it and have little or no effects on coastlines (Figure 11.8a). When the Bermuda High is strong and extensive, it may guide hurricanes along the east coast of the United States (Figure 11.8b) causing widespread death and destruction as in the New England hurricane of 1938, Diane in 1955, Donna in 1960, and Agnes in 1972. Sometimes the Bermuda High drifts southwestward toward Florida and helps direct hurricanes into the Caribbean Sea and Gulf of Mexico (Figure 11.8c). When a hurricane travels far enough to the north, then the westerly winds will push it to the northeast (Figure 10.5).

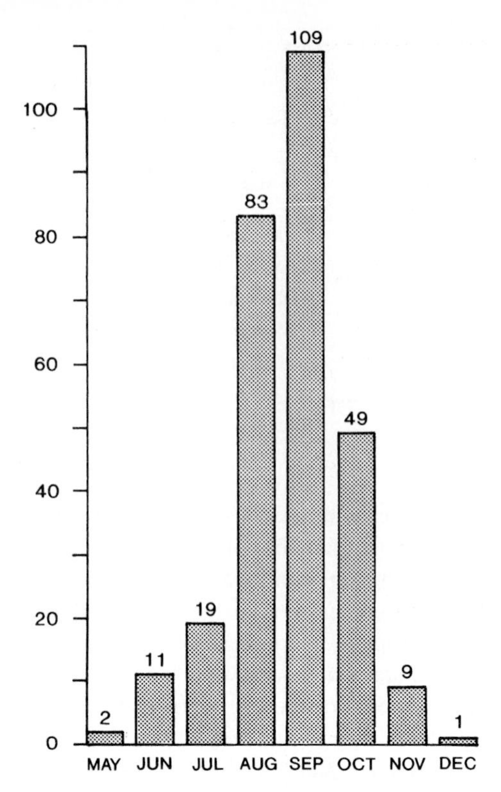

Figure 11.7 Number of hurricanes in the North Atlantic Ocean by month, 1931–1980.

Table 11.4 The Deadliest Dozen Hurricanes in Twentieth-Century United States

When	Where	Number of Deaths	Category
8 September 1900	Galveston, Texas	7,200	4
mid-September 1928	South Florida—Lake Okeechobee	1,836	4
mid-September 1919	Florida Keys/Corpus Christi, Texas	900	4
21 September 1938	New England, especially Rhode Island	600	3*
2 September 1935	Florida Keys	408	5
27 June 1957	Hurricane Audrey—Morgan City, LA	390	4
14–15 September 1944	East Coast—Virginia to Massachusetts	390	3*
21 September 1909	Grand Isle, Louisiana	350	4
17 August 1915	Galveston, Texas	275	4
29 September 1915	New Orleans, Louisiana	275	4
17–18 August 1969	Hurricane Camille—Mississippi	256	5
mid-September 1926	Miami, Florida/Alabama	243	4

*both 3s were fast-moving storms

The position and strength of the Bermuda High is part of the **North Atlantic Oscillation** (NAO), which describes the shifting of atmospheric pressures over the ocean. The NAO may strengthen and weaken on a time scale of decades, causing some coastlines to be repeatedly struck by hurricanes for a decade and then protected for another decade. For example, in the 1950s the east coast of the United States was hammered by major hurricanes, but in the 1960s and 1970s it was the Gulf Coast that was hit most frequently (Figure 11.9).

Caribbean Sea and Gulf of Mexico-type Hurricanes

Hurricanes can form above the very warm waters of the Caribbean Sea and Gulf of Mexico at the **Intertropical Convergence Zone** (ITCZ). The convergence zone occurs where the trade winds meet near the equator (Figure 10.5). The southwestward-blowing trade winds of the Northern Hemisphere (northeast trades of Figure 10.5) converge or collide near the equator with the Southern Hemisphere trade winds flowing northwestward (southeast trades). The location of the ITCZ moves with the tilt of the Earth's axis. The average position of the ITCZ is about 5°N latitude but in January it mostly lies south of the equator and by July all the ITCZ is north of the equator. Where the air flows converge, a low-pressure area or tropical depression can form with thunderstorms, a large core or center, and rising moist air that combine to create a rotating tropical cyclone that can strengthen to a hurricane, such as Mitch in 1998.

Mitch, October 1998 In the early morning hours of 22 October 1998, Tropical Depression 13 formed at the Intertropical Convergence Zone over the Caribbean Sea north of the Panama-Colombia border (Figure 11.10). The warm Caribbean water supplied so much energy that within 18 hours it was Tropical Storm Mitch, and in another 36 hours it was Hurricane Mitch. On 26 October, Mitch had grown to be one of the strongest category 5 hurricanes on record with sustained winds of 180 mph and gusts greater than 200 mph. Wind speeds remained over 155 mph for 33 consecutive hours, which is the second longest in the North Atlantic region.

Mitch was heading toward Cuba, but then turned sharply left toward Central America. As landfall for the 23 mi (37 km)-diameter eye and the category 5 winds were anxiously awaited, Mitch stalled off the coast of Honduras late on 27 October and stayed there until the evening of 29 October while its winds slowed down to tropical storm strength. At first reading this seems good. The coastline was not attacked, the powerful hurricane sat offshore while its fierce winds weakened before slowly coming onshore on 30 October. But in reality, this scenario was much worse. As the winds subsided and the central pressure increased, the massive volumes of airborne moisture spread over the land and poured down as rain. In effect, Mitch acted like a giant siphon sucking up water

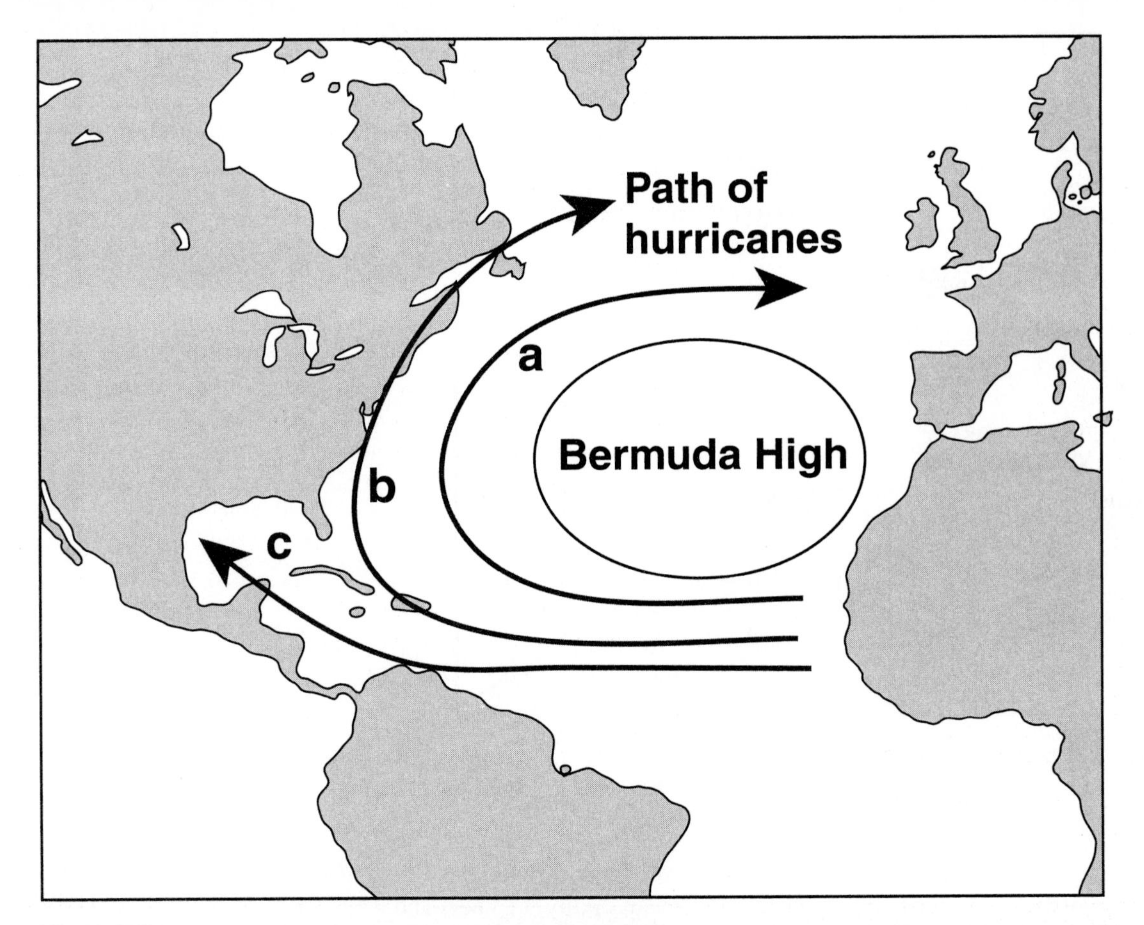

Figure 11.8 Paths of Cape Verde-type hurricanes are influenced by the size and position of a high-pressure zone, the Bermuda High. **(a)** A small Bermuda High allows hurricanes to stay over the Atlantic Ocean and miss North America. **(b)** A large Bermuda High may guide hurricanes along the east coast of the United States and Canada. **(c)** When the Bermuda High moves south, it directs hurricanes into the Caribbean Sea and Gulf of Mexico.

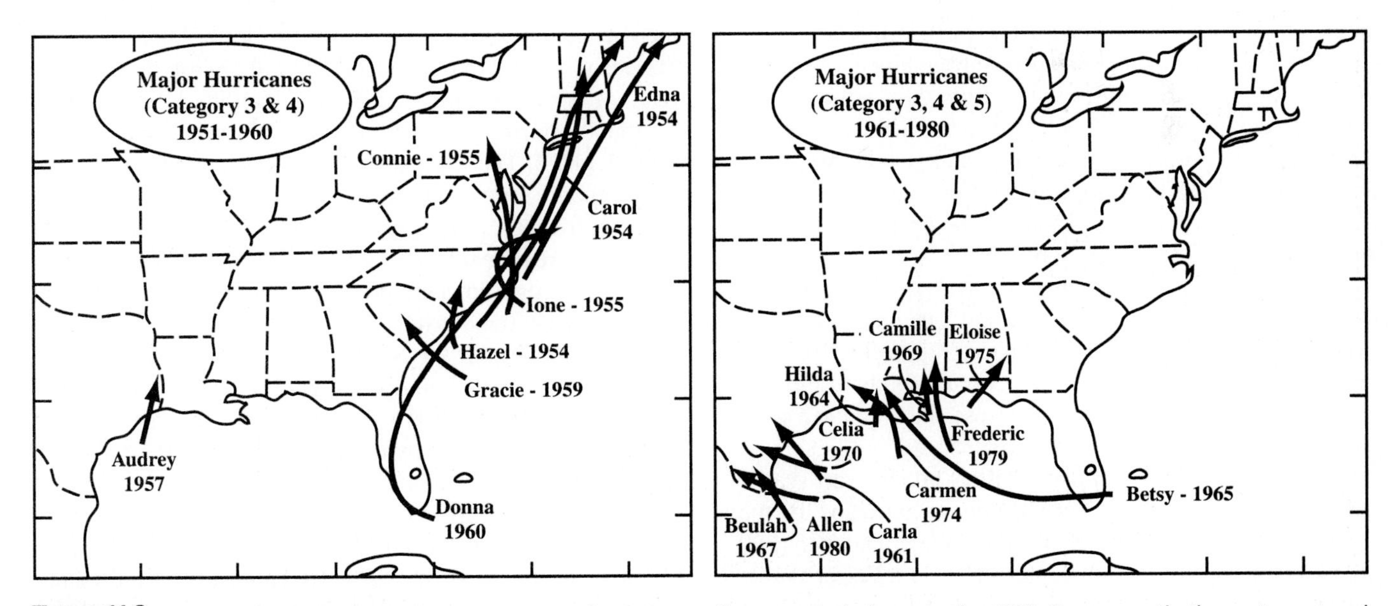

Figure 11.9 Category 3 and greater hurricanes cross the U.S. coastline at varied places. In the 1950s it was mostly the east coast and in the 1960s and 1970s it was the Gulf Coast.

Figures after Pielke and Pielke (1997).

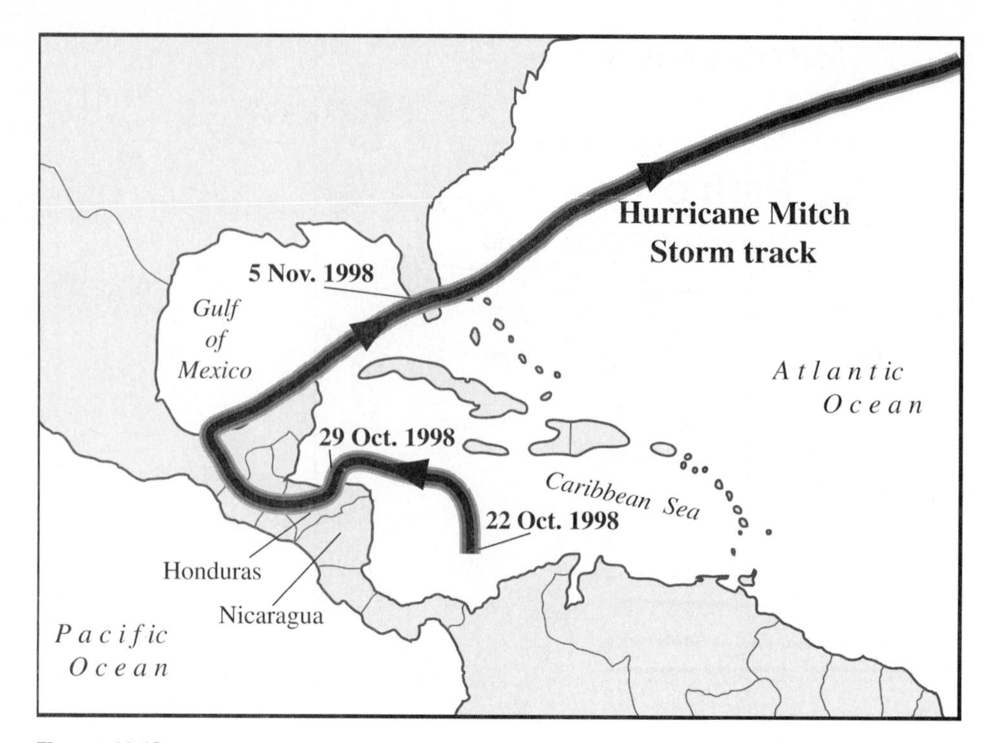

Figure 11.10 Hurricane Mitch began in the Caribbean Sea on 22 October 1998. Mitch stalled offshore from 27 to 29 October, dumping enormous volumes of rain on Honduras and Nicaragua. On 3 November, Mitch entered the Gulf of Mexico, picked up strength from the warm water, and traveled north of England on 9 November.

from the sea and dumping it onto the land, especially in Honduras and Nicaragua. Three-day rainfall totals of 25 inches were common and in some mountainous areas rainfalls up to 75 inches were estimated. Think of the problems that occur when 2 to 6 feet of rainfall must run off the land.

In Honduras, about 6,500 people were killed, 20 percent of the population was homeless, about 60 percent of roads and bridges were unusable, and 70 percent of the crops were destroyed. The President of Honduras, Carlos Flores Facusse, stated that Mitch wiped out 50 years of progress.

In Nicaragua, about 3,800 people were killed. The worst incident occurred 225 miles inland from the Caribbean Sea when the crater lake atop Casitas Volcano filled with rainwater and the crater wall failed sending lahars (mud flows) flowing 14 mi (23 km) downslope to the Pacific Ocean. Four villages were overwhelmed and about 2,000 people were buried beneath mud 6 to 18 ft (2 to 6 m) thick.

While many of the people of Central America were still fighting for their lives, Mitch moved out onto the warm water of the southern Gulf of Mexico (Figure 11.10), grew in strength up to a tropical storm, crossed southern Florida, and carried its heat and energy across the Atlantic Ocean north of the British Isles on 9 November.

During a 15-day rampage, Mitch killed over 11,000 people, making it the second deadliest hurricane in the Americas behind only the Great Hurricane of October 1780 that killed a total of 22,000 people on several Caribbean islands. As survivors worked to restore the Nicaraguan economy in the Casitas Volcano area in late 1998 and 1999, they had to cope psychologically with "the return of their dead." New rains eroding the mud deposits kept exposing the bodies of family members and neighbors.

Forecasting the Hurricane Season

Can we forecast how many Cape Verde- and Caribbean/Gulf of Mexico-type hurricanes are likely to form each year? Progress has been made. William M. Gray of Colorado State University has had reasonable success forecasting the number of named tropical storms in the North Atlantic region based on several variables. 1) When the western Sahel region of Africa (Figure 11.2) is wet, then its greater number of thunderstorms provides more nuclei for hurricanes (Figure 11.11). 2) The warmer the sea-surface temperatures, the more energy is available to help tropical depressions grow into hurricanes. 3) Low atmospheric pressure in the Caribbean region aids the formation of tropical cyclones. 4) If La Niña conditions are present in the Pacific Ocean, then west-blowing trade winds help hurricane movement over warm water. But if El Niño exists, then its east-blowing high-level winds tend to disrupt and break apart tropical cyclones. Remember that a tropical low-pressure zone can grow into a hurricane only if rapidly rising moist air releases huge quantities of latent heat into the growing storm. If upper-level winds are cutting into the tall storm clouds, they disrupt the vertically rising air and make it difficult for hurricanes to form. Most Northern Hemisphere hurricanes travel westward, whereas El Niño brings eastward-blowing winds that disrupt storms, and help make a quiet season with few hurricanes. However, if La Niña conditions are present,

Hurricanes are

fewer when

El Niño present

Cooler sea surface

Drier summer

frequent when

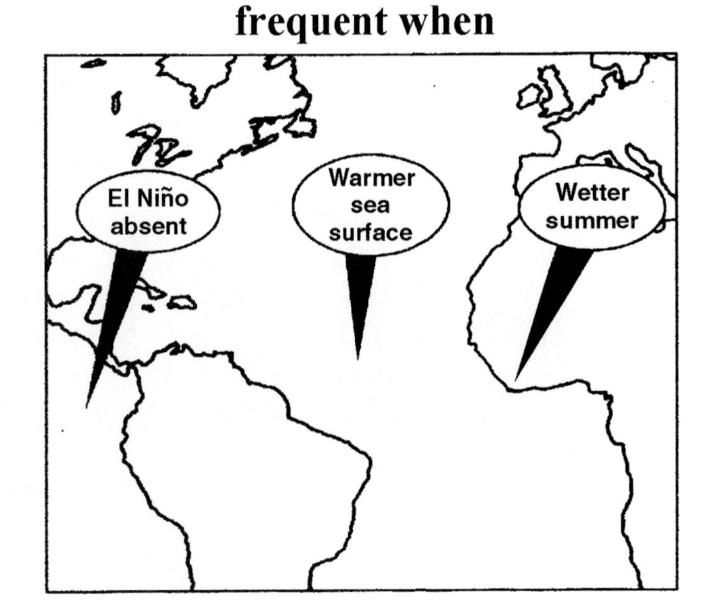

Figure 11.11 The frequency of North Atlantic region hurricanes is affected by climatic conditions.
Adapted from *New York Times,* 18 June 1997.

then westward-blowing trade winds aid the growth of hurricanes. During a La Niña in the Pacific Ocean, hurricanes may be more common and more intense in the North Atlantic Ocean, Gulf of Mexico, and Caribbean Sea.

Wherever they begin, tropical cyclones reach hurricane strength above the warm waters of the westernmost Atlantic Ocean, Caribbean Sea, and Gulf of Mexico. The annual probabilities of a hurricane over the waters of this region are sobering (Figure 11.12). Many of these hurricanes will attack the United States. The annual probabilities of hurricane landfalls on the U.S. coastline are shown in Figure 11.13.

Hurricane Damages

The catalog of twentieth-century hurricane strengths and their effects upon the United States is fairly complete. The deadliest dozen hurricanes (Table 11.4) occurred mostly in the earlier part of this century (Table 11.5). Deaths by hurricanes in the United States have dropped dramatically due to the advance warnings that are now broadcast widely before a hurricane makes landfall. Most people evacuate the low-lying areas and save themselves. Since naming of hurricanes began in 1953, only two named hurricanes have been really big killers.

Although hurricane deaths are down, the damages they cause are up (Table 11.5). An ongoing trend is for Americans to move to the coastline and to build larger and more expensive homes filled with costlier possessions. An interesting look at how expensive the future may be is gained by reevaluating old hurricanes in terms of the effects they would have on the present population and property values (Table 11.6). The destruction and deaths are brought by different aspects of hurricanes.

Table 11.5 Years of the Deadliest and Costliest Dozen Hurricanes

Years	Deadliest	Costliest
1900–1915	4	0
1916–1930	3	0
1931–1945	3	1
1946–1970	2	5
1971–1985	0	2
1986–1996	0	4

Table 11.6 Ten Historic Hurricanes, and the Damages They Would Have Caused in 1995

When	Where	Damage in Billions	Category
1926	Miami, Florida/Alabama	$72.303	4
1992	Andrew—Florida/Louisiana	33.094	5
1944	Southwest Florida	16.864	3
1938	New England	16.629	3
1928	South Florida—Lake Okeechobee	13.795	4
1965	Betsy—Florida/Louisiana	12.434	3
1960	Donna—Florida/eastern U.S.	12.048	4
1969	Camille—Mississippi	10.965	5
1972	Agnes—Florida/eastern U.S.	10.705	1
1955	Diane—Northeastern U.S.	10.232	1

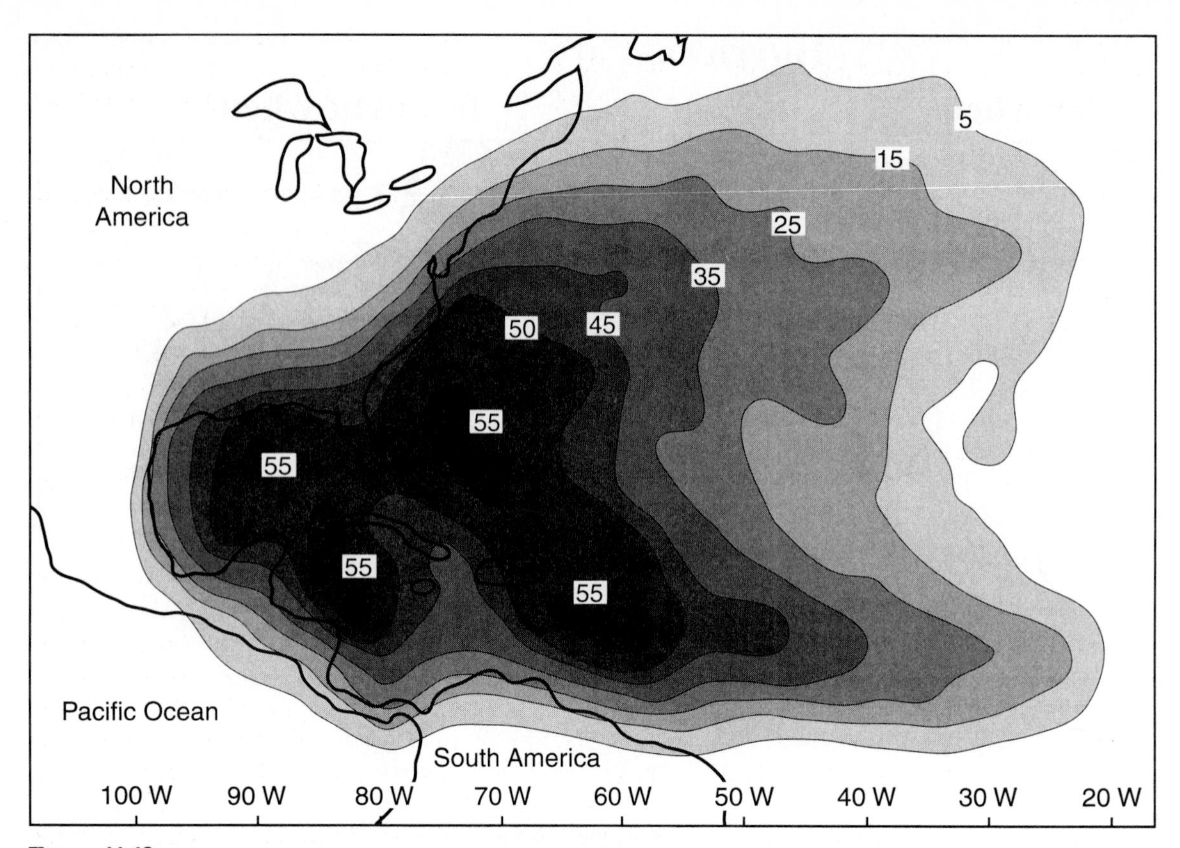

Figure 11.12 Annual probability of a named storm in each area during the June to November season for tropical storms and hurricanes, 1944–1999. Map after Chris Landsea of NOAA.

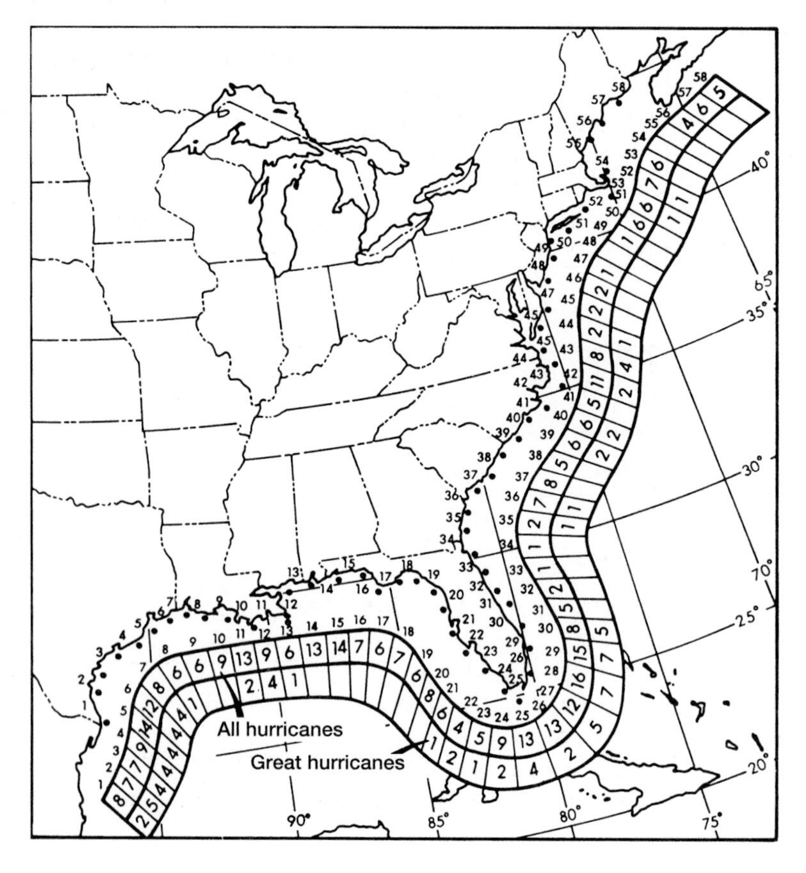

Figure 11.13 Percent probability that a hurricane (74 mph or faster) or a great hurricane (125 mph or faster) will occur in a given year along 80 km (50 mi)-long segments of the U.S. coastline.

After Simpson and Lawrence, 1971.

Storm Surges Most U.S. hurricane deaths have been associated with sea surges occurring when a hurricane moves on land. A large mound of seawater builds up beneath the eye of a hurricane for two primary reasons. 1) The eye is a very low-pressure zone, so local sea "level" rises higher. 2) Even more important are the winds racing into the chimney of the eye, pushing seawater into a tall mound (Figure 11.14). When the mound reaches shore, it is not as a wave but as a big surge of water that pours onto land, raising sea level and temporarily moving the shoreline many miles inland. On top of this elevated sea level come the large waves blown by the hurricane winds.

Why do impacting waves kill so many people? A cubic yard of water weighs 1,685 pounds (760 kg), almost a ton, and water is almost incompressible. Being hit by a wall of water is not much different from being hit by a solid mass.

Hurricane Camille is one of three category 5 hurricanes to hit the United States in the twentieth century. Camille brought winds gusting over 200 mph that hit Mississippi in 1969 with a surge of 7.3 m (24 ft); this caused many of the 256 fatalities. Fatal sites included the three-story brick buildings of the seaside Richelieu apartment complex where, instead of evacuating, 21 party-hearty people held a "hurricane party" to celebrate the event—it was the last party for 20 of them.

Remember that over half of the area of New Orleans, a major U.S. city, is at or below sea level (Figures 11.15 and 8.40). The city sits on land that is subsiding 5 mm per year. There is an at-risk population of about one million people. The hurricane that brings a surge of 6 m (20 ft) height into New Orleans may set the record for dollar damages by a storm.

Heavy Rains A hurricane holds massive amounts of water vapor in its cloud mass (Figure 11.16). After moving on land, no more water vapor is fed into the hurricane and its energy decreases. But there remains aloft a massive volume of atmospheric water that can cause severe flooding. Notice in Table 11.6 that two of the ten costliest hurricanes were only category 1 events. Their heavy damages were inflicted by rainfall and flooding away from the coast.

Nelson County, Virginia, 1969, Mudflows and Debris Avalanches Folks in the hilly Blue Ridge province of central Virginia awaited the remains of Hurricane Camille on 19 August 1969, not fully realizing just how much rain would fall and how the hillsides would respond. In 7.5 hours, beginning about 8:30 P.M., up to 710 mm (28 in) of rain fell. Hill slopes became so saturated they almost exploded, failing in long, narrow mudflows and debris avalanches that raced downhill carrying soil, gravels, trees, and water in rapid surges.

Soil on the slopes is 1 to 3 ft thick. After the slopes absorbed all the water they could hold, sections of soil simply flowed off in individual masses with average volumes of 88,000 ft^3, corresponding to about 4,000 tons. An eyewitness described how saturated ground started oozing slowly downhill, then a whole section of hillside suddenly broke free with a loud noise and flowed rapidly downhill. Other eyewitnesses described it as if waves of water squeezed earth and trees out of the hill, which then moved rapidly downslope, leaving behind a slide track. Residents reported that mudflows broke free often during the night, each giving off a loud noise reminding them of a squadron of propeller airplanes warming up or cannonballs rolling down a bowling alley.

The mudflows and debris avalanches set loose by the dying hurricane were principal agents in destroying about 150 homes and other buildings, 120 miles of roads, 150 bridges, and hundreds of cars and trucks, and burying 25,000 acres of cropland. They also killed 150 people, with 125 of the fatalities occurring in Nelson County. Despite the heavy rains and floods, very few people drowned. Virtually all of the fatalities occurred during the long, noisy night when mudflows and debris avalanches hit with battering impacts, causing multiple fractures of bones and skulls.

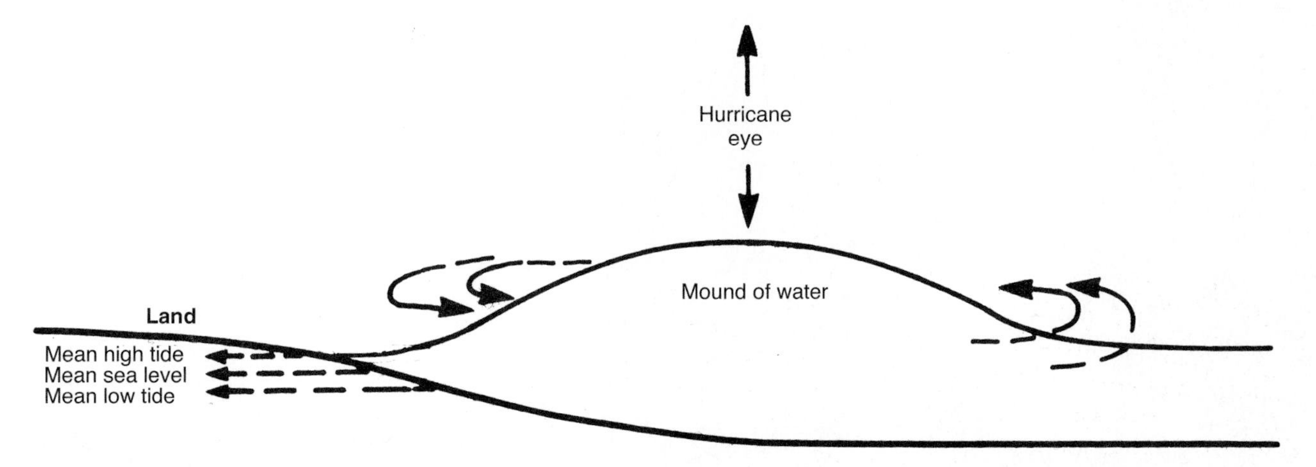

Figure 11.14 Winds rushing to and up a hurricane eye pile up a massive mound of seawater. When reaching land, the water mound rushes onshore as a surge.

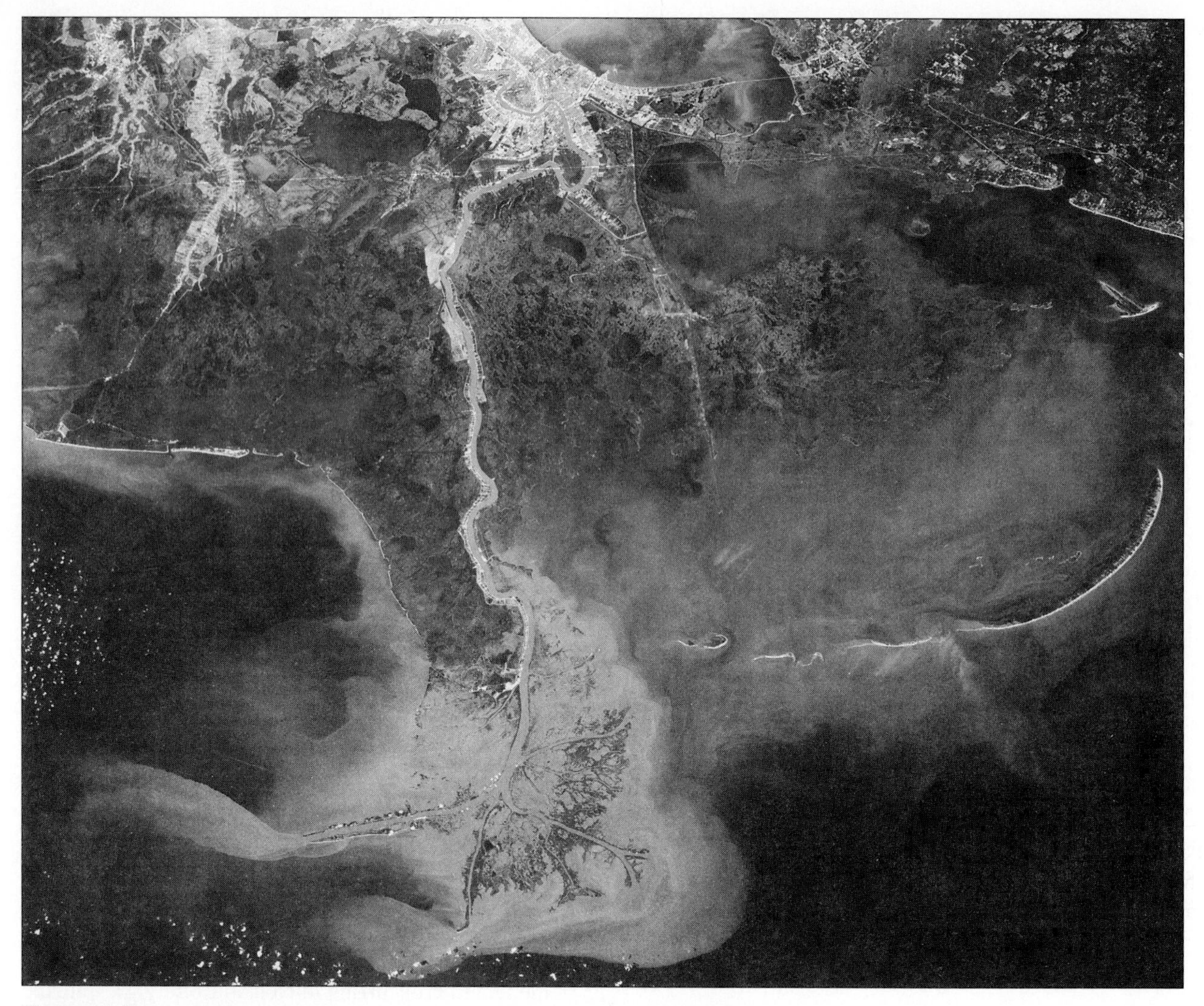

Figure 11.15 Mississippi River delta and the low-lying city of New Orleans (top center). A hurricane storm surge would flood much of the city. Photo courtesy of NASA.

Figure 11.16 Hurricane Opal on 4 October 1995 bringing winds and rains to the Florida panhandle, Alabama, Georgia, and the Carolinas. Damage exceeded $3 billion and 27 people died. Photo courtesy of NOAA.

Hurricanes and the Gulf of Mexico Coastline

Galveston, Texas, September 1900

On 8 September 1900, the deadliest natural disaster in U.S. history struck Galveston, Texas. Galveston is built on a low-lying island, a sandy barrier beach (Figures 11.17 and 11.18). Behind the sandbar island lies Galveston Bay, where trading ships made Galveston the wealthiest city in Texas early in the twentieth century. In 1900, the 38,000 residents were given warning of a possible hurricane, and many thousands evacuated the island.

The category 4 hurricane arrived in late afternoon. A high tide and the hurricane surge combined to flood the highest point on the island to a depth of 1 ft. Moving on top of this elevated sea level were storm waves blown by 120 mph winds. No place was safe. Wooden buildings were destroyed quickly. Even many of the big brick buildings fell to the high winds and ferocious waves that used the debris of other broken buildings and ships as battering rams to beat down their sturdy walls. Where to find refuge? Many people crammed into the Bolivar lighthouse, sitting on the laps of strangers on the curving metal staircase. They were jammed so close together that no one could move; there was no water to drink and no facilities for relief. Surrounding them was an air of fear permeated with the stench of human waste excreted where people sat or stood for the many hours that the 30 ft-high waves kept them confined. When finally they could open the massive door, the scene they saw was one of smashed buildings and boats, and thousands of bodies.

The morning of 9 September brought pleasant weather, but half of the buildings were destroyed and most of the rest were heavily damaged. There was no water, no food, no electricity, and no medical supplies; all bridges to the mainland were down, and all boats wrecked. The survivors were marooned. But mainland Texans came as quickly as possible to help in the overwhelming task of cleaning up. The 6,000 decaying human bodies presented a serious problem—the spread of disease. With much unhappiness, thousands of bodies were barged out to sea and dumped to avoid an epidemic. However, the tides and waves carried the floating bodies back to shore. The survivors had to pile up wood from wrecked buildings and build funeral pyres to consume the bodies.

Afterwards, the city constructed a 3 mi long seawall, 17 ft high and 16 ft wide at its base (Figure 11.17). Sand was brought in to elevate the island. Then the city rebuilt and began again. Nonetheless, another hurricane arrived on 17 August 1915, claiming another 275 lives.

Gulf of Mexico Coast Example: Texas

The Gulf of Mexico and Atlantic Coasts are passive continental margins, not active tectonic plate edges. There are over 16,000 km (10,000 mi) of coastline dominated by barrier sand masses along the shoreline from Long Island, New York, south to Florida and then west and south to Mexico. Common features along this coastline are river-cut lands now drowned by elevated sea levels, gently sloping sea floors, barrier islands of sand, and mostly lower tidal ranges.

The Texas coastal region has a long (591 km or 367 mi) and peaceful shoreline with little waves that simply invite

Figure 11.17 An early version of the Galveston seawall. Waves reflecting off the seawall have carried away beach sand.
Photo by W. T. Lee, U.S. Geological Survey.

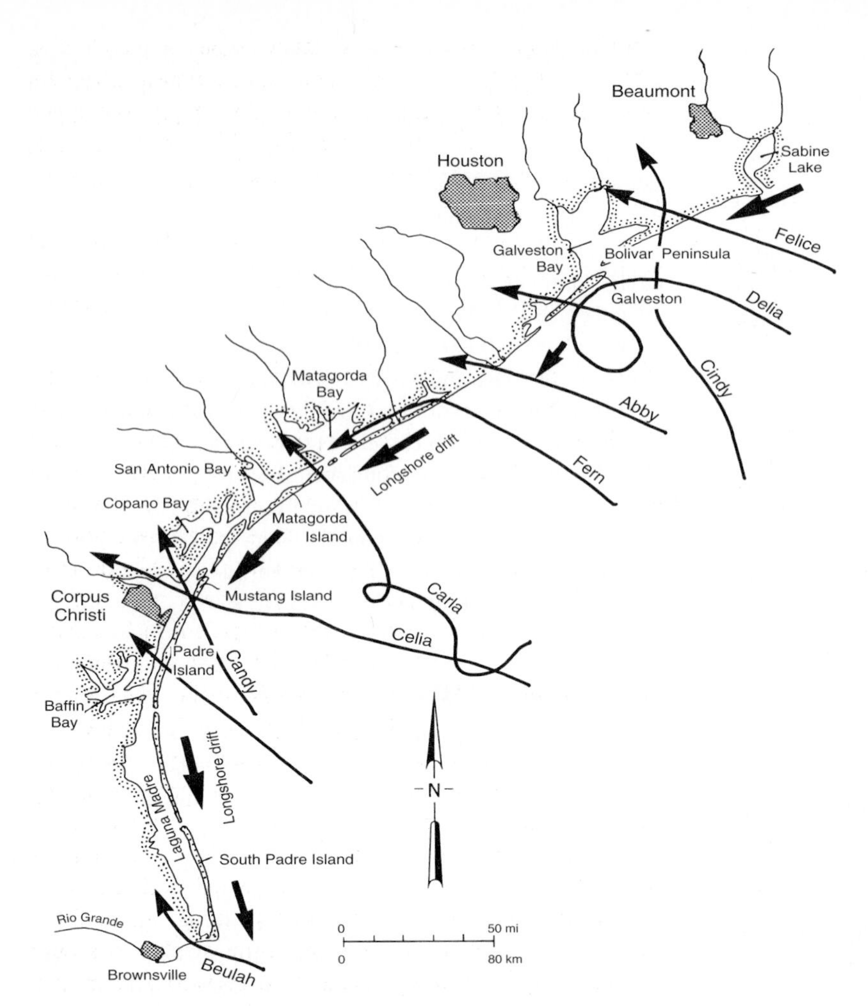

Figure 11.18 The Texas coastal zone. Longshore transport builds sandy barrier bars, separating shallow sea floor from lagoons and bays. Note the paths of hurricanes in 1960s–1970s.

Table 11.7 Some Texas Coastal Facts

Number of hurricane landfalls, 1900–1982	29	
Area below 6 m (20 ft) elevation (subject to hurricane-surge flooding)	15,031 km^2	(5,787 mi^2)
Length of bay-lagoon shoreline	1,771 km	(1,100 mi)
Length of Gulf of Mexico shoreline	591 km	(367 mi)
Length of Gulf shore with erosion over 3 m/yr	76 km	(47 mi)
Length of Gulf shore with erosion over 1.5 m/yr	81 km	(50 mi)
Length of Gulf shore with erosion up to 1.5 m/yr	167 km	(104 mi)

Source: R. A. Morton and J. H. McGowen. (1980). *Guidebook 20.* Austin: Bureau of Economic Geology, University of Texas.

people to stay. Many homes and businesses have been built along it. The Texas coastline is dominated by a southerly transport of beach sand that has constructed picturesque barrier islands and peninsulas (Figure 11.18). Behind the barrier sand masses lie an additional 1,770 km (1,100 mi) of coastline along the inner and outer margins of the lagoons and bays.

The bad news for the Texas shoreline is hurricanes. Texas has an immense area of low-elevation coastal land: over 15,000 km^2 (5,787 mi^2) lie less than 6 m (20 ft) above sea level (Table 11.7). The surges and high waves accompanying hurricanes wreak havoc on low-lying areas. The paths of ten hurricanes that hit the Texas coast between 1960 and 1979 are drawn on Figure 11.18. During the past century, major events striking the Texas coastline occurred, on average, each year. Virtually every coastal structure in Texas will experience a hurricane during its lifetime (Figure 11.13). In earthquake country, a few generations may pass between major seisms, so most people are spared the trauma. However, in hurricane country, most people who live near the shoreline will experience a storm's fury; few are spared.

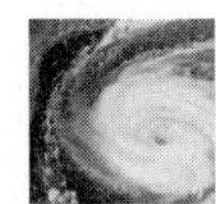

Hurricanes and the Atlantic Coastline

Hugo, September 1989

Charleston, South Carolina, has experienced many disastrous events. On 12 April 1861, the Confederate army began bombarding Charleston's Fort Sumter and started the Civil War, which ultimately left Charleston in ruins. On 31 August 1886, a major earthquake severely damaged the city (see Chapter 5). Then, at about midnight on Thursday, 21 September 1989, Hugo came to town as the tenth strongest hurricane in the twentieth-century United States.

Hugo brought a 5.2 m (17 ft)-high surge to Fort Sumter and 135 mph winds; it was a category 4 hurricane. The move onto land cut Hugo off from its warm-water energy supply. With decreasing energy, the storm adopted a curving path with a prominent northward hook that led past the west side of Charlotte, North Carolina, at 6 A.M. Friday at tropical-storm intensity. Hugo then moved up the Appalachian Mountains and lost its status as a tropical weather system by 6 P.M. Friday near the junction of West Virginia, Ohio, and Pennsylvania. The remains of Hugo were rains that crossed New York into Canada.

Although Hugo was a powerful hurricane, only 11 lives were claimed; however, property damages exceeded $7 billion. Again, here is the twentieth-century trend—decreasing deaths, increasing damages.

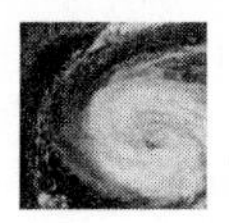

The Evacuation Dilemma

Thanks to advance warning, the number of people killed by hurricanes in the United States has dropped dramatically in the twentieth century (Table 11.5). Satellite photos show a hurricane is coming to town and people evacuate inland to safety before it makes landfall. This way of saving lives is threatened as a new migration within America has brought 41 million people to live in counties on the Atlantic Ocean or Gulf of Mexico coastline. This move-to-the-coast population growth has been aided by the knowledge that Federal disaster assistance is ready to help those people hit by a hurricane. With the financial risks reduced, more and more people build within the hurricane zone.

An evacuation dilemma is here. Population growth is far faster than the building of new roads and bridges. A September 1999 warning in South Carolina that Hurricane Floyd was approaching resulted in a massive gridlock that tied up the highway system and stopped people from leaving. A potential catastrophe was avoided this time when Floyd hit North Carolina instead.

Consider the difficulty of the evacuation process. For example, it is estimated that evacuating New Orleans could take 72 hours (3 days). Do we know 3 days in advance where a hurricane will strike? No, the detailed path of a hurricane is still unpredictable. The evacuation problem is widespread. Predicted evacuation times for other areas include 50–60 hours for Ft. Myers, Florida; 40–49 hours for Ocean City, Maryland; and 30–39 hours for Miami/Fort Lauderdale and Cape May County, New Jersey.

If the population cannot run away from an oncoming hurricane, what can be done? If you can't run far and high to escape, then you will have to run short and safe to local shelters built strong enough to withstand the hurricane attack and high enough to avoid flooding by the storm surge.

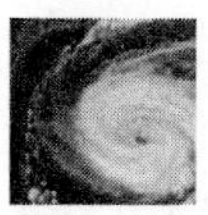

Reduction of Hurricane Damages

The costliest disaster in U.S. history occurred in 1992 when Hurricane Andrew tore through South Florida with sustained winds of 155 mph ripping apart 130,000 homes and causing $30 billion in damages. Is it just a fact of life that a powerful hurricane hitting a populated area will cause this much dollar damage? Or can we avoid much of the damage? Better planning and design will allow us to avoid most of the damages, and the need grows greater each year as more people move near the coastline.

Building Codes

After Hurricane Andrew, Florida established a Building Codes Commission to recommend ways to reduce damages from hurricanes. Florida's new statewide building code took effect in 2002, ten years after Hurricane Andrew. Stringent new regulations governing building safety apply to mapped areas that experience winds faster than 120 mph or areas that lie within one mile of the coast where wind speeds can reach 110 mph.

Hurricane winds push against the outside of buildings. The wind energy needs to be passed from the roof, down through the walls, and to the ground to prevent damage or destruction.

Roofs When hurricane winds destroy a building, a common first step is lifting off the roof. Then with building walls standing exposed and less supported, the winds proceed with their destruction. The American Society of Civil Engineers and the Institute for Business & Home Safety have named protecting the roof from uplift the highest priority need for new designs, materials, and regulations in order to save buildings during hurricanes. How can this goal be achieved? Improvements have been proposed. 1) Better design of buildings. For example, eliminate or strengthen eaves that project out from roofs and make it easier for winds to lift off roofs. 2) Strap roofs to walls. Inside the attic where the roof meets the walls, add numerous hurricane straps to help hold the roof to the walls. The straps are heavy belts of material similar to those used on suitcases or backpacks. Each strap is wrapped around a roof rafter or truss and continues wrapping onto a heavy wood stud of a wall; each hurricane strap is fastened by numerous nails. 3) Ban the common practice of using rapid-fire staples to secure thin little asphalt roofing sheets onto plywood. Hurricane winds easily strip off these lightweight flexible materials and thus gain entry to the house.

Impact of Wind-borne Debris Hurricane winds pick up and hurl debris that breaks through windows and pierces through entry and garage doors. Once an opening exists in a building, the winds will come inside and intensify their destruction. What can be done to prevent penetration? All exterior glass windows and sliding-glass doors must be made of shatter-proof glass or be protected by shutters. If the building does not have built-in shutters, then plywood should be precut to fit over all windows and sliding-glass doors.

Protecting your house against wind-borne debris also means eliminating debris. Pick up and store all outside items that could be used by the wind to attack your house, such as patio furniture, trash cans, barbecues, awnings, potted plants, and toys. Shrubs and trees in your yard can be pulled out of the ground and flung against your house. It is best to landscape with plant species native to hurricane areas because they are less likely to be uprooted by winds.

Land-Use Planning

Decisions made about land use before development takes place can prevent a lot of damage. Think of the destruction that could be avoided if cities and counties designated that low-lying coastal land be used for parks, farm fields, golf courses, nature preserves, or other uses where flooding will not create disasters. At the same time, the higher and more protected land could be zoned for house and building construction.

Inland movement of barrier ≈ 1000 d
Barrier beach
Former position of barrier
Lagoon
Sea level
Sea level
Depth increase = d
Former sea level

Figure 11.19 Sea level rise over nearly flat ground has profound effects. A sandy barrier bar may migrate inland 1,000 times the increase in water depth.

Coastal Development Restrictions

People enjoy beaches; they are aesthetic sites long enjoyed for vacations. So many people enjoy the beach that a building boom is underway along the Atlantic Ocean and Gulf of Mexico shorelines. Many thousands of expensive new homes are crowding the shoreline waiting for the inevitable hurricane attack of winds, storm surges, waves, and heavy rains. A recent study for FEMA estimates that during the next 60 years, 25 percent of houses within 500 feet of the shoreline will fall into the water unless mitigating actions are taken such as adding sand to beaches, adding riprap, or building hard walls.

Figure 11.20 Aerial view of Ocean City, Maryland in 1977. A hurricane surge and wind-driven waves will be an expensive event here.

Photo by R. Dolan, U.S. Geological Survey.

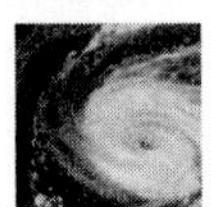

Global Rise in Sea Level

Another problem facing shoreline residents is the global rise in sea level, averaging about 1 ft per century. One foot may not sound too threatening but consider its effect on low-lying lands such as the Gulf and Atlantic coasts (Figure 11.19). A 1 ft rise in sea level may cause a beach to move inland 1,000 ft in some areas. The continuing buildup of housing along the coastline is just asking for trouble (Figure 11.20).

What are coastal residents doing about rising sea level and hurricane surges and waves? A common response is to construct a seawall for protection. Does building a seawall solve the problem? Every action has multiple reactions (Figure 11.21). A beach is an equilibrium system. The beach also has feedback mechanisms to maintain its equilibrium. A natural beach is flattened by tall ocean waves, spreading water over a broad area analogous to a stream overflowing its banks. Building a seawall disturbs the equilibrium (Figure 11.21b).

After a wall is built, the winds can no longer build sand dunes behind the beach. In a natural setting, sand dunes are redistributed by storm waves, thus dissipating some wave energy. But with a seawall, there is less sand to absorb wave energy. Instead, ocean wave energy is bounced back from the seawall, eroding the beach and steepening the offshore sea bottom. The now-deeper water allows bigger waves to attack, and they do so across a smaller beach. With time, waves hitting the seawall erode around the ends of the wall and undermine the main mass; larger waves break over the top of the wall (Figure 11.21c).

What is the typical response to wall failure? Build a bigger wall (Figure 11.21d). The Galveston seawall and the buildings it protects now stick farther into the Gulf each year as waves striking around the ends of the wall erode farther landward each year. The beach in front of the Galveston seawall is steeper and narrower than the beaches at either end. The best solution may be to leave natural processes alone and to adapt human activities to their constraints.

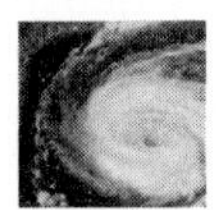

Hurricanes and the Pacific Coastline

Each year about 15 percent of Earth's tropical cyclones of hurricane strength form in the eastern Pacific Ocean, mostly offshore from southern Mexico/Guatemala/El Salvador (Figure 11.6). There are about 25 percent more hurricanes

(a) Before the wall

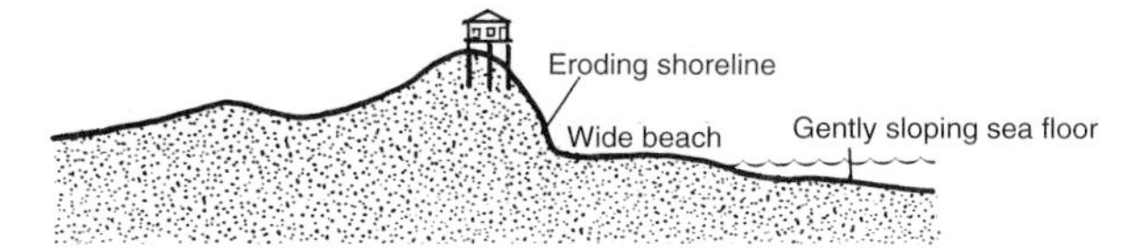

(b) The wall

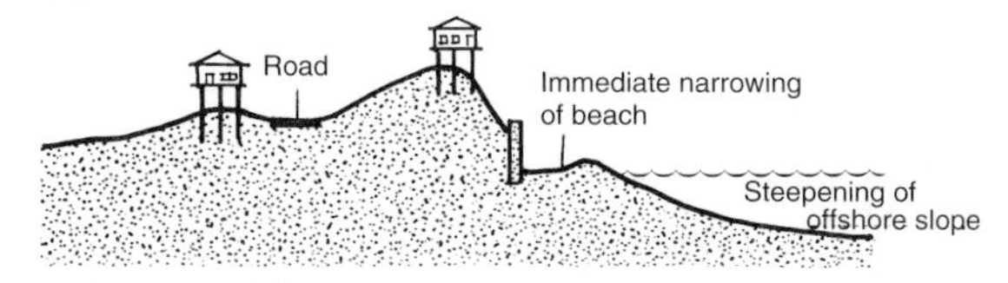

(c) Later: 2-40 years

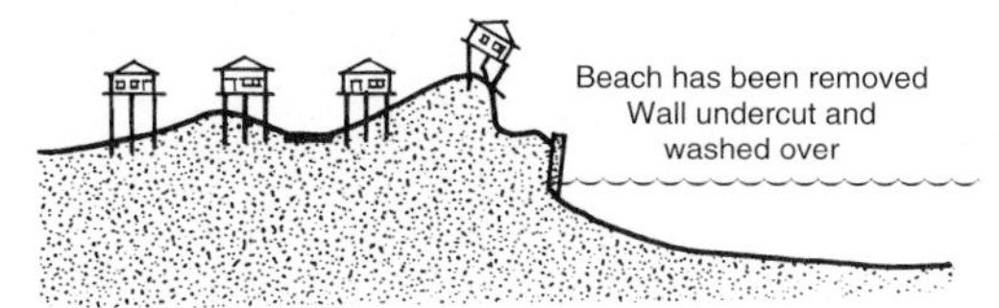

(d) Later yet: 10-60 years

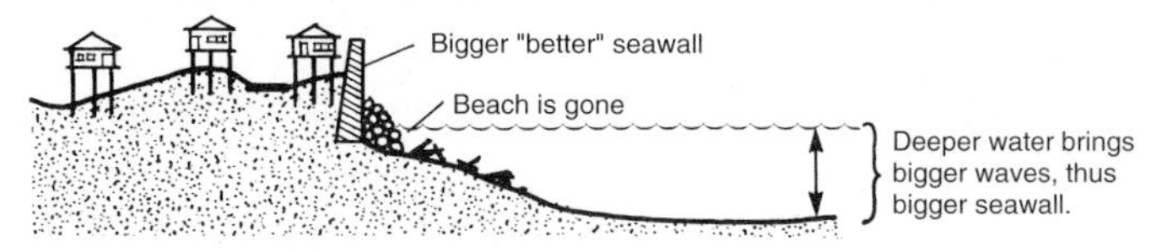

Figure 11.21 The shoreline and the seawall. **(a)** *Before the wall.* People attracted to coastal zone ignore erosion and build houses and businesses. **(b)** *The wall.* Behind the "protection" of the wall, development increases. But waves rebounding off the wall erode the beach and deepen the coastal waters. **(c)** *Later.* Beach sand is removed by waves, which then work to undermine the wall. **(d)** *Later yet.* Citizens band together and buy more "protection" with a bigger seawall. But a bigger wall reflects even bigger waves, and the beach disappears again. And on it goes. . . .

per year in the eastern Pacific Ocean than in the North Atlantic/Caribbean Sea/Gulf of Mexico. Why don't Pacific Ocean hurricanes strike the West Coast of the United States as often as Atlantic Ocean hurricanes do the East Coast? First, the trade winds blow most of the hurricanes westward out into the Pacific Ocean. Second, there is a marked difference in seawater temperatures. Along the eastern United States, the northward-flowing Gulf Stream current brings warm water from the Gulf of Mexico up along the East Coast, while the West Coast is bathed by cold water of the California Current coming down from Alaska (Figure 9.11). The cold-water current acts as a hurricane defense line. Even in the summer, water temperatures off California are usually only 64° to 69°F. The cold water drains the energy out of any hurricanes that dare to move across it.

Not all of the eastern Pacific Ocean hurricanes are blown westward out into the open ocean. Some stay nearshore and cause big trouble.

Pauline, October 1997

On Sunday, 5 October 1997, a low-pressure trough above the Pacific Ocean off Mexico grew to a tropical depression and began drifting eastward (Figure 11.22). On Monday the 6th, it grew into Tropical Storm Pauline and later in the day an eye developed showing that Pauline had matured to hurricane status. Meanwhile, a strong high-pressure system over the southeastern United States diverted Pauline's track to the northwest along several hundred miles of Pacific Ocean coastline (Figure 11.22). On Wednesday the 8th, category 4 Hurricane Pauline hit the sparsely populated coastline in the State of Oaxaca with 133 mph winds but began losing strength as it traveled over land, parallel to the coast in the State of Guerrero. When Pauline reached Acapulco in the early morning hours of Thursday the 9th, the wind speeds were down to tropical storm velocities and Pauline no longer had enough energy to support all its airborne moisture, so it fell as rain. And rain it did, almost 14 inches in 4 hours in the mountains next to Acapulco, a big rich city in the large but poor State of Guerrero.

The business of Acapulco is tourism. People come from around the world to enjoy the deep-blue bay backed by mountains. The tourists are served by Mexicans including many living in precariously built communities high in the steep hills. To find flat spots for houses they even build on the bottoms of the gullies that carry off the seasonal rainfall. The intense rains of Pauline washed sands and boulders down the steep slopes in powerful floods and debris flows that destroyed cardboard and adobe dwellings while killing 230 people. Virtually all the death and destruction took place in the poor communities on the steep slopes. The resort hotels and tourists along the seashore experienced no major problems.

Most of the hurricanes that form offshore from Mexico and Central America are blown westward into the Pacific Ocean (Figure 11.6). Some of them visit Hawaii.

Iniki, September 1992

The Hawaiian Islands lie between 19° and 22°N latitude, on the northern edge of the warm, hurricane-generating waters of the Pacific Ocean. Tropical storms form to the south of Hawaii and then follow the right-hooking, clockwise paths that the Coriolis effect produces in the Northern Hemisphere (Figure 11.6). Some of the north-hooking storms grow into hurricanes.

On Friday, 11 September 1992, the fast-moving category 4 Hurricane Iniki roared across the western side of Kauai, "the Garden Isle." Iniki means "sharp and piercing," and that well describes the damage done by its 130 mph (210 km/hr)-sustained wind speeds, with wind gusts in excess of 160 mph (257 km/hr). No buildings escaped damage, over 6,000 utility poles snapped like matchsticks, and even bark was stripped from many trees. Kauai has a population of 52,000, and 80 percent of the economy is derived from

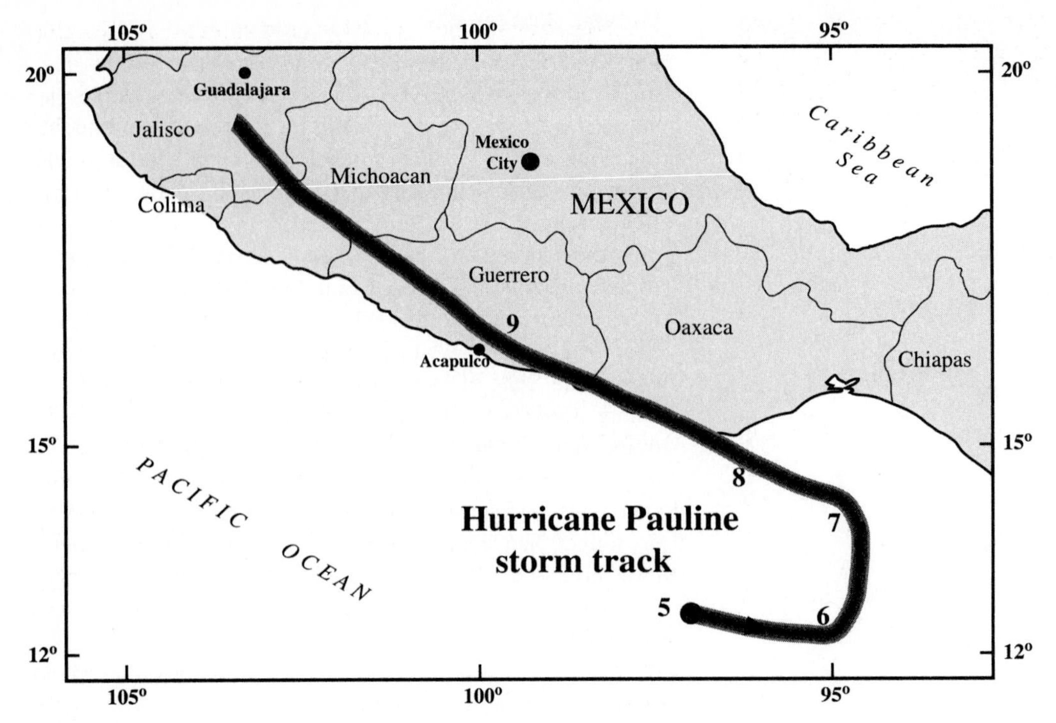

Figure 11.22 The path of Hurricane Pauline from 5 to 9 October 1997 (numbers indicate dates in October). The worst damages resulted from runoff of heavy rainfall down steep mountain slopes behind Acapulco.
Data from NOAA's National Hurricane Center.

the year-round population of 20,000 tourists. Although only two people were killed, the island economy suffered over $2 billion in damages. Iniki was reminiscent of Hurricane Iwa, which ran across Kauai on 23 November 1982. Iwa killed one person and left $234 million in damages.

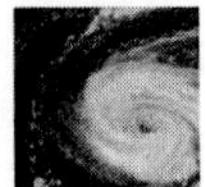

Cyclones and Bangladesh

The most deadly cyclones in the world ravage Bangladesh. In the twentieth century, seven of the nine most deadly weather events in the world have been cyclones striking Bangladesh. The country sits mostly on sediments eroded from the Himalaya Mountains and dumped into the Bay of Bengal as the delta of the Ganges and Brahmaputra rivers (Figure 11.23).

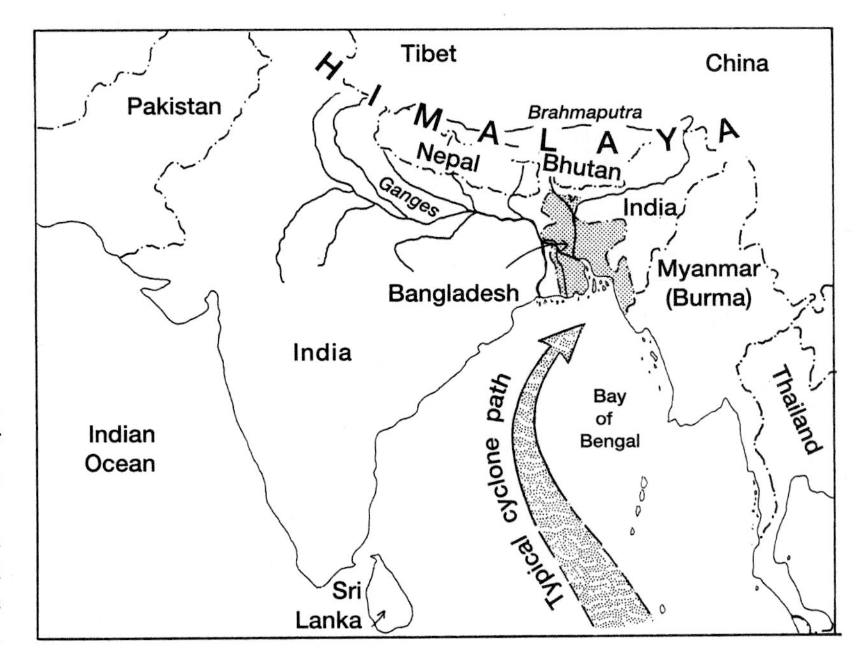

Figure 11.23 Bangladesh sits largely on the low-lying delta of the Ganges and Brahmaputra Rivers built into the Indian Ocean. Cyclones commonly move up the Bay of Bengal into Bangladesh.

The nation has densely populated a comparatively small area (Table 11.8). Since Bangladesh has a rapidly growing population and scarce land and food, it is little wonder that many millions of people are driven to the rich delta soils that yield three rice crops per year. The delta country is low-lying, most being a foot or less above sea level; over 35 percent of Bangladesh is less than 6 m (20 ft) elevation.

Bangladesh is a nation of water. Over 20 percent of the entire country is submerged beneath river floods in an average year; in 1988, 67 percent of the country was covered by river-flood waters. Then come the cyclones that bring surges of seawater of 6 m (20 ft) height, which can flood 35 percent

Table 11.8 **Comparative Statistics of Bangladesh**

	Area (km²)	Population (2001)	Median Household Income
Bangladesh	143,998	131,269,860	$500
Wisconsin	145,436	5,401,906	$39,800
Iowa	145,753	2,923,179	$35,427

Source: U.S. Census Bureau.

of the nation. It is little surprise that the national flower is the water lily.

Bangladesh has a 575 km (357 mi)-long coastline shaped like a funnel that catches the cyclones roaring up and over the warm waters of the Bay of Bengal. About five cyclones per year enter the Bay of Bengal both before (April–May) and after (October–November) the southwest monsoon season.

On 12–13 November 1970, during the high tides of a full Moon, a cyclone arrived with a surge of 7 m (23 ft) height and winds of 255 km/hr (155 mph). The tall waves drove into the low-lying delta land, killing about 400,000 people and as many large farm animals. On 30 April 1991, a cyclone packing winds of 235 km/hr (145 mph) unleashed a surge of 6 m (20 ft) height into Bangladesh, drowning 140,000 people and 500,000 large farm animals, and leaving 10 million people homeless.

The population of Bangladesh is projected to double in about 30 years, forcing millions more into the delta to await the frequent cyclones.

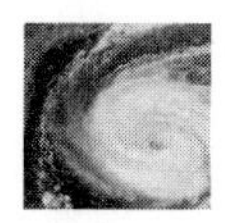

Coastline

The coastline is hit hard by hurricane waves, but it also is damaged by other waves. As ever greater numbers of people move to the coast and build communities and cities, they must learn that the sandy coast is a natural defense system helping protect them against wave attack. Virtually the entire U.S. coastline is losing beach sand (Figure 11.24). To protect our lives and property we must understand how sand comes and goes from the shore, and how waves work.

Waves in Water

Stream flow occurs when gravity literally pulls water masses downhill. Waves do not travel across lakes and oceans in a manner analogous to the downhill flow of stream water. Waves are actually pulses of energy that move through a

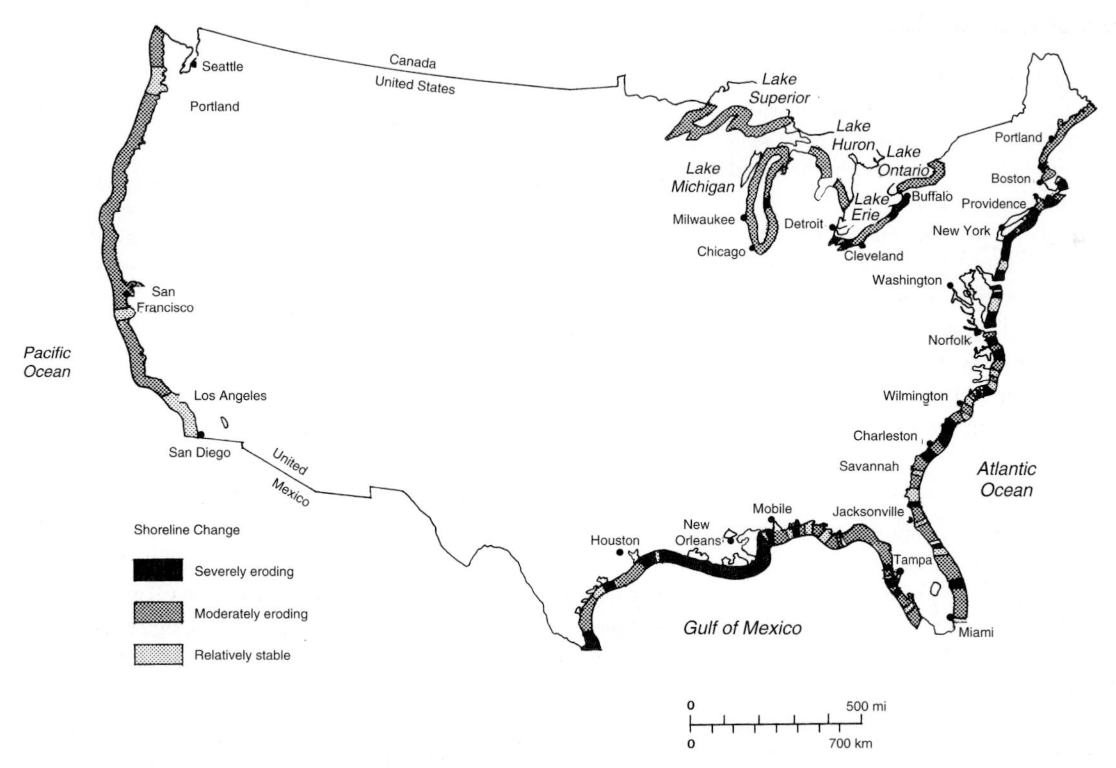

Figure 11.24 Beaches along U.S. shorelines are suffering erosion in varying amounts.
Source: S. J. Williams, et al., in *U.S. Geological Survey Circular 1075.*

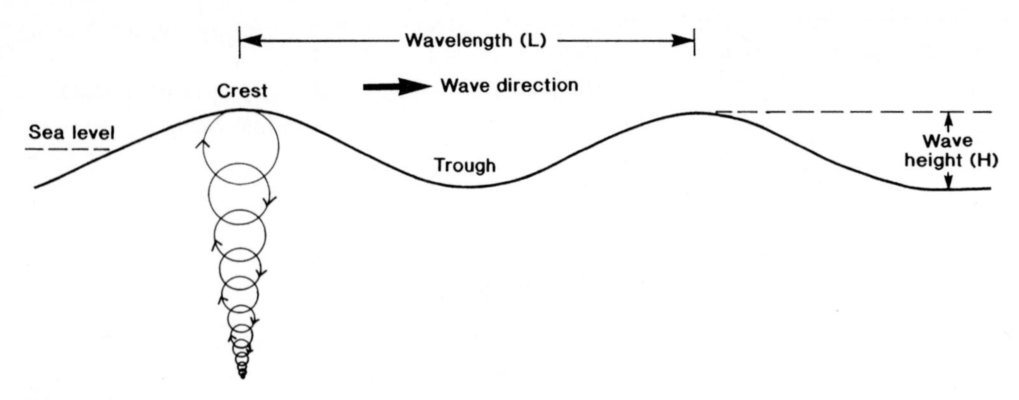

Figure 11.25 Waves are energy fronts passing through water, causing water particles to rotate in place. Rotational movement becomes insignificant at depth about one-half of the wavelength.

water mass, causing water particles to rotate in place, similar to the passage of seismic waves (Figures 3.18 and 11.25). You can feel the orbital motion within waves by standing chest-deep in the ocean. An incoming wave will pick you up and carry you shoreward, then drop you downward and back as it passes. At the water surface, the diameter of the water-particle orbit is the same as the wave height. The diameters of water orbits decrease rapidly as water deepens; wave orbital motion ceases at a depth of about one-half of the **wavelength.**

Waves vary from tiny ripples to large tidal bulges to monster tsunami generated by subsea earthquakes, but the rotational motion of water within a wave is quite similar no matter what its height or origin. Most waves are created by the frictional drag of wind blowing across the water surface. A wave begins as a tiny ripple. Once formed, the side of a ripple increases the surface area of water, allowing the wind to push the ripple into a higher and higher wave. As a wave gets bigger, more wind energy is transferred to the wave. How tall a wave becomes depends on: 1) the velocity of the wind, 2) the duration of time the wind blows, 3) the length of water surface **(fetch)** the wind blows across, and 4) the consistency of wind direction. Once waves are formed, their energy pulses can travel thousands of kilometers away from the winds that created them.

Rogue Waves

An ocean is such an extensive body of water that different storms are likely to be operating in different areas. Each storm creates its own wave sets. As waves from different storms collide, they interfere with each other and usually produce a sea **swell** that is the result of the constructive and destructive interference of multiple sets of ocean waves (Figure 11.26a). However, every once in a while, the various waves will become briefly synchronized with their energies united to form a spectacular tall wave, the so-called **rogue wave** (Figure 11.26b). The moving waves quickly disunite

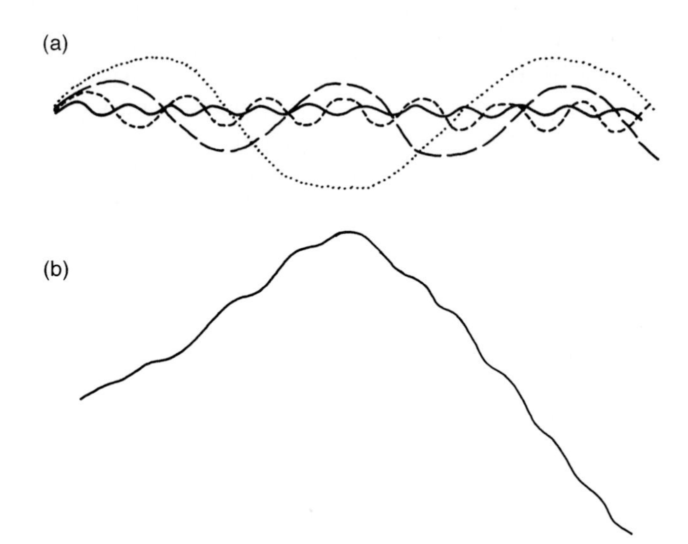

Figure 11.26 Waves on sea surface. **(a)** At any time, there usually are several different storms, each producing its own waves of characteristic wavelength. Different wave sets usually interfere with each other. **(b)** On rare occasions, the various wave sets combine to produce an unexpected giant—a rogue wave.

and the short-lived rogue wave is but a memory. But if a ship is present at the wrong time, a disaster may occur.

On 3 June 1984, the three-masted *Marques* was sailing 75 miles north of Bermuda when two rogue waves quickly sent the ship under, drowning 19 of the 28 people on board. During World War II, the *Queen Elizabeth* was operating as a troop transport passing Greenland when a rogue wave hit, causing numerous deaths and injuries. In 1987, the recreational fishing boat *Fish-n-Fool* sank beneath a sudden "wall of water" in the Pacific Ocean near a Baja California island.

On occasion, rogue waves strike the shoreline and carry people away from the beach. On 4 July 1992, a rogue wave 5.5 m (18 ft) high rose out of a calm sea at Daytona Beach,

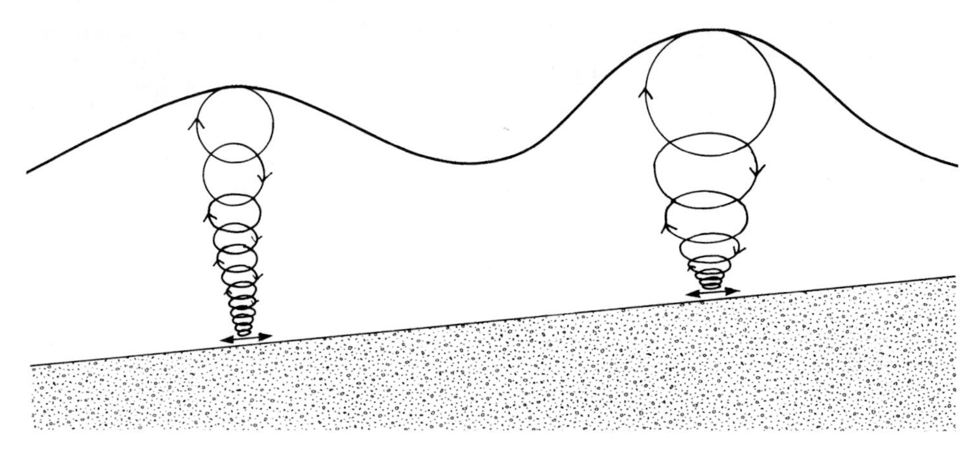

Figure 11.27 As a wave moves into shallower water, it rises higher. Circular rotating water touches bottom, causing a flattening into a back-and-forth motion.

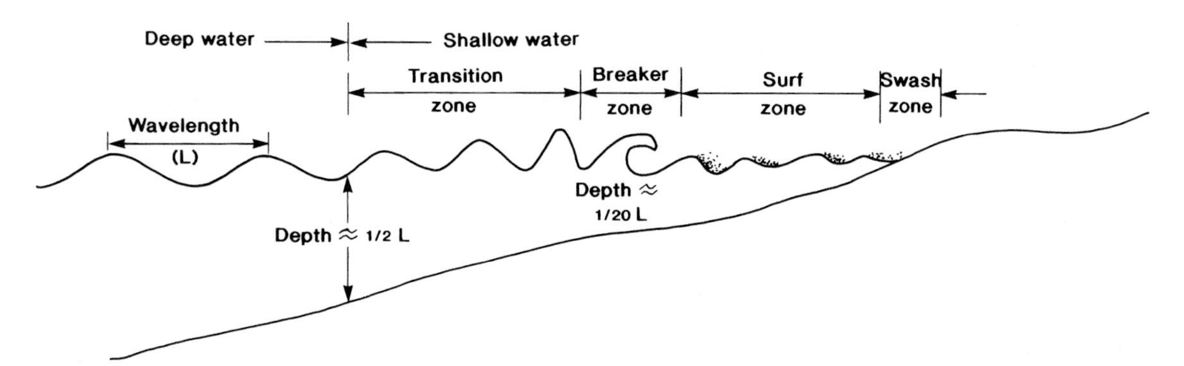

Figure 11.28 Schematic cross section of deep-water waves entering shallow water. Wavelengths decrease and wave heights increase, leading to pitching forward as breakers.

Florida, crashed ashore, and smashed hundreds of cars parked on the beach, causing injuries to 75 of the fleeing people.

Rogue waves have been measured at 34 m (112 ft) height. The problems they present also include the steepness of the wave front descending into the wave trough. A small, short boat is maneuverable and in good position to ride over the rogue wave, as long as it does not get hit sideways and rolled, or tossed from the front of one wave onto the back of the next wave. Large, long ships face the problems of either being uplifted at their midpoint, leaving both ends suspended in air, or of having both ends uplifted with no support in their middle. Either case creates severe structural strains that break some ships apart.

Waves on the Coastline

The coastline is the unique boundary on Earth where the land, ocean, and atmosphere all meet. It is the coastal zone that receives the full fury of waves.

Why a Wave Breaks

Waves undergo changes when they move into shallow water—water with depths less than one-half their wavelength. Wave friction on the floor of the shallow ocean interferes with the orbital motions of water particles, thus waves begin slowing (Figure 11.27). Friction with the bottom flattens the circular motions of the water into elliptical and horizontal movements.

As waves slow down, their wavelengths decrease, thus concentrating water and energy into a shorter length, resulting in waves growing higher. When the wave height-to-wavelength ratio (H:L) reaches about 1:7, the wave front has grown too steep, and it topples forward as a breaker (Figure 11.28). Note that the 1:7 ratio is reached by changes in both wave height and wavelength; wave height is increasing at the same time that wavelength is decreasing.

The depth of water beneath a breaker is roughly 1.3 times the wave height as measured from the still-water level. At this depth, the velocity of water-particle motion in the wave crest is greater than the wave velocity, thus the faster-moving wave crest outraces its bottom and falls forward as a turbulent mass.

Sidebar

Deep-Water Wave Velocity, Length, Period, and Energy

Waves moving through water deeper than one-half their wavelength are essentially unaffected by friction with the bottom. The waves move as low, broad, evenly spaced, rounded swells with velocities related to wavelength by:

$$V_w = 1.25 \times \text{the square root of } L$$

where V_w equals wave velocity and L equals wavelength. A swell with a wavelength of 64 m would have a velocity of the square root of 64 (i.e., 8) times 1.25, or 10 m/sec (22.4 mph). The equation is telling us that wave velocity in deep water depends upon the wavelength—as wavelength increases, so does velocity.

The period (T) is the amount of time it takes for two successive wave crests to pass a given point. Since the distance between successive wave crests is the wavelength, there must be a relationship between period (T) and wavelength (L). This relationship may be defined by:

$$V_w = \text{distance traveled/time} = L/T$$

which may be simplified to:

$$L = 1.56\ T^2 \text{ (in meters)}$$

As a rule of thumb, the velocity of waves in miles per hour may be estimated as 3.5 times the wave period in seconds. For example, waves with a period of 10 seconds are moving about 35 mph.

Higher-velocity waves carry more energy, but how much more? Wave energies can be computed with the following relationship:

$$E_w = 0.125 \text{ rho g } H^2 L$$

where E_w equals wave energy, rho equals density of water, g equals gravitational acceleration, H equals wave height, and L equals wavelength. Some representative values computed from this equation are listed in Table 11.9. Notice that doubling the wavelength will double the wave energy, but that doubling the wave height will quadruple the wave energy.

Table 11.9 Some Ocean-Wave Energies

Wave Period (T)	Wavelength (L)	Wave Height (H)	Energy in Joules
10 seconds	156 m	1 m	2.39×10^5
		2 "	9.57 "
		3 "	21.54 "
14.1 seconds	312 m	1 m	4.79×10^5
		2 "	19.15 "
		3 "	43.08 "

Summer Versus Winter Beaches

The Northern Hemisphere summer is a time of warmth and vacations. People flock to the beaches to lie on the sand, bask in the sunshine, and play in the ocean waves (Figure 11.29). But at the same beach during the winter, the conditions are quite different. Not only is the air and water colder and less inviting, but the beach sand may be gone (Figure 11.30). What happened to the sand? It moved offshore, but it will return during the next summer.

During the summer, the Northern Hemisphere is tilted toward the Sun, and weather conditions in the North Pacific and North Atlantic Oceans are relatively calm. With reduced temperature contrasts between the north polar region and the equator, the weather systems have lower wind speeds, which produce ocean waves of shorter wavelength and height. Summer waves at the beach are smaller, but they are more numerous. The abundant waves separated by short wavelengths act like bulldozers that push the offshore sand up across the shoreline to build a sandy beach (Figure 11.29).

During the winter, the Southern Hemisphere is tilted toward the Sun, and the north polar region gets very cold. The increased temperature contrast between the North Pole and the equator causes winds to blow faster and stronger. Winter storms are more energetic, and the ocean waves they create are taller and spread farther apart. What happens to the beach when it is attacked by breakers with longer wavelengths? Large breakers rush onto the beach, erode the beach sand, and carry it seaward with the backwash, leaving

Figure 11.29 Summer conditions at Boomer Beach in San Diego, California, September 1980. Small, short-length waves have piled the beach high with sand. View north. Photo by John S. Shelton.

Figure 11.30 Winter conditions at Boomer Beach, February 1980. Bigger, longer-length waves have removed beach sand, exposing underlying boulders. Photo by John S. Shelton.

the beach with less sand (Figure 11.30). Because of the longer wavelengths in winter, much of the beach sand will be dragged offshore to sit in submarine sandbars until the shorter-length waves of the next summer push the sand back onto the beaches.

A relationship exists between sandy beaches and wave lengths. Short-length waves push sand onto the beaches, whereas long-length waves pull sand offshore.

Beach Sand as Coastal Protection A sandy beach is not only good for recreation, it also is a tremendous natural barrier that absorbs the energy of breaking waves. Visualize waves packed full of energy by an intense oceanic storm. The energy-packed waves may carry their intensity for thousands of miles before rearing up and dumping it on the shoreline. What better system to absorb this wave energy than a sandy beach? The waves kick around the sand grains, knocking them every which way, but what is the harm? Sand provides an innocuous way to absorb the ocean's fury. But in winter, the sand is reduced or gone from many beaches. What absorbs the energetic attack of the bigger winter waves? Sea cliffs, houses, roads, or whatever is on the coastline.

Wave Refraction

When waves reach shallow water, their velocity becomes controlled by water depth. Waves slow markedly as water depth decreases and bottom friction increases, irrespective of their lengths or heights. Many shallow coastal regions vary in depth, causing incoming waves to vary in speed along different portions of their crests. Because of varying topography and angle of wave approach, waves bend as they approach the shore. The bending process is known as **wave refraction** (Figure 11.31a).

The wave segments that first reach shallow water slow down and bunch up (converge); the wave portions still in deeper water continue to roll ahead faster, thus stretching waves as they spread out to cover a wider area (Figure 11.31b). The refraction of waves results in headlands being hit by a concentrated attack of higher-energy waves while the embayed portions of coastline receive smaller waves.

Longshore Drift

Even where the coastline is straight, waves typically approach it at an angle (Figure 11.32). The effects of bottom friction and shallow water slow down the near-shore end of a wave, thus giving the deeper part of the wave a chance to catch up, but it rarely is able to do so.

When waves arrive at an angle, they run up the beach at an angle and drag the sand grains with them. But the return flow (backwash) is essentially perpendicular to the shoreline (Figure 11.33). The process creates a **longshore current** of water, moving along the coast and dragging along the beach sand in longshore transport.

You have probably experienced **longshore drift** if you have gone to the beach, laid down your towel and belongings, and then gone into the waves to play. When you exited the water to return to your towel, was it in front of you? Probably not; it was likely to be lying some distance away, either up or down the beach. What happened? Your body was moved along the coastline by longshore drift in the same process as the sand grains. Longshore drift is well illustrated in the movie called *The Beach: A River of Sand.* Beach sand flows with longshore drift analogous to a river; the river banks are the offshore edge of the breaker zone and the upper limit of water run-up (swash) on the beach.

(a)

(b)

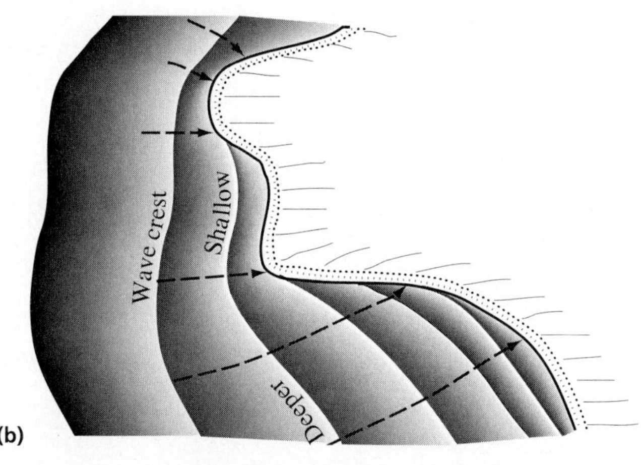

Figure 11.31 (a) View north of wave refraction around False Point in San Diego, California. (b) Wave refraction around a headland. The portion of a wave that first enters shallow water slows down and converges on the headland. The wave portion still in deeper water races ahead and stretches, thus reducing its impact on the shore.

Photo by Pat Abbott.

Figure 11.32 View of groins built out from the shoreline to trap and hold beach sand, Ship Bottom, New Jersey. What is the dominant direction of longshore drift? What would a hurricane do to this coastline?
Photo by John S. Shelton.

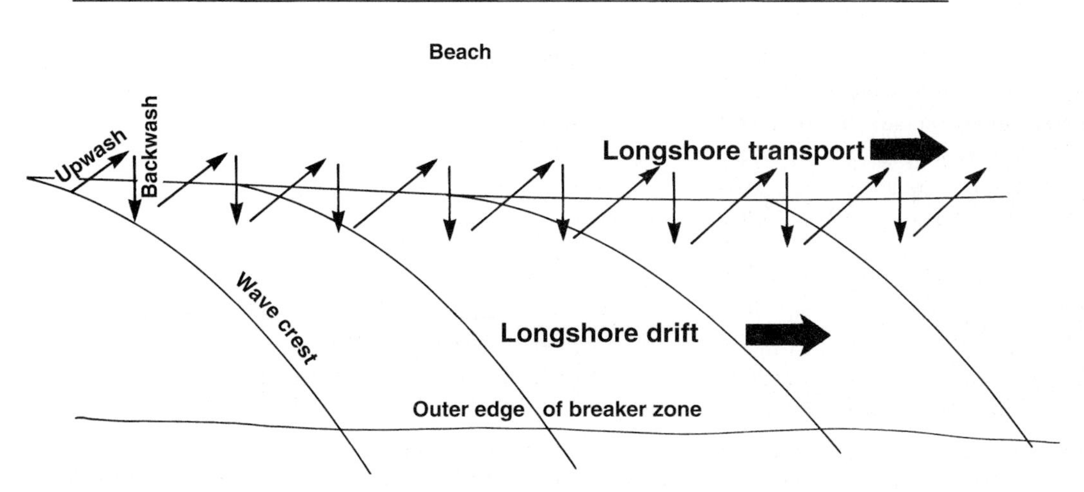

Figure 11.33 Schematic map of longshore drift. Waves run up the beach at an angle but return to sea as perpendicular backwash. Net effect is a longshore transport of water and sand along the coast.

Submarine Canyons

The rising and falling of sea level has caused many coastal canyons to be alternately exposed on land and then submerged beneath the sea as **submarine canyons.** Many of these canyons began forming during times of low sea level when rivers cut downward and outward to a distant, now submerged shoreline. When sea level rose again and drowned the canyons, they continued to grow through submarine erosion processes, such as **turbidity currents,** debris flows, and landslides. Some of these canyons are gigantic features, and many have very steep, even overhanging, walls.

One of the best studied submarine canyon systems lies just offshore from the Scripps Institution of Oceanography in San Diego, California (Figure 11.34). A broad sandy beach diminishes and stops because the beach sand pours down into the La Jolla submarine canyon. The beach sand loss to deep water averages 340,000 m^3/yr; it is a permanent loss.

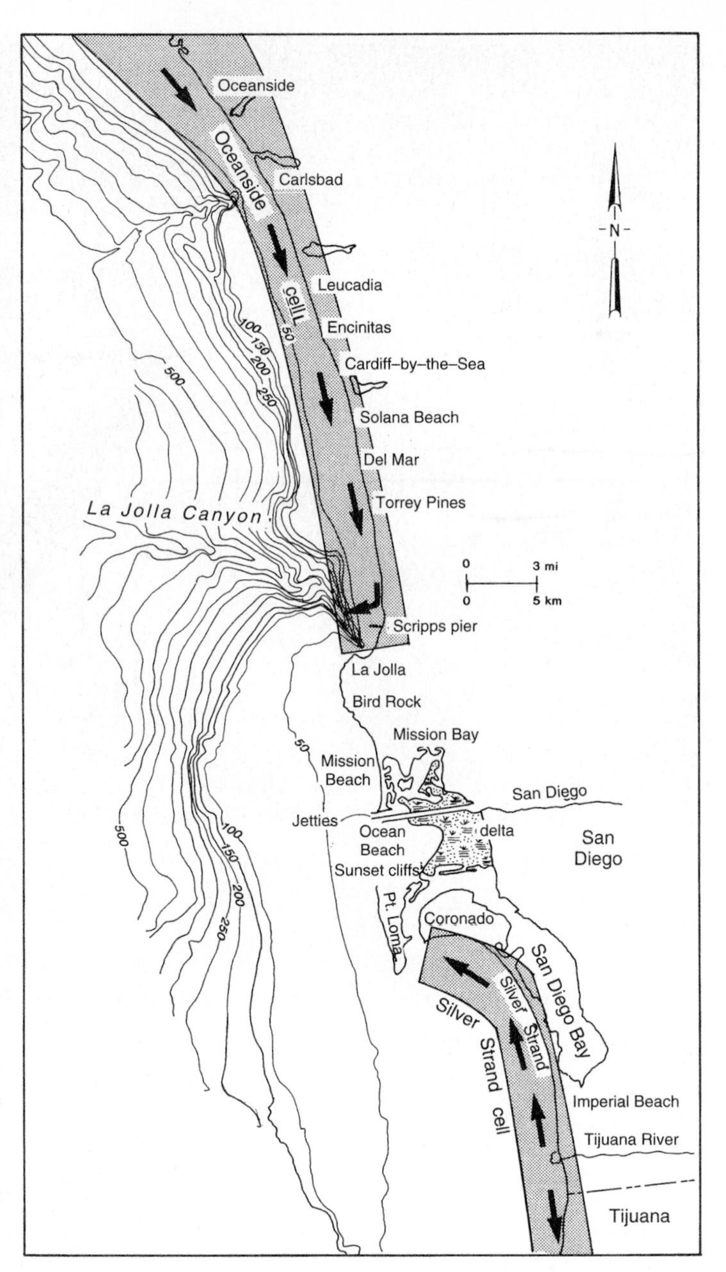

Figure 11.34 Map of coastal San Diego County, California. The Silver Strand cell is made of Tijuana River sand that moves north in longshore drift pushed by summer waves from hurricanes off Central America. The Oceanside cell is beach sand moving south pushed by waves from the north Pacific Ocean. The sand flow ends where it pours into a submarine canyon—La Jolla Canyon.

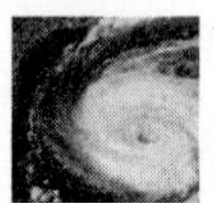

Human Effects on the Coast

People have always liked living along the coast. The climate is tempered by the large heat reservoir of the ocean, there is the additional source of food, and the recreational and aesthetic opportunities are special. But the coastal zone also presents problems, such as hurricane surges, large waves, and cliff collapses. The problems have inspired attempts to control the coastal environment; the works have had mixed success.

Dams

Despite the immense volume of ocean water, there commonly is not enough water in the coastal zone for people to drink. Thus, many dams have been built across rivers leading to the coast. The reservoirs impound freshwater for home and farm use, and the dams allow falling water to create the kinetic energy that is converted to electricity.

But dams not only trap a stream's water, they also catch its sand (Figure 11.35). The sand being transported by streams is on its way to the coast to become the next generation of beach sand. But when beaches are cut off from their sand supply, they shrink and disappear, thus depriving the coastlines of their most effective protection against storm waves.

Dams and their reservoirs are not the only human threat to beach sand supply. Sand is needed to make concrete and glass. Without sand, there would not be the houses, office buildings, roads, bridges, and other structures that mark civilization. Most sand-mining operations have moved into valleys to dig up river sands; these sands were waiting in storage for a big flood to carry them to the beach. Without them, how will the beaches be supplied with their next generation of sand?

Cliff Protection

The views from coastal cliffs are among the most enticing in the world. Sea-cliff properties are in such heavy demand that they sell at premium prices (Figure 11.36). But sea cliffs retreat under the attack of large waves and dump the cliff-edge buildings onto the beach. What can be done? Keep buildings away from the cliffs. This, however, is easier said than done.

Figure 11.35 A dam traps sand as well as water. Malibu Canyon dam was built in 1925 in the Santa Monica Mountains, California. Within 13 years, the reservoir was filled with sand.
Photo by John S. Shelton.

Figure 11.36 Condominiums built to edge of retreating sea cliff, Solana Beach, California.
Photo by Pat Abbott.

Most homeowners who can afford beach-cliff property also have the money to buy protection. Typically they prefer a barrier of heavy rocks and concrete (Figure 11.37). But these barriers are no match for the power of storm waves at work over time. If the waves do not destroy a protective wall or mass of riprap by frontal assault, they may erode underneath and around the sides, undermining the barrier and helping break it apart. Barriers also reflect waves hitting them, and the rebounding water surges powerfully erode the beach, thus removing protective sand.

The long-term picture is an unstoppable retreat of sea cliffs under the attack of ocean waves. Humans can delay the inevitable, but they cannot stop it. Anything built on the sea cliffs must be thought of as a temporary structure; it will be destroyed.

Groins

Because beach sands are disappearing in so many places, various structures are built to trap and retain sand. One of the most popular techniques is to place sets of short, elongate masses perpendicular to the coastline, the so-called **groins** (Figure 11.32). Groins interfere with longshore drift, causing sand to deposit on the up-drift side; but erosion takes place on the down-drift side. Emplacement of groins usually must be accompanied by artificial replenishment of sands.

Other problems with groins include the erosion of sand by storms coming from other directions and their tendency to direct **rip currents** offshore, carrying sand beyond the breaker zone and thus out of the longshore drift system.

Figure 11.37 Along Sunset Cliffs, the city of San Diego built cement walls and dumped riprap to "protect" the sea cliffs, then billed the homeowners. Meanwhile, a charming cove has been obliterated.
Photo by Pat Abbott.

Jetties

Even larger masses called **jetties** are built perpendicular to the coastline to create inlets to harbors and channels for boat passage. The trick is to design jetties of the right length to create a large enough tidal prism (in-and-out volume of seawater) to naturally scour the channel and keep it open. Of course, this means the jetties will extend beyond the breaker zone and will interfere with the longshore drift system (Figure 11.38). Beach sand will build up in large volumes on the up-drift side, while significant amounts of erosion will occur on the down-drift side.

Mission Bay, San Diego, California In the 1940s, development began to turn Mission Bay into a world-class aquatic recreation site. Shallow areas were dredged, creating deeper water. Dredged sands were repositioned, making long sandy shorelines. As part of the process, two jetties of riprap were extended into the Pacific Ocean to create a deep-water inlet, allowing boats to enter and leave Mission Bay (Figure 11.34).

But what happens to Mission Beach sand carried south by longshore drift? It stacks up against the northern jetty, making an extensive sandy zone at South Mission Beach. Meanwhile, Ocean Beach and Sunset Cliffs to the south of the jetties are cut off from longshore-transported sand, thus getting increased beach and sea-cliff erosion instead. Many of the once picturesque sandy coves along Sunset Cliffs are now covered with riprap and backed by cement walls (Figure 11.37); their recreational and aesthetic values have dropped markedly.

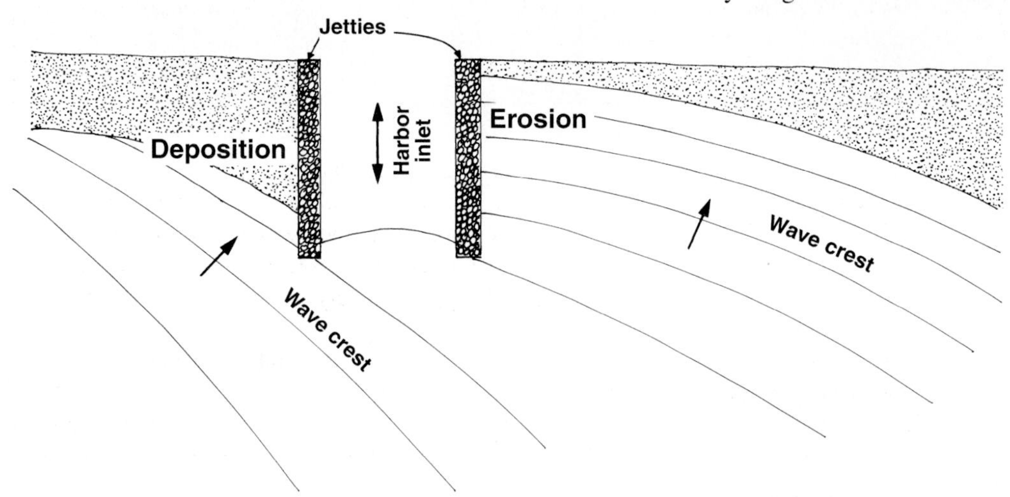

Figure 11.38 Jetties of riprap are extended beyond the surf zone, allowing boats to get in and out of a harbor. Longshore-transported sand accumulates on the up-drift side, erosion occurs on the down-drift side

Breakwaters

Another class of structures, known as **breakwaters,** are built to protect shorelines or harbors from wave attack. Breakwaters may either be attached to, or detached from, the shoreline. The goal of providing boats a safe haven from heavy waves is reasonable, but preventing waves from hitting the shoreline also cuts off the energy that drives the longshore-transport system (Figure 11.39). The sand, deprived of the energy that moves it, just stops its travels and fills in the area behind the breakwater. To keep the harbor or sheltered area open, a permanent dredging operation must be set up to move the sand back into the longshore transport system down-drift from the breakwater.

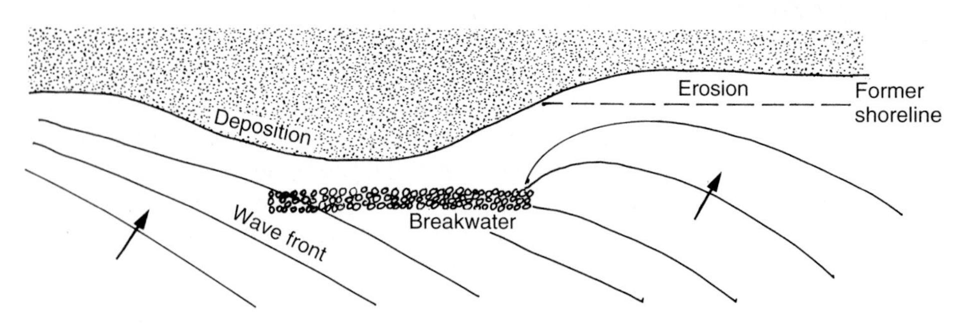

Figure 11.39 A breakwater is built parallel to shore to provide protection from powerful waves. But it also stops waves from hitting the beach, thus shutting down the longshore-transport system.

Sidebar

You Can Never Do Just One Thing

The Mission Bay jetties caused unintended problems. Building jetties to allow boats to use the ocean and bay is fine in itself, but the total effects of a construction project must be evaluated in advance. The jetties are great for boat users, but homeowners and beach lovers in Ocean Beach and Sunset Cliffs pay a high price in terms of beach loss, aesthetic depreciation, and accelerated erosion of beach cliffs. To slow the cliff erosion, the city of San Diego emplaced riprap and built cement walls (Figure 11.37), then assessed Ocean Beach and Sunset Cliffs property owners for the cost of the construction. But the walls and riprap were largely needed because the jetties stopped the longshore drift, thus reducing the protective strip of beach sand.

The planning process ignored a law of nature—you can never do just one thing. You cannot build a jetty and have only one effect. Every action has multiple reactions. When planning an action, you must remember that it will have several reactions and then evaluate the pluses and minuses in advance to see if going ahead with the project makes sense. It must be demonstrated that the positive effects will dominate.

Garrett Hardin has stated this principle another way: there is no such thing as side effects. We all know the term **"side effects,"** but the English language is playing a trick on us here. We learn a term, then incorporate it into our thought processes and use it as if it describes reality. But consider the example of taking a prescription drug: the drug is taken for a reason, and you are supposed to be aware of the side effects. But this is misleading; there are no side effects—every action has multiple reactions. Take a drug (action) and you will experience several effects (reactions). There are no side effects, just a group of reactions to evaluate. Do the benefits of taking the drug outweigh the negative consequences? If so, then proceed with the action.

Summary

Hurricanes affecting the United States usually begin as late summer storms over Africa or the Caribbean Sea/Gulf of Mexico. Cyclonic storms in west Africa move onto the Atlantic Ocean in the equatorial region and are blown westward by the trade winds. Moving over warm water (at least 80°F), surrounded by unstable, warm, and humid air, and with weak upper-level winds, a cyclonic circulation can build toward hurricane strength (74 mph and higher). Surface winds converge at the core (eye) of the hurricane and race upward, releasing tremendous quantities of latent heat that add energy to the hurricane. Nearing the Western Hemisphere and moving into higher latitudes, hurricane paths veer right (Coriolis effect) and commonly hit the United States. Much damage and many deaths can result when the large mound of seawater beneath the hurricane eye surges on land at heights over 20 ft.

Ocean waves do not travel as a physical mass like streams on land. Rather, ocean waves are pulses of energy moving through the water body, causing water particles to rotate in place, similar to the passage of seismic waves.

Most waves are created by winds blowing across the water surface. The height of a wave depends on wind velocity, length of time the wind blows, length of water the wind travels across, and consistency of wind direction. On rare occasions, the various waves moving through the ocean will briefly synchronize and produce rogue waves up to 34 m (112 ft) high.

The distance between successive waves is the wavelength, and the time for two waves to pass a common point is the period. Ocean or lake waves cause orbital motion in water to depths of one-half wavelength. Velocity of waves in miles per hour is approximately 3.5 times the wave period in seconds.

When a wave moves into shallow water, its base is slowed by friction with the sea floor. When waves slow, their lengths decrease and their heights grow. When the height-to-wavelength ratio reaches about 1:7, a wave topples forward as a breaker.

In summer, waves have shorter heights and lesser lengths, causing sand to be pushed onto beaches. During winter, the backwash from taller waves, separated by greater wavelengths, drags beach sand offshore to be stored as submarine sandbars. Beach sand provides excellent protection for coastlines under heavy wave attack. Loose sand grains absorb tremendous amounts of wave energy, then quickly fall back into place.

A wave approaching a coastline that has different water depths will refract or bend as the wave portion in shallower water slows more than the portion in deeper water. When waves strike the beach at an angle, they create a longshore current that carries beach sand along the coastline as a "river of sand." Many beaches end where the sand pours into submarine canyons and flows downslope into deep water.

Humans have major effects on the coast. We build dams across rivers that prevent sand from reaching the beaches and mine tremendous quantities of sand to make concrete and glass. This results in reduced volumes of beach sand, which allows greater wave attack on sea cliffs, where expensive buildings commonly are built. Humans build groins, jetties, and breakwaters to try to control waves and sand movement, and we place riprap and concrete walls to try to stop wave attack.

Each human-built structure triggers multiple reactions, including some negative effects. It is a law of nature that every action has multiple reactions. But many beach construction projects are done with only one purpose in mind, and then people are surprised at the "side effects" that occur. For example, concrete seawalls built to protect against wave attack also reflect waves seaward, eroding the beach sand as they return to the sea. Then with beach sand gone, waves attack the ends and bottom of the wall, causing it to fail.

Terms to Remember

breakwater 330
fetch 322
groin 329
hurricane 302
Intertropical Convergence Zone 308
jetty 330
longshore current 326
longshore drift 326
North Atlantic Oscillation 308
rip current 329
rogue wave 322
side effect 331
submarine canyon 328
surge 302
swell 322
tropical cyclone 302
tropical depression 304
tropical disturbance 304
tropical storm 302
tropical wave 307
turbidity current 328
typhoon 306
wavelength 322
wave refraction 326

Questions for Review

1. Rank the following in order of increasing strength: hurricane, tropical depression, tropical storm, tropical disturbance.
2. Explain the sequence of events that turns an African storm into a hurricane hitting the United States.
3. What factors control the path of a Cape Verde-type hurricane from Africa to North America?
4. Draw a cross section through a hurricane and explain how it operates. Label the internal flow of winds. Be sure to explain the eye.
5. Explain how latent heat helps hurricanes grow.
6. What is the most common month for hurricanes to strike the United States? Why?
7. Draw a cross section showing how a hurricane produces a sea surge that floods the land. How high can surges be?
8. Why do hurricanes strike the Gulf and East Coasts of the United States but not the Pacific Coast?
9. Compare a tornado to a hurricane: Which has the most total energy? Which has the highest wind speeds?
10. Compare water movement in a stream with that in an ocean wave.
11. Draw a cross section that defines wavelength, period, and water motion at depth.
12. Draw a cross section and explain why and how an ocean wave breaks as it nears the shoreline.
13. How are rogue waves created?
14. As wavelength increases, what happens to wave velocity? Which change will cause ocean waves to be more powerful—doubling their wavelength or their height?
15. Draw cross sections and explain the differences in beach-sand volume between summer and winter.
16. Discuss the role of beach sand in providing protection from strong waves.
17. Draw a map of a peninsula jutting into the ocean. Diagram the refraction of ocean waves hitting the shoreline.
18. Draw a map and explain how the longshore current works.
19. Draw a map showing the effects of a jetty on longshore-drifted beach sand. Explain what happens on both the up-drift and down-drift sides of the jetty.
20. Draw a map showing a breakwater and its effect on longshore-drifted beach sand.
21. Draw cross sections and explain the effects of a seawall on beach sand.

Questions for Further Thought

1. What can be done to strengthen buildings enough to withstand hurricanes?
2. Should anything be done to prevent the loss of longshore-drifted beach sand that pours into submarine canyons and is permanently lost to beaches?
3. Should buildings be allowed next to the coastline on Atlantic and Gulf of Mexico barrier-island beaches? Should federal relief funds (taxpayer monies) go to homeowners who suffer losses from hurricane surge and waves?
4. Can placing riprap and building cement walls prevent the retreat of sea cliffs being attacked by heavy waves?
5. Does the term "side effects" describe the way nature works?
6. If global ocean temperature rises, what effect will it have on hurricane strength?
7. Many hurricanes form north of the equator in the North Atlantic, Caribbean Sea, and Gulf of Mexico. Why don't hurricanes form south of the equator in the South Atlantic Ocean? Can a hurricane form at the equator?
8. Compare how many hours in advance we know where a hurricane will strike with how many hours it takes to evacuate cities along the Atlantic and Gulf Coasts.

Suggested Readings and References

American National Red Cross. (1995). *Preparing Your Home for a Hurricane.*

Brown, Joseph. (1989, April). Rogue waves. *Discover,* 47–52.

Dolan, R., Godfrey, P. J., and Odum, W. E. (1973, March–April). Man's impact on the barrier islands of North Carolina. *American Scientist,* 61, 152–62.

Emanuel, Kerry A. (1988, July–August). Toward a general theory of hurricanes. *American Scientist,* 76, 370–79.

Kaufman, Wallace, and Pilkey, Orrin H., Jr. (1983). *The Beaches Are Moving.* Durham, N.C.: Duke University Press.

Leatherman, Stephen P. (1991). Coasts and beaches. In *The Heritage of Engineering Geology: Centennial Volume 3.* Boulder, Colo.: Geological Society of America.

Living with the shore series, published by Duke University Press, Durham, N.C. Individual volumes cover: Alabama and Mississippi; California; Chesapeake Bay and Virginia; Connecticut; North Carolina; East Florida; West Florida; Georgia; Lake Erie; Long Island; Louisiana; Maine; New Jersey; Puget Sound; South Carolina; and Texas.

McHarg, Ian L. (1969). *Design with Nature.* Garden City, N.Y.: Natural History Press.

Morton, Robert A., and McGowen, Joe H. (1980). *Modern Depositional Environments of the Texas Coast.* University of Texas, Bureau of Economic Geology Guidebook 20.

Morton, Robert A., Pilkey, O. H., Jr., Pilkey, O. H., Sr., and Neal, W. J. (1983). *Living with the Texas Shore.* Durham, N.C.: Duke University Press.

Pielke, Jr., Roger A., and Pielke, Roger A. (1997). *Hurricanes:Their Nature and Impacts on Society.* New York: John Wiley & Sons.

Williams, S. J., Dodd, K., and Gohn, K. K. (1991). *Coasts in Crisis.* U.S. Geological Survey Circular 1075.

Videos

The Beach: A River of Sand. (1966). Encyclopedia Britannica (21 min.).

The Beaches Are Moving: The Drowning of America's Shoreline. (1990). Environmental Media (55 min.).

Hurricane Force: A Coastal Perspective. (1994). U.S. Geological Survey (29 min.).

Waves on Water. (1965). Encyclopedia Britannica (16 min.).

Hurricane! (1989). BBC-TV for NOVA (60 min.).

Cyclone! (1995). National Geographic Society (60 min.).

Raging Planet: Hurricane. (1997). Discovery Channel (50 min.).

Chapter 12

Floods

It is I against the world; and when the world has accidents on it[illegible]

Rainfalls vary in intensity and duration, and so does the volume of rainwater that runs across the land. When rains are heavy, floods can result (Figure 12.1). No matter where you live—be it the tropics, the plains, the desert—floods occur. Within a human lifetime, everyone will have a flood pass near them. Many people have stories to tell of their flood experiences, and the tales of the truly largest floods are passed on to succeeding generations.

What has happened in the past provides the best forecast for future events. Within small drainage basins, brief, localized downpours can cause fast-moving but short-lasting maximum floods, such as on the Guadalupe River of central Texas (see Chapter 10). Within large drainage basins, like the Mississippi River, maximum floods result from widespread rains that last for many weeks, producing high waters that also last for many weeks.

The largest flood known in an area is likely to be exceeded someday by a larger one, even in an area with a long history. For example, Florence, Italy, has been described as Europe's greatest cultural storehouse, the site of amazing artistic and scientific accomplishments in the thirteenth through sixteenth centuries. The people of Florence knew their largest flood; it occurred on 4 November 1333, when the River Arno inundated the city to 4.22 m (14 ft) depth. The flood is memorialized by several artistic creations. However, 633 years later, to the day, the River Arno flowed through the city at 6.2 m (over 20 ft) depth, causing heart-wrenching destruction and damage to ancient and priceless paintings, sculptures, tapestries, books, maps, musical and scientific instruments, and more.

The 1966 events started on 3 November in the River Arno headwaters, when enough rain fell to equal one-third of the average annual rainfall. Upstream dams could not hold all the water; the river raged through villages, killing people, ripping apart buildings, and bursting open drums of oil. Some of the treasures of Florence were smashed, some were buried beneath the 500,000 tons of mud deposited by the flood, and many more were coated with a smelly, oily slime. Will an even-larger flood arrive in Florence's future?

Even though Florence has good records of its flood history, it still is hit with nasty surprises. The United States, unlike Europe, does not have lengthy historical records to guide development, and floods are big killers

Figure 12.1 Missouri River flood of 1993 surrounds the state capitol in Jefferson City, Missouri.
Photo courtesy of Missouri Highways and Transportation Department.

Sidebar

A Different Kind of Killer Flood

The fifteenth of January 1919 was unusually hot in Boston. As the temperature climbed higher, the pressure of 27 million pounds of expanding molasses was too much for its heated tank to hold at the Purity Distilling Company. Steel bolts popped with a sound like gunfire, and the steel panels of the tank burst apart, releasing a flood of crude molasses; 2.3 million gallons flowed forth as a 9 m (30 ft)-high brown wave. The flood knocked down supports for the elevated train, pushed houses off foundations, and smothered employees at the Public Works Department, as well as teamsters and their horses unloading goods at the Boston & Worcester and Eastern Massachusetts railroads (Figure 12.2). As the molasses flowed, it cooled and congealed, holding people so tightly in its sticky grasp that rescue workers needed hours to free them. The great molasses flood of 1919 killed 21 people and injured 150. Flood threats are not always obvious.

Figure 12.2 Damage in Boston from the molasses flood of 15 January 1919.

(Table 12.1). To better understand why flood waters claim so many lives, it is useful to understand how rivers and streams operate.

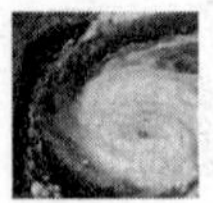

How Rivers and Streams Work

A river is simply a large stream. Streams reveal much about their behavior when examined over their total length.

Longitudinal Cross Section of a Stream

A cross-sectional plot of a stream's bottom elevation versus the stream's distance from source yields a fairly consistent and revealing relationship (Figure 12.3). Using exaggeration of the vertical scale to emphasize the relationship, the bottom profile is seen to be relatively smooth and concave upward with a steeper bottom or slope (higher **gradient**) near the stream source and a flatter bottom (lower gradient) near the stream mouth. Figure 12.3 shows the Arkansas River, but by changing the scales of elevation and length, this longitudinal cross section could serve for virtually any stream in the world.

The lessening of gradient in a stream's lower reaches is partly due to the limitations of **base level**—the level below which a stream cannot erode. For many streams, base level is the ocean, but base level for the Arkansas River is where it joins the Mississippi River. For a small stream, base level may be a lake or pond into which the stream drains.

Streams have similar longitudinal cross sections whether they run through the tropics or the deserts, whether they are long or short, and whether they run through hard or soft rocks. The worldwide similarity of bottom profiles of streams implies that some equilibrium processes must be at work.

Table 12.1 U.S. Flood Deaths and Damages by Decade, 1960–1999

	Decade	Deaths	Damages (in billions of 1997$)
	1960–1969	1,297	21.808
	1970–1979	1,819	48.887
	1980–1989	1,097	33.222
	1990–1999	992	55.586
Totals	40 years	5,205 deaths	$159.503 billion
	Average Deaths per Year = 130		
	Average Damages per Year = $3.99 billion		

Data from National Weather Service, NOAA.

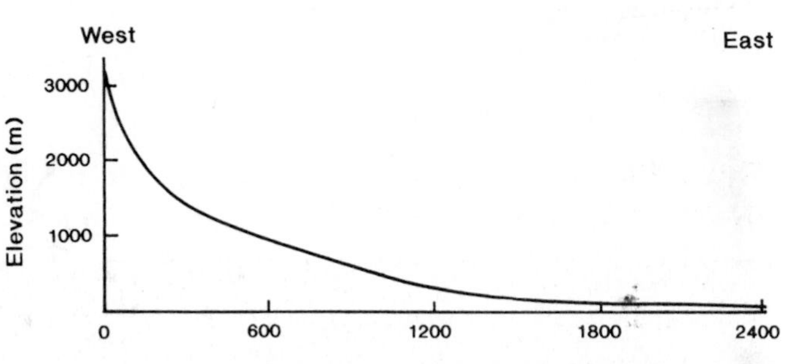

Figure 12.3 Longitudinal cross section of the Arkansas River bottom from its origin in the mountains of Colorado to its mouth at the Mississippi River. Note the extreme exaggeration of the vertical scale necessary to emphasize the concave-upward profile of the bottom.
Source: Henry Gannett, U.S. Geological Survey.

The Equilibrium Stream

Numerous factors interact to make streams seek **equilibrium,** a state of balance where a change causes compensating actions. To grasp the fundamentals of how streams work, a few key variables must be understood: 1) **discharge**—the volume of water, 2) available sediment—the amount of sediment waiting to be moved, 3) gradient—the slope of the stream bottom, and 4) channel pattern—the **sinuosity** of the stream path. Streams occupy less than 1 percent of the land surface but convey the rainfall runoff (discharge) from all the land and carry away loose sediment (**load**).

The U.S. Geological Survey maintains more than 7,000 stream-gauging stations that measure streamflow. Some of these stream gauges have been operating for more than a century. Each stream-gauging station measures water depths, channel width, and water velocity allowing calculation of the discharge or flow volume. The greater the discharge, the greater the load of sediment carried. Both discharge and available sediment are independent variables, i.e., the stream has no control over how much water it will receive or how much sediment is present. Nonetheless, a stream's task is to move the sediment present with the water provided. How can a stream accomplish this task? Excesses in discharge or load are managed by changing dependent variables, such as gradient and channel pattern.

Case 1—Too Much Discharge If a stream has too much water, it will flow more rapidly and energetically. The move away from equilibrium triggers responses to correct the imbalance. 1) Some of the excess energy is used in eroding the stream bottom (Figure 12.4). 2) The sediment picked up by erosion adds to the load carried by the stream, thus consuming more of the excess energy. 3) Notice in Figure 12.4 that the slope of the stream bottom is lowered by erosion, reducing the vertical drop downstream and causing slower and less energetic water flow.

The stream also responds by increasing the sinuosity of its channel pattern through **meandering** (Figure 12.5). A meandering stream cuts into its banks, thus using some of its

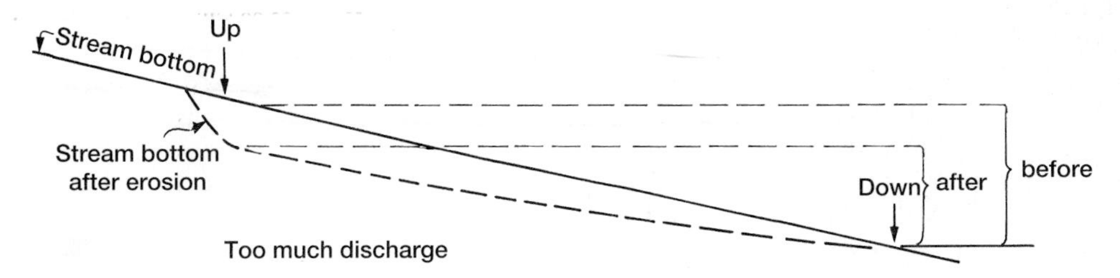

Figure 12.4 Schematic cross section of a stream with too much discharge. The excess water erodes the bottom, flattening the gradient and thus slowing water flow.

Figure 12.5 Meandering channel pattern of the Deschutes River, south of Bend, Oregon. View downstream to the north.
Photo by John S. Shelton.

excess energy to erode and transport sediment. Notice how the meandering pattern lengthens the flow path, lowering the stream's gradient and thus slowing water flow (Figure 12.6).

Case 2—Too Much Load If a stream is choked with sediment and has insufficient water to carry it away, this also triggers negative feedback. The excess sediment builds up on the stream bottom, increasing the gradient and causing stream water to flow faster and thus have more load-carrying capacity (Figure 12.7). Channel pattern responds by straightening to shorten the flow distance and increase gradient. The straighter stream still contains excess sediment, causing the water to pick its way through as a **braided stream** (Figure 12.8).

Another "too much load" situation for a stream is the presence of a lake. For example, if a landslide dams a stream, it adds excess load that the stream will attempt to carry away. The stream will gradually fill in the lake basin with its load of sediment until flow can reach the dam (Figure 12.9). When the stream is able to flow rapidly over the steep-gradient face of the dam, it does so with heightened erosive power, allowing it to carry away the landslide dam as well as the sediment fill in the lake. In a geologic sense, lakes are temporary features that streams are striving to eliminate.

Graded Stream Theory All streams operate in a state of delicate equilibrium maintained by constantly changing the gradient of the stream bottom, thus sustaining a **graded stream.** Every change in the system triggers compensating changes that work toward equilibrium. A typical stream has too much load and too little discharge in its upstream portions, thus it maintains a braided channel pattern there. In its downstream segments, the typical stream has too much discharge, finer sediments in its load, and less friction, thus it runs in a meandering pattern there. Streams also change their equilibrium states from one season to another and in response to global changes in sea level and to tectonic events. That most streams have the same longitudinal cross section is a testimony to the effectiveness of their negative-feedback mechanisms.

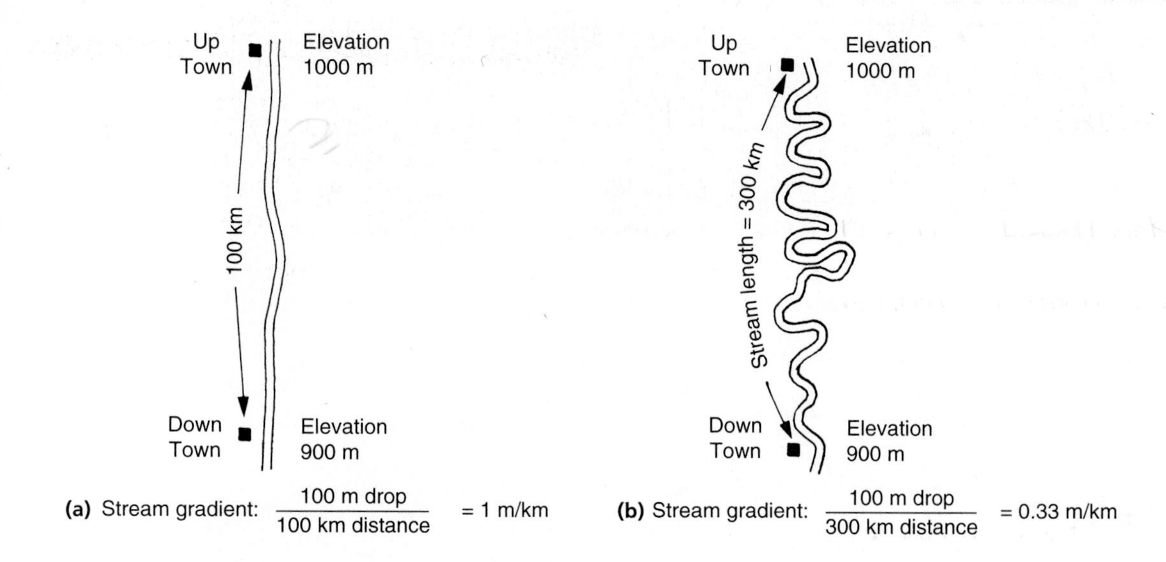

Figure 12.6 (a) Straight stream is the shortest path between two points. (b) Meandering channel lengthens stream, thus reducing gradient and slowing water flow.

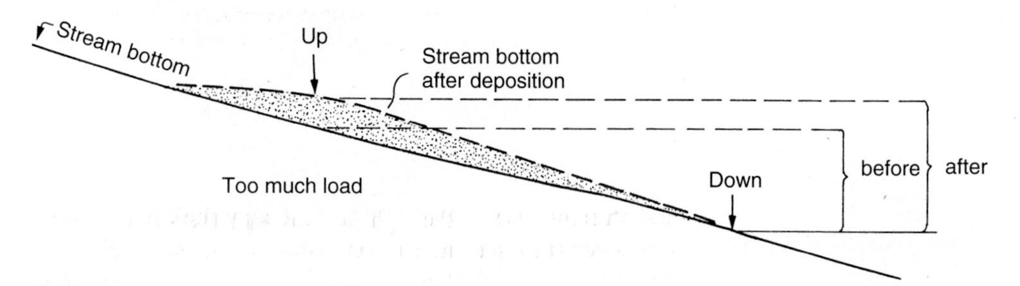

Figure 12.7 Schematic cross section of a stream with too much load. Excess sediment is dropped on the stream bottom, increasing the gradient and thus speeding water flow.

Figure 12.8 A braided channel pattern occurs as water flows through excess sediment within a fairly straight valley. View upstream on the San Juan River near Shiprock, New Mexico.
Photo by John S. Shelton.

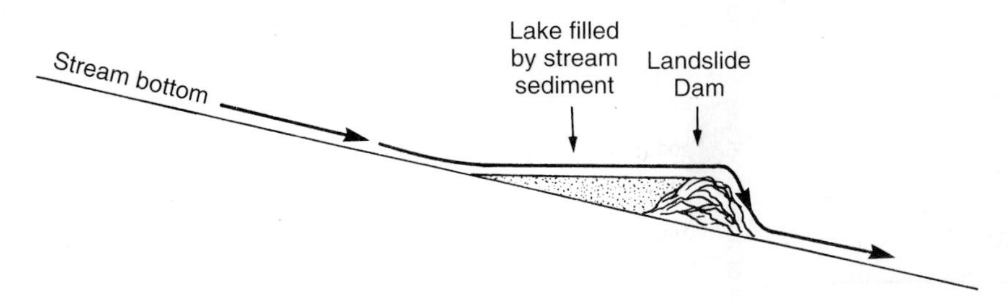

Figure 12.9 Schematic cross section of a landslide-dammed valley. Over time, a stream fills the lake with sediment, then flows across the infilled lake. Flow down the dam's steep face is at high velocity, causing erosion and transport of dam and lake sediment.

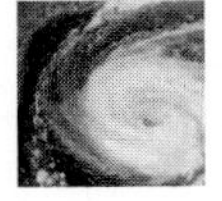

The Floodplain

Floodplains are the floors of streams during floods (Figure 12.10). Streams build floodplains by erosion and deposition, and streams reserve the right to reoccupy their floodplains whenever they see fit. Humans who decide to build on a floodplain are gamblers. They may win their gamble for many years, but the stream still rules the floodplain, and every so often it comes back to collect all bets.

Sidebar

Feedback Mechanisms

Many systems display either negative or positive **feedback. Negative feedback** occurs in self-regulating systems and works to maintain a system in equilibrium. In the case of a stream, when too much water pours into the channel, it triggers negative-feedback responses where increased erosion lowers the gradient to slow the water flow and maintain equilibrium.

Positive feedback is also known as the "vicious cycle." It occurs where one change leads to more of the same, and the whole system changes dramatically in one direction. In a human sense, positive feedback can have positive results if, for example, the system is your money in the bank; the interest you earn gathers more interest, and the system (your money) grows rapidly. Positive feedback can have negative results if, for example, you have credit-card debt and the resultant interest charges add to your debt causing you to pay interest charges on earlier interest charges. Positive feedback can make you rich or poor depending on whether you choose to invest your money or charge your debts.

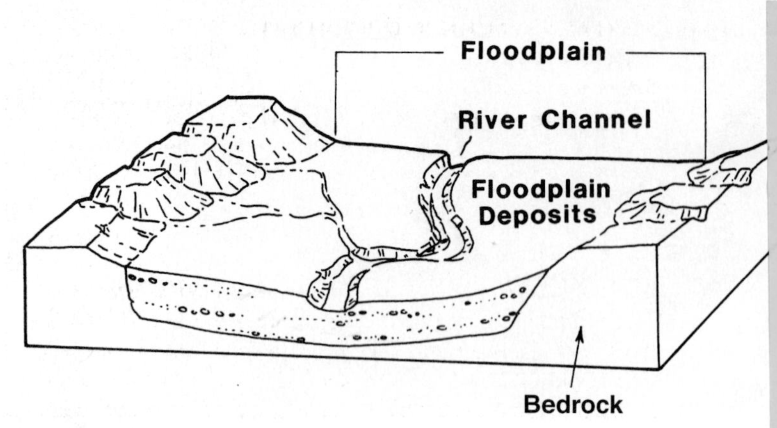

Figure 12.10 Floodplains are stream floors during floods.

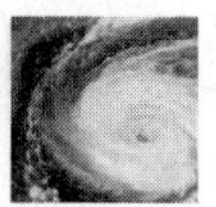

Flood Frequency

Everyone living near a stream needs to understand the frequency with which floods occur. Small floods happen quite frequently, every year or so. Large floods return less often, every score of years, century, or longer. Statistically speaking, the larger the floods, the longer are the recurrence times between each. A typical analysis of flood frequency involves a plot of historic flood sizes versus time (Figure 12.11). Flood-discharge volumes are plotted on the vertical axis, and the **recurrence intervals** are plotted on the horizontal axis in years between floods of each size. The longer the historical record of floods in an area, the more accurately the curve can be drawn. With a flood-frequency curve, the return times of floods can be estimated. For example, in Figure 12.11, move upward from 100 years, intercept a curve, then read to the left to obtain the expected flood size. The U.S. Federal Emergency Management Agency (FEMA) uses the 100-year flood in establishing regulatory requirements.

Discharge (1000's of cubic feet per second)
260
240
220

Figur[illegible] [illegible]
Johns[illegible]

Sidebar

Constructing Flood-Frequency Curves

Floods are random events. It is not possible to predict just when a flood will occur or what its discharge will be. But we need to know how often a given-size flood may occur in order to intelligently develop land. The process used is statistical. If enough runoff records exist for a river, then probable flood frequencies can be estimated.

A relatively simple method for constructing a flood-frequency curve begins by determining the peak discharge for each rainfall year. Ignoring the chronologic order of the rainfall years, rank each annual flood in order from biggest discharge (= 1), second biggest (= 2), etc., on down to the smallest. In order to plot a curve such as Figure 12.11 for a river of interest, you will first calculate recurrence intervals for each year's maximum flood using the formula

$$\text{Recurrence interval} = \frac{(N + 1)}{M}$$

where N = the number of years of flood rec[illegible] merical rank of each year's maximum flood c[illegible] culating a recurrence interval for each year, l[illegible] years) on the horizontal axis (which usually is [illegible] as in Figure 12.11); then move upward until r[illegible] priate discharge value. Stop and plot a point [illegible] section of the recurrence interval and discha[illegible] [illegible] plotting a discharge versus recurrence interval point for each year, draw a best-fit line through your plotted points.

As an example of calculating flood frequency, let's use the flood record of the Red River of the North at Fargo, North Dakota (Figure 12.12). For 115 years of records (1882–1997), the 12th highest discharge is 11,200 cubic feet per second on 3 April 1994. The recurrence interval for a flood of this size is 116/12 = 9.67 years, or about once every 10 years. The probability of a given flood occurring in any one year is the reciprocal of the recurrence interval. The 1994 flood has an annual probability of 1⁄10 or about 10 percent.

How valuable are flood-frequency curves? Their reliability is directly related to the number of years of flood records; the longer the record, the better the flood-frequency curve. In this method, the recurrence interval for the largest flood on a river is the most suspect point; it is [illegible] a sample population of one. Ther[illegible] available to help plot flood-[illegible] the Red River of the North [illegible] 8 April 1997. Inspection of [illegible] as a recurrence interval of [illegible] ility of this flood is 1⁄116 or [illegible] Notice that the flood-recurrence interval read from Figure 12.12 is longer than that calculated by the formula.

Individual flood-frequency curves must be constructed for each stream because each stream has its own characteristic floods. A flood-frequency curve should serve as the basis for designing all structures built on a floodplain and determining where buildings should be located for the highest probability of safety. Planners can decide what size flood (how many years of protection) to accommodate when determining how the land is to be used.

When designing roads, bridges, and buildings, it is seductive to consider only the smaller floods and save large amounts of money on initial construction costs. However, these initial savings are eaten away by higher maintenance and repair costs. In the long run, it is commonly cheaper to build with respect for large floods; this not only saves money in the future but also eliminates much of the human suffering that occurs when homes and other buildings are flooded.

A designer needs to know the likelihood of a given size flood occurring during the expected usage time of a structure. Flood frequency also can be expressed as the statistical probability of stream discharges of a given size arriving in any year or number of years (Table 12.2). The bigger the flood, the longer the return period and the smaller the probability of experiencing it any one year. Statistically, the 100-year flood

Table 12.2 Yearly Probabilities of Floods (Percentage Chance of This Size Flood Occurring)

Return Period (Years)	Any 1 Year	10 Years	25 Years	50 Years	100 Years
2	50%				
	40				
	30				
	25				
5	20				
	15	80	99		
10	10	65	94	99.9	
20	5	40	71	90.5	
50	2	18	40	63	86
100	1	9.6	22	39	63
200	0.5	5	12	22	39
500	0.2	2	5	9.5	18
1,000	0.1	1	2.5	4.8	9.5
2,000	0.05	0.5	1.2	2.3	5
5,000	0.02	0.2	0.5	1	2
10,000	0.01	0.1	0.25	0.5	1

Source: B. M. Reich. (1973). *Water Resources Bulletin*, 9, 187–88.

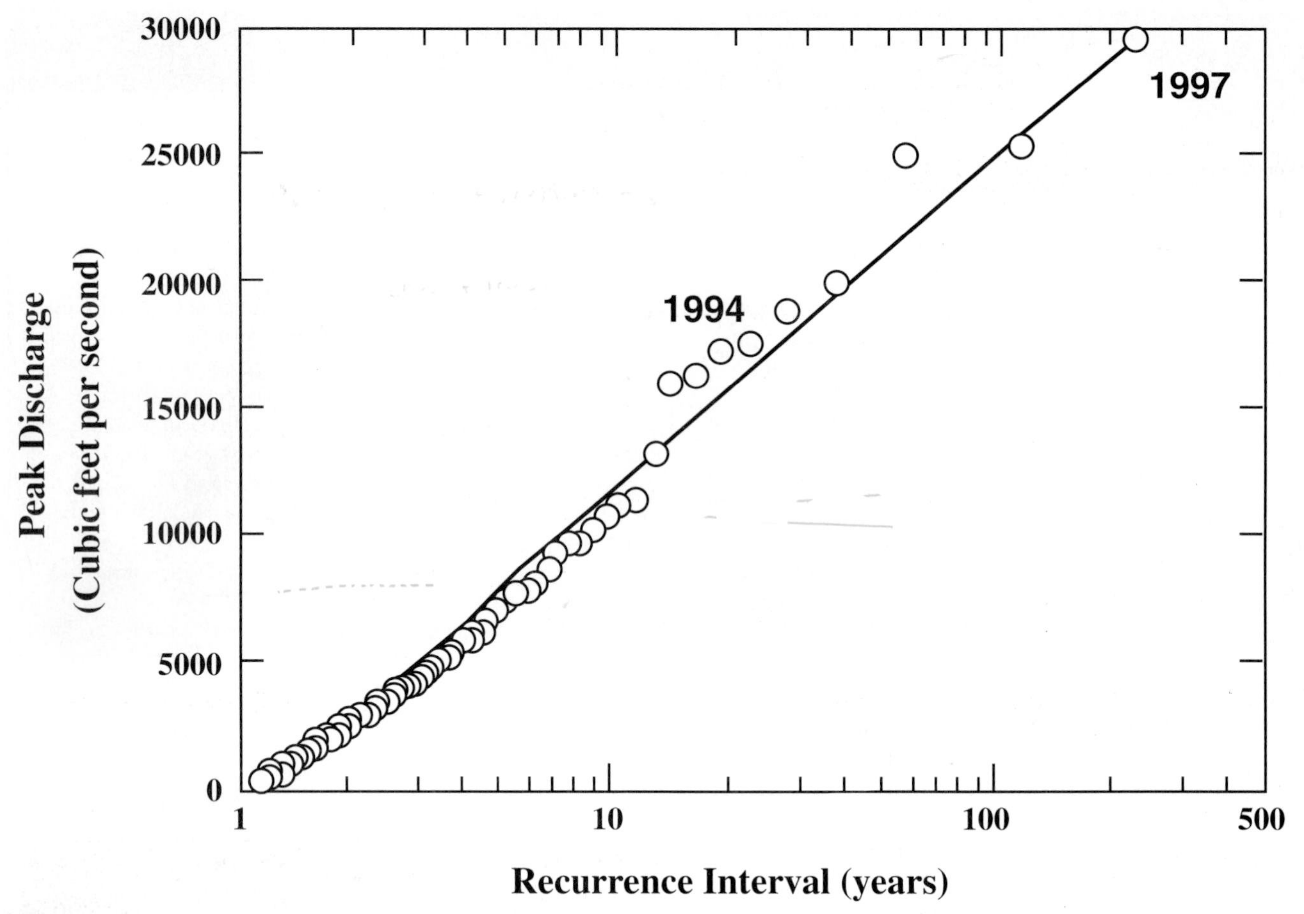

Figure 12.12 Flood-frequency curve for the Red River of the North at Fargo, North Dakota, for years 1882–1997. Data from National Water Data Storage and Retrieval System.

has a 1 percent chance of occurring any year. What is the probability that a 100-year flood will occur once in 100 years? The obvious answer of once is wrong; from Table 12.2, the probability is only 63 percent. No flood has a 100 percent chance of occurring.

One must distinguish between yearly versus cumulative probability. In cumulative probability, the longer one waits for a 100-year flood, the more likely it will occur. Nevertheless, the yearly probability of a flood is the same for any year regardless of when the last flood occurred. The confusion that commonly arises when hearing of the "100-year flood" has led some people to stop using the term and to replace it with "1%-chance flood." A 1%-chance flood is a flood event that has a one percent chance of occurring or being exceeded in any given year.

By similar analysis, even though a "150-year flood" may occur one year, it is still possible for another of the same size to come again in the following year, or even in the same year. For example, in 1971, the Patuxent River between Baltimore and Washington, D.C., had a flood that was 1.6 times bigger than its calculated 100-year flood. The next year, in 1972, the Patuxent River conveyed a flood that was 1.04 times as big as its 100-year flood. A 100-year flood, or any other size flood, is a statistically average event that occurs by chance, not at regular intervals. As the adage states: "Nature has neither a memory nor a conscience."

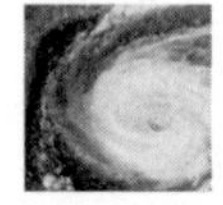

Flood Styles

Killer floods are unleashed by several phenomena. 1) A local thundercloud can form and unleash a flash flood in just a few hours. 2) Abundant rainfall lasting for days can cause regional floods that last for weeks. 3) The storm surges of tropical cyclones flood the coasts. 4) The break-up of winter ice on rivers can pile up and temporarily block the water flow, and then fail in an ice-jam flood. 5) Short-lived natural dams made by landslides, log jams or lahars fail and unleash floods. 6) Human-built dams and levees do fail causing voluminous floods.

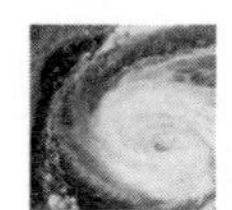

Flash Floods

Large convective thunderstorms can build up in a matter of hours and quickly set loose the terrifying walls of water known as flash floods. Steep topography helps thunderstorms

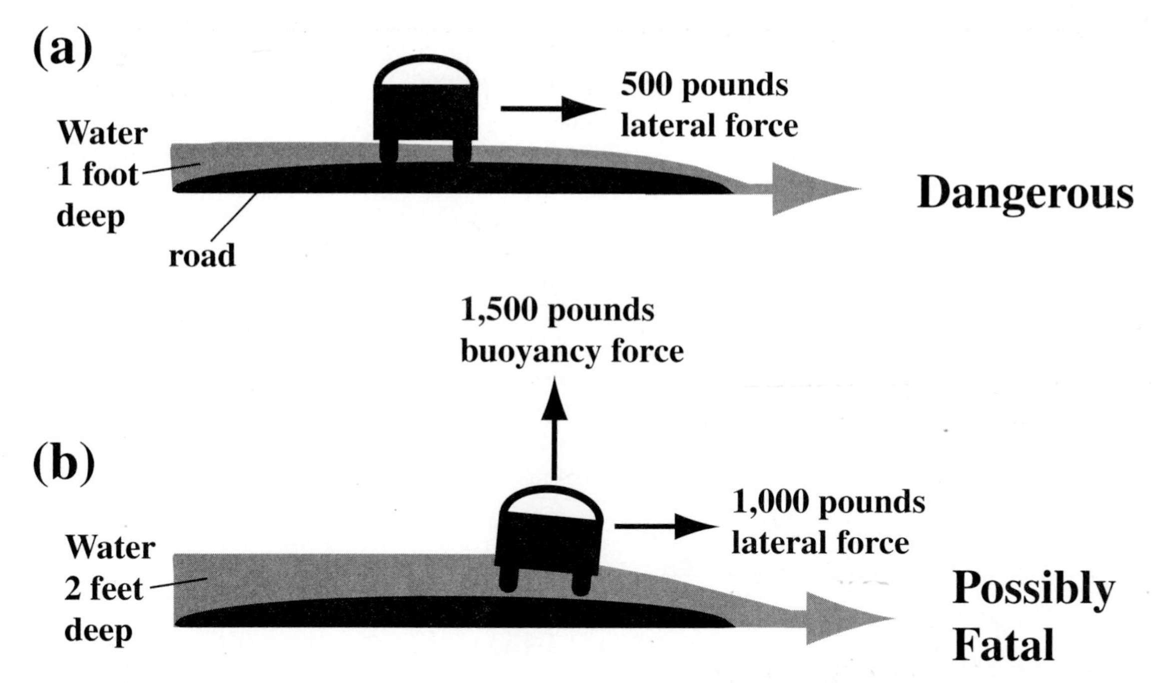

Figure 12.13 Do not drive through a flood. **(a)** Flood water 1 foot deep exerts 500 pounds of lateral force. **(b)** Flood water 2 feet deep both buoyantly lifts and laterally pushes a car.
Modified from U.S. Geological Survey Fact Sheet 024–00.

build and then provides the rugged valleys to channelize the killer floods.

Most flood-related deaths in the United States are caused by flash floods. About 50 percent of these deaths are vehicle related. Not enough people appreciate what a shallow-water flood can do to a car (Figure 12.13). Flowing water about one foot (0.3 m) deep exerts about 500 pounds (225 kg) of lateral force. If two feet (0.6 m) deep water reaches the bottom of a car, there will be a buoyant uplift of about 1,500 pounds (680 kg), which allows the 1,000 pound (450 kg) lateral force to push or roll the car off the road. Many automobile drivers and riders die in flood water that is only two feet deep.

Antelope Canyon, Arizona, 1997 The plateau country of the southwestern United States is world renowned for the canyons cut into it, for example, Grand Canyon, Glen Canyon, and numerous tributary canyons leading to the Colorado River. Some of these tributary canyons are wonder-inspiring, narrow clefts. Imagine walking down a slot canyon so narrow that you can simultaneously touch one wall with each of your hands even though the walls rise near-vertical for 100 feet above your head (Figure 12.14). The experience of walking down these slot canyons is so magical that tourists from around the world come to hike between the exquisitely beautiful, sculpted walls with their pink and orange colors; the setting is like a natural cathedral. Now imagine a localized, late afternoon thundercloud 11 miles away; too far away for you to see it or hear it. What can happen in the slot canyon? On 12 August 1997, in Antelope Canyon a flash flood roared down canyon as an 11 feet-high wall of water that picked up 12 hikers and tumbled them down canyon as helpless viewers looked down at them from the canyon rim. The natural cathedral turned into a death trap; only one hiker survived.

Figure 12.14 Arrow Canyon (upper right to left center) in Nevada is a slot canyon with steep walls that allow no escape for hikers caught in a flash flood.
Photo by John S. Shelton.

Big Thompson Canyon, Colorado, 1976 Saturday, 31 July 1976 was the eve of the one-hundredth birthday of Colorado statehood and the start of a three-day centennial weekend. The weather was pleasant, and about 3,500 people were staying in or passing through the recreational area of Big Thompson Canyon about 50 miles northwest of Denver, near Estes Park, in the Front Range of the Rocky Mountains. About 6 P.M., strong low-level winds from the east pushed a moist air mass upslope into the mountains. The unstable air kept rising, powered by heat released from condensing water vapor. Late-afternoon cloud building is a common phenomenon and usually concludes with upper-level winds pushing the thundercloud eastward over the plains. On 31 July, the mid- and upper-level winds were weak, the thunderstorm remained stationary, and a "cloudburst" ensued (Figure 12.15). The slightly tilted updraft structure allowed rain to fall profusely. From 7:30 to 8:40 P.M., rainfall was as heavy as 19 cm (7.5 in); in four hours, it equaled a typical year's total.

Rain runoff from the steep, rocky slopes fed a flash flood that roared down the canyon with an initial wall of water reaching 6 m (20 ft) high in the narrows at the eastern end of Big Thompson Canyon. The flood crest moved 23 ft/sec (15 mph) through the entire canyon, which did not allow much time to spread warnings. Over 400 automobiles were on Highway 34 within the canyon. Drivers were presented with a quick choice—either abandon their cars and run upslope or stay inside and try to outrace the flood. Those who abandoned their cars spent an uncomfortable night on the rain-swept canyon walls; those who stayed with their

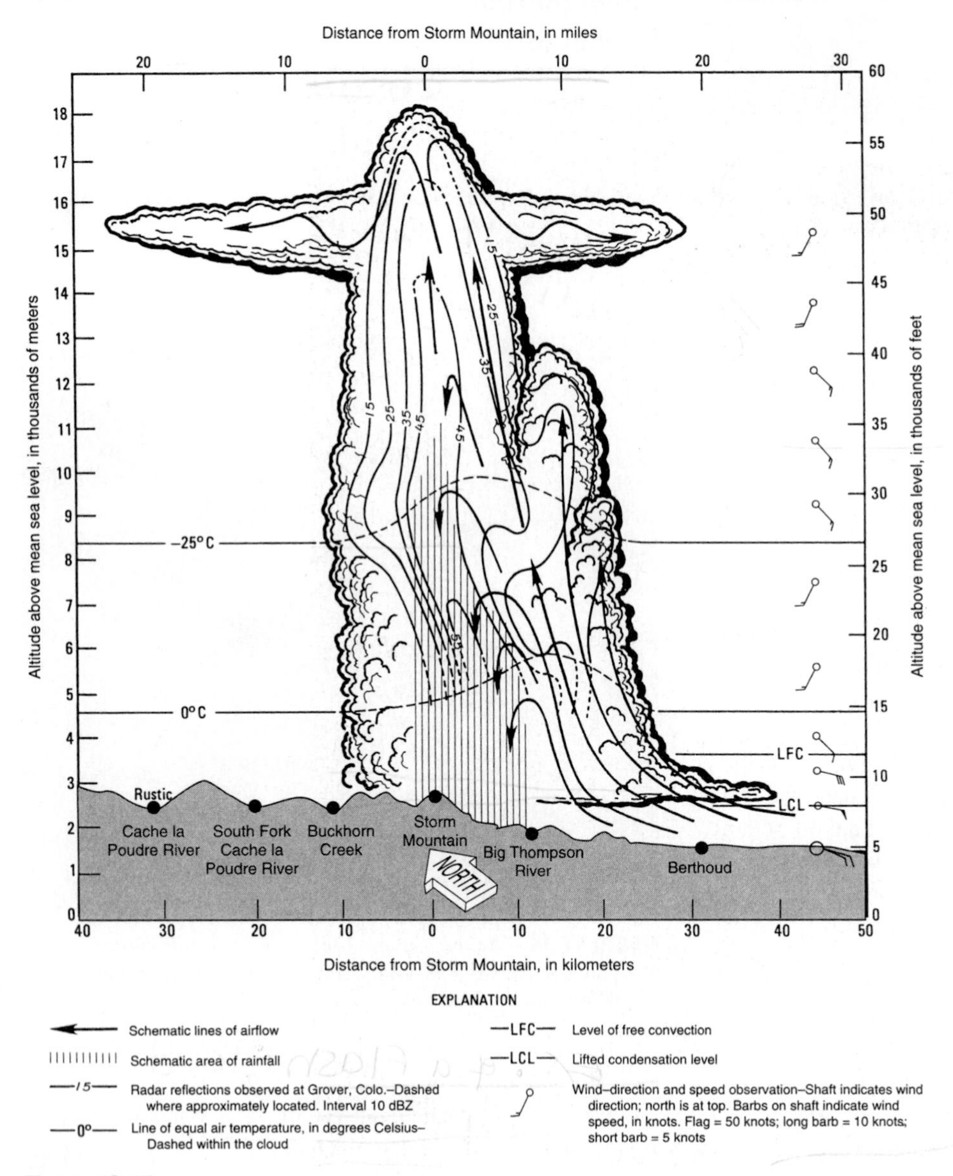

Figure 12.15 Model of the thunderstorm that fed the Big Thompson Canyon flood. Drawing depicts conditions at 6:45 P.M., 31 July 1976.

Source: U.S. Geological Survey Professional Paper 1115.

cars died. The new road signs now advise: "Climb to safety! in case of a flash flood."

At 6 P.M. on 31 July 1976, the Big Thompson River flowed with 137 ft³/sec of water; at 9 P.M., the flow was 31,200 ft³/sec. Flow volume was 3.8 times greater than the estimated 100-year flood. The flash flood killed 139 people and 6 more were never found; 418 houses were destroyed, along with 52 businesses and more than 400 cars, and damages totaled $36 million (Figure 12.16).

Rapid Creek, Black Hills, South Dakota, 1972 The Black Hills are one of the most beautiful sites in the United States; they host nearly 3 million tourists a year. At the foot of the Black Hills sits Rapid City, first settled south of Rapid Creek in 1876. The early inhabitants were wary of the Rapid Creek floodplain, and the wisdom of their caution was borne out by a large flood in 1907. As the city's population grew, the peaceful, meandering stream became a magnet that induced development on the floodplain. In 1952, Pactola Dam was built 16 km (10 mi) upstream to provide flood protection and a reserve water supply for Rapid City. The dam eliminated most small floods, giving people a false sense of security. By 1972, the floodplain was host to numerous houses, mobile-home parks, shopping centers, car lots, and other urban structures serving residents in a city approaching 50,000 population.

On Friday, 9 June 1972, southeast winds bringing moist air from the Gulf of Mexico met a cold front coming from the northwest. Under conditions similar to those at Big Thompson Canyon, the moist air turned upward to build 16 km (10 mi)-high thunderheads that remained stationary due to weak upper-level winds. Shortly after 6 P.M., heavy rain

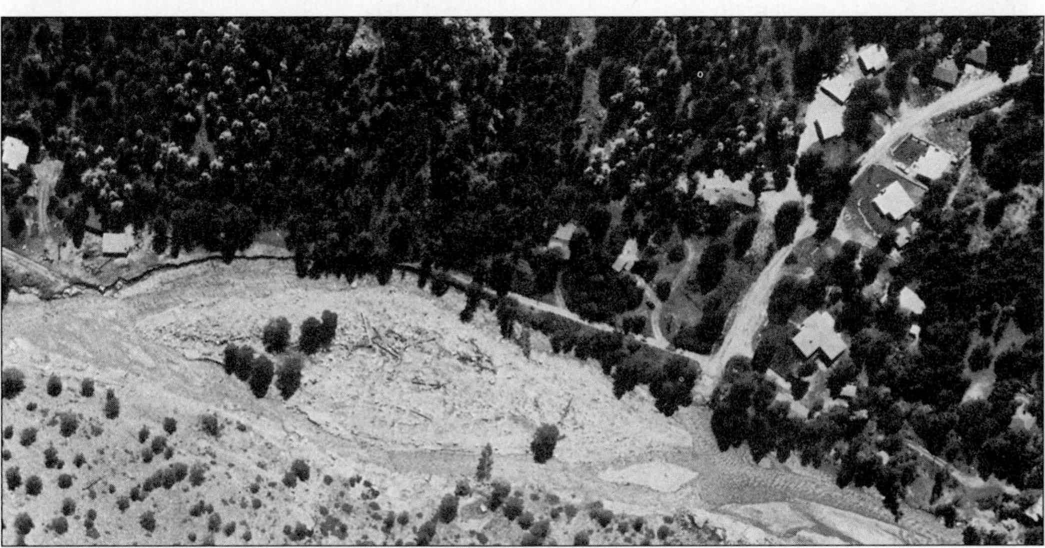

Figure 12.16 The community of Waltonia before and after the 1976 flood in Big Thompson Canyon, Colorado.
Source: U.S. Geological Survey Professional Paper 1115.

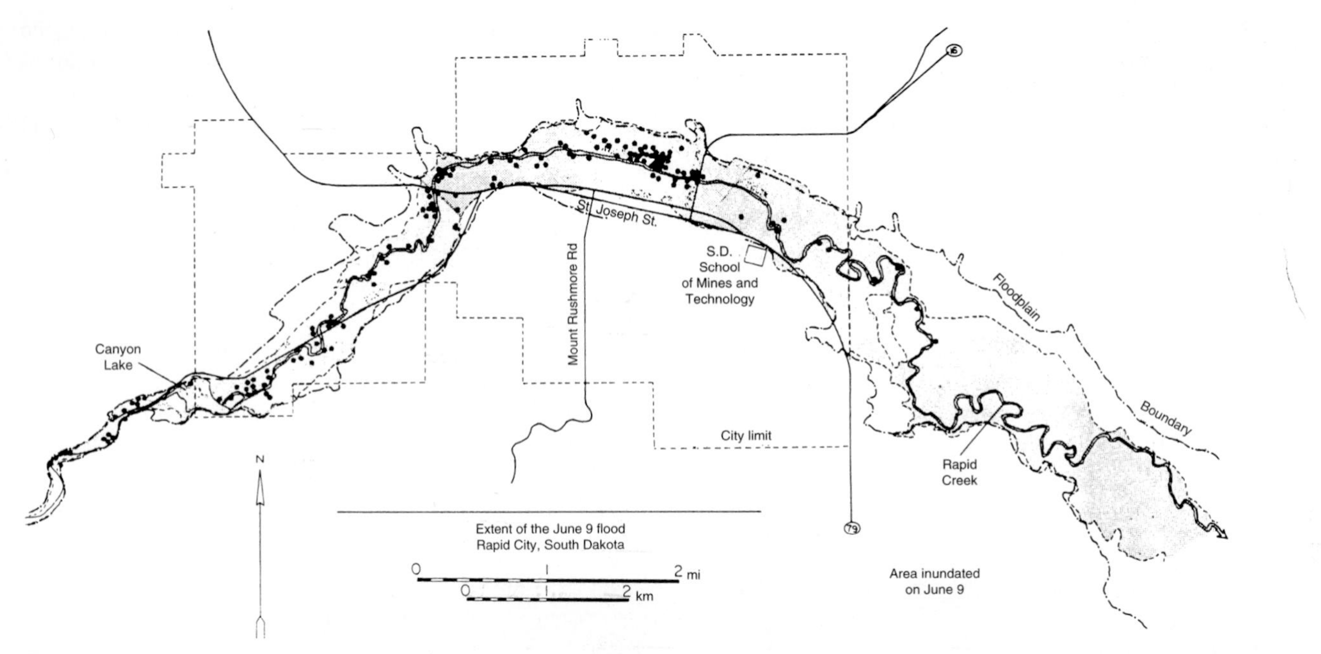

Figure 12.17 Map of Rapid City, South Dakota, showing area inundated by flood of 9 June 1972. Black circles mark locations of bodies found after the flood.

began to fall; up to 38 cm (15 in) fell in less than six hours, but most of the rain fell downstream from Pactola Dam. Rain runoff filled Canyon Lake, built on the western edge of Rapid City. The spillway at Canyon Lake Dam became plugged by automobiles and house debris, causing the lake level to rise an additional 3.6 m (12 ft). Then the dam failed at about the same time as the natural flood crest arrived, thus unleashing a torrent of water on Rapid City (Figure 12.17). The river reoccupied its floodplain with vengeance, leaving destruction totaling $664 million in 2002 dollars. Floods in the region killed 238 people, mostly in Rapid City, and destroyed 1,335 homes and 5,000 automobiles.

This time the lesson was learned by many. Canyon Lake Dam and many bridges were redesigned and rebuilt to prevent debris accumulations. The floodplain belongs to the river, and dams provide only a false sense of security. The portion of Rapid Creek floodplain inundated in 1907 and 1972 was declared a floodway, and rebuilding was not permitted. Even most buildings that survived the flood were moved out of the floodway. In their place lies an 8 km (5 mi)-long greenway featuring a golf course, picnic areas, bike and jogging paths, recreation areas, ponds and ice-skating rinks, low-maintenance grasslands, and an area reseeded with native vegetation: in short, the floodway is now being used for activities that will not be harmed by the occasional flood. A Rapid City slogan is: "No one should sleep on the floodway." As long as this policy is maintained, the next big flood will not cause deaths, destruction, and heartache, and there will be no federal relief funds wasted on rebuilding in the same floodplain only to await the next killer flood.

Regional Floods

Regional floods are different from flash floods. High waters may inundate an extensive region for weeks, causing few deaths but extensive damages and severe tests of human endurance. Regional or inundation floods occur in large river valleys with low topography when prolonged, heavy rains result from widespread cyclonic systems.

In the United States, about 2.5 percent of the land is floodplain and home to about 6.5 percent of the population. Floodplains contain much valuable property that is periodically flooded.

Red River of the North

The Red River of the North is unusual because it flows northward (Figure 12.18), draining parts of South Dakota, North Dakota, and Minnesota before flowing into the Canadian Province of Manitoba. Floods are likely here each spring. As described by John McCormick of *Newsweek,* "On the northern plains, nature is less an enemy than a sparring partner, trading rounds in a grudge bout that never ends." In 1997, nature won the round as flood levels set records throughout the region. A Presidential Disaster Proclamation was declared for all 53 counties in North Dakota.

Why are floods so common along the Red River of the North? Several factors combine to create this situation. 1) The Red River valley is geologically young, only about 9,000 years old, and has not carved a deep valley. 2) The gradient or

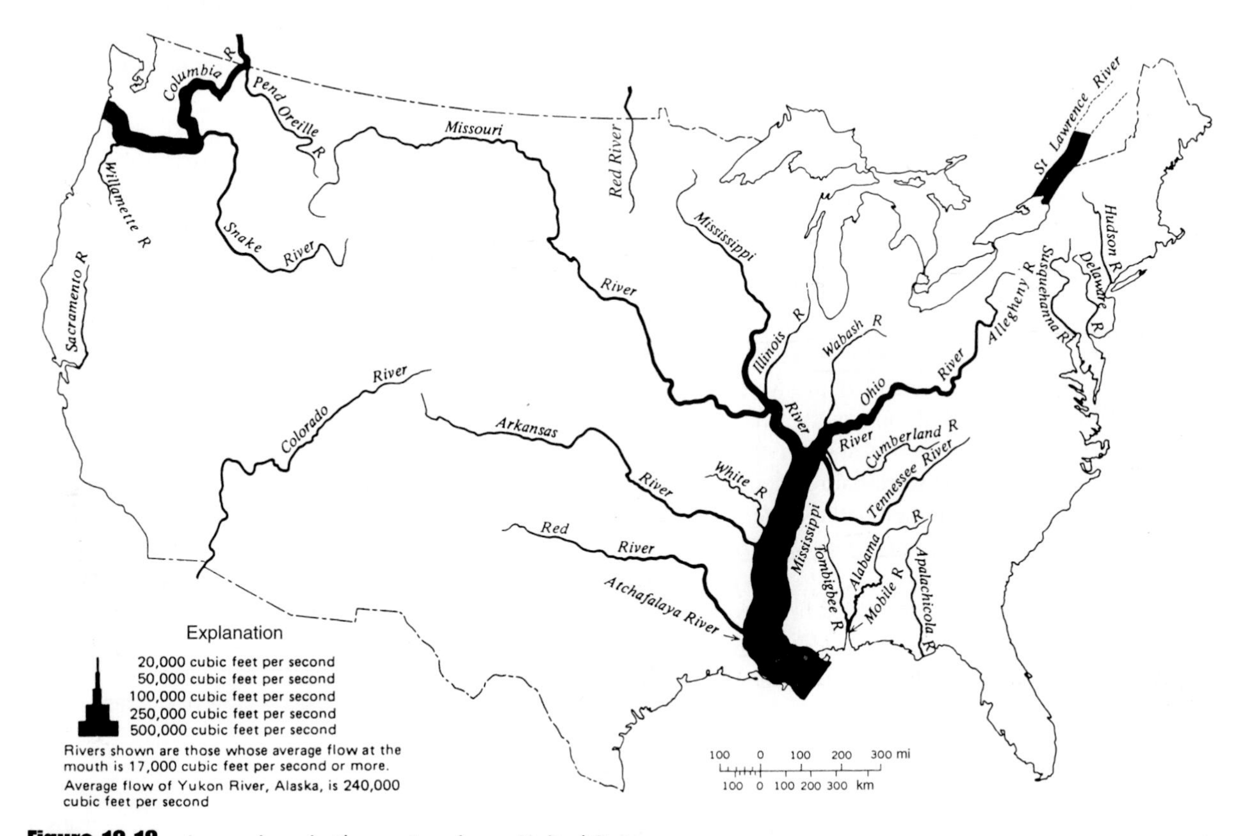

Figure 12.18 Large rivers in the conterminous United States.
Source: K. T. Iseri, and W. B. Langbein, "Large Rivers of the United States," in *U.S. Geological Survey Circular 686,* 1974.

slope of the river bed is very low, averaging only a 5-inch drop in elevation per mile (8 cm/km). The gradient is only 1.5 inches/mile (2.4 cm/km) just south of the Canadian border. The nearly flat river bottom causes slow-flowing water that tends to pool into a broad and shallow lake during high-water flow. 3) River flow increases as winter snow melts. The meltwater runs northward into still frozen lengths of the river, where ice jams up and impedes water flow causing floods.

In 1997, several variables combined to unleash record floods. 1) Fall 1996 rainfall was about four times greater than average. 2) Winter 1996 freezing temperatures began earlier than normal thus freezing the water saturated in the soil. 3) Winter 1996–97 snowfalls were 3 to 3.5 times greater than average, bringing and storing more moisture in the region. 4) Spring 1997 began cold including a blizzard on 4–6 April that brought 10 to 12 inches (25–30 cm) of snow and record low temperatures that delayed melting and draining. 5) Finally, flooding began as a rapid rise in air temperature melted snow and soil ice, sending water flowing.

On 8 April 1997 the average temperature was 9°F (−13°C), but on 18 April the average was 58°F (14°C). Snow melted, soil ice melted, and the ground everywhere was covered by overland waterflow that overwhelmed the water-transporting ability of the flat-bottomed Red River. As the flood waters slowly flowed northward, they progressively inundated farmland and towns, including the incongruous sight of a flooded downtown Grand Forks with its buildings on fire. In North Dakota, the flood damages exceeded $1 billion, forced the evacuation of 50,000 people, destroyed potato, sugar beet, and grain crops, prevented planting seeds for the next crop, and drowned farm animals including 123,000 cattle. In Manitoba, the flood damages exceeded $815 million and forced the evacuation of 25,000 people.

How common is a flood this size? The Red River flood-frequency curve (Figure 12.12) indicates the 1997 flood has a recurrence interval of about 225 years. But there are not enough controls on where to plot the largest event on a flood-frequency curve. Other estimates of the recurrence interval of the 1997 flood range from 300 to 500 years.

Mississippi River System

The greatest inundation floods in the United States occur within the Mississippi River basin, the third largest river basin in the world; it drains all or part of 31 states and two Canadian provinces (Figure 12.18). Of the 28 biggest rivers in the United States, 11 are part of the Mississippi River system. In the lower reaches of the river, the average water flow is 645,000 ft^3/sec (18,250 m^3/sec); this may be increased fourfold during an inundation flood.

Some Historic Floods New Orleans was founded in 1717 in the lower Mississippi River basin. It experienced its first large flood in the same year. The response to the flood in 1717 was the same as today—build **levees** to keep the waters away.

In 1879, the Mississippi River Commission placed a major emphasis on building levees, yet the 1882 flood broke through the levees in 284 places. By 1926, there were 2,900 km (1,800 mi) of levees averaging 6 m (20 ft) in height. But the 1927 floods breached the levees in 225 places, inundating 50,000 km^2 (19,300 mi^2) and drowning 183 people.

After 1927, more levees were built, major dams were constructed on tributary rivers to capture flood waters, Mississippi River meanders were straightened to shorten the river by 270 km (170 mi), and diversions were built to send flood waters out into uninhabited areas. Despite all these efforts, in the spring of 1973, the Mississippi River system flooded along 1,930 km (1,200 mi) of rivers in 10 states, again inundating 50,000 km^2, reaching record flood heights at numerous sites, and staying above flood stage for 97 days at Chester, Illinois, 77 days at St. Louis, and 63 days at Memphis.

The Great Midwestern Flood of 1993 The summer of 1993 saw the biggest flood in 140 years of gauged measurements for the upper Mississippi River basin (Table 12.3). Notice how late in the year the 1993 peak flood occurred. A wet winter and spring passed into an even wetter summer for Iowa and parts of eight other upper Midwest states, causing record flood levels on the lower Missouri and upper Mississippi Rivers (Figures 12.1, 12.18, 12.19). High-water levels began in April and continued into August. Some towns had more than 160 consecutive days of flooding. At the end of August, the upper Mississippi River basin had endured record high floods yet this flood water mass did not significantly affect the lower Mississippi River basin because the input from the Ohio River flow was low (Figure 12.18).

Weather Conditions Why were the 1993 floods so big? Is the weather changing? The weather pattern was remarkably similar for the big floods of 1927, 1973, and 1993. In each case, the preceding autumn and winter were wet, thus leaving the ground saturated as the new year began. In each of the years, the polar jet stream bent southward, forming huge troughs of low pressure that attracted cyclonic systems into the Mississippi River basin to drop their moisture. In 1993, a cold-air mass over Greenland and a high-pressure ridge over the northeastern United States resulted in large-scale bends in the polar jet stream (Figure 12.20). The high-pressure ridge over the eastern United States produced

Table 12.3 Top Ten Mississippi River Floods at St. Louis, 1861–2002

Date	Discharges in Cubic Feet/Second
1993 August	1,030,000
1903 June	1,019,000
1892 May	926,500
1927 April	889,300
1883 June	862,800
1909 July	860,600
1973 April	852,000
1908 June	850,000
1944 April	844,000
1943 May	840,000

Source: Illinois State Water Survey Miscellaneous Publication 151. (1994). Champaign, Ill.

Figure 12.19 Missouri River floods Highways 54 and 63 in Missouri in 1993.
Photo courtesy of Missouri Highways and Transportation Department.

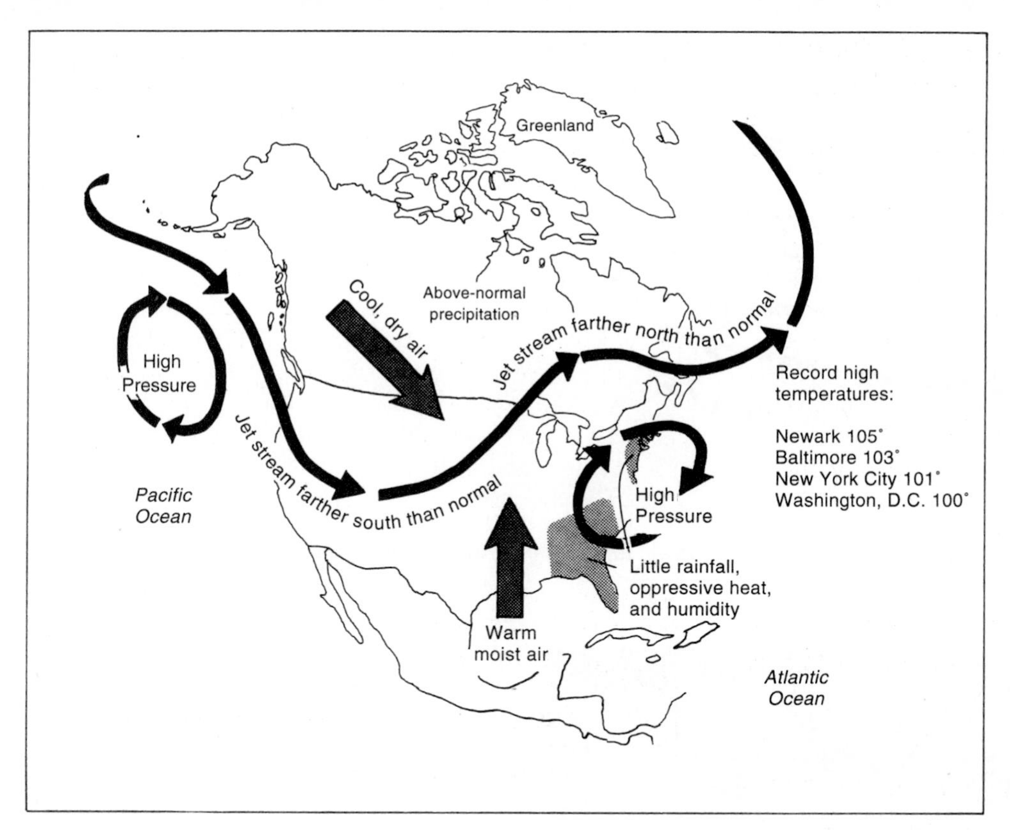

Figure 12.20 A typical weather map for the United States in July 1993.

record-breaking high temperatures. At the same time, the low-pressure trough over the Mississippi River basin drew moist air from the Gulf of Mexico into contact with cold, polar air masses, resulting in long-term heavy rainfall. The persistence of the polar jet stream pattern caused storm after storm to dump its water in the nation's heartland, producing high flood levels that went on week after week. Was this a new and unusual weather pattern? No, it is a reasonably common occurrence.

Role of Levees Are river flood heights increasing with time? At St. Louis, Missouri, the high water mark (not the same as flood volume) set in 1927 was broken in 1973 and then was exceeded twice in 1993. Were human-built levee systems part of the problem? The long systems of levees have transformed the Mississippi River and its tributaries into restricted ribbons of water. Higher levees create higher river levels. Rivers in flood are blocked by levees from flowing laterally to spread out their water; instead, they are confined and forced to rise vertically until ultimately overtopping the levees. Estimates suggest the 1993 flood at St. Louis would have crested 4 m (13 ft) lower if it had occurred 100 years earlier, before all the levees were built.

As levees become saturated, the river finds weak spots, compromising the levees by wave attack, erosion by overtopping, failing by slumping, and undermining by piping (Figure 12.21). The flood water that escapes confinement by levees spreads out, inundating farms and buildings (Figure 12.22). In 1993, 1,083 out of 1,576 levees in the upper Mississippi River system were overtopped or damaged.

Behind the seeming protection of high levees, more and more people have moved their homes and businesses onto the floodplain. But this land has only been borrowed from the river. In 1993, flood waters reoccupied more than 20 million acres (Figure 12.23). The entire state of Iowa was declared a federal disaster area, as were sections of eight other states—North Dakota, South Dakota, Minnesota, Wisconsin, Illinois, Missouri, Nebraska, and Kansas. Flooding killed 48 people (Figure 12.24), completely submerged 75 towns, destroyed or damaged 50,000 homes, closed 12 commercial airports, shut down four interstate highways (29, 35, 70, and 635), and even carried away more than 700 coffins from a graveyard in Hardin, Missouri (Figure 12.25). Damages totaled about $12 billion, making this flood one of the most expensive disasters in U.S. history.

China

The Huang (Yellow) River is reputed to have killed more people than any other natural feature; it is also known as the "River of Sorrow." In the lower 800 km (500 mi) of its course, it flows over floodplain and coastal-plain sediments. Attempts to control river flow and protect people and property go back

How leeves fail?

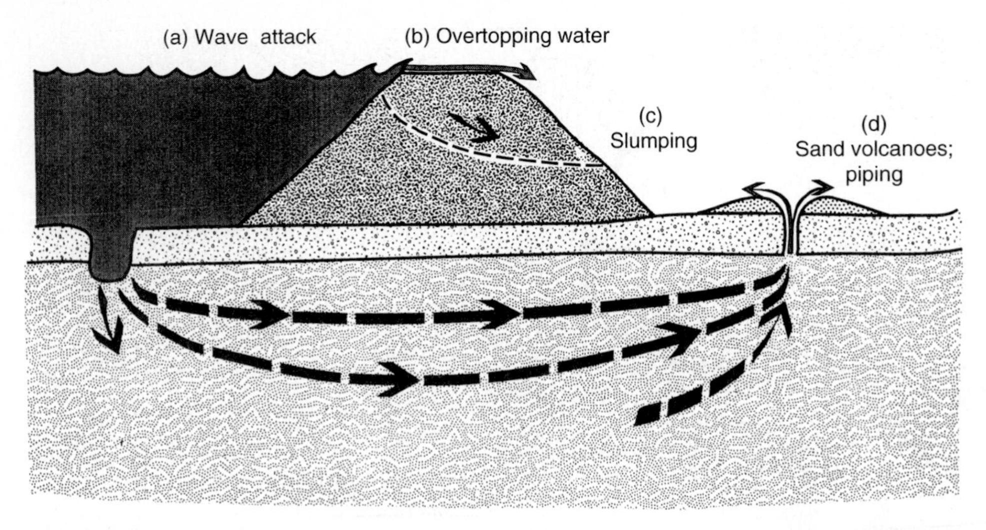

Figure 12.21 Levees are attacked by several processes: **(a)** wave attack; **(b)** erosion by overtopping; **(c)** slumping of the levee mass; and **(d)** subsurface erosion (piping).

Figure 12.22 Breach in Kaskaskia Island levee, Illinois, August 1993.
Photo courtesy of U.S. Army Corps of Engineers.

at least as far as the channel dredging of 2356 B.C.E. Levees are known to have been constructed since at least 602 B.C.E. Have these efforts controlled the river and protected the people? Not well enough. In the last 2,500 years, the Huang River has undergone ten major channel shifts that have moved the location of its mouth as much as 1,100 km (over 680 mi) (Figure 12.26).

When a river deposits sediment on its channel floor, its channel-bottom elevation grows higher. Over time the channel bottom may build to an elevation higher than its adjoining floodplain. When major floods occur and water overtopping banks and levees flows to lower elevations outside the channel, the river may adopt a new lower-elevation course and abandon its old channel; this is the process of **avulsion.** In 1887, the Huang overtopped 22 m (75 ft)-high banks, "discovered" lower elevations, and began flowing south to join the Yangtze River. The 1887 floods drowned people and crops, creating a one-two punch of floods and famine that were responsible for over 1 million deaths.

In 1938, the Yellow River levees were dynamited in the war with Japan, resulting in another million lives lost to flood and famine. Today, the riverbed is 20 m (65 ft) higher than the adjoining floodplain! The Huang River "wants" to change course, but the Chinese keep building levees to make it stay where it does not want to be. How long can the river be confined?

Figure 12.23 Flood water from the Missouri River closed this St. Charles, Missouri road to cars, 21 July 1993.
Photo by Tom Dietrich, courtesy of NOAA.

Figure 12.25 Flood water dug up coffins and carried them down river in Hardin, Missouri, in 1993.
Photo courtesy of Missouri Funeral Directors Association.

Figure 12.24 Typical flood damage in Missouri, 1993.
Photo by Mark Schreiber, Missouri Department of Corrections.

The strategy of building structures to make a river flow where it does not want to go is not unique to China. The Mississippi River has long wanted to abandon its course (avulse) and flow down the shorter and steeper path to the ocean offered by the Atchafalaya River. The U.S. Army Corps of Engineers works hard to make the Mississippi River stay put and continue flowing through the important cities of Baton Rouge (the capital of Louisiana) and New Orleans (a major U.S. port). It takes a huge commitment of time, money, and effort to control a river's course. The Chinese have tried to control the Huang for well over 4,000 years and have not been successful. Will Americans be able to control the Mississippi River indefinitely? Donald Worster in *Under Western Skies* said:

> Human domination over nature is quite simply an illusion, a passing dream by a naive species. It is an illusion that has cost us much, ensnared us in our own designs, given us a few boasts to make about our courage and genius, but all the same it is an illusion.

Societal Responses to Flood Hazards

People like to be near rivers. They provide food and drink, business and transportation, arable land and irrigation, power, and an aesthetic environment. But being near rivers also means being subjected to floods. Human responses to flood hazard have been in two main categories: structural and nonstructural.

Structural responses include constructing dams to trap flood water, building levees along rivers to contain flood water inside a taller and larger channel, engineering projects seeking to increase the water-carrying ability of a river channel via straightening, widening, deepening and removing debris, and short-term actions such as sandbagging.

Nonstructural responses include more accurate flood forecasting through use of satellites and high-tech equipment, zoning and land-use policies, insurance programs, evacuation planning, and education.

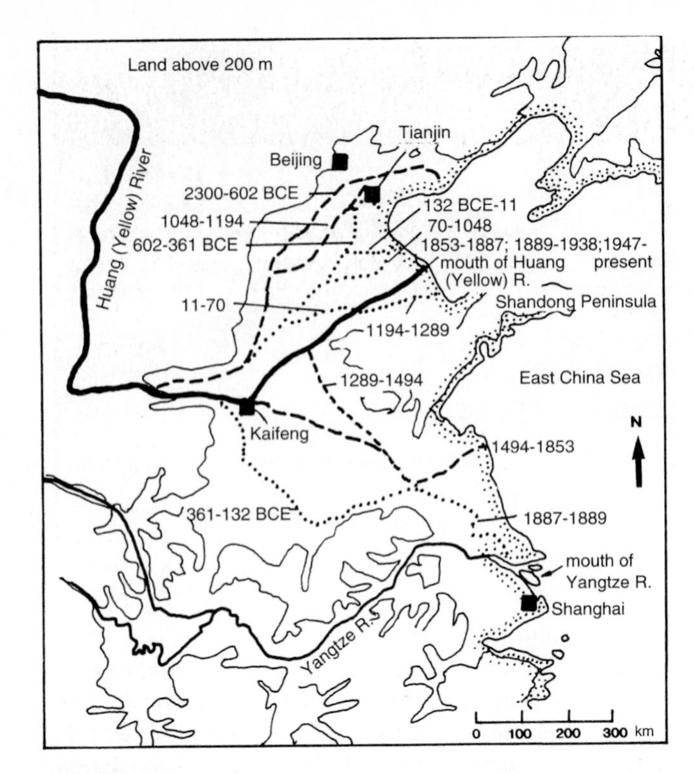

Figure 12.26 Locations of Huang (Yellow) and Yangtze Rivers. Data from Czaya, 1981.

Dams

The construction of dams to create large reservoirs is a common method of providing flood control. The presence of a dam gives a feeling of protection that invites many people to settle in the "protected" floodplain lying downstream. Yet dams do not provide flood control. They offer some flood protection—if their reservoirs are not filled to capacity or if they do not fail.

All dams have life spans limited by the durability of their construction materials and style, and the rate at which stream-delivered sediment fills in their reservoirs (Figure 11.35). Despite all the massive dams and extensive reservoirs, major floods still occur downstream due to overtopping (e.g., Rapid City, South Dakota) and to heavy rains that fall below the dam (e.g., Guadalupe River, Texas, and Big Thompson Canyon, Colorado). Then there are major killing floods unleashed by failed dams. In the United States alone, hundreds of dams have failed (Table 12.4).

The problems with dam integrity are not all solved. In 1981, the U.S. Army Corps of Engineers studied 8,639 dams and judged that 2,884 of them were unsafe.

Table 12.4 Some U.S. Dam Failures

Date	Place	Deaths
16 May 1874	Connecticut River near Williamsburg, Massachusetts	143
31 May 1889	Little Conemaugh River, Johnstown, Pennsylvania	over 2,100
27 Jan 1916	Otay River, San Diego	22
12 Mar 1928	St. Francis dam, north of Los Angeles	420
14 Dec 1963	Baldwin Hills, Los Angeles	5
26 Feb 1972	Buffalo Creek, West Virginia	118
5 Jun 1976	Teton River, Idaho	14

Levees

Some people suggest that the costs of building more levees and dams may be higher than the value of the buildings they are protecting. Plus the presence of structures advertised to control floods creates a sense of security that stimulates more development of the floodplain; development that inevitably will be flooded. They recommend lowering or removing levees along some farmland to increase wetland habitat and restore floodplain. Allowing floods to spread out and dissipate their energies over a wide expanse of land would also lower flood heights in the levee-protected major cities and towns.

On the other side, proponents of levees say that the billions of dollars in flood damages would have been many billions more without the levees. The debate on the value of the levees will go on for generations. How many miles of levees? How high? Where should they be located? Should taxpayers provide the funds to replace flood losses for people who build in the floodplain? Some towns have voted to move in their entirety to sites above the floodplain, and the distribution of some federal funds has been restricted to people who will either move to higher elevations or raise the floor level of their homes.

A study by Criss and Shock shows that peak floodwater heights in the last 150 years have increased 2 to 4 m (6 to 13 ft) for the same water volume in upper Mississippi River sections with levees and engineered channels. Meanwhile, there has been no increase in flood heights per water volume on the mostly nonengineered upper Missouri River. Where human engineering has deepened and narrowed rivers, there commonly are higher flood crests than in nonengineered sections. For example, the Mississippi River flood in St. Louis in 1903 was almost identical to the 1993 water volume, but in 1903 the river crested at 11.6 m (36 ft) height while in 1993 it reached 15.1 m (50 ft).

Sandbagging

When a big flood is on the way, a common response is to quickly build temporary levees using hastily filled bags of sand and mud weighing about 35 pounds apiece. During the 1993 flood, an estimated 26.5 million sandbags were filled and set in place. It has been suggested that the physical protection provided by sandbagging is no more important than the therapeutic value derived from citizens actively involved in helping lessen the disaster.

Forecasting

Thanks to technologic advances, our growing knowledge of weather and floods allows better forecasts of the time and height of regional flood waters. These forecasts have significantly reduced the loss of life. But it is interesting to note the twin trends of better forecasting and engineering offset by ever-greater dollar losses during big floods. We know more, yet suffer greater damages.

Zoning and Land Use

A standard approach to lessening flood losses is to ban building on the floodplain covered by the 100-year flood. This policy was adopted by the National Flood Insurance Program in the early 1970s, was issued as Executive Order 11988 in 1977, and has been used by the Federal Emergency Management Agency (FEMA) since 1982. Notice in Table 12.2 that the 100-year flood has a 1 percent chance of occurring in any year, a 9.6 percent chance of happening each 10 years, and a 22 percent chance every 25 years. Although adoption of the 100-year flood standard does not prevent structures from being flooded, it does discourage some construction from occurring in foolish places.

Insurance

Flood insurance has been available to farmers and townspeople through the National Flood Insurance Program since the 1950s, but it has not been popular. For example, when the Red River of the North flooded Grand Forks, North Dakota, in 1997, only 946 out of more than 10,000 households were covered by flood insurance. Four months before the flood, FEMA spent $300,000 on a media campaign warning citizens of the ominous snow-melting conditions in 1997. The FEMA ad campaign motivated only 73 Grand Forks homeowners to buy flood insurance.

Why is flood insurance purchased by such small percentages of people? This may be due to the knowledge that politicians love the opportunity to provide federal dollars to help disaster victims. For example, the U.S. Congress passed a bill providing $6.3 billion to aid people hit by the 1993 flood in the upper Midwest.

Presidential Disaster Declarations

Under the Disaster Relief and Emergency Assistance Act, federal disaster relief is provided to states and communities if they receive a Presidential Disaster Declaration (PDD). Declarations are made at a president's discretion but a criterion for issuing a PDD is "a finding that the disaster is of such severity and magnitude that effective response is beyond the capabilities of the state and the affected local governments." The average annual number of flood-related PDDs of seven presidents has varied considerably (Table 12.5). What influences the declaration of a flood disaster? Is it: 1) severity of flood damages, 2) the increase in population and wealth over time, and/or 3) political concerns? After controlling for flood damages, there remain statistically significant differences between presidents in the number of disasters declared.

Table 12.5 Flood-Related Presidential Disaster Declarations

President	Fiscal Years	Disaster Declarations (Average per year)	Flood Damages (Average per year in millions of 1995 dollars)
Johnson	1965–68	11.8	1,681
Nixon	1969–74	27.2	4,469
Ford	1975–76	26.0	5,370
Carter	1977–80	20.0	3,478
Reagan	1981–88	14.5	3,440
Bush	1989–92	22.3	1,469
Clinton	1993–97	32.3	7,553

Source: Downton, M. W. and Pielke, Jr., R. A. (2001) *Natural Hazards Observer,* v. xxv, no. 4.

Urbanization and Floods

As more people move near rivers and streams, they encounter unexpected problems. Yet, some of the problems are due to human activities that increase both flood heights and frequencies.

Hydrographs

A **hydrograph** plots the volume of water or stream-surface height versus time, recording the passage of water volumes flowing downstream (Figure 12.27). There is a lag time for rainwater to flow over the ground surface and reach a stream channel, but stream surface height usually rises quickly once surface runoff reaches a channel, i.e., the rising limb of the hydrograph is steep. When a flood crest passes downstream, stream level does not fall as rapidly as it rose. This is due to the stream being fed water by underground flow of rain that soaked into the ground and moved slowly to the stream, i.e., the falling limb of the hydrograph has a gentle slope.

The flood hydrograph in Figure 12.27 is typical of rural, unurbanized areas. But what happens in an urban setting? Humans cover much of the ground with houses and other buildings, pave the ground for streets and parking lots, and build storm-sewer systems to take rainwater runoff directly to streams. Covering the ground with an impervious seal

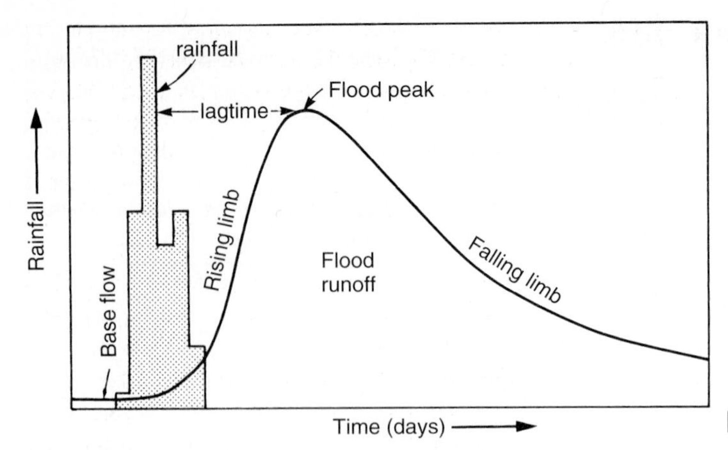

Figure 12.27 A hydrograph charts flood runoff. Commonly, stream flow rapidly increases from surface runoff, as shown by a steep rising limb reaching a peak flow. From the peak, discharge decreases slowly as infiltrated rain flows underground and feeds the stream.

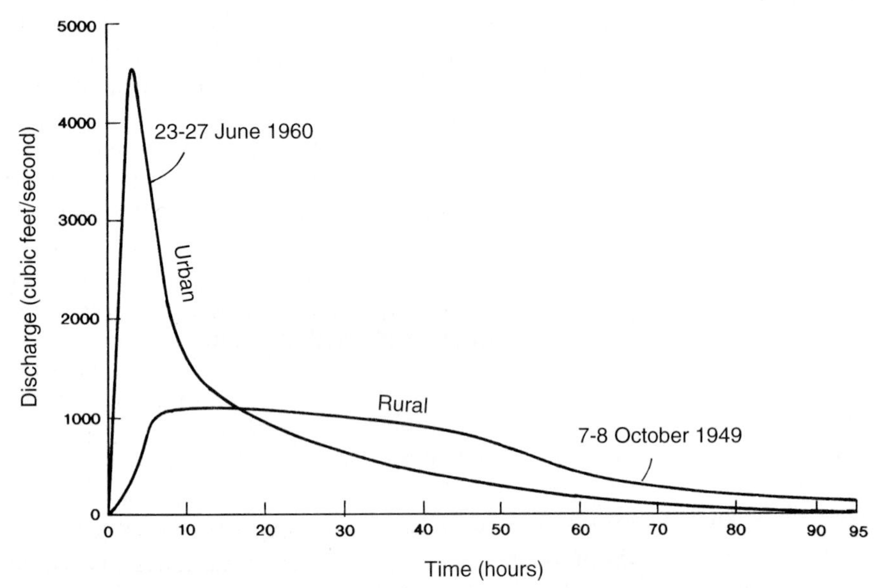

Figure 12.28 Flood hydrographs for similar-sized runoffs in Brays Bayou in Houston, Texas, before and after urbanization. In the rural setting, much of the rainfall soaked into the ground and moved slowly to the bayou. After urbanization, the roofs and pavement covering the land prevent infiltration, and most rain quickly runs over the surface to the bayou, creating much higher floods.

Source: K. Young, *Geology: The Paradox of Earth and Man.* Boston: Houghton Mifflin, 1975.

prevents rainwater from soaking into the ground and causes it to flow rapidly across the surface, thus reaching the stream ever-more quickly.

Figure 12.28 shows flood hydrographs from similar-size rainstorms in Brays Bayou in Houston, Texas, both before and after urbanization in the drainage basin. The rainstorm of October 1949 mostly soaked below the surface and flowed slowly underground to feed the stream running through the bayou. Following urbanization, the rainstorm of June 1960 produced a flood hydrograph with a very different shape. This is a proverbial good news–bad news situation for city dwellers. The good news is that the urban flood lasted only 20 percent as long; the bad news is that it was four times higher (Figure 12.28). The roofing and paving that accompany growing urbanization cause many areas to receive more frequent floods.

Flood Frequencies

Another way of looking at flood runoff within urban areas is to see how urbanization affects the frequency of floods. Roofing and paving the ground increase the surface runoff of rainwater, thus causing higher stream levels in shorter times, i.e., runoffs become flash floods. Figure 12.29 shows the effects of building storm sewers (percent of area sewered) and of roofing and paving (percent impervious). For example, notice on the vertical axis that a discharge of 100 ft^3/sec occurred about every four years in the rural setting but now happens about three times per year after urbanization.

Channelization

Humans try to control flood waters by making channels: 1) clear of debris, 2) deeper, 3) wider, and 4) straighter. A typical channelization project involves clearing the channel of trees, debris, and large boulders to reduce channel roughness and then increasing the channel capacity by digging it deeper, wider, and straighter by creating shortcuts across meander bends. All these activities make it easier for water to flow through the channel. There are over 60,000 km (37,000 mi) of channelized streams in the United States.

Think about channelization in terms of graded-stream theory. Most steps in the process push the system further into the "too much discharge" case. Straightening the channel increases the gradient of the stream bottom, making the water flow faster. Clearing the channel of obstructions reduces friction, also making water flow faster. These actions trigger a stream's negative-feedback mechanism, making it try to slow down by eroding its bottom and banks to pick up load and try to return to equilibrium by decreasing its gradient.

The Extreme Approach: Los Angeles In response to flood damage, Los Angeles not only cleared, straightened, and deepened its river channels, it also lined them with concrete to further reduce friction and speed up flow. As long as flood volumes stay smaller than channel capacity, then there are

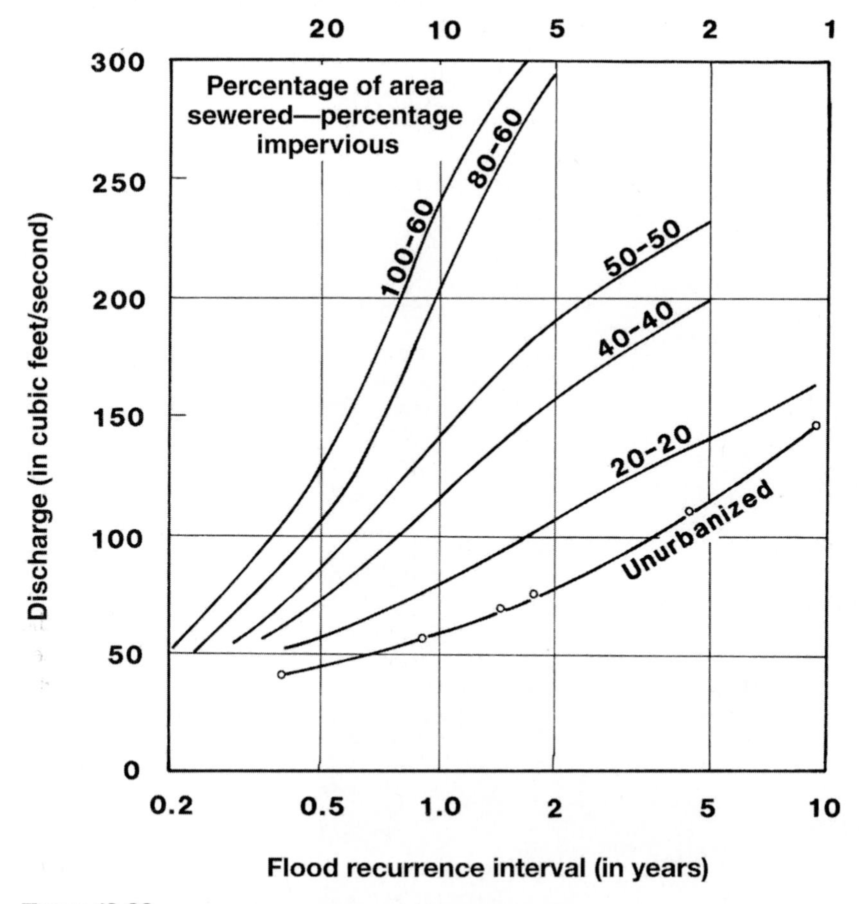

Figure 12.29 Flood-frequency curves for small drainage basins in various stages of urbanization. Floods occur more frequently as storm sewers, roofs, and pavement increase.

Source: L. B. Leopold, "Hydrology for Urban Land Planning" in U.S. *Geological Survey Circular 554,* 1968.

Figure 12.30 The concrete-lined channel of Forester Creek, El Cajon, California.

Photo by Pat Abbott.

no urban floods. But feel sorry for the person who falls into the waters charging through these channels; television news has shown people drowning in the racing water despite the frantic efforts of would-be rescuers.

It should also be noted that concrete-lined channels obliterate the habitat of all riverine plants and animals (Figure 12.30). The "soul" of a community is less well served by a concrete ditch than a tree-lined stream.

The Binational Approach: Tijuana and San Diego Rivers commonly are used as boundaries between nations. But the boundary between westernmost Mexico and the United States is drawn as a straight line, cutting through the 6,039 km^2 (2,325 mi^2) drainage basin of the Tijuana River: 80 percent in Baja California, 20 percent in California. The river's lower reaches pass through urban Tijuana, Mexico (population over 1.5 million). Water crosses the border, enters the San Diego region (population 3 million), and runs its last few miles to the Pacific Ocean.

The two countries agreed on a Los Angeles-style project to cement the river channel to the sea. Mexico carried out its side of the agreement, but then the environmental ethic arose with power in California and the cement-lined channel project was blocked in the United States. The result is shown in Figure 12.31, where the large concrete channel in the city of Tijuana sends high-velocity floods charging into the unprepared farms and subdivisions of southernmost San Diego.

The Uncoordinated Approach: San Diego The channelization style of the lower San Diego River has changed with the political winds. In the late 1940s, the U.S. Army Corps of Engineers was left alone to design a flood-control channel for the river mouth. Its channel is 245 m (800 ft) wide, has a natural bottom over which ocean tides roll in and out, and has walls of large boulders (**riprap**). The channel will handle a flow of 115,000 ft^3/sec, which is estimated to recur about every 400 years (Figure 12.32).

San Diego's population swelled after World War II, and the lower river valley, called Mission Valley, started to boom. Pushy hotel builders leaned on a weak city government in the late 1950s to gain permission to build on the floodplain next to the edge of the natural channel. The channel is about 7.5 m (25 ft)

Figure 12.31 View southeast from San Diego, California, into Tijuana, Mexico. Flood waters race through the cement-lined channel in Tijuana (top center of photo), blasting into the farms and subdivisions in southernmost San Diego, January 1978.
Photo courtesy of *San Diego Union*.

wide, is naturally vegetated, and has a capacity of about 8,000 ft^3/sec (Figure 12.33).

Mission Valley kept growing to become almost a second downtown. Ever-bigger developers were attracted to the area. In the 1980s, the developers, interested in maximizing buildable acreage, were pitted against citizen groups that envisioned a vegetated, meandering flood-control channel that would serve as a community park. The resulting compromise favored the developers, who built a 110 m (360 ft)-wide, concrete-lined channel able to hold 49,000 ft^3/sec.

The century's biggest flood on the San Diego River discharged 72,000 ft^3/sec in January 1916. Awaiting the next major flood is a concrete-lined channel of 49,000 ft^3/sec capacity, which empties into a natural channel that will hold 8,000 ft^3/sec, which in turn feeds into a 115,000-ft^3/sec-capacity channel (Figure 12.32). The planning process does have its flaws.

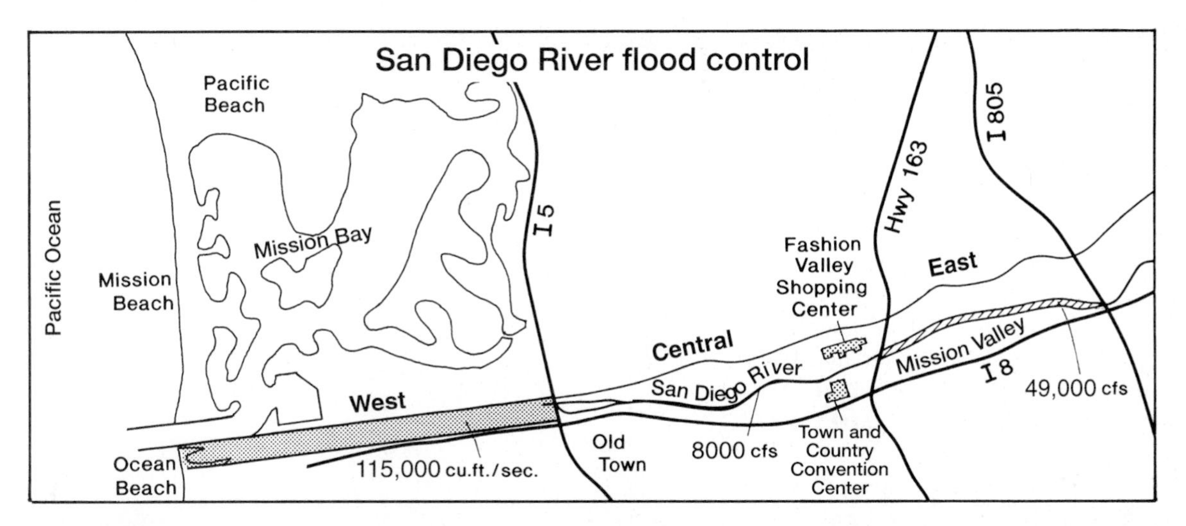

Figure 12.32 Flood channelization styles in the lower San Diego River have varied through time: in the late 1940s, 800 ft wide; in the late 1950s, 25 ft wide; and in the 1980s, 360 ft wide. All channel segments feed directly into one another. See photo in Figure 12.33.

Figure 12.33 Looking west down the lower San Diego River during a 12,000 ft^3/sec flood in January 1979. Note the 800 ft-wide channel segment in the west (top of photo) and the central segment width of 25 ft, indicated by arrows above airplane strut. Town & Country Hotel is in lower left, and Fashion Valley shopping center (number one in sales in San Diego) is in lower right. See map in Figure 12.32. Photo courtesy of *San Diego Union.*

The Hit-and-Miss Approach: Tucson In the northern reaches of the great Sonoran Desert lies Tucson, Arizona, where annual rainfall averages only 11.14 inches. The September of 1983 was the second rainiest in Tucson history; it was capped off by the arrival of Tropical Storm Octave, which dropped 17.1 cm (6.71 in) of rain.

Rillito Creek cuts across Tucson, collecting much of the urban runoff and feeding the Santa Cruz River. On 1 October 1983, the Santa Cruz River at Congress Street was discharging 1,490 m³/sec (52,700 ft³/sec) of flood water, causing major damage in the urban area. When the waters abated, 13 people were dead. Based on the entire flood history of Tucson, the Federal Emergency Management Agency had estimated the 100-year flood to be 30,000 ft³/sec. But the 1983 flood was 1.76 times bigger. Was this size flood really that rare an event? Looking at the twentieth-century flood record in Tucson shows that six of the seven largest floods occurred between 1960 and 1983. This coincides with growth in population from 265,700 in 1960 to 603,300 in 1984. All the paving and roofing that comes with urbanization may be increasing flood sizes. Computations of 100-year flood size based only on the post-1960 flows yields an estimate of 2,700 m³/sec (96,000 ft³/sec); this tripling of expectations accounts for the effects of urbanization on runoff.

Desert floods are different. Most damage is due to bank erosion, not inundation. Stream channels are cut into loose, sandy sediments, forming weak banks that easily crumble. The critical floods here are not those with an urban-style, high-water peak of short duration (Figure 12.28) but longer-duration floods (flat-topped flood hydrograph) that have time to soak the dry stream bed and banks, thus freeing later flood waters to concentrate their energies on erosion. Flood erosion changes channel location both by cutting new channel segments via overbank flow and by meander migration where the banks erode rapidly. Since the 1940s, some stream banks have eroded laterally more than 300 m (1,000 ft). One big storm and erosive flood can change stream-channel positions dramatically. This means the defined 100-year floodplain moves with it. The usual static definition of a 100-year floodplain does not mean much when the stream and floodplain migrate widely.

The 1983 flood showed the problems that come with building protective walls in hit-and-miss fashion. When the stream reaches the end of a protective wall, erosion is concentrated (Figure 12.34). When the stream is confined between walls, it increases its velocity; the added power is used in bank erosion when the protective walls are left behind (Figure 12.35). A stream is a system with delicately triggered feedback mechanisms; it must be treated as a whole or not at all. Hit-and-miss bank protection does not work.

Floods are not inviting to tourists. Evaluate the truth in the message sent after the flood by the Metropolitan Tucson Convention and Visitors Bureau to the national media:

Figure 12.34 View west of destruction of Riverview Condominiums by bank erosion along Rillito Creek in Tucson, Arizona, on 1 October 1983. Note concentrated erosion at end of "protective" retaining wall.

Photo by Peter L. Kresan, courtesy of Geophoto Publishing Company.

Figure 12.35 Bank erosion in Pantano Wash in Tucson, Arizona, removed 30 mobile homes and their spaces, 1 October 1983. View to northwest.

Photo by Tad Nichols, courtesy of Geophoto Publishing Company.

"The 100-year flood has come and gone, so, by all rights, Tucsonans should enjoy another century of great southwestern weather."

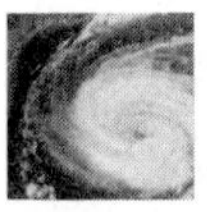

The Biggest Floods

Ancient Tales of Deluge

In almost every part of the world, tales are told of ancient deluges far greater than any seen in modern times. In India, it is said that Vishnu, the god of protection, used one of his ten lives to save Mother Earth from a great flood. In China, they celebrate Yu the Great, who helped protect the people from the overwhelming floods of the Yellow River. American Indian origin mythologies begin with an Earth completely flooded. In Babylonia, clay tablets record the Gilgamesh Epic that tells of the great flood where Utnapishtim built an ark that sailed for seven days and seven nights and saved his kin and cattle. The Genesis book of the Bible tells of Noah building an ark to save his family and pairs of all the animals from 40 days and 40 nights of rain that covered the world with water for 150 days. These sagas tell of events that occurred around 6,000 B.C.E. to 1,000 B.C.E. Were floods larger in those times? Or are these tales of rare events with very long recurrence intervals—the "1,000-year" inundation floods?

The flood of the Gilgamesh Epic, and possibly the Bible, may have occurred within the Tigris and Euphrates Rivers (in modern-day Iraq), which flow across an extensive, low-lying plain to enter the Persian Gulf (Figure 12.36). The setting is similar to that of the Mississippi, Huang (Yellow), and other long rivers with their inundation floods. Long-duration rains pouring onto the mountains of Iran and Turkey shed runoff, creating massive flood crests that inundate the floodplains of the lower Tigris and Euphrates Rivers. It is on these fertile floodplains where people congregate to reap the agricultural rewards. A long-lasting flood would submerge the "whole world" of their existence.

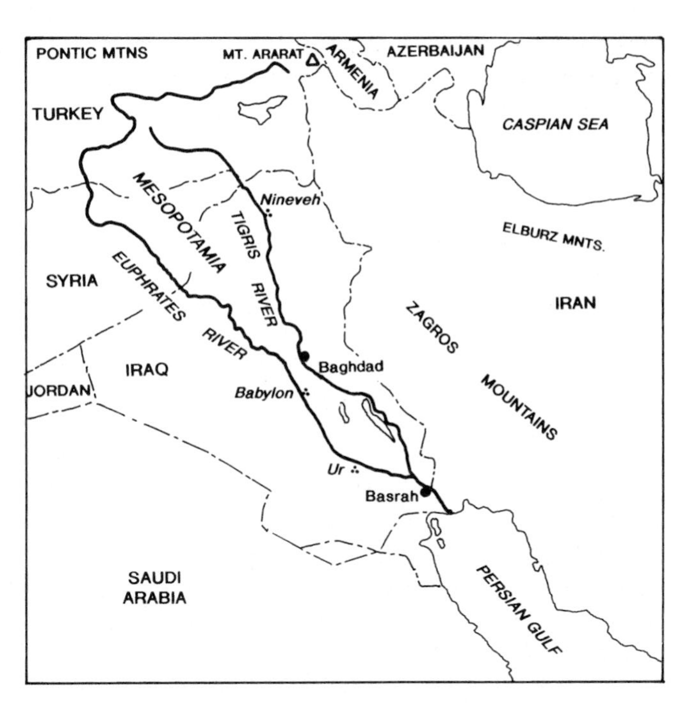

Figure 12.36 Map of the Euphrates and Tigris river plains, the site of the Babylonian (Gilgamesh Epic) and possibly the Hebrew (Noah's Ark) flood tales. The lower floodplain receives long-lasting inundation floods from extended heavy rains in the mountains of Turkey and Iran.

Black Sea Deluge Another overwhelming flood occurred when the Black Sea filled to its present level. During the last glacial advance, when global sea level was lowered, the Black Sea (Figure 12.37) was a freshwater lake much smaller than today's sea. The freshwater surface was about 150 m (500 ft) lower than today's sea surface; it was isolated from the higher elevation saltwater of the Mediterranean Sea by the land flooring the Bosporus Strait which stood high and dry. The lands around the freshwater-filled Black Sea were plowed, farmed, and irrigated by some of the earliest human farmers.

As the global glaciers began their rapid melting and retreat, world sea level began rising rapidly. When Mediterranean seawater flooded the Bosporus Strait, it reached a 140 m (460 ft) drop down to the Black Sea about 7,600 years ago. Imagine the waterfall produced by Mediterranean seawater falling into the Black Sea basin. The flowing water deepened the Bosporus Strait and roared forth as a waterfall equal to 200 Niagara Falls as more than 50 km^3 (12 mi^3) of Mediterranean seawater poured into the Black Sea each day. Visualize the impact on the lives of the farming communities around the Black Sea as their freshwater lake turned salty and the water level rose rapidly, maybe 6 in (15 cm) per day, flooding their lands. Rising water covered over 100,000 km^2 (about 40,000 mi^2) of land. This event caused a mass migration of displaced farmers forced to emigrate to Central Europe, Egypt, and Mesopotamia carrying tales of a great flood whose waters rose day after day for months and inundated their world. How has history passed the news of this event on to us? It has been suggested that this is the flood described in the Gilgamesh Epic and as Noah's flood in Genesis.

Ice-Dam Failure Floods

The biggest floods known on Earth occurred during the melting of the continental ice sheets. Glacial meltwater tends to pond in front of glaciers due to downwarped land (isostatic adjustment) and dams of stagnant ice and glacial débris. Thousands of lakes formed along the glacial front in different locations at different times. The sudden failure of ice dams still occurs today and has been observed and photographed, for example, at ice-dammed Strandline Lake in Alaska.

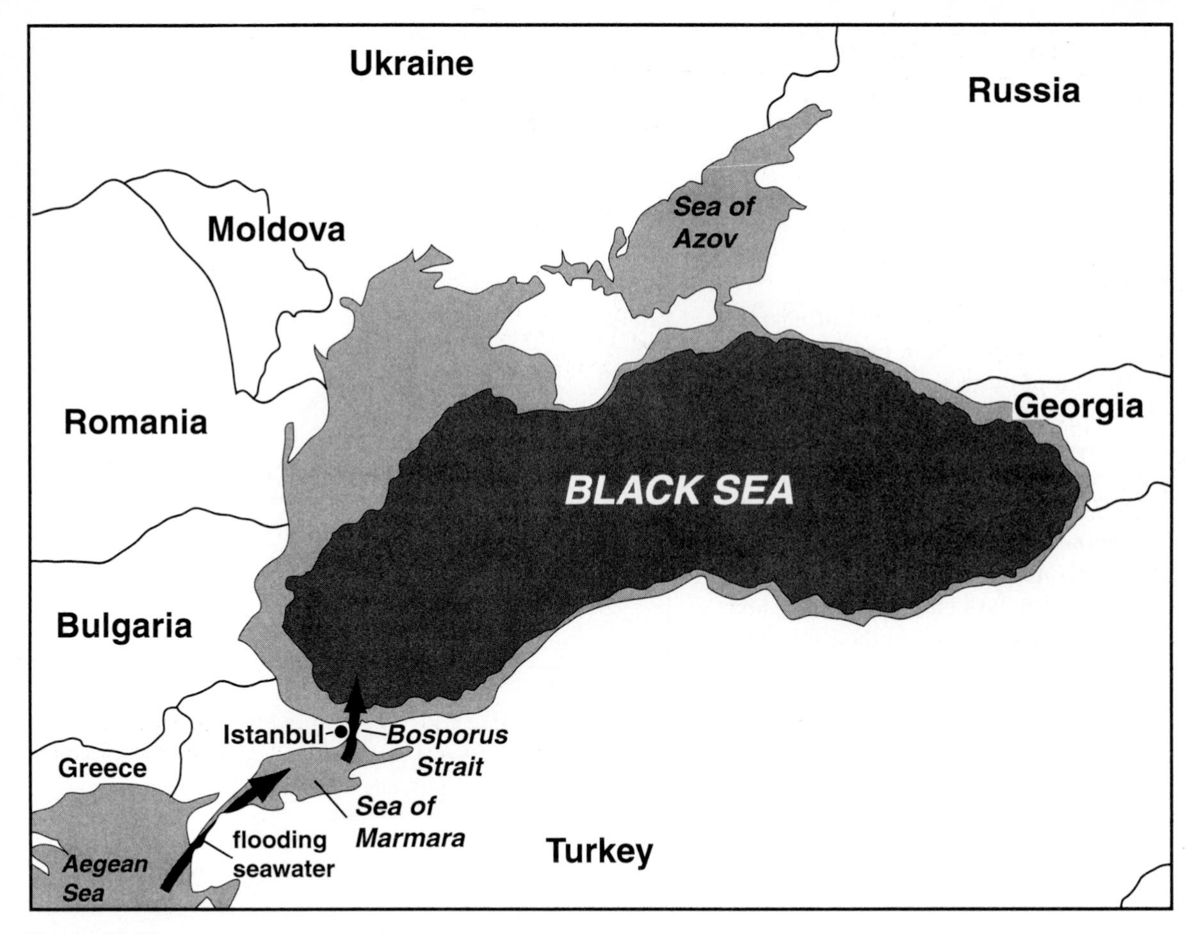

Figure 12.37 The Black Sea was a freshwater lake (dark gray) during the last glacial advance. When glaciers began melting, sea level rose (Figure 9.21). About 7,600 years ago, sea level was high enough to flow through the Bosporus Strait, causing the Black Sea to grow rapidly in size (medium gray). Untold thousands of people were forced to migrate to new areas.

When ice dams in front of the largest glacial lakes failed, stupendous floods resulted whose passage is still recorded in lake sediments; by countryside stripped of all soil and sediment cover; by high-elevation flood gravels; by an integrated system of braided channels (a mega-braided stream); by abandoned waterfalls; by high-level erosion; and by large-scale sediment deposits. The most famous of the ice-dam failure floods are preserved in the "channeled scablands" topography in southeastern Washington. Here, the underlying Columbia River flood basalts were swept clean of overlying sediments, while river valleys were cut into the plateau, creating a maze of gigantic channels between scoured flatlands.

Some of the biggest floods burst forth from the widespread glacial Lake Missoula, which was impounded by a glacial lobe lying across Clark Fork and acting as a dam (Figure 12.38). Ancient wave-cut shorelines testify to a 610 m (2,000 ft)-deep lake. When the ice dam failed, a water volume of 2,500 km^3 (600 mi^3) was released in a flood that lasted up to 11 days. For comparison, the flood volume was more than five times greater than the volume of Lake Erie. Discharge is estimated to have been greater than 13.7 million m^3/sec (484 million ft^3/sec). Floods moved at velocities greater than 30 m/sec (67 mph).

The immense floods flowed southwestward across the Columbia River plateau, eroding channels and stripping the overlying sediments off an area of 7,250 km^2 (2,800 mi^2) to create the channeled scablands topography. The scablands landscape is marked by scoured bedrock and immense former waterfalls, such as Dry Falls in Grand Coulee, Washington (Figure 12.39). The massive floods carried boulders over 10 m (33 ft) diameter for miles. Flood waters moved gravelly sediments in giant ripples up to 15 m (50 ft) high with distances between their crests of 150 m (500 ft) (Figure 12.40).

The failure of an ice-dammed lake can send so great a volume of water running over and eroding the land that it can reorganize the paths of rivers. In North America, the meltwater floods varied in their paths among the following: the Mississippi River to the Gulf of Mexico, the St. Lawrence and

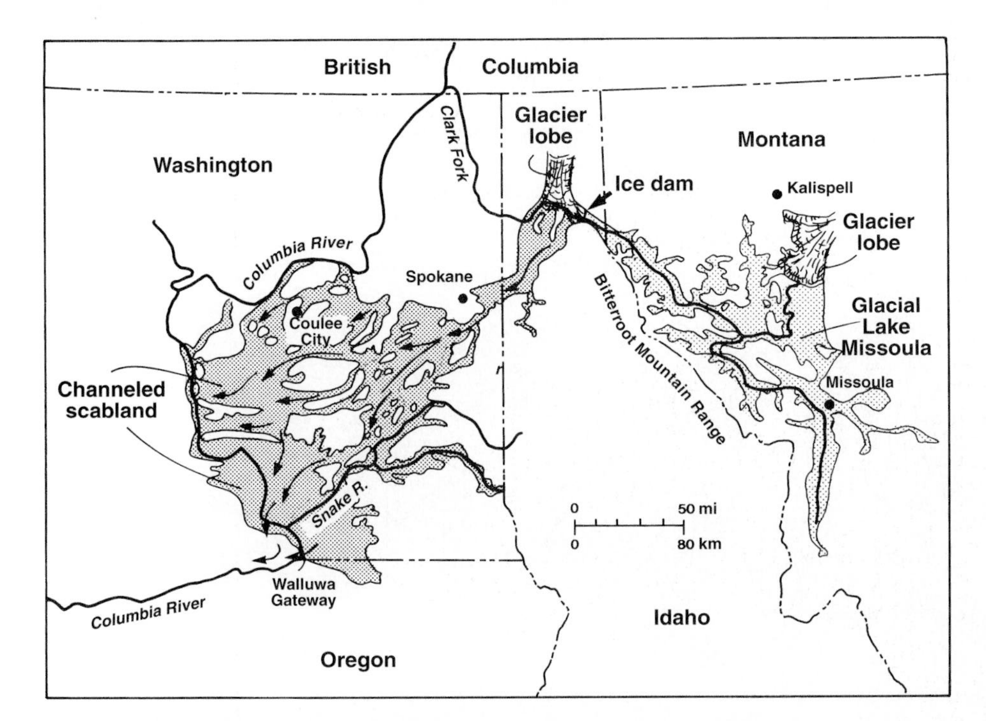

Figure 12.38 The "channeled scabland" is a landscape intensely scoured by stupendous floods that ran southwesterly across Washington. Flood waters were unleashed when ice dams across Clark Fork in Idaho failed catastrophically and glacial Lake Missoula drained rapidly.

Figure 12.39 The Dry Falls in Washington are a 125 m (400 ft)-high cataract that operated when glacial Lake Missoula was rapidly emptying. Note the scoured surface above the falls where flood waters eroded all sedimentary cover to expose the Columbia River basalts. See the giant plunge pools where waterfalls eroded big holes. In the background lies Coulee City.

Photo by John S. Shelton.

Figure 12.40 Flood waters crossing the channeled scablands moved gravelly sediment in massive "ripples" that traveled toward top of photo. The megaripples are about 15 m (50 ft) high with wavelengths of 150 m (500 ft). View is down the modern Columbia River near Trinidad, Washington.

Photo by John S. Shelton.

Hudson Rivers to the North Atlantic Ocean, the Mackenzie River to the Arctic Ocean or via the Hudson Strait to the Labrador Sea.

The discharge of a huge volume of cold, low salinity glacial meltwater could change the global circulation of deep water through the world ocean (Figure 9.3). The deep-ocean circulation is driven by regional differences in heat and salinity of ocean water and this could be changed by a huge influx of cold fresh meltwater. A change in ocean circulation would in turn make changes in global climate. At 12,900 years ago, climate cooled about 5ºC (9ºF) in the event known as the Younger Dryas (Figure 9.24). This dramatic plunge back into colder temperatures is thought to have been caused by a gigantic meltwater flood through the St. Lawrence River into the North Atlantic Ocean.

The outbursts from glacial meltwater lakes created the largest known floods in Earth history. In the span of a few thousand years, the massive continental ice sheets melted in gigantic quantities, raising sea level by some 130 m (425 ft) (Figure 9.25).

Summary

Floods usually are produced by heavy rains. An intense thundercloud may unleash a fast-moving but short-lasting flash flood in a local stream. Widespread rainfall over weeks may produce inundation floods on many streams that last for weeks.

Streams are equilibrium systems. Streams convey a volume of water (discharge) and carry sediment (load) toward the ocean. To balance relative differences of discharge versus load, streams vary the slope of their bottom (gradient) and channel pattern (e.g., meandering or braided).

If a stream has excess discharge, and thus energy, it erodes its bottom and banks, producing a meandering course that decreases its gradient and slows the water flow. If a stream valley has excess sediment available, it deposits load on its bottom and straightens its valley, thus increasing its gradient and speeding the water flow. Both mechanisms are examples of negative feedback, where a change in a system provokes other changes that tend to cancel the first change and restore equilibrium.

Floodplains are flattish areas used as stream floors during floods. Small floods happen frequently; large floods happen uncommonly. Flood-frequency curves are produced by plotting flood volumes versus recurrence interval (average number of years between floods of same size). A common standard for design near streams is protection from the 100-year flood. But 100-year floods are only statistical approximations; two 100-year floods can happen on the same stream in the same year.

Streams routinely adopt new courses in a process called avulsion. Humans build levees to try and hold streams in place. For example, the Mississippi River would have changed its course in the 1930s if the U.S. Army Corps of Engineers had not dredged the channel bottom and built bigger levees.

In a rural setting, a heavy rain supplies water to a stream quickly by overland runoff and slowly by underground seepage and flow. After urbanization occurs, pavement and roofs seal off most of the ground, increasing the amount of surface runoff. The result is more high-water levels (peak floods) than a region's preurbanization history predicts.

Channels are constructed to try to control floods. Channelization involves debris clearing, deep digging, widening, and straightening. All these actions cause faster and more powerful water flow.

The biggest floods known occurred thousands of years ago during retreat of continental glaciers. Dams of ice occasionally failed catastrophically, releasing gigantic volumes of lake water that overwhelmed landscapes and probably changed deep ocean circulation that in turn triggered climate changes.

Terms to Remember

avulsion 350
base level 336
braided stream 338
discharge 336
equilibrium 336
feedback 340
floodplain 339
graded stream 338
gradient 336
hydrograph 353
levee 349
load 336
meander 337
negative feedback 340
positive feedback 340
recurrence interval 340
riprap 355
sinuosity 336

Questions for Review

1. Explain what is meant by negative feedback. Is negative bad?
2. Explain what is meant by positive feedback. Is positive good?
3. Draw a map showing the channel pattern of a stream having excess discharge. Explain the feedback processes involved.
4. Geologically speaking, lakes are temporary. Draw a cross section of a lake and use stream feedback processes to explain how lakes are removed.
5. Draw a cross section and explain floodplains. Label where it is safe to build.
6. Draw a flood-frequency curve and explain its use in planning.
7. Draw a cross section through a levee and explain the processes pushing it toward failure.
8. Describe the history of levee control of the Huang River in China. Has it been a success?
9. Draw a hydrograph record for a two-week-long flood in a rural setting. What controls the shape of the hydrograph, i.e., what is happening on the rising and falling limbs? Now draw the hydrograph resulting from the same-size rainstorm after the land has been urbanized.
10. What activities are typically involved when humans modify stream channels to provide flood "control"? Use equilibrium stream processes (graded stream theory) to explain the changes in stream flow characteristics after channelization.
11. Explain the steps leading to the largest floods of the last 20,000 years.

Questions for Further Thought

1. Make a list of appropriate uses for a floodplain. Make a list of inappropriate uses.
2. Should the federal government (i.e., taxpayers) provide disaster-relief funds to home and business owners to rebuild their flood-ravaged properties on the floodplain?
3. How often does a 100-year flood occur on a stream?
4. For the stream nearest you, obtain the annual records of largest floods throughout its recorded history and draw a flood-frequency curve. How large a flood is likely to occur every 10, 50, and 100 years?
5. Draw a cross section and explain why avulsion occurs.
6. Can the United States control the course of the Mississippi River indefinitely?
7. Why are some urban areas receiving more frequent floods and higher peak flows than at any time in their history?

Suggested Readings and References

Abbott, Patrick L. (1991). Flood control in the lower reaches of the San Diego River. In *Earthquakes and Other Perils* (189–94). San Diego: San Diego Association of Geologists.

Baker, Victor R. (1973). *Paleohydrology and sedimentology of Lake Missoula flooding in eastern Washington.* Geological Society of America Special Paper 144.

Colman, Steven M. (2002). A fresh look at glacial floods. *Science,* 296, 1251–52.

Costa, John E., and Baker, Victor R. (1981). *Surficial Geology.* New York: John Wiley & Sons.

Criss, Robert E., and Shock, Everett. (2001). Flood enhancement through flood control. *Geology,* 29, 875–78.

Iseri, Kathleen T., and Langbein, W. B. (1974). *Large rivers of the United States.* U.S. Geological Survey Circular 686.

Kresan, Peter L. (1988). The Tucson, Arizona flood of October 1983. In *Flood Geomorphology* (465–89). New York: John Wiley & Sons.

Leopold, Luna B. (1968). *Hydrology for urban land planning.* U.S. Geological Survey Circular 554.

Leopold, Luna B. (1974). *Water, A Primer.* New York: W. H. Freeman.

McPhee, John. (1989) *The Control of Nature.* New York: Farrar, Straus and Giroux.

National Oceanographic and Atmospheric Administration. (1994). *The Great Flood of 1993.* Silver Spring, MD.

Perry, Charles A. (2000). *Significant floods in the United States during the 20th century.* U.S. Geological Survey Fact Sheet 024–00.

Rahn, Perry H. (1984). Flood-plain management program in Rapid City, South Dakota. *Geological Society of America Bulletin,* 95, 838–43.

Ryan, W. B. F., et al. (1997). An abrupt drowning of the Black Sea shelf. *Marine Geology,* 138, 119–26.

Ryan, W. B. F., and Pitman, W. (1998). *Noah's Flood.* New York: Simon and Schuster.

U.S. Geological Survey. (1975). *The 1973 Mississippi River basin flood.* U.S. Geological Survey Professional Paper 937.

U.S. Geological Survey. (1979). *Storm and flood of July 31–August 1, 1976, in the Big Thompson and Cache la Poudre river basins, Larimer and Weld Counties, Colorado.* U.S. Geological Survey Professional Paper 1115.

Videos

Flood! (1996). NOVA/WGBH (60 min.).

The Great Floods. (1994). Northwest Interpretive Association (14 min.).

After the Flood. (1994). The Learning Channel (50 min.).

The Awesome Power. (1989). National Oceanographic and Atmospheric Administration (15 min.).

The Earth Revealed—Rivers. (1992). Annenberg/CPB (30 min.).

The Earth Revealed—Running Water. (1992). Annenberg/CPB (30 min.).

Erosion: Leveling the Land. (1964). Encyclopedia Britannica (14 min.).

Seeking Noah's Flood. (1999). BBC/The Learning Channel (45 min.).

Chapter 13

Fire

> The moving finger writes; and having writ,
> Moves on; nor all your piety nor wit
> Shall lure it back to cancel half a line,
> Nor all your tears wash out a word of it.
>
> —Omar Khayyam, c. 1100, *Rubaiyat*

Fire is so familiar; it is both friend and foe, slave and master. Fire is a natural force, yet it was used by humanity's ancestors many hundreds of thousands of years ago. Later, humans developed the ability to artificially generate fire. The control and use of fire has been a major factor in the development of the human race. The control of fire for warmth allowed humans to migrate into cold climates and build diverse and successful civilizations. Fire for cooking greatly increased the number of foods and improved their taste, ease of digestion, nutrition, sanitation, and preservation. Fire has long been used to drive game out of hiding during hunting and to scare away predatory animals during the night. Fire aided agriculture by clearing the land of brush and creating fertilizer with its ashy residue. With the help of fire, humans have been able to expand farmland and pasture against both climatic and vegetational gradients.

The possibilities of fire have stimulated creative thinking, which in turn has spurred human development. One invention has followed another. The use of fire to harden materials led to pottery, cookware, weapons, and more. The ability to produce ever-higher temperatures led to the smelting and use of metals. The use of fire provided the benefits of sterilization, which advanced public health. Fire controlled inside machinery supplies the energy that underlies our civilization. The heat from the burning of oil, coal, and natural gas is converted into the electrical and mechanical energy that powers our industries, the lighting and heating of our homes, and our ability to travel quickly to any point in the world. As a rough estimate of the benefits of fire, through it, each American today has available for her or his personal use the energy equivalent of owning 100 slaves in the past.

The destructive use of fire shifted from trying to control wild animals to trying to dominate other humans. Fire passed from the individual torch to a method of destroying whole cities. For example, Troy was so obliterated by fire that its very site remained unknown to the world for nearly 3,000 years.

Another destructive use of fire has been to deny enemies their prize. For example, on 12 September 1812, Napoleon and the French army reached the hills outside of Moscow and looked down on its green, blue, and gold domes. Upon entering the great city, the French found it to be largely deserted with fires burning throughout. For six days, the fires raged until 90 percent of the city was incinerated. The Russians chose to destroy their own heritage rather than let it aid their French conquerors. Denied the support of a conquered populace, Napoleon began his disastrous retreat from Russia during winter's harsh conditions. Napoleon won the military battles but lost the war.

New destructive uses of fire were conceived with time and technological innovation. In the twentieth century, "fire" was

Table 13.1 U.S. Wildland Fire Causes in 2000

		Number of Fires	Acres Burned
Natural Cause			
Lightning		18,417 (15%)	4,826,643
Human Cause		104,410 (85%)	3,595,594
Arson	(26%)		
Equipment	(10%)		
Juveniles	(4%)		
Smoking	(4%)		
Campfires	(3%)		
Railroads	(3%)		
Other & Unknown	(50%)		

Source: National Interagency Fire Center.

Table 13.2 Civilian Fire Deaths and Damages in the United States (1978–1999)

Year	Fire Deaths	Damage to Structures (in billions of dollars)
1999	3,570	$11.482
1998	4,035	11.510
1997	4,050	12.940
1996	4,990	12.544
1995	4,500	11.887
1994	4,275	12.778
1993	4,635	11.331
1992	4,730	13.588
1991	4,465	11.302
1990	5,195	9.495
1989	5,410	9.514
1988	6,215	9.626
1987	5,810	8.504
1986	5,850	8.488
1985	6,185	7.753
1984	5,240	7.602
1983	5,920	6.320
1982	6,020	5.894
1981	6,700	5.625
1980	6,505	5.579
1979	7,780	4.851
1978	8,130	4.008

Source: U.S. Census Bureau. *Statistical Abstracts of the United States—1996 and 2001.*

Table 13.3 Where People Died in U.S. Fires in 1982

Residences	82%
one- and two-family homes (66%)	
apartments (14%)	
hotels or motels (12.5%)	
other (7.5%)	
Nonresidential buildings	4.3%
Highway vehicles	9.6%
Other vehicles	2%
Other	2.1%
	Total 100%

Source: U.S. Census Bureau. (1983). *Statistical Abstract of the United States—1982.*

packaged into bombs that could be dropped from airplanes. During World War II, entire cities were ignited as thousands of tons of bombs created massive **firestorms,** killing tens of thousands of people in such cities as Hamburg and Dresden.

Fire has always been an important part of the natural world. For example, about 1,800 thunderstorms are in action around the Earth each hour, and their lightning bolts start many fires. Humans have brought fire from the natural world into the cultural world, expanding its role. We control fire for our benefit, but we pay the price when this control is lost. In the United States today, the abundant strikes of lightning start about 15 percent of all fires. Humans are the main source of fire ignition in the United States (Table 13.1). More of the price paid for the cultural use of fire is presented in Tables 13.2 and 13.3.

What Is Fire?

Fire is the rapid combination of oxygen with carbon, hydrogen, and other elements of organic material in a reaction that produces flame, **heat,** and light. In effect, fire is the **photosynthesis** reaction in reverse.

In the photosynthesis reaction, plants take in water (H_2O) and carbon dioxide (CO_2) and use solar energy to build organic material, their tissue. Oxygen is given off as a by-product of the reaction. The molecules of plants are tied together by chemical bonds between atoms. These bonds store some of the Sun's heat as **potential energy.** An example of the photosynthesis reaction is:

$$6\ CO_2 + 6\ H_2O + \text{heat from the Sun} \rightarrow C_6H_{12}O_6 + 6\ O_2$$

The organic molecule ($C_6H_{12}O_6$) in this equation is glucose, which approximates that of cellulose, the main component of wood.

In the fire reaction, plant material is heated above its point of **combustion,** and oxygen begins combining rapidly with the organic material. The old chemical bonds between carbon and hydrogen are broken, new bonds form between carbon and oxygen and between hydrogen and oxygen, and the stored energy is given off as heat during the fire. The fire equation is identical to the photosynthesis equation, except that it runs in the opposite direction:

$$C_6H_{12}O_6 + 6\ O_2 \rightarrow 6\ CO_2 + 6\ H_2O + \text{released heat}$$

In effect, the solar energy stored by plants during their growth is returned to the atmosphere during fire.

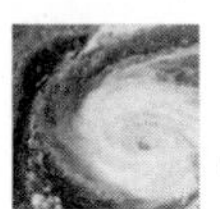

The Need for Fire

Through photosynthesis, plants grow and collectively build large volumes of trunks, branches, leaves, needles, grasses and such. The mass of organic material is recycled by the combined effects of slow decomposition through rotting and digestion plus the rapid burning through wildfire. Decomposition requires warmth and moisture to operate efficiently. In a tropical rain forest with abundant warmth and moisture, rotting can decompose the dead vegetation and recycle the nutrients for the production of new plant material via photosynthesis; there is no need for wildfire to recycle materials (except in cases such as after a hurricane destroys large numbers of trees). In the deserts, there is little moisture for either plant growth or decomposition, so fire does not need to be frequent.

In many environments, fire is necessary to recycle nutrients and regenerate plant communities. Fire-dependent ecosystems include grasslands, seasonal tropical forests, some temperature-climate forests, and the Mediterranean-climate shrublands.

In the Mediterranean climates such as in the Californias, Australia, and South Africa, the wet winters are too cold and the warm summers are too dry for rotting to recycle the products of photosynthesis, consequently wildfire must be frequent. In these areas, fire is necessary for the health of plant communities. Many of the plant species must have the smoke and/or heat of fire to germinate their seeds, to control parasites, and to influence insect behavior. Thus fire is necessary in Mediterranean-type climatic zones; it is nature's way of cleaning house. But this built-in need for fire should serve as a warning to anyone planning to build her or his house in this fire-dominated plant community.

We earlier viewed plate tectonics as causing the buildup of stresses that are released during earthquakes; by analogy, photosynthesis causes the buildup and storage of potential energy in plants that is released during fires.

The Fire Triangle A fire may begin only when **fuel,** oxygen, and heat are present in the right combination. These three critical components are referred to as the fire triangle (Figure 13.1). Oxygen makes up 21 percent of the atmosphere, so as long as a steady supply of air is available then heat and fuel are the most important factors. Heat during summers and droughts both warms up and dries out vegetation making it easier for a lightning strike to ignite a fire. Given the common presence of oxygen and heat, the occurrence of fires is mainly limited by the amount of fuel available.

The fire triangle also is useful for visualizing how to fight a fire. Remove one of the sides of the fire triangle, and the fire collapses. Firefighters spray water on a fire to reduce its *heat.* Air tankers drop reddish-orange viscous fluids in front of a fire to coat the unburned vegetation and block *oxygen* from contacting the plant. Firefighters commonly bulldoze and remove vegetation to eliminate *fuel.* Fuel also may be reduced by setting **backfires** (Figure 13.2). A backfire is lit by firefighters in

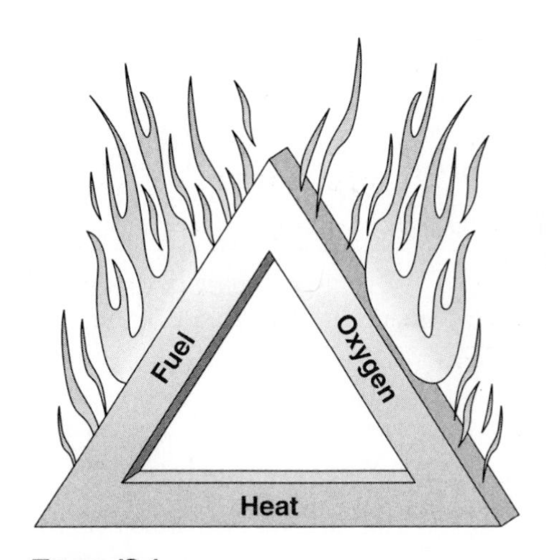

Figure 13.1 The fire triangle. When all three sides are present, a fire results; eliminate one side and the fire cannot burn.

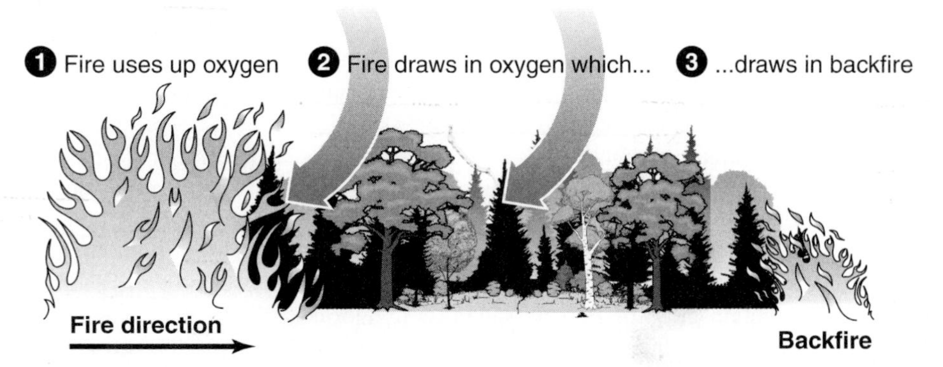

Figure 13.2 How to fight a fire with fire. Light a backfire in front of an advancing wild fire. As the wildfire draws in oxygen to continue burning, it also draws in the backfire—thus eliminating fuel in front of the wildfire.

Sidebar

The Burning of Rome, 64 C.E.

A famous fire of the ancient world consumed Rome, the capital of the most powerful empire of its time. The popular myth about the fire tells of a vain and bored Emperor Nero who fiddled while Rome burned. As usual, the truth, as best as we can know it, is a bit different from the myth. On 19 July 64, fire broke out in the Circus Maximus and was swept by strong winds through the small, closely packed buildings along the narrow, winding streets of the city. Upon hearing news of the fire, Nero left Antium and returned to Rome. For six days and seven nights, the fires moved up and down the hills of Rome until only four of its 14 districts were unscathed (Figure 13.3). Gone were mansions, shrines, and temples built over the centuries; also destroyed was Nero's palace on the Palatine Hill.

Although no eyewitness accounts remain, both Tacitus and Suetonius, writing two and three generations later, describe the actions of Nero that provoked the rage of the populace. Nero is possibly the only ruler of a major empire who considered himself to be primarily a singer and stage performer. During the emotional time of the long-lasting fire, Nero donned his singer's robes, played his lyre, and sang a lengthy song of his own composition called "The Fall of Troy." Expressing his emotions musically during the tragic fire was perceived as callousness, and stories spread until Nero was even rumored to have started the fire. In the absence of contemporary manuscripts, the truth shall probably never be known.

After the fire, Nero quickly rebuilt the Circus Maximus, religious shrines, and other public facilities. He personally paid to have debris from the fire cleared and provided bonuses to building-site owners who quickly completed houses and blocks. For the reconstruction, Nero ordered measures to reduce the probability of future city-destroying fires. Rome was rebuilt with broad streets, spacious and detached houses, and heavy usage of massive stone; households were required to have fire-fighting apparatus in an accessible place. This story from 2,000 years ago is amazingly similar to the events in the news today.

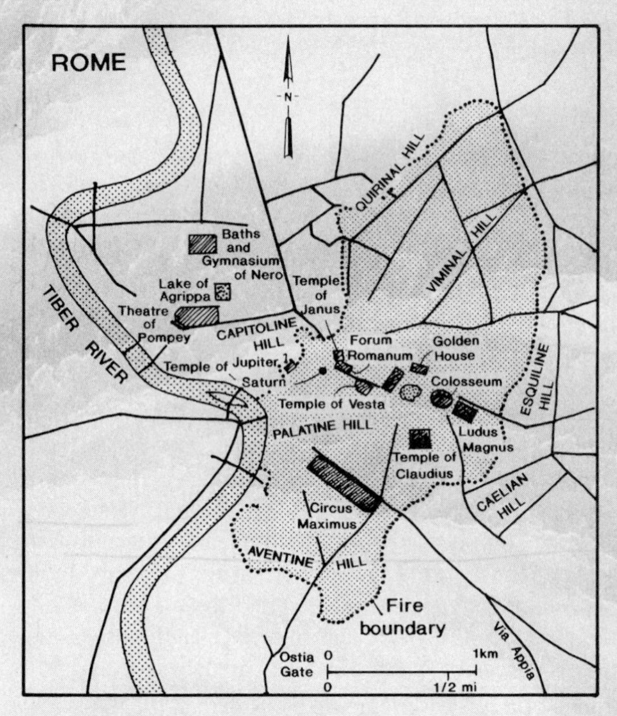

Figure 13.3 The fire of 64 C.E. burned most of Rome. Flames began near the Circus Maximus and raged for a week on the seven hills of Rome.

front of the advancing wildfire. As the wildfire draws in oxygen it also draws in the backfire thus eliminating fuel in front of the advancing wildfire.

Fuel Categories Any combustible material is fuel. Common categories of fuel include grasses, shrubs, trees, and **slash,** the organic debris left on the ground after logging or windstorms. As the human population spreads into wildlands, houses have become a fifth category of fuel.

Grasses have broad, exposed surfaces resulting in fast-spreading fires. Grass fires commonly move about 4 mph (6.5 km/hr) with flames about 8 ft (2.5 m) high. Examples include the fountain grass of Hawaii, sawgrass of Florida, prairies of the central United States, and annual grass understories in forests.

Shrubs are only 2 to 12 feet high but their loose layering allows easy burning. The most intense fires occur in shrubs containing high contents of natural oils. Shrub fires commonly move about 8 mph (13 km/hr) with flames about 50 ft (15 m) high. Examples include the Mediterranean-type chaparral of California, snowberry beneath Douglas Fir forests, palmetto understory in Florida, and shrublands regenerated after a fire or grading of the land.

Trees and forests are affected by the amount of organic litter or slash on the ground beneath them. If litter is scarce, fires pass through quickly and do little harm to trees. If litter is thick and dry, fire will burn hot and slow and kill the trees. If the forest has an understory of litter and shrubs, they may act as ladders leading the fire from the ground to the treetops where the fire then moves as crown fires.

The presence of **ladder fuels** allows small ground fires to quickly carry upward into tall trees and create major wildfires. Vegetation of varying heights makes it easy for fire to climb from grasses to shrubs to trees similar to climbing up

Sidebar

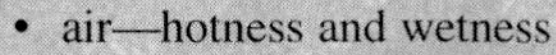

An Ancient View of Fire

The Greeks developed influential theories concerning fire. The synthesis by Aristotle in the fourth century B.C.E. was taken seriously for nearly 2,000 years. In this view, all matter on Earth was made of varying proportions of four elements: air, earth, fire, and water. The behavior of matter varied according to the relative abundances of two opposing qualities: hotness versus coldness and wetness versus dryness. The four elements combined these qualities as follows:

- air—hotness and wetness
- earth—coldness and dryness
- fire—hotness and dryness
- water—coldness and wetness

The elements were not thought by the Greeks to be identical with everyday materials but rather were essences or purer forms. Their relative qualities may be plotted against temperature and humidity (Figure 13.4).

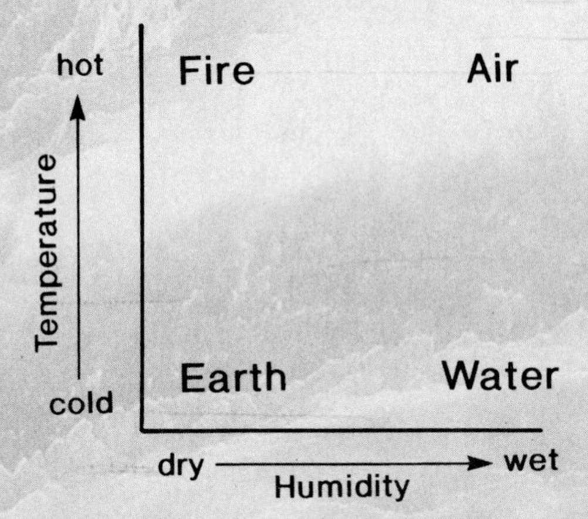

Figure 13.4 The four elements of Aristotle.

the rungs of a ladder (Figure 13.5). How can the ladder fuel threat be reduced? Cut and prune vegetation to create vertical separations between layers.

With the emergence of humans and our ability to both make fires and to fight the flames, the old natural balance has changed. No longer does fuel supply simply accumulate until ignited by a lightning strike. Now humans extinguish many small fires causing the buildup of tremendous volumes of dead-plant fuel, which can ignite and cause fires bigger than we can handle. Humans interfere with the natural cycle of plant growth and fire, but we don't control it.

Figure 13.5 Ladder fuels. Where vegetation of varying heights exist together then fire can climb from grasses to tree tops similar to climbing up the rungs of a ladder.

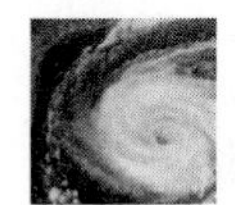

The Stages of Fire

Before a fire breaks out, a *preheating* phase occurs where water is expelled from plants, wood, or fossil fuels by nearby flames, drought, or even a long summer day. The water in wet wood has such a high capacity to absorb heat that the wood becomes extremely difficult to ignite. To burn, wood not only needs to be dry, but its temperature must be raised considerably. For example, the cellulose in wood remains stable even at temperatures of 250°C (480°F); however, by 325°C (615°F), it breaks down quickly, giving off large amounts of flammable gases.

The thermal degradation of wood involves the process of **pyrolysis.** During pyrolysis, the chemical structure of solid wood breaks apart and yields flammable hydrocarbon vapors along with water vapor, tar, and mineral residues. If oxygen is present when the temperature is raised, then the pyrolized gases can ignite and combustion begins.

In *flaming combustion,* the pyrolized surface of the wood burns fast and hot; this is the stage of greatest energy release in any fire. The released heat is carried via **convection** through air flowing up and away from the fire (Figure 13.6); heat is transferred in **radiation** as electromagnetic waves or particle waves; in **conduction** the heat energy is transmitted downward or inward through physical contact; and in **diffusion** heat is in particles that move from hotter to cooler areas (Figure 13.7).

Figure 13.6 Heat rises upward in convection. Hot gases are less dense than the surrounding air; they expand and rise buoyantly.
Drawing by Jacobe Washburn.

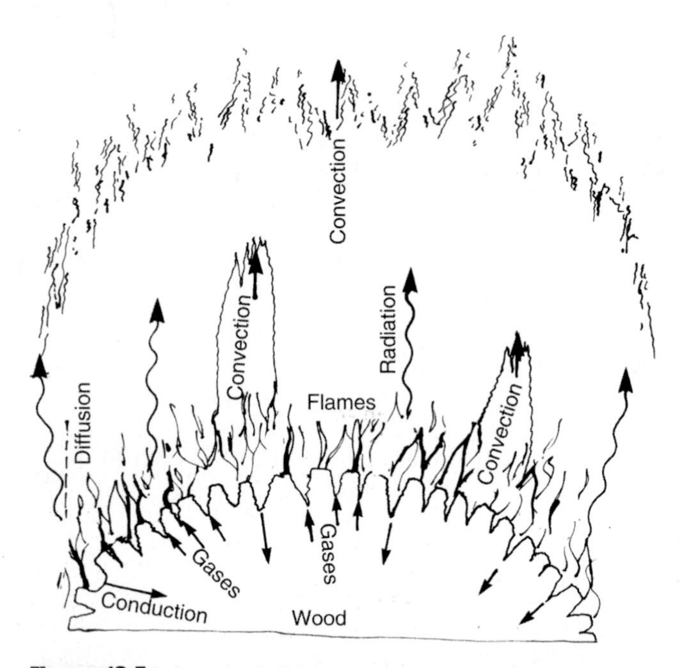

Figure 13.7 Schematic cross section of a burning log. Heat moves inward by conduction, decomposing the cellulose and lignin of wood into gases (pyrolysis) that move through cracks to fuel flames at the log surface. Heat flows outward by: 1) diffusion of particles from hotter to cooler areas, 2) radiation from flames and from hot surfaces, and 3) convection of hot, lightweight buoyant gases that rise upward.

The phases of a fire are well-known to anyone who starts logs burning in a campfire or fireplace. The preheating phase is usually accomplished by lighting newspaper or kindling with a match to dry out the logs and raise their surface temperature. When kindling burns beneath logs and is supplied with an efficient air flow carrying abundant oxygen, then the cellulose and lignin in the logs decompose and give off gases in the process of pyrolysis. As the logs heat, they expand in volume and form cracks that release gases to the surface to feed the flames. Where cracks do not form, the wood will "pop" and throw out sparks and embers.

Because of the poor conductance of heat in wood (Figure 13.8), the interior of a log remains below the combustion point even when the exterior is engulfed in flames. For a log to burn completely, there must be enough outside heat (from

Figure 13.8 Wood conducts heat poorly; metal conducts heat readily.
Drawing by Jacobe Washburn.

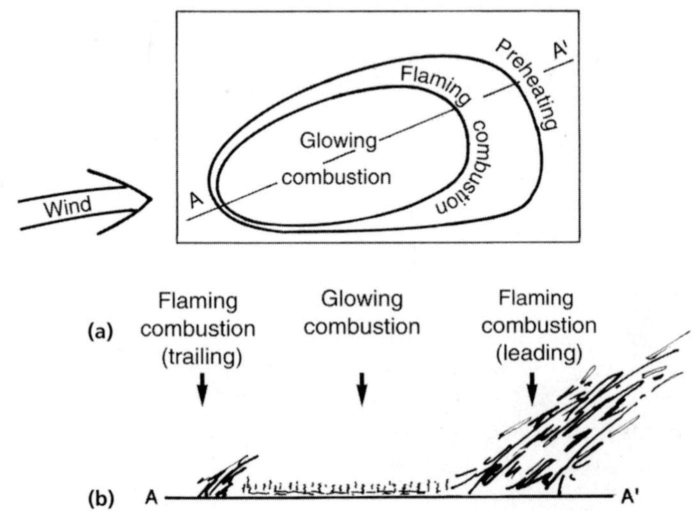

Figure 13.9 The stages of fire are shown in horizontal or map view **(a)**. Line AA[1] on the map is shown in vertical slice or cross-section view in **(b)**.

embers below, adjoining burning wood, etc.) conducting into the log to continue the pyrolizing process; it is the gases from pyrolysis that fuel the surface flames that slowly eat into the heart of the log. The radiant and convective heat that keeps the logs hot enough to continue burning is also the heat that warms us.

In nature, the flaming combustion stage involves a flaming front that passes by and leaves behind *glowing combustion.* In a fireplace, the active flames disappear, but the wood surface glows. In each case, the wood now burns more slowly and at a lower temperature as the fire consumes the solid wood instead of pyrolized gases. The process is a slower **oxidation** of the charred remainders left by the flames.

All the stages of a fire occur simultaneously in different areas of a wildfire (Figure 13.9). The character of a wildfire and the area it covers depend on several factors described below.

The Spread of Fire

Wildfires occur in different styles: 1) they may move slowly along the ground with glowing combustion playing an important role, 2) they may advance as a wall of fire along a flaming combustion front, or 3) they may race through the treetops as a crown fire.

The spread of fire depends on: 1) fuel—the types of plants or other material burned, 2) weather—especially the strength of winds, 3) topography—the shape of the land, and 4) behavior within the fire itself.

Fuel The energy release in a fire strongly depends on the chemical composition of the plants and organic debris. For example, the eucalyptus family of trees and shrubs has a high oil content that favors easy ignition and intense heat of burning. In the world before global travel and trade, eucalyptus was confined to Australia, but it has been exported in abundance to many areas where it thrives like a native plant. The eucalyptus fire hazard is now a significant threat in southern California, North Africa, India, and the Middle East.

Wind Many of the worst fires in history were accompanied by strong winds. The wind brings a continuous supply of fresh oxygen, distributes heat, pushes the flames forward, and bends them toward preheated plants and other fuel. If winds are absent, a vertical column of convected heat dominates, and the fire may move very slowly (Figure 13.10a, type I). If winds are fast, they push the fire front rapidly ahead and prevent a vertical convection column from forming (Figure 13.10f, type VI). Strong, gusty winds and fire tornadoes both pick up flaming debris and burning embers and drop them onto unburned areas, starting new blazes.

Topography The topography of the land has numerous effects on fire behavior. Before a fire, the topography sets up microclimates that result in different plant communities that will burn with different intensities. Wind blowing over rugged topography develops turbulence. Canyons with steep slopes and dense vegetation cause high levels of radiant heat that consume virtually all the organic matter in the canyon (Figure 13.10g, type VII). Fire burns faster up a slope because the convective heat rising from and above the fire front dries out the upslope vegetation in an intense preheating phase; in effect, a chimney is created up the slope, allowing fire to move quickly (Figure 13.11).

Fire Behavior The strength of a fire is partly created by its own actions. The vast quantities of heat given off by a fire create unstable air. Heat-expanded air is less dense and more buoyant; thus, it rises upward in billowing convection columns (Figure 13.10a–e, types I–V). The rising columns of hot, unstable air may spin off fire tornadoes. Winds sucked into the base of a fire tornado bring oxygen to feed the flames, while huge quantities of heat race up the column, venting above as in a megachimney. Fire tornadoes are commonly 10 to 50 times taller than their diameters, and they may spin at speeds of 250 km/hr (over 150 mph). Fire tornadoes may carry fiery debris called **firebrands** and drop it miles away, starting new fires.

A fatal example showing the strength of a fire tornado occurred on 24 April 2000 near Winkler in the province of Manitoba, Canada. A fire burning 90,000 bales of flax straw spun off a fire tornado that lifted a man and his pickup truck and carried them 165 feet away before dumping their remains onto a field.

Fire Weather

Fire hazards are greatest in those regions with the biggest differences between their wet and dry seasons. Problems occur when a rainy season promotes rapid plant growth that builds a large biomass, only to be followed by prolonged dry weather. A drought dehydrates plants, both living and dead, making them easier to burn. During a drought, if high winds with low humidity start blowing, small fires may be spread into major blazes. And big fires rarely occur alone. A dry, windy weather pattern affects a large region, and major fires commonly break out in bunches. Major fires are of great significance; more than 95 percent of all burned area is caused by only 2 to 3 percent of fires.

Winds of Fire

Winds causing fire problems may be large-scale associated with the movement of air-mass fronts (see Chapter 10) or small-scale local winds controlled by temperature and topography differences.

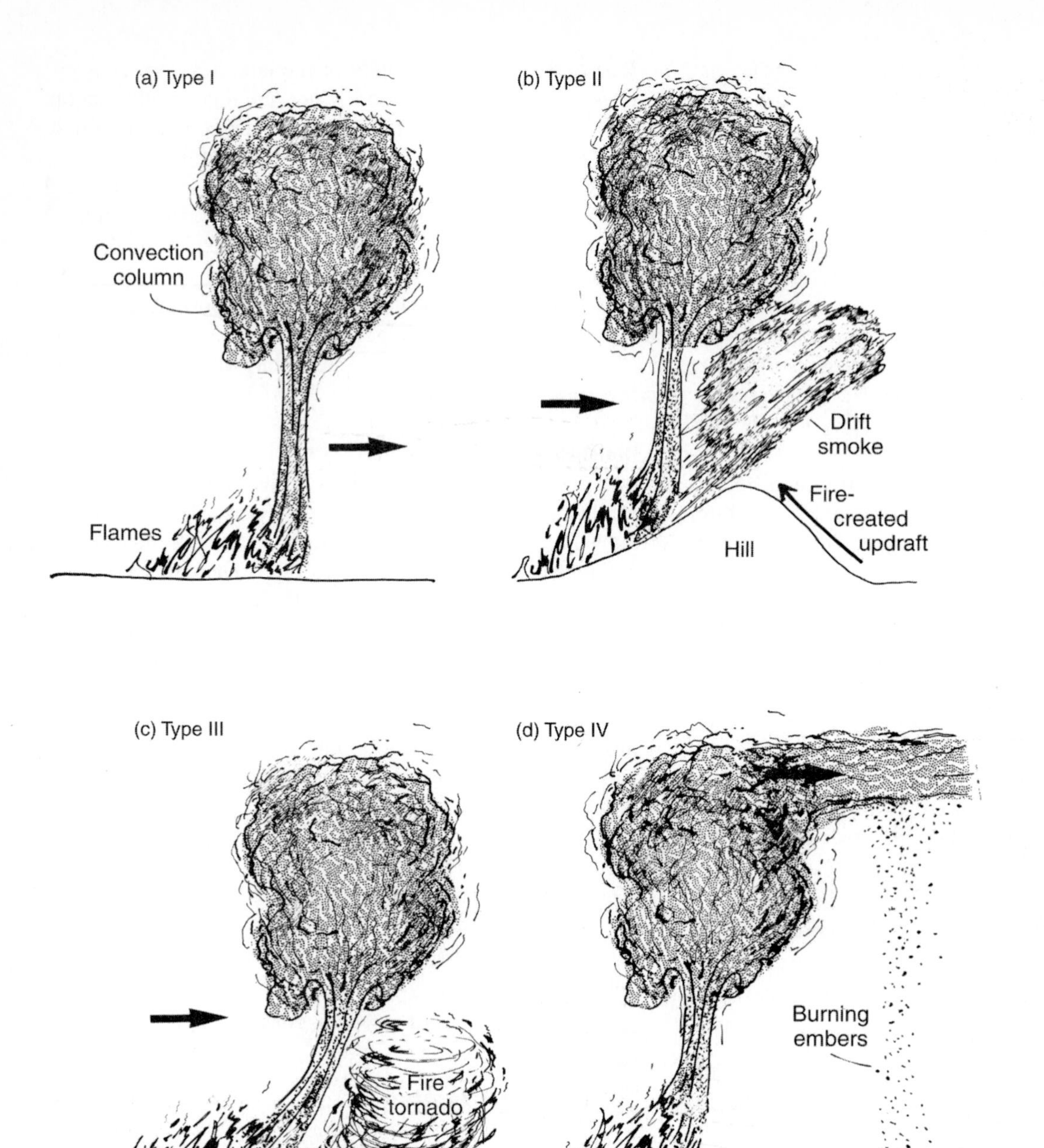

Figure 13.10 Some fire types. **(a)** Type I has a tall convection column and weak winds that push the column at a moderate rate of spread. **(b)** Type II has a tall convection column that moves rapidly upslope. **(c)** Type III has a powerful convection column pushed by surface winds; fire tornadoes are spun off. **(d)** Type IV has a convection column distorted above the ground surface by strong winds; glowing embers are dropped.

Cold-front Winds Fire problems increase when a cold front blows in at its typical speeds of 20–30 mph. As the cold front approaches, wind speeds increase and gusty wind conditions usually persist for hours after it passes. In winter the cold fronts commonly bring rain but in summer the cold fronts are usually dry.

Foehn Winds Fierce winds occur when a high-pressure air mass spills over a mountain range and descends as warm, dry wind toward a low-pressure zone. **Foehn** is a German word applied to winds flowing down from the Alps. In the western United States, foehn winds commonly occur in September through April, when a strong, high-pressure system stagnates over the Great Basin. Winds blow when a low-pressure zone sits on the other side of the mountains and a pronounced pressure gradient exists between the atmospheric air masses, causing air flow across the mountains. This is dry air flowing down from high elevations; it warms adiabatically as it descends, and it flows fast, with speeds commonly at 40 to 60 mph (65 to 100 km/hr) with intervals up to 100 mph (160 km/hr).

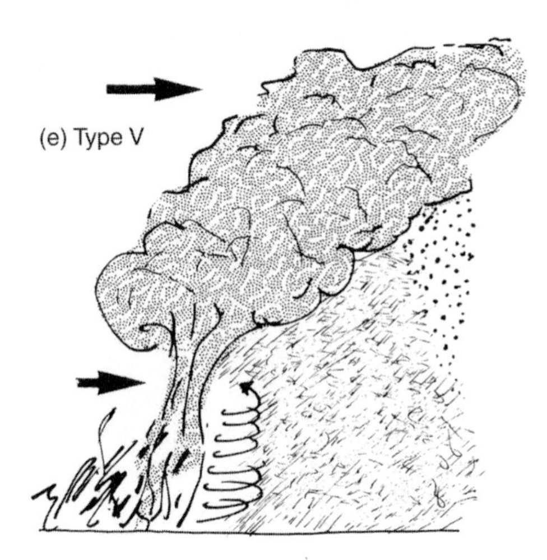

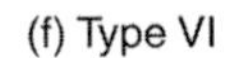

(f) Type VI

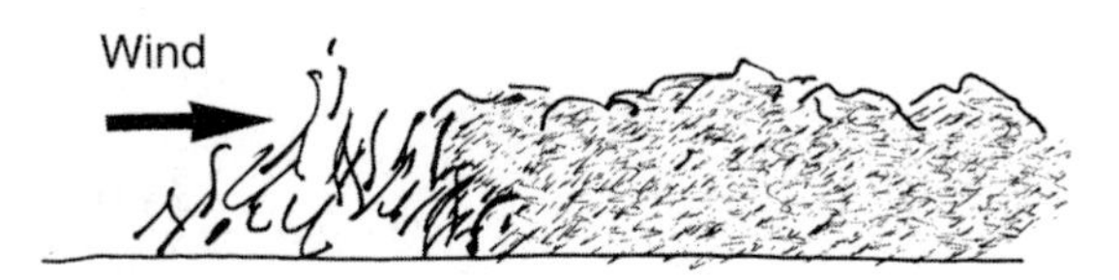

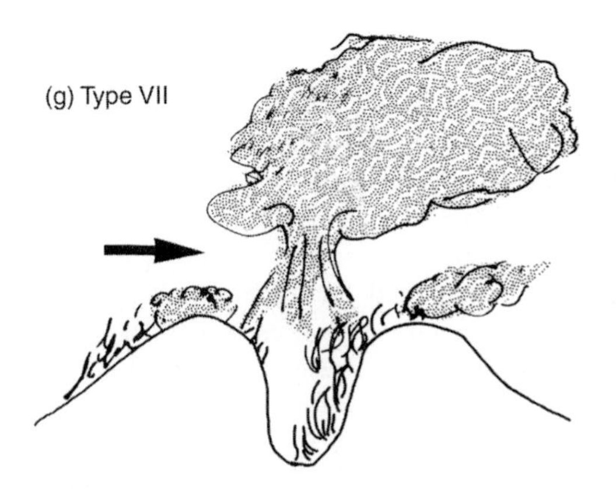

Figure 13.10 —Continued. **(e)** Type V has a convection column bent by winds of differing speeds. **(f)** Type VI is a wind-driven fire that spreads rapidly; the winds are too strong for a convection column to function. **(g)** Type VII shows the effects of topography.

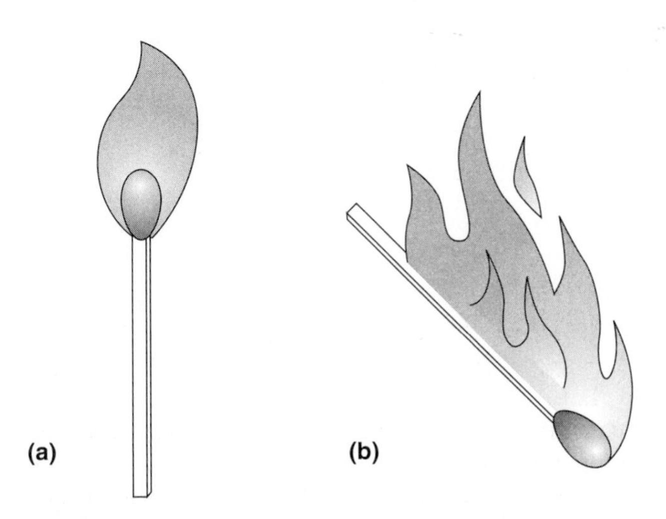

Figure 13.11 How fires burn on slopes. **(a)** An upright match burns slowly downward, similar to fire on top of a hill. **(b)** A sloping match burns rapidly upward, similar to fire up a hill.

If a low-pressure zone sits east of the Rocky Mountains, then the foehn winds flow east and are called Chinooks (derived from an American Indian word meaning "snow eater"). Chinook winds can raise surface temperatures by 40°F and drop humidities to 5 percent within a few minutes (Figure 13.12). Along the Pacific Coast, the foehn winds are known by several names, including Diablo (Spanish for "devil") in the San Francisco Bay region and Santa Ana in southern California. A strong air flow from a high-pressure area over the Great Basin may push to the southwest along the ground surface and extend over the Pacific Ocean (Figure 13.13). The strong air flow pushes moisture-laden Pacific air away and dries out the vegetation. The rough topography in California creates turbulence in the air flow, spawning eddies of variable sizes. Hot, dry winds racing over dehydrated California vegetation produce some of the worst fire-weather situations in the world.

Local Winds Several local wind patterns are factors in fire behavior: sea breezes, land breezes, slope winds, and valley winds.

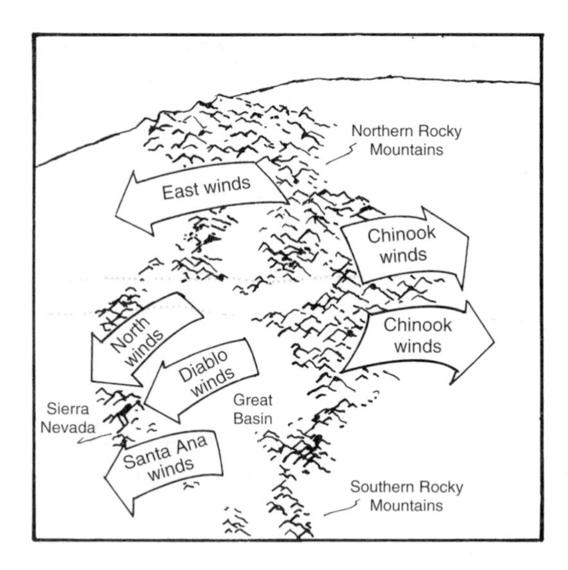

Figure 13.12 Foehn winds have local names. In the western United States, they form when high-pressure air over the Great Basin spills over the mountains as warm, dry, high-speed winds.

Source: M. J. Schroeder, and C. C. Buck, "Fire Weather," 1970. U.S. *Department of Agriculture Handbook* 360.

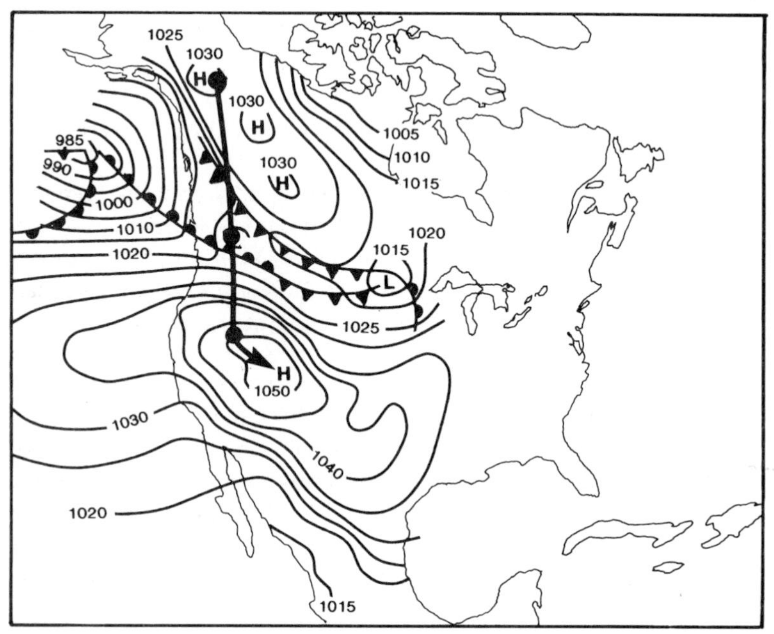

Figure 13.13 A high-pressure center over the Great Basin creates a marked gradient in air pressure to the southwest, producing strong Santa Ana winds in southern California.

Source: M. J. Schroeder, and C. C. Buck, "Fire Weather," 1970. *U.S. Department of Agriculture Handbook* 360.

Sea breezes are a daily condition when warm air over the land rises, and cooler air over the ocean flows in to replace the risen air. Sea breezes usually begin in the midmorning to early afternoon and then cease after sunset. Their on-shore wind speeds of 10 to 30 mph reach a maximum during the hottest part of the day.

Land breezes are the opposite of sea breezes. After sunset, the land cools more rapidly than ocean and lake surfaces. During the night, the cooler air over the land flows seaward at typical velocities of 3 to 10 mph until shortly after sunrise.

Slope winds flow upslope and downslope during the day. As the ground heats, the warmed and buoyant air flows upslope with some turbulence. As the ground cools at night, the cooled nighttime air flows downslope but slower and smoother than the upslope winds of the daytime.

Valley winds are similar to slope winds. During the day, the warmer air in canyons and valleys flows upvalley. During the night, the cooler air flows down to valley bottoms and then flows down the valley axis.

The regional and local wind behaviors strongly influence the behavior of fires as is shown in examples from the Great Lakes region and California.

Great Lakes Region

In the late 1800s, the Great Lakes region was heavily forested, dominantly by pine trees. The settlers in the region both logged the forests for timber and clear-cut the native vegetation to expose ground for farming. The widespread removal of forests left abundant woody debris (slash) on the ground, and farmers routinely used fire to clear land for grazing and plowing. Every day, numerous small fires burned. Many individual fires burned for weeks at a time. In the late 1800s, the fire practices of rural America were similar to the slash and burn clearing of forests occurring in Brazil today. Nearly half of the U.S. forest fires in 1880 were started by people clearing land (Table 13.4). Compare the causes of fire in 1880 (Table 13.4) to those in 2000 (Table 13.1).

The summer and early autumn of 1871 was a time of drought, of excessive dryness. Persistent high-pressure zones over the region blocked moisture-bearing winds from the Gulf of Mexico. The long spell of dry weather sucked the moisture out of logging slash and reduced the water content in forest trees. On the evening of 8 October 1871, gale-force winds blew in from the southwest, and the region's abundant small fires blew out of control from the Ohio Valley to the Great Lakes states and the High Plains; farms burned in Indiana, towns were incinerated in Michigan, Wisconsin, North Dakota, and South Dakota, and the great city of Chicago was gutted by flames.

Peshtigo, Wisconsin Small fires were burning most days in late nineteenth-century Wisconsin. When strong winds arose, they could whip small fires into large blazes, including the deadliest forest fire in U.S. history. Around 9 P.M. on 8 October 1871, the townspeople of Peshtigo heard an approaching roar and saw the southwestern sky glowing red

Table 13.4	Causes of U.S. Forest Fires in 1880
Total fires = 2,983	
Clearing of land	45.2%
Hunters	21.1%
Locomotives	17.0%
Malice	8.8%
Campfires	2.4%
Indians	1.9%
Smokers	1.2%
Lightning	1.1%
Others at less than 1% each include Prospectors, Coal pits, Woodcutters, Carelessness, Travelers, Spontaneous combustion	

Source: Sargent, C. S. (1884) *Forests of North America.* U.S. Government Printing Office.

with flames and smoke. The fire did not advance slowly but instead raced forward through the dried-out treetops as a crown fire. Tornadoes of fire in the treetops dropped enveloping flames to the ground below. The fire front was 15 miles wide and moved at high speed; this did not leave many options for people living in wooden houses and lacking fast transportation to escape. The wide fire front raced 40 miles northward along the northwest shore of Lake Michigan. After it passed, there were 1,152 known dead.

Chicago, Illinois In 1871, Chicago was home to 334,000 people. The city was already an important center for grain and livestock markets, railroad and farm machinery, and financial institutions. But most of Chicago was built with little thought of fire hazard. Not only were most buildings made of wood, but many of the sidewalks also were constructed with resin-rich pine wood.

At about 9:30 P.M. on 8 October 1871, a fire broke out in Patrick O'Leary's barn. The same gale-force winds from the southwest that were fanning the Peshtigo inferno picked up the flames from O'Leary's barn and pushed them rapidly to the northeast. In the beginning, no major problem was foreseen; the fire was 2 miles from the river, across which lay the heart of Chicago. But the wind-driven fire moved swiftly through the flammable businesses in its path—numerous stables, lumberyards, grain stores, distilleries, and coal yards all fed the flames. In less than three hours, the fire reached the river, and then the gusty winds quickly carried blazing debris across the river, dropping it on dry, wooden tenement buildings. Flames raced through the densely packed dwellings and litter-strewn alleys as fast as an adult can walk. Two journalists from the *Chicago Daily Tribune* wrote this eyewitness account:

> Flames would enter at the rears of buildings, and appear simultaneously at the fronts. For an instant the windows would redden, then great billows of fire would belch out, and meeting each other, shoot up into the air a vivid, quivering column of flame, and poising itself in awful majesty, hurl itself bodily several hundred feet and kindle new buildings. . . . The whole air was filled with glowing cinders, looking like an illuminated snowstorm. Interspersed among these cinders were larger brands, covered with flame, which the wind dashed through windows and upon awnings and roofs, kindling new fires. Strange fantastic fires of blue, red and green, played along the cornices of the buildings. On the banks of the river, red hot walls fell hissing into the water, sending up great columns of spray and exposing the fierce white furnaces of heat which they had enclosed.

In 27 hours, 300 people were dead, 3.3 mi^2 of the city were incinerated, 96,000 people were homeless, and the heart of the city lost its great hotels, banks, theaters, the Opera House, City Hall, and most of the water works. The Patrick O'Leary family lost their barn, but their house was undamaged.

California

Many people move to California because the weather is warm all year; it is said of California that "the four seasons do not exist there." Others say California does have four seasons—flood, drought, fire, and earthquake. A few weeks of rain sets plants into a fast-growth mode. Then, months of drought kill the annual plants and dehydrate the perennials, setting the stage for fire.

Much of California is covered by a group of perennial evergreen shrubs known as **chaparral.** The chaparral species dominate hillsides throughout the state. These plants are one of the most flammable groups in the world, and it is a plant community whose generations are in equilibrium with fire. The warmth of the California climate promotes plant growth, but the cold and wet winter months plus the many hot and dry months each year do not allow large populations of decomposer organisms to flourish. So, the chaparral community has evolved to let fire take the place of decomposer organisms; it is fire that consumes organic debris and recycles nutrients. After a fire destroys the surface vegetation, shoots spring forth from the subsurface parts of many chaparral plants, beginning the next crop of plants. Many of the chaparral plants encase their seeds in shells that open only after the intense heat of fire destroys the outer coat thus freeing the seed to later absorb water and germinate. Some other plant seeds respond to the smoke of a fire. For example, nitrogen dioxide (one of the gases in smoke) induces complete germination of some seeds in as short a time as 30 seconds. Thus, the very fire that burns away a dense plant cover releases and germinates seeds on ground made bare by the same fire. The brush removal by the fire also allows the seedlings to get the sunlight they need to grow.

Many hillsides in California and Baja California are covered by dense growths of chaparral. Figure 13.14 shows impenetrable thickets of stiff chaparral shrubs just waiting to

Figure 13.14 Dense impenetrable thickets of highly flammable chaparral cover the hillsides along the dry Santa Margarita River in northern San Diego County.
Photo by Pat Abbott.

ignite and burn. Chaparral on the adjacent land in Fallbrook, California, ignited in February 2002 and burned over 5,000 acres, destroying two fire engines and 30 luxury homes including the home of singer Rita Coolidge.

Oakland and Berkeley Hills Rising behind the cities of Berkeley and Oakland are steep hills where million-dollar homes command sweeping views over San Francisco Bay. The hills are covered with trees and shrubs that allow the homeowners to "live in the woods" and "get away from it all." However, these wildland-urban interfaces have terrible threats from fire.

A five-year-long drought in the late 1980s dried shrubs and trees in the hills. A December 1990 freeze killed many ornamental plants, only to be followed by a rainy March that set grasses into rapid growth. Then drought quickly returned for the rest of 1991. The volume of dead and dehydrated vegetation was dangerously high. On Saturday, 19 October 1991, a fire of suspicious origin started near the top of the hills. Firefighters fought the blaze and controlled it in three hours with 5 acres burned. The fire site was in an area of Monterey pine trees with a 1-ft-thick bed of pine needles and organic debris known generically as **duff.** Firefighters know duff as a problem; it burns on top but smolders below the surface, where oxygen is in short supply. Water extinguishes the surface flames, but it also can combine with ash to form a crust over smoldering combustion. The firefighters left early Saturday evening, planning to return on Sunday morning to "mop up" the smoldering duff.

However, on Sunday, the Diablo winds (Figure 13.12) also showed up at 10:45 a.m., bringing low humidity, a 96°F day, and wind speeds averaging 17 mph with gusts up to 25 mph. The Diablo winds picked up sparks from the duff and started a fire in a steep-walled canyon. Fire loves steep slopes. By 11:15 A.M., the flames had raced up the slopes of the confining canyon and leapt over the rim. The blaze was already out of control (Figure 13.15). Flames in grass ignited the fuel-rich chaparral, which fed the fire into the low branches of Monterey pine and eucalyptus trees, which in turn sent flames climbing up the "fuel ladder" to set off a crown fire that moved through the tree tops at nearly 4 mph. Erratic Diablo winds pushed the fire in changing directions. The flames reached 2,000°F and quickly grew to firestorm status, creating their own winds, towering convection column, and fire tornadoes. In its first hour of existence, the fire consumed 790 homes. Winds grew to speeds of 20 mph with gusts up to 50 mph, pushing flames across an unbroken chain of dry vegetation and combustible homes (Figures 13.16 and 13.17). Burning embers carried by the winds started new blazes. By early evening, the winds died down to 5 mph, reversed their direction, and slowly pushed the flames back over the burned area.

The winds spread the fire and pushed the flames where they pleased all day long. But then, as the winds died down in the early evening, firefighters finally had a chance to control the flames. During the day, 25 people were trapped and killed, 150 were injured, and 2,449 single-family dwellings were destroyed, along with 437 apartments and condominiums. In all, 1,600 acres were burned, and property damages exceeded $1.5 billion.

Southern California Southern California has a long dry season, chaparral vegetation, and foehn winds—southern California is a land that was born to burn. When conditions are ripe for foehn winds, southern California all too commonly is ablaze with numerous fires. On Wednesday, 27 October 1993, Santa Ana winds blew in and fires broke out (Figure 13.18). Wind-felled power lines created sparks that the winds fanned and carried as flames far and wide through the flammable chaparral (Figure 13.19). Fires also began when the winds picked up flames from a transient's campfire; several more were set by arsonists. The Santa Ana winds died down on Friday, 29 October, allowing firefighters to gain control of some fires, but not all.

Then, on Tuesday night, 2 November 1993, resurging winds hit speeds up to 50 mph, and the flames took off for another two days (Figure 13.20). Fires burned clear to the ocean despite the best efforts of firefighters; surfers rode waves in front of bushes burning on the beach.

The massive firestorms spun off tornadoes that sent embers leaping large canyons in a single bound to start new blazes. The flames consumed whatever they wanted. Firestorms are not stopped by humans; they go as long as the wind blows. When

Sidebar

The Winds of Madness

The Santa Ana winds of southern California not only push firestorms, they also affect people's moods and behaviors. In his fictional story *Red Winds*, Raymond Chandler wrote:

> "There was a desert wind blowing that night. It was one of those hot, dry Santa Anas that come down through the mountain passes and curl your hair, and make your nerves jump and your skin itch. On nights like that, every booze party ends in a fight, meek little wives feel the edge of the carving knife and study their husbands' necks. Anything can happen."

What is it about Santa Ana winds that affect some people? It is the high wind speeds, extra low humidity, and electrically charged air carrying 7 to 9 times the normal level of positive ions. During "the winds of madness," records of public agencies show increases in domestic violence and household mishaps. There are common complaints of allergic reactions and migraine headaches, and the suicide rate is said to double.

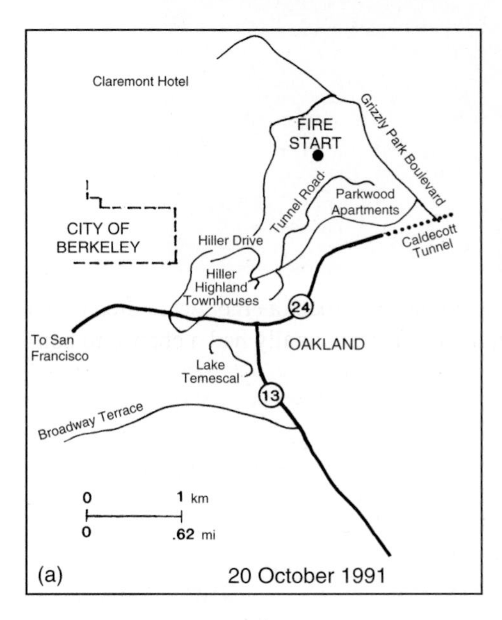

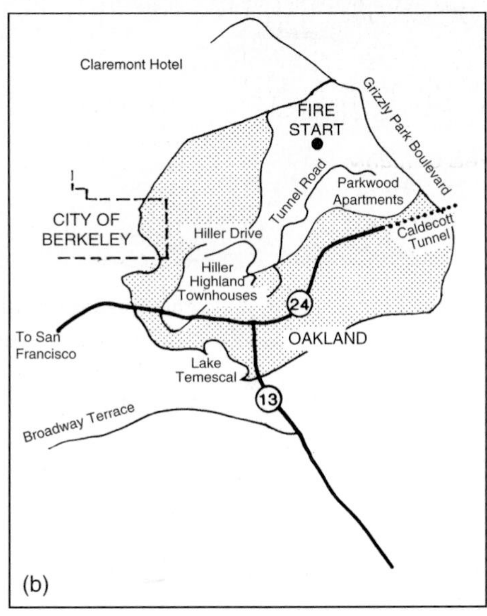

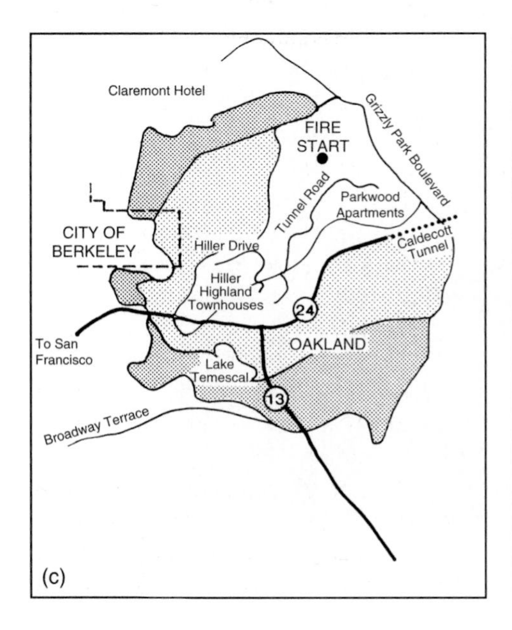

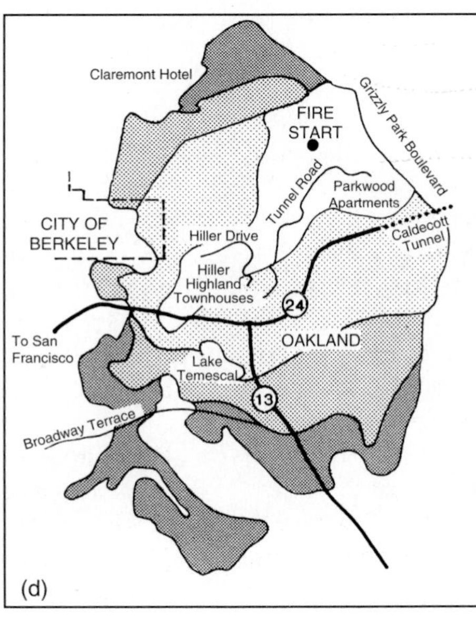

Figure 13.15 The burn history of the Oakland/Berkeley Hills fire of 20 October 1991. Area burned by: **(a)** 11:30 A.M.; **(b)** 1 P.M.; **(c)** 3 P.M.; **(d)** 5 P.M.

Source: The Oakland/Berkeley Hills Fire. Courtesy of The National Fire Protection Association, Quincy, Mass.

Figure 13.16 Firestorm advances through the Berkeley Hills, consuming vegetation and houses, on 19 October 1991.
Photo by Jose Aguirre.

Figure 13.17 Firestorm advances toward the University of California, Berkeley campus.
Photo by Jose Aguirre.

these fires finally ended, 3 people were dead, 200 were injured, 1,150 homes were destroyed, and damages exceeded $1 billion. Over 215,000 acres (over 300 mi²) were charred. Despite the horror of this event, it is not a rare one. Every year brings the possibility of similar conflagrations.

The fire peril is evident enough that a few people intelligently design their houses to withstand fires (Figure 13.21). The choice is yours—either pay higher construction costs to fireproof your home, or pay the entire cost to rebuild your home, and life, after a fire.

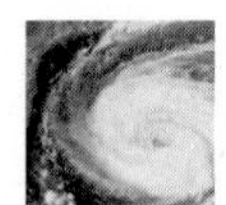

Home Style and Fire

Not all of the death and destruction from fires can be called a "natural disaster." Poor decisions on landscaping and home construction style and materials are partly to blame (Figure 13.22). Many homes are made of wood or roofed with wooden shake shingles. Wooden decks extend out over steep slopes and help concentrate heat, igniting the houses.

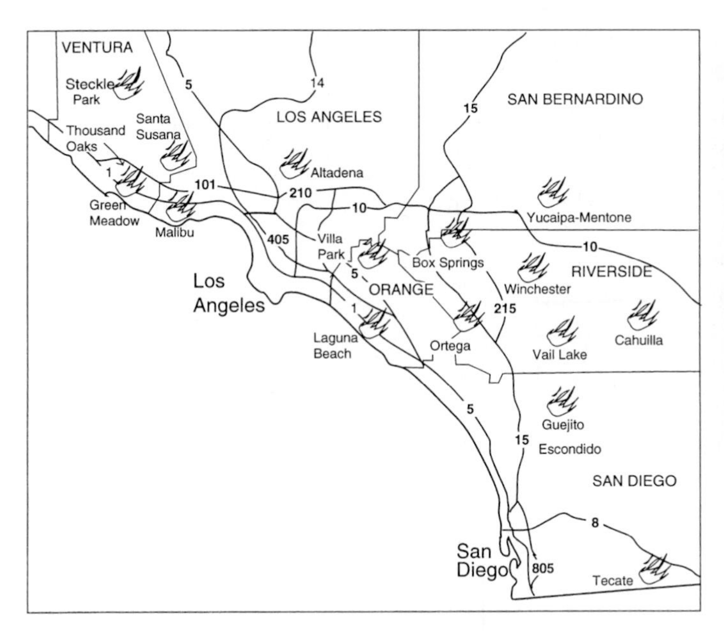

Figure 13.18 Sites of some large wildfires that raged between 27 October and 4 November 1993. When the Santa Ana winds blow, the fires occur in bunches.

Figure 13.20 Firestorm racing through chaparral-covered hillslopes as a wall of flame up to 14 m (45 ft) high.

Photo courtesy of the *Los Angeles Times.*

Figure 13.19 Convection column pushed by Santa Ana winds during the Malibu fire.

Photo by Richard Derk, courtesy of the *Los Angeles Times.*

Figure 13.21 Houses designed to withstand fire remain unharmed while their million-dollar neighbors are nothing but ashes, near Las Flores Canyon in Malibu.

Photo by Patrick Downs, courtesy of the *Los Angeles Times.*

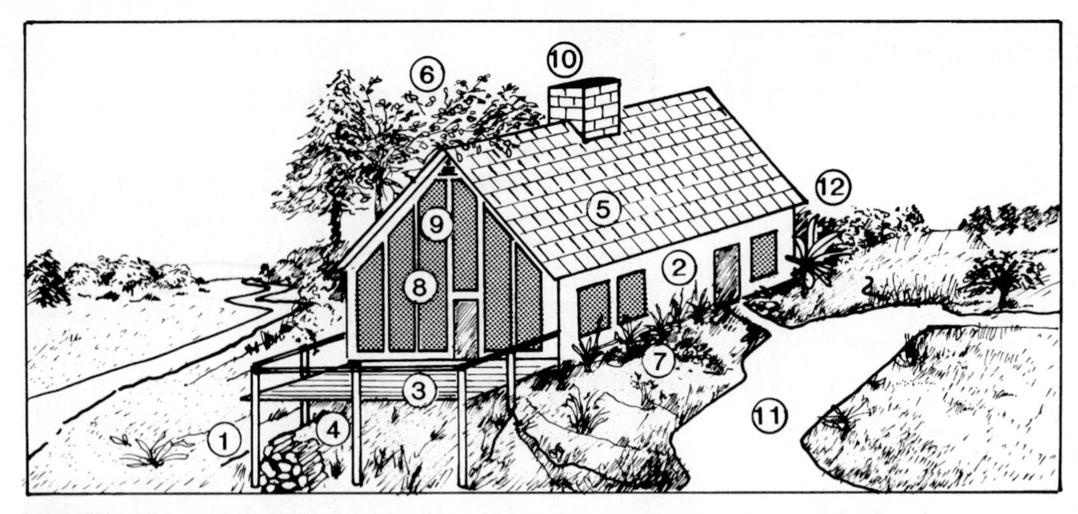

Figure 13.22 Twelve mistakes, or how to sacrifice your house to the fire gods. 1) House is located on a slope. 2) House is made of wood. 3) Wooden deck hangs out over the slope. 4) Firewood is stored next to the house. 5) Roof is made of flammable wood shingles. 6) Tree limbs hang over roof. 7) Shrubs continue up to house. 8) Large, single-pane windows face the slope. 9) Unprotected louvers face the slope. 10) No spark arrester is on top of chimney. 11) Narrow road or driveway prevents access of fire trucks. 12) Wooden eaves extend beyond walls.

Source: The Oakland/Berkeley Hills Fire. Courtesy of The National Fire Protection Association, Quincy, Mass.

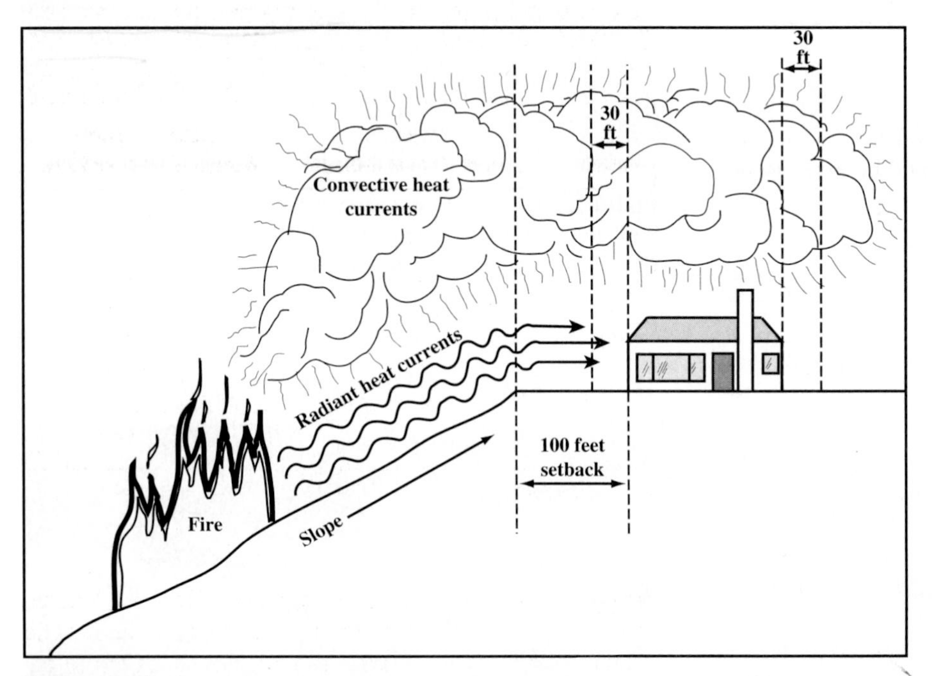

Figure 13.23 Buildings at the top of a slope need setbacks to help avoid the increased heat flow by convection and radiation. The defensive space includes a vertical zone around the building that has no tall trees.

Natural and planted vegetation commonly continue from the wildland right up to houses or drape over the roofs. All of these flammable materials act to convey fire into and through houses.

The houses that flames passed by commonly had clay- or concrete-tile roofs, stucco exterior walls, double-pane windows, few overhanging roofs or decks, and fire breaks of cleared vegetation extending about 30 feet from the house. Houses built on slopes need even larger areas of cleared space (Figure 13.23). If you plan to build a house in the woods or along the urban-wildland interface, the decisions you make about construction materials and landscaping may well determine whether your house will end up as fuel for flames or remain your home (Table 13.5). Even little decisions about landscaping can determine whether your house burns or endures (Figure 13.24).

Table 13.5 How to Protect Your House from Wildfire: A Must-do List

Roof. The roof is the most vulnerable part of your house because of wind-blown burning embers. You must build or re-roof with lightweight materials that will not burn. Remember to remove branches hanging over your roof and sweep off accumulated leaves, needles, and other plant debris.

House. Build or remodel with fire-resistant materials. The ultimate foolishness is to build a wooden house among trees and shrubs designed to burn.

Decks. The undersides of above-ground decks, balconies, and eaves must be covered with fire-resistant materials.

Windows. Use only double- or triple-paned windows to reduce the potential of breakage during a fire.

Flammables. Be sure to place natural-gas tanks and firewood piles at least 30 feet from your house.

Trees. Hold back fire from your house by 1) reducing the number of trees in densely wooded areas and 2) cutting off branches within six feet of the ground so grass fires cannot climb up to the treetops.

Yard. Create a defensible space by 1) replacing fire-loving plants with fire-resistant plants and 2) removing all dry grass, dead brush, and leaves at least 30 feet from your home, or 100 feet if you are on a slope (Figure 13.23).

The American Society of Civil Engineers' list of ten most-wanted changes to reduce losses from natural hazards includes one on wildfires. The highest priority should be given to developing fire-resistant materials for the building envelope.

Fire Suppression

In August 1910, over three million acres of forest in the Bitterroot Mountains of Idaho and Montana burned in a firestorm known as The Big Blowup. The fire destroyed towns, killed 85 people, and provoked the U.S. Congress into authorizing expenditure of federal money to fight forest fires. And so began the policy of aggressively suppressing forest fires with well-trained armies of professional fire fighters.

During the twentieth century, people were taught to hate fire and to stamp it out quickly. Fire suppression tactics and equipment improved during the century, resulting in dramatic reductions in the number of acres burned (Table 13.6). Over the decades, forests were transformed by the fire-suppression practices. Forests that once held 30 big trees per acre were accustomed to having ground fires move quickly through grasses and thin litter on the forest floor without harming most of the big trees. After years of limiting fires, some of these forests now support 300 to 3,000 trees per acre plus an understory of shrubs. When lightning ignites fires in these dense forests the flames burn slow and hot, killing the big trees. The widespread recognition of the fire problem for dense and crowded forests led to a dilemma. Should the dense forests be thinned? Should natural fires be allowed to burn? Yes answers to these questions led to the following events.

Figure 13.24 Right versus wrong is illustrated above a busy street in San Diego. Non-thinking homeowner on the left has dead grasses on a slope leading up to dead trees and an overhanging wood patio. Thinking homeowner on the right has non-burning succulent ice plant on the slope leading up to a concrete-block wall.

Photo by Pat Abbott.

Table 13.6 U.S. Wildland Fires by Decade

Decade	Number of Fires (Annual Average)	Number of Acres Burned (Annual Average)
1920–29	97,599	26,004,567
1930–39	167,277	39,143,195
1940–49	162,050	22,919,898
1950–59	125,948	9,415,796
1960–69	119,772	4,571,255
1970–79	155,112	3,194,421
1980–89	163,329	4,236,229
1990–99	106,393	3,647,597

Source: National Interagency Fire Center.

Yellowstone National Park

Yellowstone is our oldest national park; it was authorized on 1 March 1872. The park averages about 15 fires per year started by lightning. This is not an unusual number of fires, considering that the Earth as a whole is estimated to have about 8 million lightning discharges per day.

Not all lightning strokes have the same fire-igniting potential. The most capable are "hot strokes," which are cloud-to-ground discharges of high amperage and longer duration.

About one in 25 lightning strokes matches these characteristics. Whether a fire is ignited depends on what the lightning hits. A hot stroke may start a fire if it strikes kindling, such as dry grass, rotten wood, or "organic dust" mechanically blasted into the air by the force of the lightning strike. Lightning has started fires in settings ranging from the tundra of Alaska, the forests of North America, and the marshes of the Great Lakes to the swamplands of the South, the chaparral of California, and the grassy deserts of Arizona.

The policy of Yellowstone Park from the 1880s to the 1970s was to extinguish all fires as soon as possible. But fire is a natural process. So the question arose, "who can better manage these wildlands—humans or nature?" Nature became the answer in the 1970s, so the policy was changed to one of putting out human-caused fires but letting natural fires alone to run their course. Lightning-caused fires burn the forest irregularly; they cause formation of a mosaic of meadows, burned-over ground, and forests ranging from young to mature to old growth. The different areas provide a diversity of environments, supporting various communities of plants and animals.

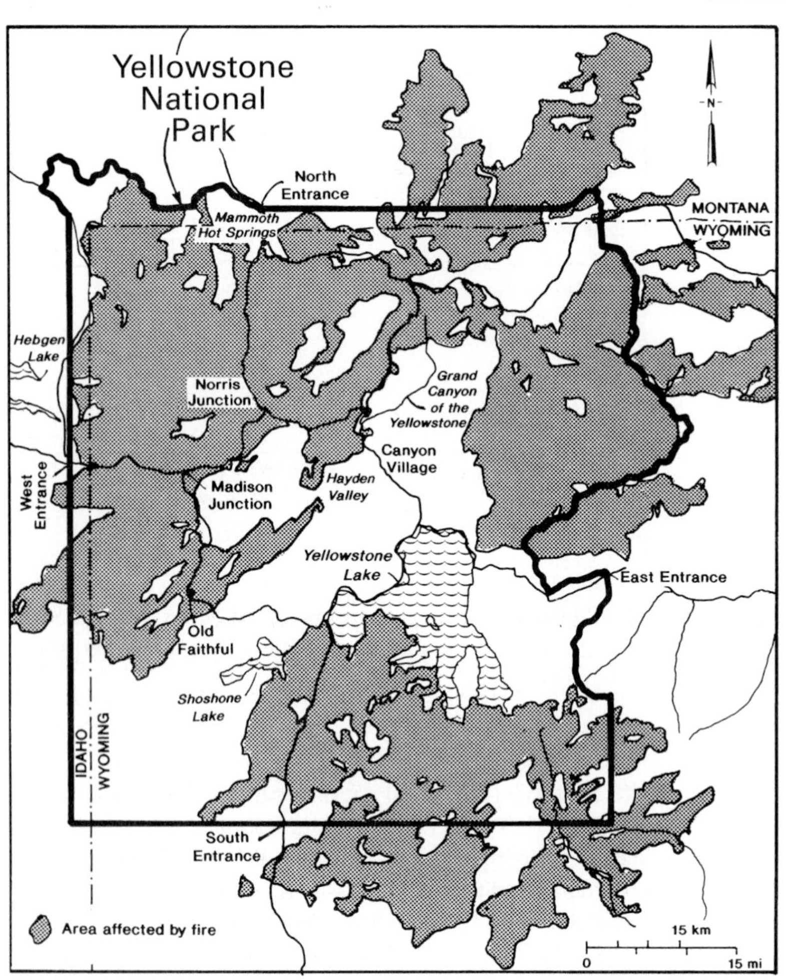

Figure 13.25 Area burned by the Yellowstone fires of 1988.

Following the policy change, between 1976 and 1987 there were 235 lightning fires. A typical fire burned about 100 acres, only eight fires burned more than 1,000 acres, and the largest burned 7,400 acres. The policy change was judged to be a great success. Fires were occurring in moderate amounts and were promoting ecological diversity. Then came 1988.

The winter of 1987–88 was a dry one. Lightning started fires as usual in June 1988, but this year, no rains followed as in the previous six June–July intervals. By late July, over 17,000 acres had burned, and the decision was made to extinguish the blazes and suppress any new fires regardless of their origin. But the situation had added complexities: 1) many forest stands had been killed by infestations of mountain pine beetles, 2) nine decades of fire suppression had created an extensive buildup of dead wood on the ground, and 3) moisture levels in dead wood had dropped from the usual 15 to 20 percent down to 2 to 7 percent.

As firefighters struggled to control the situation, the weather worsened as a wave of high temperatures arrived with sustained high winds. In the 24 hours of 20 August 1988, more square miles of Yellowstone Park burned than the total area burned in any preceding decade.

Despite the continuing efforts of firefighters, the blazes did not weaken until the arrival of mid-September snows, and the flames did not quit until winter conditions prevailed in November. The weather controlled the flames that humans could not. At the fire's conclusion, 1.4 million acres lay burned, including almost half of Yellowstone Park (Figure 13.25). In the previous 116 years, a total of only 146,000 acres had burned.

Ten Years Later The fires of 1988 drastically changed the look of Yellowstone National Park. Thousands of acres of charred trees held the bodies of 269 elk, 9 bison, 6 black bears, 4 deer, and 2 moose. The fires killed trees but opened land to increased sunlight and nutrients that brought forth grasses, wildflowers, and shrubs. Trees are recolonizing the burnt ground with seedlings of lodgepole pine, Engelmann spruce, subalpine fir, and Douglas fir. Standing dead trees continue to fall, and thus enrich the soil as fungi, bacteria, beetles, ants, and other organisms decompose the tree remains. The pre-1988 Yellowstone is gone, but the living and functioning ecosystem is going through many interesting changes and time will allow a return to the past.

California versus Baja California: Pay Now or Pay Later

Many people hold the view that humanity is separate from the environment, that humans are sup-

Table 13.7 Chaparral Areas Burned, 1972–1980

	Total Area (thousands of hectares)	Area Burned (thousands of hectares)	% Area Burned	% Burned after September 1	Number of Fires
Southern California	2,019	166	8.2	72	203
Baja California	1,202	95	7.9	20	488

Source: R. A. Minnich (1983).

posed "to be fruitful, multiply, and subdue the earth." Many seek to control nature for their own benefit. This includes building houses wherever it strikes their fancy. Fires? No problem, we will just extinguish them before they cause any damage. But what is the long-term effect of short-term suppression of fires? An interesting study by geographer Richard Minnich worked on this question using Landsat imagery ("photos" from satellites) to determine the number and size of all chaparral fires in southern California and northern Baja California from 1972 to 1980. The life cycle of the chaparral plant community takes it through a sequence of fire susceptibility. Younger plants do not burn easily, but after 40 years of growth, there is an increased proportion of dead-plant material which acts as fuel aiding fire to burn readily and intensely.

In the United States, fires are fought energetically and expensively. The goal is to not let fire interfere with human activities, no matter how much money it costs to suppress it. In Mexico, fires are simply allowed to burn with little or no human interference. The fire histories of the chaparral areas in the United States versus those in Mexico show some interesting differences (Table 13.7).

In the United States, fires that break out during the cooler, wetter months are quickly extinguished. Thus, most of the chaparral is allowed to grow older and more flammable. Then, when the hot and dry Santa Ana winds come blowing (usually in September, October, and November), firestorms are unleashed that firefighters are powerless to stop, and great numbers of acres burn. Southern California has fewer fires but more large ones.

In Mexico, the fires are smaller because older chaparral is commonly surrounded by younger, more fire-resistant plants. This distribution creates an age mosaic that mixes volatile older patches with younger, tougher-to-burn growths. Fires are more numerous in Baja California, but they are smaller and more of them occur during the cooler, wetter months.

For the 1972–80 period, the percentage of chaparral acreage (1 acre equals 0.405 hectare) burned in the United States and Mexico was about the same. This is despite the enormous expenses and valiant efforts made in the United States to suppress fire. The U.S. fire-control efforts seem to have reduced the number of fires but not the amount of acreage burned. In southern California, the monster firestorms pushed by Santa Ana winds burn tremendous numbers of acres, killing people and destroying thousands of buildings.

Table 13.8 U.S. Wildland Fires in 2000

Number of Fires	122,827
Average Year (1990–1999)	106,393
Acres Burned	8,422,237
Average Year (1990–1999)	3,786,411
Buildings Burned	861
Cost of Fire Fighting	$1.36 billion
On the busiest day—29 August 2000	
84 large fires were burning on 1,642,579 acres in	
16 states	
28,462 people were fighting fire using	
1,249 fire engines	
226 helicopters	
42 air tankers	

Source: National Interagency Fire Center.

The Western and Southern United States in 2000

The year 2000 fire season began early in Florida; it began on January 1. When summer arrived, the conditions for fire in the West and South were ominous: hot weather, low relative humidities, wind, absence of monsoon rains in the Southwest, and a potent source of ignition in lightning accompanying dry thunderstorms. The long dry summer of 2000 provided a reminder of the problems created by 90 years of forest-fire suppression. Lightning strikes throughout the western states ignited major fires that burned millions of acres including many of the same acres in the Bitterroot Mountains that burned in The Big Blowup of 1910 (Table 13.8). Despite the valiant efforts of firefighters, the fires were not extinguished until weather fronts brought welcome rains. The debate about the proper role for humans in the natural forest and fire system will continue for many years. What is the best approach? Logging and mechanical thinning, or controlled burns, or just let nature take its course?

La Niña A La Niña condition (see Chapter 9) existed in the Eastern Pacific Ocean during 1999 and 2000, and it helped set the stage for the fires of 2000. La Niña brings a pool of cool ocean water to the Americas, leading to drier than average weather in the southern tier of states from Florida to California. The Pacific Northwest receives a great deal of rain, but the other western states receive thunderstorms that commonly are dry. The dry thunderstorms bring lightning that starts fires, but there are no accompanying rains to extinguish the blazes. In 2000, the fires began early in spring and did not quit until rain and snow began falling in autumn.

Prescribed Fires

One solution to the problem of dense forests and shrublands is for trained people to deliberately set fires, prescribed fires, at times carefully selected for low wind speeds, low temperatures, humid air, good soil moisture, and other factors that limit the size and power of fires. Prescribed fires have become a widely implemented program. From 1995 to 2000, more than 31,000 prescribed fires were set by federal agencies (Table 13.9). The hope is that a prescribed fire will be a controlled fire that burns only the desired area. But with so many fires being set by so many people, mistakes will happen.

Los Alamos, New Mexico, May 2000 On 4 May 2000, National Park Service personnel set what they hoped would be a controlled fire at Bandelier National Monument to clear an understory of shrubs that could provide fuel for a wildfire. The next day, high winds whipped the prescribed fire into a wildfire that consumed 50,000 acres of national forest, 235 houses in Los Alamos, displaced 25,000 people, damaged or destroyed 115 buildings in Los Alamos National Laboratory and came frighteningly close to the hazardous materials sites at the nuclear weapons research facility.

The prescribed fire caused the very disaster it was supposed to prevent. The federal government has paid 455 million dollars in compensation to victims of the fire. Critics say that the fire-setting personnel were careless and incompetent. Proponents say a lightning strike would have started the same fire anyway. And so, the debate rages on. Shall we practice fire fighting or fire lighting?

Table 13.9 U.S. Government Prescribed Fires

Year	Acres Burned	Costs
2000	1,077,314	>$455,000,000
1999	2,240,105	$99,104,000
1998	1,889,564	$70,793,000
1997	1,601,158	$36,146,000
1996	915,163	$29,550,000
1995	918,300	$20,448,000

Source: Bureau of Land Management, Bureau of Indian Affairs, Fish and Wildlife Service, National Park Service, Forest Service.

Australia

Australia is bedeviled by bushfires, major conflagrations swept by high winds through the eucalyptus-dominated land. Most eucalyptus trees and bushes have highly flammable oil in their branches. It makes good wood for your fireplace, but in a natural setting the trees may be heated so hot that their sap boils and whole trees explode in flame. This presents obvious problems for houses and towns in the path of a eucalyptus-fueled bushfire. But it is not a problem for the plant species. Most eucalyptus varieties are designed to burn hot and fast; it is then that their fruit opens and releases unburned seeds to germinate in the fire-cleared ground. The fast-moving fires may burn down trees, but many of the resilient plant species simply send up shoots from their stumps or roots and grow to full-size trees again in several years.

The worst bushfires in history correlate with drought and wind. Many of these weather episodes occur when an El Niño–Southern Oscillation circulation system is operating in the oceans (Figures 9.28 to 9.30). When the ocean waters off Australia are cooler, evaporation and precipitation are reduced. The lessened rainfall creates drought that can extend through an entire Southern Hemisphere winter. Bushfire conditions may start with a high-pressure system off southern Australia. As the high-pressure zone moves slowly eastward, it induces air masses over the central desert to blow southeastward from the interior toward the heavily populated southeastern region of Australia. As the winds descend from the high desert, they warm in foehn-wind fashion, producing temperatures around 40°C (104°F), humidities of less than 20 percent, and speeds of at least 60 km/hr (37 mph).

The El Niño–Southern Oscillation of 1982–83 was particularly strong, and the Australian summer was the driest in recorded history. On Ash Wednesday, 16 February 1983, strong foehn-type winds from the interior were reinforced by the jet stream in the upper atmosphere. The winds blew more than 60 km/hr (40 mph), and temperatures were about 40°C (104°F). Adelaide, the capital of South Australia, and Melbourne, the capital of Victoria, were ringed by fires pushed by the southeastward-blowing winds.

In midafternoon, a cold front swept through the area, dropping temperatures 10°C (50°F). However, the cooler winds did not cool off the fires but instead changed their di-

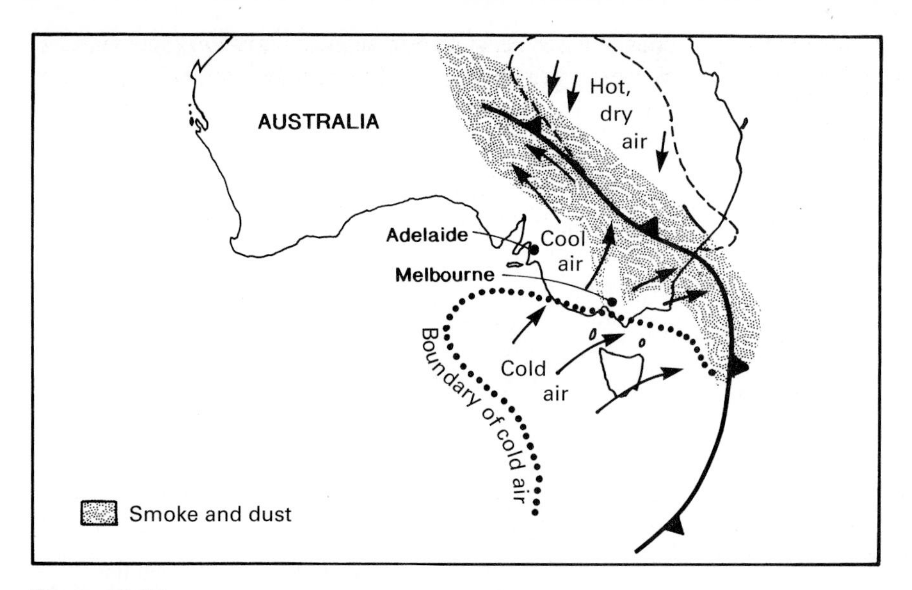

Figure 13.26 Weather systems at noon, 17 February 1983. The hot, dry air mass that spawned the foehn winds has been pushed northeastward, along with the flames, by the advancing cold front.

rection of movement to the northeast and increased their speed (Figure 13.26). The winds were blowing 70 km/hr (45 mph) with gusts up to 170 km/hr (105 mph), and the fire front advanced at speeds up to 20 km/hr (over 12 mph). The firestorms were so strong that their winds snapped tree trunks.

When the fires were finished, 76 people lay dead, 3,500 had suffered injuries, and 1,700 houses and 20 towns were gone. The livestock toll was also great as more than 300,000 sheep and 18,000 cattle perished along with uncounted numbers of native animals.

The bushfire hazard is ever present. In the second week of January 1994, the hot, dry, fast winds returned to char more than 1.2 million acres, again leaving dead bodies and incinerated homes behind.

The Similarities of Fire and Flood

Stephen J. Pyne has pointed out that floods can serve as a metaphor for fire. Fires and floods seem so different, yet they have some general characteristics in common.

Both fire and flood are closely related to weather, plant cover, and topography.

Both fire and flood are at their strongest when atmospheric conditions are extreme. Fast, dry winds push flames, while heavy rains feed floods.

Both fire and flood move across the landscape and through human developments as waves of energy. A fire front is a wave of chemical energy released from temporary storage in organic matter. A flood crest is a wave of mechanical energy unleashed when the potential energy of high topographic position is converted to the kinetic energy of motion.

Both fire and flood become more turbulent the faster they move and the bigger they grow.

Both fire and flood can be described by their size and frequency. As a first approximation, the bigger the fire or flood, the longer the return time until the next big one. Both fire and flood are effectively understood as 50-year, 100-year, or other recurrence-time events.

Both fire and flood are aggravated by human activity. Fires are made more intense by buildings placed in dense growths of plants and by our habit of quickly suppressing small fires, thus allowing organic debris to build into large masses that provide fuel for gigantic firestorms, such as in Yellowstone National Park in 1988. Floods are made more destructive by buildings placed on floodplains and by levees built to protect against floods that inadvertently cause the record-high flood levels, such as in the Upper Mississippi River Valley in 1993.

Summary

Fire is a natural force used by humans for purposes both good and bad. Controlled use of fire provides energy for industry, travel, plus lighting, heating, and cooking in our homes. But fire also kills and destroys; in the United States, from 1978 to 1999, fire caused a yearly average of 6,100 deaths (82 percent in residences) and over $6 billion in property damage.

Fire is the rapid combination of oxygen with carbon, hydrogen, and other elements of organic material in a reaction that produces flame, heat, and light. In effect, fire is photosynthesis run in reverse.

Before a fire starts, a preheating stage occurs where water is expelled from wood, plants, or fossil fuels by flames, drought, or hot weather. When temperatures exceed 300°C, wood breaks down and gives off flammable gases in the process of pyrolysis. If oxygen is present, pyrolized gases can ignite and combustion begins.

In the flaming combustion stage, the pyrolizing wood burns hot. Released heat keeps the wood surface hot through conduction, diffusion, radiation, and convection. After the active flames pass, a glowing combustion stage exists where fire slowly consumes the solid wood.

The spread of fire depends on several factors. 1) Types of plants or fuel burned; some plants, such as eucalyptus and chaparral, ignite easily and burn intensely. 2) Strong winds bring oxygen and push flames forward. 3) Topography has numerous effects. For example, before a fire, it helps control plant distribution, and during a fire, a steep slope acts like a chimney that fire races up. 4) Fires heat air, which rises buoyantly, creating its own winds.

Fire threats are greatest in areas with big contrasts between wet and dry seasons. A wet season triggers voluminous plant growth. Then dry conditions dehydrate plants, making it easier for ignition to occur. Add some strong winds (e.g., foehn winds), and major fires break out, usually in numerous places.

Buildings can be constructed to withstand fire. Traditional structures can be made safer by eliminating flammable vegetation and woodpiles near them, avoiding wood-shingle roofs and overhanging wood balconies or decks, and using double-pane glass in windows and doors and spark arresters on chimneys.

Fires cannot be prevented, only deferred. We pay a price for fire suppression. Allowing natural fires to burn helps prevent buildups of extensive debris that can fuel a firestorm during heavy winds. Large firestorms are essentially unstoppable. We must choose between lots of little fires or a few monstrous blazes.

Terms to Remember

Questions for Review

1. How many people die on average each year in fires in the United States?
2. Write an equation that describes fire.
3. Compare fire to photosynthesis.
4. Explain the process of pyrolysis.
5. Explain the differences between conduction, diffusion, radiation, and convection.
6. What is the difference between flaming combustion and glowing combustion?
7. Will a typical wildfire burn faster upslope or downslope? Why?
8. Explain how a wildfire can create its own winds.
9. Draw a cross section and explain how foehn winds form. Why are they typically hot and dry?
10. What difficulties are presented when duff catches fire?

Questions for Further Thought

1. Explain how to build a campfire. Explain the fire processes occurring at each stage of your campfire.
2. Make a detailed list of actions you could take to make your current residence safer from fire.
3. If you were designing your dream house on your dream lot, what features could you incorporate into the house and landscape design to better protect your house from destruction by fire?
4. Evaluate the wisdom of quickly suppressing a local wildfire during a time of cold weather with low wind speeds.

Suggested Readings and References

Albini, Frank A. (1984, November–December). Wildland fires. *American Scientist,* 72, 590–97.

Jeffery, David. (1989, February). Yellowstone, the great fires of 1988. *National Geographic,* 175, 255–73.

Lyons, John W. (1985). *Fire.* New York: Scientific American Books.

Minnich, Richard A. (1983, March 18). Fire mosaics in southern California and northern Baja California. *Science,* 219, 1287–94.

National Fire Protection Association. (Undated). *The Oakland/Berkeley Hills Fire.* Boise: NFPA.

Pyne, Stephen J. (1982). *Fire in America: A Cultural History of Wildland and Rural Fire.* Princeton, N. J.: Princeton University Press.

Pyne, Stephen J. (1995). *World Fire: The Culture of Fire on Earth.* New York: Henry Holt and Co.

Pyne, Stephen J., Andrews, P. L., and Laven, R. D. (1996). *Introduction to Wildland Fire,* 2nd edition. New York: John Wiley & Sons.

Romme, William H., and Despain, Don G. (1989, November). The Yellowstone fires. *Scientific American,* 261, 37–46.

Rossotti, Hazel. (1993). *Fire.* New York: Oxford University Press.

Schroeder, Mark J., and Buck, Charles C. (1970). *Fire Weather.* U.S. Department of Agriculture Handbook 360.

Voice, M. E., and Gauntlett, F. J. (1984). The 1983 Ash Wednesday fires in Australia. *Monthly Weather Review,* 112, 584–90.

Videos

Fire in America. (1981). National Fire Protection Association (30 min.).

Yellowstone Aflame. (1991). Finley-Holiday Film Corporation (30 min.).

Yellowstone's Burning Question. (1989). Nova—PBS (60 min.).

Wildfire. (1990). PBS Video (60 min.).

Why America Burns. (1982). Nova/WGBH-Boston (57 min.).

Chapter 14

The Great Dyings

History fades into fable; fact becomes clouded with doubt and controversy; the inscription molders from the tablet; the statue falls from the pedestal. Columns, arches, pyramids, what are they but heaps of sand; and their epitaphs, but characters written in the dust?

—Washington Irving, 1820, *The Sketch Book*

Recorded human history tells of horrifying natural disasters. Individual earthquakes and floods have each killed in excess of three-quarters of a million people, single cyclones have drowned more than one-half million humans, and individual volcanic eruptions have killed tens of thousands of people. Yet these disasters etched into the historic record seem almost insignificant when compared to the great dyings documented in the fossil record. The human tragedies wrought by natural disasters involve individuals of a **species,** not the entire species or millions of species. Several times in the last 544 million years, many, or even most, of the species on Earth became extinct in a geologically short time. These great dyings or mass **extinctions** are the biggest natural disasters known to have occurred on Earth.

Early Understanding of Extinctions and Geologic Time

In 1786, the French paleontologist Georges Cuvier proved that extinction of species had occurred. He demonstrated that the skeletons of mammoths (Figure 14.1) were distinctly different from those of living elephant species and yet mammoths no longer live; thus, mammoths had gone extinct.

Cuvier also was impressed with the profound changes in the sedimentary rock record. Fossils made abrupt first appearances, were found in abundance in overlying rock layers, but then were not found in higher rock layers (Figure 14.2). Cuvier recognized that entire communities of plants and animals died out in geologically short lengths of time.

Figure 14.2 also illustrates Steno's 1669 Law of Superposition, which explains that younger layers of sediment are deposited on top of older layers. Thus, in a nonoverturned sequence of sedimentary rocks, the lowest layer is the oldest, and each bed encountered moving up the sedimentary section is progressively younger.

In 1799, the English civil engineer William Smith published a list describing the sequence of sedimentary **strata** and its contained fossils in the area of Somerset, England. He indicated clearly that specific intervals of sedimentary strata could be recognized over broad areas by the unique assemblages of fossils they contain. The principle was called the **Law of Faunal Assemblages;** it

Figure 14.1 Once there were many species of mammoths, now there are none.

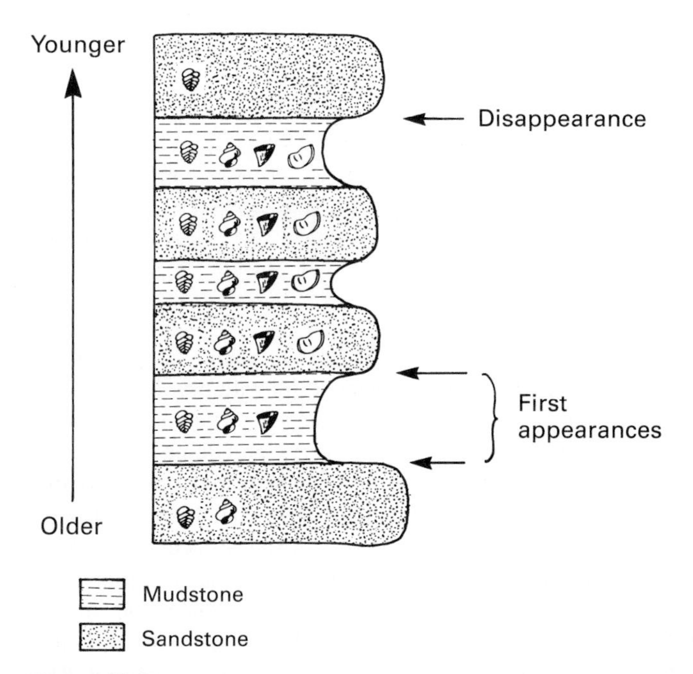

Figure 14.2 A vertical sequence of sedimentary rock layers illustrates the principles of superposition (the lower the layer, the older) and faunal succession (older forms of life die out and new forms develop).

explains that strata of like age can be recognized by the like assemblage of fossils they contain.

The Laws of Superposition and Faunal Assemblages meld together in the observation that fossils collected from older rock layers (lower by superposition) are older and more different from present-day organisms than are fossils collected from younger rock layers (higher by superposition). The fact that older fossils are more unlike living organisms implies that old forms of life have died out and that new forms of life have developed. This relationship is known as the **Law of Faunal Succession.**

William Smith used his principle of faunal assemblages to make geologic maps of England and Wales that were published in 1815. Geologists throughout the world reacted to the work of Smith, Cuvier, and others by recording the order of sedimentary strata and the sequence of their contained fossils. All over the world, sedimentary strata were classified and subdivided in vertical columns based on their differing fossil assemblages. The sequences of sedimentary rocks and fossils in different parts of the world cover different intervals of Earth history. Where fossil assemblages in different areas were found to be the same, the rocks that contain the fossils were considered to be the same age. By 1841, the co-occurrences of fossils had been compiled to erect a standard geologic column for the world; this was an early version of the geologic time scale based on fossil succession that we use today (Figure 14.3).

Beginning in the twentieth century, geologists have added quantification to the fossil record using dates determined from radioactive elements in igneous rocks associated with fossiliferous sedimentary rocks. The ongoing refinements of the geologic time scale involve geologists and **paleontologists** studying rocks and fossils throughout the world.

Brief History of Life

The fossil record contains the appearances of millions of life forms (Figure 14.3). There have been some remarkable changes in life on Earth. As described by T. A. Conrad: "Race after race resigned their fleeting breath—The rocks alone their curious annals save."

At 3.85 billion years ago, **archaea** were alive, reproducing and evolving. They obtained energy for their lives by breaking the chemical bonds within molecules such as CO_2, H_2O, and N_2. Archaea are one of the three major branches of life, along with bacteria and eukarya (which includes plants and animals). Archaea today are found down to 3.5 km (2.3 mi) below the surface under pressures over 200 atmospheres and at hot springs and deep-ocean spreading centers with temperatures up to 113°C (232°F), but they are killed by oxygen. These organisms may be relatives of the earliest life forms on Earth.

Over 3.5 billion years ago, photosynthetic bacteria were removing some of the abundant carbon dioxide (CO_2) from Earth's atmosphere and combining it with water using the Sun's energy. The process of photosynthesis gives off oxygen (O_2) and thus began the radical transformation of the Earth's atmosphere that ultimately led to its present composition (Table 9.1).

Organisms of these early times reproduced by simple division of their cells. Thus, if a **gene** had an advantageous

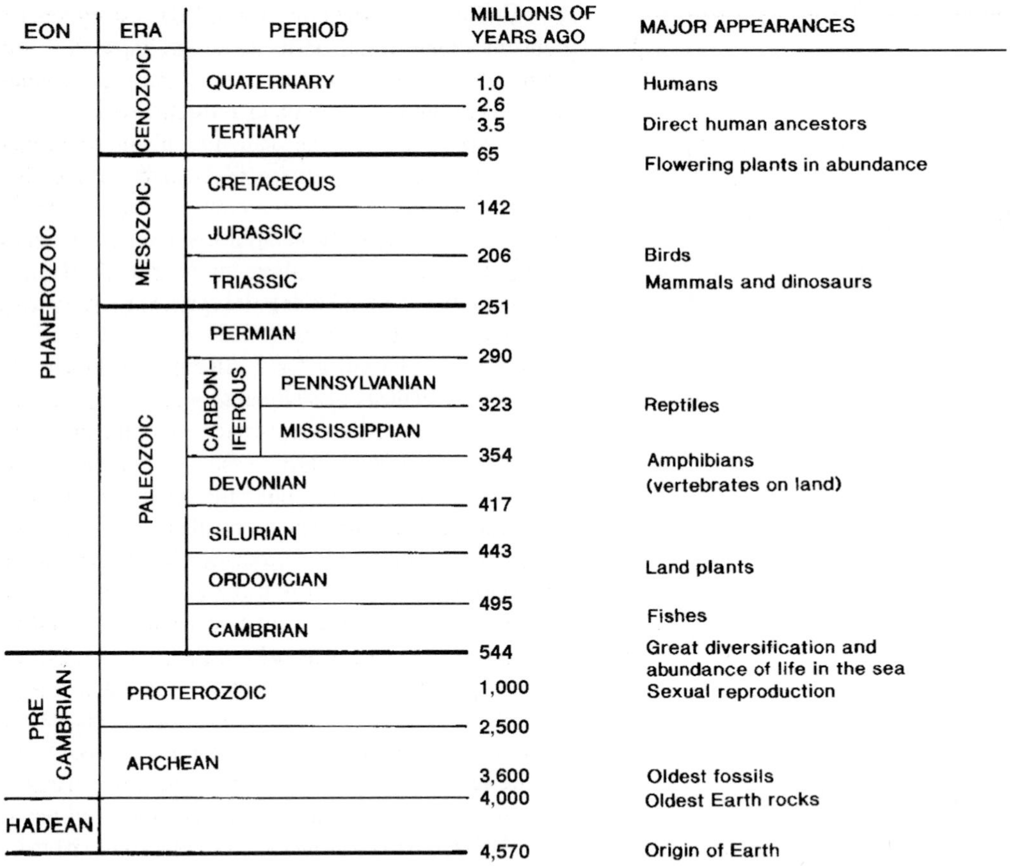

EON	ERA	PERIOD	MILLIONS OF YEARS AGO	MAJOR APPEARANCES
PHANEROZOIC	CENOZOIC	QUATERNARY	1.0	Humans
			2.6	
		TERTIARY	3.5	Direct human ancestors
			65	Flowering plants in abundance
	MESOZOIC	CRETACEOUS		
			142	
		JURASSIC		
			206	Birds
		TRIASSIC		Mammals and dinosaurs
			251	
	PALEOZOIC	PERMIAN		
			290	
		CARBONIFEROUS: PENNSYLVANIAN		
			323	Reptiles
		CARBONIFEROUS: MISSISSIPPIAN		
			354	Amphibians (vertebrates on land)
		DEVONIAN		
			417	
		SILURIAN		
			443	
		ORDOVICIAN		Land plants
			495	Fishes
		CAMBRIAN		
			544	Great diversification and abundance of life in the sea
PRE CAMBRIAN	PROTEROZOIC		1,000	Sexual reproduction
			2,500	
	ARCHEAN		3,600	Oldest fossils
			4,000	Oldest Earth rocks
HADEAN			4,570	Origin of Earth

Figure 14.3 Geologic time scale based on superposition of sedimentary rock layers and the irreversible succession of fossils. Numerical ages were measured on igneous rocks found in association with fossil-bearing sedimentary rocks.

mutation, it was limited to a single line of offspring. By 1 billion years ago, sexual reproduction had appeared. With the arrival of sex, cells could share and mix genetic material and thus speed up evolutionary changes by thousands of times.

During the Earth's early history, organisms lived primarily below the surface in the oceans and within sediments and rocks. As oxygen (O_2) given off during photosynthesis built up to a large volume in the atmosphere, some was altered to ozone (O_3), thus building a shield from the Sun's lethal ultraviolet radiation. With a protective atmospheric shield, more multicellular life could come out into the open. Marine rocks from 670 million years ago contain the oldest known fossils of multicellular animal life, including nearly all the major body plans that exist today.

About 544 million years ago, life on Earth began a 40 million-year-long burst of remarkable evolutionary change. From this time onward, the fossil record improves because many of the new groups of organisms began creating hard parts, such as shells, that preserve well as fossils. The initial development of hard parts has been referred to as "the world's first arms race" wherein organisms covered their bodies with armor, possibly as protection against predators.

Moving on through time (up the sedimentary rock sequence), the Earth has continued to fill with new forms of life (Figure 14.3). For example, the waters became home to fishes, while plants moved out of the water and spread across the lands. Later on, vertebrate animals appeared on land—amphibians, then reptiles, and finally the mammals and birds. The **evolution** and increasing diversity of species is continuing.

Today, life is almost everywhere—in waters ranging from 0°F to above boiling, on land, in the air, in the soil, in ice, around deep-sea volcanic vents, within the pores of rocks buried several kilometers deep, and inside other organisms. The overall trend has been an increase in the diversity and abundance of life, but there have been major setbacks and reorganizations. Figure 14.4 plots the number of families of marine animals with hard skeletons that have lived during the last 600 million years. Over time, there has been an obvious growth in the number of families; this growth reflects the increase in diversity and abundance of marine life. However, there also have been marked die-offs in which more than 50 percent of the families became extinct within relatively short time intervals of several million years or less. To better understand these dramatic extinction events, it is necessary to look at events on the species level, which is where the real action takes place.

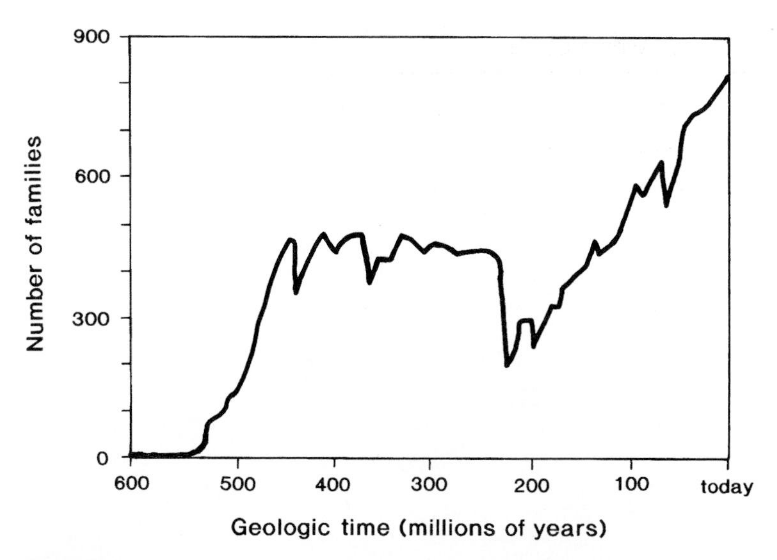

Figure 14.4 Number of families of marine animals with mineralized hard parts versus geologic time. The overall increase in families with time is interrupted by extinction events.

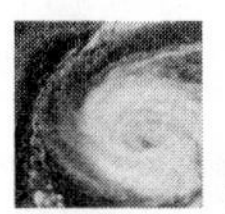

Species and the Fossil Record

The classification of life is a topic being hotly debated today. However, the basic terminology laid out by the Swedish botanist Linnaeus (Carl von Linne, 1707–1778) provides us with terms still widely used. He organized life into kingdoms and then subdivided successive categories down to species (Figure 14.5). Overviews of life history tend to focus on larger groupings in the hierarchy, such as the families in Figure 14.4. However, it is at the species level where both evolution and extinction occur.

A species is a population of organisms so similar in life habits and functions that they can breed together freely and produce reproductively viable offspring. They are reproductively isolated by such differences from other species. Members of a species share a common pool of genetic material (**genome**). A species may migrate over a broad geographic area, causing reproductive isolation between local populations in widely separated areas. As genes mutate and are recombined in sexual reproduction, geographically segregated populations within a species may begin to develop differences in behavior and anatomy that over time produce enough changes to create new species. This is the process of evolution wherein changes in the genetic pool of a population lead to the origins of new species: as defined by Charles Darwin, descent with modification. A species is the smallest biologically real and distinct unit of individuals that share a common ancestor not shared with other organisms. Each species is unique; each new species is never entirely the same as any previous one.

Species are subjected to many environmental changes. Oceans recede and expose land, climates change from hot to cold and from wet to dry, volcanoes spew out great volumes of magma and gases and then quiet down—the physical, chemical, and biological conditions on Earth change both locally and globally. Some species are flexible enough to handle environmental fluctuation, but other species are not

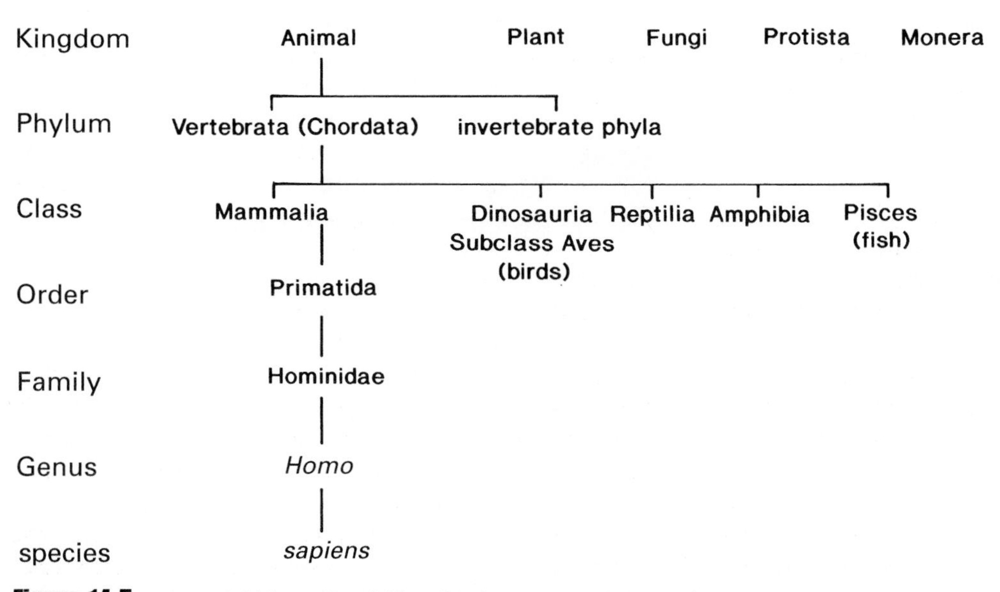

Figure 14.5 A partial hierarchy of life using humans as an example.

and thus become extinct singly, in relay, or in groups. Extinctions of species are an ongoing fact of life on Earth.

Causes of extinction are complex and vary among events, but they combine to create a relatively constant background level of extinction; there are always some species going extinct. Again, the fossil record speaks clearly: each species is nonrecurring; once a species dies out, it never reappears. Most of the species that have ever lived on Earth are now extinct; in fact, over 99.9 percent of all plant and animal species that have ever lived are extinct. At present, species diversity on Earth is estimated to range from 40 to 80 million; but consider that these large numbers are less than 0.1% of the species in Earth history.

It is increasingly noted that extinctions clear out living space (habitats or "ways of living"). Extinctions are like evolutionary foreclosures that provide new opportunities for different organisms to evolve new ways of life, to occupy vacated **niches** in the environment, and in so doing, create even more new niches. For example, the extinction of most dinosaur species (except birds) 65 million years ago, opened up opportunities for ways of life that mammals adopted and filled during the following several million years. The constant elimination of old species and refilling of their vacated spaces in the environment by new species has created an incredible variety of life forms during Earth's history, increasing diversity fourfold since Cambrian time. Major groups of new organisms have usually arisen, not as better competitors that eliminated already living species, but as groups of opportunists that extended into vacant environments and adopted new ways of life.

Background extinctions occur at a moderate rate and continuously open up a limited number of opportunities for the origin of new species and new niches. Mass extinctions are relatively uncommon events; they cause worldwide elimination of numerous species, thus opening up a wealth of niches for new organisms to fill by the evolutionary process. Following a mass extinction, life on Earth takes on a different appearance as the surviving species are joined by numerous new species in the following several million years.

Figure 14.6 A tropical coral reef.

The Tropical Reef Example

The tropical oceans today are host to massive **reefs** built by scleractinian corals along with other framework-building species. Reefs have porous, wave-resistant frameworks, creating shelters (new niches) occupied by numerous other species, both within the reefs themselves and in the quiet, protected areas behind the reefs (Figure 14.6).

The fossil record of reefs tells a varying story. Reefs have waxed and waned through time, and the cast of characters has changed. At times in the past, tropical oceans were home to well-developed reefs, but there also have been long intervals when no reef-framework builders existed in the world and other lengthy times when organisms were merely evolving reef-building abilities (Figure 14.7). Each time a mass extinction eliminated the reef-builders of its day, there followed a long interval of time before other creatures were able to fill the environmental void. Each time reefs reappeared in the world, they were built by a different group of organisms.

Mass Extinctions During Phanerozoic Time

Extinction occurs on the species level. When so many individuals of a species die that reproduction fails, then the continuity of their kind is stopped, and extinction occurs. The average "life span" of a species is about 4 million years. Because there are so many millions of fossil species, plotting all of their extinctions through geologic time is difficult. However, a major effort was carried out by American paleontologist John Sepkoski at the next hierarchial level—the genus. The times of extinction were plotted for about 25,000 genera of marine invertebrates and protozoa (Kingdom Protista). The study involved plotting a tremendous amount of extinction data calculated against geologic time as:

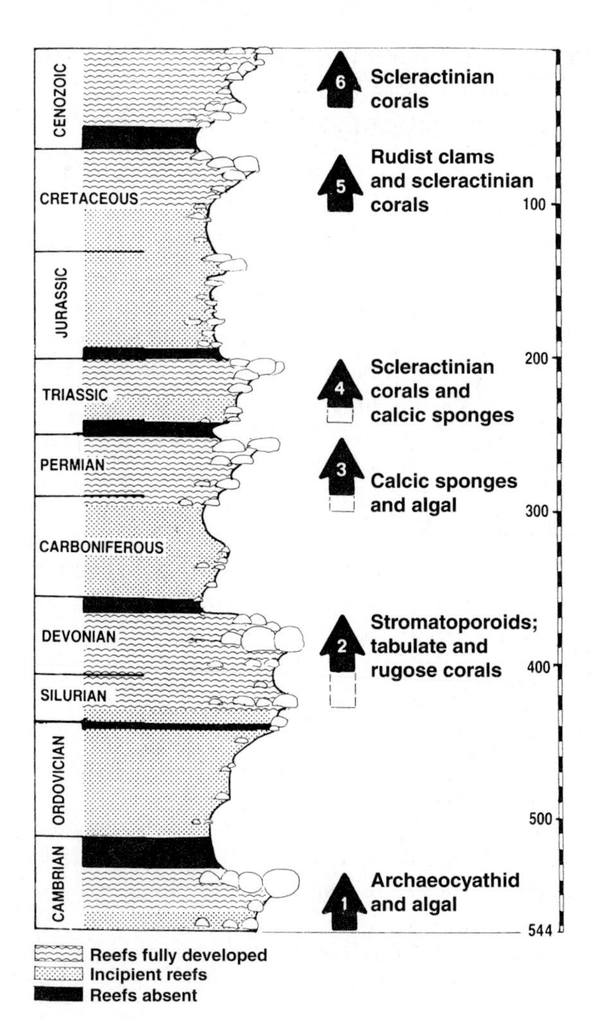

Figure 14.7 Geologic column showing reef-building organisms and their times. Six major successions are shown terminated by mass extinctions. Species are different in each reef phase.

$$\% \text{ extinction} = \frac{\text{number of generic extinctions}}{\text{number of genera alive at that time}}$$

The results are preliminary but fascinating (Figure 14.8). The plot of generic extinctions versus time produces a highly variable, jagged line that obscures trends. But note that the level of background extinctions was around 50 percent in the early going (during Cambrian time) when so many new life forms appeared and their extinctions were common. Ignoring the peaks and valleys of the solid line, it can be seen that background extinctions have declined with time to around 5 to 10 percent in more recent (Late Cenozoic) time.

Against this declining rate of background extinction, there are marked spikes on Figure 14.8 that identify mass extinctions; each spike records a significant increase in the number of extinctions of organisms that occurred in a geologically short length of time.

The geologic time scale can be redrawn to emphasize the times of mass extinctions (Figure 14.9). The biggest extinction events coincide fairly well with divisions in the nineteenth-century geologic time scale, but new perspectives are also needed. A new approach to assessing the frequency of mass extinctions is used by American paleontologist Dave Raup (Figure 14.10). The analysis follows the same logic used in analyzing the size and frequency of floods (Figure 12.11). Remember the approach for floods: yearly flood volumes are plotted against the historic time record to produce a flood-frequency curve, allowing estimates of how often certain size floods will occur. The same approach is used to forecast mass extinctions: numbers of extinctions are plotted against time producing an extinction-frequency curve (Figure 14.10). The curve allows estimates of how often a certain size mass extinction might occur.

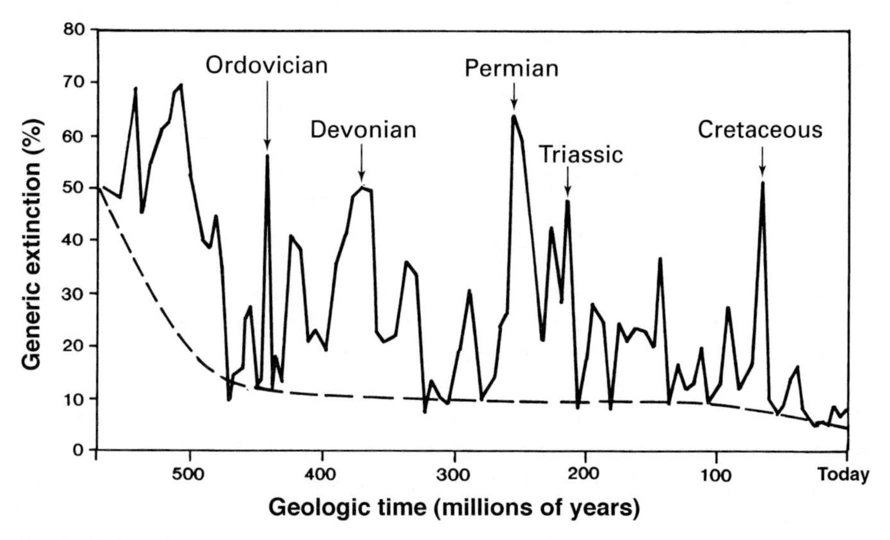

Figure 14.8 Extinction patterns of genera of marine invertebrates and protozoans versus time. The percentage of extinctions was calculated by dividing number of extinctions by number of genera alive at that time. Heavy dashed line represents background extinctions.

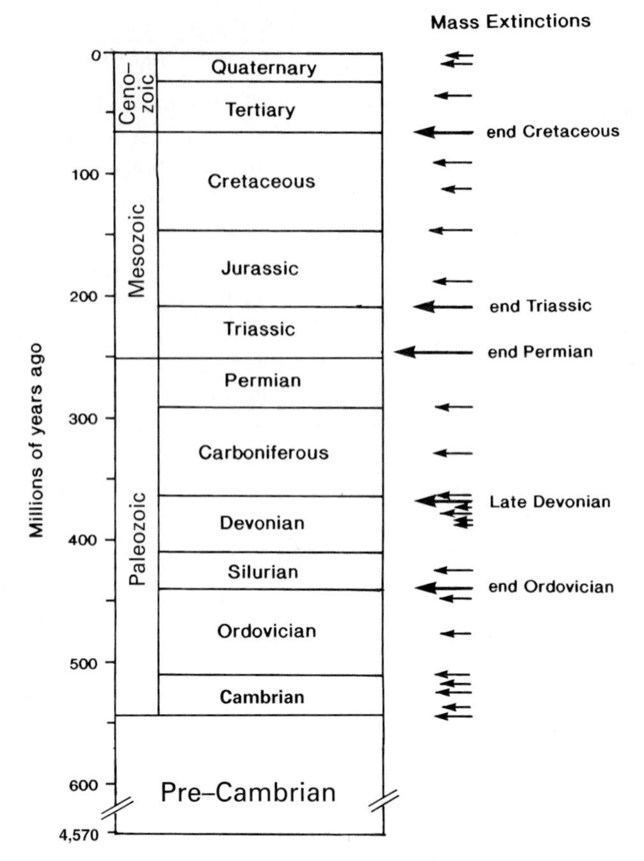

Figure 14.9 Geologic time scale showing mass extinctions. The bigger the extinction, the bigger the arrow.

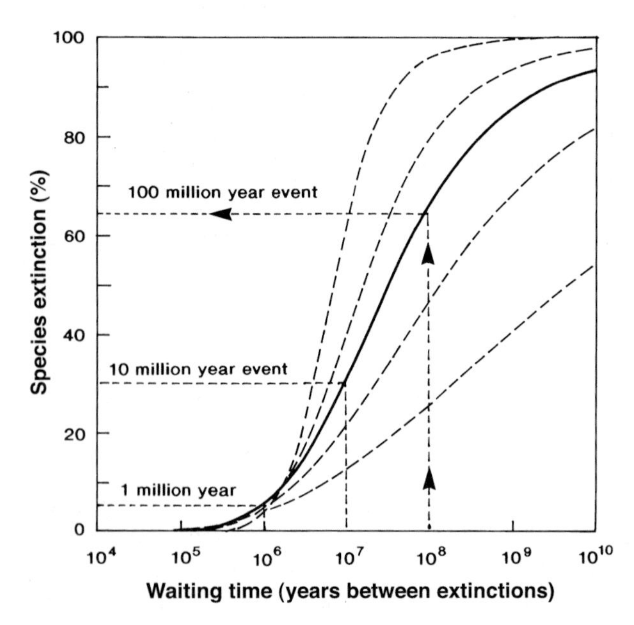

Figure 14.10 Extinction-frequency curves. The dark curve is a best-estimate curve. For example, approximately every 100 million years (10^8 years), up to 65 percent of species may die out.

Possible Causes of Mass Extinctions

Natural disasters take large numbers of human lives; just hearing the news about the devastation they wreak shocks our senses. Yet earthquakes, hurricanes, floods, and such cause only local mass mortality. What are the powerful events that drive large numbers of species into extinction all around the Earth? There are numerous possible causes of mass extinctions, including the longer lasting and more far-reaching effects of plate tectonics, voluminous volcanic outpourings, climate changes, asteroid and comet impacts, and biologic processes.

Plate-Tectonic Causes

The rate of seafloor spreading varies, the number and size of continental landmasses differ over time, and the rise and fall of sea level changes the percentage of land versus sea. Today, the oceans cover about 71 percent of the Earth's surface, and land comprises 29 percent. Figure 14.11 shows that the world shoreline lies on top of the gently sloping continents. A sea-level drop could increase the percentage of land on Earth to upwards of 40 percent, or a sea-level rise could drop the percentage of land to around 17 percent. Changes of these magnitudes have happened in the geologic past, and they have had significant effects on life.

Changes in Seafloor Spreading Rates The volume of magma rising up through spreading centers varies over geologic time. During times of more rapid spreading, the volcanic mountain chains at the spreading centers/ridges greatly increase in mass and volume with new rock that retains warmth and buoyancy (Figure 14.12). The increase in volume of oceanic crust has an effect on the oceans similar to you dunking your body in a bathtub full of water. What happens in each case is that displaced water spills over the sides—of the bathtub onto the floor, or of the ocean basins onto the continents.

For example, during mid-Cretaceous time, from 110 to 85 million years ago, the tempo of seafloor spreading is thought to have greatly increased. A large volume of new oceanic crust can cause the oceans to spill over onto the continents and flood the interior of North America (Figure 14.13). At that time, global sea level was over 200 m (660 ft) higher than today, the area of shallow seas was almost doubled, the amount of exposed land was severely reduced, and world climate was warmer due to the abundance of shallow seas. Many of the species that flourished under these conditions were severely tested when the oceans pulled back from the continents: the area of shallow seas was sharply reduced, the land area and its interconnections grew, and climates changed.

Sea-Level Changes The water that falls as snow to build glaciers comes from the oceans via evaporation. The bigger the glaciers, the lower the sea level. Today, about 25 million km³

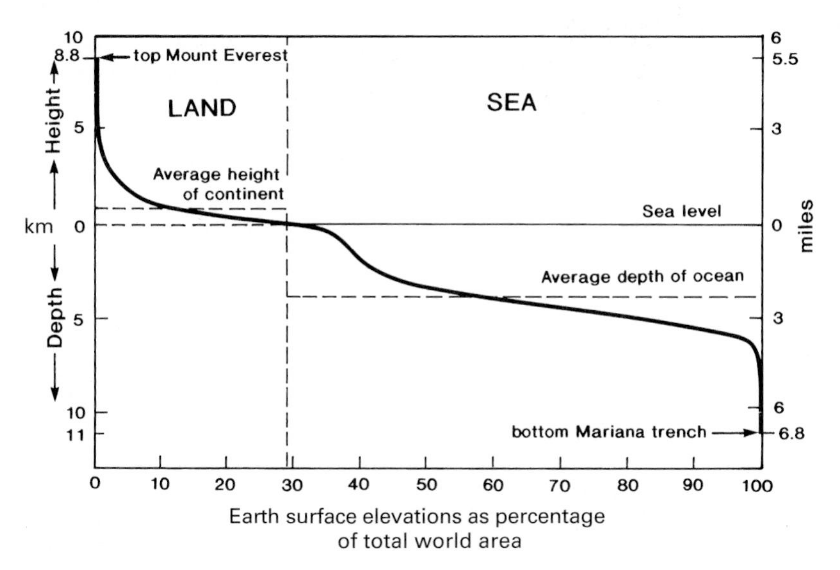

Figure 14.11 Elevations of land and depths to sea floor as percentages of Earth's surface area.

From P. J. Wyllie, *The Way the Earth Works.* Copyright © 1976 John Wiley & Sons, Inc., New York. Reprinted with permission of John Wiley & Sons, Inc.

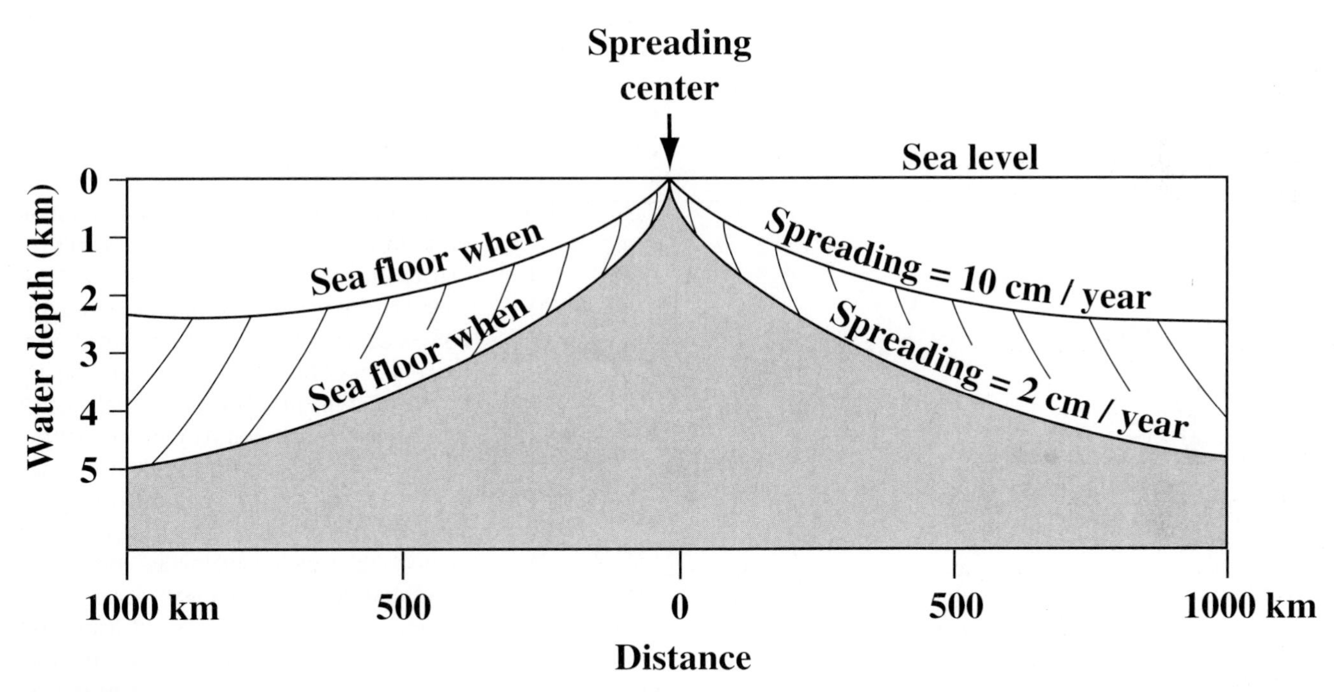

Figure 14.12 Comparative volumes of volcanic mountains at spreading centers. When spreading rates are higher, the greater volume of volcanic rock causes ocean water to spill out of the ocean basins and over the edges of the continents.

(6 million mi^3) of ice ride on the continents, mostly on Antarctica and Greenland. If all of this ice were to melt, world sea level would rise about 70 m (230 ft).

Just 20,000 years ago, during the most recent expansion of the continental glaciers (Figure 9.22), about three times as much ice (75 million km^3 or 18 million mi^3) sat on the continents. To build glaciers this voluminous requires that sea level be drawn down about 140 m (460 ft) from today's levels. The total change in sea level from the glacial peak of 20,000 years ago to a world free of continental glaciers would see a sea-level swing of about 210 m (690 ft); this would greatly change the percentages of land versus sea (Figure 14.11).

Remember that world sea level can also change by over 200 meters via changes in rates of seafloor spreading. The sea-level changes due to the presence or absence of

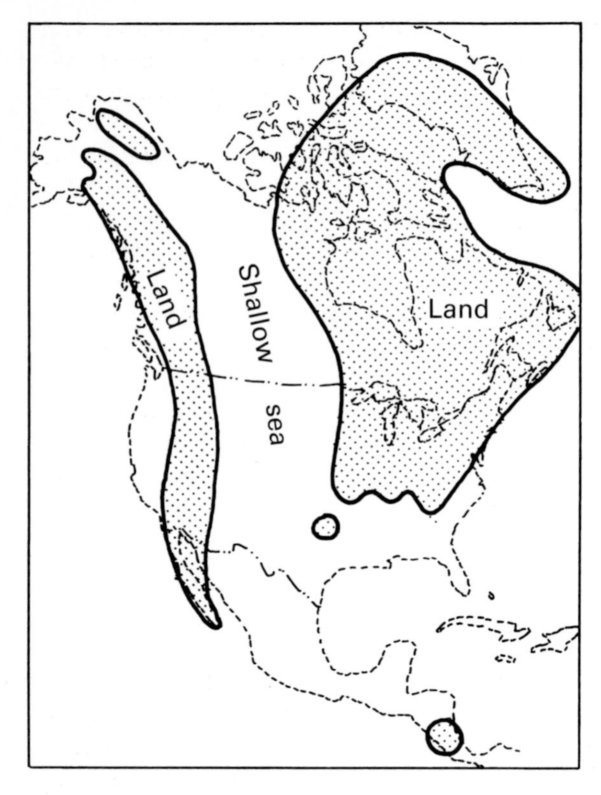

Figure 14.13 The proportions of land and sea in North America 100 million years ago.

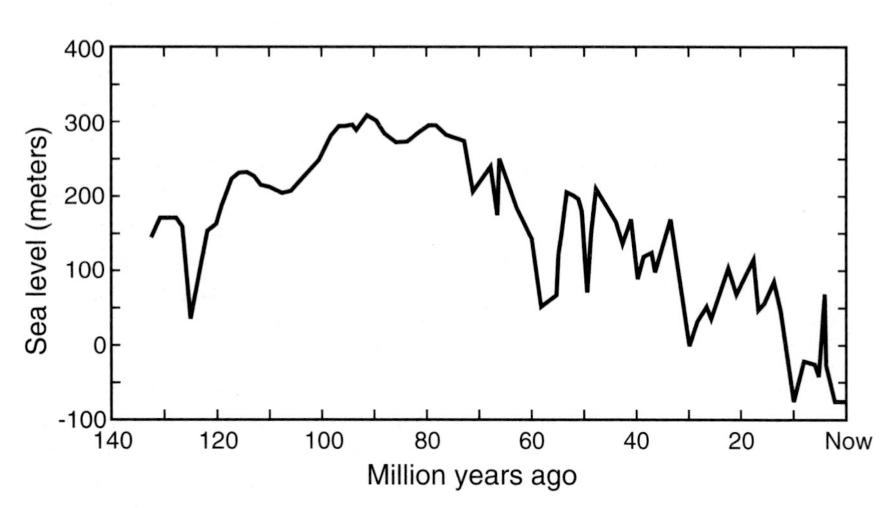

Figure 14.14 Sea-level changes in the last 133 million years. Based on data from Haq et al., 1987.

continental glaciers and the variations in seafloor spreading rates operate independently. They can cancel each other out, or they can combine to cause great rises and falls of sea level that put major stresses on life. The result of their changes is a world with ever-changing sea levels (Figure 14.14).

Numbers and Sizes of Continents In Late Permian—Early Triassic time (about 260 to 240 million years ago), the continents were combined into the supercontinent Pangaea (Figure 2.16). By 200 million years ago, Pangaea was being dismembered by the seafloor spreading that has produced the numerous continents of today. The breakup of the supercontinent Pangaea greatly lengthened the world's shoreline, reduced the areas of climatically harsh continental interiors, and made other habitat changes with which species either coped or went extinct (Figure 2.17).

A large, combined landmass will have a lesser number of species than several smaller, isolated landmasses. Today, for example, the isolated landmasses of Australia, New Zealand, Madagascar, and Africa all have very different species from each other and from the rest of the world. If these four landmasses were recombined, there would be numerous extinctions as the different species were thrown into competition for the shared food supply and living space.

Continental Position and Glaciation The drifting of continents plays a role in the climatic conditions on Earth. It is only when large landmasses have moved near the North or South Poles that enough falling snow can be caught and accumulated into the massive ice sheets that bury entire continents and plunge the Earth into an Ice Age (e.g., Figure 9.22). During an Ice Age, climatic extremes affect the majority of the Earth with colder and warmer intervals alternating as glacial ice sheets advance and retreat (Figures 9.19 and 9.24). Major shifts in climatic belts place great stresses on many species.

Volcanic Causes

In the past, immense volumes of basaltic lava flowed out and covered millions of square kilometers of the Earth in geologically brief time spans. Eruptions of voluminous magmas are fed by deep-rooted, narrow plumes of hotter rock that have risen up through the mantle (hot spots), melted further by decompression as they neared the surface, and then welled out onto the surface as lavas similar in composition to spreading-center lavas. Magma pours out onto stable tectonic plates as continental **flood basalts** and onto the ocean floor, forming oceanic plateaus (see Chapter 6). For example, the Ontong Java oceanic plateau was created about 120 million years ago when 36 million km^3 of lava poured forth in less than 3 million years to cover an area equivalent to two-thirds of Australia. It is estimated that world sea level rose about 10 m (33 ft) due to this volcanic outpouring alone. The rate of lava emission at Ontong Java was equivalent to the present annual outpourings of magma from *all* the world's spreading centers combined.

Changes in Atmospheric Composition An outpouring of flood basalt is accompanied by massive volumes of gases. In subsea eruptions, the oceans absorb and dilute some of the gases; however, ocean-water acidity and oxygen concentrations can change. With continental flood basalts, all the gases are pumped into the atmosphere, and they may be greenhouse gases. Some of the warmest climatic intervals in Earth history have accompanied flood basalts. For Ontong Java, average annual temperatures may have increased up to 13°C (23°F). Temperature increases this large create extreme stresses on some species.

Climate Change Causes

Climates change for numerous reasons (see Chapter 9). Each single change triggers a complex network of both negative and positive feedback responses. From the volcanism example above, the following sequence of responses could occur: 1) concentrated volcanism emits tremendous volumes of gases, 2) the composition of the atmosphere changes, 3) the altered atmosphere changes the heat balance of the Earth via the greenhouse effect, and 4) the global climate undergoes significant changes. Some of the volcanic gases may dissolve in the ocean and change its composition at the same time that increased climatic warmth raises the temperature of the ocean. The oceanic changes then trigger more atmospheric and climatic changes, etc. You can never do just one thing. Knock over one domino, and many more will fall. Every action has multiple reactions.

The climatic history of the Earth shows many changes (Figure 9.9). Change is difficult to handle for many species. For example, many forms of life are unable to cope with cooling and drying. But cope they must or else face extinction.

Ocean Composition Causes

The world ocean is chemically connected to its dissolved salts and bottom sediments, the continents, and the atmosphere. Numerous buffer systems using negative feedback work to maintain a dynamic equilibrium composition for the ocean. However, on some occasions, the buffering systems are overcome. The resultant changes in ocean composition can be lethal to some species.

Today, the oceans are well stirred by the worldwide movements of water layers of differing densities (Figure 9.3). Near the poles, the cold surface waters of the seas are so dense that they sink and flow through the deep oceans. The deep waters are well oxygenated, and the ocean floors are rich with life.

However, during some of Earth's warmest climatic intervals (Figure 9.9), the polar waters were too warm to sink. At these times of lessened circulation of water, the decay of organic matter on the ocean bottoms robbed the bottom waters of their oxygen. The decay process uses oxygen to destroy organic matter, with carbon dioxide being given off in the reaction. The process is photosynthesis in reverse and is analogous to our human digestive tract; we inhale oxygen to burn (digest) our food and then exhale carbon dioxide as a waste product. During portions of Late Devonian (375 to 360 million years ago), Pennsylvanian/Permian (323 to 251 million years ago), Jurassic (170 to 160 million years ago), and Cretaceous time (124 to 83 million years ago), warm waters at the bottom of the ocean became depleted in oxygen (became **anoxic**), and bottom-dwelling species were pushed into extinction.

Salinity changes can be lethal also. For example, during intervals of glacier retreat in an Ice Age, the catastrophic emptying of glacial meltwater lakes (Figures 12.38 to 12.40) could quickly cover the surface of the ocean with cold freshwater (low salinity). The cold freshwater is lethal to many sea-surface species used to warmer temperatures and higher salinities. A different example occurred when the northward tectonic movement of Africa closed the Mediterranean Sea from the Atlantic Ocean during parts of Late Miocene time (7 to 5.3 million years ago). Evaporating water from the sealed-off Mediterranean basin increased the water salinity to crisis levels for most species.

Extraterrestrial Causes

Life on Earth is also subjected to bombardment by space debris (Figure 14.15). Large bodies, such as asteroids and comets, hit the Earth with tremendous force (see Chapter 15). When a 10 km (6 mi)-diameter object strikes the Earth, it is likely to cause such effects as wildfires, acid rain, tsunami, and a huge dust cloud that blocks sunlight and creates weeks or months of dark winter. When the dust settles, voluminous gases may remain aloft, creating a greenhouse rise in temperature. Events such as these would be traumatic for much of the life on Earth.

Earth is constantly being bombarded by cosmic rays from outer space and by the subatomic debris emitted from the incinerator that is our Sun. The influx of high-energy radiation and tiny particles varies as the Sun's intensity changes or when a **supernova** (stellar) explosion occurs at astronomically close distances. The Earth's magnetic field provides a protective envelope that diverts or entraps most of the incoming subatomic particles, but the strength of the magnetic field varies markedly over time. During intervals of a weakened magnetic field, the amount of radiation received increases significantly. It has been hypothesized that during these times, the cosmic-ray bombardment causes increases in life-altering genetic mutations that could deliver lethal doses of radiation. As appealing as the cosmic-ray hypothesis sounds, no correlation to the fossil record has yet been demonstrated.

Biologic Causes

The diversity of species on Earth is so great that it cannot yet be defined. Estimates of the number of species alive at the present time range from 40 to 80 million. Despite the abundance of life, it is difficult for a species to persist through time. Some species go extinct every day for biologic reasons, such as low population size, reduced geographic area, competition, and predation including epidemic disease.

Species-Area Effects Studies of isolated environments, such as islands, have shown a strong relationship between the

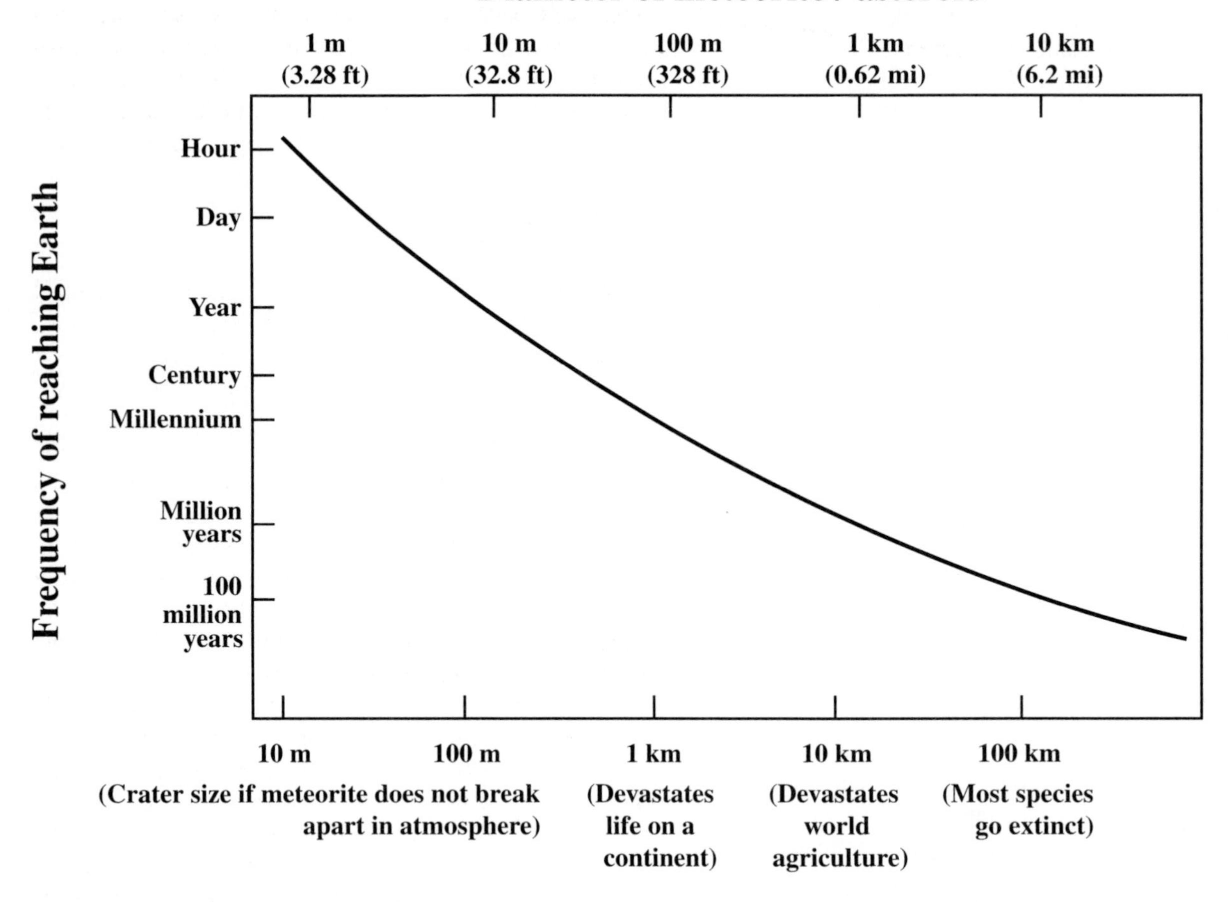

Figure 14.15 Relationships between meteorite/asteroid diameters, crater sizes, and their effects on life.

number of species present compared to the geographic area occupied (Figure 14.16).

Random Extinction As the number of individuals in a species's population decreases, they become subject to extinction via gambler's ruin. The probability theory of gambler's ruin can be illustrated by following the fortunes of a pile of money played by a gambler (Figure 14.17). The amount of money goes up and down, along an unpredictable path called a random walk. No matter how big the pile of money grows, it is never big enough, because a downward trend can start at any time and the money pile will go to zero. The random walk has no memory; there is no force that will cause it to return to its starting point. It is inevitable that a random walk will reach zero, the absorbing boundary. When that time comes, and the money is gone, the game is over.

A random walk also describes the existence of a species. The number of individuals in a species may grow and the species may spread over a wide area. But there is no abundance or distribution level that can ensure permanent survival of a species. As changes restrict the geographic area within which a species lives and the number of its individuals decreases, then extinction is probably near. Extinction is another absorbing boundary from which there is no return.

Predation and Epidemic Disease A species may be driven into extinction by excessive **predation.** The predators do not have to do the whole job, just drive the population of a species to a low enough level that the gambler's ruin scenario can complete the extinction. Predators may be large carnivores, or more likely they are small life forms, such as viruses (epidemic disease is a form of predation). In today's world, the predator responsible for most biologic extinctions is us, *Homo sapiens,* which means "wise or knowing man."

Multiple Causes of Mass Extinction

All of the above-mentioned factors contribute to extinctions along with still other factors not described. Any one of these factors operating alone can cause the local stress that drives one or a few species into extinction. This is the process of ongoing disappearances that make up background extinction. However, to wreak the havoc that ends the existence of numerous species around the world probably requires two or more of the mentioned causes to occur either simultaneously or closely spaced in time and on a global scale. The best way

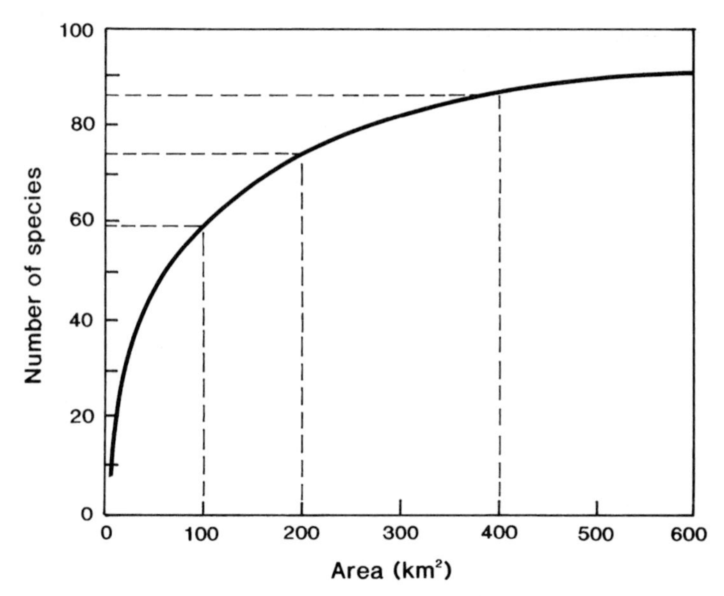

Figure 14.16 A species-area curve showing the numbers of species on islands of increasing size. Note that doubling the area does not double the number of species.

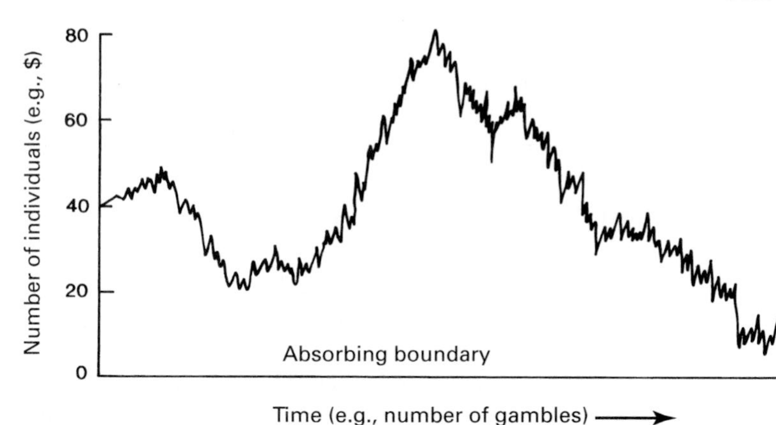

Figure 14.17 In gambler's ruin, the total number of dollars held varies after each bet. No amount of success can prevent the fact that a time will come when the money will be gone. When the money hits the absorbing boundary, there is no more. This is the same plot followed by individual members of a species—only their absorbing boundary is extinction.

to analyze the combined killing effects is through examination of specific mass extinctions.

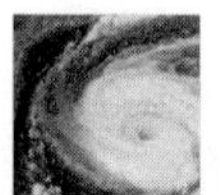

Examples of Mass Extinctions

After life made its explosive leap forward beginning about 544 million years ago (Figure 14.3), there were numerous extinction events, each of which decimated many of the species of its time (Figures 14.8 and 14.9). After each great dying, many of the world's environmental niches were vacated, thus clearing the stage for the next set of actors to appear. To illustrate the multiple causes of these great dyings, three examples from different times will be examined: 1) End Permian, 2) End Cretaceous, and 3) Quaternary.

Closing of Permian Time (Ended 251 Million Years Ago)

In Permian time, the seas were full of animals that lacked mobility; there were reef-building tabulate corals and solitary rugose corals, crinoids on stalks fastened to the sea floor, encrusting bryozoans (colonial coral-like animals), and brachiopods (shelled creatures resembling clams) lying on the sea floor. The marine animals mostly stayed in place, either waiting for prey to pass by or else quietly filtering the water for food. On land, there were amphibians as big as pigs and mammal-like reptiles larger than sheep.

Although they did not know it, most of these species were doomed as Earth's largest wave of extinction lay ahead in the closing 5 million years or less of Permian time. On land, just one of four amphibian orders and only one of 50 reptile genera survived the Permian Period. In the oceans, the die-offs were so drastic that 80 percent or more of animal species went extinct along with 70 percent of vertebrate families. Global changes were so dramatic they are described in a "horrible luck" hypothesis where a multitude of problems arose at the same time and combined their effects to trigger the mass extinctions. What were the events that combined to make life so difficult? It was a combination of several changes, each operating on a different time scale.

Formation of the Supercontinent Pangaea As Paleozoic time was drawing to a close, the continents of the world were being pushed together into one supercontinent stretching from north to south across the face of the Earth (Figure 2.16). The uniting of the continents closed most of the equatorial sea, thus severely reducing the area of shallow tropical ocean and triggering great die-offs among species that required warm and shallow marine water.

Sea-Level Fall Seafloor spreading apparently slowed its pace during Permian time. As spreading centers were fed less magma, the volcanic ridge masses shrank in size, thus increasing the capacity of the world ocean basin (Figure 14.12). With a larger ocean basin, seas retreated from the continents; sea level dropped about 200 m (660 ft); and shorelines moved seaward as much as 1,900 km (1,200 mi). After the shoreline retreat, the total area of shallow seas, whether of warm or cool water, was greatly reduced, thus pushing more species into extinction.

Climate Changes The formation of a single supercontinent affected climate. There was less shoreline, which meant that greater percentages of land were located away from the climate-moderating effects of the ocean. The continent itself was elevated above the former sea level, causing the interior landmass to suffer a dryer climate with marked seasons and severe shifts in temperature. The aridity of the interior led to the spread of deserts with wind-blown sand dunes.

Ocean Composition Changes The close of Permian time also was the close of a very long Ice Age (Figure 9.9). It has been suggested that as cold polar waters disappeared, ocean circulation slowed and much of the deep ocean became sluggish. Stagnant bottom water becomes anoxic (depleted in oxygen) thus killing many deep-water organisms. This deep water may occasionally overturn (see Lake Nyos in Chapter 7) causing shock and death to surface-water organisms.

Siberian Traps Flood Basalt As more and more species fell into extinction during these trying times, along came a possible *coup de grâce,* a finishing blow in the form of a massive volume of flood basalt extruded during a brief interval of time. In the region now known as Siberian Russia, the greatest outpouring of flood basalt on a continent occurred, apparently within a million years. During this geologically short time, up to 3,000,000 km^3 (> 700,000 mi^3) of basaltic lava flowed out to bury 3,900,000 km^2 (1,500,000 mi^2) of land. With Siberia then in northern latitudes, the lavas likely flowed out on top of permafrost thus converting frost to water vapor. Direct heating of permafrost would release tremendous quantities of methane that were frozen in hydrates. Erupted along with the lavas was a huge volume of gases, apparently including CO_2, emitted directly into the atmosphere. The abundance of atmospheric methane, carbon dioxide, and water vapor would have warmed climate worldwide by increasing the greenhouse effect. Life would have been hammered by global warming and acid rain.

Duration of the Extinction Events How long did the end-Permian extinction events take? They took much less than one million years. Newly determined radiometric ages from China indicate the climactic extinction occurred at 251.4 (+/– 0.3) million years ago.

Organic-rich sedimentary rocks in China contain a marked drop in the ratio of heavy carbon atoms to light carbon that occurred in less than 165,000 years, and maybe less than 30,000 years. The change in carbon isotopes tells of a collapse of biological productivity in a geologically short time. Were the evermore difficult living conditions on Earth accelerated by a rapid event such as the impact of a carbon-rich comet? This question is being worked on now.

Life at the End of Permian Time Many changes occurred as the Permian Period was closing: 1) tropical seas were virtually eliminated; 2) the extent of shallow marine water was reduced worldwide; 3) only one major landmass existed, which placed species-area effect pressures on terrestrial life (Figure 14.16); 4) lands were marked by great deserts; 5) deep-ocean water probably became anoxic and CO_2-rich, and finally, 6) the climate warmed, perhaps as a result of flood-basalt volcanism.

The oceans were home to abundant corals, brachiopods, bryozoans, fusulinid foraminifera, crinoids, and ammonoids. The lands were populated by diverse, large amphibians and mammal-like reptiles. All of these successful lineages were either exterminated, or their diversity was drastically reduced. The dominant life forms reached the inevitable fate indirectly described in 1751 by Thomas Gray in these lines from *"Elegy Written in a Country Churchyard":*

> The boast of heraldry, the pomp of power
> And all that wealth and fame e'er gave
> Await alike the inevitable hour,
> The paths of glory lead but to the grave.

The Permian mass extinction left an impoverished global fauna. But the characteristics of the remaining species allowed them to evolve into new life forms that spread throughout the world in Mesozoic time. The immobile organisms of the Permian ocean were replaced by increased diversities and abundances of armored snails, deep-burrowing clams, and free-swimming predators, such as cephalopods and reptiles. The hardy survivors of the Permian extinction were better suited to life in the new conditions and filled more niches in the Mesozoic ocean than had their Paleozoic predecessors. Most surviving terrestrial reptiles may have been more warm blooded and had more efficient respiratory systems. From these survivors arose the dominant terrestrial life of Mesozoic time—the fabled dinosaurs.

Close of Cretaceous Time (Ended 65 Million Years Ago)

Following the Permian extinctions, the surviving species were few in number but in the following millions of years, new species originated and the diversity and volume of life increased significantly. In the Triassic oceans, invertebrate life included abundant molluscs—bivalves (clams, oysters), snails, and ammonoids (hard-shelled, coiled squidlike animals)—as well as new types of reef-building corals. By mid-Triassic time, there were numerous vertebrate species, including bony fishes, crocodiles, turtles, frogs, rodentlike mammals, and dinosaurs. By mid-Jurassic time, some small dinosaurs evolved into birds. All of these lines of organisms were hit hard by Triassic, Jurassic, and early Cretaceous extinctions. Yet, despite these setbacks, the major lines kept evolving new species. Late in Cretaceous time, the North American heartland was covered with herds of dinosaurs in a scene reminiscent of Africa today, except that the large animals were dinosaurs rather than mammals. The North American dinosaur herds in Late Cretaceous time included herbivorous ceratopsians (analogous to rhinoceros) and noisy trumpeting duckbills (analogous to antelope and wildebeest) that were preyed upon by tyrannosaurs (analogous to lions). Overhead were large flying pterosaurs (analogous to vultures and eagles). In the oceans were large marine reptiles including ichthyosaurs (analogous to dolphins) (Figure 14.18).

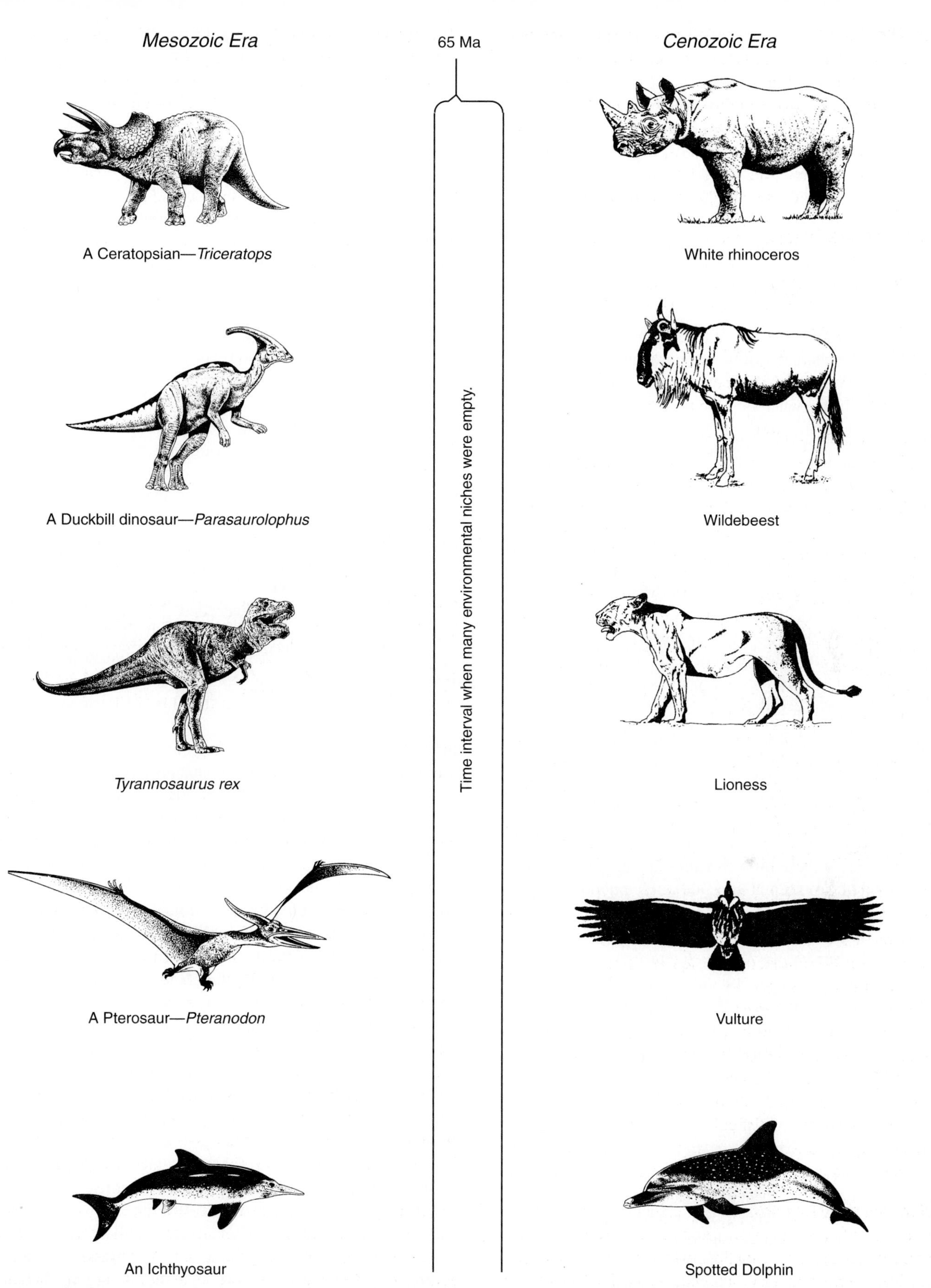

Figure 14.18 In Mesozoic time, dinosaurs thrived on land and in the air while large reptiles lived in the sea. Then, 65 million years ago—mass extinction! After a few million years, large animals again filled the same places in the environment, but this time they were mammals and birds.

Plant life had become very different in Late Cretaceous time as the flowering plants (angiosperms) had evolved. Because these hardwood trees, grasses, and other flowering plants reproduced rapidly, quickly colonized bare ground, and competed well for light and nutrients, they expanded by competitive displacement of the existing vegetation of ferns and gymnosperms (flowerless seed plants). For a gymnosperm example, conifer seeds take a long time to develop, so many niches were lost to the flowering plants. Today, the conifers mainly survive only in temperate, cold, or dry areas. At present, there are over 500 species of gymnosperms, but there are more than 200,000 species of angiosperms.

But the world was about to undergo radical changes that would reshape the character of life on Earth again. In the last several million years of Cretaceous time, many groups of plant and animal life, whether on land or in the seas, were losing species to extinction. It appears that slow-acting changes elevated the level of background extinction. Then this deteriorating situation apparently was finished off by a deadly one-two punch of volcanism and asteroid impact to end Cretaceous time.

The scorecard of die-offs includes over 35 percent of genera and 65 percent of species. In the oceans, the reptiles and ammonoids went extinct, and major die-offs occurred in species of bony fishes, sponges, sea urchins, foraminifera, snails, and clams. On land, many species of mammals and reptiles and all the dinosaurs (except birds) plunged into extinction. Land plant life took a heavy hit as shown in the fossil pollen and spore record. In the North American fossil record, fern spores increase from about 25 percent to nearly 100 percent of the fossils, suggesting a massive disturbance (burnoff?) with a slow recovery. The extinction of tiny floating organisms (plankton) in the oceans was overwhelming; the events affected both primary producers (photosynthesizers) and grazers with calcareous and siliceous skeletons. About 60 percent of these little creatures apparently died out right at the end of Cretaceous time; although individually insignificant, they summed to an immense biomass (at least 40 percent of Earth's total biomass).

With 10 million years of the Cretaceous remaining, there were numerous genera of dinosaurs; with 2 million years to go, there were markedly fewer (most of the survivors were in North America); and then there were none. Their slow decline was abruptly terminated, and these once mighty "rulers" of the Mesozoic world were reduced to a condition similar to that described by Percy Bysshe Shelley in *"Ozymandias of Egypt":*

> I met a traveler from an antique land
> Who said: Two vast and trunkless legs of stone
> Stand in the desert. Near them on the sand
> Half sunk, a shatter'd visage lies, whose frown
> And wrinkled lip and sneer of cold command
> Tell that its sculptor well those passions read
> Which yet survive, stamp'd on these lifeless things,
> The hand that mock'd them and the heart that fed;
> And on the pedestal these words appear:
> "My name is Ozymandias, king of kings:
> Look on my works, ye Mighty, and despair!"
> Nothing beside remains. Round the decay
> Of that colossal wreck, boundless and bare,
> The lone and level sands stretch far away.

What were the changes that led to the massive end-Cretaceous die-off? Long-lasting changes in sea level and climate finished by flood-basalt volcanism and asteroid impact.

Sea-Level Fall During mid-Cretaceous time sealevel stood high as the ocean flooded low-elevation portions of the continents (Figure 14.13). But during the final 18 million years of Cretaceous time, the seas retreated, climates cooled, vegetation suffered, and animal life declined.

Deccan Traps Flood Basalt The Deccan traps are the remains of a massive outpouring of basaltic lava most prominently displayed in west-central India, but much is covered by the Indian Ocean. The Indian landscape is made of thick piles of hardened lava that give a stair-step feel to the topography (*trap* is Dutch for "staircase"; *deccan* is Sanskrit for "southern"). The original area covered by Deccan flood basalt was probably about 2 million km^2 (over 770,000 mi^2), and the volume extruded probably exceeded 2 million km^3 (over 480,000 mi^3). The lavas poured forth probably in less than a million years beginning about 65.5 million years ago. The climatic effects would have been felt worldwide, similar to the Siberian traps flood basalt at the close of Permian time.

Chicxulub Impact On the Yucatan peninsula of Mexico, oil drillers in the 1950s encountered shattered rock about 2 km (1.2 mi) below the surface. Recent studies have shown that the shattered zone has a circular pattern with an inner ring of 180 km (112 mi) diameter, an outer ring of 300 km (185 mi) diameter, and a shattered zone up to 60 km (37 mi) thick. The shattered region apparently resulted from an asteroid impact that occurred 64.98 plus or minus 0.06 million years ago. The feature is called the Chicxulub impact structure, using the Mayan word for devil's tail (sanitized translation). The bull's-eye of the structure lies below the surface north of Merida, capital of the state of Yucatan.

What were the effects on life caused by this asteroid impact? They are still being determined. Despite the popularity of this event with the media and its easy description—big asteroid hits Earth and the dinosaurs all die—the magnitude of the worldwide effects continue to be debated. Impacts are the topic of Chapter 15.

The worst-case scenario for the Cretaceous impact goes like this: An asteroid of 10 km (6.2 mi) diameter plunged through the atmosphere faster than 10 km/sec (over 22,000 mph). Its impact with Earth set off a fireball of 1,000 km (620 mi) radius; its intense winds and searing heat ignited wildfires that especially affected vegetation in nearby North

America. The impact created an earthquake with magnitude up to 12 and tsunami up to 2 to 3 km (1.2 to 2 mi) high. The impact blasted a hole up to 60 km (37 mi) deep, sending off a horizontally directed base surge of melted and pulverized rock material and a plume of vaporized water and rock that shot up into the stratosphere. Some of the gases and water vapor fell as acid rain that may have destroyed the planktonic protists and algae in the surface waters of the oceans. The cloud of pulverized dust blocked the incoming Sun for months and plunged the Earth into a long, dark "winter" in which photosynthesis was virtually stopped. After the dust settled, the greenhouse-effect gases remained aloft; thus replacing the nuclear winter with a few years of elevated temperatures. Life that had survived months of "winter" freezing then had to live through years of "summer" weather; the die-offs would have been overwhelming. The scenario is easy to visualize, probably too easy, but it remains hard to rigorously demonstrate the detailed effects at this time.

When things calmed down, the surviving organisms found themselves in an emptied world where all sorts of opportunities existed to evolve new ways of life, and evolve they did. Flowering plants expanded rapidly, new forms of birds filled the air, more efficient reef-building corals seized the opportunity, and mammals took over the land vacated by the dinosaurs. Mammals had existed as long as the dinosaurs but always as little squirrel- or shrew-like creatures that kept out of the way of the dominant dinosaurs. But given the opportunity, the mammals grew into large herbivores that were preyed upon by large carnivores (Figure 14.18); they entered the seas (whales and seals) and even flew through the air (bats). The extinction event led to the evolution of new mammalian forms, including us, *Homo sapiens.*

Table 14.1 Extinctions of Large-Bodied Animals

	Genera Extinct in Last 100,000 Years	Genera Still Living	% Genera Gone Extinct
Africa	7	42	14%
North America	33	12	73
South America	46	12	79
Australia	19	3	86

Source: Martin and Klein (1989).

Table 14.2 Calibration of Human Migrations and Large-Animal Extinctions (number of years before present)

	Appearance of Humans	Concentrated Extinctions
Africa	over 2,000,000	—
Europe	over 1,000,000	12,000–10,000
Australia	~56,000	~46,000
North America	13,000	12,000–10,000
South America	13,000	12,000–8,000
Madagascar	1,500	by 500
New Zealand	1,000	900–600

Source: Martin and Klein (1989).

Quaternary Extinctions

The fossil record tells of significant die-offs of large-bodied animals in the last 1.5 million years. This has occurred during the advances and retreats of the continental glaciers during our present Ice Age (Figures 9.9 and 9.19). Are the extinctions of large-bodied animals (over 45 kg or 100 lbs) just the elevated background extinction associated with severe climatic changes? Or did other causes combine with climate change to push even more than the expected number of species over the brink into extinction? The concentrations of extinctions suggest multiple causes. What is the suspected additional cause that has spelled doom for so many large animals? *Homo sapiens.* The present Ice Age coincides with the growth and spread of the human population and the increasingly sophisticated abilities of human hunters.

The fossil record of large-bodied animal extinctions shows variations during the last 100,000 years (Table 14.1). Large animals were decimated in the Americas and Australia. African animals fared best; this may be because humans originated and evolved in Africa and the animals there had already co-existed with humans for many thousands of generations. For the other continents and islands, the extinctions appear to follow the arrival of human immigrants—wherever humans went, extinctions followed (Table 14.2) (Figure 14.19).

Were the documented climate changes enough to have caused the extinctions? Several points argue against climate acting alone. 1) Many more large-animal genera went extinct than plant genera. This is contrary to expectations: why should animals die off faster than their food supply? 2) Large mammals should not have been that much affected by climate change because they had heat-regulating bodies. 3) Retreat of glaciers in the last 11,000 years increased the amount of habitable land, which should cause an increase in large-animal species, not a decline. 4) There were no equivalent extinctions of large-bodied animals during the earlier phases of our present Ice Age.

Humans migrated to Australia 56,000 (+/– 4,000) years ago. The new arrivals found 24 genera of large-bodied animals. By 46,000 years ago, 23 of the genera (marsupials, reptiles, bird) were extinct; only a genus of large kangaroos survived. How did it happen? The extinctions occurred during a short time in all climate zones and habitats. The die-offs occurred tens of thousands of years before the extinctions in the Americas, Madagascar, or New Zealand,

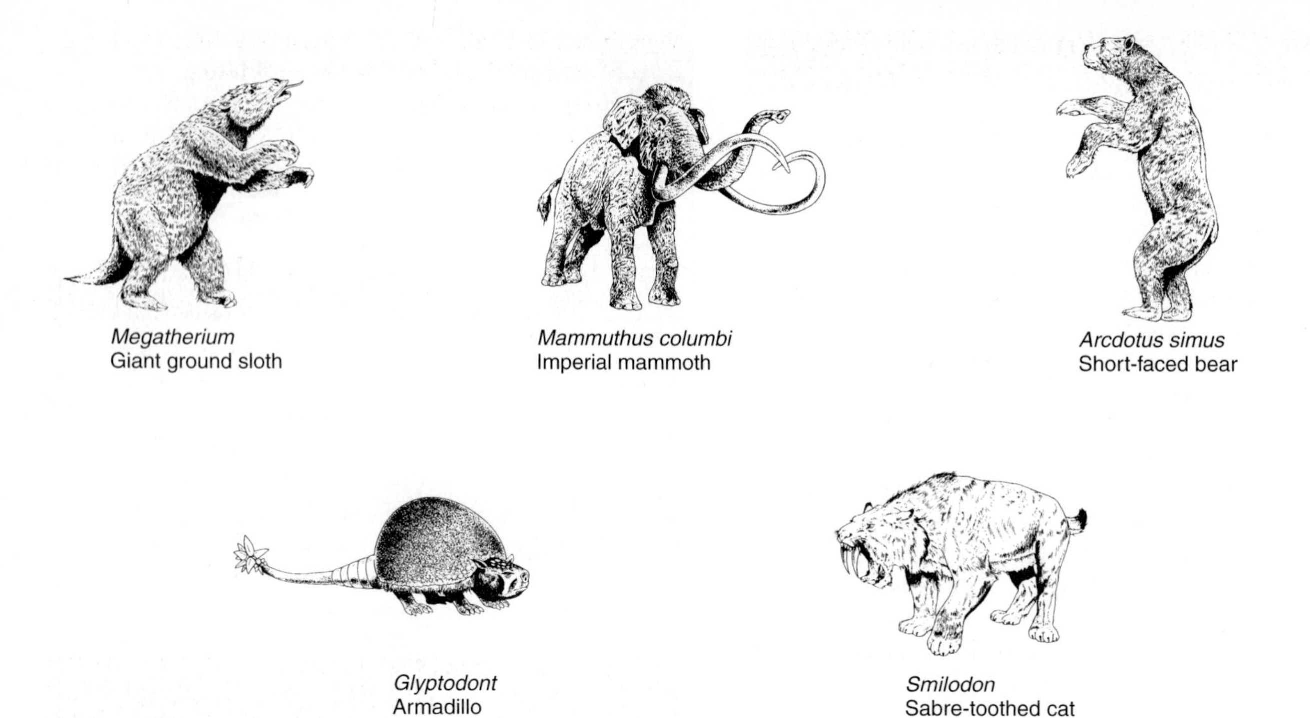

Figure 14.19 The earliest humans migrating to the Americas 13,000 years ago found many species of large land mammals—but the animals went extinct shortly thereafter.

Figure 14.20 The first humans to reach Madagascar and New Zealand found that the largest animals living there were flightless birds. The humans drove the birds into extinction.
Drawing by Jacobe Washburn.

suggesting that the Australian extinctions were a regional problem; they were not part of a global event. The most likely cause for the extinctions was humans overhunting naive, large-bodied herbivores (plant eaters), which in turn helped cause the carnivores (meat eaters), who preyed on the herbivores, to die out for lack of food.

Humans have been documented to be the exterminating agent in the extinctions on Madagascar, New Zealand, Hawaii, and Chatham Island. The first humans to reach Madagascar and New Zealand found that the largest animals living there were flightless birds. On Madagascar, the largest elephant bird species stood 11 ft (3.4 m) tall and weighed up to 1,100 lbs (500 kg). On New Zealand, the largest moa species reached 13 ft (4 m) tall but was not as heavy. Humans killed the big birds (Figure 14.20) and stole their eggs, driving them into extinction. Relatives of these extinct birds include the ostrich (up to 8 ft tall and 300 lbs) of Africa, the rhea of South America, and the emu of Australia. The appearance of humans using spears, fire, and hunting dogs was overwhelming to large animals unprepared for them. Humans did not have to do all the slaughter themselves. Just driving down the population size is enough; the gambler's ruin scenario will finish off most species.

It appears that the heavy effects of humans on the environment are *not* something that arrived with the Industrial Revolution, but rather it seems to have accompanied every human advance from toolmaking to control of fire to agriculture to taming of companion animals. The rate of human induced or related extinction has increased during the last 12,000 years. The past two centuries have seen even faster rates of extinction that, if continued for another two centuries, could equal the greatest mass extinctions of the geologic past.

Summary

Earthquakes, volcanic eruptions, hurricanes, floods, and such can each kill hundreds of thousands of people in a single event. As horrifying as these death totals are, they pale in comparison to the great dyings preserved in the fossil record. There have been brief times when most of the species on Earth became extinct. Over 99.9 percent of all plant and animal species that have ever lived are extinct.

Every year, species go extinct; this is the background level of extinction. Mass extinctions are relatively rare events wherein large percentages of Earth's species die off. The vacated niches left by extinct species are mostly filled by newly evolved species within several million years.

Global extinctions must be caused by far-reaching, but familiar, processes of plate tectonics, volcanic outpourings, climate changes, asteroid and comet impacts, and population biology. During times of rapid seafloor spreading, ocean waters can spill out onto the continents, flooding another 12 percent of the Earth (leaving 17 percent land); during times of slow spreading, ocean waters pull back into enlarged ocean basins, leaving about 40 percent of the Earth exposed as land. Changes in land-to-sea percentages have major effects on some species. As continents split apart and combine together, the number and sizes of continents change, resulting in major stresses on life—sometimes isolated, sometimes thrown together. When continents drift near the poles, major glaciations can result that drastically change climate and land-to-sea percentages.

Volcanic causes of extinctions are likely related to eruptions of flood basalt. Gases emitted during these voluminous outpourings can create a greenhouse effect and change the chemistry of the ocean and atmosphere. Climate changes by any cause make life difficult for many species.

Ocean temperature may vary significantly from ice ages to torrid ages. Ocean salinity and oxygen contents also vary; all changes place stress on some species.

Impacts of large asteroids and comets can cause mega-earthquakes, 2-mile-high tsunami, wildfires, acid rain, and huge dust clouds that block sunlight, creating a year-long winter that is followed by dust settling but greenhouse gases that remain aloft to create a year-long summer.

Biologic causes of extinction are also significant. When the population size of a species drops too low, the probability theory of gambler's ruin explains how extinction is near.

Mass extinctions where over two-thirds of the species on Earth become extinct probably have multiple causes. At the end of Permian time (251 million years ago), life was stressed by continents combining into one landmass (Pangaea), falling sea level, harsh climates in the interior of the continent, very little tropical shoreline, and the voluminous eruption of flood basalt in Siberia.

At the end of Cretaceous time (65 million years ago), life was challenged by falling sea level, the Deccan traps flood basalt in India, and the impact of a large (10 km diameter) asteroid in the Yucatan area of Mexico.

During our own Quaternary time, a mass extinction is apparently occurring. Life is stressed by advances and retreats of continental glaciers and the killing abilities of *Homo sapiens*—us.

Terms to Remember

- anoxic 397
- archaea 389
- evolution 390
- extinction 388
- flood basalt 396
- gene 389
- genome 391
- Law of Faunal Assemblages 388
- Law of Faunal Succession 389
- niche 390
- paleontologist 389
- predation 398
- reef 392
- species 388
- strata 388
- supernova 397

Questions for Review

1. Give examples of species that have become extinct. What percentage of species in Earth's history have become extinct?
2. Explain the Laws of Superposition, Faunal Assemblages, and Faunal Succession.
3. What is the oldest fossil found on Earth? How old is it?
4. Explain the concept of species.
5. Sketch an extinction-frequency curve. How is it similar to a flood-frequency curve?
6. Why does sea level change during intervals of rapid seafloor spreading?
7. How much can sea level rise and fall due to seafloor spreading? Due to continental glaciation? Are these numbers additive, subtractive, or . . . ?
8. Discuss the species-area effect. Apply it to a discussion of species diversity on Pangaea.
9. Explain the effects on world climate of: A) intervals of rapid seafloor spreading, B) continental glaciation, C) continents united into a single landmass, and D) a flood-basalt episode.
10. In what ways could a massive outpouring of flood basalt affect the existence of species on Earth?
11. In what ways can ocean composition change to affect the existence of species on Earth?
12. Explain the concept of gambler's ruin.

Questions for Further Thought

1. How does the Law of Faunal Succession provide support for the theory of evolution?
2. How does a mass extinction set the stage for increased diversity of life?

3. What types of organisms are likely to increase greatly during times of rapid seafloor spreading? What types are likely to suffer the most?
4. Are we alive in the midst of a mass extinction event? In what ways are humans playing a role in mass extinction?

Suggested Readings and References

Albritton, Claude C., Jr. (1989). *Catastrophic Episodes in Earth History.* London: Chapman and Hale.

Burney, David A. (1993, November–December). Recent animal extinctions: Recipes for disaster. *American Scientist,* 81, 530–41.

Copper, Paul. (1988). Ecological succession in Phanerozoic reef ecosystems: Is it real? *Palaios,* 3, 136–52.

Eldredge, Niles. (1991). *The Miner's Canary.* Englewood Cliffs, N. J.: Prentice-Hall.

Erwin, Douglas, H., ed. (1993). *The Great Paleozoic Crisis.* New York: Columbia University Press.

Gore, Rick. (1989, June). Extinctions. *National Geographic,* 175, 662–99.

Martin, Paul S., and Klein, Richard G., eds. (1989). *Quaternary Extinctions.* Tucson: University of Arizona Press.

Raup, David M. (1991). A kill curve for Phanerozoic marine species. *Paleobiology,* 17, 37–48.

Raup, David M. (1991). *Extinction—Bad Genes or Bad Luck?* New York: W. W. Norton.

Roberts, R. G., et al. (2001). New ages for the last Australian megafauna extinction. *Science,* 292, 1888–1892.

Sepkoski, J. J., Jr. (1982). Mass extinctions in the Phanerozoic oceans: A review. In *Geological Implications of Impacts of Large Asteroids and Comets on the Earth* (283–89). Geological Society of America Special Paper 190.

Sepkoski, J. J., Jr. (1986). Phanerozoic overview of mass extinction. In *Patterns and Processes in the History of Life.* Amsterdam: Springer-Verlag.

Sharpton, Virgil L., and Ward, Peter D. (1990). *Global Catastrophes in Earth History.* Geological Society of America Special Paper 247.

Stanley, Steven M. (1987). *Extinction.* New York: Scientific American Library.

Videos

The Death of the Dinosaur. (1993). PBS Video (60 min.; one of series of four).

Dinosaurs. (1991). Fox/Lorber (30 min. each; four programs).

Geologic Time. (1986). Encyclopedia Britannica (24 min.).

Buried in Ash. (1994). Nova/Nebraska ETV (60 min.).

Mammoths of the Ice Age. (1994). Nova/PBS (55 min.).

Chapter 15

Impacts with Space Objects

The universe as Hardy understood it was governed, not by a benevolent god, but by mindless and indifferent chance. All living things were subject to the same injustices, and man and nature could therefore reveal the same philosophic themes, could stand as metaphors for each other. There was one natural world, and man was simply one of the unfortunate creatures in it.

—Samuel Hynes, 1967, Introduction to Thomas Hardy, 1872, *Under the Greenwood Tree*

Earth moves rapidly through space occupied by comets and asteroids traveling on different paths (Figure 15.1). Sometimes the paths cross and all life on Earth feels the effects. In his 1882 novel *Two On A Tower,* Thomas Hardy described comets as "Members of the solar system, these dazzling and perplexing rangers, the fascination of all astronomers, rendered themselves still more fascinating by the sinister suspicion attaching to them of being possibly the ultimate destroyers of the human race."

Could a comet or asteroid impact destroy the human race? The statistics in Table 15.1 suggest that it is possible.

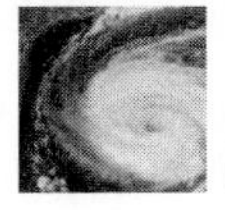

Impact Scars

A good place to see impact scars made by collisions with space debris is the surface of the Moon (Figure 15.2). In its first few hundred million years of existence, the Moon was a violent place as millions of objects slammed into it. The intense bombardment apparently occurred as a sweeping up of debris left over from the formation of the planets. The Moon's surface still displays tens of millions of ancient impact craters, some with diameters of hundreds of miles (Figure 15.2). By 3.9 billion years ago, the heavy barrage of space debris had died down.

Flood basalts poured forth on the Moon from about 3.8 to 3.2 billion years ago. They created the dark-colored **maria** (*mare* is Latin for "sea") so prominent on the Moon's surface today. The maria have relatively few impact scars on them, thus providing evidence that the period of intense bombardment was over before 3.8 billion years ago. Impact craters on the lunar maria were made by collisions with asteroids and comets in the same process that continues today.

For over 3 billion years, the Moon has been essentially "dead"; it is an orbiting museum showing only the scars of its ancient past. The Moon has no plate tectonics, no water or agents of erosion, and no life. About the only event that disturbs the cemetery calm of the Moon is the occasional impact of an asteroid or comet.

Why are impact craters so common on the Moon but so rare on Earth? The Moon is geologically dead, so impact scars remain. But the Earth is dynamic: it destroys most of the record of its past. Plate-tectonic movements consume impact scars during subduction (Figure 2.18) and crumple them during continent

Figure 15.1 Comet Hale-Bopp with a 30 million mile-long tail streaks above Joshua Tree National Monument, California on 18 April 1997.

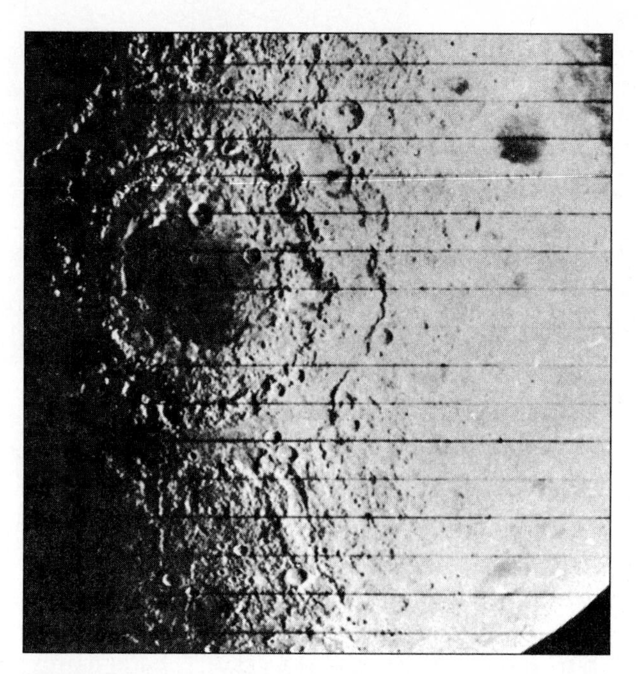

Figure 15.2 The Moon's surface is ancient and pockmarked by numerous impact craters. Notice the large Orientale multiring basin. Its outer ring is 1,300 km (more than 800 mi) across; on Earth, it could stretch from the Great Lakes to the Gulf of Mexico. Photo courtesy of NASA.

Table 15.1 Frequency of Globally Catastrophic Impacts

Average interval between impacts	500,000 years
Annual probability a person will be killed	1/500,000
Assumed fatalities from impact	1/4 of human race
Total annual probability of death	1/2,000,000

Source: D. Morrison (1992).

collisions. The agents of erosion work to erase all impact craters on Earth (Figure 1.19). Nevertheless, some impact scars have avoided destruction, especially the geologically younger ones (Figure 15.3). The American geologist Robert Dietz calls impact scars **astroblemes,** or literally, star wounds (the Greek word *astron* means "star" and *blema* means "wound" from a thrown object).

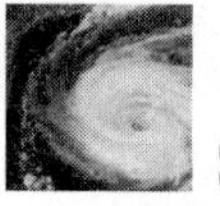

Sources of Extraterrestrial Debris

Space debris that collides with the Earth comes primarily from fragmented asteroids and secondarily from comets. The pieces of asteroids and comets that orbit the Sun are called **meteoroids.** When meteoroids blaze through the Earth's atmosphere as a streak of light or **shooting star,** they are referred to as **meteors.** The objects that actually hit the Earth's surface are called **meteorites.** The main types of meteorites are either "irons" (metallic) or "stones" (rocky). Although most space objects that reach the Earth's atmosphere are "stones," they are not very abundant on the ground. Stony meteorites are less commonly collected because: 1) they break up more readily while passing through the Earth's atmospheric filter, 2) those that reach the ground are weathered and destroyed more rapidly, and 3) they are not as easily recognized as "irons." Thus, most of the collected meteorites are "irons."

Asteroids

The Solar System has nine planets in orbit around the Sun. The four inner planets are small, close together, near the Sun, and rocky. The outer planets are mostly larger, spaced far apart, lie at great distances from the Sun, and are composed mainly of hydrogen and helium gas surrounding rocky cores; they commonly are orbited by icy moons and

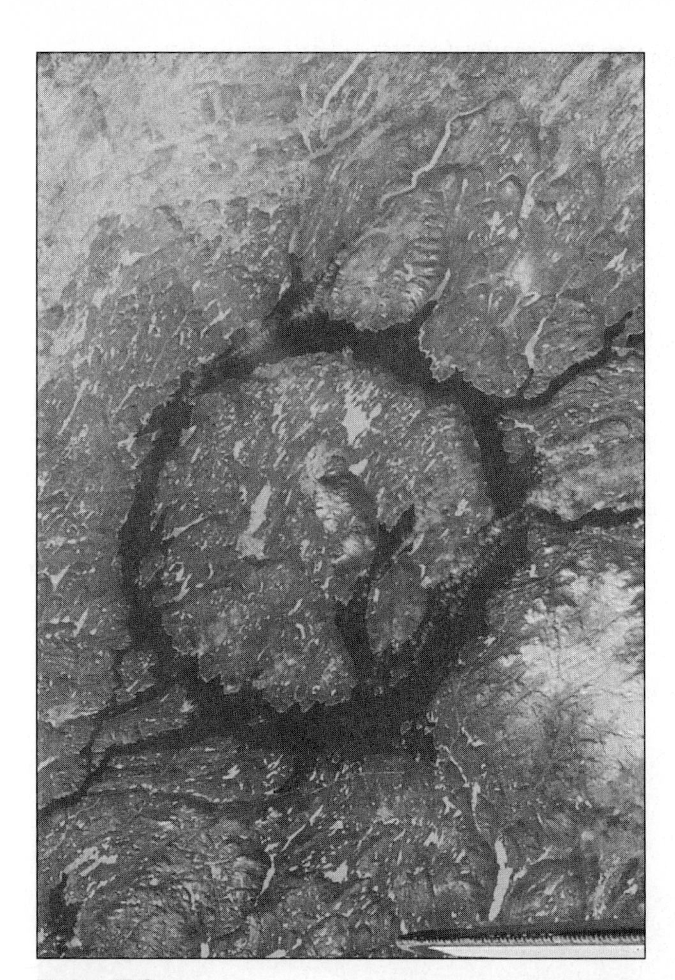

Figure 15.3 The Manicouagan impact crater formed about 214 million years ago in Late Triassic time, in northern Quebec, Canada. It is 75 km (45 mi) across but probably exceeded 100 km before glacial erosion stripped away its upper levels.
Photo courtesy of NASA.

rings of icy debris with compositions of water (H_2O), ammonia (NH_3), carbon dioxide (CO_2), and methane (CH_4). Between the inner and outer planets lie the asteroids, a swarm of small (under 1,000 km diameter) rocky, metallic, and icy masses (Table 15.2).

Meteorites appear to come from the inner part of the Solar System and especially from the asteroid belt (Figure 15.4). Asteroids are small bodies orbiting the Sun. The three largest asteroids make up about half the combined total mass of all asteroids; they are Ceres, Pallas, and Vesta, with respective diameters of 933, 523, and 501 km. There are more than 200 asteroids with diameters greater than 100 km, about 1,000 with diameters greater than 30 km, and another million with diameters over 1 km. If all the asteroids were brought together they would make a planet about 932 miles in diameter, that is less than half the diameter of our Moon (Table 15.2).

The asteroids lie mostly between Mars and Jupiter in a zone where a tenth planet might have been expected to form. Many asteroids are similar to the ingredients from which the planets were assembled via low-velocity collisions. However, the asteroids were apparently too strongly influenced by the gravitational pull of Jupiter and thus have been unable to combine, or recombine, and form a planet. The gravitational acceleration caused by the massive planet Jupiter creates asteroid velocities that are too fast and individual collisions that are too energetic to allow the asteroids to collide, unite, and stick together to form a planet.

Table 15.2 Bodies in the Solar System

	Diameter (miles)	Specific Gravity (water = 1)	Gravity (Earth = 1)
Sun	864,886	1.4	27.9
Inner rocky planets			
Mercury	3,024	5.4	0.4
Venus	7,522	5.3	0.9
Earth	7,918	5.5	1.0
Moon	2,160	3.3	0.2
Mars	4,200	3.9	0.4
Asteroid belts			
Outer ice and gas planets			
Jupiter	86,692	1.3	2.7
Saturn	72,352	0.7	1.1
Uranus	29,168	1.6	1.1
Neptune	28,230	2.3	1.4
Pluto	7,084	1.7	0.2

Source: R. T. Dodd (1986).

Notice that the asteroids are concentrated in belts and that there are gaps between the belts (Figure 15.4). The gaps occur at distances related to the orbit time of Jupiter. The asteroids rarely collide, but when they do, their collisions may be spectacular impacts at 10,000 mph. The force of these smashups may bump an asteroid into one of the gaps in the asteroid belt. An asteroid nudged into a gap experiences an extra gravitational acceleration from Jupiter that makes its orbital path more eccentric, and thus, it becomes more likely to collide with a planet.

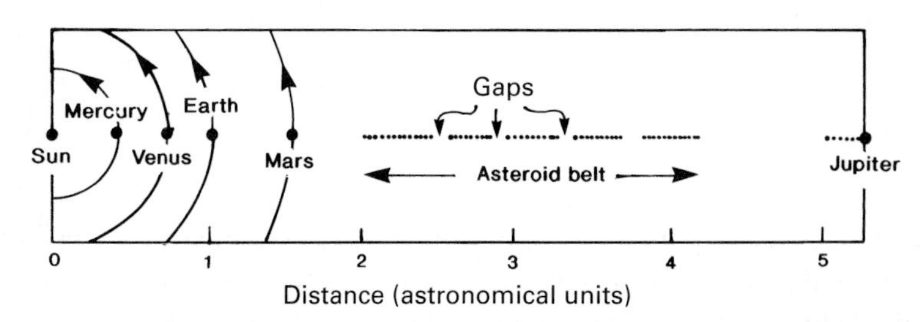

Figure 15.4 The Solar System from the Sun to Jupiter. The asteroid belt is composed of millions of rocky and metallic objects that did not combine to form a tenth planet. One astronomical unit equals 93 million miles, the distance between the Sun and Earth.

A recent photo of the asteroid Ida shows it has impact craters and its own moon, now named Dactyl (Figure 15.5). When Ida and Dactyl collide with a planet, craters will form simultaneously in two different areas. This observation explains some of the double-impact sites found on Earth (Figure 15.6).

It has been shown recently by radar data that some asteroids are not solitary, solid masses but rather are made of two or more similar-sized bodies bound together by gravitational attraction. Calculations indicate that the amount of energy needed to break up an asteroid is much less than the amount of energy required to scatter all its fragments. Thus, an asteroid may be broken into pieces during a collision, and then the pieces may be held together by gravity, creating a loose collection of rocky debris—a rubble pile.

The realization that collisions have made some asteroids into rubble piles raises interesting questions. Did some of the impact craters on Earth form within hours of each other when a multi-chunk asteroid hit the surface? One line of impact structures being investigated to see if they are the same age includes Manicouagan (Figure 15.3). Removing the effects of 214 million years of continental drift places three impact sites along a 4,462 km-long (2,766 mi) line parallel to the ancient 22.8° N latitude. Saint Martin in Manitoba, Canada (40 km diameter) lines up with Manicouagan in Ontario, Canada (100 km diameter) and both line up with Rochechouart in France (25 km diameter). If these three impacts occurred hours apart, then life on Earth must have suffered a terrible blow.

Clusters of asteroids such as the Apollos have orbits that intersect Earth's orbit (Figure 15.7). There are more than a thousand mountain-sized Apollo asteroids that could hit the Earth. Other groups of asteroids, such as the Amors, pass near Earth but intercept the orbit of Mars. The Apollo and Amor asteroid groups are sources for large asteroids and meteorites that occasionally slam into the Earth's surface. With so many asteroids whizzing about, we are lucky that space has such an immense volume and that the Earth is such a small target.

Comets

The Solar System is surrounded by about a trillion (10^{12}) comets, icy objects with orbits that take them *far* beyond the outermost planets of our Solar System. The vast and diffuse envelope of encircling comets is known as the **Oort cloud.** The planets and the asteroid belt orbit the Sun in planes, whereas the Oort cloud has a spherical distribution. The comets pulled into the inner parts of the Solar System are only a miniscule fraction of the Oort cloud. Most of the comets we see have wildly eccentric orbits that bring them in near the Sun at one end of their orbit **(perihelion),** but they swing out beyond the outermost planet at the other end of their journey **(aphelion).** A comet may travel 100,000 astronomical units away during its orbit (an astronomical unit is 93 million miles, i.e., the distance between Earth and Sun).

Comets are called "dirty snowballs" to describe their composition of ice and rocky debris. When an incoming comet passes Saturn on its journey toward the Sun, it begins to be affected by sunlight and the **solar wind** (the stream of subatomic particles flying outward from the Sun). Material from the frozen outer portion of a comet **sublimates** directly to vapor, thus liberating gases and trapped dust to form the distinctive luminous "tail" of a comet (Figure 15.8). The term *comet* is derived from a Greek word for long-haired. A comet's tail is produced by the charged particles of the solar wind acting on comet ices. The nearer an icy comet approaches the Sun, the larger its tail becomes. As the comet curves around the Sun, its tail rotates also, always pointing away from the Sun. The tail lines up with the solar wind moving outward from the Sun; the tail is not simply a vapor trail behind a comet. A comet that has lost most of its ices over time will have a dim and small tail. Despite their visibility in the sky, comets are surprisingly small; most have heads less than 15 km (10 mi) diameter.

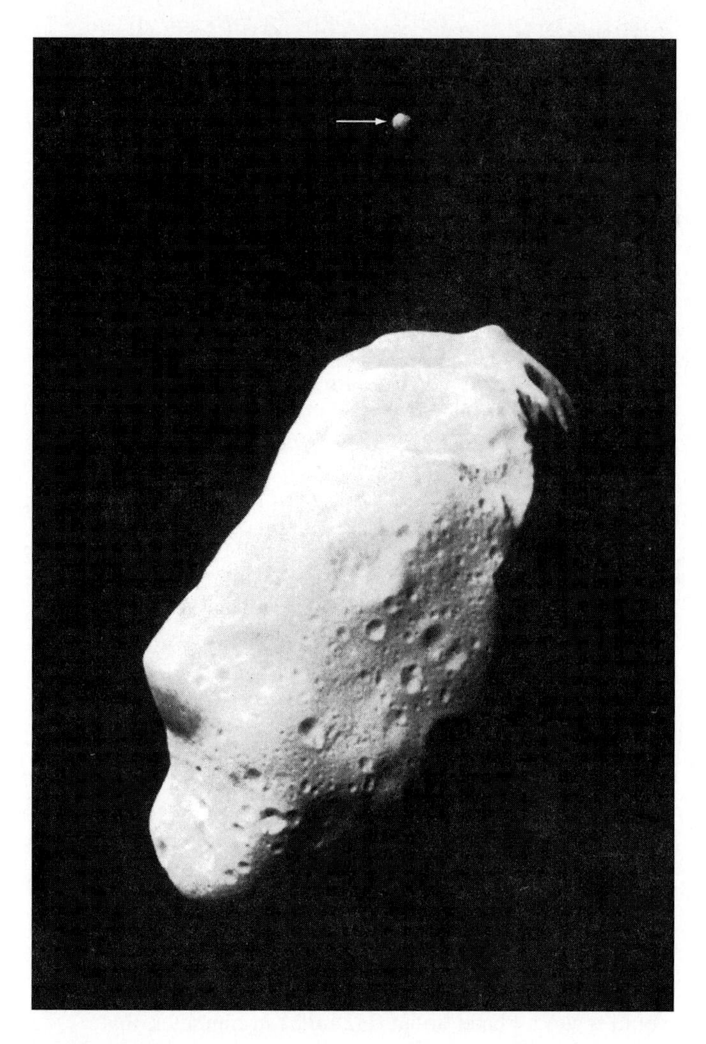

Figure 15.5 The asteroid Ida is 56 km (35 mi) long and pockmarked with impact craters. Traveling with Ida is its near-spherical moon Dactyl with dimensions of 1.2 × 1.4 × 1.6 km. Photo courtesy of NASA.

Figure 15.6 Space shuttle view of the Clearwater Lakes double impact sites, Quebec, Canada. About 290 million years ago, a two-part asteroid hit the ground. The western crater is 32 km (20 mi) across and has a central uplift. The eastern crater is 22 km (less than 14 mi) across; its central uplift is below water level.
Photo courtesy of NASA.

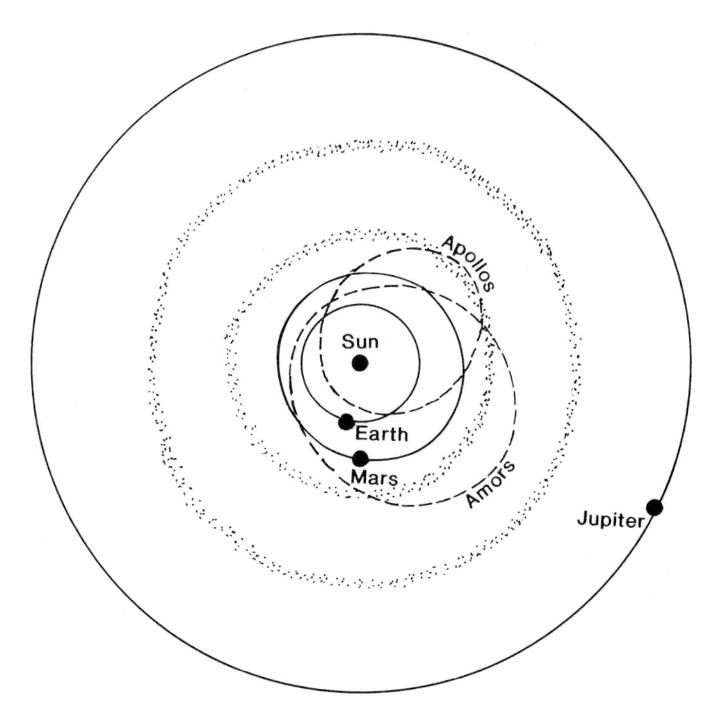

Figure 15.7 Earth's orbit around the Sun is intersected by the Apollo asteroids. The Amor asteroids cross the orbit of Mars and pass near the Earth.

Figure 15.8 Halley's comet.
Yerkes Observatory.

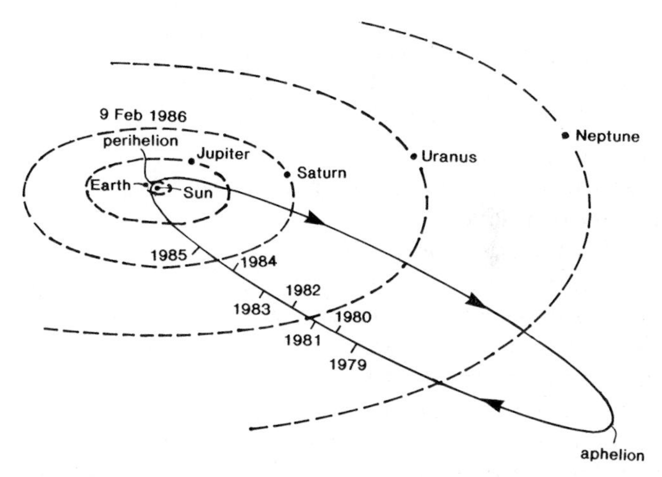

Figure 15.9 Orbit of Halley's comet during its 76-year round trip. Halley's elongate elliptical orbit is steeply inclined to Earth's orbit.

Some comets operate more directly under the gravitational influence of the Sun and Jupiter. As a comet's orbit loses eccentricity and begins passing by the Sun more often, its ice volume declines and its tail shrinks. There are at least 800 of these short-period comets with orbits of less than 200 years duration. Some short-period comets may be from orbits about the outer gas-giant planets and are thus distinct from the Oort cloud comets. If their ices are completely sublimated or melted, the remaining rocky body is quite similar to an asteroid.

The most famous of the comets is the one carrying Edmund Halley's name, the man who calculated its orbit in 1682 and predicted its return to the inner Solar System. Halley's comet travels from near the Sun (its perihelion) to beyond Neptune (its aphelion). The orbit of Halley's comet takes from 74 to 79 years, averaging 76 years (Figure 15.9).

Sidebar

Shoemaker-Levy 9 Comet Impacts on Jupiter

A once-in-a-lifetime event occurred during the week of 16–22 July 1994, as a series of comet fragments plunged into Jupiter. The comet was named after its discoverers: geologists Eugene and Carolyn Shoemaker and comet hunter David Levy. In 1992, the comet had flown too close to Jupiter, and the planet's immense gravitational attraction pulled in the comet and broke it into pieces. In 1994, the broken-up comet was stretched out like a string of beads as it again approached Jupiter. In succession, 21 large fragments plunged into Jupiter's dense atmosphere at speeds up to 60 km/sec (134,000 mph). Each impact caused: 1) an initial flash as a fragment collided with the heavy atmosphere, 2) a superheated fireball of hot gas rising upward as a plume thousands of kilometers above Jupiter's clouds, and 3) radiation as the plume crashed back down at high velocity.

The largest fragment (G) apparently was only 1 km (0.6 mi) across, yet it left an impact scar larger than the diameter of the Earth (Figure 15.10). Each impact caused a rising plume of hot gas that expanded and cooled as it rose (Figure 15.11). Although the impacting fragments were small and penetrated only into Jupiter's upper atmosphere, the impact energy released by fragment G was equivalent to about 315 million World War II atomic bombs.

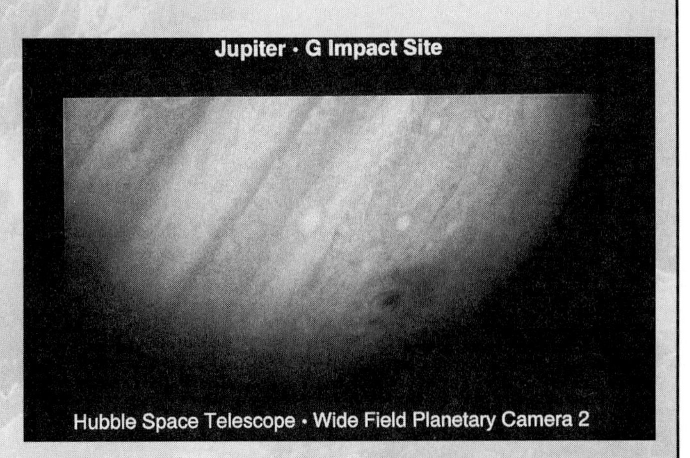

Jupiter • G Impact Site
Hubble Space Telescope • Wide Field Planetary Camera 2

Figure 15.10 Impact scar of Shoemaker-Levy 9 comet fragment G on Jupiter (in lower right of photo), 17 July 1994. To lower left of G is the smaller impact scar of fragment D.
Photo courtesy of NASA.

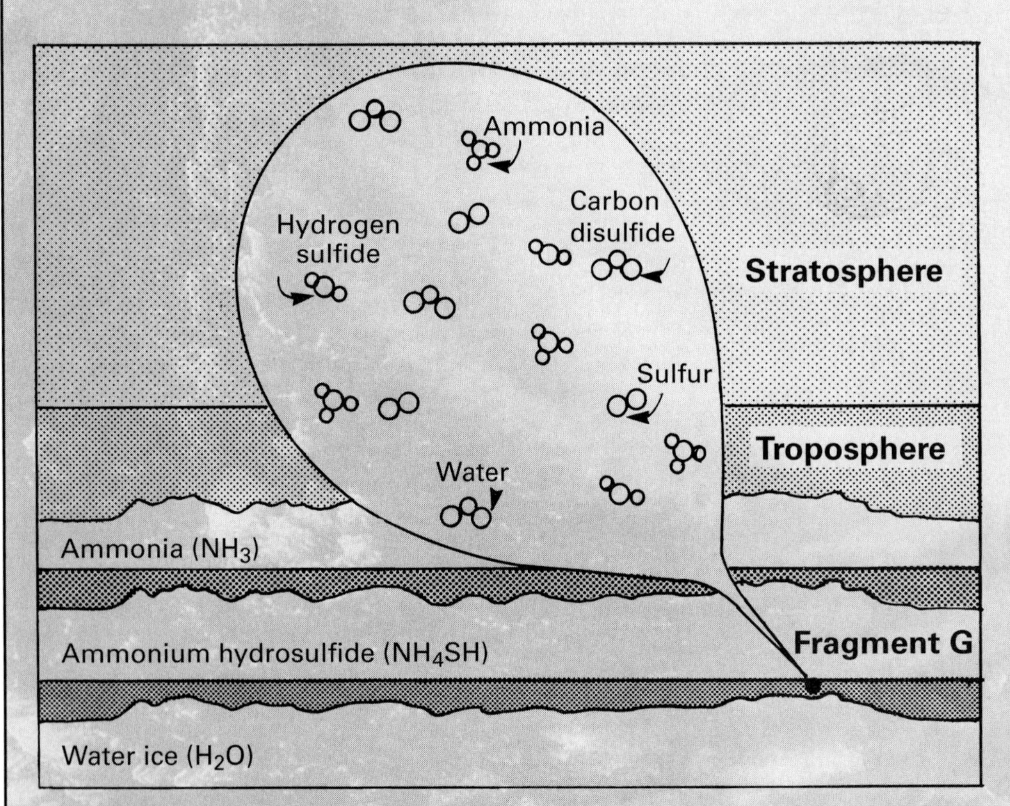

Figure 15.11 Path and impact plume of Shoemaker-Levy 9 comet fragment G plunging through cloud layers thought to make up Jupiter's upper atmosphere.

Near Neptune, the comet travels 1.5 km/sec (over 3,300 mph), but it speeds up to 55 km/sec (over 120,000 mph) as it nears the Sun due to the Sun's immense gravitational acceleration. On its round-trip journey, Halley's comet spends only about 15 months inside the orbital region of Jupiter, but this is where its size is reduced most and its tail develops and glows bright before it returns to the deep freeze of its outer orbit. Halley's comet glowed brightly during the Norman conquest of England in 1066; it passed Earth in 1835 as Samuel Clemens (Mark Twain) was born and then returned again in 1910, the year he died. On 20 May 1910, Earth passed through the tail of Halley's comet. The latest visit of Halley's comet was in 1986, and, with some luck, you will get to see it on its next visit.

The ices of comets contain carbon compounds, some of which are important building blocks of life on Earth. For example, Halley's comet contains carbon (C), hydrogen (H), oxygen (O), and nitrogen (N) in ratios similar to that in the human body. Many scientists think that the compounds used to build life were brought to Earth by comets. Are we the offspring of comets?

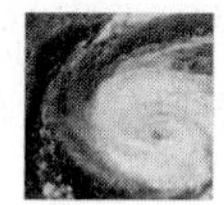

Rates of Meteoroid Influx

It is estimated that more than 100,000 million meteoroids enter the Earth's atmosphere every 24 hours. The numbers of incoming objects are directly related to their size; the smaller the meteoroids, the greater their abundance (Figure 15.12). Earth is largely protected from this bombardment by its atmosphere. At about 115 km (70 mi) above the ground, the atmosphere is dense enough to cause many meteoroids to begin to glow. A typical meteor is seen about 100 km (60 mi) above the ground and has largely or entirely vaporized before reaching 60 km (35 mi) above the surface. All this incoming debris adds from 100 to 1,000 tons of material to the Earth's surface each day.

Earth is also protected from meteoroids by the very great speeds with which they hit the atmosphere—from 11 km/sec (over 25,000 mph) to over 30 km/sec (about 70,000 mph). At high impact speeds, the low-viscosity atmosphere behaves more like a solid. Remember how hard the water in

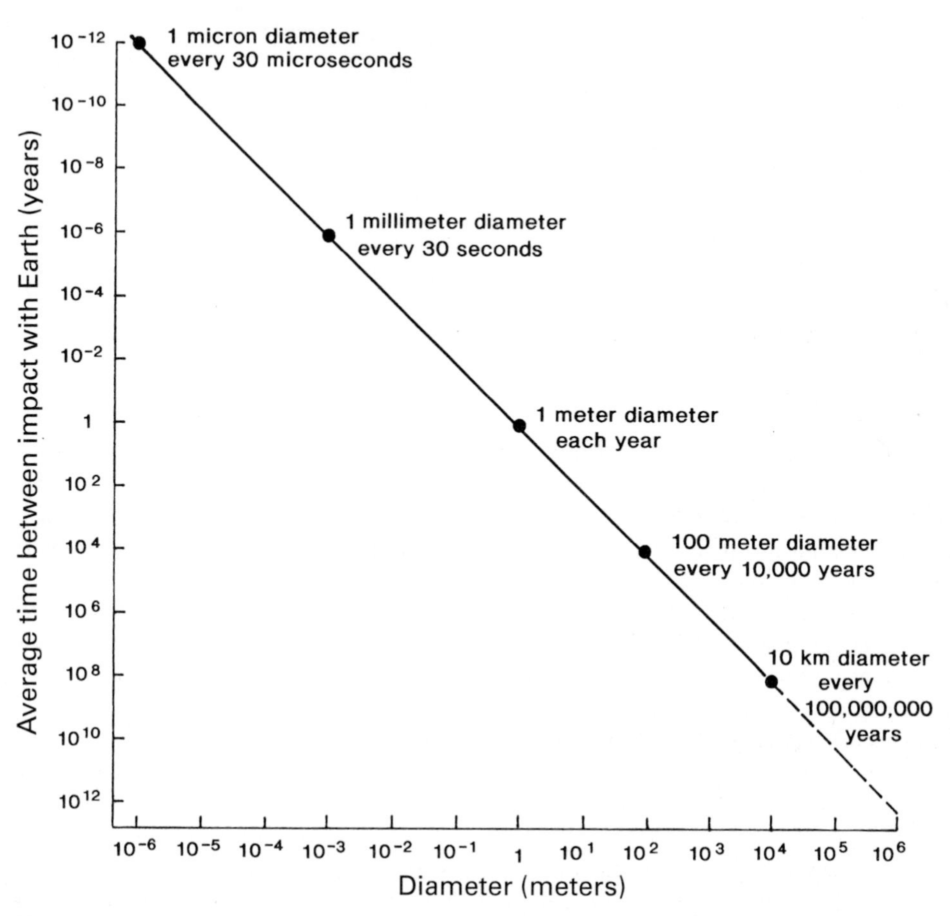

Figure 15.12 Frequency versus size of impacting space debris.

a pool or lake feels when you hit it doing a belly flop? Meteoroids hitting the atmosphere at incredible velocities experience a similar effect; they may be destroyed on impact with the atmosphere, deflected into space if their angle of approach is low enough, or slowed down due to friction. Only incoming objects weighing more than about 350 tons are big enough to be largely unaffected by the atmosphere.

Cosmic Dust

The littlest meteoroids are so small that they pass downward through the atmosphere effectively unchanged and settle onto the surface as a gentle rain. Particles with diameters around 0.001 mm have so much surface area compared to their volume that their frictional heat of passage is radiated as quickly as it develops, and they escape melting.

Shooting Stars

Incoming debris the size of sand grains, with diameters around 1 mm, typically flame out as shooting stars—flashes of friction-generated light about 60 miles above the ground that blaze for about a second. A shooting star melts in the atmosphere, and tiny droplets fall to the Earth's surface as little spheres of glassy rock.

Meteorites

Meteoroids weighing 1 gm (about 0.04 ounce) or more will pass through the atmosphere and fall onto the surface of the Earth. During their meteoric phase, the frictional resistance of the atmosphere causes their exteriors to melt. The melted surface materials are stripped away, also removing heat and thus protecting their interiors from melting. On the ground, they can be recognized as meteorites by their glazed and blackened outer crusts.

Consider a multipound meteoroid traveling many times faster than a rifle bullet. It hits the atmosphere and violently compresses the air in front of it, creating a minisonic boom. We all experience loud booms, rattling windows, and shaking houses; usually, we attribute the disturbances to airplanes (sonic booms), explosions, earthquakes, or weather phenomena. But a few of these atmospheric concussions may be caused by meteors hitting the top of the atmosphere (Figure 15.13). To be heard on the ground, an incoming meteor must be at least as big as a basketball.

On 3 October 1996, the sky had two brilliant light shows. First, a meteorite hit the atmosphere over New Mexico, lighting up the New Mexico and Texas sky before bouncing back into space. Second, after a 100-minute orbit, the meteorite was pulled back into the atmosphere over the Pacific Ocean, crossing over the California coast at Point Conception, then exploding in mid-air at least twice, sending off sound waves recorded by seismographs. The meteorite fragments apparently fell in the eastern Sierra Nevada foothills, but no pieces have been found—yet.

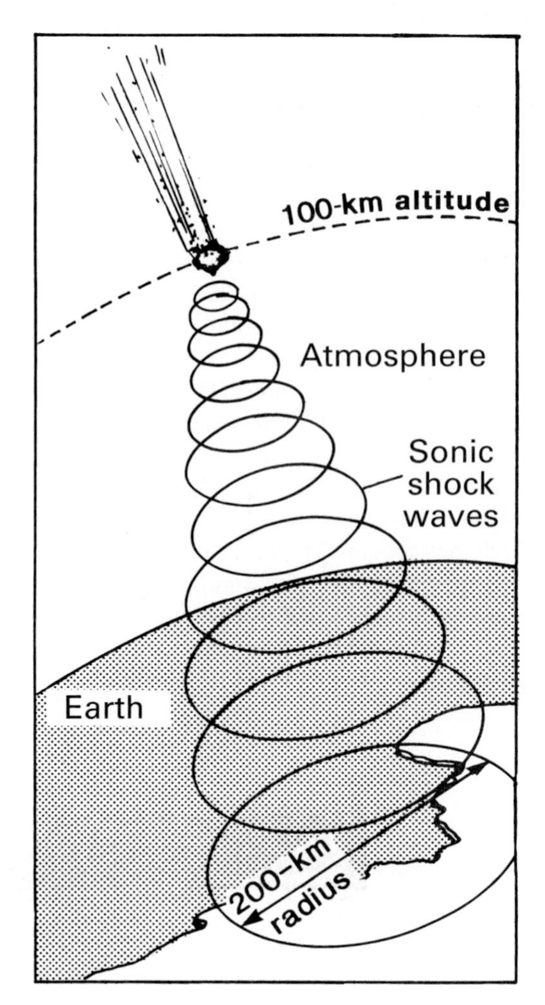

Figure 15.13 A meteor hitting the top of the atmosphere sends off sonic shock waves that might be heard on the ground.

The heat of atmospheric friction may raise the surface temperature of an incoming object to 3,000°C (about one-half the temperature of the Sun's surface). Melted surface material is stripped off to feed the glowing tail of a fireball, lighting up thousands of cubic miles of sky.

Friction with the atmosphere also slows a meteorite down; it may hit the ground at only 200 to 400 mph. In 1954, an 8.5-pound stony meteorite crashed through the roof of a woman's home in Sylacauga, Alabama, and severely bruised her hip.

On 22 November 1996, a meteorite struck near San Luis in western Honduras, excavating a 165-feet-wide crater, damaging a main highway, and sparking a fire that destroyed several acres of coffee plants. It has been suggested that some mysterious deaths may be attributable to meteorite impact.

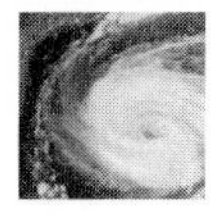

Crater-Forming Impacts

Meteoroids with weights greater than 350 tons are not slowed down much by the atmosphere. The big ones hit the ground at nearly their original speed, explode, and excavate craters. The record of crater-forming impacts on Earth is sparse. Craters are erased by erosion, consumed by subduction, mangled by continent collisions, and buried beneath younger sediments. There are 164 known impact craters including 57 in the United States and Canada (Figure 15.14).

Meteor Crater, Arizona

The world's classic meteorite crater lies on an arid portion of the Colorado Plateau in north-central Arizona about 26 km (16 mi) west of the town of Winslow. Meteor Crater, also known as Barringer Crater, is over 1 km (4,000 ft) wide, excavated nearly 185 m (600 ft) below the plateau, and

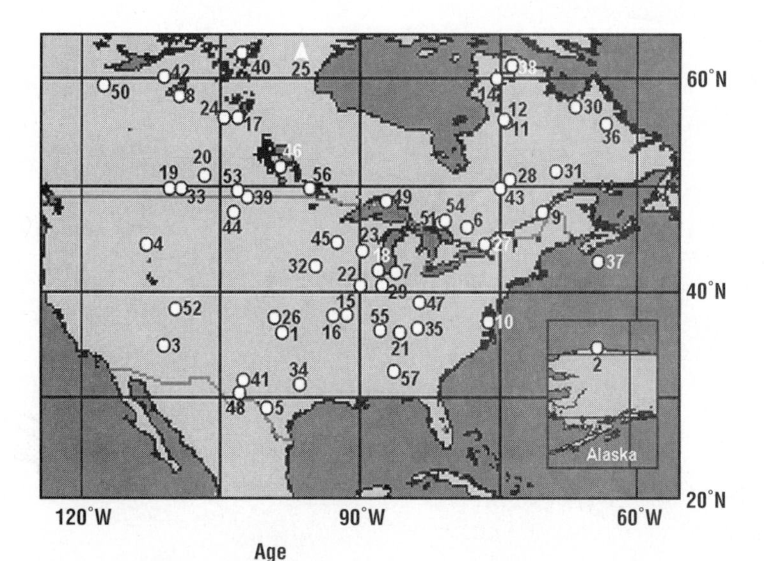

	Crater Name	Location	Diameter (km)	Age (millions of years)
1	Ames	Oklahoma	16	470 ± 30
2	Avak	Alaska	12	> 95
3	Barringer	Arizona	1.18	0.049 ±
4	Beaverhead	Montana	60	0.003
5	Bee Bluff	Texas	2.4	~ 600
6	Brent	Ontario	3.8	< 40
7	Calvin	Michigan	8.5	396 ± 20
8	Carswell	Saskatchewan	39	450 ± 10
9	Charlevoix	Quebec	54	115 ± 10
10	Chesapeake Bay	Virginia	90	342 ± 15
11	Clearwater East	Quebec	26	35.5 ± 0.3
12	Clearwater West	Quebec	36	290 ± 20
14	Couture	Quebec	8	290 ± 20
15	Crooked Creek	Missouri	7	430 ± 25
16	Decaturville	Missouri	6	320 ± 80
17	Deep Bay	Saskatchewan	13	< 300
18	Des Plaines	Ilinois	8	99 ± 4
19	Eagle Butte	Alberta	10	< 280
20	Elbow	Saskatchewan	8	< 65
21	Flynn Creek	Tennessee	3.8	395 ± 25
22	Glasford	Illinois	4	360 ± 20
23	Glover Bluff	Wisconsin	8	< 430
24	Gow	Saskatchewan	5	< 500
25	Haughton	Nunavut	24	< 250
26	Haviland	Kansas	0.01	23 ± 1
27	Holleford	Ontario	2.35	< 0.001
28	Ile Rouleau	Quebec	4	550 ± 100
29	Kentland	Indiana	13	< 300
30	La Moinerie	Quebec	8	400 ± 50
31	Manicouagan	Quebec	100	214 ± 1
32	Manson	Iowa	35	73.8 ± 0.3
33	Maple Creek	Saskatchewan	6	< 75
34	Marquez	Texas	12.7	58 ± 2
35	Middlesboro	Kentucky	6	< 300
36	Mistastin	Newfoundland/Labrador	28	36.4 ± 4
37	Montagnais	Nova Scotia	45	50.50 ± 0.76
38	New Quebec	Quebec	3.44	1.4 ± 0.1
39	Newporte	North Dakota	3.2	< 500
40	Nicholson	Northwest Territories	12.5	< 400
41	Odessa	Texas	0.16	< 0.05
42	Pilot	Northwest Territories	6	445 ± 2
43	Presqu'ile	Quebec	24	< 500
44	Red Wing	North Dakota	9	200 ± 25
45	Rock Elm	Wisconsin	6	< 505
46	Saint Martin	Manitoba	40	220 ± 32
47	Serpent Mound	Ohio,	8	< 320
48	Sierra Madera	Texas	13	< 100
49	Slate Islands	Ontario	30	~ 450
50	Steen River	Alberta	25	95 ± 7
51	Sudbury	Ontario	250	1850 ± 3
52	Upheaval Dome	Utah	10	< 170
53	Viewfield	Saskatchewan	2.5	190 ± 20
54	Wanapitei	Ontario	7.5	37.2 ± 1.2
55	Wells Creek	Tennessee	12	200 ± 100
56	West Hawk	Manitoba	2.44	100 ± 1.5
57	Wetumpka	Alabama	6.5	81 ± 1.5

Figure 15.14 Impact crater locations, sizes, and ages in Canada and the United States.

Map courtesy Planetary and Space Science Centre, University of New Brunswick. www.unb.ca/passc/ImpactDatabase

Figure 15.15 View northwest of Meteor Crater, Arizona. Notice the upturned rock layers in the crater rim, the little hills of ejected debris surrounding the crater, and the individual blocks of resistant rock (e.g., limestone) strewn about the plateau.
Photo by John S. Shelton.

surrounded by a rock rim rising 30 to 60 m (100 to 200 ft) above the countryside (Figure 15.15).

What is the evidence demonstrating that the crater formed by meteorite impact? 1) The crater is steep-sided and closed; 2) the rim of surrounding rock was created by uplifting the horizontal sedimentary-rock layers of the region and tilting them away from the crater; 3) little hills of rock outside the crater rim are inverted piles of the rock sequence exposed in the crater walls; 4) huge blocks of limestone are strewn around outside the crater; 5) the crater floor holds a 265 m (870 ft) thickness of shattered rock; 6) numerous pieces of nickel-iron metallic meteorite with a combined weight of nearly 30 tons have been collected in the area; and 7) several features indicate the occurrence of high temperature and pressure, such as unusual varieties of quartz (the minerals **coesite** and **stishovite**), cooled droplets of once-melted metal, fused masses of sand grains, and **shatter cones** (a structure of cones inside of cones) that form under pressure.

There also is negative evidence that argues against other processes being responsible for the crater. 1) There is no volcanic material within or nearby, and thus, the crater is not a volcanic-explosion pit. 2) There are no solutional features to argue for a solution-collapse process of subsidence similar to sinkholes.

When all the evidence is considered, the words of Sherlock Holmes in "The Adventure of the Bruce-Partington Plans" apply: "Each fact is suggestive in itself. Together they have a cumulative force." In sum, the evidence at Meteor Crater, Arizona, is overwhelming; the site has become the most photographed meteorite crater in the world.

Meteor Crater formed about 50,000 years ago when a nickel-iron metallic meteorite came blazing through the atmosphere. The meteorite had a diameter of about 30 m (100 ft), weighed around 110,000 tons, and hit the ground traveling about 20 km/sec (45,000 mph). The enormous energy of impact was largely converted into heat, which liquefied about 80 percent of the meteorite and the enveloping ground in less than a second. Impact craters are almost always circular because the meteorite explosion upon impact creates shock waves that excavate the crater. About 100 million tons of rock were pulverized in this Arizona event, generating about double the energy released by the Mount St. Helens volcanic eruption (Figure 7.10a–f). The shock wave leveled all trees in the region, wildfires broke out, and dust darkened the sky.

When the surface features are as clearly evident as they are at Meteor Crater, Arizona, there is little debate about the origin of a crater. But how are geologically ancient meteorite-impact craters recognized? What clues remain after the obvious topographic features have been removed by erosion or buried beneath later sediments? Some ideas have been generated during the worldwide search for evidence to explain the great dying of organisms that occurred 65 million years ago at the close of Cretaceous time.

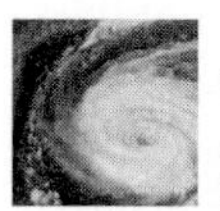

The Cretaceous/Tertiary Boundary Event

To learn what happened at the close of Cretaceous time, one should follow the advice spoken by Sherlock Holmes in "The Problem of Thor Bridge": "If you will find the facts, perhaps others may find the explanation." To find the facts of 65-million-year-old events means to examine rocks of this age.

The modern search began near Gubbio, Italy, where the Cretaceous/Tertiary boundary is well exposed. The latest Cretaceous (K) rocks are limestone loaded with fossil foraminifera (microscopic animal protists) whose individuals had diameters up to 1 mm. The earliest Tertiary (T) limestone above it contains an impoverished and markedly changed fossil assemblage. Between the limestones lies a 1 cm-thick clay layer that marks the Cretaceous/Tertiary (K/T) boundary. Does the K/T boundary clay layer hold facts that might explain the events of its days? The late Luis Alvarez, a Nobel prize-winning physicist, and his geologist son Walter focused their research group's efforts on this topic. A fact determined in their investigation is the high percentage of the element iridium in the clay layer, an enrichment about 300 times greater than the normal abundance. Here is a fact begging for an explanation.

Iridium is a siderophile, or iron-loving element, whose concentration on Earth parallels that of iron. Most of Earth's iridium lies deep in its iron-rich core; it migrated with iron to the core during the time of early heating when the Earth separated into layers of different density (Figure 2.2). But the K/T boundary clay layer, which is found in many places around the world, holds an estimated one-half million tons of iridium. How did this layer become so enriched in iridium? Luis Alvarez reasoned that since meteorites are enriched in iridium, a 10 km (over 6 mi)-diameter asteroid could have supplied the volume of iridium estimated to be present in the K/T boundary clay.

In a condensed version, here is a popular theory of our times: an asteroid with a diameter equal to the height of Hawaii (from sea floor to peak of Mauna Kea) slammed into the Earth 65 million years ago; the impact caused a great dying among life worldwide, including the extinction of dinosaurs (excluding birds), and left its incriminating fingerprint as iridium in a global clay layer. The theory is intriguing, easy to grasp, and beguilingly simple to accept. But for a theory to gain widespread approval in the scientific community, it must explain all relevant facts and allow predictions to be made. If these predictions later become supported by facts, then a theory gains wider acceptance. Thus, before the K/T impact theory could gain wide acceptance in the scientific community, many more facts needed to be discovered to verify predictions made by the theory.

Evidence of the K/T Impact Once the K/T theory was proposed, it excited scientists worldwide, and the search for facts shifted into high gear. Researchers around the world began examining the K/T boundary clay layer for other evidence of Earth-like versus meteorite-like components. 1) The clay layer was found on the continents, thus ruling out the possibility that the iridium enrichment was due simply to a change in ocean composition. 2) The K/T boundary clay minerals have a different composition from clays in the limestone layers above and below it; they might be explained by a mixture of one part asteroid to ten parts Earth crust. 3) Quartz grains are present with shocked crystal structures, indicating a short and violent impact. 4) Sand-sized spherules of minerals are present suggesting a melting and resolidification. 5) Ratios of the radioactive element rhenium to its decay product osmium are similar to those in meteorites and are quite different from the ratios in Earth surface rocks. 6) Abundant microscopic diamonds, found in some meteorites, occur in the K/T boundary clay layer. 7) Carbon-rich grains with "fluffy" structures indicative of fire are abundant in the K/T boundary clay layer.

Site of the K/T Impact The facts from the K/T boundary clay layer compelled more scientists to agree that a massive impact had occurred, but even more evidence was needed. If an asteroid slammed into Earth some 65 million years ago, then where was the impact site? Could it be found? Or had it been: 1) subducted and destroyed? 2) buried beneath a continental glacier? 3) hidden under piles of sediment on land or sea floor? 4) covered by flood basalt? 5) eroded and erased from the face of the Earth? 6) crunched into oblivion by a continent collision? Geologists searched for impact scars worldwide, but some were too small, while others were too old or too young.

Then related evidence began to focus the search. On Haiti was found a 65.01 (+ or – 0.8) million-year-old sedimentary rock layer containing shocked quartz grains and

1 cm-diameter glassy spherules formed from melted rock. In Cuba, a thick, chaotic sedimentary deposit with huge angular blocks was found. In northeastern Mexico, similar particles were found in a thick bed of sediment containing land debris, ripple marks, and other features interpreted as a tsunami deposit; this bed is 65.07 (+ or – 0.1) million years old. Similar but thinner deposits were found in the K/T boundary position in the banks of the Brazos River in Texas and in coastal deposits in New Jersey and the Carolinas. The sedimentary features suggested a Caribbean region impact, but where?

The excitement of the search led to the Yucatan Peninsula of Mexico, where the Mexican national petroleum company (PEMEX) had drilled exploratory wells in the region of Merida. At depths of 2 km, the PEMEX well bores had encountered a 90 m (300 ft)-thick zone of shattered rock containing shocked quartz grains and glassy blobs of once-melted rock. On the ground surface lie solution pits (sinkholes) aligned in a circular pattern. Geophysical measurements show circular patterns of gravity and magnetic anomalies suggesting a circular disturbance at depth. A seismic survey reveals a raised inner ring of 80 km (50 mi) diameter and an outer ring of about 195 km (120 mi) in diameter.

The above data all help define the Chicxulub structure of 64.98 (+ or – 0.06) million years age. The evidence continues to mount that a massive asteroid slammed into the shallow, tropical sea 65 million years ago.

Angle of Impact An asteroid can hit the Earth at angles ranging from 90° (vertical impact) to 0° (grazing blow). At Chicxulub, the subsurface features measured by gravity and magnetism show some opening to the northwest, like a horseshoe (Figure 15.16a). This asymmetry could be the result of the asteroid coming in from the southeast (Figure 15.16b), hitting and excavating a deep crater which shallows to the northwest. An oblique impact of 20 to 30° would concentrate its energy into vaporizing surface rocks and making a mammoth vapor/dust cloud. This impact scenario would cause great grief to life in North America by spraying the continent with a high-speed vapor cloud hot enough to ignite plants.

The impact was so great that its effects were not just regional but would have been felt worldwide; they probably played a significant role in the great dying that marked the end of Cretaceous time.

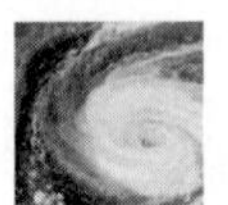

Problems for Life from Impacts

What does life have to tolerate when a massive asteroid slams into the land? There are many difficult conditions on both regional and global scales. 1) The impact certainly created an earthquake of monumental magnitude along with numerous gigantic aftershocks. Seismologist Steven M. Day has assumed the magnitude of the K/T earthquake can be estimated by scaling up from the energy released in nuclear explosions. Extrapolating upward from an atomic bomb blast of magnitude 4 leads to a K/T impact earthquake of magnitude 11.3 (see Figure 3.25). 2) Wildfires would rage regionally, or even globally. A recent study suggests that the K/T impact ejected so much hot debris into the atmosphere

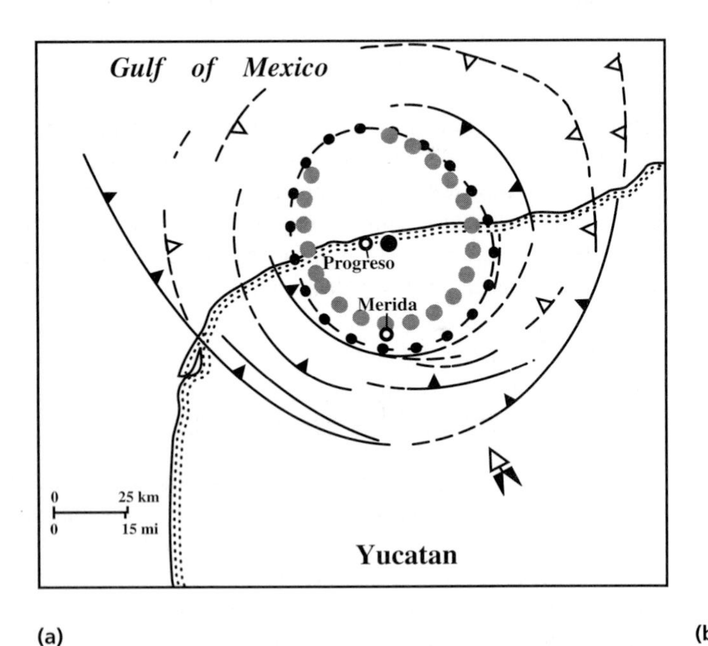

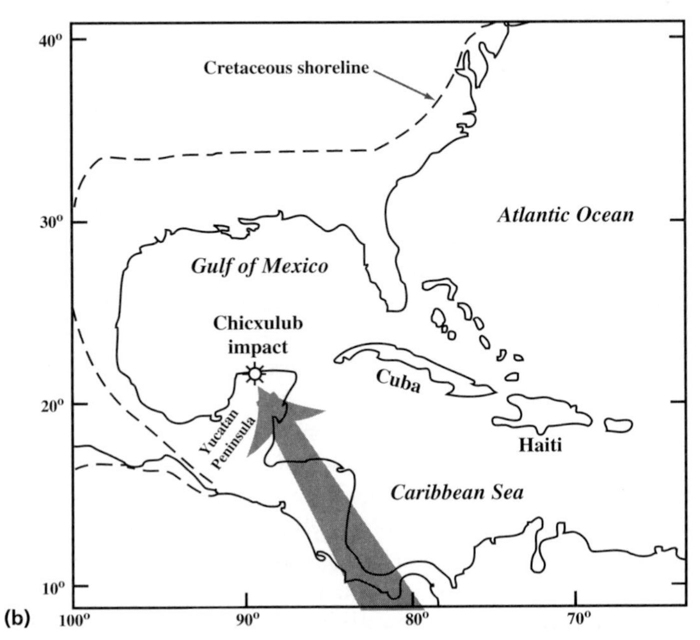

Figure 15.16 **(a)** A buried impact crater is shown on the tip of the Yucatan Peninsula. Notice how the gravity data appear to open to the northwest suggesting that the asteroid came from the southeast. **(b)** An approximate path of the K/T asteroid. Its impact would have sent a superhot vapor cloud over North America.

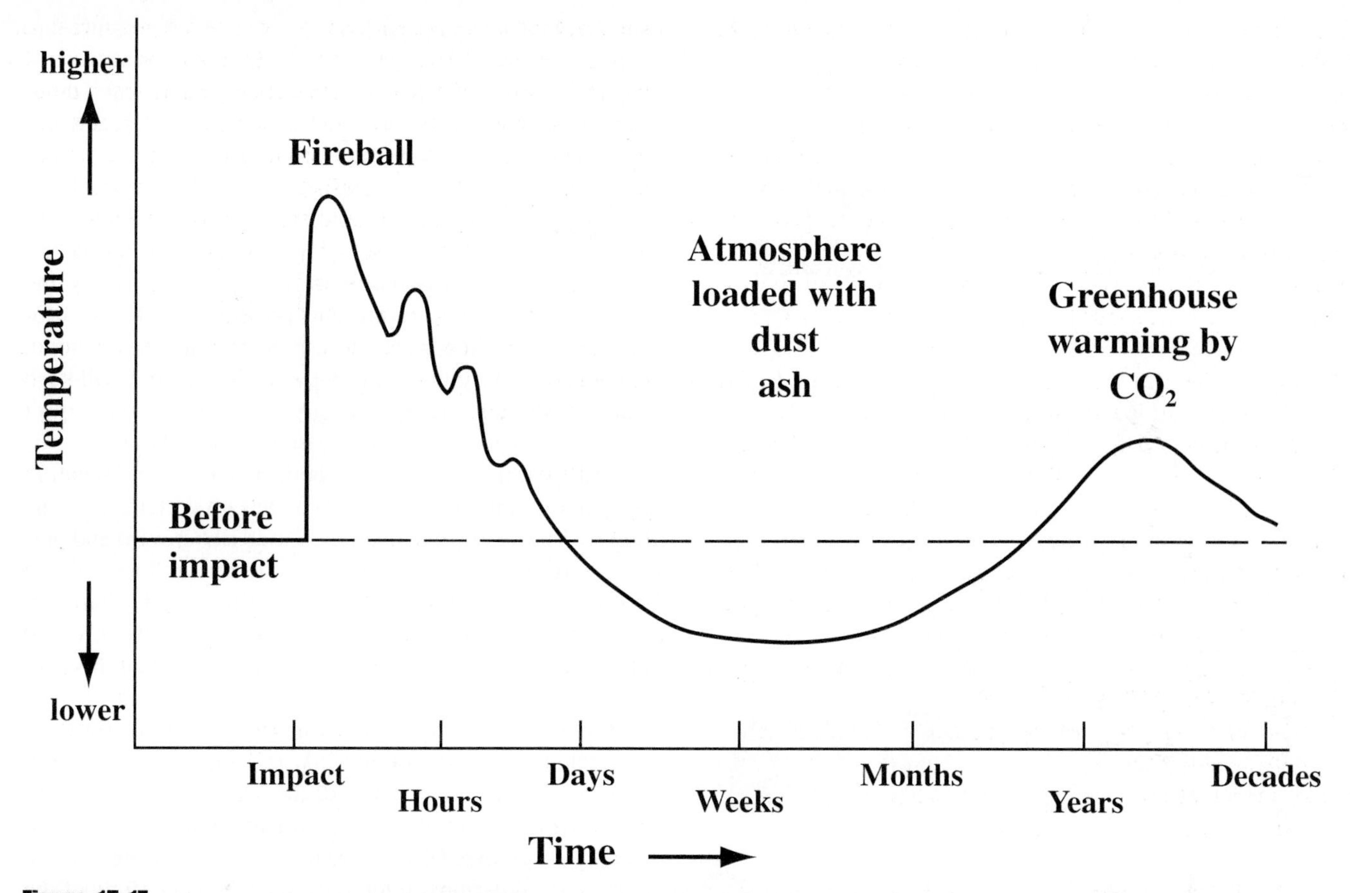

Figure 15.17 Impact of the K/T asteroid had marked effects on Earth's surface temperatures. First, there was a fireball and hot gases that lasted for many hours. Second, temperatures dropped to wintry conditions as airborne dust and soot blocked much incoming sunlight for several months. Third, after the dust settled, CO_2 remained aloft creating a greenhouse effect that lasted for years.

After David A. Kring, August 2000, "Impact events and their effect on the origin, evolution and distribution of life." *GSA Today,* v. 10, no. 8, p. 4.

that it caused massive wildfires that consumed much of the vegetation in North America, the Indian subcontinent, and the equatorial region of the world. 3) Huge amounts of nitrogen oxides in the atmosphere would fall as acid rain and acidify surface waters. 4) Dust and soot in the atmosphere would block sunlight and turn day into night, thus making photosynthesis difficult and plunging much of the world into dark wintry conditions for weeks to several months. 5) After the atmospheric dust settled, the water vapor and CO_2 remaining in the atmosphere would lead to global warming for years (Figure 15.17).

What additional insults does life have to survive after an oceanic splashdown of a 10 km-diameter asteroid? 1) Tsunami up to 300 m (1,000 ft) tall. 2) A bubble of steam up to 500 km^3 (120 mi^3) volume that blows into the upper atmosphere carrying Earth rock and asteroid debris. 3) Another problem occurred at Chicxulub where the K/T asteroid landed in shallow, tropical marine water underlain by limestone ($CaCO_3$). The impact must have vaporized enormous quantities of limestone, thus increasing atmospheric CO_2, maybe by an order of magnitude. After the winter-causing asteroidal dust settled, the added CO_2 in the atmosphere could have elevated the Earth's climate into global-warming conditions. Average temperatures in the world may have risen 10°C (18°F), and life on Earth would have been forced to endure the shift from an "extracold winter" to an "overly hot and long summer."

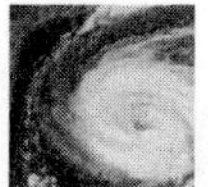

The Crater-Forming Process

The amount of energy released by an asteroid or comet impact depends upon its speed and size. Incoming objects in excess of 350 tons may impact at speeds as high as 30 km/sec (70,000 mph). The impact of smaller meteorites creates *simple craters* with raised rims and concave bottoms lacking central uplifts, such as Meteor Crater, Arizona (Figure 15.15).

The impact of larger meteorites (Figure 15.18a) forms *complex craters* with central uplifts and collapsed outer

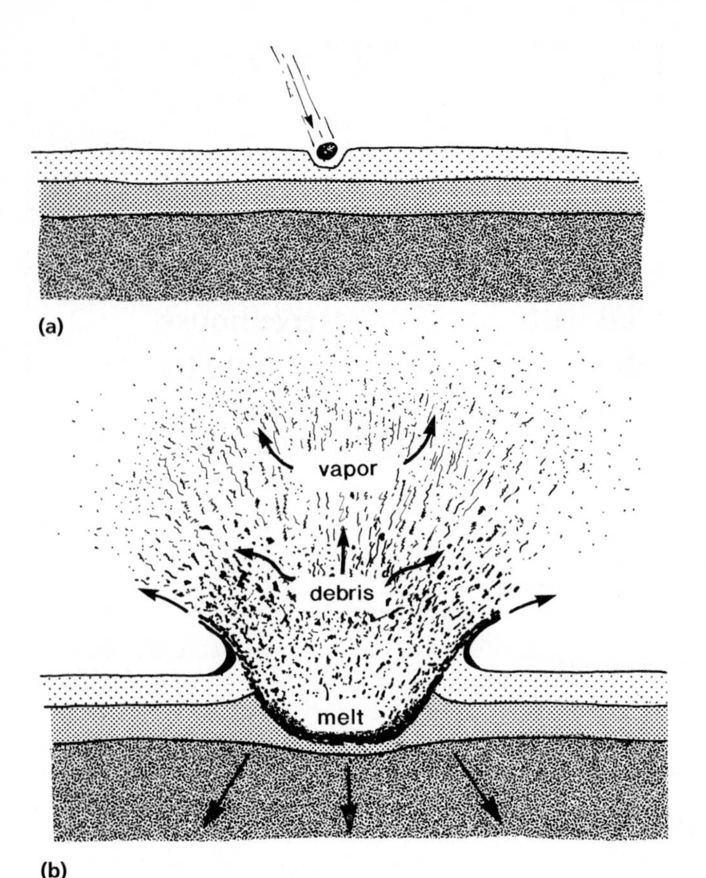

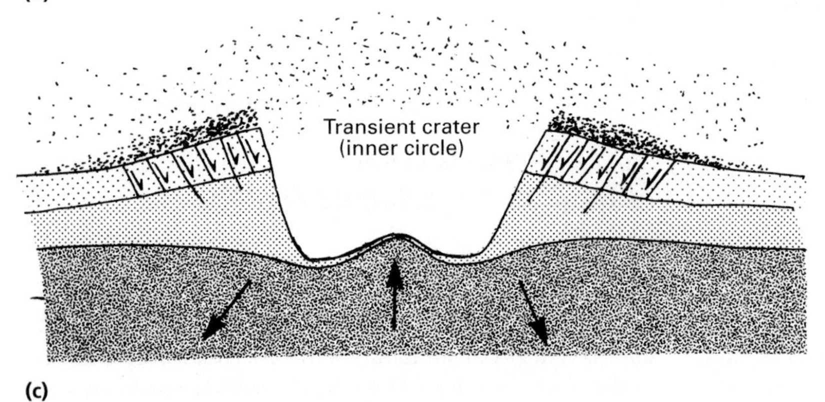

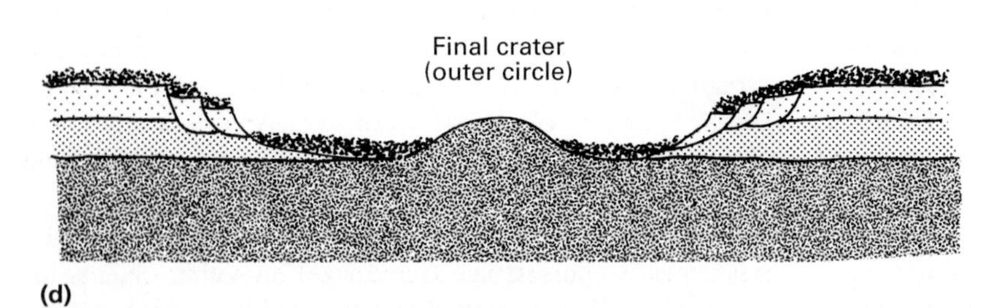

Figure 15.18 Impact of a meteoroid. **(a)** An incoming meteoroid heavier than 350 tons may be moving faster than 30,000 miles per hour. **(b)** The impact shock causes such high temperatures and pressures that most of the meteoroid and crater rock are vaporized and melted. **(c)** The release wave following the shock wave causes the center of the floor in the transient crater to rise. **(d)** The fractured walls fail and slide into the crater, creating a wider and shallower final crater.

rims. Large impacts generate so much heat and pressure that much of the asteroid and crater rock is melted and vaporized (Figure 15.18b). In an instant, temperatures may reach thousands of degrees centigrade, and pressure may exceed 100 gigapascals (more than the weight of a million atmospheres). The shock wave pushes rocks at the impact site downward and outward in rapid acceleration of a few kilometers per second. The rocks in the crater and the debris thrown out of the crater are irreversibly changed by the short-lived high temperature and pressure. Rocks are broken, melted, and vaporized; new minerals, such as diamond, are created; and a common mineral such as quartz will have its atomic structure transformed by the high-pressure impact into its high-density form as the mineral stishovite.

Still within the initial second, a release or dilatation wave follows into the Earth and catches up with the accelerating rocks, causing a deflection of material upward and outward, forming a central uplift on the crater floor (Figure 15.18c). The crater that exists in this split second is transient and soon to be enlarged. An example of a transient crater may be the 135 km-diameter circle of the Chicxulub structure in Yucatan.

As the crater is emptied of vaporized and pulverized asteroid and rock, the fractured walls of the transient crater fail and slide in toward the center of the crater (Figure 15.18d). This is the final enlarged crater with an upraised center, surrounded by a circular trough and then by an outermost fractured rim. The outer circle of the final crater may have a diameter 100 times wider than the crater is deep.

The seismically indicated 195 km-diameter circular anomaly around the Chicxulub structure is suggested to be an outer circle of a final crater; geophysical imaging indicates a central peak of 80 km diameter. The Manicouagan impact crater shows an upraised central area surrounded by a circular trough (Figure 15.3). An outer circle of a final crater may have existed at higher elevations at Manicouagan but has been eroded away by post-impact continental glaciation. Similar features are seen on other planets as well. The Yuty crater on Mars (Figure 15.19) has a well-developed central peak surrounded by a circular trough. Apparently, subsurface ice deposits were melted at Yuty, yielding muddy, liquefied ejecta that flowed over the adjacent area. Similar features would form on Earth if impact occurred on the frozen ground of Siberia, northern Canada, or Alaska.

The impact process may be visualized in miniature using a falling drop of water hitting a still body of water. At the point of impact, water springs upward, ripples and troughs surround the impact, and a spray of fine water shoots upward and outward. Although the water quickly

Figure 15.19 Yuty crater on Mars has a well-developed central peak and surrounding circular trough.
Photo courtesy of NASA.

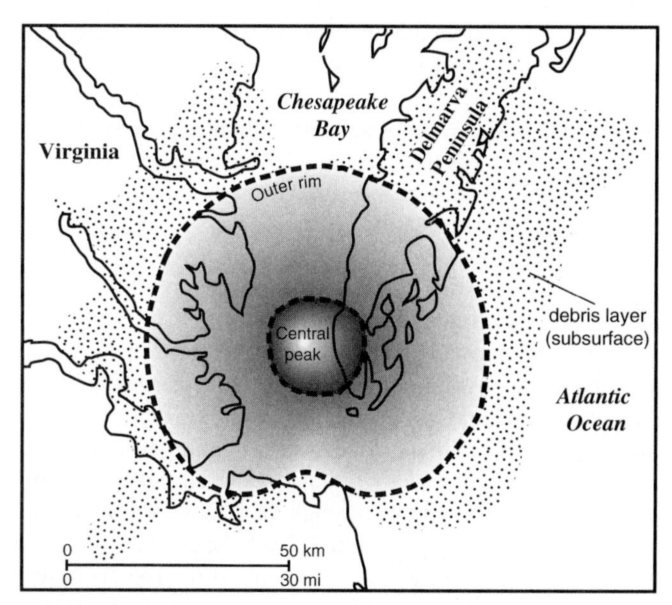

Figure 15.20 Outline of the buried impact crater in lower Chesapeake Bay, Virginia.

returns to its normal still condition and retains no evidence of impact, rocks altered by impact remain in their broken, melted, and shocked states for the rest of their existence.

Another way of visualizing the impact process is by imagining a bullet fired into soft sand. The bullet hits the sand, then fragments and explodes, creating a circular crater. The explosion creates the crater so it is circular whether the bullet path is vertical or at a shallow angle to the surface. After the event, what evidence remains? There is a circular crater with small fragments of metal that are difficult to find.

Impact Origin of Chesapeake Bay

Drilling bore holes and analyzing seismic lines, it has recently been shown that Chesapeake Bay is the site of a 90 km (56 mi)-diameter crater formed by asteroid impact 35.5 million years ago (Figure 15.20). The crater has a 25 km (16 mi)-diameter central peak surrounded by a circular trough or moat up to 400 m (>1,300 ft) deep and is 30 km (19 mi) across. The shape of the impact structure is similar to Yuty crater (Figure 15.19). The impact crater was a topographic low spot that rivers flowed toward, thus forming the network of river valleys that are today drowned by the Atlantic Ocean to form Chesapeake Bay.

Geologists have long puzzled over the source of **tektites,** glassy spherules formed by in-air cooling of impact-melted rock. The North American tektite field consists of glassy spherules spread over the southeastern United States, Gulf of Mexico, and Caribbean Sea area. Now we know that an asteroid heading south 35.5 million years ago slammed into Virginia and showered tektites over a 9,000,000 km^2 area south of the impact site.

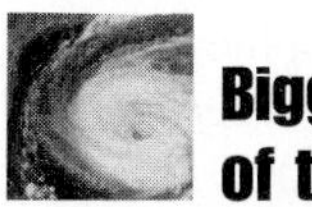

Biggest Event of the Twentieth Century

Tunguska, Siberia, 1908

The morning was sunny in central Siberia on 30 June 1908. Then, after 7 A.M., a massive fireball came streaking in from the east. It exploded about 8 km (5 mi) above ground in a monstrous blast heard 1,000 km (over 600 miles) away. No humans lived immediately under the blast point, but many reindeer did, and they died. A man 60 km (37 mi) away was enveloped in such a mass of heat that he felt his shirt almost catch fire before an air blast threw him 2 m (7 ft). People and horses 480 km (300 mi) from the explosion site were knocked off their feet. From 650 km (400 mi) away, visible in bright sunlight, a huge column of fire rose 20 km (12 mi) high. The ground shook enough to be registered on seismometers in Russia and Germany. Barometric anomalies were recorded as the air blast traveled twice around the world. In Sweden and Scotland, an extraordinarily strong light appeared in the sky about an hour after sunset; it was possible to read books by this light until after 2 A.M.

Scientists around the world speculated on what had happened, but it was years before an expedition went to the remote area to search for evidence of the event. Near the Tunguska

River, the forest in an area greater than 1,000 km² (over 400 mi²) was found to have been knocked down and destroyed; many trunks were charred on one side. Over a broader area exceeding 5,000 km² (2,000 mi²), 80 million trees were down, and many others had tilted trunks, broken branches, and other signs of disturbance. But there was no impact crater or even broken ground. The relative lack of facts led unrestrained minds to invent all sorts of wild stories. It was not until 1958 that scientists returned to the site and collected little globules of once-melted metal and silicon-rich rock.

Several important facts must be explained. There was an intense bluish-white streak in the sky, a horrendous explosion, a searing blast of heat, blasts of air that encircled the globe, a brilliant sunset and bright night, yet no impact crater, only little globules of melted material. So what happened? A meteoroid racing through the atmosphere broke up and exploded about 8 km (5 mi) above the ground. It either was a fragment of an icy comet about 50 m (165 ft) in diameter, or it was a large, stony meteorite about 30 m (100 ft) in diameter. The object was traveling about 15 km/sec (33,000 mph) when it disintegrated in a spectacular mid-air explosion. If it had been a metallic body, it would almost certainly have slammed into the ground. But comets and stony meteorites are weak bodies traveling at outrageous speeds, and the resistance of the Earth's atmosphere is so strong that they typically break apart. At the end of June 1908, the comet Encke was passing by Earth; one of its fragments is the likely culprit for the Tunguska event.

The Tunguska comet explosion rocked a sparsely inhabited area and devastated a forest. Imagine if it had exploded over a city like Washington, D.C., obliterating its buildings and people (Figure 15.21). The broader area of knocked-down trees was big enough to include Baltimore.

Tunguska

Washington, D.C.

0 10 40 km

0 25 mi

Figure 15.21 The Tunguska event in 1908 devastated more than 400 mi², knocking down most trees. Had this comet exploded over Washington, D.C., it would have been one of the major events in history.

How common are these Tunguska-like events? Are such events frequent enough for humans to be concerned about?

Biggest "Near Events" of the Twentieth Century

On 22 March 1989, the asteroid 1989FC with a diameter of about 500 m (0.3 mi) crossed Earth's orbit at almost the wrong moment; Earth was at the spot six hours earlier. The asteroid missed us by less than 700,000 km (400,000 mi). Had a collision occurred on land, the impact would have created a crater about 7 km (4.4 mi) across. Was this close call a freak occurrence? Apparently not: several thousand such bodies are in Earth-approaching orbits.

On 19 May 1996, a 150 m (500 ft)-wide asteroid missed Earth by only 453,000 km (<281,000 mi). This asteroid is four times wider than the meteorite that excavated Meteor Crater, Arizona (Figure 15.15).

In March 1998 the media widely and excitedly reported that an asteroid labeled 1997XF11 might hit Earth in the year 2028. In the same year two big-budget Hollywood movies were released sensationalizing the effects of collision with a comet in *Deep Impact* and with an asteroid in *Armageddon.* In order to communicate calmly about the threat of comet and asteroid impacts, Richard Binzel developed the Torino Scale (Table 15.3), which assesses the threat on a scale of 0 to 10.

For smaller objects, the number of near misses is surprisingly high. Detailed telescopic examination has shown that up to 50 house-sized bodies pass between the Earth and Moon each day. Should we worry about the consequences of impacts from speeding bodies of 50 m (165 ft) diameter? Apparently not, because the Earth has its own defense system against small bodies—its atmosphere. When comets and stony meteorites traveling at 50,000 km/hr (over 30,000 mph) hit the atmosphere, the great strains break most of them into smaller pieces that burn up explosively. Most of the flameouts occur 10 to 40 km (6 to 25 mi) above the surface, which is too high for significant damage to occur on the ground. However, most iron meteorites are internally strong enough to stay intact as they pass through the atmosphere and they do hit the ground. Luckily, iron meteorites are relatively uncommon.

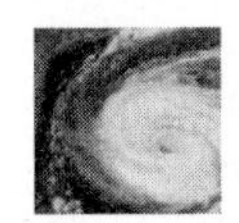

Frequency of Large Impacts

How often do impacts of large bodies occur? This question is hard to answer looking at our planet because of the continuous recycling of Earth's surface materials by plate tectonics and their destruction by weathering and the agents of

Table 15.3	The Torino Scale
Assessing Comet and Asteroid Impact Hazards	
Events with no likely consequences (White zone)	
0	No collision hazard, or object is small.
Events meriting careful monitoring (Green zone)	
1	Collision is extremely unlikely.
Events of concern (Yellow zone)	
2	Collision is very unlikely.
3	Close encounter with >1% chance of local destruction.
4	Close encounter with >1% chance of regional devastation.
Threatening events (Orange zone)	
5	Significant threat of regional devastation.
6	Significant threat of global catastrophe.
7	Extremely significant threat of global catastrophe.
Certain collisions (Red zone)	
8	Collision will cause localized destruction (one event each 50 to 1,000 years).
9	Collision will cause regional devastation (one event each 1,000 to 100,000 years).
10	Collision will cause global catastrophe (one event each 100,000 years).

erosion. It is easier to answer this question by looking at the long-term record of impacts preserved on the dead surface of the Moon and then extrapolate the results back to Earth. The dark volcanic maria on the Moon formed after the few hundred-million-year period of intense asteroidal bombardment over 3.9 billion years ago. The basalt-flooded maria cover 16 percent (6 million km^2) of the Moon's surface; they formed by about 3,200 million years ago. The maria are scarred by five craters with diameters greater than 50 km (over 30 mi) and another 24 craters with diameters between 25 to 50 km (15 to 30 mi). This averages to one major impact somewhere on the maria every 110 million years.

Applying these impact rates to Earth generates the following numbers: Earth's surface area is more than 80 times the area of the lunar maria, so it would have had more than 80 times as many impacts, i.e., about 2,400 impacts leaving craters greater than 25 km (15 mi) diameter. Land comprises about 30 percent of the Earth's surface, so about 720 of these craters should have formed on land. More than 160 craters have been discovered so far, but most of them are less than 25 km diameter. Most of the missing craters have probably been destroyed or buried.

The odds are extremely small that a large asteroid will hit the Earth during your lifetime. However, so many people will be killed when a big space object does hit that it skews the probabilities. Statistically speaking, every individual has a greater chance of being killed by a comet or asteroid than of winning a big jackpot in a lottery! The probabilities of death by meteoroid impact were indirectly assessed in the words of paleontologist George Gaylord Simpson: "Given enough time, anything that is possible is probable." Because the risks from large meteoroid impact are high, they should be of concern to humans. American astronomer David Morrison has described the Earth as a target in a cosmic shooting gallery of high-speed asteroids and comets (Figure 15.22). The situation has been evaluated for defensive actions we humans might take.

Figure 15.22 The Earth presents a small target in the cosmic shooting gallery.
Photo courtesy of NASA.

A Defense Plan

The problems presented by meteoroid impacts were addressed at National Aeronautics and Space Administration (NASA) international workshops in 1991–92, as requested by NASA and the U.S. Congress. The conclusions are presented in a report entitled *The Spaceguard Survey.* The risks analyzed were those presented by space objects with diameters greater than 1 km (0.62 mi) (Tables 15.1 and 15.4). About 90 percent of the potential impactors are near-Earth asteroids or short-period comets; the other 10 percent are intermediate- or long-period comets (greater than 20-year return times).

There are over 2,000 Near-Earth Objects (NEOs), and about 25 to 50 percent of them will eventually hit the Earth. However, the average interval of time between impacts exceeds 100,000 years. The risk of any individual in the United States being killed via the impact of an asteroid with a diameter of 1 km or greater in the next 50 years is around one chance in 20,000. The surprisingly high risk is an artifact of the calculations—a tremendous number of people killed in an impact (e.g., 1.5 billion) compared to a long return time between events. The calculations show that your risk of death from meteoroids is about 150 times greater than your odds of dying from botulism, 50 times higher than that from fireworks, and 3 times greater than that from tornadoes. Some more of *The Spaceguard Survey* findings are summarized in Table 15.5.

Table 15.4 Frequency of Impacts and Annual Risks of Death

For Tunguska-sized Events:	
Average interval between impacts	300 years
Average interval for populated areas only	3,000 years
Average interval for urban areas	100,000 years
Average interval for U.S. urban areas only	1,000,000 years
Total annual probability of death	1/30,000,000

Source: D. Morrison (1992).

Table 15.5 Odds of Dying in the United States from Selected Causes

Cause of Death	Odds of Happening
Motor vehicle accident	1 in 100
Murder	1 in 300
Fire	1 in 800
Firearms accident	1 in 2,500
Electrocution	1 in 5,000
Asteroid or Comet Impact	1 in 20,000
Airplane crash	1 in 20,000
Flood	1 in 30,000
Tornado	1 in 60,000
Venomous bite or sting	1 in 100,000
Food poisoning by botulism	1 in 3,000,000

Source: D. Morrison (1992).

Can we do anything about this threat? Or must we just sit back fatalistically and say, "It will happen if it is meant to be"? The first step in a plan to protect ourselves is to locate the near-Earth objects (NEOs), determine their orbits, and learn which ones present immediate threats. In 1998, the U.S. Congress authorized $40 million for NASA to find 90 percent of the near-Earth asteroids (NEAs) greater than 1 km (0.62 mi) diameter by 2008. The search is being conducted by six international observatories and their success is impressive and improving (Figure 15.23). By October 2002, they had discovered 2,072 NEOs; 635 are NEAs and 465 of them have been classified as potentially hazardous asteroids (PHAs). So far, only one has a chance of hitting Earth soon, and it is a 1-in-300 chance for an impact in 2880.

What can be done if we discover a large NEO on a path to impact the Earth? We could take an appropriate engineering action. Suggested ways of changing an NEO to alter its collision course include: 1) blowing it apart with a nuclear explosion, 2) attaching a rocket engine that could drive it away, 3) using a big mirror to focus sunlight that would vaporize rock, and 4) scooping rock (mass) and tossing it away.

Is a large impact too unlikely an event to take seriously? Arthur C. Clarke said: "We tend to remember only the extraordinary events, such as the odd coincidences; but we forget that almost every event is an odd coincidence. The asteroid that misses the Earth is on a course every bit as improbable as the one that strikes it." Are we willing to make a commitment to an Earth defense system? Or is the situation too low a priority for dollars when competing with crime, AIDS, welfare, and other issues?

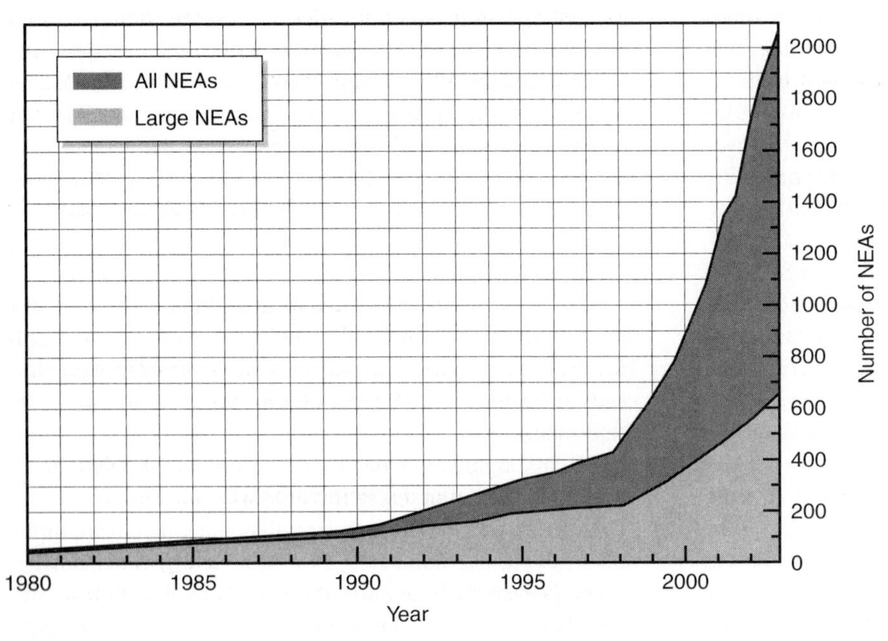

Figure 15.23 Number of near-Earth asteroids (NEAs) by year of discovery. Total NEAs are dark gray; large NEAs (diameter > 1 km) are light gray.
Alan Chamberlin, NASA's Near-Earth Object Program Office

Summary

Space debris colliding with Earth are primarily rocky (stones), metallic (irons), or ices. Metallic or stony bodies are called meteorites if they land on the surface of the Earth; icy masses are comets. Irons are the most common form found by collectors because they pass through the atmosphere more readily, are more resistant to weathering, and are more easily recognized. Impact scars are hard to find on Earth because they are destroyed by plate-tectonic actions and the agents of erosion. Conversely, the geologically dead Moon is a museum of impact scars.

Asteroids abound in a Sun-orbiting belt between Jupiter and Mars. The strong gravitational pull of the massive planet Jupiter sends collided fragments outbound on collision courses with nearby planets, including Earth.

Comets surround the Solar System in a vast envelope known as the Oort cloud. Some comets are pulled inward close to the Sun and may collide with the Sun or a planet, as in July 1994, when the Shoemaker-Levy 9 comet collided with Jupiter.

About 100,000 million meteoroids enter Earth's atmosphere every 24 hours. Most are small and burn up due to atmospheric friction before reaching 35 miles above ground. Up to 1,000 tons of material are added to Earth's surface each day.

Meteoroids hit Earth's atmosphere at speeds from 25,000 to 70,000 mph; some are destroyed on impact, some are deflected into space, and others are slowed by friction as they pass through the atmosphere. Meteoroids weighing more than 350 tons are slowed little by the atmosphere and may hit the ground at high speeds, explode, and excavate craters.

The impact of a large asteroid generates such tremendous heat and pressure that much of the asteroid and crater rock is vaporized. Rocks in the crater, and debris thrown out of the crater, are broken, melted, and vaporized; minerals have new high-pressure atomic structures and include forms such as diamonds. As the initial or transient crater is emptied, the crater bottom rebounds upward, and the fractured walls slide inward toward the crater center, forming a final, enlarged crater. The crater may be 100 times wider than deep.

Life on Earth is subjected to great stress by an impact. A large asteroid impact can generate an earthquake over magnitude 11; cause widespread wildfires; create nitrous oxides in the atmosphere that fall as acid rain; and place dust in the atmosphere that blocks incoming sunlight to create "winter." After the dust settles, water vapor and CO_2 remain in the atmosphere, causing a global-warming "summer." Additionally, if the impact occurs in the ocean, tsunami of 1 to 3 km height can occur.

About 50 house-size bodies pass between the Earth and Moon each day, but the Earth is protected by its atmosphere. It is difficult to estimate how often large bodies impact Earth because plate tectonics and erosion destroy the evidence. However, looking at the Moon's surface indicates Earth receives a crater greater than 15 mi diameter every 1.33 million years.

Statistically, your risk of being killed by a meteoroid impact is greater than that of dying by flood or tornado. We have the ability to locate incoming meteoroids and could possibly change their paths by sending explosives or a rocket out to redirect them.

Terms to Remember

aphelion 410
astrobleme 408
coesite 416
maria 407
meteor 408
meteorite 408
meteoroid 408
Oort cloud 410
perihelion 410
shatter cone 416
shooting star 408
solar wind 410
stishovite 416
sublimation 410
tektite 421

Questions for Review

1. Why does the Moon display impact scars so well?
2. Why are impact scars relatively rare on Earth?
3. Distinguish between a meteor, a meteoroid, a meteorite, an asteroid, and a comet.
4. Why are metallic meteorites so commonly collected?
5. Why did the asteroids of the asteroid belt not assemble into a tenth planet?
6. Why does a comet's tail glow brighter as it nears the Sun? Why does a comet's tail point away from the Sun?
7. How much space debris is added to the Earth each day? How big must a meteoroid be to pass through the atmosphere with little slowing?
8. Draw a series of cross sections showing what happens when a 10 km-diameter asteroid hits the Earth at 20,000 mph.
9. Make a long list of the evidence you could collect to demonstrate that a specific area was the site of an ancient asteroid impact.
10. Describe the sequence of life-threatening events that occur when a 10 km-diameter asteroid slams into Earth.
11. Explain the Torino Scale of impact threats.

Questions for Further Thought

1. Extrapolating impact rates from the Moon, about 720 craters with diameters greater than 15 mi should have formed on land on Earth; only about 120 have been found so far. Might some of the missing impact sites have been big and caused mass extinctions? How could you proceed scientifically to investigate this possibility?
2. Should the United States spend the money and effort to develop engineering devices that could land on large asteroids and comets and change their courses away from hitting the Earth?
3. It is proposed that we send rockets or explosives to divert incoming large asteroids or comets. Might this action just shatter the incoming object into many devastating impactors? Or cause the object to hit Earth on a more direct path?

Suggested Readings and References

Binzell, R. P., Barucci, M. A., and Fulchignoni, M. (1991, October). The origins of the asteroids. *Scientific American,* 88–94.

Chapman, Clark R., and Morrison, David. (1989). *Cosmic Catastrophes.* New York: Plenum Press.

Chapman, Clark R., and Morrison, David. (1994, 6 January). Impacts on the Earth by asteroids and comets: Assessing the hazard. *Nature,* 367, 33–39.

Cowen, R. (1994, 30 July). Comet impact poses intriguing riddles. *Science News,* 146, 68–69.

Davies, J. K. (1986). *Cosmic Impact.* London: Fourth Estate, Ltd.

Dodd, Robert T. (1986). *Thunderstones and Shooting Stars.* Cambridge: Harvard University Press.

Grieve, Richard A. F. (1990, April). Impact cratering on the Earth. *Scientific American,* 66–73.

Melosh, H. Jay. (1988). *Impact Cratering: A Geologic Process.* New York: Oxford University Press.

Morrison, D. (1992). *The Spaceguard Survey: Report of the NASA International Near-Earth Object Detection Workshop.* Pasadena: NASA/Jet Propulsion Lab.

Morrison, David, and Chapman, Clark R. (1990, March). Target Earth: It will happen. *Sky and Telescope,* 261–65.

Poag, C. W., et al. (1994). Meteoroid mayhem in Ole Virginny. *Geology,* 22, 691–94.

Verschuur, G. L. (1996). *Impact! The Threat of Comets and Asteroids.* New York: Oxford University Press.

Weaver, Kenneth F. (1986, September). Meteorites, invaders from space. *National Geographic,* 170, 390–418.

Wetherill, G. W., and Shoemaker, E. M. (1982). Collision of astronomically observable bodies with the Earth. In *Geological Implications of Impacts of Large Asteroids and Comets on the Earth* (1–13). Geological Society of America Special Paper 190.

Videos

The Doomsday Asteroid. (1995). NOVA/BBC-TV/WGBH (60 min.).

Asteroids: Deadly Impact. (1997). National Geographic Society (60 min.).

Comet Halley. (1985). WETA-Washington (60 min.).

The Miracle Planet—The Third Planet. (1994). Nova/KCST (60 min.).

Planet Earth—Tales from Other Worlds. (1986). WQED-Pittsburgh (60 min.).

Powers of Ten. (1968). Pyramid Film & Video (8 min.).

Chapter 16

Population Growth

> Most men will not swim before they are able to.
>
> —*Novalis* (Friedrich von Hardenberg, 1772–1801)

> Naturally they won't swim! They are born for the solid earth, not for the water. And naturally they won't think. They are made for life, not for thought. Yes, and he who thinks, what's more, he who makes thought his business, he may go far in it, but he has bartered the solid earth for the water all the same, and one day he will drown.
>
> —Herman Hesse, 1929, *Steppenwolf*

Today, the world population of humans is in the midst of a breathtaking growth. The greater the number of people (Figure 16.1), the greater the need for land to live upon, to farm, and to mine. And so, more and more people migrate into hazardous locations. They live and farm on the slopes of active volcanoes, in the lowlands of river floodplains, and along hurricane-prone coastlines. How have the numbers of people grown so large? The present situation can best be appreciated by examining the record of population history.

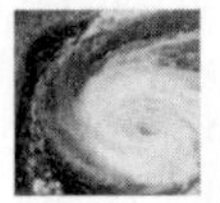

Overview of Human Population History

The most difficult part of human history to assess is the beginning because there are no historic documents and the fossil record is scanty. Nonetheless, the present pace of population change is so great that it overwhelms the small-scale changes of the distant past, thus leaving a clear picture. For the sake of mathematical

Figure 16.1 The number of people on Earth continues to grow.
Photo courtesy of Pat Abbott.

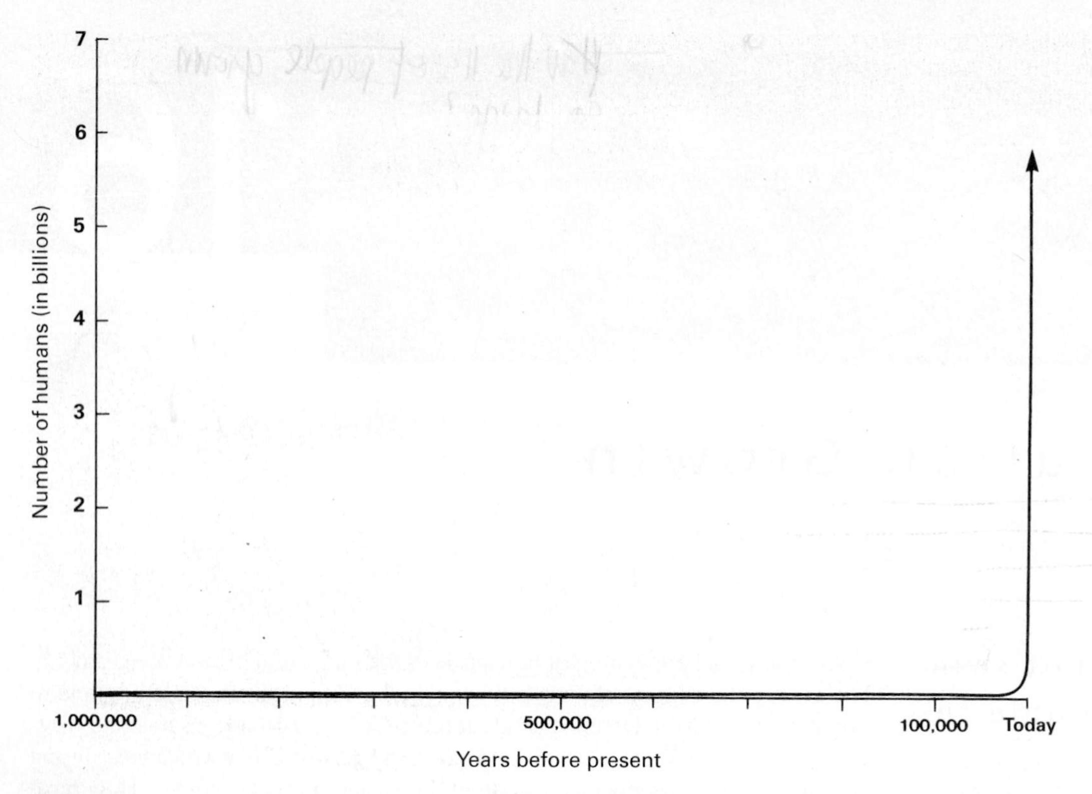

Figure 16.2 The human population in the last 1 million years.

analysis, we can hypothesize a start a million years ago with a few thousand people; the rate of population growth and number of people alive early in human history were so small that they cannot be plotted accurately on the scale of Figure 16.2. The growth from a few thousand people a million years ago to over 6.2 billion people in the year 2002 requires an average growth rate of only 0.02 percent per year, but an average growth rate is misleading.

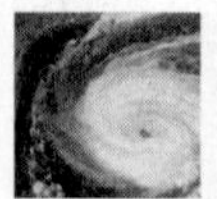

The Power of an Exponent on Growth

The most stunning aspect of Figure 16.2 is the peculiar shape of the human population curve; it is nearly flat for most of human time and then abruptly becomes nearly vertical in recent years. In the last two or three centuries, birth rates have changed little, while death rates have plunged. The marked upswing in the curve of Figure 16.2 shows the result of the **exponential growth** of the human population. Possibly the least appreciated concept of present times is what a growth-rate exponent does to the size of a population over time. Exponential growth moves continuously in ever-increasing increments; it leads to shockingly large numbers in surprisingly short times. Probably our most familiar example of exponential growth occurs when interest is paid on money.

It can be difficult to visualize the results of exponential growth when it is expressed only as a percentage over time, such as 0.02 percent growth of the human population in a million years or 7 percent interest on your money for 50 years. It is commonly easier to think of exponential growth in terms of doubling times—the number of years required for a population to double in size given an annual percentage growth rate. A simple formula, commonly called the rule of 70, allows approximation of doubling times:

$$\text{Doubling time (in years)} = \frac{70}{\text{\% growth rate/year}}$$

Learning to visualize annual percentage growth rates in doubling times is useful whether you are growing your money in investments or throwing it away by paying interest on debts (especially at the high rates found with credit-card debt). Dividing interest rates into 70 is an easy habit to acquire (Table 16.1).

The Last 10,000 Years of Human History

The long, nearly flat population curve in Figure 16.2 certainly masks a number of small-scale trends, both upward and downward. The fossil record is not rich enough to plot a detailed record, but surely at times when weather was

Sidebar

Interest Paid on Money: An Example of Exponential Growth

Compare the growth of money in different situations (Figure 16.3). If $1,000 is stashed away and another $100 is added to it each year, a linear growth process is in operation. Many of the processes around us can be described as linear, such as the growth of our hair or fingernails.

If, on the other hand, another $1,000 is stashed away but this time earns interest at 7 percent per year and the interest is allowed to accumulate, then an exponential growth-rate condition exists. Not only does the $1,000 earn interest, but the interest from prior years remains to earn its own interest in compound fashion.

Notice that an exponential growth curve has a pronounced upswing or J shape. A comparison of the linear and exponential curves in Figure 16.3 shows that they are fairly similar in their early years, but as time goes on, they become remarkably different. The personal lesson here is to *invest money now.* Smaller amounts of money invested during one's youth will become far more important than larger amounts of money invested later in life. Individuals who are disciplined enough to delay some gratification and invest money while they are young will be wealthy in their later years. Albert Einstein described compound interest, the exponential growth of money, as one of the most powerful forces in the world.

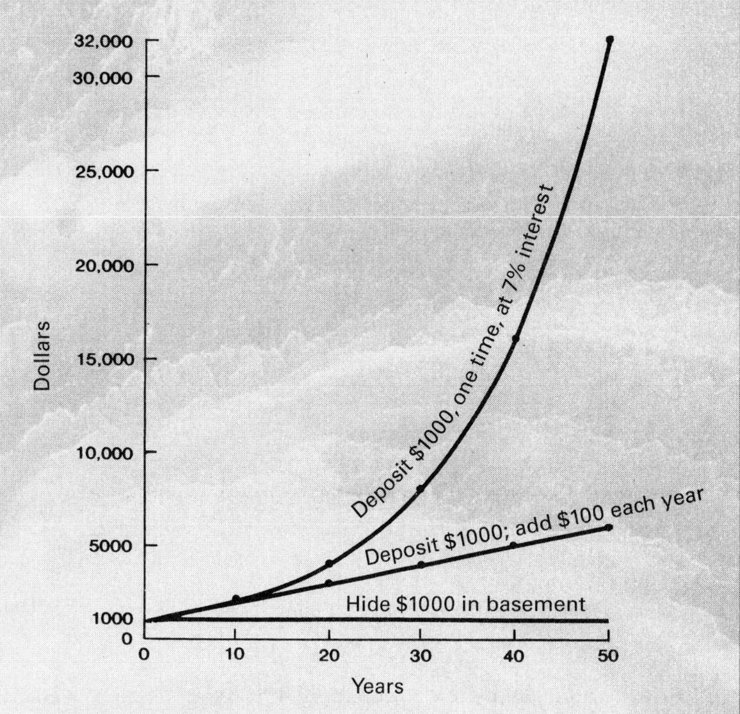

Figure 16.3 Amounts of money versus time. Compound interest (exponential growth) produces truly remarkable sums if given enough time.

Here is a riddle that illustrates the incredible rate of exponential growth; it shows the significance of doubling times in the later stages of a system. Suppose you own a pond and add a beautiful water lily plant that doubles in size each day. If the lily were allowed to grow unchecked, it would cover the pond in 30 days and choke out all other life forms. During the first several days, the lily plant seems small, so you decide not to worry about cutting it back until it covers half the pond. On what day will that be?

Table 16.1 Doubling Times at Some Common Percentage Rates

Growth Rate (% per Year)	Doubling Time (Years)
0.02	3,500
0.5	140
1	70
1.4	50
2	35
5	14
7	10
10	7
17	4

pleasant and food from plants and animals was abundant, the human population must have risen (Figure 16.4). Conversely, when weather was harsh, food was scarce, and diseases were rampant, the human population must have fallen.

The nearly flat population growth curve began to rise more rapidly by about 8,000 years ago when agriculture became established and numerous species of animals were domesticated (Figure 16.5). The world population is estimated to have been about 8 million people by 10,000 years ago. After the development of agriculture and the taming of animals removed much of the hardship from human existence, the population growth rate is likely to have increased from about 0.0015 percent to 0.036 percent per year, yielding a net gain of 360 people per million per year. This heightened rate of population growth probably raised the number of humans to about 200 million alive at 2,000 years ago.

Ex.

a result of change ↑ pop

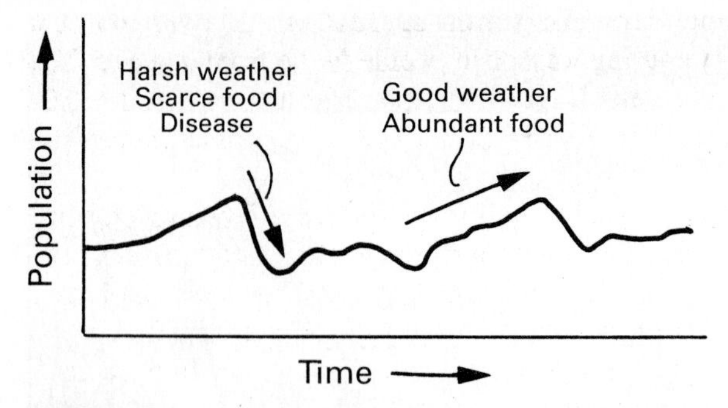

Figure 16.4 The nearly flat curve of human population size in Figure 16.2 would show considerable fluctuations on a larger-scale plot. Good weather and plentiful food cause upsurges in population; bad weather, disease, and scarce food cause downswings in population.

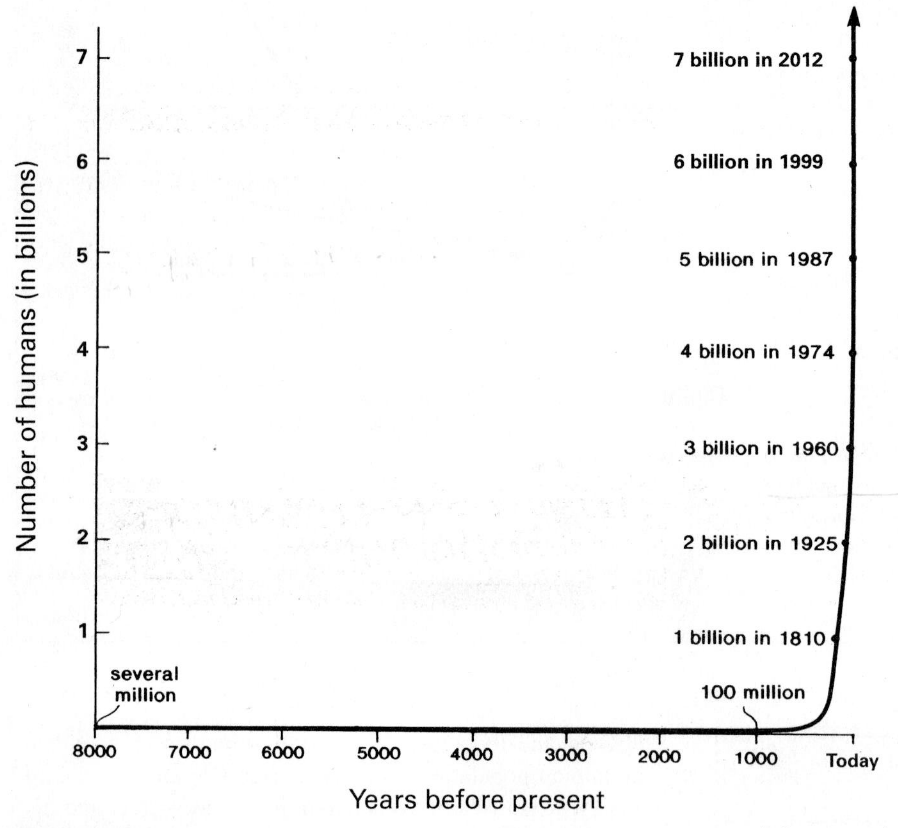

Figure 16.5 The human population of the last 8,000 years shows a slight upswing following the agricultural revolution and explosive growth following the scientific-medical revolution.

As humans continued to improve their ability to modify the environment with better shelter and more reliable food and water supplies, the world population grew at faster rates. From about 1 C.E. to 1750, world population grew to about 800 million. Growth was at an average rate of 0.056 percent per year, or another 560 people were added per million per year.

The eighteenth century saw many of the intellectual advances that set the stage for the present phase of cultural change. At long last, the causes of many diseases were being recognized, and the principles of public health were being established. Advances in the medical world greatly improved the odds for the survival of individual humans through their reproductive years. No longer were many mothers and great numbers of children dying during childbirth and infancy. Throughout the history of the human race, high rates of birth were required to offset high rates of infant mortality and thus maintain a viable-sized human population.

The eighteenth century saw death rates drop dramatically, but birth rates remained high and population doubling times dropped dramatically, thus population size soared. About 1810, the human population reached 1 billion; by 1925, it had grown to 2 billion; in 1960, it reached 3 billion; by 1974, it was 4 billion; by early 1987, it was 5 billion; in 1999, it reached 6 billion, and is heading toward 7 billion in 2012 (Figure 16.5). Notice the continuing decline in the number of years it takes for a net gain of another 1 billion people on Earth; the effect of exponential growth is racing ahead.

 The Human Population Today

At present, the world population is growing at about 1.3 percent per year for a doubling time of 53 years. The 1.3 percent gain is a net figure derived by measuring the birth rate (**fertility** rate) and subtracting the death rate (**mortality** rate). Even after subtracting all the human lives lost each year to accidents, diseases, wars, and epidemics such as AIDS, the human population still grows by almost 80 million people per year. Each year, the world population increases by about the total population of the Philippines or Germany.

Many people complain that there are too many people in the world today; they say there is too much crowding into cities, too much crime, pollution, illegal migration, disease, and other evils. Yet others say there are not too many people in the world. From a simplistic approach, using mathematics only (and ignoring biological and environmental realities), you can calculate that if 6 billion people stood shoulder-to-shoulder with each person having her or his own 3-by-1-ft space (0.28 m^2), the entire world population would fit inside a square fence less than 42 km (about 26 mi) on a side. From numbers like these, some argue that the world seems almost uninhabited, and thus, there is plenty of room for more people. Of course, this simple calculation is not realistic since it ignores water and food supply, energy resources, living space, waste disposal, etc.

What does the future hold? Nobody knows for sure, but let us continue with mathematics and push the exponential growth model into the future to see what happens. The land surface area of the Earth is $1,488.5 \times 10^{11}$ m^2; this is a large enough area to have 529 trillion humans (529,000 billion people) stand shoulder-to-shoulder (again all these standing humans are simply theoretical; this is not biologically or environmentally realistic). At today's 1.3 percent growth rate per year, the human population would reach 529 trillion in less than 900 years.

If the same 1.3 percent rate of human population increase continued, the volume of human flesh would about equal the volume of the Earth in less than 2,000 years from now; this is less time than the religions of Judaism and Christianity have existed. This shocking scenario of the human future is based simply on mathematical extrapolation of current population trends into the future. Does trend have to be destiny? Will religious, political, and cultural traditions change before population size grows beyond the **carrying capacity** of the Earth? This question is difficult to answer.

Carrying Capacity

Calculations of Earth's carrying capacity for humans vary markedly. However, biologists studying carrying capacity of the environment for individual species of mammals, birds, and other animals find that population size is regulated by the resources available. When a resource such as available food increases, a feeding population grows in size. If a food resource decreases due to drought, disease, etc., the population dependent on that food dies back and decreases in size. Thus it is commonly assumed that populations of most animals are at or near their carrying capacity.

Ireland in the 1840s provides a human example of carrying capacity. The European explorers of the 1500s brought the potato back from South America. The potato is a highly nutritious food. A diet of potatoes, milk (from animals fed potatoes), and greens makes a whole diet. An acre of potatoes could feed an Irish family of six for a year. In Ireland, the potato was the wonder crop that allowed a children-loving population to grow explosively. In 1841, Ireland's population had grown well past 8 million with nearly half the people surviving wholly or largely on potatoes. In 1845, heavy spring rains aided growth and spread of a fungal infestation, the potato blight, which caused potatoes to rot in storage. But no other food source existed; when the potatoes rotted there was no substitute food. Malnourishment became common. Then the winter of 1846–47 hit with unusual severity causing weakened people to die from diseases. The toll was severe; a million people died and another 1.5 million emigrated. During their travel to the United States and Canada, 1 in 7 of the emigres perished.

The carrying capacity of Irish land increased for humans when the potato arrived. Potato plants covered the lands, even extending into bogs and up steep mountain slopes. The human population fed by the increased resource grew rapidly but when the potato resource dropped suddenly, so did the human population.

Thus, the laws governing carrying capacity apply to the human animal as well. Some view the Polynesian history of Easter Island (Rapa Nui) as a small-scale example of what the exponential growth of humans is likely to do to the whole Earth.

Easter Island (Rapa Nui)

The history of human population on Easter Island (Rapa Nui) is instructive for the whole Earth. Easter Island is a triangular-shaped volcanic mass with an area of about 165 km^2 (64 mi^2). It lies over 2,000 km (more than 1,200 mi) east of Pitcairn Island and over 3,700 km (more than 2,200 mi) west of Chile (Figure 16.6). Easter Island is isolated; it

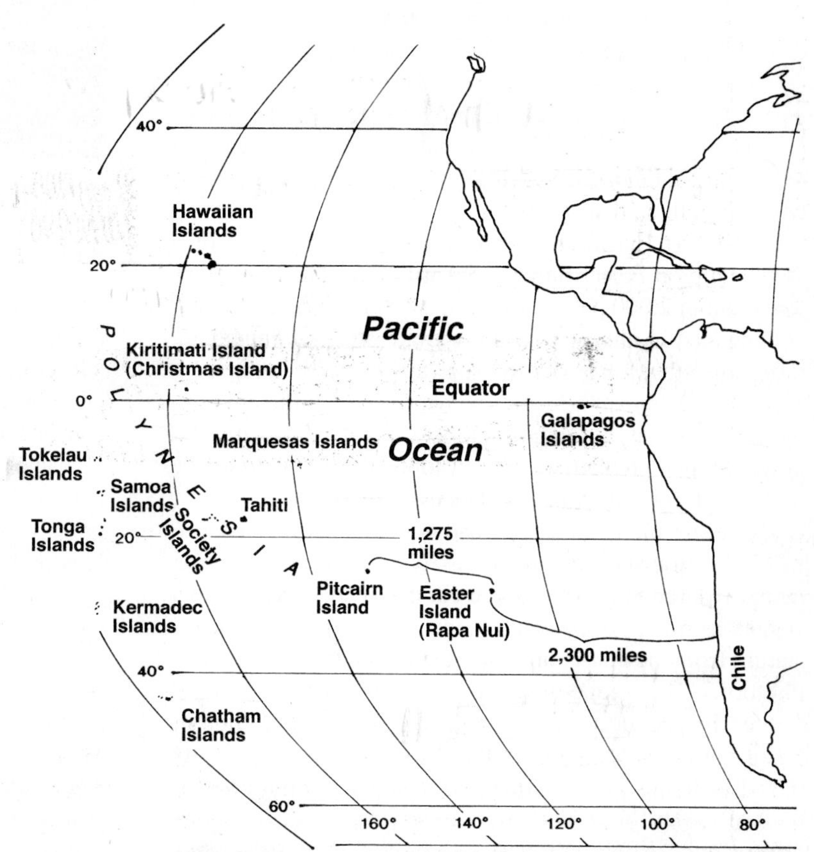

Figure 16.6 Easter Island (Rapa Nui) is an isolated outpost of Polynesian civilization nearly lost in the vast Pacific Ocean.

has high temperatures and humidity, poorly drained and marginal soils, no permanent streams, no terrestrial mammals, about 30 native plant species including trees in locally dense growths, few varieties of fish in the surrounding sea, and year-round water available only in little lakes within the volcano caldera.

Early in the fifth century C.E., seafaring Polynesian people arrived on Rapa Nui with 25 to 50 settlers. They were part of the great Polynesian expansion outward from southeast Asia that led them to discover and inhabit islands from Hawaii in the north to New Zealand in the southwest and to Rapa Nui in the southeast. The wide-ranging voyagers colonized islands over a Pacific Ocean area more than twice the size of the United States.

The colonizers of Rapa Nui brought chickens and rats, along with several of their food plants. The climate was too severe for most of their plants except the yam. Their resulting diet was based on easily grown chickens and yams, and housing was fashioned using wood from native trees; the people had lots of free time.

The islanders used their free time to develop a complex social system divided into clans that practiced elaborate rituals and ceremonies. Their customs included competition between the clans in shaping and erecting mammoth statues. The statues were carved out of volcanic rock using obsidian (volcanic glass) tools. The statues (moai) were more than 6 m (20 ft) high, weighed about 15 tons apiece, and were erected upon ceremonial platforms (ahu) (Figure 16.7).

The peak of the civilization occurred about 1550 C.E., when the human population had risen to about 7,000; statues numbered more than 600, with half as many more being shaped in the quarries. But from its peak, the civilization de-cline [illegible]

[illegible] out trees, there was little fuel for cooking or to ward off the chill of colder times. Without trees, there was increased soil erosion and less agricultural production. Without trees, there were no canoes and thus lesser amounts of fish were caught. Without canoes, there was no escape from the remote and

Figure 16.7 Rapa Nui inhabitants spent much of their energies creating giant statues (moai).
Photo courtesy of Pat Abbott.

isolated island. As food resources declined, the social system collapsed, and the statue-based religion disintegrated. Clans were reduced to warfare and cannibalism in the struggle for food and survival.

The competition between clans had been so consuming that the health of the environment was not considered and a price was paid: the human population on the island collapsed. Easter Island is one of the most remote inhabited areas on Earth, a tiny island virtually lost in the vast Pacific Ocean. When problems set in faster than the Rapa Nui customs could solve them, there was no place to turn for help, no place to escape. The carrying capacity of the land had been exceeded and the human population paid the price. What lesson does Easter Island have for the whole world? The Earth is but a tiny island lost in the vast ocean of the Universe (Figure 16.8); there is no realistic chance of escaping to another hospitable planet.

The Easter Island example raises interesting philosophical questions. How fast can human value systems change to meet

Figure 16.8 The Earth is an isolated outpost nearly lost in the vast "ocean" of the Universe.
Photo courtesy of Pat Abbott.

the mathematical reality of exponential growth of a human population? If all the people on Earth had to face the Easter Island situation, could our collective customs change fast enough to handle a human population growing out of control?

It can be difficult to live during an emotional time of major change and still remain intellectually detached enough to recognize and cope with the ongoing major changes. How does one distinguish between fundamental truths and the mere customs of the past? How can one separate important principles from the fads of the day?

There is no question that the human population of the world is changing drastically and with exponential rapidity (Figures 16.2 and 16.5), yet many political, religious, and social institutions deny that a problem exists. Is mathematical reality being set aside due to the unpopularity of its message that birth rates must decrease? Another scenario dealing with the future of the human population follows.

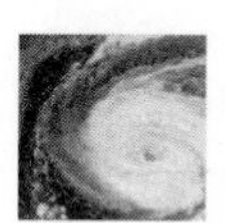

Demogra[illegible] Transition Model

To understand the hist[illegible] people like to look fo[illegible] that the growth in human popula[illegible] steady rise over time but rather follows cycles. In this view, a major technological advance occurs, and it is followed by an increase in population up to a new and higher level. The technological advance is said to increase the carrying capacity of the Earth for humans, and thus, the population rises to use the new opportunity. In his book *Wealth of Nations,* published in 1776, Adam Smith said, "Men, like all other animals, naturally multiply in proportion to the means of their subsistence."

Three major technological changes are usually described to explain the long-term increase in the human population worldwide:

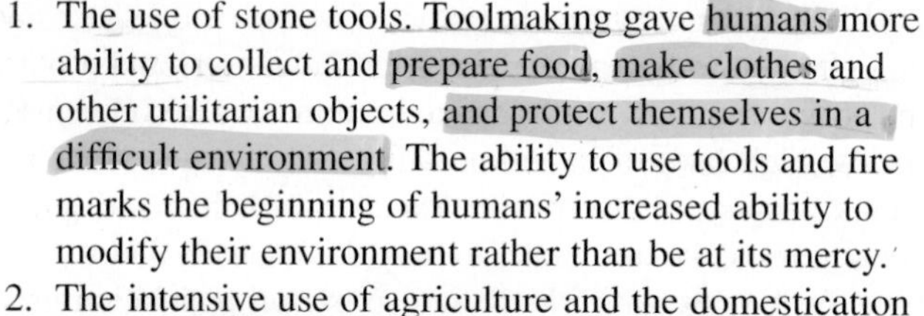

1. The use of stone tools. Toolmaking gave humans more ability to collect and prepare food, make clothes and other utilitarian objects, and protect themselves in a difficult environment. The ability to use tools and fire marks the beginning of humans' increased ability to modify their environment rather than be at its mercy.
2. The intensive use of agriculture and the domestication of animals was well under way by 8,000 years ago. As humans gained control of their food supply, population could rise to higher levels.
3. The scientific, medical, and industrial revolution gave humans some degree of control in sustaining individual lives, and thus, population has grown explosively to previously unthinkable levels.

It is thought that following each of the first two revolutions, the human population increased rapidly and then leveled off at a new and higher carrying capacity (Figure 16.9). The process is thought to have progressed from: 1) discovery of an important technology, to 2) rapid growth of human population, to 3) overshooting the carrying capacity, and finally, to 4) a decline to and stabilization at a population level supportable by the new carrying capacity. The switch from rapid population growth to a stable population appears to occur during a **demographic transition** that is benign, i.e., without negative effects. However, it must be pointed out that the levelings of population shown on Figure 16.9 are schematic, conceptual lines—they are not plots of real data.

Note also in Figure 16.9 that the vertical axis plot of population uses a **logarithmic** scale that compresses numbers, thus changing the shapes of the curves and distorting the real picture. The greater the number on a logarithmic scale, the greater is the compression of the numbers, and the more the real events are visually distorted. Conversely, the dramatically different shape of the population curve in Figure 16.2 is an actual plot of real data on the human population.

The demographic transition model is based on the recent experiences in the economically wealthy, Western nations.

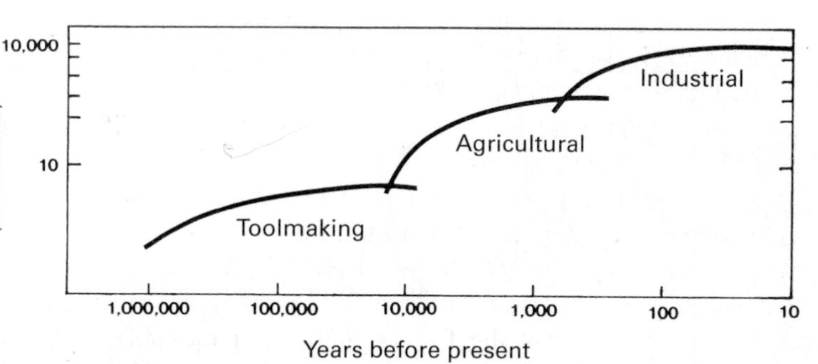

Figure 16.9 The growth of the human population can be viewed as having been due to revolutions in toolmaking, agriculture, and science and medicine. Each major leap forward caused an upsurge in population. The leveling of population shown after each revolution is an artistic interpretation, not a plotting of real data on population size.

Up through the seventeenth century, a woman had to bear several children to have a few survive to adulthood and replace the prior generation. Births had to be numerous to compensate for the high rates of infant mortality. Beginning in the eighteenth century, discoveries in science, medicine, and public health caused the death rates to drop dramatically. During this time, birth rates stayed high, and so overall population grew rapidly (Figure 16.10). As time passed and people realized that most of their children would survive to greater ages, birth rates dropped and population stabilized at a new and higher level.

The demographic transition is described as taking place in three steps:

1. Before the transition: high death rates are offset by high birth rates to maintain a population.
2. During the transition: low death rates couple with continuing high birth rates, causing population to soar.
3. After the transition: low death rates combine with low birth rates to achieve a stable population at a significantly higher level.

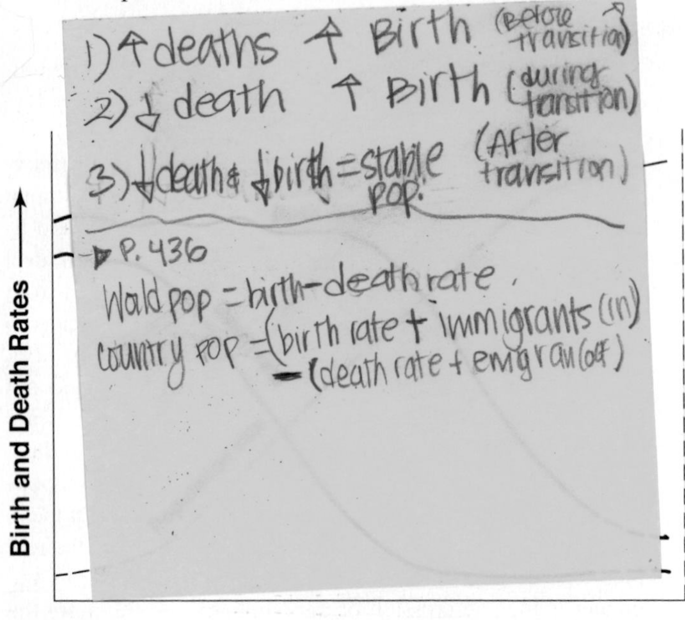

Figure 16.10 The demographic transition. First, the death rates decrease due to implementation of better public health procedures, nutrition, and medical intervention. When succeeding generations become convinced that their children will survive, then birth rates drop and overall population size stabilizes at a new, higher level.

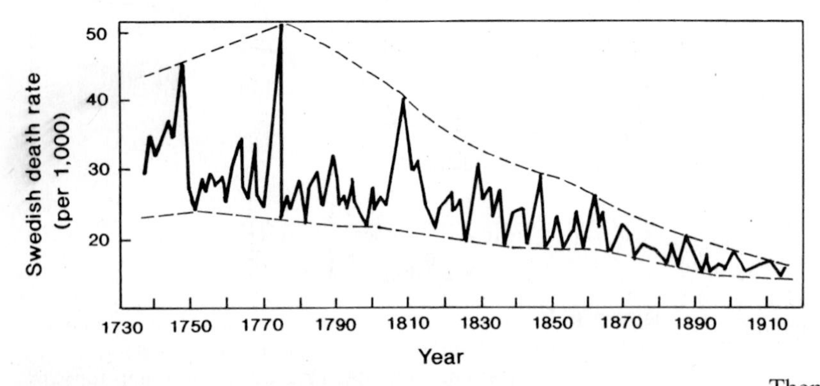

Figure 16.11 Mortality rates in Sweden. Notice that over time, there are decreases in both average rate of mortality and year-to-year oscillations in mortality.

During the eighteenth through twentieth centuries, Europe experienced the demographic transition thusly:

- Average life expectancies increased from 25–35 to 75–80 years.
- Average number of children borne per woman dropped from five to less than two.
- Annual rates of births and deaths dropped from 3–4% down to 1%.

As an example, the mortality rates in Sweden dropped in both average numbers and in amounts of oscillation from year to year (Figure 16.11). The change to stable and relatively predictable death rates prompted people to bear fewer children.

The demographic transition model has applied mostly to the economically wealthy countries of today. Some long-term population histories show changes over time in areas such as Mesopotamia. They can be quite instructive.

Population History of Mesopotamia

The Euphrates and Tigris Rivers bring large volumes of water long distances to the flat desert lowlands of their distal reaches (Figure 16.12). The availability of a water supply in a warm area with flood-deposited sediments for soils has attracted human agricultural development for over 6,000 years. The historical and archaeological records have been examined to reconstruct a population history for this region that today is part of Iraq. By 4100 B.C.E., an agricultural society worked the region using gravity-powered canals to supply irrigation systems. For the following 2,200 years, the population size varied in the short term but overall increased from about 25,000 to 630,000 as farming and towns spread through the river basins (Figure 16.13).

After this long interval of irrigation farming, much of the land suffered from environmental abuse that created areas of water-logged soils, regions with salt-poisoned soil, and many stretches of canals choked by sediments. The decrease in agricultural productivity saw the population mostly decline for 1,500 years; wars and disease also took their tolls, and groups migrated out of the region.

Then the Persians conquered the lowlands and instituted centralized control and a sophisticated irrigation system. The population grew for a millenium—from about 200,000 people in 450 B.C.E. up to 1.5 million by 550 C.E. The population remained relatively stable for a few centuries.

After the Arabs invaded and took over, the population dropped markedly for about 500 years, hitting a low of 140,000 around 1300 C.E. The population stayed relatively low until twentieth-century medical practices allowed the number of people to grow so rapidly that the only way to plot it on Figure 16.13 is by changing the vertical scale. The population currently is racing forward at 3.9 percent per year, despite a murderous war with Iran (1980–1988), genocide practiced by the Baghdad regime against Shiite Muslims in the south and Kurds in the north, and the Persian Gulf War of 1992.

How can the population cycles on Figure 16.13 be interpreted? Have there been two benign demographic transitions already and a third yet to occur? Or does the human population simply grow out of control until some negative events knock it back down to a lower level?

The best information we have available to analyze are the population records of other species. When weather is favorable, food supplies are large, and predators and disease are not significant, animal populations grow rapidly. Greater numbers of offspring survive and reproduce, thus creating positive exponential growth, i.e., positive feedback to population size. When greater amounts of resources are available, the carrying capacity of the environment is increased, and animal populations increase.

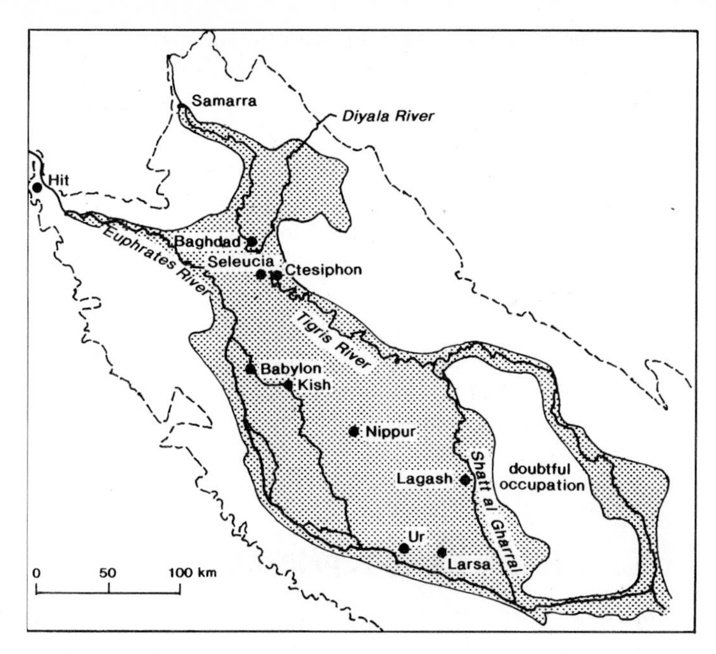

Figure 16.12 Map of the lower stretches of the floodplain of the Euphrates and Tigris Rivers of Mesopotamia. Today, the region is a part of Iraq.

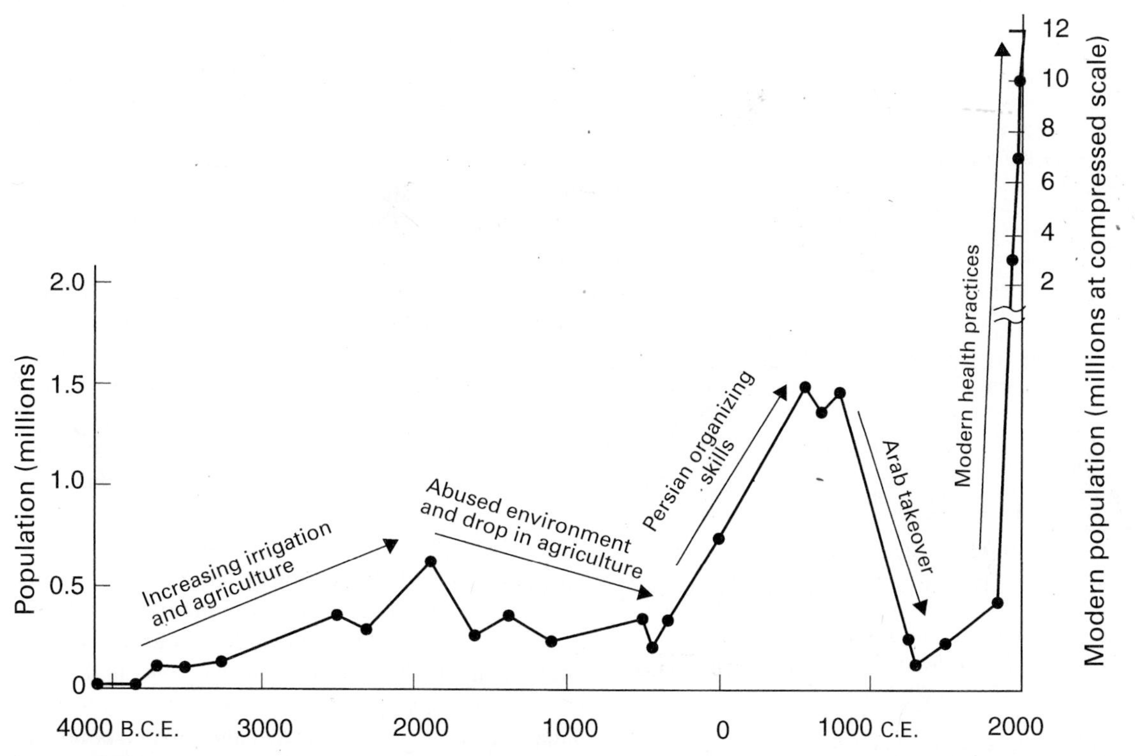

Figure 16.13 Population history of the Euphrates-Tigris river lowlands of Mesopotamia. Notice that to fit the twentieth-century population explosion onto the page, the vertical scale had to be reduced.

But when climate deteriorates, food supplies shrink, and disease strikes, then animal populations collapse in mass die-offs. Lesser numbers of surviving animals means lesser numbers of offspring, thus creating negative exponential growth, i.e., negative feedback to population size. When the carrying capacity of the environment decreases, population sizes also decrease.

Setting technology aside, humans do not differ from other species in their relationship to the environment. We increase the carrying capacity of the environment by using tools, developing agriculture, implementing public health practices, and building an industrial society on an energy base of fossil fuels. When events occur such as fossil fuel depletion, viral epidemics, and wars, then the carrying capacity of the environment is reduced, and the human population will drop correspondingly.

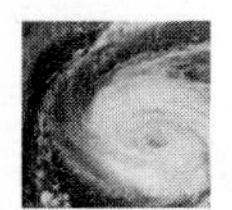

World Population Data

The growth rate of the world population equals the birth rate minus the death rate. The growth rate for a country or region equals the birth rate and the number of immigrants (incoming) minus the death rate and the number of emigrants (outgoing). Excess people can migrate from one impacted region to another less crowded or more amenable site; but there is no escape from the Earth, and the population total only grows at present.

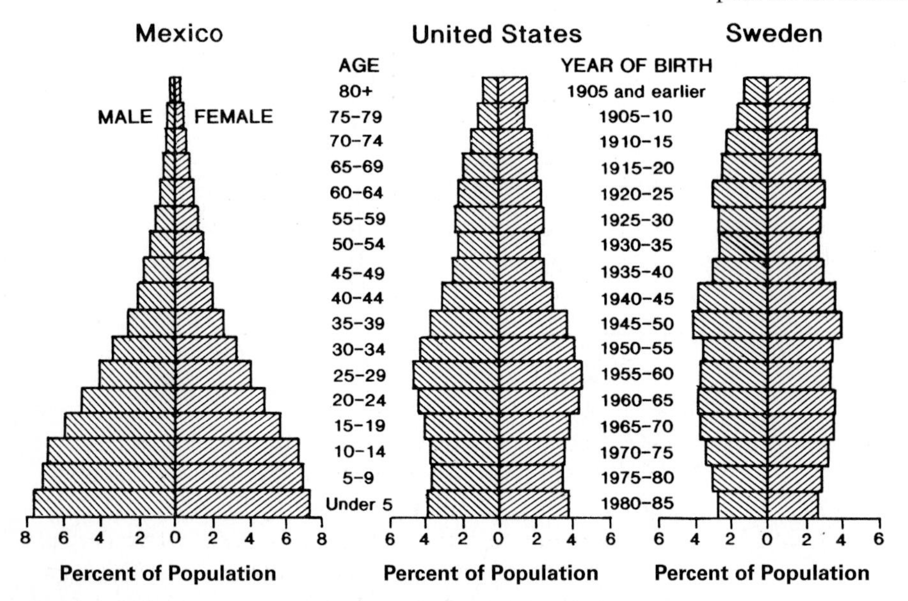

Figure 16.14 Population pyramids of three countries in 1985. Sweden is quite stable, with about equal percentages of people in all age groups. Mexico is bottom heavy, with greater percentages of young people who will grow up, reproduce, and add greatly to the population size. The United States pattern is not so regular; it is affected by large numbers of immigrants.
Data from Population Reference Bureau, 1985 estimates.

Some of the economically wealthy nations have undergone demographic transitions and, like Sweden, have reached population stability (Figure 16.14). Other of the more-developed nations have undergone a demographic transition like that of the United States but continue to admit large numbers of immigrants with high birth rates. Most of the less-developed countries have low death rates plus high birth rates, creating a bottom-heavy population pyramid, as in Mexico (Figure 16.14). The large number of young people in a country like Mexico means that there will be rapid population growth in the future even if birth rates are held low. As of mid-2002, world population data varied greatly from one region to another (Table 16.2).

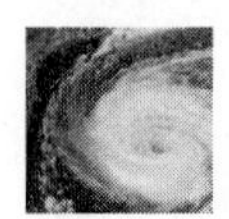

Future World Population

Today, most of the more-developed countries have gone through demographic transitions; they have low death rates and low birth rates. But the less-developed countries have low death rates and high birth rates; will they go through demographic transitions? In demographic transition theory, both mortality and fertility decline from high to low levels because of economic and social development. It is widely spoken and written that the world does not have a population problem, that a benign demographic transition will occur. But is this simply an assumption? A hope that the recent demographic history of a few nations will be the inevitable path for all nations, thus solving a difficult situation that few leaders are willing to address? A demographic transition appears to have begun in many less-developed countries even without benefit of strong economic development, good education, and high living standards.

The Population Reference Bureau estimates of the rates of world population growth were 1.8 percent in 1990, dropping to 1.6 percent in 1997, falling to 1.4 percent in 2000, and to 1.3 percent in 2002. What is causing this decrease in fertility? It appears to be due largely to urbanization and increased opportunities for women. At the beginning of the twentieth century, less than 5 percent of people in less-developed countries lived in cities; but by the year 2005, the majority of people will live in urban areas (Table 16.3). This is a change from farmer parents wanting many children to work in the fields and create surplus food, to city parents wanting fewer children to feed, clothe, and educate. Urban women have greater access to education, heath care, higher incomes, and family-planning materials. When presented with choices, many women

Table 16.2 World Population Data, Mid-2002

	Population (Millions)	Birth (Rates/1,000)	Death (Rates/1,000)	Yearly Growth %	Doubling Time (in Years)	Projected Population in 2050 (Millions)
World	6,215	21	9	1.3	53	9,104
More-developed countries	1,197	11	10	0.1	809	1,231
Less-developed countries	5,018	24	8	1.6	43	7,873
Less-developed countries (excluding China)	3,737	27	9	1.9	36	6,479
Africa	840	38	14	2.4	29	1,845
Asia	3,766	20	7	1.3	53	5,297
Europe	728	10	11	–0.1	—	651
North America	319	14	9	0.6	124	450
Latin America	531	23	6	1.7	41	815
Oceania	32	18	7	1.0	70	46

Source: World Population Data Sheet. (2002).

Table 16.3 Data Influencing Future Population, Mid-2002

	Percent of Population of Age <15	Percent of Population of Age 65+	Percent Urban (Cities >2000 People)	Percent of Married Women Using Modern Contraception (Pill, IUD, Condom, Sterilization)
World	30	7	47	55
More-developed countries	18	15	75	58
Less-developed countries	33	5	40	54
Less-developed countries (excluding China)	36	4	41	41
Africa	43	3	33	20
Asia	30	6	38	59
Europe	17	15	73	52
North America	21	13	75	71
Latin America	32	6	75	62
Oceania	25	10	69	56

Source: World Population Data Sheet (2002).

choose to have fewer children and to bear them later. Both of these choices lower the rate of population growth.

In the last 50 years of the twentieth century, population grew from about 2.5 billion to over 6 billion, an increase of 3.5 billion people. Even with the recent decreases in fertility rates, the population explosion is not over. A growth rate of 1.3 percent per year will cause the world population of humans to exceed 9 billion by the year 2050 (Table 16.2), an increase of another 3 billion people within 50 years. An important factor in estimating future growth is the age distribution of the population (Table 16.3). A significant percentage of the population today is less than 15 years old meaning their prime years for childbearing lie ahead. The century from 1950 to 2050 will probably witness the world population grow from 2.5 billion to 9 billion people.

The number of births per woman has a dramatic effect on human population growth. Starting in the year 2000 with a world population in excess of 6 billion people, look at 3 scenarios for population size in the year 2150 based on births-per-woman. 1) If women average 1.6 children, world population drops to 3.6 billion. 2) If women average 2 children, population grows to 10.8 billion. 3) If the present average of 2.6 children per woman continues, population will grow to 27 billion. The difference between a world population of 3.6

billion or 27 billion rests on a difference of only 1 child per woman. What will the women of the world choose?

The subject of population has become almost taboo; it is difficult for anyone to address when its very mention brings scathing attacks from some religious leaders and cries of racism from some political activists. Nonetheless, exponential growth is in full swing, and it is a mathematical reality. As Aldous Huxley said nearly a century ago, "Facts do not cease to exist because they are ignored."

No increasing exponential growth rate of population can continue for long on an Earth of finite size and resources. If humans do not regulate population size, then nature will step in with powerful negative feedbacks of social chaos, famine, disease, and death. Endless growth is the ideology of the cancer cell, and cancer kills its host. Human populations cannot grow endlessly.

Summary

The curve describing the history of human population growth is flat to gently inclined for several hundred thousand years, then it rises rapidly in the last three centuries. In the past, women bore numerous children, but many died, so overall population growth was slow. With the arrival of the scientific-medical revolution and the implementation of the principles of public health, human population has soared. Birth rates remain high in much of the world, even though death rates have plummeted. Population reached 1 billion in 1810, 2 billion in 1925, 3 billion in 1960, 4 billion in 1974, 5 billion in 1987, 6 billion in 1999, and is heading for 7 billion in 2012.

A steeply rising growth curve is exponential; for population, more people beget ever-more people. One way to visualize exponential growth is by using doubling times, the length of time needed for a population to double in size. Doubling times can be approximated by the rule of 70:

$$\text{Doubling time (in years)} = \frac{70}{\%\ \text{growth rate/year}}$$

At present, after subtracting deaths from births, world population increases 1.3 percent per year for a doubling time of 53 years. Extrapolating a 1.3 percent growth rate into the future yields humans standing shoulder-to-shoulder covering the face of the Earth in less than 900 years; or continuing for another 1,200 years longer, the volume of human flesh would equal the volume of the Earth. Obviously, these numbers are biologically and environmentally impossible. Something must change.

Much hope is placed in the demographic transition model, which holds that economic wealth, combined with knowing that one's children will survive, leads to dramatic drops in birth rates. This model holds for some wealthy countries. Now some less-developed countries are experiencing drops in birth rates presumably due to urbanization and more choices for women.

Population growth patterns for other animal species are related to the carrying capacity of the environment, i.e., how many can be supported. When climate is favorable, food is abundant, and predators are insignificant, then carrying capacity increases and population size increases to the maximum possible. When carrying capacity decreases, populations collapse in mass die-offs.

Humans have increased the carrying capacity of the environment through technological innovation. But the ultimate carrying capacity of the Earth is finite, and human population must level off at some finite number.

Terms to Remember

carrying capacity 431
demographic transition 433
exponential growth 428
fertility 430
logarithmic 433
mortality 430

Questions for Review

1. What are the population doubling times given these annual growth rates: Africa, 2.4%; world, 1.3%; Europe, –0.1%?
2. What is the size of the world population of humans today? Extrapolating the current growth rate, what will the population be in 100 years? In 200 years? Are these large numbers environmentally realistic?
3. Draw a curve showing the world population of humans in the last 10,000 years. Why has the curve changed shape so dramatically?
4. Explain the concept of exponential growth. Use positive feedback in an example in your answer.
5. Explain the concept of carrying capacity for a species. What negative-feedback processes might limit the numbers of a species?
6. Explain the demographic transition model.

Questions for Further Thought

1. What is the carrying capacity of the Earth for humans, i.e., how many humans can the Earth support? What factors are most likely to slow human population growth?
2. Assess the likelihood that benign demographic transitions will occur in poor African nations.
3. Compare the rate of change of human populations to the rate of change in religious and cultural institutions. Can religious and cultural institutions change fast enough to deal with world population growth?

4. Evaluate the suggestion that the overpopulation problem on Earth can be solved by colonizing other planets.
5. Which is a bigger number—the age of the Earth in years or the number of humans alive today?
6. Is a nation's destiny determined by its demography?

Suggested Readings and References

Bongaarts, John, and Bulatao, Rodolfo (eds.). (2000). *Beyond Six Billion: Forecasting the World's Population.* Washington, D.C., National Academy Press.

Cohen, Joel E. (1995). *How Many People Can the World Support?* W. W. Norton and Co.

Demeny, Paul. (1990). *Population.* In *The Earth as Transformed by Human Action* (41–54). Turner, B. L., II, et al., eds. Cambridge University Press.

Hardin, Garrett. (1993). *Living within Limits.* New York: Oxford University Press.

Haupt, Arthur, and Kane, Thomas T. (1998). *Population Handbook.* Population Reference Bureau.

Keyfitz, Nathan. (1989, September). The growing human population. *Scientific American,* 261, 119–26.

Livi-Bacci, Massimo. (1989). *A Concise History of World Population.* Cambridge: Blackwell.

Ponting, Clive. (1991). *A Green History of the World.* New York: St. Martin's Press.

Weeks, John R. (1992). *Population.* Belmont: Wadsworth.

Whitmore, Thomas M., and others. (1990). Long-term population change. In *The Earth as Transformed by Human Action* (25–39). Turner, B. L., II, et al., eds. Cambridge University Press.

World Population Data Sheet. (2002). Washington, D.C.: Population Reference Bureau.

Videos

Human Population Growth. (1977). Dallas County Community College (29 min.).

Increase and Multiply? (1988). Better World Society (55 min.).

The People Bomb. (1992). Cambridge Educational (90 min.).

Population Story: Collision with the Future? (1988). Encyclopedia Britannica (24 min.).

The Tragedy of the Commons. (1971). Biological Science Curriculum Studies (23 min.).

What Is the Limit? (1988). National Audubon Society (23 min.).

Glossary

A

aa Lava flow with a rough, blocky surface.

acceleration 1) To cause to move faster. 2) The rate of change of motion.

acceleration due to gravity The acceleration of a body due to the Earth's gravitational attraction, expressed as the rate of increase of velocity per unit of time (32.17 ft/sec/sec at sea level and 45° latitude).

acoustic fluidization A theorized process where sound waves trapped inside a dry, fallen mass lessen internal friction to enable fluidlike flow.

acre foot A measure of water volume where 1 acre of surface is covered 1 ft deep. An acre is about 90 percent of the area between the goal lines on a football field.

actualism Using the actual processes operating on Earth today to interpret the past; not inventing unrecognized processes to explain the past.

adiabatic process The change in temperature of a mass without adding or subtracting heat. Examples are cooling with expansion and warming upon compression.

aerosol A suspension of fine solid or liquid particles in air.

albedo The reflectivity of a body; for the Earth, how much solar radiation is reflected back to space.

alluvial fan A gently sloping cone of sediment formed where a stream leaves the hills or mountains and flows out onto a plain or valley.

amplitude The maximum displacement or height of a wave crest or depth of a trough.

andesite A volcanic rock named for the Andes Mountains in South America. It is intermediate in composition between basalt and rhyolite and commonly results from melting of continental rock in basaltic magma.

anoxic Depleted of oxygen.

anticline A fold where rock layers are compressed into a convex-upward pattern.

anticyclone A region of high atmospheric pressure and outflowing air that rotates clockwise in the Northern Hemisphere.

aphelion The point in the orbit of a body that is farthest from the Sun.

archaea An ancient branch of life whose species can thrive under high pressures and temperatures; they derive their energy by breaking the chemical bonds of inorganic molecules.

ash Fine pyroclastic material less than 4 mm in diameter.

asteroids Stony or metallic masses that orbit around the Sun.

asthenosphere The layer of the Earth below the lithosphere in which isostatic adjustments take place. The rocks here deform readily and flow slowly.

astrobleme An ancient impact site on Earth usually recognized by a circular outline and highly disturbed, shocked rocks.

atmosphere The gaseous envelope around the Earth, chiefly of nitrogen and oxygen. The average weight of the atmosphere on the Earth's surface is about 14.7 pounds/in^2.

avalanche A large mass of snow, ice, soil, or rock that moves rapidly downslope under the pull of gravity.

avulsion An abrupt change in the course of a stream and adoption of a new channel.

B

backfire A fire deliberately set to consume fuel in front of an advancing wildfire in order to stop it.

barometer An instrument for measuring atmospheric pressure.

barrier island An elongated sandbar parallel to the land but separated by a lagoon or marsh.

basalt A dark, finely crystalline volcanic rock typical of low-viscosity oceanic lavas.

base isolation Protecting buildings from earthquakes by isolating the base of the building from the shaking ground via rollers, shock absorbers, etc.

base level The level below which a stream cannot erode, usually sea level.

base surge A cloud of volcanic gas and suspended debris that flows rapidly over the ground.

bauxite Aluminum ore formed as a residual soil in the tropics where warm, heavy rainfalls dissolve and carry off most other materials.

B.C.E. Before the Common Era. Equivalent to B.C.

bedding The layering of rocks, especially sedimentary rocks.

bedrock Solid rock lying beneath loose soil or unconsolidated sediment.

bentonite A general term for clays derived from volcanic ash; these clays have extreme swell-shrink properties.

blind thrust A reverse fault of shallow inclination that does not break the surface.

blizzard Strong cold winds filled with snow.

body wave Seismic waves that travel through the body of the Earth, e.g., primary and secondary waves.

Bowen's reaction series The order of crystallization of common minerals from a cooling magma.

braided stream An overloaded stream so full of sediment that water flow is forced to divide and recombine in a braided pattern.

breakwater An offshore structure built parallel to the coastline to provide shelter from wave attack.

brittle Behavior of material where stress causes abrupt fracture.

buffer A mixture of substances in solution that neutralize changes, thus acting to maintain an equilibrium composition. For example, the world ocean has several buffer systems.

buoyancy The quality of being able to float, usually on water or rock.

C

caldera A large (>2 km-diameter), basin-shaped volcanic depression, roughly circular in map view, that forms by a piston-like collapse of a cylinder of overlying rock into an underlying, partially evacuated magma chamber.

calorie The amount of heat required to raise the temperature of 1 gram of water 1 degree centigrade at a pressure of 1 atmosphere.

carbonic acid A common but weak acid (H_2CO_3) formed by carbon dioxide (CO_2) dissolving in water (H_2O).

carnivore A flesh-eating animal.

carrying capacity The maximum population size that can be supported under a given set of environmental conditions.

cavern A large cave.

C.E. Common Era. Equivalent to A.D.

centigrade A temperature scale that divides the interval between the freezing and boiling points of water into 100°. Conversion from the Fahrenheit scale is by $C = 5/9 (F - 32)$.

chaparral A dense, impenetrable thicket of stiff shrubs especially adapted to a dry season about six months long; abundant in California and Baja California. Fire is part of the life cycle of these plants.

chemical weathering The decomposition of rocks under attack of base- or acid-laden waters.

chondrite A stony meteorite characterized by presence of small rounded grains or spherules.

cinder cone Steep volcanic hill made of loose pyroclastic debris.

clay minerals Very small (under 1/256 mm) minerals with sheet- or booklike internal crystal structure. Many varieties absorb water or ions into their layering, causing swelling or shrinking.

climate The average weather conditions at a place over many years.

coal A sedimentary rock made largely of plant remains. Coal is readily combustible.

coesite A very dense mineral of SiO_2; a polymorph of quartz created under a pressure around 20,000 atmospheres.

cohesion A mass property of sediments where particles cohere or stick together.

combustion Act of burning.

comets Icy bodies moving through outer space.

compaction The decrease in volume and porosity of a sediment via burial.

composite volcano A volcano constructed of alternating layers of pyroclastic debris and lava flows. Syn. stratovolcano.

compression A state of stress that causes a pushing together or contraction.

condensation The change of state of a substance from vapor to liquid.

conduction Transfer of heat downward or inward through material by communication of kinetic energy from particle to particle.

conglomerate A sedimentary rock dominated by gravel (pebbles, cobbles, boulders).

continent Lower-density masses of rock, exposed as about 40 percent of the Earth's surface: 29 percent as land and 11 percent as the floor of shallow seas.

continental drift The movement of continents across the face of the Earth, including their splitting apart and recombination into new continents.

convection A process of heat transfer where hot material at depth rises upward due to its lower density while cooler material above sinks because of its higher density.

convergence zone A linear area where plates collide and move closer together. This is a zone of earthquakes, volcanoes, mountain ranges, and deep-ocean trenches.

core The central zone or nucleus of the Earth about 2,900 km (1,800 mi) below the surface. The core is made mostly of iron and nickel and exists as a solid inner zone surrounded by a liquid outer shell. The Earth's magnetic field originates within the core.

Coriolis effect Moving objects experience the Earth move out from beneath them; in the Northern Hemisphere, bodies move toward their right-hand sides, while in the Southern Hemisphere, they move toward their left.

crater An abrupt basin commonly rimmed by ejected material. In volcanoes, craters form by outward explosion, are commonly less than 2 km diameter, and occur at the summit of a volcanic cone. Similar rimmed basins form by impacts with meteorites, asteroids, and comets.

creep The slow, gradual, more or less continuous movement of ice, soil, and faults under stress.

cross section A two-dimensional drawing showing features in the vertical plane as in a canyon wall or road cut.

crust The outermost layer of the lithosphere, composed of relatively low-density materials. The continental crust has lower density than oceanic crust.

crystallization The growth of minerals in a fluid such as magma.

curie A measure of radioactivity equal to 3.7×10^{10} disintegrations per second.

Curie point The temperature above which a mineral will not be magnetic.

cyclone A region of low atmospheric pressure and converging air that rotates counterclockwise in the Northern Hemisphere.

D

daylighted bedding Rock layers that dip less than the slope of a hill, thus the ends of the beds "see daylight." These layers are prone to mass movement due to lack of support of their ends.

debris Any accumulation of rock fragments; detritus.

debris flow Loose sediment plus water that is pulled downslope directly by gravity.

decompression melting The most common process creating magma is by reducing pressure on hot rock, not by adding more heat.

delta The mass of sediment brought by a river and built outward into a standing body of water.

demographic transition The change from a human population with high birth rates and high death rates to one with low birth rates and low death rates.

demography The statistical study of populations.

density The mass per unit volume of a substance.

derecho Winds that blow straight ahead.

dew point temperature The air temperature where the relative humidity of an air mass reaches 100 percent and excess water vapor condenses to liquid water.

dielectric constant A measure of the displacement currents occurring after applying an electric field.

diffusion Intermingling movement caused by thermal agitation with flow of particles from hotter to cooler zones.

dike Magma injected as tabular masses into older rocks.

dip The angle of inclination measured in degrees from the horizontal.

dip-slip fault Faults where most of the movement is either up or down in response to pushing or pulling.

discharge The volume of water flowing in a stream per unit of time.

divergence zone A linear zone formed where plates pull apart as at a spreading center.

ductile Behavior of material where stress causes permanent flow or strain.

ductility The ability to change shape markedly without breaking; "plastic" behavior.

duff A mat of organic debris in which fire can smolder for days.

E

earthquake The shaking of the Earth by seismic waves radiating away from a disturbance, most commonly a fault movement.

earthquake weather A common misconception; there is no connection between weather and earthquakes. Weather occurs at the Earth's surface and earthquakes occur deep below the surface.

eddy A circular moving water current; a whirlpool.

elastic Behavior of material where stress causes deformation that is recoverable; when stress stops, the material returns to its original state.

element Distinct varieties of matter; an atom is the smallest particle of an element.

El Niño A climate pattern that occurs every 2 to 7 years when the trade winds relax and warm ocean water in the equatorial Pacific Ocean flows to the west coast of North America.

embayment An indentation of the shoreline; depressed land near the mouth of a river.

energy Capacity for performing work.

epicenter The point on the surface of the Earth directly above a fault movement (i.e., earthquake location).

equilibrium A state of balance in a system; a condition in which opposing processes are so balanced that changes cause compensating actions.

erosion The processes that loosen, dissolve, and wear away earth materials. Active agents include gravity, streams, glaciers, winds, and ocean waves.

escape tectonics Collision of continents may cause large areas to move away from the pressure, such as Turkey moving west away from Arabia, and Indo-China squeezing eastward away from India.

evolution The change of life forms (species) over time.

exponential growth Growth in a compound fashion that, given time, leads to incredible numbers.

extinction The die-off or elimination of a species.

F

Fahrenheit A temperature scale where the boiling point of water is 212° and the freezing point is 32°. Conversion from the centigrade scale is by $F = 9/5 (C + 32)$.

failed rift Site of a spreading center that did not open far enough to create an ocean basin.

fall A mass moving nearly vertical and downward under the influence of gravity.

fault A fracture or belt of fractures where the two sides move past each other.

feedback A change in a system that provokes further changes. See **positive feedback** and **negative feedback.**

fertility The ability to produce offspring; the proportion of births to population.

fetch The length of water surface the wind blows across to create waves.

fire The rapid combination of oxygen with organic material to produce flame, heat, and light.

firebrand Burning debris such as branches and embers that are lifted above the fire and carried away to possibly start new fires.

firestorm A fire of large enough size to disturb the atmosphere with excess heat, thus creating its own winds.

fissure A narrow parting or crack in rock.

flank collapse A catastrophic event where the side of an oceanic volcano falls into the sea.

flood basalt Tremendous outpourings of basaltic lava that form thick, extensive plateaus.

floodplain The nearly flat lowlands that border a stream and act as the stream bed during floods.

flow A mass movement where the moving body of material behaves like a liquid.

foehn A warm, dry wind on the leeside of a mountain range. Pronounced *foon.*

fold Beds of rock that have been bent and warped. See **anticline** and **syncline**.

footwall The underlying side or block of a fault.

force Mass times acceleration.

forces of construction Land-building processes of volcanism, seafloor formation, earthquakes, continent collisions fueled by Earth's internal energy.

forces of destruction Land-destroying processes such as erosion and landsliding fueled by Earth's external energy sources of Sun and gravity.

fossil Evidence of former life, including bones, shells, teeth, leaves, footprints, etc.

fracture A general term for any breaks in rock. Fractures include faults, joints, and cracks.

fracture zone Major lines of weakness in oceanic crust; former transform faults.

freezing rain Supercooled rain that turns to ice when it touches objects such as trees and powerlines.

frequency Number of events in a given time interval. For earthquakes, it is the number of cycles of seismic waves that pass in a second; frequency = 1/period.

friction The resistance to motion of two bodies in contact.

front A boundary separating air masses of different temperature or moisture content.

fuel Any substance that produces heat by combustion.

fusion Act of melting.

G

gene The fundamental unit in inheritance; it carries the characteristics of parents to their offspring.

genome The common pool of genetic material shared by members of a species.

geothermal energy Energy derived from subsurface water heated to high temperatures by nearby magma.

geyser A hot spring that gushes magma-heated water and steam.

glacier A large mass of ice that flows downslope or outward due to the internal stresses caused by its own weight.

glass Matter created when magma cools too quickly for atoms to arrange themselves into the ordered atomic structures of minerals. Most glasses are supercooled liquids.

global climate model (GCM) A three- and four-dimensional computer model of Earth's atmosphere that simulates global climates produced by varying temperature, rainfall, atmospheric pressure, winds, and ocean currents.

global positioning system (gps) Accurate measurement by satellite of monitored ground sites.

global warming potential (GWP) The ability of a greenhouse gas to trap heat in the atmosphere as compared to CO_2.

Gondwanaland A southern supercontinent that included South America, Africa, Antarctica, Australia, New Zealand, and India during Jurassic time.

graded stream An equilibrium stream with evenly sloping bottom adjusted to efficiently handle water flow (discharge) and sediment (load) transport.

gradient The slope of a stream channel bottom; change in elevation divided by distance.

granite A quartz-rich plutonic rock.

gravel Sediment pieces coarser than 2 mm diameter, including granules, pebbles, cobbles, boulders.

gravity The attraction between bodies of matter.

great natural disaster A disaster so overwhelming that outside assistance is needed to handle the rescue and recovery for the region.

greenhouse effect The buildup of heat beneath substances such as glass, water vapor, and carbon dioxide that allow incoming, short-wavelength solar radiation to pass through but block the return of long-wavelength reradiation.

groin A low, narrow barrier built perpendicular to shore to slow the longshore transport of sand.

groundwater The volume of water that has soaked underground to fill fractures and other pores; it flows slowly down the slope of the underground water body.

H

Hadley cell A thermally driven atmospheric circulation pattern where hot air rises at the equator, divides and flows toward both poles, and then descends to the surface at about 30° latitude north and south.

hail Precipitation of hard, semispherical pellets of ice.

half-life The length of time needed for half of a radioactive sample to lose its radioactivity via decay.

halite Table salt; a mineral made of sodium chloride (NaCl) formed from evaporating seawater.

hangingwall The overlying side or block of a fault.

harmonic tremors Nearly continuous, small earthquakes created by underground magma on the move.

haze Fine dust, smoke, water and salt particles that reduce the clarity of the atmosphere.

heat The capacity to raise the temperature of a mass, expressed in calories.

heat capacity The amount of heat required to raise the temperature of 1 gram of a substance by 1°C.

herbivore A plant-eating animal.

hertz One hertz (Hz) equals one cycle per second.

hot spot A place on Earth where a plume of magma has risen upward from the mantle and through a plate to reach the surface.

humidity A measure of the amount of water vapor in an air mass.

hurricane A large, tropical cyclonic storm with wind speeds exceeding 74 mph; called a typhoon in the western Pacific Ocean and a cyclone in the Indian Ocean.

hydrograph A plot of water volume, height, flow rate, etc., with respect to time.

hydrologic cycle The solar-powered cycle where water is evaporated from the oceans, dumped on the land as rain and snow, and pulled by gravity back to the oceans as glaciers, streams, and groundwater.

hydrosphere The waters of the Earth, including the oceans, lakes, and rivers.

hypocenter The initial portion of a fault that moved to generate an earthquake. Hypocenters are below the ground surface; epicenters are placed above them on the surface.

I

igneous rock Rock formed by the solidification (crystallization) of magma.

impermeable Impervious; the condition of rock that does not allow fluids to flow through it.

inertia The property of matter by which it will remain at rest unless acted on by an external force.

insolation Amount of solar radiation received at any area on Earth.

Intertropical Convergence Zone The zone where collision occurs between the trade winds of the Northern and Southern Hemispheres.

inversion layer An atmospheric layer in which the upper portion is warmer or less humid than the lower.

ion An electrically charged atom or group of atoms.

isostasy The condition of flotational equilibrium wherein the Earth's crust floats upward or downward as loads are removed or added.

isotope Any of two or more forms of the same element. The number of protons is fixed for any element, but the number of neutrons in the nucleus can vary, thus producing isotopes.

J

jet stream Fast-moving belts of air in the upper troposphere that flow toward the east.

jetty A structure built perpendicular to the shore usually to protect the entrance to a harbor.

joint A fracture or parting in rock.

jokulhlaup Glacial outburst flood.

K

kinetic energy Energy due to motion. See **potential energy.**

L

ladder fuel Vegetation of varying heights in an area that allow fire to move easily from the ground to the tree tops.

lahar A volcanic mudflow composed of unconsolidated volcanic debris and water.

La Nada A climate pattern that occurs when seawater temperatures in the tropical eastern Pacific Ocean are neither excessively warm nor cool, but instead are neutral.

La Niña A climate pattern that occurs when cooler than normal seawater exists in the tropical eastern Pacific Ocean.

latent heat The energy absorbed or released during a change of state.

latent heat of condensation The heat released when vapor condenses to liquid. For water, the heat release is about 600 calories per gram.

latent heat of fusion Water releases about 80 calories per gram when it freezes. In reverse, ice absorbs about 80 cal/g when it melts.

latent heat of vaporization Water absorbs about 600 calories per gram when it evaporates. This stored heat is released during condensation.

latitude Reference lines that encircle the Earth parallel to the equator. The equator is 0° latitude and other lines are proportioned up to 90°N (North Pole) or 90°S (South Pole). Perpendicular to **longitude.**

Laurasia A northern supercontinent that included most of North America, Greenland, Europe, and Asia (excluding India) from about 180 to 75 million years ago.

lava Magma that flows on the Earth's surface.

lava dome A mountain or hill made from highly viscous lava, which plugs the central conduit of volcanoes.

Law of Cross-cutting Relationships A feature (rock body, fault, erosion surface) is younger than any rock body it cuts across.

Law of Faunal Assemblages Similar assemblages of fossil organisms indicate similar ages for the rocks that contain them.

Law of Faunal Succession Fossil organisms succeed one another in a definite and recognizable order.

Law of Original Continuity A water-laid sediment body continues laterally in all directions until it thins out due to nondeposition or butts against the edge of the basin of deposition.

Law of Original Horizontality Sediments are deposited in nearly horizontal layers.

Law of Superposition In a sequence of sedimentary rock layers, the oldest layer is at the base, and ages are progressively younger toward the top.

left-lateral fault A strike-slip fault where most of the displacement is toward the left hand of a person straddling the fault.

levee A natural or human-built embankment along the sides of a stream channel.

lifting condensation level The altitude in the atmosphere where rising air cools to saturation (100 percent humidity) and condensation begins.

lightning A flashing of light as atmospheric electricity flows between clouds or between cloud and ground.

limestone A sedimentary rock composed mostly of calcium carbonate ($CaCO_3$), usually precipitated from warm saline water. Limestones on continents may later be dissolved by acidic groundwater to form caves.

liquefaction The temporary transformation of water-saturated, loose sediment into a fluid, typically caused by seismic waves.

lithosphere The outer rigid shell of the Earth that lies above the asthenosphere and below the atmosphere and hydrosphere.

Little Ice Age A colder interval between about 1400 to 1900 C.E. with renewed glaciation in the Northern Hemisphere.

load The amount of material moved and carried by a stream.

loess Extensive deposits of wind-blown fine sediment (silt, very fine sand, clay) commonly winnowed from glacially dumped debris.

logarithm The exponent of that power of a fixed number that equals a given number.

longitude Reference lines connecting the North and South Poles. The line running through Greenwich, England is the 0° line; all other lines are counted away toward either 180°W or 180°E.

longshore current Waves in the breaker zone hit the beach at an angle causing a "current" to flow down the beach.

longshore drift Beach sediment (and bathers) are carried along the shoreline if waves are striking the beach at an angle.

M

magma Molten or liquid rock material. It crystallizes (solidifies) on the Earth's surface as volcanic rock and at depth as plutonic rock.

magnetic field A region where magnetic forces affect any magnetized bodies or electric currents. Earth is surrounded by a magnetic field.

magnetic pole The point where the Earth's magnetic field flows back into the ground. Currently, this point is near the North Pole.

magnetism A group of physical phenomena associated with moving electricity.

magnitude An assessment of the size of an event. Magnitude scales exist for earthquakes, volcanic eruptions, hurricanes, and tornadoes. For earthquakes, different magnitudes are calculated for the same earthquake when different types of seismic waves are used.

mantle The largest zone of the Earth comprising 83 percent by volume and 67 percent by mass.

map A two-dimensional representation showing features in a near-horizontal surface, such as the ground.

marble Metamorphosed limestone.

maria Dark, low-lying areas of the Moon filled with dark volcanic rocks.

mass A quantity of material.

mass movement The large-scale transfer of material downslope under the pull of gravity.

Maunder Minimum A cooler interval between 1645 to 1715 C.E. when astronomers noted minimal sunspots on the Sun's surface.

meander The curves, bends, loops, and turns in the course of an underloaded stream that shifts its bank erosion from side to side of its channel.

Medieval Maximum A relatively warm interval in the Northern Hemisphere between about 1000 to 1300 C.E.

mesosphere The atmosphere layer above the stratosphere and below the thermosphere.

metamorphic rock A former igneous or sedimentary rock whose mineralogy, chemistry, and texture have been changed due to high temperature and pressure.

metamorphism The changes in minerals and rock textures that occur with the elevated temperatures and pressures below the Earth's surface.

meteor The light phenomena that occur when a meteoroid enters Earth's atmosphere and vaporizes; commonly called a shooting star.

meteorite A stony or metallic body from space that passed through the atmosphere and landed on the surface of the Earth.

meteoroid A general term for space objects made of metal, rock, dust, or ice.

methane A gaseous hydrocarbon (CH_4).

methane hydrate An ice-like deposit in deep-sea sediments of methane combined with near freezing water.

microburst Sudden strong downrushes of wind and water from a thundercloud.

mineral A naturally formed, solid inorganic material with characteristic chemical composition and physical properties that reflect an internally ordered atomic structure.

mitigation Actions taken by humans to minimize the possible effects of a natural hazard.

monsoon The season of heavy rains in southern Asia.

mortality Death rate; the proportion of deaths to population.

N

natural disaster An event or process that destroys life and/or property.

natural hazard A source of danger to life, property, and the environment.

negative feedback Occurs in equilibrium systems where one change triggers another change that tends to negate the initial change and restores equilibrium.

neotectonics The study of the youngest faults and tectonic movements.

niche A site in the environment where an organism or species can successfully exist.

normal fault A dip-slip fault where the upper fault block has moved downward in response to tensional stresses.

North Atlantic Oscillation (NAO) A shifting of atmospheric pressures over the North Atlantic Ocean occurring on a multi-year time scale.

nuclear fission Splitting the nucleus of an atom with resultant release of energy.

nuclear fusion Combining of smaller atoms to make larger atoms with resultant release of energy.

nuée ardente A turbulent "glowing cloud" of hot, fast-moving volcanic ash, dust, and gas; a pyroclastic flow.

O

obsidian Dark volcanic glass.

Oort cloud A vast and diffuse envelope of comets surrounding the Solar System.

order A unit in taxonomy that contains one or more families; its rank is below class.

oxidation Combination with oxygen. In fire, oxygen combines with organic matter; in rust, oxygen combines with iron.

ozone A gaseous molecule composed of three atoms of oxygen.

P

pahoehoe Lava flow with a smooth, ropy surface.

paleontologist One who studies the fossils of animals, plants, and other life forms.

paleontology The study of fossils and the evolution of life through time.

paleoseismology The study of prehistoric earthquakes.

Pangaea A supercontinent that existed during Late Paleozoic time when all the continents were unified into a single landmass.

Panthalassa A massive, single ocean that occupied 60 percent of Earth's surface in Late Paleozoic time.

perihelion The point in the path of an orbiting body that is closest to the Sun.

period The length of time for a complete cycle of seismic waves to pass; equals 1/frequency.

permeability The capacity of a porous material to transmit fluids.

photosynthesis The process where plants produce organic compounds from water and carbon dioxide using the energy of the Sun.

pillow lava Lava cools underwater into pillow-shaped masses.

piping Formation of conduits due to erosion by water moving underground.

plate A piece of lithosphere that moves atop the asthenosphere. There are a dozen large plates and many smaller ones.

plateau An elevated, extensive tract of land; a tableland.

plate tectonics The description of the movements of plates and the effects caused by plate formation, collision, subduction, and slide past.

Plinian eruption The eruptive phase where an immense column of pyroclastic debris and gases are blown vertically to great heights.

plume An arm of magma rising upward from the mantle.

plutonic rock Rock formed by the solidification of magma deep below the surface. Name is from Pluto, the Greek god of the underworld.

polymorph The characteristic of a chemical substance to crystallize in more than one mineral structure.

pore An opening or void space in soil or rock.

pore-water pressure Pressure buildup in underground water that offsets part of the weight (pressure) of overlying rock masses.

porosity The percentage of void space in a rock or sediment.

positive feedback Occurs in nonequilibrium systems where one change triggers more changes in the same direction.

potential energy The energy a body possesses because of its position; for example, a large rock sitting high on a steep slope. See **kinetic energy.**

power The rate of work.

predation Killing and eating other organisms.

primary or P wave First seismic wave to reach a seismometer. Movement is by alternating push-pull pulses that travel through solid, liquid, and gas.

pumice Volcanic glass so full of holes that it commonly floats on water.

pyroclastic Pertaining to magma and volcanic rock blasted up into the air.

pyroclastic flow A high-temperature, fast-moving cloud of fine volcanic debris, steam, and other gases.

pyroclastic surge A variety of pyroclastic flow with higher steam content and less pyroclastic material. Surges are lower density, more dilute, high velocity, and may flow outward in a radial pattern.

pyrolysis Chemical decomposition by the action of heat.

Q

quarry An openly mined area on the surface

quartz A resistant, rock-forming mineral made of silicon and oxygen (SiO_2). The most common mineral found as a sand grain.

quick clay A clay that loses nearly all its shear strength after being disturbed.

quicksand Loose sand partially supported by upward-flowing water to create a semiliquid mass into which heavy objects sink.

R

radiation Heat emitted as rays.

radioactive elements Unstable elements containing excess subatomic particles that are emitted to achieve smaller, stable atoms.

radioactivity The breakdown of unstable atomic nuclei by emission of particles or radiation. The decay process produces smaller atoms and gives off heat.

radiometric dating The determination of ages using measured laboratory amounts of: 1) radioactive parent atoms and 2) decay-product atoms, then 3) using half-lifes to compute lengths of time of radioactive decay.

recurrence interval The average time interval between floods or earthquakes of a given size.

reef An organism-built structure or current-deposited mound of $CaCO_3$ material (limestone).

resonance The act of resounding, ringing. A vibrating body moves with maximum amplitude when the frequency of seismic waves is the same as the natural frequency of the body.

resurgent caldera A large topographic depression formed by piston-like collapse of overlying roof into a magma chamber with a later central uplift of the caldera floor.

resurgent dome The uplifted floor and mass of magma in the center of a large volcanic caldera.

return period Amount of time between an event of a given size.

reverse fault A dip-slip fault where the upper fault block has moved upward in response to compressional stresses.

rhyolite A volcanic rock typical of continents. Typically forms from high-viscosity magma.

ridge The volcanic mountain ranges that lie along the spreading centers on the floors of the oceans.

rift The valley created at a pull-apart zone. Term commonly used to describe the valley that occurs along the axis of the volcanic mountain ranges of seafloor spreading centers.

right-lateral fault A strike-slip fault where most of the displacement is toward the right hand of a person straddling the fault.

rip current A strong current flowing seaward from the shore; erroneously called riptide.

riprap Large, irregular boulders placed to slow the attack of erosion.

rock A solid aggregate of minerals.

rogue wave An unusually tall wave created when several wave systems briefly and locally combine their energies.

rotational slide A downward-and-outward movement of a mass on top of a concave-upward failure surface.

S

salinity The total quantity of dissolved salts. For seawater, salinity is about 35 parts per thousand.

salt dome An upward-risen plug of salt that behaves like a viscous fluid when the high pressures of deep burial cause the salt to deform and flow.

sand Sediment grains with diameters between 1/16 and 2 mm.

scarp A steep, clifflike face or slope.

scoria Basaltic rocks with numerous holes formed by gases escaping from magma.

scoria cone A small cone or horseshoe-shaped hill made of pyroclastic debris from Hawaiian- or Strombolian-type eruptions. They commonly occur in groups.

seafloor spreading Where tectonic plates pull apart, magma wells up and solidifies to create volcanic mountains, which in turn are pulled apart as new ocean floor.

seamount Submarine hills or mountains, typically former volcanoes.

secondary or S wave Second seismic wave to arrive at the seismometer. Movement occurs by shearing particles at right angles to travel path. S waves move through solids only.

sediment Fragments of material of either inorganic or organic origin. Sizes are gravel (over 2 mm), sand (2 to 0.0625 mm), silt (0.0625 to 0.0039 mm) and clay (less than 0.0039 mm). A mixture of silt and clay equals mud.

seiche An oscillating wave on a lake or landlocked sea that varies in period from a few minutes to several hours. Pronounced saysh.

seism Earthquake.

seismic-gap method Earthquakes are expected next along those fault segments that have not moved for the longest time.

seismic moment A measure of earthquake size that involves amount of movement on the fault, the shear strength of the rocks, and the area of fault rupture.

seismic wave A general term for all waves generated by earthquakes.

seismogram The record made by a seismograph.

seismograph An instrument that records vibrations of the Earth.

seismology The study of seismic waves generated by earthquakes.

seismometer An instrument that detects Earth motions.

shatter cone Distinctively grooved and fractured conical fragments of rock.

shear The failure of a body where the mass on one side slides past the portion on the other side.

shield volcano A very wide volcano built of low-viscosity lavas.

shooting star Tiny space particles (about 1 mm diameter) that burn up with a flash of friction-generated light in the Earth's atmosphere.

side effect A misleading term that directs attention away from the concept that every action has multiple reactions.

sill A tabular body of igneous rock intruded between/parallel to rock layers.

silt Sediment grains with diameters between 1/16 and 1/256 mm.

sinkhole A circular depression on the surface created where acidic water has dissolved limestone.

sinuosity The length of a stream channel divided by the straight-line distance between its ends.

slash Debris such as logs, branches, needles left on the ground by logging or high winds.

slate Mud changed to hard rock by the high temperatures and pressures of metamorphism.

sleet Precipitation of fine icy particles formed as frozen rain.

slide A gravity-pulled mass movement on top of a failure or slide surface; a landslide.

slip The actual displacement along a fault surface of formerly continuous points.

slump A landslide above a curved failure surface.

slurry A highly mobile, low-viscosity mixture of water and fine sediment.

soil The surface layers of sediment, organic matter, and decomposing bedrock.

solar radiation Energy emitted from the Sun mostly in the infrared, visible light, and ultraviolet wavelengths.

solar wind The outflow of charged particles (ions) from the Sun.

species Organisms similar enough in life functions to breed freely together.

specific gravity The ratio of the density of a material to the density of water.

specific heat The amount of heat required to increase the temperature of 1 gram of a substance by 1°C.

spreading center The site where plates pull apart and magma flows upward to fill the gap and then solidifies as new lithosphere/ocean floor.

spring A place where groundwater flows out onto the surface.

stishovite A high-pressure, extremely dense mineral made of SiO_2. It is a polymorph of quartz produced by shock metamorphism.

strain A change in form or size of a body due to external forces.

strata Sedimentary or volcanic rock layers with distinct physical or paleontological characteristics. Singular form is *stratum.*

stratigraphic sequence A stack of rock layers accumulated over time.

stratosphere The stable atmospheric layer above the troposphere.

stratovolcano A composite cone built of layers of both lava and pyroclastics.

stress External forces acting on masses or along surfaces; forces include shear, tension, and compression.

strike The compass bearing of the trend of a rock layer as viewed in the horizontal plane.

strike-slip fault Faults where most of the movement is horizontal or slide past in character.

sturzstrom Long-runout movements of huge masses at great speeds.

subduction The process of one lithospheric plate descending beneath another one.

sublimation Changing from solid to gas without passing through a liquid phase.

submarine canyon An underwater canyon cut into the continental shelf.

subsidence Downward movement of the ground either slowly or catastrophically.

supernova The cataclysmic eruption of a star that releases tremendous quantities of energy.

superposition, law of The principle that successively younger rock layers are deposited on top of lower, older layers.

surface tension The attractive force between molecules at a surface.

surface wave A class of seismic waves that travel along the surface only, e.g., Love and Rayleigh waves.

surge A large mound of seawater that builds up within the eye of a hurricane and then spills onto the land.

swell One of a series of regular, long-period, somewhat flat-crested waves that travel outward from their origin.

syncline A fold where rock layers are compressed into a concave-upward position.

T

tectonic cycle New lithosphere forms at oceanic volcanic ridges, the lithospheric plates spread apart to open ocean basins, and then the oceanic plates are reabsorbed into the mantle at subduction zones.

tectonics The deformation and movement within the Earth's outer layers.

tektites Rounded pieces of silicate glass formed by impact melting.

tension A state of stress that tends to pull the body apart.

thrust fault A reverse fault where the upper fault block is pushed up a shallow-dipping fault surface.

thunder The sound given off by rapidly expanding gases along the path of a lightning discharge.

thunderstorm A tall, buoyant cloud of moist air that generates lightning and thunder usually accompanied by rain, gusty winds, and sometimes hail.

tides The alternate rising and falling of land and water surfaces due to the gravitational pulls of the Moon and Sun.

topography The shape of the Earth's surface both above and below sea level.

topple A large rock mass that has fallen over.

tornado Spinning funnels of wind with rotating wind speeds up to 300 mph.

trade winds Drying winds that blow toward both sides of the equator; from 30° North latitude, they move to the southwest, and from 30° South latitude, they blow to the northwest.

transform fault A strike-slip fault that connects the ends of two offset segments of plate edges such as spreading centers or subduction zones.

translational slide A mass that slides downward and outward on top of an inclined planar surface.

trench The elongate and narrow troughs where ocean water can be more than twice as deep as usual. Trenches mark the downgoing edges of subducting plates.

triple junction A place where three plate edges meet.

tropical cyclone Any weather system formed over tropical waters that rotates counterclockwise in the Northern Hemisphere.

tropical depression A tropical cyclone with wind speeds less than 39 mph.

tropical disturbance A low-pressure system in the tropics with thunderstorms and weak surface wind circulation.

tropical storm A tropical cyclone with wind speeds between 39 and 74 mph.

tropical wave Surface low-pressure systems over northwest Africa that move westward within the trade winds. Above warm Atlantic Ocean water they may grow into tropical storms or hurricanes.

tropopause The top of the troposphere.

troposphere The lowest layer of the atmosphere, 11 miles thick at the equator to 5 miles thick at the poles.

tsunami Giant, long-period sea waves caused by oceanic disturbances, such as fault movements, volcanic eruptions, meteorite impacts, and landslides.

turbidite A rock layer that grades from coarse sediment at the bottom to fine at the top. It is deposited from underwater density (turbidity) current events.

turbidity current A bottom-flowing current of high density due to sediment load, colder temperature, or high salinity.

typhoon A large, tropical cyclonic storm with wind speeds exceeding 74 mph; called a hurricane in the Western Hemisphere.

U

Uniformitarianism The concept that the same laws and processes operating on and within the Earth throughout geologic time are the same laws and processes operating today.

V

vaporization Act of conversion to gas.

viscosity The property of material that offers internal resistance to flow, its internal friction. The lower the viscosity, the more fluid the behavior.

viscous Ease of flow. The more viscous a substance, the less readily it flows.

volatile Substances that readily become gases when pressure is decreased, or temperature increased.

volcanic plug Vertical mass of igneous rock that cooled inside the central conduit of a volcano.

volcanic rock Rock formed by solidification of magma at the Earth's surface.

volcano An opening of the Earth's surface where magma has poured or blown forth, typically creating hills or mountains.

vortex A whirling body of water or air whose circular motion creates a vacuumlike effect in the center.

W

wavelength The distance between two successive wave peaks, or troughs, in seismic waves or ocean waves.

wave refraction The bending of waves. A segment of ocean wave in shallower water will slow while the segment in deeper water continues to race ahead, thus causing a bend in the wave. A seismic wave will bend when it passes into rocks with different physical properties.

weather The state of the air at a place with respect to hot or cold, wet or dry, calm or storm.

weathering The surface processes that physically disintegrate and chemically decompose rock to produce soil and sediment.

work Distance times force, where force equals mass times acceleration.

Credits

Chapter 1

Table 1.7: Source: Data from Hubbert (1971); **Table 13.5:** Source: National Interagency Fire Center; **Fig. 1.10:** Drawings by Jacobe Washburn; **Fig. 1.20:** Reprinted with permission from *Global Change in the Geosphere-Biosphere.* Copyright © 1986 National Academy Press, Washington, D.C.

Chapter 2

Fig. 2.10: Reprinted from *Deep Sea Research.* Copyright © 1966, 13:435, with kind permission from Elsevier Science Ltd., The Boulevard, Langford Lane, Kidlington OX5 1GB, UK; **Fig. 2.13:** From T. Utsu, "Seismological Evidence for Anomalous Structures of Island Arcs with Special Reference to the Japanese Region" in *Review of Geophysics and Space Physics,* 9:389, 1971. Reprinted by permission of the author.

Chapter 3

Fig. 3.24: Reprinted by permission of the California Division of Mines & Geology; **Fig. 3.32:** Reprinted by permission of the California Division of Mines & Geology.

Chapter 4

Fig. 4.29: Source: U.S. Geological Survey; **Fig. 4.32:** Reprinted by permission of the California Division of Mines & Geology; **Fig. 4.34:** Source: Data from EERI (Earthquake Research Institute) "Scenario for a Magnitude 7.0 Earthquake on the Hayward Fault," EERI, Sept. 1996; **Fig. 4.43:** Source: C. Bayarsayhan, et al., "1957 Gob8-Altay, Mongolia Earthquake as a Prototype for Southern California's Most Devastation Earthquake" in *Geology,* Vol. 24, No. 7, 1996.

Chapter 5

Fig. 5.4: From F. E. Wallace, "Active Faults, Paleoseismology, and Earthquake Hazards in the Wester U.S." in *AGU Earthquake Prediction,* D. W. Simpson and P. G. Richards, eds., 1981, pp. 209–216. Reprinted by permission; **Fig. 5.5:** From A. Holmes, *Physical Geology.* Copyright © 1965 Nelson (A division of Stanley Thornes (Publishers) Ltd). Reprinted by permission; **Fig. 5.9:** From R. B. Smith and W. J. Arabasz, "Seismicity of the Intermountain Seismic Belt" in *GSA, Neotectonics of North America,* 185–228, fig. 1, 1991. Reprinted by permission; **Fig. 5.11:** Source: D. P. Schwartz and K. J. Coopersmith, "Seismic Hazards: New Trends Using Geologic Data" in *Active Tectonics,* 1986; **Fig. 5.13:** From W. S. Baldidge and K. H. Olsen, "The Rio Grande Rift" in *American Scientist,* 77:240–247, 1989. Reprinted by permission of Sigma XI, Scientific Research Society; **Fig. 5.14:** From D. B. Slemmons, "Chapter 1 Introduction" in *GSA, Neotectonics of North America.* Reprinted by permission of the Geological Society of America; **Fig. 5.27:** Source: Moore and Kirby, *Journal of Geophysical Research,* Vol. 69, 1964, U.S. Geological Survey.

Chapter 6

Fig. 6.6: From S. L. deSilva, "Altiplano-Puna Volcanic Complex of the Central Andes" in *Geology,* 17: 1102–1106, 1989. Reprinted by permission of the author; **Fig. 6.16:** From A. Holmes, *Physical Geology.* Copyright © 1965 Nelson (A division of Stanley Thornes (Publishers) Ltd.) Reprinted by permission; **Fig. 6.18:** From A. Holmes, *Physical Geology.* Copyright © 1965 International Thomson Publishing Services LTD. Reprinted by permission; **Fig. 6.22:** Reprinted from *Volcanoes of the Earth,* Second Revised Edition, by Fred M. Bullard. Copyright © 1984. By permission of the University of Texas Press.

Chapter 7

Fig. 7.6: Source: H. Lansford, "Vulcan's Chimneys: Subduction Zone Volcanism" in Mosaic, March/April, 12:46–53, 1981; **Fig. 7.10:** Source: H. Lansford, "Vulcan's Chimneys: Subduction Zone Volcanism" in MOSAIC, March/April, 12:46–53, 1981; **Fig. 7.14;** Reprinted from *Volcanoes of the Earth,* Second Revised Edition, by Fred M. Bullard. Copyright © 1984. By permission of the University of Texas Press; **Fig. 7.19:** From D. R. Crandell, et al., "Catastrophic Debris Avalanche from Ancestral Mt. Shasta Volcano, California" in *Geology,* 12, March 1984. Reprinted by permission.

Chapter 8

Fig. 8.2a: From *Landslides and Related Phenomena* by S. F. Sharpe. © 1938 by Columbia University Press. Reprinted with permission of the author; **Fig. 8.2b:** Reuse of block diagram from *Landslides and Related Phenomena* from C. F. S. Sharpe, Falls Church, VA; **Fig. 8.5b:** Reprinted with permission from E. B. Eckel, *Landslides and Engineering Practice.* Copyright © 1958 National Academy Press, Washington, D.C; **Fig. 8.23:** From G. A. Kiersch, "Vaiont Reservoir Disaster" in *Geotimes,* May–June, 1965. Reprinted by permission of American Geological Institute; **Fig. 8.24:** From G. A. Kiersch, "Vaiont Reservoir Disaster" in *Geotimes,* May–June, 1965. Reprinted by permission of American Geological Institute; **Fig. 8.29:** Reprinted by permission of the California Division of Mines & Geology; **Fig. 8.33:** From Gilluly, Waters, and Woodford, *Principles of Geology,* 4th edition, 1975. Reprinted by permission; **Fig. 8.43:** From C. W. Kreitler, "Faulting and Land Subsidence From Groundwater and Hydrocarbon Production, Houston-Galveston, Texas" in *Bureau of Economic Research* Note 8, 1978. Reprinted by permission of Bureau of Economic Geology, University of Texas at Austin.

Chapter 9

Fig. 9.3: From E. Broecker, "Will Our Ride Into the Greenhouse Future Be a Smooth One?" in *GSA Today,* Vol. 7, No. 5, 1997, Geological Society of America. Reprinted by permission of the author; **Fig. 9.9:** From L. A. Frakes, *Climate Throughout Geological Time.* Copyright © 1979 Elsevier Science Publishing. Reprinted by permission of the author; **Fig. 9.13:** From J. C. Crowell, "Gondwanan Glaciation, Cyclothems, Continental Positioning and Climate Chante" in American Journal of Science, Vol. 278, 1978. Reprinted by permission of *American Journal of Science;* **Fig. 9.15:** From J. C. G. Walker and L. C. Sloan, "Something is Wrong with the Climate" in *Geotimes,* June: 16–18, 1992. Copyright © 1992 American Geological Institute. Reprinted by permission; **Fig. 9.24:** Data from S. B. Johnsen and H. B. Clausen, "Irregular Glacial Interstadials Recorded in a

New Greenland Ice Core" in *Nature,* 359:311, 1992; **Fig. 9.30:** From Edward Bryant, *Natural Hazards.* Copyright © 1991 Cambridge University Press. Reprinted by permission.

Chapter 10

Fig. 10.6: From Edward Bryant, *Natural Hazards.* Copyright © 1991 Cambridge University Press. Reprinted by permission.

Chapter 11

Fig. 11.6: Data from Gordon E. Dunn and Banner I. Miller, *Atlantic Hurricanes,* Revised Edition, Louisiana State University Press; **Fig. 11.19:** From R. A. Morten, et al., *Living with the Texas Shore,* fig. 2.3, p. 15. Copyright © 1983, Duke University Press. All rights reserved. Reprinted with permission.

Chapter 12

Fig. 12.17: From P. H. Rahn, "Flood-Plain Management Program in Rapid City, SD" in *GSA Bulletin,* 95:838–843, 1984. Copyright © 1984 Geological Society of America. Reprinted by permission of the author; **Fig. 12.32:** From P. L. Abbott, "Flood Control in the Lower Reaches of the San Diego River" in *Envrironmental Perils, San Diego Region, 1991.* Copyright © 1991 Patrick L. Abbott.

Chapter 13

Fig. 13.7: Drawing by Jacobe Washburn; **Fig. 13.11:** Source: M. J. Schroeder and C. C. Buck, "Fire Weather," 1970. *U.S. Department of Agriculture Handbook* 360; **Fig. 13.25:** From M. E. Voice and F.J. Gauntlett, "The 1983 Ash Wednesday Fires in Australia" in *Monthly Weather Review.* Copyright © 1984 American Meteorological Society. Reprinted by permission; **Fig. 14.4:** Source: Data from J.J. Sepkoski, Jr., "Mass Extinctions in the Phanerozoic Oceans" Geological Implications of Impacts of Large Asteroids, GSA Spec. Paper 190, 1982, fig. 2, p. 285; **Fig. 14.7:** From Paul Cooper, "Ecological Succession in Phanerozoic Reef Ecosystems" in *Palois,* 3:136–152, fig. 4, p. 147, 1988. Reprinted by permission of SEPM (Society for Sedimentary Geology); **Fig. 14.10:** From David M. Raup, "A Kill Curve for Phanerozoic Marine Species" in *Paleobiology,* 17:37–48, 1991. Reprinted by permission; **Fig. 14.12:** Source: Data from R. A. Press and R. Siever, *Earth,* 4th edition, 1986, W. H. Freeman and Company.

Chapter 15

Fig. 15.16a: Source: Data from P. H. Schulz and S. D'Atondt, "Createaceous–tertiary (Chicxulub) Impact Angle and It's Consequences" in *Geology,* Vol. 24, 1996; **Fig. 15.16b:** Source: Data from R.K. Olsson, et al., "Ejecta Layer at the Cretaceous Tertiary Boundary Bass River, NJ" in *Geology,* Vol. 25, 1997; **Fig. 15.20:** Data from Poag, et al., U.S. Geological Survey, "Meteroid Mayhem in Ole Virginny: Source of the North American Tektite Strew Field" in *Geology,* Vol. 22, August 1994.

Chapter 16

Fig. 16.9: From T. M. Whitmore, *The Earth as Transformed by Human Action* edited by B. L. Turner II, et al., 1991. Copyright © Cambridge University Press; **Fig. 16.11:** From K. F. Helleiner, "The Population from the Black Death to the Eve of the Vital Revolution" in E. E. Rich and C. H. Wilson (eds.), *The Cambridge Economic History of Europe, Vol. 4, The Economy of Expanding Europe in the Sixteenth and Seventeenth Centuries, 1967.* Reprinted with permission of Cambridge University Press; **Fig. 16.12:** From T.M. Whitmore, *The Earth as Transformed by Human Action* edited by B. L. Turner II, et al., 1991. Copyright © Cambridge University Press.

Chapter 1

1.1: AP Photo / Paras Shah; **1.2:** AP Photo / La Prensa Grafica; **1.3:** © *Reuters* / Zohra Bensemra / Getty Images; **1.6:** Photo by Wesley Bocxe / Liaison / Getty Images

Color Tip-Ins

TI1: Photo by Peter W. Weigand; **TI2a:** © Reuters / Corbis–Bettmann; **TI2b:** © AP / Wide World Photos; **TI3a:** © Kevin West / Gamma Liaison; **TI3b:** © Kevin West / Gamma Liaison; **TI4:** Ed Youmans; **TI5:** Photo by Peter W. Weigand; **TI6:** Weatherstock Inc. / International Stock; **TI7&8:** © AFP / Corbis; **TI9:** Photo by Larry Mayer / Getty Images; **TI10:** © Photo Researchers, Inc.

Chapter 10

10.1: © Stone / GettyImages; **10.30:** © W. Faidley / Weatherstock

Index

A

B

C

D

E

F

G

H

I

J

K

L

M

N

O

P

Q

R

S

T

U

V

W

Y

Z